The Printing Ink Manual

The Printing Ink Manual

FIFTH EDITION

Edited by

R.H. Leach, R.J. Pierce
E.P. Hickman, M.J. Mackenzie and H.G. Smith

BLUEPRINT
An Imprint of Chapman & Hall
London · Glasgow · New York · Tokyo · Melbourne · Madras

Published by Blueprint, an imprint of Chapman & Hall, 2–6 Boundary Row, London SE1 8HN, UK

Chapman & Hall, 2–6 Boundary Row, London SE1 8HN, UK

Blackie Academic & Professional, Wester Cleddens Road, Bishopbriggs, Glasgow G64 2NZ, UK

Chapman & Hall Inc., One Penn Plaza, New York NY10119, USA

Chapman & Hall Japan, Thomson Publishing Japan, Hirakawacho Nemoto Building, 6F, 1-7-11 Hirakawa-cho, Chiyoda-ku, Tokyo 102, Japan

Chapman & Hall Australia, Thomas Nelson Australia, 102 Dodds Street, South Melbourne, Victoria 3205, Australia

Chapman & Hall India, R. Seshadri, 32 Second Main Road, CIT East, Madras 600 035, India

First edition 1961
Reprinted 1963 (with revisions)
Second edition 1969
Third edition 1979
Reprinted 1984
Fourth edition 1988
Reprinted 1988, 1989, 1991
Fifth edition 1993

Typeset in Meridien 10/11 by Falcon Graphic Art, Wallington, Surrey

Printed in Great Britain by TJ Press (Padstow) Ltd, Padstow, Cornwall

ISBN 0 948905 81 6

A catalogue record for this book is available from the British Library

Library of Congress Cataloging-in-Publication data available

∞ Printed on permanent acid-free text paper, manufactured in accordance with the proposed ANSI/NISO Z 39.48-1992

Contents

Preface

The first edition of the *Printing Ink Manual* was published by the Society of British Printing Ink Manufacturers in 1961 to fill the need for an authorative textbook on printing technology, which would serve both as a training manual and a reliable reference book for everyday use. The book soon became established as a standard source of information on printing inks and reached its fourth edition by 1988. This, the fifth edition, is being published only five years later, so rapid has been the development in technology.

The objective of the *Printing Ink Manual* remains unchanged. It is a practical handbook designed for use by everyone engaged in the printing ink industry and the associated industries. It provides all the information required by the ink technician for the day-to-day formulation of printing inks. It supplies the factory manager with details of the latest equipment and manufacturing methods, including large-scale production, and gives guidance on achieving quality assessment and total quality management specifications. Care has been taken to maintain the value of the *Manual* for training both technical personnel and others who require some knowledge of inks. Readers with little scientific knowledge will not find difficulty in using the *Manual*, but sufficient chemistry and physics have been included to provide an explanation of the underlying principles and theories governing the behaviour of inks for use by the advanced technologist. Suppliers of raw materials, substrate manufacturers, printers and print users will find the book a valuable source of information.

Both environmental issues and quality management concepts have grown in significance and are having a marked effect on ink development and manufacture. These trends are covered in detail, incorporating the latest practices in a continuously changing situation. Many chapters have been written by authors working for international companies, ensuring that the contents will include the widest international practices. Legislation, mainly European and from the United States, together with specifications set by world-wide end-users have established printing ink as a truly international product and the *Manual* will consequently be of help to a very wide readership.

This edition contains many significant changes. The lithographic chapter has been expanded to cover in more detail cold-set, heat-set, sheet-fed paper and metal-decorating inks; dry-offset inks have now been allocated a separate chapter and have been divided into inks for plastics and two-piece inks. Analysis has been separated from quality control and both are covered in individual chapters. A new chapter has been introduced dealing with ink-jet inks. All the remaining chapters have been reviewed, updated and expanded where needed.

R.H. Leach
R.J. Pierce

ACKNOWLEDGEMENT

The editors wish to acknowledge the valuable help from P.W.G. Seabrook, Director of the Society of British Printing Ink Manufacturers. Without his organizational skills their tight target schedule could not have been met.

The editors and authors

Editors

Dr R.H. Leach	Joint Editor-in-Chief
Mr R.J. Pierce	Joint Editor-in-Chief
Mr E.P. Hickman	Technical Consultant, Manders Printing Inks
Mr M.J. Mackenzie	Technical Director, Usher-Walker Printing Inks Ltd
Dr H.G. Smith	Technical Director (Oil Inks), BASF Coatings + Inks Ltd

Authors

Chapter 1	Dr R.H. Leach	
	Mr R.J. Pierce	
Chapter 2	Mr J.W. Birkenshaw	PIRA International
Chapter 3	Mr J.F. Brown	Mander-Kidd (UK) Ltd
	Mr K. Lau	Datacolour International
Chapter 4	Mr M. Clayton	Sun Chemical Ltd
Chapter 5	Mr G.B. Burdall	Usher-Walker plc
Chapter 6	Mr G.B. Burdall	Usher-Walker plc
	Mr D. Owen	Sun Chemical Ltd
	Dr R. Paradine	Carnaud Metalbox Technology plc
Chapter 7	Mr J.E. Kenee	Can Print Ltd
	Mr P. O'Flynn	Coates Lorilleux Ltd
Chapter 8	Mr B. Hancock	BASF Coatings + Inks Ltd
Chapter 9	Mr F.C. Wyatt	Coates Lorilleux Ltd
Chapter 10	Mr M.J. Barker	Dane and Company Ltd
Chapter 11	Mr I. Hargreaves	Sun Chemical Ltd
Chapter 12	Dr A. Hudd	Domino Amjet Ltd
Chapter 13	Mr M.J. Heath	Coates Lorilleux Ltd
Chapter 14	Mr G.A. Tabbernor	Manders Oil Inks Ltd
Chapter 15	Mr P. Simpson	Manders Oil Inks Ltd
Chapter 16	Mr D. Easterby	Coates Lorilleux Ltd
Chapter 17	Mr J.L. Orpwood	Coates Lorilleux Ltd
Index	Mr P. Greene	PIRA International

List of figures

List of tables

Foreword

The first edition of *The Printing Ink Manual* was published in 1961. Since then it has been constantly up-dated to reflect changes in raw materials, manufacturing and printing processes to meet the new challenges of the packaging and publishing industries. It has clearly become an internationally recognized authority on all aspects of printing ink.

It is with this formidable background that I have pleasure in welcoming the publication of the fifth edition of *The Printing Ink Manual*. It has maintained its traditions and in particular has recognised the significant changes which have occured since the fourth edition published five years ago.

The editors and authors are experts in their fields from international companies and I congratulate them on their knowledge and presentation.

The European ink industry will find the fifth edition essential to all involved in and practising printing ink. Students, raw material suppliers, schools and printers will also find *The Printing Ink Manual* an important reference book.

Hans A. Lentze
Secretary General
European Confederation of Paint, Printing Ink and
Artists' Colours Manufacturers' Associations (CEPE)

CHAPTER 1
The nature of printing ink

1.1 A BRIEF HISTORY

Throughout history little has been said or written about the influence of ink in spite of its ancient origin, yet from the moment writing was conceived, so was ink.

The story of printing ink began with the invention of the printing process. The earliest evidence of printing came from the Far East, centuries after the development of writing. Records suggest that the Chinese were using hand-carved wooden blocks for reproduction in AD 251. Writing ink was adapted for the purpose, composed of lamp-black and gum dissolved in water. The blocks were carved with the design in relief, a technique of printing which remained unique for the next 1300 years and which became identified as the letterpress process following the invention of other printing processes.

Writing at this time was mostly carried out with brushes, though there is some evidence of the use of quills. The ink was made up as dry tablets from a mixture of gum and lamp-black. When writing, ink was applied with a wet brush, when printing, similar ink was applied by using a wet brush to ink the wooden blocks.

Printing did not reach Europe until the Middle Ages. The demand for books, religious tracts and the works of poets and writers grew rapidly at this period and copyists were overwhelmed with the demand. Hence the search for an alternative, more rapid, method of making copies. One other element was necessary to make the change possible, the introduction of paper, Up to this time, manuscripts had been written on parchment, prepared by hand from sheepskin or goatskin. The production of sheets of parchment in sufficient quantity to allow printing was very expensive. However, in the 12th century, a new 'parchment' appeared in Italy, introduced from Arab countries, namely paper. Over the next 200 years both the quality and availability of paper had increased to such an extent that by the late 14th century it was rapidly replacing parchment for most purposes, except for legal documents. Paper was easily adapted for printing and through the 14th century copies were made more and more by a simple printing technique. Whole pages of a book were cut out in relief from a wooden block using one block per page. Illustrations were cut from separate blocks. The blocks were inked with the water-soluble inks which had been used for writing the original manuscripts.

This method of working was cumbersome and led to the search for a form of movable type. Attempts had been made in China as early as AD 1041, when Pi Sheng is said to have experimented with separate characters from moulded clay. Later, more durable type was made from wood. However, the Chinese attempts were abortive because of the immense number of characters required. It was in the 15th century that serious experiments began in Europe which culminated in the work of Gutenberg. Johann Gutenberg, who was a goldsmith, successfully developed a process of casting metallic type from meticulously prepared moulds producing characters of excellent design and detail. Once the moulds had been prepared any number of characters could be cast, all of the same high standard. Gutenberg's outstanding achievement was the printing of the 42-line Bible in 1456. The quality of the reproduction was highly recognized for the beauty of the type, the layout of the page and the intense blackness of the print. That the revolution from written manuscript to printed book was so successful owes very much to the brilliant craftsmanship of Gutenberg and his assistants.

The ink played a considerable part in this success. The blackness of the print was striking, the prints made from wooden blocks using water-type inks being brown and washy by comparison. An entirely new ink preparation had been used. Gutenberg found that the water-based ink was completely unsuitable for use on the new metallic type. Not only had there been difficulty in completely wetting the type with ink, but the very fluid nature of the water mixture resulted in poor reproduction of the characters because the ink did not remain on the surface of the metal and print sharpness was lost. A new ink had to be devised because water could no longer form the basis of the ink, and it was the printer who had to help solve the problem. Gutenberg in Germany, and later Caxton in England, with others, examined alternative materials and developed their own inks. The precise details of the composition of these inks remain a mystery to this day, although their basic composition is known.

It is one of the coincidences of history that exciting developments were taking place at that time in a not unrelated craft, and this gave Gutenberg the solution to his problem. Until the 15th century, the medium that artists were using for painting pictures was tempera. The pigments were ground with water and yolk of egg was added to provide a binder. The paint was usually applied to an absorbent surface such as plaster or wood. The medium was difficult to use as it dried rapidly, leaving little time to change or correct the work. Artists were therefore ready to try any means of overcoming such a restriction and the answer was found in drying oils.

Paints made from linseed oil to which litharge was added as a drier, were used in Roman times, a technique lost to Europe until it re-appeared in the 15th century. Artists learned how to process linseed oil and walnut oil by heat, producing thin or thickened oil as required. The new medium allowed artists to change their methods of painting and to produce quite different results. They had acquired a new and exciting medium. It is interesting to note that Holbein the Younger (1497–1543) painted his early works in tempera but used oil paint for his later remarkable portraits.

Gutenberg experimented with oil-based mediums and found them ideal for printing from metal. Using the painters' materials, Gutenberg developed his own formulations and produced inks of very high quality. Whilst the precise composition of these inks is not known, some information is available on the materials in use at the time. These included linseed oil, walnut oil, turpentine, rosin, pitch, Venice turps, lamp-black and vermilion.

The practice of treating linseed oil by heat was well established by the end of the 15th century and was further refined during the 16th century by printers for the production of ink. Raw linseed oil on heating could be clarified. Mucilage separated and was removed by throwing bread into the hot oil to which mucilage adhered. Litharge was added producing a quick drying unbodied oil. Prolonged heating at higher temperatures gave rise to bodied oils, the degree of bodying being determined by the length of string that could be drawn from the cold varnish. Temperature was gauged by adding an onion to the hot oil and noting the violence of the frothing that occurred. Varnishes were further modified by the addition of rosin.

By the beginning of the 17th century water-based printing inks were obsolete and all inks were being made from drying oils and resins, the formulations of which changed little for the next 300 years. When stocks ran low, the printer and his employees took a day off and adjourned outside the city walls to an open space, where they set up the pots in which the oil was to be heated. Heating and bodying the oil took a considerable time, so the event was enlivened with copious draughts of schnapps supplied by the Master. It was customary, too, to eat the charred bread scooped from the surface oil and many praised its health-giving properties. Grinding in the pigment undoubtedly took place the following day when both oil and heads had cooled. Printers continued to make their own ink until the end of the 18th century, when independent ink manufacture began.

By the late 1700s the demand for print in all forms had increased to such an extent that the printer found inkmaking an irksome, time-wasting task. This encouraged the start of the independent inkmaker. It also opened the way to a much greater use of colour. Coloured inks had been little used since the printer found them much more difficult to make.

Coloured pigments were minerals as used by the artist painter. Not only were these difficult to grind to a fine enough powder, but each colour required its own proportions of varnishes. The independent ink maker made it his task to produce a workable coloured ink and demand for coloured inks increased rapidly.

Relief printing, i.e. letterpress, remained the only method of printing text commercially until the early part of the 19th century, although engraving was used for fine pictorial work from the 17th century. The invention in the 1790s of the lithographic process by Senefelder greatly increased the scope of the printer and enabled him to produce fine illustrations.

During the remainder of this century many excellent illustrations in a number of colours were produced by lithography. This process created a challenge for the inkmaker. Inks had to be modified and more carefully formulated to resist the water successfully. They continued, however, to be based on linseed oil and rosin with the addition of waxes and other additives which appeared to improve the printing qualities.

Initially, printing was by direct lithography, the print being taken directly from the inked lithographic stone. By the end of the 19th century, offset lithography largely replaced direct lithography, though direct lithography continued to be used for printing large hoarding posters until the end of the 1930s. As less ink was transferred to the paper by offset lithography, it was necessary to formulate inks of appreciably greater colour strength.

However, by the end of the century, as offset lithography expanded, the new chemical industry associated with coal tar made possible the introduction of new synthetic pigments with much finer texture and greater colour strength. Over the next sixty years the range of synthetic pigments expanded to embrace a complete spectrum of colours, allowing the ink maker useful flexibility in formulation.

By the end of the 19th century two other printing processes had been invented: aniline and photogravure. They both required quick-drying inks based on volatile solvents. The aniline process, so-called because it made use of the new water- and alcohol-soluble aniline dyestuffs, was essentially an adaption of the relief letterpress process, and is now known as the flexographic process. The aniline dye solutions contained tannic acid to reduce the water solubility on a print. The main source of alcohol was from the decomposition of molasses. Photogravure was an intaglio process initially used for the continuous tone reproduction of illustrations and the source of fine art prints for many years. It is now known more briefly as the gravure process. Early photogravure inks were based on varnishes made by dissolving rosin derivatives in toluene and xylene solvents obtained from coal tar. The composition of letterpress and lithographic inks changed little until the introduction in the 1920s of the first synthetic resins, derivatives of phenol–formaldehyde polymers.

During the next decade considerable research into the chemistry and physics of polymers and colloidal systems was initiated both in universities and industrial laboratories. To exploit this new knowledge printing ink companies engaged chemists and physicists and established research departments. There was close collaboration between the printing ink industry and the universities, which led to a greater understanding of the underlying principles of ink behaviour and laid the foundation for the rapid expansion in ink technology brought about by the development of the petrochemical industry at the end of the 1940s. Many new pigments, polymers, solvents and additives were produced, which the ink chemist was able to adapt and incorporate into ink. This evolution signalled the rapid end of the traditional formulations, which had changed little over so many centuries.

From 1950 onwards the ink industry was further transformed by the tremendous expansion brought about both by the birth of the packaging industry and new technology used in the publishing and printing of magazines. The craft aspect of printing was changed to a highly sophisticated technology which could only be satisfied by an equally developed ink industry. It is ironic, however, that in the 1990s much R + D effort is being spent on water-based inks, which is how the inkmaker started some 1000 years ago; although, now of course, for 20th-century environmental reasons.

1.2 THE NATURE OF PRINTING INKS

Printing inks are coloured liquids or pastes, formulated to transfer and reproduce an image from a printing surface. They are used mainly to convey a message, provide protection, but they also can give a decorative effect to the substrate to which they are applied. Printing inks are used on a wide range of papers, boards, plastics, glass and textile surfaces in flat and, in some instances, preformed shapes. Lacquers or overprint varnishes are uncoloured forms of printing inks and can be used to give added gloss and protective properties to the print and substrate. Printing inks are applied by five major printing processes which will be covered later. The film thickness will depend upon the process used which is usually between 2 and 3 μm.

Printing inks consist of dispersions of insoluble colourants or solutions of dyes in a varnish or vehicle so that the resulting combination forms a fluid which will distribute and transfer on the printing press. The vehicle must then dry and bind the colourant to the substrate under the press running conditions. Both the colourants and vehicles will determine the end-behaviour of the print with respect to the specific resistance properties. In summary, an effective printing ink must meet both press and print requirements.

This chapter will discuss how each of the fundamental components of an ink, that is to say colourants, vehicles and their constituents and additives, affect printing and print properties. Individual chapters in this manual will deal with each of these requirements in detail.

1.3 VISUAL CHARACTERISTICS OF INKS

The visual characteristics of an ink are recognized in terms of its colour, its intensity, transparency or opacity and gloss, and these properties are largely determined by the quantity and nature of the colourant used (pigment or dye) specifically related to the vehicle system used.

The colour of inks

Colour is a complex concept and is determined in the following interrelated ways. First, the 'hue' of a colour indicates what kind of colour it is: for example, red, blue or green. Secondly, the 'strength' of a colour is measured by its intensity or saturation. The third property is its 'purity' which indicates how bright or dark it is.

There are many factors which affect each of these properties to one degree or another. The chemical nature of the colourant has a fundamental effect on its hue but so do particle size and shape surface characteristics and the amount of colourant contained in a given printed ink film. In general, the higher the concentration of a particular colourant the greater the strength of the colour, although, in most cases, an optimum concentration occurs, beyond which there is no further increase in strength.

Ink vehicles consist of resins and oils or solvents and these can alter the hue and purity of a particular colourant. This can be due to the colour of the vehicle itself or because different ink vehicles develop different colours when dispersing the same colourant, owing to the wetting or chemical relationship between the vehicle and the colourant. Selected additives to aid dispersion may improve the quality of the colour.

The choice of ink vehicle is also important in order to keep the colourant particles apart and well dispersed; otherwise, settling or flocculation may occur, resulting in loss of strength. The strength of an ink as a print is also dependent upon the substrate on which the ink is printed. For example, the same ink on an absorbent or non-absorbent paper will show different strengths. Similarly, differences will be seen between inks drying at substantially different rates on the same surface due to different penetration rates. For these reasons it is necessary for an ink maker to take into consideration the substrate when formulating an ink of required shade and colour strength.

The transparency and opacity of printing inks

Printed inks have varying degrees of opacity or transparency depending upon the amount of light which is transmitted through, or reflected from, the surface of the ink. The choice of the colourant and the degree to which it has been dispersed are major influences on this property. The more opaque colourants have a greater tendency to reflect and refract light due to the particle size and refractive index of the colourant.

Thus, in the case, for example, of titanium dioxide pigment the high refractive index and particle size effectively reflects and refracts and, therefore, scatters light of all visible wavelengths making it one of the most opaque pigments used in printing inks.

Soluble dyes, on the other hand, have a relatively low refractive index and small particle size and give highly transparent ink films.

The ink vehicle can influence the opacity of an ink by both its pigment-dispersing properties and its own refractive index. The substrate on which the ink is printed will affect the degree to which light is reflected back through the ink film and, therefore, will influence the opacity of the print. In this respect, for example, gloss or matt papers, transparent film and metallic surfaces will give considerable differences in opacity properties.

The gloss of printing inks

The gloss of an ink is a measure of its ability to reflect incident light and depends to a large extent on whether or not the ink forms a smooth film on the surface of the substrate. Generally, the higher the ratio of vehicle to colourant the more gloss is obtained. Gloss level can be influenced by the nature of the vehicle itself and its interaction with the colourant which give both effective dispersion and a continuous film when printed. Gloss will also depend upon the substrate printed; penetration will reduce gloss, as will insufficient wetting or affinity between the ink and the substrate. In addition, higher gloss can be achieved if the adverse effect of the substrate is minimized by the application of thicker films of ink which will improve flow-out and surface continuity.

Most inks contain significant amounts of solvents which are used either to dissolve the resins in the vehicle systems or to adjust the viscosity of the ink for printing requirements. The solvent choice is also important in controlling gloss as it can affect pigment dispersion, hold-out, the percentage of resins in the ink at its printing viscosity and the ink-flow characteristics. If the solvent is not, for example, effective in dissolving resins, a poor flowing, less glossy ink can result. Also, if the solvent either evaporates or penetrates the substrate before flow out of the ink occurs, then reduced gloss will be obtained.

Additives, although they are present in the inks in small amounts for specific ink film properties, can have a disproportionate effect on gloss. For example, plasticizers can often improve gloss by aiding flow-out, whereas waxes used to improve rub resistance form an uneven profile of high refractive index, scattering incident light which creates a matt finish.

1.4 THE NATURE OF PRINTING INKS AS DETERMINED BY THE PRINTING PROCESS

The composition of a printing ink and its physical properties are determined greatly by the nature of the printing process used to transfer an image from the printing surface to the substrate. There are five basic ink types which will be considered.

Lithographic and letterpress inks

When compared with the fluid nature of flexographic and gravure inks, lithographic and letterpress inks are much higher in viscosity and can be 'buttery' in consistency. For this reason they are often called paste inks. This increase in viscosity is necessary for correct ink transfer on the press roller distribution system and to reproduce the image surface.

Because of the large surface area of the roller system and the relatively long time the ink is exposed, lithographic and letterpress inks are formulated on slower-evaporating solvents. These solvents have to be balanced in drying rate between duct and press stability and drying on the substrate. Newspaper inks, as an extreme example, are formulated to have almost infinite press stability but this leads to slow drying on the paper.

In the relevant chapters of this manual it will be seen that the letterpress process makes fewer demands on the range of raw materials used in inks compared with the lithographic process because of the water used in the latter process, and greater consideration has to be given to the choice of pigments, vehicles, additives and overall rheological requirements.

Flexographic and gravure inks

Inks which are printed by the flexographic and gravure printing processes are characterized by their very fluid nature and are generally termed liquid inks.

Fluidity of gravure inks is necessary to allow the recessed cells of the gravure cylinder to be filled rapidly. Similarly, fluidity is required for the flexographic process as the inking system depends upon an engraved cylinder metering the ink on to a raised rubber image. Both inking systems involve only a short distance and time between the ink duct and the printing surface and, therefore, highly volatile solvents are used to allow the print to dry as quickly as possible without the inks drying in the duct or on the press inking system.

Liquid inks are mainly used for reel-to-reel printing of a wide range of substrates such as paper, films and foils; they can also be used for sheet-fed printing of rigid surfaces such as board. They dry mainly by evaporation of the solvent but in some instances, such as in the case of water-based inks printed on paper and board, they also dry by penetration.

In both printing processes it is ideal to have inks with maximum strength and minimum viscosity and, therefore, careful pigment selection relative to the particular resin systems has to be made. Many ink makers use both a wide range of pigments to satisfy the number of substrates to be printed and the resin systems needed for adhesion and fast press speeds.

Screen inks

In screen printing the ink is forced through the open areas of a stencil supported on a mesh of synthetic fabric stretched across a frame. The ink is mechanically forced through the mesh onto the substrate underneath by drawing a squeegee across the stencil.

The inks need to remain fluid and coherent on the mesh before printing, transfer through the mesh without sticking it to the substrate, flow to make the ink film continuous and so avoid the pattern of the mesh but maintain the image, and then dry rapidly once applied to the substrate. Suitable inks have a viscosity intermediate between liquid flexographic and gravure inks, and paste lithographic and letterpress inks.

Screen inks can be formulated to dry by a variety of methods. These depend both on the type of substrate such as paper, board, plastic, glass

and metal and on the chemistry required to meet wide end-use specifications. The most widely used technique, however, is evaporation, although consideration has to be given to ink screen stability and drying rate on the substrate.

1.5 METHODS OF DRYING PRINTING INKS

Printing inks are liquids of various viscosities which, when applied to a substrate by means of a printing press, must be converted to a solid structure which will allow the print to be handled and meet end-use specifications. This, ideally, must be achieved as quickly as possible within the restraints of the printing process itself. This change of state is referred to as ink drying and it can come about by physical and chemical means or a combination of both. The ink maker's task is to produce carefully balanced inks which ensure that both high quality printing and production speeds are maintained along with satisfactory drying rates.

Absorption drying

An ink dries by absorption when it is applied to an absorbing surface such as newsprint or corrugated board. The ink structure is obtained mainly by penetration of liquid components into the substrate leaving a dry-to-handle ink on the surface. All inks will have an absorption mechanism when applied to an absorptive substrate, even if the eventual main drying procedure is by an additional method.

Oxidation drying

An ink dries by oxidation when oxygen in the atmosphere chemically combines with drying oil components of the vehicle system, converting the ink to a semi-solid or solid form. This method is accelerated by the use of catalysts such as metallic salts and, where appropriate, heat. Oxidation drying is often used in combination with absorption drying; for example, in sheet-fed paper or board printing where penetration of some of the liquid components occurs, leaving the remainder of the ink to reach a hard condition on the surface by oxidation.

The speed of oxidation can also be affected by other factors such as the type of pigments used, fount solutions, the use of anti-oxidants and the substrate.

Evaporation drying

Many inks dry by the physical removal of volatile solvents in the ink, leaving resins and other materials to bind the pigments to the substrate. The rate of drying depends upon the evaporation rate of the solvents selected, the affinity and retention of the solvents by the resins and the ambient and additional means of applying energy to evaporate the

solvent. It is important to ensure that solvents do not evaporate so quickly on the press that they cause instability but that they dry quickly on the substrate and meet print requirements as soon as possible under the press running conditions.

Chemical drying

Oxidation drying is a chemical method of drying but there are other reaction methods used to convert a liquid into a solid. For example, there are acid-catalysed two-pack systems which cross-link to produce a solid and more stabilized systems used for metal decorating two- and three-piece inks which dry by chemical reaction and thermal polymerization. All chemical drying systems are accelerated by the use of heat. In general, these systems are used when high levels of adhesion, heat resistance, or chemical resistance are needed.

Radiation-induced drying

There are several forms of radiation used to dry inks. These include ultra-violet radiation, infra-red, electron beam and, experimentally, microwave and radio frequency.

Each type of radiation determines the chemistry of the ink vehicle which is used in printing ink.

Ultra-violet curing inks contain vehicle systems which, in the presence of both photo-initiators and the appropriate wavelength ultra-violet (UV) light, produce a solid by free radical reaction. This reaction can occur very quickly to give prints which are immediately dry at the end of the press. UV drying is used in lithographic carton printing, for example, and has replaced, in many instances, slower oxidation drying.

Short and medium wavelength infra-red is used as an efficient means of providing heat and is used mainly to cause evaporation of solvents. It can also be applied in lithographic printing to assist penetration and to accelerate the oxidation of conventional quick-setting inks.

Electron beam drying inks are similar in composition to UV drying inks except that the use of high-energy electrons to bring about a free radical reaction makes photo-initiators unnecessary. This method of drying is used in web printing of cartons and paper coatings.

Microwave and radio frequency radiation have been examined for their uses in drying inks which contain a high proportion of polar solvents or water. Heat is generated by molecular rotation which causes evaporation to give a solid film.

1.6 THE ADHESIVE NATURE OF PRINTING INKS

One of the fundamental functions of an ink is that it should adhere to the substrate and should remain there for the lifespan of the print. Colourants can affect the adhesive properties of an ink. Their influence depends upon the compatibility with the vehicle system, the type of pigments and their

percentage, and the degree of dispersion in the final ink.

It is, however, the vehicle system used in the ink which largely determines its adhesive properties. On absorbent substrates, adhesion is influenced by the degree of vehicle penetration when a layer of relatively unbound pigment can be left on the surface to powder off. On non-absorbent substrates such as films or foil, adhesion is primarily controlled by the film-forming ability of the resin and the molecular affinity for the substrate. The choice of resin for a specific substrate is, therefore, extremely important.

Solvents can also affect adhesion of an ink by two means. First, solvents assist the wetting and flow-out of ink to give a continuous film which is essential for good adhesion. Secondly, solvents can penetrate the substrate, such as PVC, for example, to cause softening of the surface to assist physical and chemical bonding.

Inkmakers use, in some circumstances, adhesion promoters in small amounts to assist the compatibility between the substrate and the ink to give improved chemical bonding.

1.7 THE RESISTANCE PROPERTIES OF PRINTED INKS

In supplying products, inkmakers must not only consider all the printing conditions which have to be met for the successful production of the print itself, but they also have to formulate a product taking into consideration the formation of the final printed article and its end use. For example, there would be no purpose in producing print which restricted packaging speeds or which was, perhaps, softened by the product being wrapped. Clearly, inkmakers must be aware of end requirements for the print and that specifications must be agreed between the inkmaker and the printer to meet the end properties. Often, effective laboratory simulation tests are devised to reproduce actual large-scale processes and resistances.

It is the principle of BS5750 (ISO9000) that formal specifications are agreed between the supplier and the user and that these specifications are met for all deliveries. This concept applies to the raw materials used by ink makers so that the 'right first time' objective is more likely to be achieved in the production, delivery, application and end use of the print.

Nearly all printed substrates have to meet at least one critical specification such as the following.

Lightfastness

The lightfastness of a print will depend upon a number of factors and, if not adequate, the colour will fade, become dirtier or change shade and, eventually, the colour can disappear altogether. These factors are the exposure conditions, time of exposure, substrate and ink film thickness, but primarily the colourants used.

Colourants can be put into broad classes of lightfastness and specified

according to the blue wool scale. A few pigments, including carbon black and ultramarine, and some inorganic pigments such as iron oxides can be considered permanent, but the remainder are, eventually, fugitive.

In general, the lightfastness of a pigment decreases with its degree of dilution by other pigments, especially white. Highly concentrated coloured inks have better resistance to light because of the self-masking action of the pigments. Hence, inks of pastel shades are generally not as lightfast as the stronger colours. Also, the ink film thickness affects lightfastness with thick films being more lightfast due to greater reflection or light scattering.

The precise lightfastness requirements need to be agreed for the final product and it is as essential not to overspecify as to underspecify. In general, greater resistance is achieved by the use of more expensive colourants.

Heat resistance

Heat resistance of a print may be required for pasteurization conditions or heat-sealing operations involved in carton or bag production. It is essential that the vehicle system is not softened or that the colourant is not thermally decomposed or sublimed to give a loss of strength or shade change during the manufacture or treatment of the finished product.

Abrasion resistance

Most, if not all, printed articles are subject to different forms of rub and abrasion during their life cycle. All types of inks are formulated with this in mind. The abrasion resistance of a print will depend upon the ratio of the vehicle to colourant and, in some instances, upon the dispersion method used in making the ink. In other words, better dispersion leads to less abrasion.

However, for optimum abrasion resistance, it is necessary to include surface and slip compounds in inks such as waxes which tend to migrate to the surface of the dried print. There are occasions when a print can have too much slip, leading to stacking problems, so ink formulators must be careful in their judgements in the use of additives.

Product resistance

Ink, in its many applications in the packaging field, has to be able to withstand contact with the substances packaged. Colourants, for example, vary widely in their resistance to acids, alkalis, oils, fats, detergents and other substances and must, therefore, be selected with the end use of the ink in mind. Resins also have differing resistances or solubilities and must be tested carefully in combination with the colourants to ensure that the appropriate resistance of the print will be obtained. Also, where food is involved, ink components must be chosen to avoid odour and taint problems.

Weathering

An ink may have to withstand attack from physical and chemical forces when exposed to the weather or chemical elements such as sea water. Colourants must be chosen to resist these environmental attacks so that there is the least effect on the print as considerable fading can occur. Attempts to correlate accelerated weathering tests with actual conditions are continually being made but, in general, they can only be used as guidelines.

BIBLIOGRAPHY

(1961) *The Printing Ink Manual*, 1st edn, W. Heffer & Sons Ltd, Cambridge, Chapter 1.

Bloy, C.H. (1967) *History of Printing Ink, Balls and Rollers*, 1440–1850, Evelyn Adams and Mackay Ltd, London.

Febvre, L. and Martin, H.-J. (1984) *The Coming of the Book* Verso Editions, London.

CHAPTER 2
Printing processes

Worldwide, the printing industry, together with its associated supplier industries, represent an important sector in all economies. In the developed economies printing and its related industries are likely to rank in the first ten industry sectors by size. When all branches of printing are combined with those of publishing and the related industries of packaging, paper and board making, ink making and machinery manufacturing, they represent a sector equal to such industries as aerospace and automobile manufacturing.

In the 16th and 17th centuries one could talk of the printing industry in precisely defined terms. Business was highly regulated and there existed only one process of printing. Today the picture is very different. The role of the printer is changing and it is not only in terms of technology. The relationship between the printer and his customer, particularly in publishing, is evolving under the impact of new communications technology. The newspaper sector has undergone a revolution in the last 25 years and this continues today. The packaging printer is now part of a sophisticated multidisciplined distribution business. This fast changing environment causes the printer to make heavy demands on his supplier industries, and not least the manufacturers of printing ink.

The main printing processes are:

- lithography, which uses a flat printing plate. The image area is ink-receptive, while non-printing areas are wetted by water and repel ink;
- flexography, which is a relief printing process. It is a rotary process using rubber or photopolymer plates and an inking system suited to low-viscosity water and solvent-based inks. The image area is raised above the non-printing areas, and is inked from an 'anilox' roller;
- gravure, where the image is sunk into the plate surface. The entire plate surface is flooded with ink and the excess removed with a doctor blade. Ink-filled cells remain and transfer the ink to the substrate;
- letterpress, which is a relief printing process. The image area is raised above the non-printing areas, is inked by rollers and pressed into contact with the substrate;

- screen printing, which employs a stencil principle. A rubber squeegee is used to push ink through the stencil on to the substrate;
- Ink-jet printing, in which a computer-controlled stream of ink drops is projected at high speed onto the substrate;
- toner printing systems such as electrophotographic printing, in which light is used selectively to discharge an electrostatic field, thus forming an electrostatic latent image. The image is 'developed' using toner of appropriate electrical charge, which is then transferred to the substrate and fused by heat or other means.

There are other minor processes such as intaglio or recess printing and collotype which are rarely used except for special purposes.

2.1 THE OFFSET LITHOGRAPHIC PROCESS

Lithography is a planographic process. The image and non-image areas are in the same plane on the printing image carrier. Chemical treatments are used to ensure that ink adheres to some areas and not others. That is, the effect of the treatment is to create areas of different surface energy on the plate. This results in image areas that are water repellent and ink accepting, and non-image areas that are water accepting.

Thus, the plate is usually damped before it is inked (although there is one type of newspaper press where this is reversed). Water forms a film on the water-accepting areas, but contracts into tiny droplets on the water-repellent areas. When an inked roller is passed over the damped plate, it is unable to ink the areas covered by a water film – the non-image areas – but it pushes aside the droplets on the water-repellent areas – the image areas – and these ink up. (It should be noted that there is also a 'waterless' process. This also uses the principle of differential surface energy, but relies on silicon-based compounds to form the ink-rejecting, non-image areas.)

The process is called offset lithography because the inked image on the plate does not print directly on to paper, but is first 'offset' onto a rubber blanket, and thence transferred to paper. This process was first used for printing on tinplate, to overcome the difficulty of printing directly from metal onto metal. Originally, lithography was direct, but the image on the plate was rapidly worn away by contact with paper. The direct process is, however, used for the printing of some newspapers.

The printing unit

Litho presses may be sheet or web fed. Several different designs exist for sheet-fed presses, the basic one being shown in Fig. 2.1. This shows a single-colour press consisting of a plate, offset or blanket cylinder and impression cylinder. These three printing cylinders, together with the inking and damping systems, make up the printing unit.

The plate is clamped round the plate cylinder. At each revolution it is contacted first by the damping rollers, then by the inkers. The blanket is

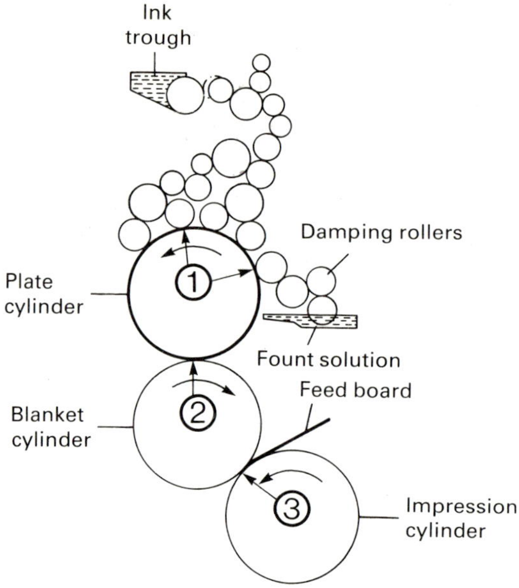

Fig. 2.1 Single-colour rotary offset machine: (1) damping and inking; (2) plate/blanket nip; (3) blanket/paper nip.

clamped round the blanket cylinder, and the paper passes between the blanket and impression cylinders, but does not contact the plate.

Press configurations

Sheet-fed presses can be made to print from one to six colours. The common types are one-, two-, four-, and five-colour, with three-, and six-colour available for special requirements. Perfecting presses are also available.

Figures 2.2 and 2.3 show how two-colour and perfecting presses are built up. A four-colour press can either consist of two, two-colour units linked together, or a number of single colour units placed in line, see Fig. 2.4. By assembling basic printing units together any number of colours can be printed. A combination of single-, two- and four-colour presses, plus a perfector, can be used to print efficiently any number of colours (by using different machines to print successive colours).

All sheet-fed machines have a sheet-feeding mechanism attached to the back of the press and a pile delivery unit attached to the front. To complete the picture, a sheet-fed printing press must have a method of transporting the sheet from feeder to printing unit, around the printing cylinders and finally, to the delivery pile (Fig. 2.4).

The small offset press

In some ways, these machines are the big brothers of duplicating machines. Their size is generally in the A4–A3 range. Because they are intended for use in in-plant printing works and other small companies,

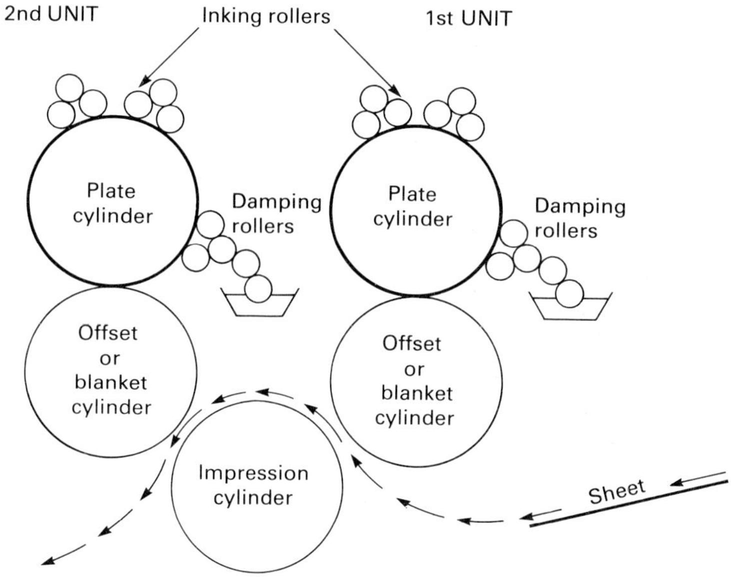

Fig. 2.2 Two-colour rotary offset machine.

they are simple machines to operate, and yet, with care, can handle a wide range of work. Some machines are highly automated to optimize their use for short runs and incorporate features for automatic plate change, wash up and start up and can be programmed to print from several plates in succession without any manual intervention.

Larger sheet-fed presses

Heavy offset machines take the larger sheet sizes up to a maximum of around A0. They have a much heavier construction, are more highly specialized, and are usually bought with a particular type of work in mind. Whereas small offset machines can be operated by relatively unskilled staff, these machines require much greater knowledge and training.

Web-offset presses

Three types of web-offset press are in use, the most common being blanket-to-blanket perfectors with the web going either vertically or horizontally through the units. Secondly, there are common-impression units, and finally the three-cylinder principle press.

Blanket-to-blanket press

In these the blanket cylinder acts as the impression surface for the printing on the opposite side of the web (Fig. 2.5). Any number of identical units may be mounted in-line to form a single printing press, although five is normally the maximum. A five-unit press enables two webs to run, with black both sides on one, and four colours both sides on

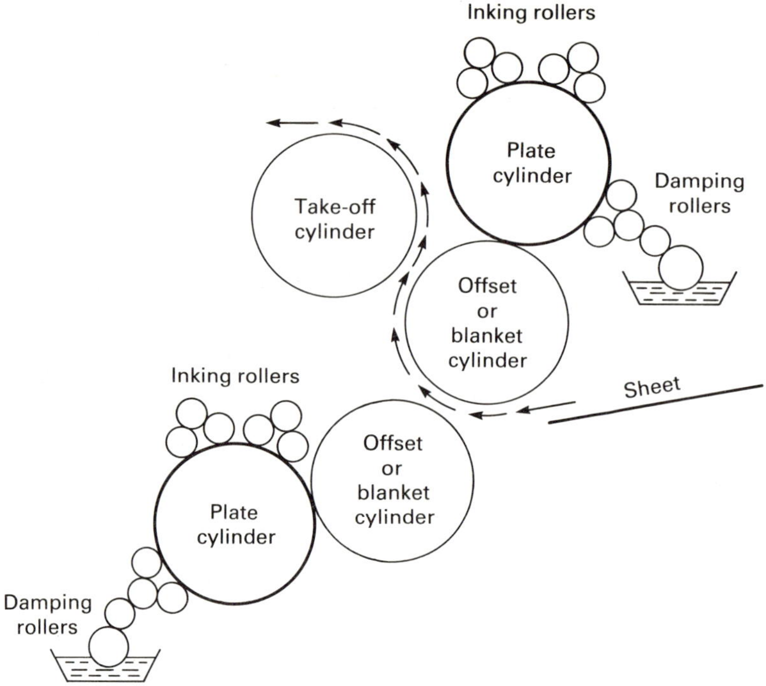

Fig. 2.3 Blanket-to-blanket perfector press.

the other. Other combinations can be used if needed. The vertical web-fed printing unit is mostly found in newspaper plants. On the other hand, commercial web-offset printers find the horizontal web blanket-to-blanket press the most suitable.

Common-impression drum presses

In these, up to five plate/blanket units are arranged circumferentially around a single-impression drum of large diameter and consequently only one side of the web is printed in one operation (Fig. 2.6).

Three-cylinder presses

This type of press is the least popular but is used for packaging, continuous stationery or small jobbing work (Fig. 2.7).

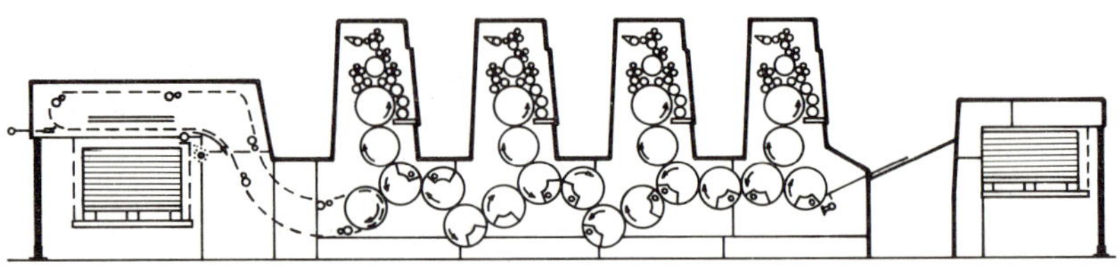

Fig. 2.4 A typical complete four-colour press.

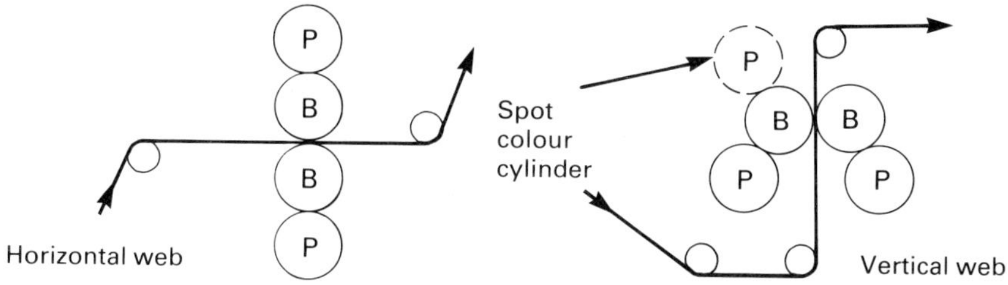

Fig. 2.5 Blanket-to-blanket perfector press units.

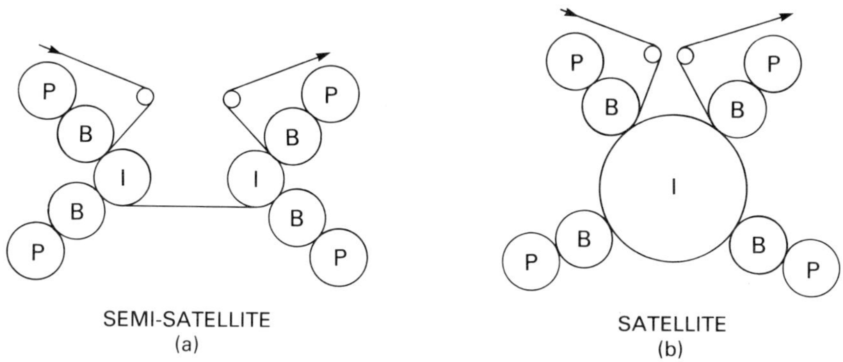

Fig. 2.6 Web printing units with common-impression cylinders: (a) semi-satellite; (b) satellite.

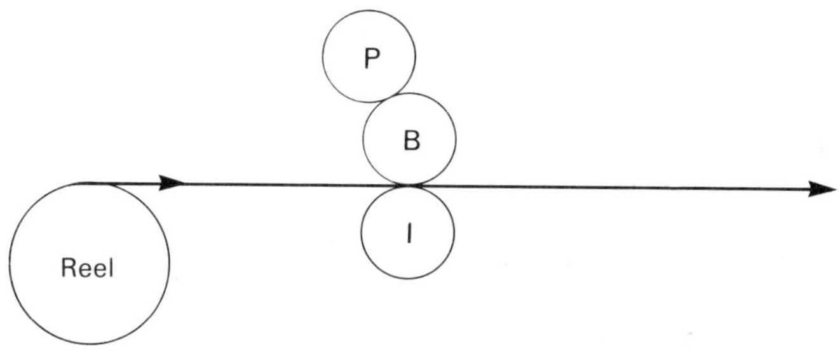

Fig. 2.7 Three-cylinder principle.

The damping system

A damping system supplies a carefully controlled amount of water, or fountain solution to the plate. There are various forms of damping units, for example a fountain reservoir and rollers, rotary brushes, sprays and canvas flaps.

One of the most common systems consists of a reservoir for the fountain solution, and five rollers. The fountain roller is partly immersed in the reservoir (called the fountain). As this rotates it carries a film of water up to the second roller and the distributor. The distributor then applies the fountain solution to two plate-damping rollers, and has a reciprocating action as well as rotational motion. The amount of water fed to the plate dampers is governed by how long the feed roller stays in contact with the fountain roller. The two main functions of the fountain solution are:

- to keep non-image areas clean and free from ink;
- to help minimize the amount of water necessary to obtain a clean print and so assist in the maintenance of ink/water balance.

When a plate leaves the plate-making department, it should have on its surface a fine film of gum arabic or similar desensitizer. This makes the plate surface extremely hydrophilic (water attracting). If the plate remained in this condition, it would be possible to run with water only in the fountain. However, the gum film is very fragile and is easily worn away by the contact of roller and the blanket. When this occurs the non-image areas tend to accept very minute amounts of ink which print as 'scum', a light tint over non-image areas.

The desensitized layer must therefore be continually replaced if scumming is to be avoided: this is achieved by using a desensitizer. It is the desensitizer which is the main active component of any fountain solution.

The fountain solution is usually made up from a concentrate, supplied ready to mix with water. A basic formulation would consist of gum arabic, phosphoric acid, and water, which produces an acid solution. Some of the fountain solution concentrates are formulated to give an alkaline solution, whereas others may be slightly acidic or neutral. Generally, fountain solutions consist mainly of water (98%) while the remaining components enhance its interactions with the printing plate. However, fountain solutions containing isopropyl alcohol (IPA) are very popular, in which case between 15 and 25% IPA may be used.

The following components may be contained in fountain solutions.

- Film-forming agents – which maintain the water-receptive nature of the non-image area while running, and protect these areas during press stops so that a quick recovery is achieved when the run is continued. Historically, gum arabic has been used for this purpose since it is highly water soluble and forms a non-porous film over the non-image area. The film formed has a high affinity for the aluminium oxide on the surface of the non-image areas of the plate rendering them extremely hydrophilic. The gum arabic only adheres to the non-image areas and is repelled by the image areas. The main disadvantage of gum arabic is that it is incompatible with IPA and therefore synthetic alternatives are now widely used.
- Desensitizers – in addition to the film-forming agents, desensitizers are also used which react chemically with the non-image areas to increase their hydrophilic nature. Desensitizers are the main active component of fountain solutions and provide a continuous means of preventing scumming compared with gum arabic which is gradually worn away during plate use. Examples of desensitizing compounds are fluorides, phosphates, nitrates and organic compounds such as phosphate esters.

- A buffering system – required to maintain the pH of the fountain solution. It is common practice to use salts which can also act as desensitizers, hence a citrate or phosphate system may be used.
- Surfactants – their principal role is to lower the surface tension of the solution. This allows a thin film of fount to wet the surface of the plate rapidly and aid absorption of water into the ink. These properties are intended to be similar to those imparted by IPA. Two types of surfactants are used, anionic or block copolymers. In order to prevent scumming it is necessary to reduce the surface tension of the fount to a similar value to the surface tension of the ink, that is approximately $37 \times 10^{-3} Nm^{-1}$.
- Humectants – non-drying materials which leave a thin wet film on the plate surface during press stops, and thereby aid a rapid recovery. Typical humectants are glycols or polymeric surfactants.
- Sequestering agents – which deal with excessive water hardness.
- Biocides – to prevent the growth of fungi and bacteria.
- Corrosion inhibitors – to protect the metal parts of the press.
- Anti-foams – to counteract the foaming effect of surfactants.
- Hydrated magnesium sulphate (Epsom salts) – prevent casein scumming which may be encountered when printing some coated papers which contain casein. If this is not fully hardened, it may remain soluble in water and can therefore be removed by the fountain solution. The casein in the fountain solution causes the non-image areas to become sensitized and print as an overall background scum. On a multi-colour machine the scum is generally most severe on the later units. The solubility of the casein in the fount solution is reduced by adding magnesium sulphate.

The addition of alcohol to the fountain solution has become very popular, particularly with the advent of vibratorless dampening systems. It is claimed that a higher rate of evaporation of moisture from the plate surface between dampening and inking gives less emulsification, improved drying of inks, quicker ink/water balance and easier maintenance of this balance. Other claims made are that less fluff and debris is picked up from the plates and better trapping of ink on the paper is obtained. IPA certainly does lower surface tension, and induces an increased evaporation rate, but a complete understanding of why it works so well is not yet available.

However, there are a number of drawbacks to the use of IPA such as the extra cost. It is necessary to use refrigerated circulation tanks, measuring devices and mixers to maintain the alcohol content. Also there is growing concern over the evaporation of the alcohol into the atmosphere. As a result many 'alcohol substitutes' have been produced, based usually on surfactants. However, while some of these work reasonably well, none appears to perform quite the same as alcohol systems.

Many printing problems are associated with the pH of the fountain solution. High pH values can cause excessive ink/water emulsification problems, while low pH values can cause ink drying difficulties, loss or sharpening of image on the plate and poor definition of dots. The ideal running pH is generally regarded as being between 5 and 7.

The offset blanket

In litho printing it is the blanket which transfers ink to the substrate, and consequently much of the character of the printed image is a result of the properties of the blanket. The blanket consists of a backing fabric made of cotton plies or manmade fibres, which are woven and laminated in thin layers, and then coated in a synthetic rubber or similar compound to give a smooth surface and overall even caliper. Modern blankets, especially if of the so-called compressible type, have a far more complex internal structure than older conventional ones, which gives improved rolling characteristics and better dot definition. This is commonly achieved by incorporating a layer of very small air or gas bubbles in the structure of the blanket, so enabling the surface to deform without causing it to be stretched which would lead to slur and other defects.

The surface of the blanket is made of a synthetic rubber such as nitrile with blends of polysulphide rubber to improve surface characteristics. Polyurethane compounds are used to improve sheet release properties. Other ingredients are added such as plasticizer, sulphur, zinc oxide, fillers which toughen and harden, and pigments to give the required colour. These are chosen to produce a blanket with appropriate hardness, solvent resistance, and resilience properties.

Blankets come in many varieties to suit different materials or inks, e.g. conventional inks, UV-cured inks, heat-set inks, certain solvent-based inks, varnishes and lacquers. Hardness recommendations are generally as follows although personal preferences count a great deal especially if compressible blankets are used:

Soft blankets, 70–74 degrees Shore hardness
- uncoated paper
- large solids
- polyurethane
- low speed machines
- proofing presses
- metal

Medium blankets, 74–76 degrees Shore hardness
- the normal blanket suited to a wide range of substrates
- coated or uncoated papers
- metal
- plastic sheets

Hard blankets, 76–82 degrees Shore hardness
- used on high speed presses (e.g. newspapers) and web-offset
- coated papers
- foil and plastic material.

Compressible blankets can be used for most types of work. It should be noted that a packing used under blankets to bring them to the correct height on press. The hardness of this packing has an effect on the printing characteristics.

Some small offset machines use self adhesive blankets of approximately 1 mm calibre, which obviate the need for punching, barring and lock-up. In recent years the increased use of 'no-pack' blankets has taken place in

America and Europe. These are considered to be simpler to install as they remove the need for underpacking.

Automation of lithographic printing presses

In the last few years there have been great strides forward on the part of the press manufacturers to incorporate features which reduce make-ready time. In the main these have been made possible by application of computer technology, which enables a sequence of actions to be programmed, recorded data to be used for certain press settings, and calculations to be made, with consequent alteration to press settings. The most common examples are:

- automatic blanket washing – of all units simultaneously;
- automatic ink duct setting. This works in conjunction with a plate scanner – a device which scans the printing plate prior to mounting on the press to determine the ink demand for each strip of the plate running in the machine direction. The width of a strip corresponds to the ink duct zone width on the press. Data from the plate scanner is recorded on disc or tape, subsequently fed into the press console and the ink ducts set accordingly. The method is not perfect, but for typical jobs with reasonable image coverage, will produce passable prints almost from the start of a run (Fig. 2.8);
- automatic plate changing. New plates are pre-loaded into a holding system on each printing unit before the current job is finished. At changeover, a programmed set of operations begins involving blanket washing, removal of the old plates, and loading of the new ones. Under demonstration conditions, a complete change can be accomplished in under 10 minutes;
- automatic register control. On web presses, this has been commonplace for decades, but not on sheet-fed ones. This has changed recently, with the introduction of systems especially adapted to use on sheet-fed machines. A typical example uses a portable unit rather like a densitometer to measure the register marks in the corners of a sheet. The portable unit is placed back on the press console, the data transferred, and automatic changes made to the press to achieve correct register.

These features might make up 30% of the cost of a new press but because of the reduced downtime that results, and consequent increased output from a press, are usually considered well worthwhile. In addition, modern presses are adopting digital control systems which besides controlling the press, provides remote diagnostics of many aspects of the press functions.

Lithographic platemaking

A wide variety of plates are used for offset-litho printing. The choice of plate is governed primarily by the length of run required, and also by cost and technical factors such as time and complexity of preparation.

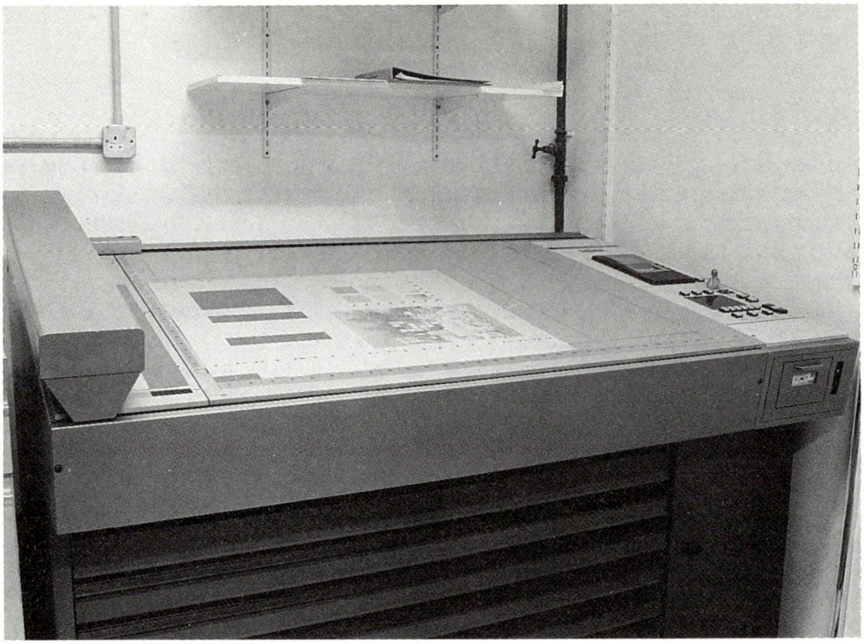

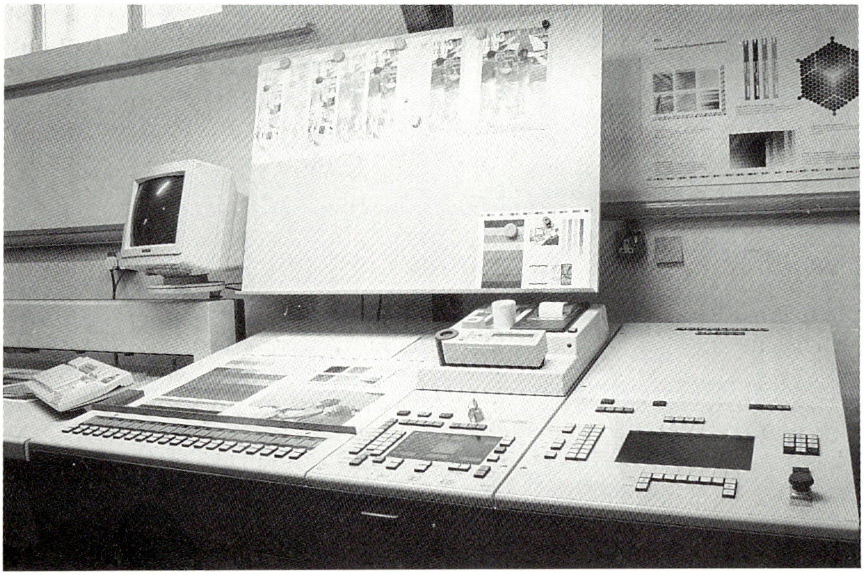

Fig. 2.8 (a) A plate scanner. (b) Press console with tape reader and ink duct controls. (Courtesy of PIRA)

A light-sensitive coating is the basis of most lithographic platemaking, enabling the image to be formed on the plate photographically. Plates are supplied ready coated by the manufacturer – these are called pre-sensitized.

Pre-sensitized surface plates

These plates are the most common now in use. They are produced using a number of different base materials including plastic, paper and aluminium. Various surface treatments can be applied to aluminium to modify its performance as a non-image metal, the most common being that of graining. The detailed nature of the grain depends on the method of producing it. Many plates are now available which have a hard porous layer formed in the aluminium by anodizing. These plates give increased wear resistance with easier handling on the press.

A light-sensitive coating is applied to the base material by the plate manufacturers. Several types of pre-sensitized coating are now used, the two main ones being diazo and photopolymer. These can be either positive or negative working.

For negative working plates the coating becomes insoluble during exposure to light and forms a bond with the plate base, unexposed (non-image) areas are removed with a developing solvent. With positive working plates the coating becomes soluble upon exposure and is removed during subsequent development. In both cases the coating which remains forms the ink-receptive image area. Production of the plate is fairly simple but some special equipment is required. The film negative or positive is positioned on the plate and held in contact in a vacuum frame during its exposure to UV light. After exposure the plate is developed, usually in an automatic plate processor which applies developer and cleans the plate. Print runs of up to 250 000 impressions are possible from certain of these plates. Some can have their press life extended by baking the plate image after processing, making runs of up to 1 million impressions possible.

Other types of plate

Wipe-on plates

These are supplied uncoated, the light-sensitive resin coating is applied by the platemaker prior to use. The main advantage to be gained compared with pre-sensitized plates is their low material cost. The press life is not normally as good, 100 000 impressions being the maximum obtainable

Deep-etch plates

Deep-etch plates are so-called because the image forming layer is very slightly below the surface of the non-image base metal. They are positive working and require long processing times. Their use has declined sharply.

Multi-metal plates

These have both the image and non-image areas composed of metal. A grease-receptive metal (e.g. copper or brass) is used for image areas, while metals which have an affinity for water (e.g. aluminium, stainless steel and chrome) are used for non-image areas. Only two metals are used for the lithographic surface, but a third metal may be used as a base support.

Multi-metal plates give high quality and a life of up to 2 million impressions. They are expensive but are suited to long runs or repeat work.

Electrostatic imaged plates

These employ photo-conductive materials to produce the image. Photoconductors have the ability to hold an electrostatic charge until exposed to light.

When the charged photo-conductive coating is exposed to an image the charge remains in image areas (black) but is dissipated in non-image areas (white). The coating is then developed with toner (powder or liquid) which has an electrostatic charge of the opposite polarity.

Print runs of up to 100 000 impressions are possible on aluminium-backed plates and makes them suitable for some newspaper printing. The quality of reproduction will depend to a great extent on the equipment which is used. The process is not normally suitable for large areas of solid or halftone screen rulings greater than 34 lines/cm (85 lines/in).

Chemical diffusion transfer plates

Silver salts which form the image are transferred from one surface to another (paper negative to plate) under the pressure of rollers in the presence of developer into which the salts dissolve. Diffusion transfer negatives are available in two speeds, slow for contacting and fast for projection.

The process is primarily intended for text and line copy although screen rulings up to about 40 lines/cm can be reproduced. Print runs of up to 40 000 are possible from metal plates.

Photodirect plates

These employ a dimensionally stable paper base in roll or sheet form, impregnated with plastic on which is coated an emulsion composed of three layers:

- pre-fogged top layer with moisture-attracting characteristics but also possessing an affinity for litho inks when developed;
- a middle layer of light-sensitive silver halides in gelatine;
- a bottom layer containing a developer which will bring out the latent ink affinity of the pre-fogged top layer.

During exposure, light reflected from the white areas of the original passes through the top layer without affecting it, producing a latent image in the light-sensitive layer. No light is reflected from the image, therefore the emulsion is unaffected.

Using this system it is possible to produce a large number of plates quickly at very low cost. They are only suitable for print runs up to

10 000 therefore their ideal application is work where only short print runs are required from a large number of different masters.

Direct image plates

With these the image is applied by direct contact to the plate base, which may be aluminium, plastic or paper, using special crayons, inks, pencils or typewriter ribbons. Printing by letterpress or litho with a suitable ink may also be used to apply the image to the plate base.

The plates are cheap and quick to produce but are only suitable for short runs up to about 5000 impressions, where good print quality is not of prime importance.

The driographic plate

This is the plate used in the waterless process. The non-image areas of the plate are composed of a silicone rubber coating which has the ability to prevent the special inks used sticking to its surface. Plates are available, pre-sensitized and either negative or positive working.

In addition to the advantage of being able to print without water, the process is capable of printing exceptionally sharp dots with very good colour strength. The silicone rubber surface is rather fragile and care is required to avoid scratching. A temperature-controlled inking system is required to maintain the correct ink rheology.

A similar type of plate is made on the relatively new (and at present unique) Heidelberg GTO-DI press. In this case the plate is made by an electro-erosion process that takes place on the press. Image quality is lower than from a conventional plate but perfectly satisfactory for certain classes of work.

Laser exposed plates

The use of these is not yet widespread but is expected to grow rapidly. It has been possible for some while to expose paper and polyester plate materials in some laser imagesetters, thereby removing the need for intermediate films – Purup imagesetters, for example, have proved particularly suitable for this. However, two metal plates are now available and others are under development which have been specially developed for laser exposure – that is, they have high light sensitivity to give reasonably short exposure times, and spectral sensitivity matched to argon lasers. The following table summarizes the current plate types and laser systems available or in development.

Coating	Sensitivity (mJ/cm^2)	Laser	Power (mW)	Wavelength (nm)
Silver polymer	2	argon	5	488
Silver	7	argon	5	488
Electrophotographic	11	argon	10	488
		HeNe	10	633
		diode	10	780
Photopolymer	100	argon	50	488

For large size plates the availability of suitable imagesetters is still limited but this is expected to change shortly.

Platemaking control

For consistent results it is necessary to control image transfer from film positive or negative to plate. Some light spread always occurs when platemaking, resulting in undercutting or sharpening of the image. Thus, when working with positive film images, the effect will be to reduce slightly the size of the dots or line widths. In practice this effect is used to reduce the problems of dust spots and film or tape edges. Negative film images lead to some dot growth.

Control can be achieved by making use of special test elements which are printed down on every plate along with the rest of the working image.

Available control strips make use of three types of elements that are suitable for controlling image transfer.

- Micro-lines – Brunner, Fogra and Ugra – these are very sensitive, the resolving power of the plate needs to be established for correct interpretation.
- Small dot patches – Brunner, Fogra, Ugra and Gretag – these are less sensitive and less dependent on plate resolution.
- Continuous tone wedge – Gretag, Ugra and Stouffer – these provide a good indication of the level of, or change in exposure, but do not really indicate differences in image transfer.

Control in colour printing

In lithographic printing in particular, various aids have been developed to assist the press operator to maintain a given press condition so that consistent colour work is produced. In general terms, similar or equivalent techniques are applicable to other printing processes but have not been so well developed, and so the following description refers mainly to lithographic printing.

In litho printing in order to achieve consistent results when colour printing it is necessary to monitor and control the following:

1. dot gain;
2. solid density;
3. trapping of one colour on another;
4. grey balance (which should be correct if 1–3 are correct);
5. register.

Dot gain is the term used to describe the change in dot size which occurs between the image on the film positive or negative and that on the final print. There are a number of contributory factors to this change in size:

- the image transfer from film to printing plate (hence the need for control of platemaking procedures, use of light-integrating meters to control exposure, and use of control elements such as the UGRA wedge);
- the image transfer from plate to rubber blanket;
- the image transfer from blanket to paper.

The last two occur on the printing press and are in turn influenced by characteristics of the ink (especially viscosity), the blanket and paper. Dot gain is an inherent characteristic of lithographic printing (as is squash in letterpress and flexo) and not a fault. However, excessive and especially uncontrolled dot gain is bad practice.

Dot gain is normally measured using a densitometer to measure the density of tint patches of known original dot area. Density can be approximately converted to percentage dot area using the Murray–Davies equation:

$$\text{Area} = \frac{1-10^{D\text{t}}}{1-10^{D\text{s}}}$$

where D_t = density of tint and D_s = density of the corresponding solid.

Dot gain is then the measured area on the print minus the original area on the film. If this is done for a whole range of original dot sizes and the measurements plotted graphically the type of diagram shown in Fig. 2.9 results. The curve is intuitively of this general shape since a dot of 0% cannot grow and neither can a dot of 100% grow any larger. Clearly dots in the 50% region to start with have the greatest potential for growth, and on coated paper figures of 18% are quite typical. Thus a 50% dot on film can be expected to grow to a 68% dot on paper. On newsprint, apparent dot gain of around 30% is common. The curve is not generally smooth or symmetrical. Figure 2.10 shows a more realistic typical case, the kinks being caused by the halftone screen dot join up. It is important to measure and control dot gain. If it is known that dots are going to grow by a consistent amount between film and print, then a corresponding allowance can be made on the films, i.e. dot sizes are reduced on the films by an amount equivalent to the growth which will occur when printing. In the last few years a number of 'standards for lithographic printing' have been published, the single most significant component of which is the specification of dot gain. These specifications give the dot gain which should be induced in a proof so that it will be a reasonable representation of the average result from the production run. Making proofs with these amounts of dot gain can be difficult but special inks are available which will give the desired results on standard proofing presses.

A great step forward has been the progressive development of photo-mechanical proofing systems of which the DuPont Cromalin and 3M Matchprint products are the most well established, both based on photopolymer systems, and available to work from positive or negative separations. More recently, systems employing electrographic principles such as the Kodak Signature and Stork systems have been introduced, and the latest developments are systems which proof from digital image data, rather than from the separation films. The advantage of all these systems is that they provide a controllable and reproducible proof. The difficulty with the earlier proofing systems is that their capability to match different press characteristics is limited, and so they have been designed to produce a proof in the mid range of dot gain, say around 18% on a 50% tint. Some adjustment to this is possible, but not easily to simulate dot gains around 10% or less (which may be encountered on a newish

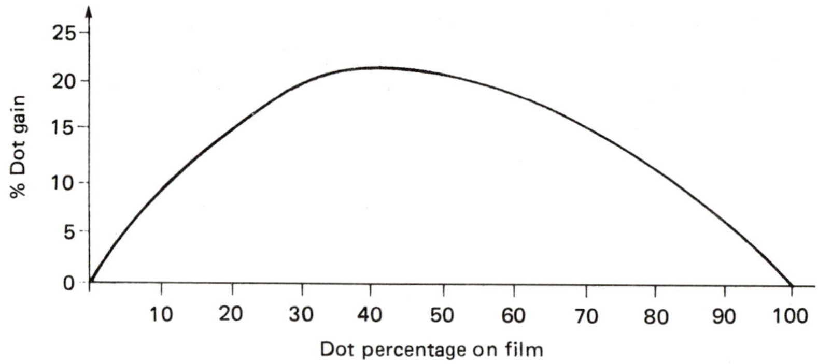

Fig. 2.9 Dot gain characteristic curve.

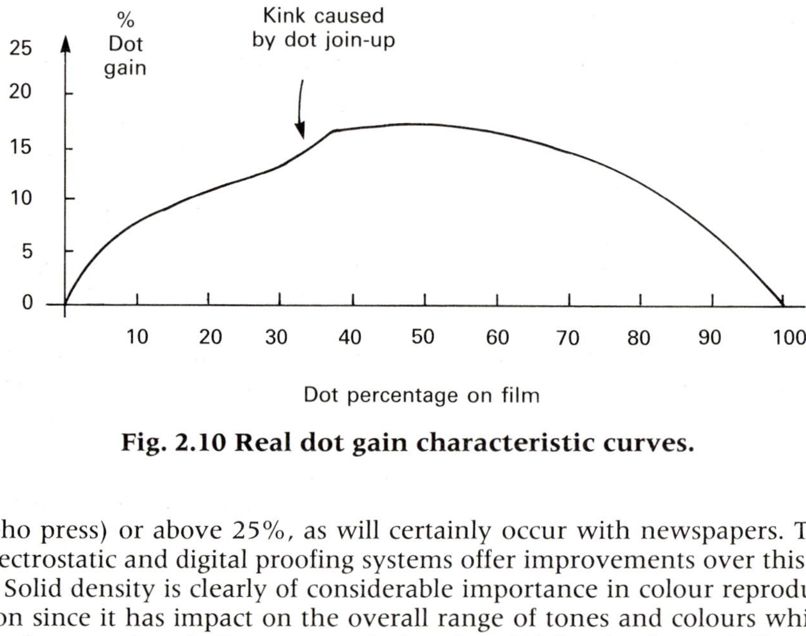

Fig. 2.10 Real dot gain characteristic curves.

litho press) or above 25%, as will certainly occur with newspapers. The electrostatic and digital proofing systems offer improvements over this.

Solid density is clearly of considerable importance in colour reproduction since it has impact on the overall range of tones and colours which can be reproduced. However, variations in solid density do not influence the appearance of the printed result to the same degree as variations in dot gain. There are various techniques for determining the optimum solid density for a given press/ink/paper combination (based, for example on maximizing print contrast, i.e. the optimum solid density is taken to be that which gives the maximum difference between a tint and the solid. Such methods taken alone, however, can be misleading). But the preferred approach now is simply to follow the recommendations of one of the recognized standards.

The trapping or transfer of one ink on to another, whether it be wet on dry or wet on wet, is seldom 100%. That is to say that for a given ink film thickness, ink transfer to unprinted paper is greater than to a previously printed ink film. This, in itself, does not present too great a problem. Difficulty is encountered, however, if:

- the trapping characteristics do not remain consistent;
- attempts are made to match prints on presses where the trapping characteristics are very different;
- the ink sequence is changed.

The most common example of this occurs when trying to match on a multi-colour production press (wet on wet), prints produced on a single-colour proof press (wet on dry).

Some of the factors affecting trapping are:

- ink film thickness;
- area coverage;
- ink rheology;
- properties of the paper, e.g. absorbency;
- time interval between printings;
- wet on wet or wet on dry.

Greys in a colour reproduction are made up from the correct proportions of yellow, magenta and cyan. Clearly, if one of these changes significantly relative to the other two a colour shift will occur – greys will no longer be grey, and in fact most colours in the picture will be incorrect. So maintaining this correct balance is a key task for the press operator.

Various faults can also occur, the most common being slur and doubling, which superficially appear as excessive dot gain. Slur is caused by a movement of the sheet or web relative to the blanket in the printing nip and distorts the shape of dots in one direction. It is generally caused by slightly incorrect packings under the rubber blanket. Doubling occurs in wet on wet printing when, in a subsequent printing unit, dots of the previous ink colour become transferred back on to the blanket, then back to a sheet slightly out of register. Slur and doubling must be eliminated before reliable dot gain measurements can be made.

To assist the press operator in controlling these various aspects of litho colour printing, colour bars have been developed (Plate 6). These contain elements which in the main can be checked visually, and show up the above faults particularly well. They are independent of the job being printed and so provide a constant reference point from one job to the next. It is common practice to include such colour bars on all proofs, where they are also checked rigorously using a densitometer, and where possible on the production run. This is not always possible in the latter case but usually some reduced version can be fitted in somewhere. This is better than no control element at all.

Successful colour printing depends on good transfer of information about the printing process to those involved in making colour separations. There are various well established techniques for measuring the printing characteristics of a printing press in such a way that the information is useful to a colour scanner operator.

In the halftone process the colour which is visually perceived from a combination of process colours can be influenced by the register of these colours to one another. In practice, however, these differences are only slight and do not generally have a significant influence on the final reproduction. What is of greater relevance is the effect which register has on the resolution and detail of the reproduction. The tolerance will vary depending on the class of work but will normally be in the region of ±0.05 mm between colours.

All of the press settings which control the passage of the sheet/web through the press will affect register, as will the condition and properties of the stock on which printing is taking place.

Ambient conditions can also contribute to the variations in register. Many web presses are fitted with automatic register controls which maintain register to close tolerances.

Ink duct pre-setting and control

Maintaining solid density within suitable tolerances on the run is not too difficult, particularly if a densitometer is used. However, in colour reproduction, control of mid tones is vital and more difficult. Again, a densitometer can be used to make measurements of the colour control bar but the time involved becomes excessive in relation to the press output. Consequently efforts have been made to automate some of this measurement and control task.

A typical litho ink duct consists of a roller with a blade pushed against it to meter the ink (Fig. 2.11). The pressure of the blade is controlled by an array of ink duct keys spaced at intervals of a few centimetres along its length. Control is achieved in two ways. Overal ink level is controlled by the angle that the ink duct roller moves through every time the oscillating transfer roller comes into contact with it. Control across the width of the press is achieved by using the ink duct keys. The objective in setting these is to match the ink transferred in each strip corresponding to each duct key with the demands of the image in that strip. This clearly varies from one job to the next and accurate setting is a quite difficult task and almost always involves an element of trial and error at the start of a run. For this reason, plate scanners have been developed (see Fig. 2.8(a)).

Attempts have also been made to mount densitometers on presses to measure the printed density of each colour across the width of the sheet and automatically adjust the ink duct. This has proved more difficult, partly at least because the printed density is also altered by the level of damping. Independent damp control systems have been developed based either on gloss measurement of the wet plate surface or infra-red absorption, but these have not been widely adopted.

Substrates and inks

The majority of sheet-fed litho inks dry by oxidation, aided by absorption on porous materials. When printing on non-absorbent substrates or where very rapid drying is required, it is necessary to use special equipment and inks which enable the drying to be accelerated. Heat-set inks are used on a large number of web-offset presses. In these inks the vehicle consists of resin dissolved in a solvent, and drying takes place principally by evaporation.

In printing with heat-set inks, the printed web is passed through driers

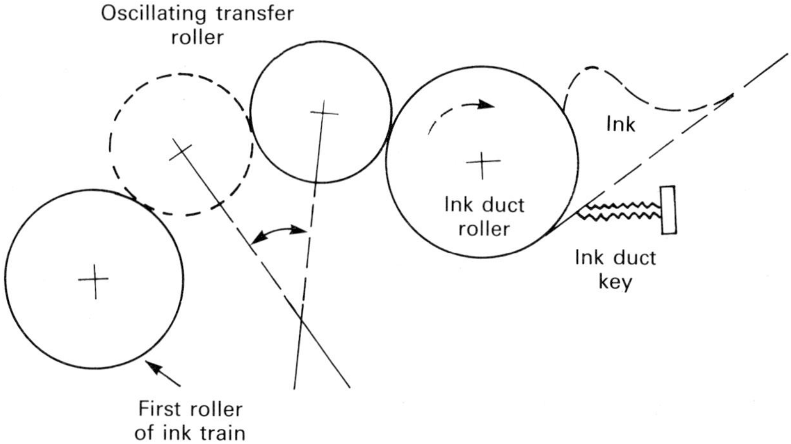

Fig. 2.11 Cross-section through simple ink duct on a sheet-fed press.

which originally used gas flames playing directly on the web. Now, however, forced hot air driers are more frequently used. These must raise the temperature of the web enough to cause evaporation of the solvent, leaving only the resin to bind the pigment into a film and to the paper. From the drier the web passes on over chill rollers to cool it, after which it can be cut, folded and delivered. The cooling serves to set the ink and is necessary to prevent set-off or marking in the delivery.

Ultra-violet (UV) curing inks

The UV ink drying system involves specially formulated inks which contain a photo-initiator. After printing, the sheet is exposed to UV radiation from lamps housed within the press, this causes a chemical reaction to take place, producing a completely dry ink film virtually instantaneously. This method of drying is used on some web-offset presses, but is most commonly found on sheet-fed presses printing cartons, where it permits cutting and creasing operations to be carried out immediately after printing. Some offset-litho presses printing tin-plate and plastics also use this method for drying.

Infra-red radiation

This is also used to dry inks. Again, specialized drying units are required but infra-red drying differs from UV curing in that specialized ink systems are not required. Infra-red drying is best considered simply as a machine improvement that generates heat in the print and accelerates the normal drying process.

2.2 THE FLEXOGRAPHIC PROCESS

Flexography is a process in which the printing image stands up in relief. A liquid ink is used which may be solvent-based, and dries mainly by solvent evaporation. Water-based inks are also widely used, and UV-cured systems are being introduced.

A low printing pressure is essential to the process because of the combination of very fluid inks and soft, flexible printing plates that are used. The process has several distinctive features.

- Liquid inks are used that dry rapidly by solvent evaporation, thus enabling fast printing speeds to be achieved on non-absorbent materials such as films and foils.
- 'Soft' and flexible relief printing plates are employed that can be mounted and registered on a plate cylinder away from the printing press. Proofs can also be obtained. Individual plates can easily be changed or repaired, and a portion of a plate can be removed to enable items such as price or expiry date to be changed.
- The application of ink to the surface of the printing plate is by means of a screened (Anilox) roller. The result is a simple ink feed system that consists of not more than two rollers, or perhaps a single roller and doctor blade(s).
- Although most flexographic printing is reel to reel, the machines enable changes in the print repeat length to be made simply.

The printing unit

The printing unit consists of three basic parts (Fig. 2.12):

- the inking unit;
- the plate cylinder;
- the impression cylinder.

The function of the inking system is to meter out a fine and controlled film of liquid ink, and apply this to the surface of the printing plate. It

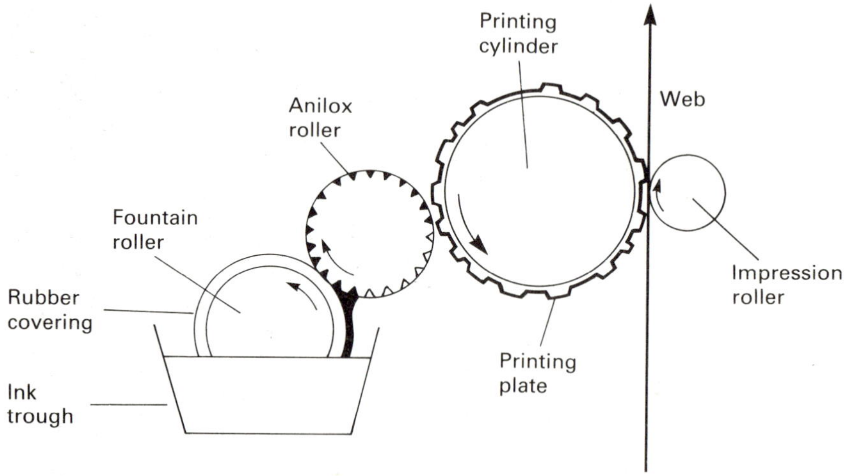

Fig. 2.12 Flexographic printing unit – ink doctoring by roller.

typically consists of an ink trough, a rubber-covered fountain roller, and a screened (Anilox) inking roller into which cells of uniform size and depth are engraved. The fountain roller lifts ink to the nip position, where it is squeezed into the cells in the screened inking roller and by a shearing action is removed from the roller surface. The ink in the cells is then transferred to the surface of the printing plates. To regulate ink film thickness in printing, screened ink rollers are available which have screens of from 40 to 200 cells/cm. These may be engraved or etched metal or ceramic. The engraved cells are generally square in shape (although many other shapes are available now) with sloping side walls.

A particularly effective and controllable method of manufacturing anilox rollers is to laser-engrave a ceramic coating. This provides not only a very hard wearing roller but also great flexibility in the control of screen ruling, cell shape and cell profile – the latter being especially significant since it greatly affects ink release characteristics. On laser-engraved ceramic rollers the cell count and pattern is regular and a rounded cell profile is produced allowing ink transfer volumes to be predicted more accurately. The cells produced are capable of carrying 15% more ink than conventional rollers which in turn means that a 15% finer screen can be used. The angle of the cell wall is made more acute (26° instead of 45°) to reduce the effect of wear on its ink carrying capacity (Fig. 2.13).

When printing halftones, the cells per centimetre of the anilox roller needs to be about 3.5 times the halftone screen ruling. The number of cells and their size regulate the volume of ink transferred. Further regulation of the ink is achieved by varying the surface speed of the

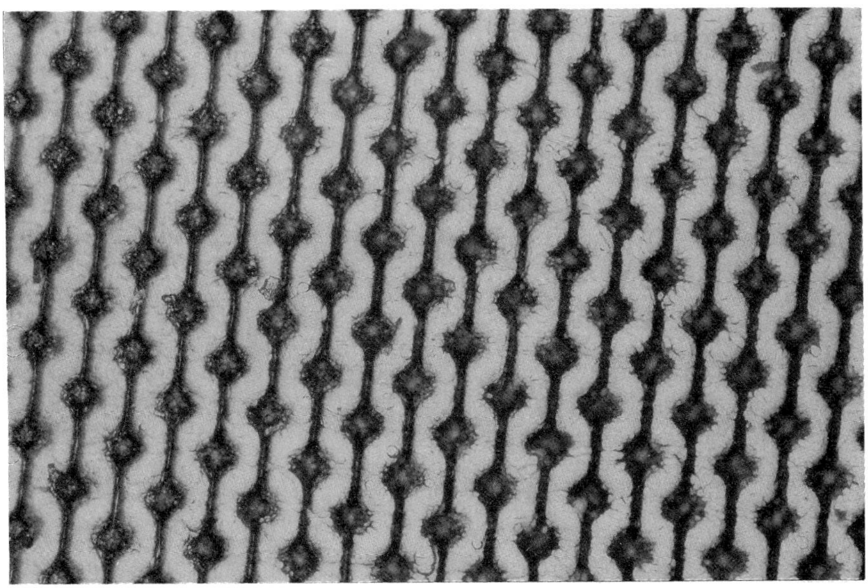

Fig. 2.13 Surface of laser-engraved ceramic anilox roller. (Courtesy of PIRA)

fountain roller, altering the pressure between the fountain roller and screened roller, and also altering the hardness of the rubber covering on the fountain roller. Despite these controllable factors it is still the basic characteristic of the anilox roller which determines the ink supply to the plate.

The anilox roller is a crucial factor in achieving good-quality flexo printing, and yet there is still much to be learnt about it. Ceramic rollers and chrome plated rollers have very different ink release characteristics, and supply different amounts of ink even when all other factors such as cell volume and screen ruling are the same (presumably because of different surface energies, etc.). Compensating for wear is a matter requiring careful attention, but in this respect ceramic rollers are a great improvement, lasting up to eight times as long as their predecessors.

For high-quality flexographic printing, reverse angle doctoring of the anilox roller has been introduced (Fig. 2.14). Steel blades were originally used but have largely given way to nylon, polyester or ultra-high molecular weight polyethylene (of thickness 0.004–0.012 in) which give less anilox wear. On some presses an ink chamber formed by a double doctor system is used which has the benefit of providing a totally enclosed inking system (Fig. 2.15). Reverse angle doctoring is not speed dependent to the same extent as the other methods, which is a distinct advantage.

The plate cylinder is usually made from steel. The printing plates, which have a thickness of up to a few millimetres are secured to the cylinder with two-sided self-adhesive material.

The impression cylinder is also made from steel. The substrate passes between the plate and impression cylinders, which generate light printing pressure. The ink is transferred from the cells in the screened ink roller to the plate surface, and then to the substrate, during which it reaches virtually a uniform film. For high-quality flexographic printing the components of the printing unit must be engineered to very tight

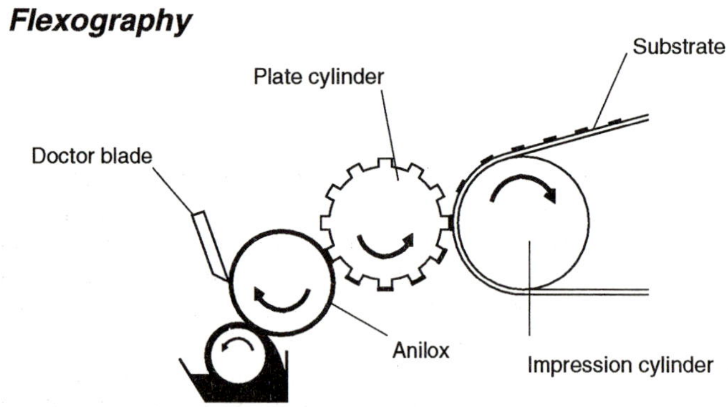

Fig. 2.14 Ink doctoring using a reverse angle doctor blade. (Courtesy of PIRA)

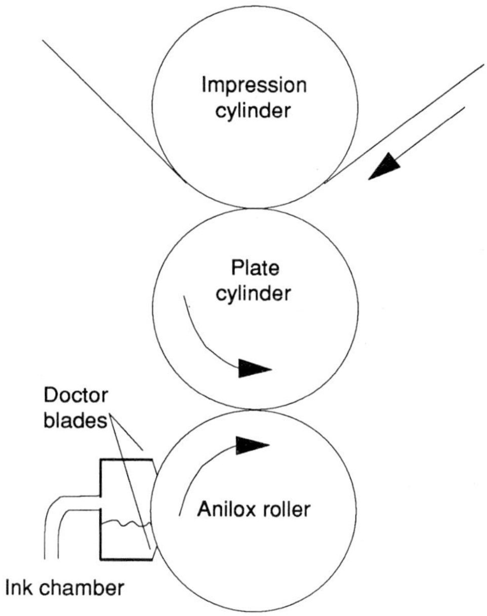

Fig. 2.15 An enclosed ink chamber system. (Courtesy of PIRA)

tolerances (measured in tenths of thousandths of an inch). The ability to manufacture to these standards is one of the factors which has contributed to the growth in flexographic printing, and its use for higher-quality products than was previously possible.

Press configurations

There are three basic types of configuration of flexographic presses:

- stack;
- common-impression;
- in-line.

The first two are the more common.

The stack press consists usually of two or three integral printing units arranged in vertical formation, and this is repeated to provide the number of printing units required. The stack press is mostly used for printing on paper of all grades, and is less suitable for printing on film. This machine enables reverse side printing on the web. Each printing unit has its own integral impression cylinder. The web enters the first upper unit and passes downwards through the units, where its direction is reversed to travel upwards towards the drying equipment. Where there is a second bank of printing units the web will travel through those to the drying equipment. Figure 2.16 shows that the web travels directly from one printing unit to the next being supported only from the non printed side, thus allowing six or more colours to be printed. Electronic register controls can be added to this type of press (and in-line) to enable higher-quality colour work to be produced (Fig 2.17).

The common-impression machine (Fig. 2.18) consists of a large cylinder around which are arranged either four or six printing units. The cylinder is very accurately made from steel. Usually the web enters the top unit on one side of the cylinder, travels to each unit with the cylinder, and emerges from the top unit on the opposite side of the cylinder. With the web wrapped around most of the impression cylinder well controlled register between colours, even with thin gauge extensible films, can be achieved. Most multi-colour work on extensible film that requires precise register is printed on common-impression machines. To improve further the capability in this area, temperature controlled drums are now used controlled to ±0.5°C.

The in-line machine which is a less popular configuration, consists of printing units arranged in horizontal formation, with the impression cylinder situated below the web, thus providing easy access to the plate cylinder. The web passes through each printing unit in a horizontal path. This configuration is used mostly for printing on lightweight board and less flexible materials that cannot be wrapped around rollers.

Many products printed by flexography are required in reel form for subsequent processing, and so machines provide suitably versatile winding equipment. Where necessary the printed web can be chopped into sheets in register. The machine also provides facilities for supporting and controlling the supply reel during unwinding. A variety of equipment is available for both manual and automatic splicing, and also infeed control.

An ink drying system, which usually blows hot air on to the web, is situated on top of the machine. This is to ensure that the inks dry rapidly enough to enable printing and rewinding at a suitable speed without

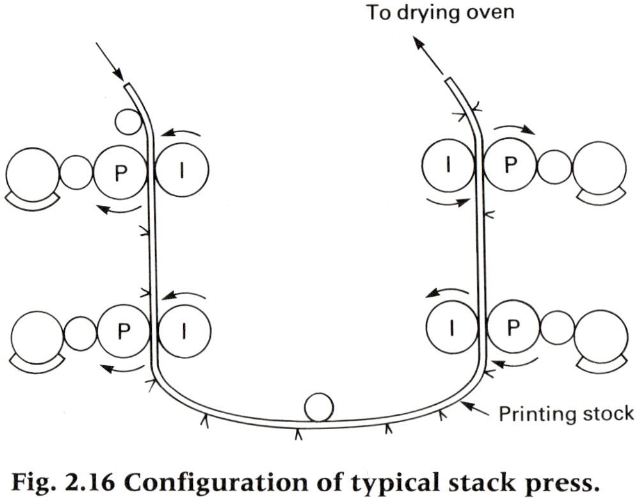

Fig. 2.16 Configuration of typical stack press.

Fig. 2.17 A stack press. (Courtesy of PIRA)

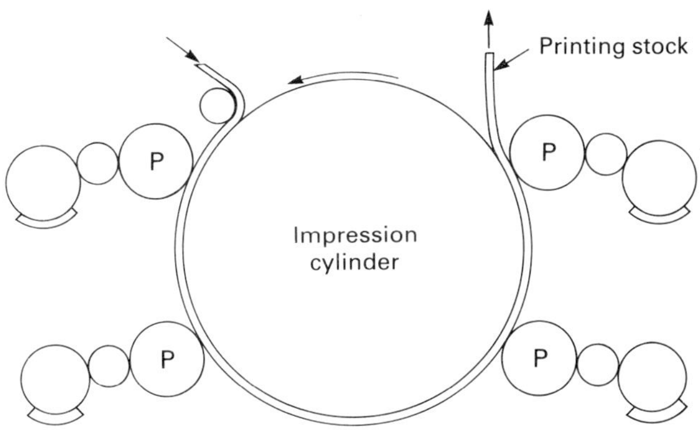

Fig. 2.18 Configuration of typical common-impression press.

set-off and sticking. Alternatively, heated drums may be used for the same purpose.

Most machines feature rewinding equipment which employs either a centre or surface winding principle. Much development work is currently taking place in connection with flexographic printing. The process has been adapted to newspaper printing, electronic pre-setting and control systems added, and new types of ink (such as water based and UV-cured systems) and inking systems are being introduced.

Flexographic platemaking

There are several kinds of image carrier in flexography:

- the traditional rubber plate;
- photopolymer plates;
- laser-engraved rubber plates or rubber rollers.

Rubber plates

The rubber plate is essentially a duplicate plate obtained from an original. The original may consist of various type elements such as, hand-set foundry type, Monotype or Linotype to provide the type display, and zinc plates to provide the line and halftone image. These are assembled and locked up in a steel chase for moulding. Alternatively, the type matter can be set on phototypesetting equipment to provide a negative and line image negatives combined with this to produce a one-piece zinc plate of the entire printing image. A one-piece plate can also be electronically engraved, which can be originated by hand-drawn artwork or photographic print. This equipment will engrave line and halftone images. A mould is required to produce a rubber plate. The mould is made by pressing a heat-setting matrix material against the original under heat and pressure in a hydraulic moulding press. The thermosetting resin in the matrix material is initially softened by the heat to take the impression and then the material becomes hard and rigid in approximately 8 min at a temperature of 150°C.

The rubber stereo is made by placing a layer of uncured rubber over the mould and pressing these together under heat and pressure. The grades of rubber used are usually cured in approximately 8 min at a temperature of about 150°C. The raised areas of the stereo, which are the image areas, correspond to the image areas of the original. The back of the stereo has to be ground to achieve a suitable thickness. Rubber plates are still widely used, especially for line work, and in situations which demand unusual solvents in the inks. Rubber is available in a range of hardnesses to suit the substrate.

Photopolymer plates

There are various photopolymer plates suitable for flexographic printing. These plates are made directly from photographic negatives. They are available in various thicknesses, hardness and with different backing materials. Some plates are made from two or more layers, providing a hard thin surface, and more resilient supporting layers. Ultra-thin (less than 1.7 mm) plates are said to give improved consistency and lower dot gain.

The plates consist basically of a light-sensitive photopolymer (which may be liquid or solid before exposure), which when exposed to UV light through a photographic negative polymerizes and becomes resistant to the washing-out medium. The non-image areas of the plate which have not been exposed to UV light and are therefore not polymerized or resistant, are removed in a plate-processing machine. On completion of the plate processing, the plate is usually ready for printing. The washing-out or removal of the non-image area of the plate is done by the use of either solvent, an aqueous solution or air. These plates have suitable thickness accuracy, and do not require back grinding. Plate technology has advanced greatly in recent years, the main features being:

- the introduction of softer plates of Durometer 30–35 Shore A hardness;
- the introduction of larger plates up to 1524 × 2032 mm (60 × 80 in);
- the use of thinner plates, below 1.7 mm (0.067 in). Recently, plates which are only 1.14 mm (0.045 in) thick including a 0.22 mm (0.0085 in) antihalation base mounted on suitable rubber 'support' rollers have been shown to produce very significant reductions in dot gain;
- the use of alternative wash-out solvents that are more environmentally friendly;
- the introduction of water-washable solid plates;
- the widespread use of liquid photopolymers;
- the introduction of an integral foam layer between the polymer and the polyester base to give increased tolerance to substrate thickness variation and any errors in the machine;
- the introduction of ozone-resistant plates;
- improved surface tension properties of the plates to give more predictable printing results.

Rubber suitable for flexographic printing can be engraved by laser techniques. The equipment will handle black and white positive copy for line work, and screened negatives or positives for halftone work. Screen rulings of 47 lines/cm (120 lines/in) are possible, and is expected to improve to 60 lines/cm. Engraving by this method can be done on either separate pieces of rubber, or rubber rollers. The ability to engrave rollers is unique, and an advantage in the printing of continuous designs.

Because flexographic printing is done from an image in relief it is essential that the shank of the image has a steep angle and is smooth. A suitable depth in the non-image area is also essential.

Plate mounting

Flexographic printing plates are usually secured to the printing cylinder by means of two-sided self-adhesive material. The plates are mounted on the plate cylinder and pre-registered in position on special equipment designed for this purpose. Several plate cylinders are normally available for one machine to enable pre-mounting of plates. This reduces the unproductive time on the machine to a minimum. A 'cushion-back' adhesive layer behind the stereo is sometimes used to compensate for any inaccuracies in the plate or press. However, in halftone printing this can lead to greater enlargement of highlight dots than would otherwise have been the case. The current best practice is considered to be to use the thinnest possible adhesive tape since the potential for introduction of thickness variation is then minimized.

Control of flexo printing

As previously mentioned, the anilox roller is a crucial element in producing good print, but assuming it has been appropriately specified and is in good condition, there are several other factors the pressman must attend to. Foremost amongst these is ink viscosity (which is, of course, temperature dependent). The correct viscosity will depend on the anilox roller, plate material and substrate, but is usually 20–25 s using a Zahn 2 cup. Viscosity is often controlled automatically by the ink circulation system, by adding more solvent when required. Of course, this also dilutes the pigment concentration in the ink, and so some other compensation may be required. It is usual to run inks with a proportion of medium in them (medium is unpigmented ink) so that the pigment concentration can be increased and decreased. For high-quality halftone work, it is good practice to run at the highest possible viscosity compatible with good ink lay. In this way dot gain is minimized, which is so important to achieve good highlight rendition.

Typical flexo press speeds are 7.5–10 m/s, although narrow web presses for label printing will commonly go much slower at around 3–5 m/s. Flexo printing normally produces much less waste than, say, litho, at around 5–8%. Much attention has been given to reducing make-ready times, with one press actually working in a non-stop fashion, the changeover occurring with the wastage of only 9 m of material.

Applications

Flexographic printing is widely used for package printing, including labels, particularly on plastic films. To print on plastic films, surface treatment is required, usually by corona discharge. For a wide range of materials this raises the surface energy to around 40 mN/m, although this value decays with time. Consequently, for repeatable results, the time between treatment and printing should be constant, and one way of achieving this is to do it on the printing machine. However, pre-treated films are available, which should be used within six months.

A wide range of paper-based products such as corrugated cardboard, for example, are also printed by this method. It is being used for some newspaper printing. The quality of flexographic printing has risen substantially in the last few years, enabling it to produce work that would otherwise have been printed gravure or litho.

2.3 THE GRAVURE PROCESS

In the gravure process the printing image is engraved into a cylinder in the form of cells which become filled with ink. Printing is achieved by passing the substrate between the gravure cylinder and an impression roller under pressure.

The printing unit

The printing unit of a gravure press consists of an ink duct in which the etched cylinder rotates in a fluid solvent-based ink. A metal doctor blade, which reciprocates from side to side scrapes excess ink from the cylinder surface. The substrate is fed from reels into a nip between the engraved cylinder and a rubber-covered impression roller which supplies the pressure needed to transfer ink from the cells to the substrate (Fig. 2.19). The printed web runs upwards through a heated drying system where the solvents are evaporated and extracted, and the ink is thus dried. In gravure printing each colour must be nominally dry before the succeeding colour is printed over it, therefore each printing unit has its own integral drying equipment. The ink which is usually stored underneath each unit, is pumped up to the ink trough and continuously circulated, and usually viscosity control is incorporated in this system. Because each printing unit has an integral drying system and impression roller, most presses consist of units arranged in line, where the web travels between units in a horizontal plane. As the impression cylinder is not gear driven, but obtains its drive through contact with the gravure cylinder, cylinders of different size can be used to provide variable print repeat dimensions within certain limits.

The inking system

The oldest and simplest method which is still widely used is to place the cylinder in the ink trough. This has its deficiencies, however, since the ink is not well agitated, solvent is readily given off and it is unsuitable for high speeds since the cylinder tends to carve a trough in the ink.

An alternative uses transfer rollers. The roller does not actually touch the cylinder – there is a gap of about 0.15 mm (0.006 in). This method was used on some gravure machines printing from plates rather than cylinders.

Enclosed duct designs were introduced in 1924 and meant that more volatile solvents could be used which contributed to a rise in press speeds. Further refinement using the weir system provides the ability to run up to about 15 000 revs/h. A circulation pump provides continuous agitation, and viscosity control is normally also a part of such a system. The whole arrangement is enclosed to avoid solvent losses. The pump takes the ink from a large tank and delivers it to a small trough in which the cylinder actually sits. Excess ink overflows through a sieve back into the large tank.

A further development of this is the spray system which can be used at even higher speeds. It is basically the same as the weir system except that the pump delivers the ink to nozzles pointing at the cylinder. It has the advantage that the cylinder surface never has the chance to dry out. Again, the system is totally enclosed.

Another system, instead of, or in addition to, spraying the cylinder, pumps the ink upwards through an open-mouthed jet directly under the doctor blade or a pre-metering blade. This is used at speeds of 25 000 revs/h.

Several variations on the above systems have also been used. (Gravure inks are described in Chapter 8.)

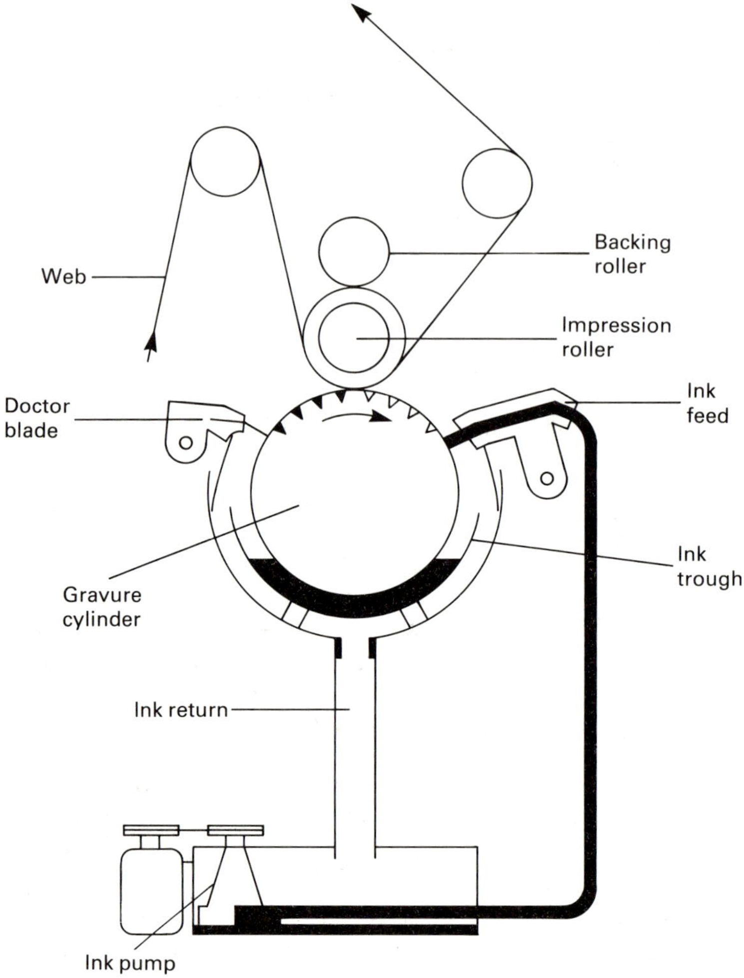

Fig. 2.19 A gravure printing unit.

Doctor blades

The function of the doctor blade is to remove surplus ink from the surface of the cylinder leaving the ink in the cells. There are many possible configurations for the doctor blade and they have an effect on the printed result. The thickness of the blade is generally 0.15–0.25 mm (0.006–0.01 in) and it is made from high-carbon flexible steel. Doctor blades are usually supported by a backing blade to give extra support (say 0.76 mm or 0.03 in thick).

Blades can be sharpened by hand but it is more usual, especially with wide presses, to use a special grinding machine, or blades which are supplied pre-sharpened. Blades are ground with a bevel edge and the angle of bevel is one of the factors influencing the printing results from the blade. A blunt blade will leave more ink in the cells than a sharp one,

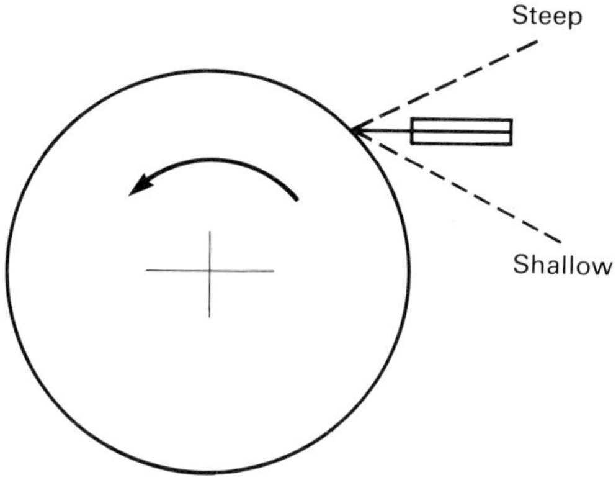

Fig. 2.20 Doctor blade angles.

and possibly not give a clean wipe of the non-image areas. Similarly, the angle which the doctor blade makes with the cylinder affects the ink left in the cells. A steep angle gives a cleaner wipe (Fig. 2.20).

Pressure must be exerted on the doctor blade or it will be forced up by the ink being pushed up underneath it. This can be done in several ways.

- Using a screw or air pressure – this has the effect of locking the blade rigidly in position. Cylinders must be perfectly cylindrical for this to work well but it is ideal for high-speed presses.
- Using weights – the advantage of this approach is that the blade will follow an uneven surface. It is used especially on plate machines.
- Using springs – this has similar advantages to weight systems.

Doctor blades are normally made to reciprocate by up to 6 cm. This gives a better wipe and disperses paper fibres which may get trapped under the blade. Blade mountings must have adjustments to cope with different sizes of cylinder and also movement for making the blade exactly parallel with the cylinder axis.

High-speed presses may be equipped with a pre-doctoring system. This has one doctor blade which allows an ink film of 0.5 mm (0.02 in) to remain, then another which performs the final doctoring. This has the advantages that the pressure on the second blade can be substantially reduced so that cylinder wear is less, and printed results are less affected by speed.

The impression roll

This has a steel core with a rubber covering 12–20 mm thick. It is a relatively hard rubber (60–100 Shore A). The rougher and more compressible the substrate, the harder the roller that is used. Plastic films are normally printed with a soft roll and relatively low impression pressure. In general, the pressure applied between the impression roll and the printing cylinder is high in relation to other processes which is one reason why the side frames of gravure presses are so substantial.

On wide printing machines, such as those used for publications, the impression roller can deflect, giving rise to low impression pressure in the centre. To overcome this, it is common to use a third roll to back up the impression roller. Another technique is to use a so-called 'flexible' roll which can be adjusted to even out the pressure across the width of the web.

Gravure printing frequently suffers from speckle, caused by individual cells not printing on 'rough' papers. In this context it is the smoothness of the substrate under pressure which matters and consequently an uncoated, but compressible paper may well print better than a coated one which has a less compressible surface. In an effort to overcome this, electrostatically assisted ink transfer was introduced. A special impression roller made with an electrically conductive rubber is used, and in the common version of this equipment the inside of the roller contains an electrical generator. The turning action of the roller causes a high voltage to be generated (e.g. 1000 V). The resulting electrical field encourages the ink to leave the cells and transfer to the paper even where contact is imperfect. Very marked improvements in the printed result are possible.

Drying system

Slow-speed sheet-fed machines may not have any drying system other than an extended delivery transport but some form of cold air blower is more normal. High-speed web machines demand more sophisticated driers between each unit. A web tends to carry a solvent layer with it, and consequently high-velocity hot air is used to break up this layer and remove the solvent. Steam-heated drums may be used to heat the web, and cold chill rollers are used on exit from the drier to cool the web. Some driers are split into sections so that highly saturated air is taken away first (to solvent recovery) while other sections recirculate the air thus conserving heat.

Press configurations

Most gravure printing is done on web-fed presses, which provide facilities for supporting and controlling the supply reel during unwinding. A variety of equipment is used for both manual and automatic splicing. An input feed control system is used to provide stability of web movement to the first printing unit.

Suitable folding equipment on the end of the machine is essential to produce a folded product such as magazines. For package printing where the printed product is required in reel form, rewinding is necessary, which employs either a centre- or surface-winding principle. Sometimes cartons are gravure printed on web-fed presses and are 'flat' die-cut in line with the printing operation. Web-fed presses will also produce a sheeted product that registers with the printed image. There are sheet-fed presses that print from solid cylinders, or plates fastened around a cylinder. Most fast running web-fed presses producing items such as magazines, cartons and packaging products on film are printed from hollow cylinders. With thin gauge film, delicate control of tension is essential, and often the tension in winding is graduated. With centre winding a lay-on roller is used to expel air between layers and most of the rollers in the machine are chevron engraved to smoothe the film in the cross direction. Electronic control of register between colours is usually employed, and scanning equipment is provided to enable inspection of the web during the printing operation.

Gravure cylinder preparation

A gravure cylinder is generally made from a steel tube (although solid ones are also used) which is copper plated by electrolytic deposition from an acid copper sulphate bath. Additions to the plating bath are used to produce a surface finish which is almost ready for subsequent processing. Since it is not possible to electroplate steel directly from such a bath, a new steel cylinder is first lightly plated either with copper from a cyanide bath or with nickel. The cylinder is then transferred to the copper sulphate bath and grown to the appropriate size.

The base cylinder is obviously costly and so is reused. This can be done by turning off the image on a lathe and then growing the cylinder back to size by electroplating more copper. A more convenient and cheaper process is to use the Ballard skin technique. In this method, the base copper cylinder is grown to a few thousandths of a inch under the required diameter, and a layer of silver is then deposited followed by the rest of the copper. After use the silver layer allows easy separation of the copper skin on top, which can be pulled off by hand. The cylinder is cleaned, a new silver layer deposited and the process repeated for the next job.

The production of a good cylinder is a very exacting process requiring close control of the constituents of the electroplating bath (which employs special additives to give an even deposit), the current during plating, and the configuration of the anodes. For mechanical engraving it is vital to obtain a deposit which has uniform properties across the entire width.

There are several different ways of producing the image on a gravure cylinder.

Conventional etching

This was the original method of producing a gravure cylinder, and is capable of producing very fine results – almost continuous tone in nature. However, it is difficult to control, and has been largely superseded.

The image to be printed is transferred to the cylinder by means of a material called carbon tissue or pigment paper. Carbon tissue is a gelatine coating containing a light yellow/brown pigment on a paper base. The coating is made sensitive to light (peak sensitivity in the UV) by immersion in a solution of potassium dichromate (2–3% by weight in water) for 2–3 min. It is dried at room temperature in contact with glass or chrome glazing plates. When dry it can be lifted off and is generally stored in a freezer for 7–10 days before use. For repeatable results the temperature, concentration and replenishment of the dichromate bath must be closely controlled.

When exposed under a positive of the subject to be printed, the gelatine coating 'hardens' (i.e. becomes insoluble) to a depth depending on the amount of light received. Thus in shadow areas very little hardening occurs but it becomes progressively deeper as the tones become lighter. A gravure positive generally has a density range of approximately 0.35–1.65. The exact densities, once chosen, are very closely adhered to.

A second exposure is made under a gravure screen (Fig. 2.21). This screen is not used to produce a halftone dot pattern, but only to split the image up into tiny cells to produce a support for the doctor blade. It consists of a regular pattern of black squares, the width of the square being 2.5–3 times the space between them (i.e. the cell:wall ratio is said to be 3:1).

The exposed tissue is transferred to the copper-plated cylinder by rolling it into contact while feeding water into the nip. The gelatine layer adheres to the copper. The base paper is drenched in industrial methylated spirits, and then soaked in hot water at 45° C until it can be lifted off.

Development of the remaining gelatine layer continues, washing away the unhardened coating, leaving a relief image of hardened gelatine corresponding to the tones of the photographic original. The cylinder is finally drenched in industrial methylated spirits again, wiped dry with a squeegee and possibly blown dry. It is allowed to adjust to room temperature (Fig. 2.22).

Etching is carried out with ferric chloride solution through this gelatine layer. The etch penetrates most rapidly through thin gelatine, giving deep cells in the shadows, and shallow cells in the highlights (Fig. 2.23).

Etching is a skilled operation. It has to begin with a strong solution of etch (conventionally measured in degrees Baume; 43–45 degrees Be would be the starting concentration) which causes the etching to start only in the shadows. The etcher will judge when to change to weaker solutions (down to perhaps 35 degrees Be) to start the etching of the mid-tones and finally the highlights. Considerable skill is needed to

Fig. 2.21 A gravure screen.

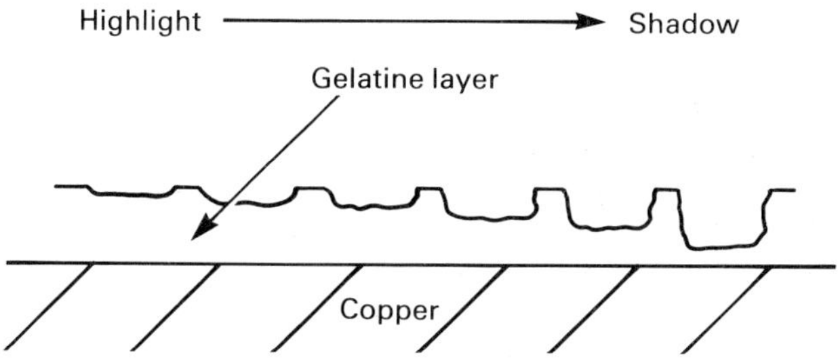

Fig. 2.22 Gelatine layer on copper cylinder after development.

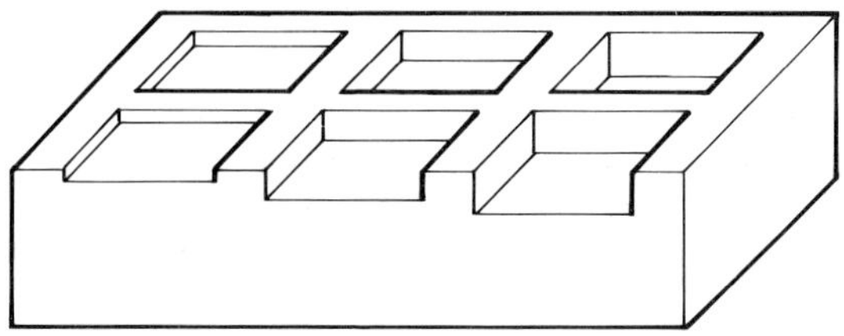

Fig. 2.23 The etched surface.

ensure that dark tones will have the correct depth of etch, just as etching starts in the highlights. At least five solutions of varying strength are used. The composition and temperatures of these solutions is critical. Usually they will have a certain amount of copper dissolved in them before use to give a smoother etch. Other additives may also be used.

The end result is a cylinder in which different tones are made up of square cells that are roughly the same area, varying only in depth. Deep cells carry more ink, and thus will print darker tones than shallow ones.

Single bath etching

In order to obtain more repeatable results than manual etching produces, single bath etching was introduced, and because only one strength of ferric chloride solution is used, is obviously easier to carry out with a machine. To be successful, times, temperatures and concentrations at all stages must be very closely controlled. Normally, a single strength solution of ferric chloride will not penetrate all the range of tones properly. However, the addition of ferric ammonium sulphate to the ferric chloride makes a single bath etch possible using a concentration of 38–40 degrees Be. In practice, completely repeatable results are almost impossible to achieve.

Cells vary in diameter with the screen size used; with a 60 line/cm screen the cells will be about 125 mm in diameter, the depth varying from about 2 μm in the highlights to 38 μm in the shadows. During etching, etching proceeds sideways as well as downwards. (There is no banking agent as in letterpress plate etching.) Consequently, the original cell:wall ratio of say 3:1 on the gravure screen changes to around 10:1 on the etched cylinder. Higher ratios are undesirable since the cell walls may collapse. The finished cylinder is usually chromium-plated for longer life.

Halftone processes

Double positive system–halftone gravure

This is similar to the conventional process except that instead of exposing the carbon tissue to a continuous tone positive and then a gravure screen, it is exposed to a continuous tone positive and a screened positive. The screened positive is made from the continuous tone positive by contact using a special contact screen and duplicating film (to yield a positive directly). The screened positive has dots of varying area, but ones which never join up completely because a cell wall pattern is always present. Etching is similar to before, using baths of ferric chloride of different strengths or a single-bath etching technique. The result is a cylinder in which different tones are made up of cells which vary in both depth and area.

The advantage of this technique is that etching is easier to control and the cylinder has greater tolerance to wear on long runs. This is because the highlight cells have smaller area but much greater depth so will not be affected by wear. Speckle can be a more prominent problem with this type of etch.

Another technique will produce cells which vary only in area and are all the same depth. Carbon tissue is not used and only a single halftone positive. A light-sensitive coating is applied directly to the cylinder, exposed to the positive and developed. Etching can be done in a one-bath etch. The Acigraf process is the most widely used process of this type although other methods based on powderless etching (as in letterpress platemaking) and electrolytic etching have been tried.

Mechanical engraving

This is a technique that was developed by the Hell organization in Germany, who have produced a range of Helio–Klischograph machines for several types of work such as consumer magazines and packaging. The main difference between the machines is that those for magazines – that is where the image is conveniently split up into pages – have several cutting heads, whereas those for packaging applications have only one. In the latter case it is quite possible that the image is continuous around and across the cylinder, and to create an invisible join between areas engraved by different cutting heads would be very difficult.

These machines, which are similar to a lathe in general layout, use a diamond stylus which is shaped in a very precise way, to engrave cells shaped like inverted pyramids, 2 to 50 μm deep. As the cylinder rotates the stylus moves in and out of the copper surface cutting between 2800 and 5000 cells per second, although 3200 cells per second is typical. For deeper cells the stylus penetrates deeper into the copper so that the area at the top of the cell becomes greater as well as the depth. Cells in adjacent rows are staggered by half a cell so that they nestle together, with a screen angle of 45°. For colour work the screen angle must, of course, be varied to avoid moiré patterning. This is done by altering the cylinder rotational speed in relation to the frequency of the cutting tool in a carefully calculated fashion. In this way, compressed or elongated cells are produced with screen angles at 30 or 60°. The composition of the copper on the cylinder is important to achieve a consistent cutting action. If the copper is too soft, the chips of copper produced by the cutting tool do not come out of the surface cleanly, and tear leaving splinters of copper on the cell edges. Variation in hardness can lead to variation in depth of cut, and hence cell volume. The best copper hardness is in the range 200 to 220 Vickers. The use of mechanical engraving has become widespread since, by comparison with etching methods, it is controllable, and also lends itself to direct output from computer pre-press systems. The inverted pyramid shape of the cells promotes good ink release and a lower cell volume can therefore print an equivalent density. Laser engraving methods have also been tried, but have so far failed to find wide acceptance.

Press control systems

In contrast to the difficulty of controlling the cylinder making processes, the printing operation is more straightforward, and particularly on web presses, automatic control systems for register, web tension and viscosity control have been available for years.

Viscosity control of the ink is more or less essential on high-speed presses. There is a definite relationship between the speed at which a press is running and the viscosity which must be used in order to achieve good printing results. If the viscosity is too high then the cell wall pattern will still be evident even in shadow tones and the print will lack smoothness. If the viscosity is too low, then bleeding will occur (spreading of the image from shadow areas).

Many modern presses are equipped with solvent-recovery systems. The hydrocarbon solvent-laden air from the driers is passed through large tanks of activated charcoal. The solvent is adsorbed on the charcoal. When saturated, another tank of charcoal is brought into line. Steam is passed through the first tank and this drives the solvent off. When condensed, the solvent and water separate and can be pumped off separately. Recovery of 75–90% is possible. In the packaging industry, the use of alcohol/ester inks precludes the use of activated charcoal and other forms of solvent recovery are used.

Web tension is now generally automatically controlled but on less sophisticated presses this is not always so. On some multiunit presses it is necessary to graduate cylinder sizes through the length of a machine so that each one is just slightly larger than the one in the previous unit. This ensures that each unit is very slightly pulling the paper from the previous unit, so satisfactory web tension is maintained.

Substrates and inks

Very fluid, solvent-based inks are used to print gravure, which dry by evaporation, leaving a dry film of resin and pigment on the substrate. (For details of the raw materials used see Chapters 4 and 8.) Evaporation is accelerated by moving air and by heat, so drying units are built on to each unit to speed up the drying.

The advantages of solvent-based inks are as follows.

- Rapid drying can be obtained on non-absorbent materials such as plastics films and metal foils.
- Converting operations can be done in-line.
- Solvents can be chosen to key on to a wide variety of different surfaces, and to dissolve film-forming resins that will impart specific properties to the inks.

However, environmental concern over releasing solvents into the atmosphere is resulting in efforts to use water based and UV-cured inks.

Applications

Gravure printing is used primarily for long runs. Many magazines in Europe and the USA are printed by gravure. The quality of the colour printing and the low waste of the printing operation are attractive features.

Much package printing of cartons, plastic films and metal foils is printed by gravure. It is attractive since the rapid ink drying makes in line finishing processes possible. Gravure is also used for a number of special purposes. Many postage stamps, reproductions of oil paintings and even some telephone directories are or have been printed by gravure.

In the field of printing magazines, mail order catalogues and travel brochures, there is strong competition between gravure and heat-set web-offset printing. No universally accurate guidelines can be given as to what is most appropriate but the following issues require consideration.

- Run length – while the economic minimum run length for gravure has been reducing, so it has also for heat-set web, and consequently it is usually the case that heat-set web is preferred for shorter runs. However, the breakeven point between the processes may be as low as 150 000.
- Print quality – on good quality paper there is little to choose between the processes. However, as the paper quality reduces, so the balance becomes in favour of gravure which will provide a good result even on uncoated papers.
- Consistency – gravure, once running well will produce a stable quality product fairly easily. The press operator has only to ensure that no faults are developing and he can be certain that a consistent product is being produced. Heat-set web-offset presses, however, require constant attention, particularly during early phases of a run when the ink/water balance is relatively unstable and temperatures are rising, giving changes in the way the ink behaves.
- Waste – paper waste levels for web-offset printing are typically 8–15%, but very much lower for gravure.
- Time to press – in general, the time from producing final page films to press running can be substantially shorter with heat-set web-offset than with gravure (minutes rather than hours).

2.4 THE LETTERPRESS PROCESS

The letterpress process requires a relief printing image carrier. That is, the areas to be printed are raised above the blank parts, i.e. non-printing areas. A film of ink, having sufficient tack and body to adhere to the raised image surface, is applied to it using rollers.

The substrate to be printed is brought into contact with the inked surface and sufficient pressure is applied to transfer the ink from the face of the image carrier to the surface of the substrate. In order to achieve a good and even transference of ink the pressure must be uniform, in proportion to the area, over the whole image surface.

Press configurations

There are three main types of letterpress press:

- platen;
- flat-bed cylinder;
- rotary.

The platen press

A platen press achieves impression by bringing two flat surfaces into contact. The forme and impression surface may be either in a horizontal or vertical plane. There are two types of platen press: the clamshell in which the platen rocks up against the forme, and the sliding platen in which the platen is first positioned parallel with the bed and then drawn up against it (Fig. 2.24).

With this method the whole of the forme is under impression at the same time. As it requires a considerable force to transfer ink from a forme to paper clearly and smoothly, it naturally follows that machines of this class are strictly limited in size (less than A3).

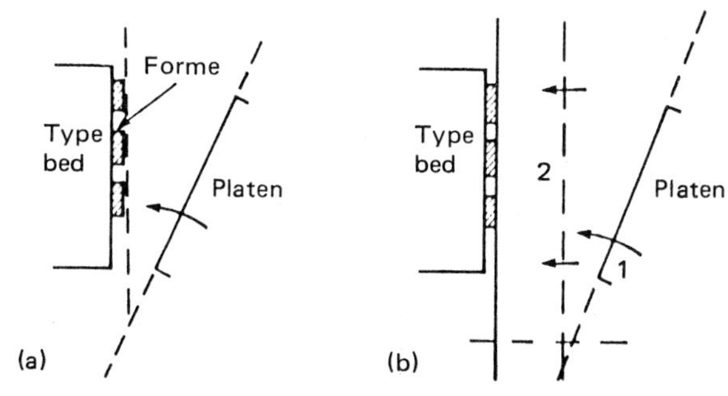

Fig. 2.24 Platen presses: (a) clamshell type; (b) sliding platen type. (1) Platen moves to vertical position, (2) parallel approach to forme.

Flat-bed cylinder press

Flat-bed cylinder machines have the forme placed on a flat bed, usually but not necessarily in a horizontal position and the paper is fed round a cylinder. The rolling action results in only a small portion of the forme being under impression at one time, so much larger machines can be successfully used. The impression cylinder usually makes two revolutions for each impression. It prints while making the first revolution and is raised during the second to clear the forme and permit inking on the return stroke. The type bed is driven back and forth by what is basically a crank mechanism. This would cause the bed to vary in speed during the impression stroke (giving changes in ink transfer) and consequently ingenious gear mechanisms are used to avoid this. Some presses only make one revolution for each impression. This is achieved by using an impression cylinder with a circumference larger than the print length. The type bed moves at high speed back to the start position during the passage of the non-printing part of the impression cylinder. This is cut away to clear the type bed, so it does not have to be continually raised and lowered (Fig. 2.25).

Rotary presses

Rotary presses use two cylinders, one carrying a curved plate specially made to fit the particular cylinder and the other having the same circumference as the surface of the plate. As both these cylinders are continuously revolving, with the paper between them, fairly high speeds of around 8 m/s can be attained.

Keyless inking

A significant variation of the basic letterpress process was developed during the 1980s and is now used for the production of some newspapers. The main difference is in the inking system. This uses an etched roller rotating in a trough of ink, the excess being removed by a reverse angle doctor blade. The ink film is transferred to the printing forme by one or two rubber rollers, typically of 30–35 Shore hardness when using hard

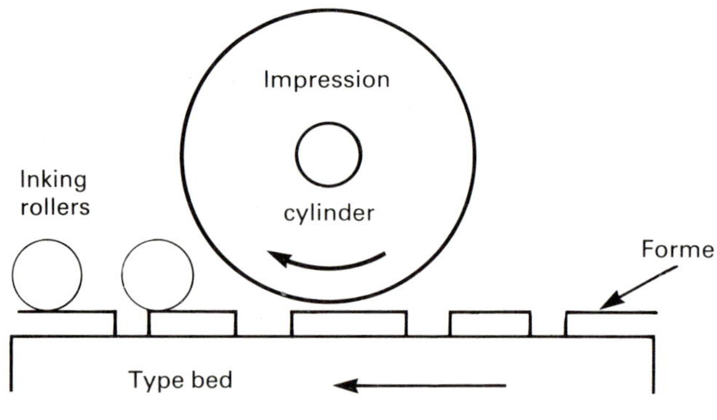

Fig. 2.25 Flat-bed cylinder press arrangement.

polymer plates. Because the ink-feed system does not have control keys to adjust the amount of ink being used, the system has become known as the 'keyless inking system'. It is sometimes referred to as 'indirect flexo' because of the use of the etched roller that is also used on flexographic presses, and because this roller is not in direct contact with the printing plate.

The main advantages of this system are:

- automatic even ink coverage which leads to reduced printed waste, and gives better solids and halftones;
- lower energy consumption due to the fewer inking rollers required;
- the lack of any adjustment to the operation when printing, eliminating the need for operator attention;
- ink mist is reduced to negligible levels due to the reduced number of inking rollers;
- the ability to benefit from modern technology inks (i.e. water emulsion and resin based) which gives sharper print, reduced rub-off and less strike through.

Letterpress forme production

Type matter for letterpress printing can be produced by two main methods.

- Line casting, in which lines of characters are cast in a lead alloy – a single line is referred to as a slug.
- Single characters, which may be assembled by hand or cast using Monotype machines.

The galleys of type from the casting machines are assembled into a chase (a metal frame) and locked into position. The result is the printing forme.

The casting of type by either method is a precision operation. Type metal is an alloy made from lead, antimony and tin, different proportions being used depending on the application. A slug is cast from 11–12% antimony, 3–4% tin, the balance being lead. The result is soft and not very hard wearing but perfectly adequate for use in making stereos for newspaper printing. Monotype characters are cast from 15–16% antimony, 6–7% tin, the rest being lead. The result is harder and more wear resistant. Still harder alloys can be made by raising the antimony and tin levels.

The composition of the lead alloy, the temperature and rate of cooling of the mould must all be well controlled in order to cast good type. Also, the dimensions of a character must be closely controlled otherwise characters may move about when printing, leading to uneven impression. Pictures are printed from plates produced by some form of etching process, and duplicate plates are often also made.

Original plates

Original plates are produced by process engraving. This is a process of photographically producing an acid resist on a metal plate and then etching away unprotected parts by means of an acid. The etched parts become lower than the protected parts which remain in relief and so become the printing part of the plate. Original plates can be classified in two groups – line and halftone.

Line plates are used for the reproduction of black-and-white line originals. The plate is made of zinc, etched in nitric acid and then mounted so that it is type high. These plates are used for advertisements in newspapers, label work, letter headings, diagrams and work involving clear lines.

Halftone plates are used to reproduce continuous-tone illustrations. The copy can be a photograph, wash drawing, water-colour or oil painting.

Halftone plates are made of copper, zinc or magnesium, and mounted on blocks to bring them to type height. The method differs from the line plate by the use of the halftone screen. Screens vary from 25 to 40 lines/cm for coarse paper, 40 to 50 lines/cm for calendered papers, and 50 lines/cm upwards for art papers. Electronic engraving methods are also available for these plates.

Plates are generally etched by what is known as the powderless etching process. A blank plate is coated with a light-sensitive resist which is exposed to a negative of the required image. The exposed image areas become insoluble. The plate is developed and baked to harden the resist. The plate is then mounted in the etching machine which uses an arrangement of sprays or paddles to throw the etching solution forcefully against the plate. Additives are present in the etchant which protect the side walls of the image as etching progresses. The rate of etch downwards is therefore greater than the rate sideways (Fig. 2.26).

Plates from original copy can also be made from photopolymers. Several suitable plate materials are available consisting of a photopolymer coated on a flexible metal backing sheet. The great advantage of these plates is their simplicity of preparation and, in particular, the repeatability of the process (something that powderless etching could never really provide). So, on photopolymer plates there is an almost exact correspondence between dot sizes on film and those on the plate which leads to much higher quality reproductions. The use of such plates has largely superseded etched metal plates. The flexibility of photopolymer plates is particularly useful as they can be used flat on a platen or flat-bed cylinder machine, or curved and used on a rotary press. One-piece photopolymer plates for rotary printing are called 'wrap-around' plates.

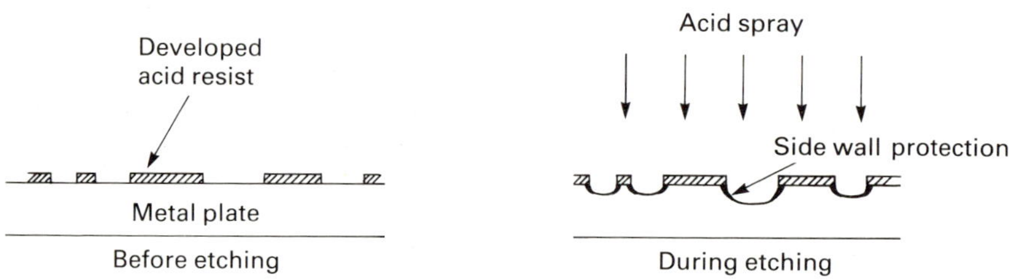

Fig. 2.26 Etching a letterpress plate.

Duplicate plates

Letterpress would be very restricted if it could only print from original type and photo-engravings. In package printing, multiples of the same subject need to be printed on the same press. This is made possible by the use of duplicate plates. There are several kinds of duplicate plates, but they are all produced from existing relief plates or complete formes by making an intermediate, called a matrix, and moulding from this with type alloy, or electroplating from it, or moulding in rubber or plastic.

Make-ready

Letterpress printing requires 'make-ready' to distribute pressure evenly over the printing plate, and to compensate for small variations in type height or plate mountings. Solids require greater printing pressure than highlights to ensure proper ink transfer and a number of 'pre-make-ready' systems have been developed to achieve this fairly easily and more quickly than the old-fashioned technique of sticking tissue paper in various thicknesses on the impression cylinder or platen. These all involve taking a print from the forme and converting this into a slight relief image by some means. One method uses a resin powder which is dusted over the freshly printed sheet and therefore sticks to the ink film. It is then heated in an oven causing the resin to fuse, swell and harden. More powder adheres to solids than highlights so the end result is that the resin is thickest in the shadow areas. This sheet is then placed on the impression cylinder or platen, in register with the forme, and is known as an overlay. Further make-ready by hand is usually necessary. Underlays are also possible in which case the additional sheet is placed under the printing plate. A good make-ready is essential for high-quality letterpress printing.

Substrates

A wide variety of substrates can be printed by letterpress provided the press and inking levels are suitably chosen. For halftone work, the screen ruling must suit the paper roughness. Thus 22 lines/cm (55 lines/in) is used with newsprint and up to 60 lines/cm (150 lines/in) for high-quality colour work. See Chapter 5 for details of the types of inks used.

Applications

Each type of machine has its particular uses, although it must be said that the use of letterpress printing has declined sharply in favour of the other processes. Platen and flat-bed cylinder machines are appropriate in the jobbing or commercial sections of the industry, owing to their flexibility. Automatic platen machines, such as the Heidelberg, are suitable for simple jobbing work such as letter headings and business cards. The feeders will deal with almost any kind of paper with equal ease. Automatic platens can also be obtained for embossing, cutting and creasing, carbon printing and numbering. Specially adapted machines are used for gold and metallic blocking.

Flat-bed cylinder machines are capable of producing a wide range of work, at speeds of about 5000 impressions/hour in the case of small presses, while large machines are slower because of the mechanical inertia involved.

Sheet-fed rotary machines give tremendous mechanical advantages over the traditional flat-bed machines and permit a greater volume of work to be produced in a given time – about three times more sheets per hour. Perfectors – that is presses which are able to print both sides of a sheet at one pass through the press – have proved to be popular for bookwork.

Web-fed rotary presses have been widely used in newspaper production owing to their high speed output. The rotary type of machine is also extensively used for the production of office stationery and continuous stationery. As the size of the page or copy must of necessity be a subdivision of the circumference of the cylinder, these machines cannot be adapted to such a wide range of jobs as other methods of printing. Narrow-width rotary presses are commonly used for label printing.

2.5 THE SCREEN PRINTING PROCESS

Screen printing (formerly called silk-screen printing) is a stencil process whereby ink is transferred to the substrate through a stencil supported by a fine fabric mesh of silk, synthetic fibres or metal threads stretched tightly on a frame. The pores of the mesh are 'blocked-up' in the non-image areas and left open in the image area. This image carrier is called the screen.

During printing the frame is supplied with ink which is flooded over the screen. A squeegee is then drawn across it, forcing the ink through the open pores of the screen. At the same time the substrate is held in contact with the screen and the ink is transferred to it. The principle is shown in Fig. 2.27.

Because of their simplicity, screens can be produced cheaply and this makes it an attractive process for short-run work. Furthermore, since the image is produced through a screen rather than from a surface the impression pressure is very low. This makes it ideal for printing on fragile boxes or awkward shapes.

Irrespective of the type of machine the printing procedure is generally the same. A working supply of ink is placed at one end of the screen and the screen is then raised so that the stock may be fed to register guides or grippers on a base. The screen is then lowered and a rubber or plastic squeegee drawn across the stencil to produce the print. Ink replenishment is undertaken as necessary.

On most flat-bed machines the base to which the substrate is applied is of a vacuum type. This prevents the stock sticking to the screen and being lifted by tacky inks.

To a certain extent the thickness of the ink film printed can be controlled by the pressure, sharpness and angle of the squeegee blade.

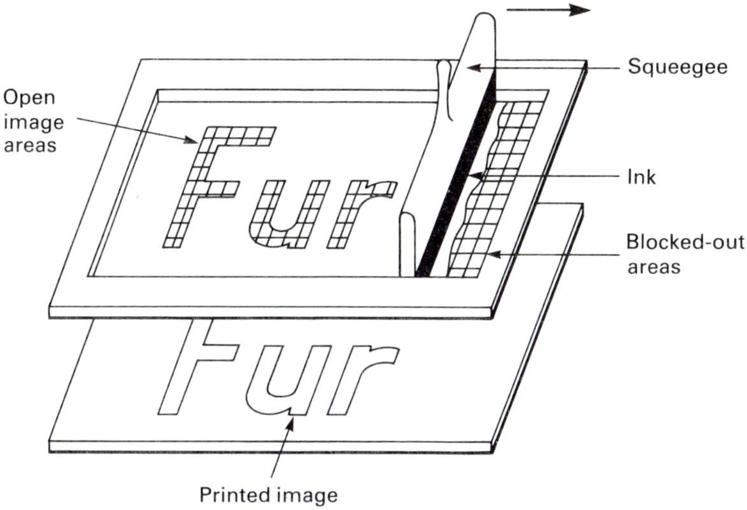

Fig. 2.27 Basic screen press arrangement.

The more upright the blade the thinner the deposit of ink. Thus, in general, fine work requires a more upright blade. However, the type of ink, stock and machine govern the blade setting also.

Press configurations

Many printing machines still consist of a simple hand-operated unit. With these the substrate is fed in by hand, ink is placed in the frame and flooded over the screen and the squeegee is then hand drawn across the screen. These can be particularly useful where very thick or thin materials, which cannot be automatically fed, are printed or where a test run of a new package is required.

Semi-automatic machines have the stock fed and taken off by hand but utilize a mechanized squeegee blade stroke.

Fully automatic flat-bed presses are also available where the substrate is fed in and taken off by means of automatic feed and delivery systems. After printing the sheets are taken through an air drier to evaporate the solvent. Prints emerge dry and can be stacked ready for further processing. Such fully automatic machines have only really become feasible since the advent of thin ink films which dry more rapidly.

In order to speed up the process further, cylinder presses have been developed (Fig. 2.28). With these presses the squeegee remains stationary and the screen, cylinder and substrate all move in unison. This permits faster operation since the paper does not have to be brought to a halt and fed into a vacuum base as with flat-bed presses. Speeds of 5000 impressions per hour or more are not uncommon with such machines, but they are not suitable for rigid materials.

One of the major advantages of the screen process is the ability to obtain prints on non-flat objects. For example, printing on bottles or

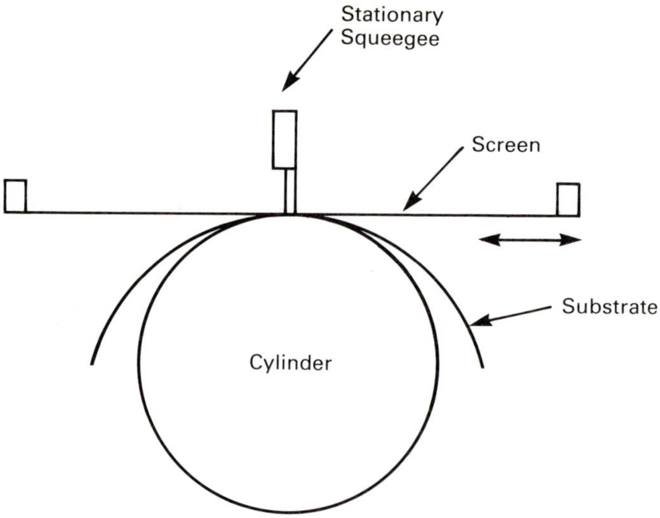

Fig. 2.28 Cylinder press.

other cylindrical objects is achieved by using a press of the cylinder type described above but the object to be printed is placed in the machine where the impression cylinder is shown. After each impression the bottle is removed and another unprinted one substituted. There are few limitations on size or shape. Special screens and jigs are produced for printing on shaped objects such as cups with handles or tapering cylinders, and screens with high elasticity combined with shaped squeegees are used for conforming to irregular objects. Print heads can also be bolted to automatic production lines, so that printing becomes a part of the total production process of such objects as filled polythene bottles.

Screen stencil manufacture

The screen consists of a mesh of threads woven together. Early screens were made of silk and each element of the mesh consisted of a number of stranded threads woven together. These 'multifilament' screens had a tendency to distort and swell, as ink was deposited between the fibres, and were difficult to clean. Because of these disadvantages, and the shortage of silk, alternative materials were sought after the war and today the materials used are almost wholly synthetic. Monofilament screens have become widely used, where each element of the mesh consists of a single strand of material. This minimizes the disadvantages described above. Typical monofilament materials used are nylon and polyester, the latter being used where dimensional stability is important. Metal meshes are also used where a long life and complete dimensional stability is required but these are considerably more expensive. Multifilament screens, usually polyesters, are occasionally used where a heavy ink film is required.

Every screen has a characteristic mesh opening and thickness. Which particular opening and thickness is chosen for a given application depends on the thickness of ink film required, the pigment particle size of the ink and the fineness of detail it is necessary to reproduce.

Mounting the screen

Prior to printing the screen has to be stretched and mounted on a frame. The correct tension is important to avoid misregister, premature wear of the stencil or splitting of the fabric. Although hobbyists and small printers may do this by hand, on small wooden frames, it is normally achieved in production by use of a pneumatic stretching frame. A number of clamps grip the screen and can be individually adjusted to achieve tension. The screen is then placed over a frame and stuck to it after which it is degreased before the stencil is applied.

Application of the stencil

For simple work a stencil may be hand-cut from a laminated film material and stuck down to the screen. Obviously this is only useful for large blocks of colour or simple geometric designs. A direct photographic stencil is produced by applying a light-sensitive coating such as a dichromated resin to the screen and drying it. (Usually a number of separate coatings are required.) Prior to exposure the coating is water soluble, but upon exposure to light it hardens. Thus the stencil is produced by coating the screen and then exposing it to a photographic positive. This hardens the coating in the non-image areas and leaves the image areas water soluble. By washing-out the screen with water the image areas will be cleared and the mesh left open.

Indirect stencils are produced using the same principle but during exposure and development the stencil is held on a temporary backing film. It is only after exposure that it is transferred to the screen, while in a moist state, and dried. The backing sheet can then be carefully removed leaving the stencil adhered to the screen.

Direct stencils have a longer life than indirect because the coating is bonded into the mesh but are much more difficult to expose (since the screen is attached to a frame during exposure). Furthermore, because the washing out action tends to follow the mesh threads, a serrated edge to the image may sometimes result. Indirect screens avoid these problems and therefore tend to produce finer detail. In addition they can be obtained pre-sensitized, so that they may be used direct from the packet without any of the mixing of chemicals and application to a base which was previously necessary.

After treatment

Following stencil production any remaining areas of the screen around the stencil have to be sealed to avoid ink penetration. A number of materials, such as lacquered tape or synthetic varnish, may be used which must be resistant to ink and yet remain flexible so that they do not crack under pressure of the squeegee.

Finally the stencil is inspected against a light box. Any areas not properly filled are then filled with an opaque filler.

Substrates and inks

Typically screen printing inks used to be very thick preparations based on oil-bound paints. They printed an ink film that literally stood out in relief on the surface. Such inks were highly suited for printing on almost any type of substrate, rigid or flexible, and it became a popular process for producing posters, display advertising, metal signs, glass and china decorating and textile printing. However, these oil-based inks dried by aerial oxidation and being thick, drying times were long. For many years work was placed on drying racks stacked on top of each other and gradually these were replaced by mechanized racking systems. Even so the production speeds achievable were very low.

However, considerable developments are taking place with UV-cured and other ink systems (refer to Chapters 10 and 11).

Applications

Since screen printing inks can be formulated to adhere to almost any surface, and the printing process itself can handle almost any substrate in a wide variety of shapes, screen printing is a very versatile process. It is used for package printing, especially of plastic containers, posters, point of sale materials and many other products. It has even been used for carpet tiles!

The process has always had limitations for four-colour process printing because of the many variables in the process and the total lack of standards. However, this is now being tackled and draft specifications have been produced.

2.6 INK-JET PRINTING

Ink-jet printing is a non-impact means of generating images by directing small droplets or particles in rapid succession on to the surface of a substrate under computer control. There are various possible ways of generating and projecting droplets but only the following appear suited to real production applications.

Continuous jet

The primary ink-jet system is the continuous jet, formed by forcing ink under pressure (about 3×10^5Pa) out of one or more small nozzles (Fig. 2.29). In these circumstances, a liquid jet tends to break up into a stream of droplets of a size and frequency determined mainly by the surface tension of the liquid, the pressure applied to it and the size of the nozzle. However, to encourage this process regular pressure pulses are applied continuously to the ink within the reservoir at a suitable frequency and then the size and spacing of droplets become much more regular. This can be done by applying a high-frequency alternating voltage (possibly up to 1 MHz) to a piezo-electric crystal that is in contact with the ink. A stream of uniformly sized and spaced droplets is generated, which is projected towards the moving web of paper at a speed of 5–20 m/s producing a continuous line of dots, each having a diameter about three times that of the droplets.

For any practical message to be generated by a single nozzle, the droplets must be individually controlled and deflected and this is done electrostatically by surrounding the ink jet in the region of droplet formation by a charged electrode. The jet becomes charged by induction and each droplet carries a charge depending upon the voltage applied to the electrode at the instant the droplet separates from the jet. The droplets then pass between oppositely charged deflector plates and are deflected by the electric field to an extent proportional to the size of the charge it carried. Characters can be generated by depositing dots within a matrix: the minimum matrix size for reasonable legibility is 7 × 5 dots. Thus by repeatedly increasing the voltage applied to the deflection electrodes in seven steps and then returning it to zero, a row of seven dots is printed on to paper. By moving the paper sideways, a parallel row of seven dots can follow. In this way, a rectangle of dots is printed. The other dots must be prevented from reaching the paper, and this is achieved by leaving them uncharged so that they are not deflected on to the paper but proceed straight to a trap where the unwanted ink is collected and can be filtered and returned to the reservoir. Hence the stepped voltage pulses applied to the charging electrode are gated by signals that define the character matrix and allow dots to be printed as needed.

The physical constraints of useful droplet size, nozzle size and surface tension of the ink mean that in practice droplet frequencies will be somewhere in the region of 100 kHz and this determines, for a single

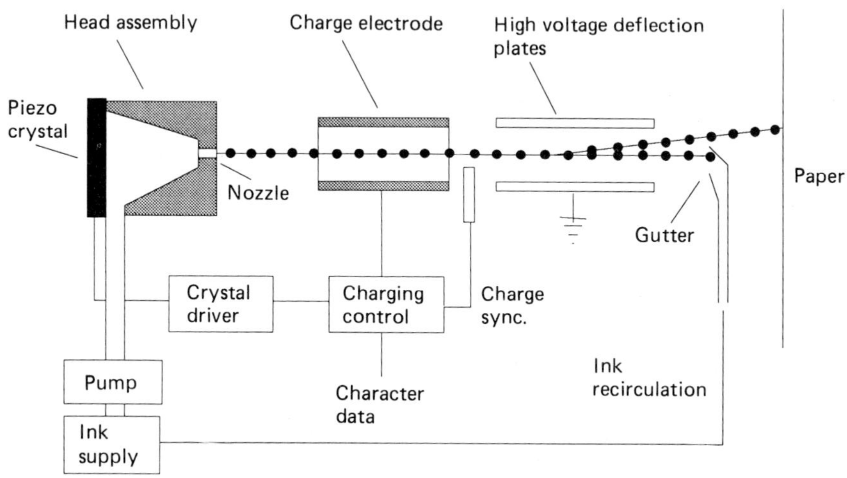

Fig. 2.29 Continuous ink-jet printhead. (Courtesy of PIRA)

nozzle, the limitations on writing speed and image definition. Thus, for 7 × 5 matrix characters, with some space between them, up to about 2000 characters per second (cps) may be printed, whereas for a 40 × 30 matrix, the speed may be under 100 cps. For in-line numbering, coding and addressing applications 7 × 5 characters from a single continuous jet are widely used and, typically, messages of 100 or 200 characters can be printed at speeds of up to 2000 cps. According to the application, this corresponds to a web speed of 5 m/s or 100 000 articles per hour.

An alternative form of continuous jet system uses an array of closely spaced nozzles, each of which serves only one line of dots. Unwanted droplets only are charged and deflected into a gutter. Since the writing droplets are not charged, electrostatic interaction between adjacent droplets is avoided and flight distances are equal. The system is therefore in some ways simpler than the deflected droplet type. The complexity comes in assembling and controlling a multiplicity of nozzles. The Mead Dijit 2800, for example, has an array of nozzles spaced at 120 per inch in a double row that may be up to 10.65 in long and print up to 512 characters.

Yet another form of continuous ink-jet printing is known as Hertz technology, which is characterized by the use of extremely small nozzles and hence very small drops (Fig. 2.30). Each spot on the paper is produced from a group of these small drops, and by varying the number of such small drops a grey level capability is made possible. Rather higher print quality can be achieved compared with other ink-jet systems but there are difficulties with such small nozzles and droplets.

At present these systems are very slow, taking nearly an hour to produce a four-colour A0 print.

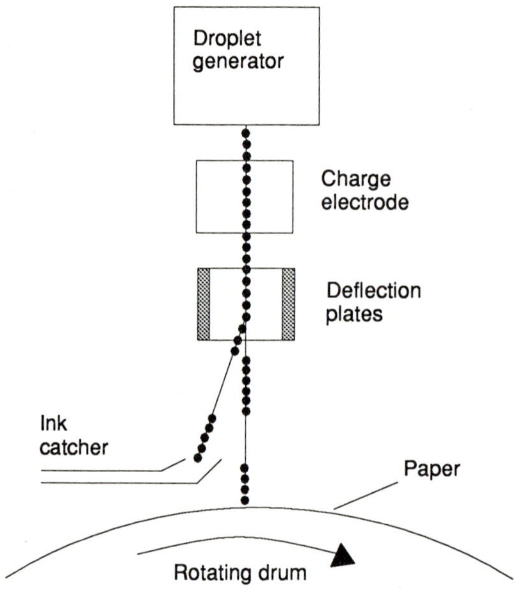

Fig. 2.30 Hertz ink-jet system. (Courtesy of PIRA)

Impulse or drop on demand

With this method, the pressure on the ink reservoir is not maintained continuously but is applied when a droplet is needed to form part of a character (Fig. 2.31). In a widely used system, an array of 12 nozzles, each actuated by its own piezo-electric crystal, is used to generate a matrix column, so that no deflection of droplets is needed to form the image. Since, also, the recovery and recirculation of non-printing droplets does not arise, the equipment can be both electrically and mechanically simpler than the other classes of ink-jet printer considered.

A variation on this type is the Canon bubble-jet (Fig. 2.32) which, instead of piezo-electric crystals, has very small heating elements behind each nozzle so that ink droplets can be ejected by rapidly formed solvent vapour bubbles. Yet another method patented by Xaar deflects the side walls of each ink chamber causing drops to be ejected.

Finally there is an impulse jet system in which the ink is solid at room temperature so that the print head has to be heated to allow the ink to melt and be printed. This provides fast drying on a wider range of substrates than conventional impulse jet systems. However, the image is in the form of a waxy solid sitting mainly on the paper surface and is prone to damage.

Drop on demand ink-jet systems are being used mainly in matrix printers but their application is not, in principle, limited to that. Examples of four-colour reproductions have been produced using the bubble-jet technique at resolutions of 16 points/mm suggesting that this may be a viable method for short runs in the future, and also for colour proofing.

Ink-jet inks

Ink-jet inks are low viscosity fluids, typically 4–30 cP and consist of a solvent blend, colourant, binder and additive.

Some five to eight solvents are typically used, a prime consideration being to control the drying time. In continuous ink-jet systems the majority of drops are not printed but recirculated. Solvent is lost during this process leading to an increase in viscosity, so the ink system must

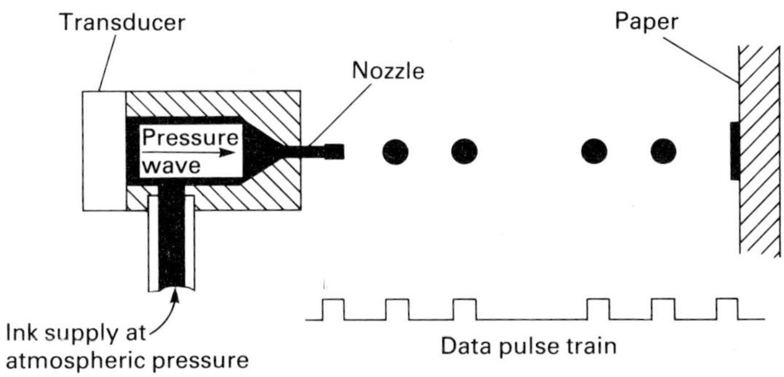

Fig. 2.31 Impulse or drop on demand printhead.

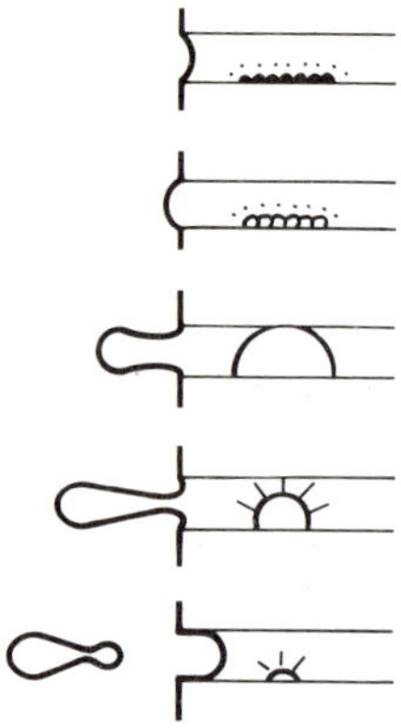

Fig. 2.32 Canon bubble-jet.

replenish the loss with an appropriate mix.

The colourants are predominantly dyes, but pigments must have a particle size less than 3 μm. Dyes need to have high solubility in the solvent system, high thermal stability and be lightfast. The binder consists of one or more polymers suitably chosen to control viscosity, promote good drop formation, bind the colourant and adhere to the surface to be printed.

Finally, various additives are included to modify flow properties, surface energy (which should be >35 mN m^{-1} to produce a well defined stream of droplets), and plasticize the binder. Also, a conducting salt may be needed to make the ink electrically conductive ($> 10^{-3}\Omega^{-1}$ cm^{-1}) – necessary for charging the droplets. Additives typically make up less than 1% of the ink but are crucial to success. (Chapter 12 gives details of formulations used.)

2.7 TONER PRINTING SYSTEMS

As with ink-jet printing systems there are many different types of toner printing systems. They are all computer driven and use optical or electrical methods to form an electrostatic latent image to which toner is attracted and subsequently transferred to the substrate. The main types are:

- electrophotography;
- ion deposition;
- electrostatic;
- magnetographic;
- electrographic.

Of these, electrophotography or xerography, as it is sometimes called, is the most widely used by far, being the basis of most photocopiers and so-called laser printers. However, the other systems have their special advantages in some situations, and may become very significant in the future.

Electrophotography

Electrophotographic printing is commonly used in photocopiers and laser printers. Some of the higher-quality machines in both categories are currently used for print runs where up to a few hundred copies are required. Electrophotography is the technique of producing an image by causing light to affect the charge distribution on a photoconductive drum or belt. The charge pattern is subsequently used to attract toner, which is then transferred to paper. The toner is then fixed by some means and forms the final image.

Figure 2.33 shows a schematic layout for an electrophotographic printer. The unit is based on a rotating drum or belt, coated with photoconductive material. The photoconductor is electrically insulating in the dark, but becomes more conductive when illuminated. Various materials are used, ranging from amorphous selenium to complex organic compounds. Each kind of material has a sensitivity which depends on wavelength, hence light sources have to be matched to the photoconductor (Fig. 2.34).

The first step in the process leading to the printed image is to generate a uniform electric charge on the dark photoconductor surface. This charge is then dissipated by exposure to light in the non-image areas. In a conventional copier, light reflected from the original is projected on to the drum. In a printer, the image is generated point by point using one of several types of imaging device. Types of imaging devices include:

- laser;
- light-emitting diode (LED) array;

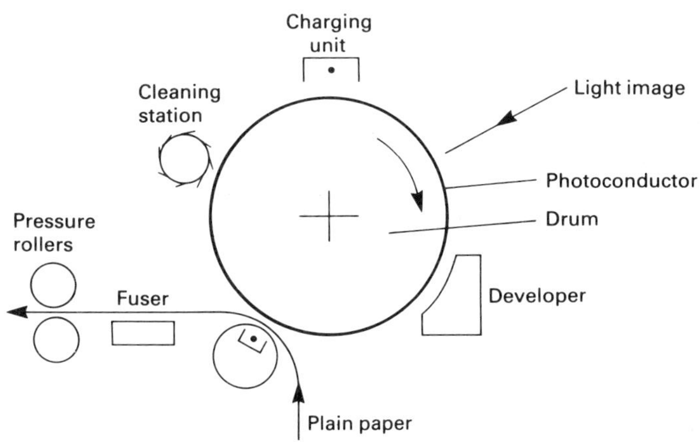

Fig. 2.33 Schematic layout of electrophotographic printer. Process steps: (1) charging; (2) imaging; (3) developing; (4) transfer; (5) fixing; (6) cleaning.

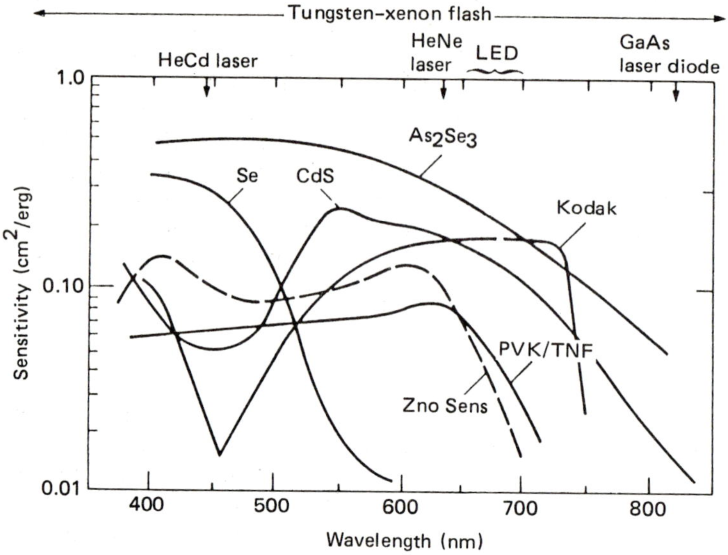

Fig. 2.34 Photoconductor sensitivity.

- liquid crystal array (LCD) (plus light source);
- magneto-optic array (plus light source);
- cathode ray tube with fibre optics.

The laser is the most common device. This is a suitable light source since it produces light which can be collimated and focused into small, high-intensity spots. Three types are in use: HeNe gas, HeCd gas and diode. A complete laser scanning sub-system includes several essential components – the laser, a scanner to sweep the beam across the drum, a modulator to turn the beam 'on' and 'off' and several lenses and mirrors.

Rather than have a single source, it is possible to have an array of light-emitting diodes. The total width of the drum is covered by several monolithic devices, each with a line of LEDs. LCDs are used as a light switch, but this method is not widely used. The other types of light source, while they have been tried, are not used in current machines.

The next step in the electrophotographic process is to develop the latent electrostatic image. The ink in this process is known as toner, of which there are various types, described later in this chapter. After development, the image exists on the photoconductive drum as a distribution of toner particles. This is transferred to paper, which is moved past the drum so that the paper speed is the same as the drum surface speed. Behind the paper, at the point of contact between drum and paper, is a corona which charges the back of the paper so that toner is attracted from the drum to the paper. This process is not 100% efficient, and some toner (up to 20%) can remain on the drum and needs to be removed subsequently.

After transfer the toner is still only held on the paper electrostatically, and so a final fixing process is required. This may be done by heat, or heat and pressure, which melts the toner and fuses it to the paper. Heat can be applied by a radiant heater, a heated roller (coated with silicone rubber or some similar material to which toner will not adhere), or some combination. Rollers will typically approach 200°C which is necessary to cause the toner to melt in the very short time (11 ms to < 1s) in which the toner is in the roller nip. There are other methods of fusing such as the use of solvent vapour or high-intensity flash.

Ion deposition

This process is very similar to electrophotographic systems except that the electrostatic latent image is formed directly and not by the use of light. The drum is a hard dielectric material, and imaging is performed by an ion source array which deposits charge directly on the drum (Fig. 2.35). The remainder of the process is similar to the electrophotographic system. The advantages are that the drum is hardwearing, and should give long life, and that fusing takes place just by pressure. This

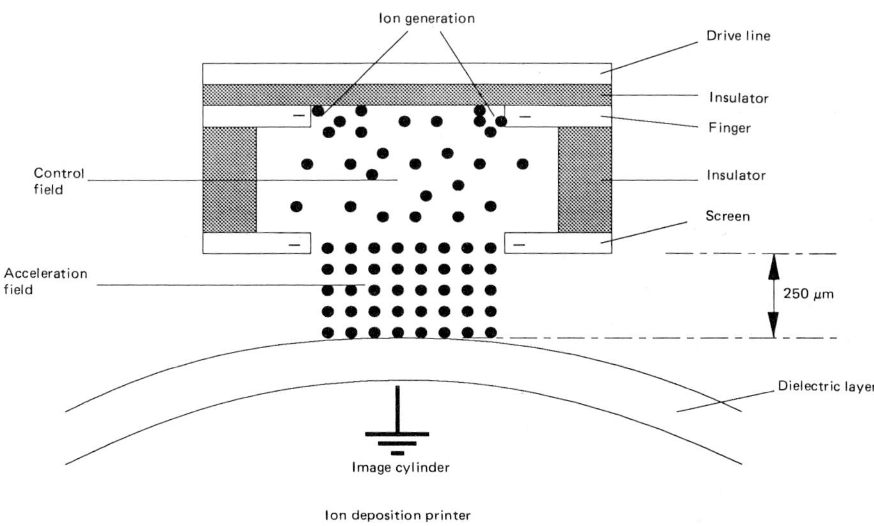

Fig. 2.35 Ion deposition printer. (Courtesy of PIRA)

is therefore, in principle, a simpler process, but it has not proved possible to obtain the quality of print really required, and so applications have been limited. One area of application is that of plastic self-adhesive labels where the lack of heat in the fusing process enables a plastic substrate to be used.

Electrostatic

In this process the paper itself acts as the dielectric medium which is charged by an electrode array over which it is passed, forming an electrostatic latent image (Fig. 2.36). The paper then moves over a toner reservoir and liquid toner is applied by spray or roller, which adheres to the charged areas. The process is relatively simple. Multiple colours can be produced by passing the paper repeatedly over the electrode array and presenting a different toner each time. Print quality has remained static for some years, apparently limited by the electrode array, but there are possible improvements in the future.

Magnetographic

This method uses a hard drum with a magnetic coating similar in properties to that on a recording tape or floppy disc. The 'image' is produced by an array of electromagnets which create magnetized areas

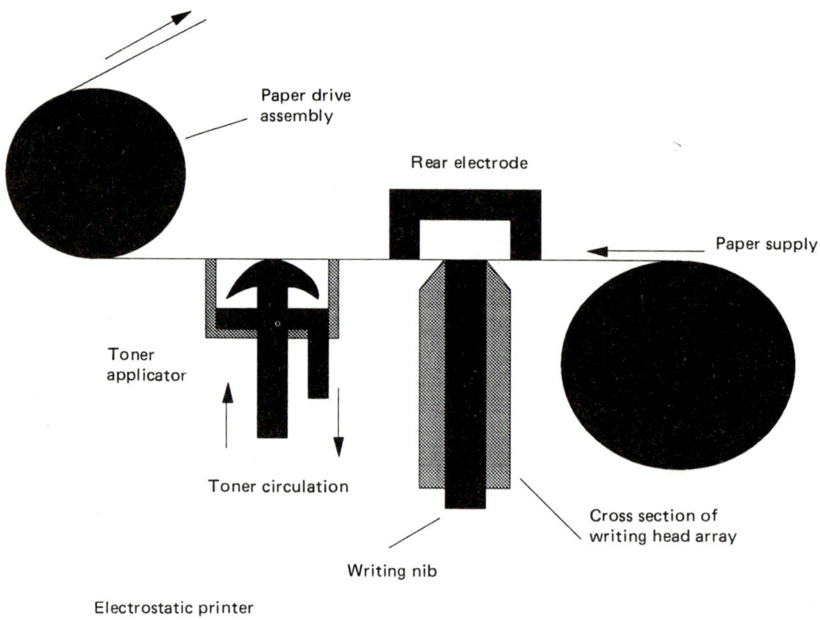

Fig. 2.36 Electrostatic printer. (Courtesy of PIRA)

on the drum. These areas are developed using magnetic toner particles. An interesting feature of this process is that the magnetic latent image is more permanent than is the case with any of the electrostatic systems, and that multiple prints can be obtained without having to re-write the image. Other aspects of the process are similar to electrophotographic systems.

While having some obvious attractions, this process also has its problems. The toners have to be magnetic, which is very difficult with transparent colours as would be needed for process colour printing. The advantage of the semi-permanent image on the drum is also a problem in that the drum is also more difficult to clean, and image quality suffers. Finally, there is the obvious difficulty of manufacturing arrays of ever smaller electromagnets, to increase the resolution.

Electrographic

In electrographic printing, a latent electroconductive image, or master, is first formed by any of a variety of methods. Then, in a separate printing cycle, multiple identical copies of the image are generated. The master-forming technologies are categorized as either physical or chemical, based on the method by which the latent image is formed. After the master image has been created, the rest of the process is very similar to electrophotographic printing.

The master-forming technologies can be classified as follows:

Physical methods
- toner fusion xeroprinting
- overcoated relief printing plate
- charge retention masters
- 'AMEN' and 'HEP'

Photochemical methods
- photopolymerization/photohardening
- photo-induced differential triboelectric potentials
- doped photoconductors
- photoconductors with switching layers
- silver masters
- electrochemography

It is beyond the scope of this publication to give a detailed account of each of these processes since none of them is used very widely at present. However, some commercial products and many patents exist from companies such as Fuji, Xerox, Olympus, Canon, DuPont, Dai Nippon Printing, Asahi Chemical, Minolta, and others. (An excellent review paper is given by D.E. Bugner, *Journal of Imaging Science*.

Toners

Toners are either in the form of a dry powder, or a dispersion in a liquid. The main features of each type are given below.

Dry powder

Two-component

This is a dry powder consisting of fine particles of toner pigment, and coarse beads (say, 200 μm diameter) known as the carrier. The toner and carrier become oppositely charged, and so the carrier beads become coated with toner.

Particle size:	5–32 μm
Resistivity:	$> 10^{13}$ Ω cm
Charge:	10–30 μC g^{-1}
Composition:	pigment, most commonly carbon black; charge control agent; 90% polymer binder or resin; surface flow and cleaning additives (i.e. silicon oil or low molecular weight polyethylene or polypropylene waxes)
Typical use:	high-speed laser printers and copiers

Conductive mono-component

The toner particles are both magnetic and conductive, and are of intermediate size. The primary advantage is the simplicity of the system but the larger particle size limits resolution.

Particle size:	5–45 μm
Resistivity:	10^3–10^{12} Ω cm
Composition:	50–70% magnetic oxide; polymer binder; carbon black; cleaning additive
Typical use:	ion deposition system

Resistive mono-component

These are similar to the conductive toners.

Particle size:	5–32 μm
Resistivity:	$>10^{13}$ Ω cm
Charge:	2–20 μC g^{-1}
Composition:	30–60% magnetic oxide; charge control agent; polymer binder; surface flow and cleaning additives
Typical use:	HP Laserjet printers

Hot fusing of dry toners requires different resin properties depending on the heating method. For radiant heat, polyester or epoxy resins are used with a molecular weight of 5000 to 50 000, and a melting point of 50–60°C. Flash fusing can cause temperatures to reach 200°C so styrene copolymers, epoxies and polycarbonates are used. For roll fusing, styrene acrylates, styrene methacrylates and styrene butadienes with molecular weights of 30 000 to 100 000 are used.

Liquid toners

These consist of a colloidal dispersion of very fine sub-micron particles in a highly insulating isoparaffin, having a boiling point of 155–210°C. A wide range of resin binders may be used.

Particle size:	0.1–2.0 μm
Concentration:	0.5–1.0%
Charge:	2000 $\mu C\ g^{-1}$
Composition:	pigments; polymers for stabilization and fusing; charge control agents; dispersant

Substrate requirements

Substrate requirements for toner printing are as much to do with paper transport requirements as to do with image quality issues. Consequently, correct grammage, thickness, bulk, stiffness, moisture content, smoothness and electrical characteristics are necessary. Smooth coated papers give the best image quality from laser printers but this is really only of any consequence if halftones are being printed. High moisture content should be avoided since this can lead to poor toner transfer (because the moisture affects conductivity), poor toner fusing or adhesion (since the energy to fuse the toner evaporates the water instead), and curl. High moisture content can occur if litho printed material is subsequently fed through a laser printer – as a result of picking up fount solution during the first printing. In some cases a dry-offset or offset letterpress process may be preferred to avoid this problem. Very rough papers may also give rise to difficulties due to the poor contact with the photoreceptor drum and fuser rolls, leading to poor image definition and inadequate toner fusing and adhesion.

Applications and systems

The majority of commercial systems employing toner printing are intended as office printing systems, and lack the output capability needed to find application within a printing industry environment. However, some exceptions to this are emerging.

The AM Graphics Electropress

This is a web-fed, high-speed (300 ft/min), 300 dpi liquid toner, two colour, 20-inch print width, electrophotographic machine with integrated variable size finishing. This machine looks much like a conventional small web-offset machine, and for some direct mail, business forms, pamphlets and similar products can do just as well, with the added feature of being able to vary every image printed. A special version is available for printing books. The manufacturers anticipate that by 1994 they will have a similar four-colour press running at 600 ft/min with an image resolution of 600 dpi.

The Xerox DocuTech System

This is based around established Xerox technology, but built into a system that perfects, collates and binds the product, all at relatively high speeds (135 A4 pages/minute), so providing another approach to satisfying the requirement for on-demand book production.

The Canon Colour Laser Copier and the Xerox 5775
These are both four-colour printing machines producing 5 and 8 A4 pages/min, respectively, at a resolution of 400 dpi. Although slow by printing industry standards, both machines are likely to find application for short run colour work, and indeed, are already being used as digital proofing systems.

2.8 OTHER PRINTING PROCESSES

There are a few other printing processes which are rare but used in special applications.

Intaglio or recess printing (commonly used on bank notes) is the process of printing from an engraved steel, copper or brass plate or cylinder. The printing of fine tapering lines and a very thick ink film are characteristics of this process (Fig. 2.37). Various press configurations are possible, but all require a very strong press frame indeed. Impression pressure is typically achieved with hydraulic rams. The inking roller is like a flexo printing cylinder, that is, it has raised parts which transfer the ink on to the image areas only. A very thick ink film is transferred so that the engraved image is completely flooded with ink. The excess ink is then wiped off by a counter rotating roller which in turn is cleaned in a bath of solvent or by some other method. Ink drying is typically by infra-red heaters.

Cylinder manufacture can be by a variety of methods including etching, but each print works tends to develop its own in-house methods.

Collotype printing reproduces illustrations in continuous tone – it is a screenless process which involves printing from a plate coated with

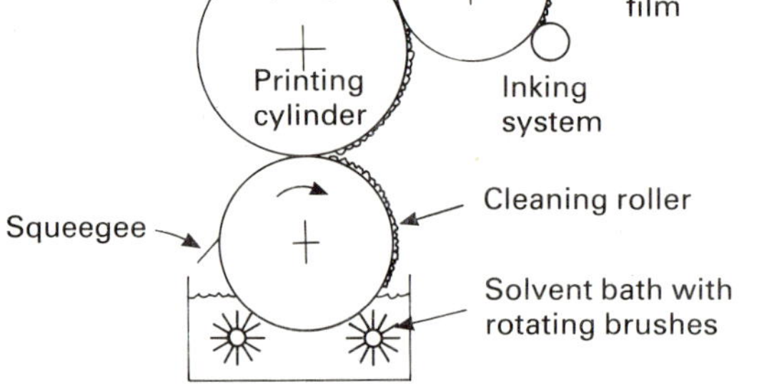

Fig. 2.37 Intaglio printing unit.

gelatine. It is a difficult process to control and only relatively short runs are possible. It is still used very rarely for fine art reproductions and until a few years ago was used for some postcards.

Tampon printing is a form of offset gravure in which a compressible (in the true sense of the word) tampon is used to transfer the image onto an irregular shaped object, or one with a recessed surface. The tampon deforms to the surface profile of the object, and because it is truly compressible does not distort the image greatly. Tampons are made in a wide variety of shapes and sizes to suit different objects.

2.9 PRINT RECOGNITION

It is not always possible to make a positive identification of the printing process used since the appearance characteristics of some processes are rather similar. However, in most cases the following features, examined under a magnifying glass, will provide a good guide:

- the edges of type characters;
- the appearance of halftone dots and solids;
- presence of embossment on either the top or reverse side of a printed sheet;
- the ink film thickness.

The kind of stock on which the print is made can modify its appearance considerably but may also be a clue to the printing process used. Also, some processes have particular areas of application and are more likely to be used for certain kinds of work. Table 2.1 shows the main characteristics of the most common printing processes.

Lithography

The main characteristics of a lithographic print are the uniform ink coverage and the absence of squash. There is no rim of ink around a type character, and even on rough paper, non-bottoming does not occur since the rubber blanket conforms well to the paper surface. Halftone dots are of the same shape and pattern as letterpress dots, but the ink coverage is uniform and the smallest dots are smaller than is possible with flexo or letterpress. The edges of dots are fairly well defined on smooth coated stocks, but have rather fuzzy edges on uncoated stocks, so the dot pattern is less distinct. Ink film thickness is around 2 μm.

There are two defects specific to offset lithography which assist in identification if they happen to be present on a print. The first is scum which occurs when the non-image areas of a plate start to accept ink for some reason. Scum appears under magnification as randomly distributed fine spots of ink. A superficially similar fault is that of tinting which occurs when the ink becomes emulsified in the water. When this occurs, a thin wash of colour appears over the whole print which under high magnification can be seen to be made up of minute droplets of ink (Fig. 2.38).

Table 2.1 Main characteristics of the printing processes

Print process	Ink film thickness (μ)	Typical halftone screen ruling (lines/cm)	Substrate types	Typical applications
Offset litho				
Sheet-fed	<2	47–80	Wide range of paper and board, plastic sheet and metal	All general print, business forms, technical documentation, promotional, magazines, credit cards
UV drying		60	Carton board	Packaging
Web-fed				
Heat-set	<2	52–69	Wide range of coated and uncoated paper	Magazines and similar format products
Cold-set	<2	25–40	Newsprint	Newspapers
Flexography				
Narrow web	0.75–2	60	Paper and plastic film	Labels, flexible packaging
Wide web	0.75–2	33	Newsprint	Newspapers
	0.75–2	52	Wide range of plastic film, paper, corrugated cardboard	Boxes and many other types of packaging, sacks
Gravure				
Large web	<6	60–80	Coated or uncoated paper	Magazines and similar products, mail order catalogues, woodgrain patterns
Smaller web	<6	60–80	Coated or uncoated paper, plastic films, board	Packaging (esp. flexible), cigarette cartons, postage stamps
Sheet	<6	150–200	Paper	Fine art reproductions

Table 2.1 Main characteristics of the printing processes *(cont)*

Print process	Ink film thickness (μ)	Typical halftone screen ruling (lines/cm)	Substrate types	Typical applications
Screen	<30	25–40	Card, fabric, wide range of plastic, shaped containers	Point of sale displays, plastic containers, labels, 'T' shirts
Ink-jet	3	4–12 dots/mm	Paper, plastic, metal, board, special coated substrates, sometimes required for best colour saturation	Sell-by-dates, mail wrappers, direct mail personalization, office stationery
Toner	<15	12–24 dots/mm	Paper	Technical documentation, short run books, reports, labels, business forms
Intaglio	<30	N/A	Paper	Bank notes, security documents
Tampon	2	40	Wide range of plastic and metal objects	Irregularly shaped objects such as bottle tops, compact discs

Offset letterpress

A process which should be mentioned here is offset letterpress (also sometimes called dry offset or letterset) which combines features of both letterpress and lithographic printing. In this process, the image is raised – as in letterpress – but is offset on to a rubber blanket before printing on to paper. Prints made by this process can easily be confused with lithographic prints, and there is no certain means of identification. It is much less common than lithographic printing and is mainly used for printing simple line, tone and solid designs, often for business forms and continuous stationery.

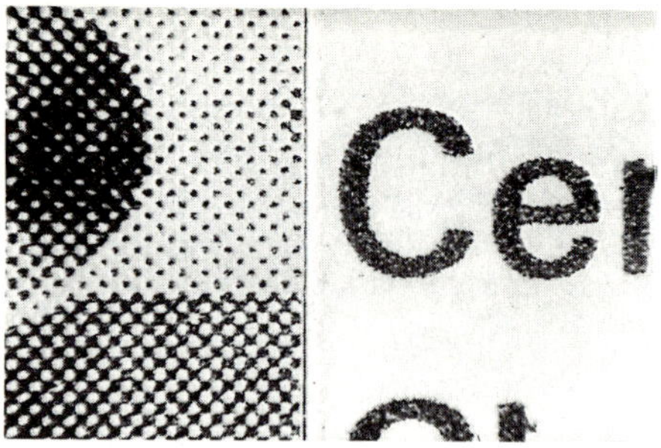

Fig. 2.38 Offset lithographic print.

Flexography

Flexography is a relief process like letterpress, but of course uses rubber plates and very fluid inks. Ink squash is therefore a significant characteristic but is kept to a minimum by using very light pressures. Other characteristics of letterpress printing are, however, absent so there should be no possibility of confusing the two processes. There is no embossment of the sheet and because of the ability of the rubber plates to conform to the surface contours of the printing stock, non-bottoming should not arise. Flexography is widely used for packaging and especially where this involves plastic films, metal foils and waxed papers. The substrate and application are therefore also good guides to flexo printing (Fig. 2.39).

Gravure

Gravure printing exhibits the characteristic cell pattern, which gives a saw-tooth edge to lines and especially type characters. However, care is needed since type is sometimes screened at a finer screen ruling than the illustrations and this reduces the sawtooth effect. Also, special techniques have been used on mechanical engraving systems which also reduce the effect, making it difficult to be sure from this aspect alone. To produce a fully covered solid, ink has to flow out of each cell and join up to

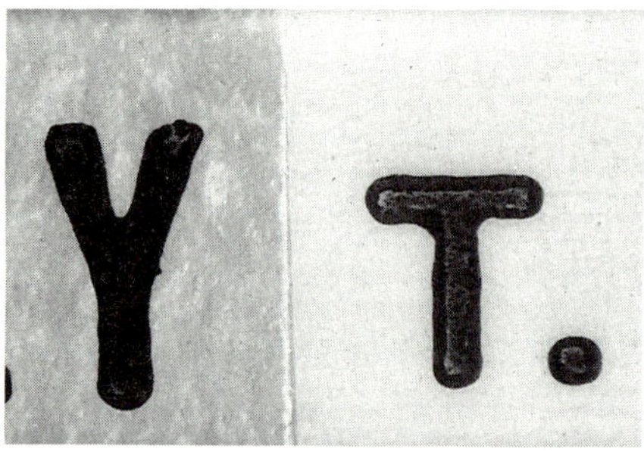

(a) (b)

Fig. 2.39 Flexographic print: (a) on paper; (b) on film.

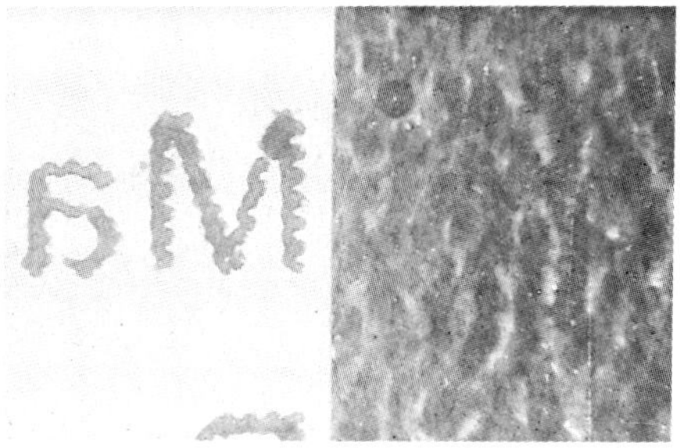

(a)

Fig. 2.40 Gravure print.

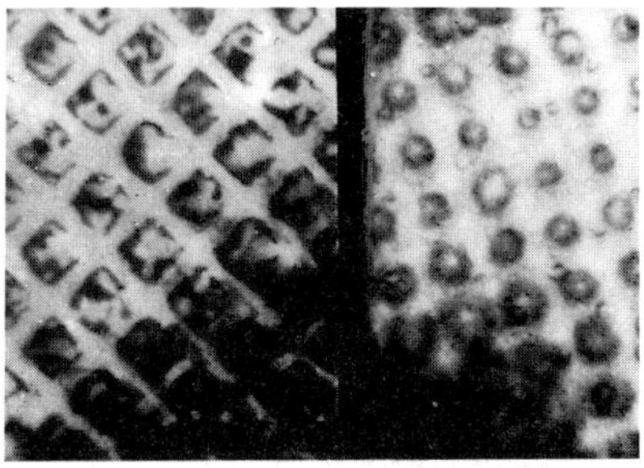

(b)

Fig. 2.41 Gravure prints: (a) conventional gravure; (b) halftone gravure.

obliterate the cell wall pattern. If this action does not go to completion, white spots can be seen on the solids corresponding to the diamond-shaped areas between groups of four cells, and sometimes white lines corresponding to cell walls. The cell wall formation can, in any case, generally be seen in the darker mid-tones where the cell join up is not complete. If the ink is too fluid, a fully covered solid is produced, but with a mottle pattern, and 'bleeding' on the trailing edge of solid areas – a ragged overflowing edge of lower density.

The different gravure processes also have different appearances. Conventional gravure produces cells of approximately the same area, but of different depth to reproduce a range of tones. Cells only print as individual squares transferred from the two back edges of the cells giving a V-shaped pattern and as the cells increase in depth, ink first transfers from all four edges to give a cell outline, this then gradually fills in until a fully-covered square is produced (Figs 2.40 and 2.41).

Invert or halftone gravure uses cells which vary in area to produce the tone range. Light tones are printed from small round cells, darker tones from larger cells similar to conventional gravure. The light to mid-tones can appear very similar to lithography, or even letterpress since the ink tends to transfer from the edges of cells, producing a rim of ink with a hollow centre – sometimes referred to as a 'doughnut'. This can be confused with 'squash'.

Two defects are particularly characteristic of gravure printing. Sometimes, thin streaks of ink will be seen running through the print. These are called 'doctor streaks' and are caused by either a nick in the doctor blade, a piece of grit lodged under the doctor blade or a score in the cylinder surface. The other defect is called 'skipping' or 'speckle'. It usually appears in the light to mid-tones and is caused by white spots where ink has failed to transfer from one or more cells. It is caused by rough or poorly coated paper and will occur wherever there is a pit in the paper surface larger than a gravure cell. In this case, the cell cannot make contact with the paper and no ink transfers.

Letterpress

Letterpress prints are identified by reverse side embossment and 'ink-squeeze' or 'squash'. The absence of reverse side embossment, it should be noted, does not mean a print is not letterpress. However, the edges of letters are particularly characteristic showing a heavy rim of ink, slightly separated from the main body of the character. The amount of squash is influenced by a number of factors. On a rough, uncoated paper, the rim of ink is less clearly defined than it is on the smoother coated paper. The printing pressure, amount of ink applied, and the material of the printing plate will also affect the degree of squash. Letterpress halftone dots are essentially circular, in the mid-tones they form a chequerboard pattern and in dark tones they join together. They exhibit squash in the same way as the edges of letters. A highlight dot often appears as a rim of ink with a hollow centre. Solid areas, particularly on uncoated papers, often show white spots on the print where the ink was unable to reach the bottom of the pits in the paper surface. This is called 'non-bottoming'. This can be minimized by increasing pressure and ink film thickness but this also increases ink squash on halftones, thereby changing tonal value. This degradation of quality is more pronounced the finer the screen, so coarse screens have to be used for halftones printed by letterpress on uncoated papers (Fig. 2.42).

Screen printing

The main characteristic of screen printing is the thickness of the ink film which may be up to 30 μm. It can often be felt by running the fingertips

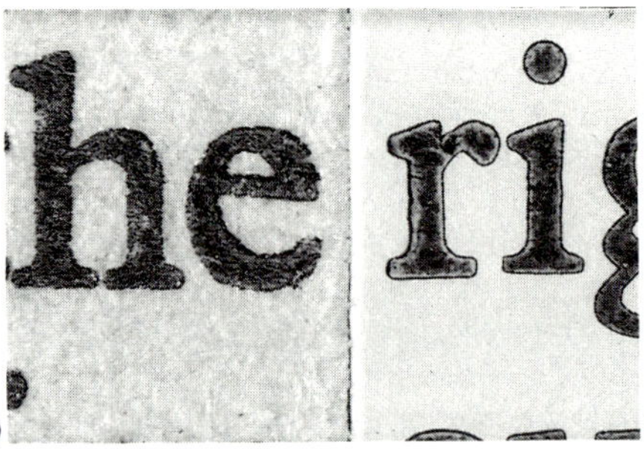
(a) (b)

Fig. 2.42 Letterpress prints: (a) uncoated paper; (b) coated paper.

lightly over the printed surface. Even with 'thin film' inks the deposit is still much thicker than that obtained from other processes. The print will also be patterned with the mesh structure of the screen fabric. This is most apparent at the edges of a print and gives an irregular zig-zag outline. Type characters also show this but generally, much care is taken to minimize it so the effect may not be obvious.

Electronic printing processes

It is difficult to give guidelines on the recognition of the electronic printing processes since they are so varied. Ink-jet processes usually use a fairly coarse matrix for type characters, and this is commonly found on packaging materials. The highest resolution ink-jet system currently works at about 8 dots/mm so resolution will always be low compared with other printing processes. Ink-jet images also tend to have a flat and matt look about them, and are low in density.

Electrophotographic prints, especially if of type only, can appear very similar to lithographic printing. However, if viewed at reasonable magnification the fine particle structure of the toner will be apparent, especially around edges, and indeed there may be a random scattering of toner particles extending into non-image areas. The typical resolution of about 12 dots/mm will only allow the reproduction of very coarse halftones (say 30 lines/inch screen). Such prints can often be easily scratched off the surface.

2.10 SUBSTRATE SELECTION

Paper requirements for printing can be classified into three main areas:

- general paper properties;
- runnability;
- printability.

General paper properties

Paper has a number of physical and chemical properties which influence printing characteristics. The main ones are:

- smoothness;
- absorbency;
- compressibility;
- surface strength;
- hygrostability;
- acidity or alkalinity;
- opacity;
- gloss;
- colour.

Optical properties that govern the visual appearance of paper – colour, opacity and gloss – will be of equal importance whatever printing process is used, as will absorbency. The level required of these properties will depend on the nature of the work being done.

The requirements of other properties – such as smoothness, surface strength, hygrostability and acidity – vary with the printing process being used.

Runnability

Most research on the suitability of paper for printing has been concentrated with its ability to produce a good quality print, but there is little point in buying paper of high printability if it is incapable of running through the press from feeder to delivery. This runnability requirement becomes of greater importance as press speeds increase. A paper that may run well can be ruined by bad handling and storage. For sheet-fed presses, good trimming, flatness and even moisture content are vital. Web printing imposes many additional requirements since web breaks are very undesirable, and current attention is turning to the elastic properties of the web since this greatly affects dimensional stability and hence register.

Printability

The printing processes have certain requirements of the substrate.

For offset litho printing, the paper must have a strong surface. If fibres come away from the surface during printing (linting), they will build up on the offset blanket, decreasing ink transfer. To return the print to normal the press has to be stopped to wash the blanket. It is impossible for the printer to maintain constant colour throughout the run.

The deposits on the blanket accept water rather than ink, produce white fibrous markings on solids, and interfere with halftone printing, causing loss of contrast and definition. The fluff deposit on the blanket will wear the plate, especially if it contains abrasive loading materials from the paper.

Coated papers for offset must be strong enough to resist the pull of the tacky ink, otherwise areas of coating may be pulled away from the surface, giving a defect known as picking.

Paper for all processes, but especially litho printing, must be treated properly or problems will occur. Creasing down the centre of the sheet, for example, happens on tight-edged paper. If a stack of paper is left standing in air that is too dry for it, the edges will lose moisture and shrink. The sheets of paper become saucer-shaped, with tight edges and a baggy middle. When this paper passes through the nip on the press, it is mangled flat and the baggy middle forms a crease.

Stacks of paper left standing in an atmosphere that is too damp will develop wavy edges, and these sheets will also crease when printed, but the creases are usually nearer the edges. Even if the distortions are not big enough to form creases, misregister will occur.

Creases are more likely in offset litho than in other methods of printing due to the overall contact of the rubber blanket, both in printing and non-printing areas.

Paper must be reasonably dimensionally stable, i.e. change as little as possible with changes in humidity. It must also have the right moisture content for the printing conditions it is to encounter. Care must be taken to protect the edges of stacks from atmospheric variations, by using moistureproof wrappers and stack covers.

The problems of casein scumming have been mentioned previously, although such coated papers can be printed without difficulty, providing the casein has been hardened so that it is insoluble in water and does not transfer to the plate. The problem is not often seen now since it is well known and therefore avoided.

This is an example illustrating an important property of all offset lithographic papers. They should not contain water-soluble materials that would adversely affect the plate if they transferred to it in the damping water. Paper coatings should be insoluble and unaffected by contact with water. If a coating swells in water, it will be weakened, and picking may occur on the second colour which is being printed on the weakened surface, even though the surface strength may be adequate when dry.

Flexographic printing prefers a smooth substrate and is extensively used on plastic films and foil. The rubber or polymer plates used to, however, provide some latitude and rough materials such as paper sacks and corrugated cardboard can be printed.

Gravure can be used to print a wide range of substrates including plastic films. Smoothness, under pressure, is perhaps the most important substrate property otherwise speckle occurs. Thus rough papers which are highly compressible may print satisfactorily while some 'higher quality' coated papers may not. The problem may be reduced by using electrostatic assist.

For letterpress printing, smoothness is the most important property. If a rough paper is used, solids will not 'bottom' properly. This can be overcome by increasing the ink film thickness but then dot squash will increase and coarser screens must be used for halftones.

Screen printing is perhaps the most tolerant of all processes and can be made to print on almost any substrate, and on many pre-formed containers, providing the ink can be suitably formulated. Electrographic printing generally has quite stringent requirements, and the faster the particular printer runs, the more exacting these requirements are. Often the properties are more to do with paper handling than printing – many of these printers heat the paper strongly to fuse the toner (e.g. 200°C) and excessive curl and brittleness must be avoided. The paper surface is, however, important. Non-coated papers tend to give better image adhesion since the toner can key on. Difficulties can be experienced in trying to print over pre-printed areas (e.g. tinted areas on forms) – image adhesion is poor and tends to produce a form of set-off.

Ink-jet printing can be used on a wide range of substrates but since a very fluid ink is used, the final dot size printed is affected by the wetting characteristics of the substrate's surface. Quite objectionable dot spread can occur in bad cases.

Substrates should be selected not just to suit the characteristics or requirements of the printing process but also those of any subsequent processing stages and the end use of the product. Surprisingly, this is frequently overlooked. Thus, if a job is to be laminated after printing, the substrate and inks must be appropriately chosen (and an excess of anti-set-off spray powder avoided). Prints on matt coated papers tend to have poor rub resistance, and the resulting product will be disappointing if the finishing processes involve rough handling. Much packaging work involves not only coating or laminating after printing, but cutting and creasing. Cracking of folds is a frequent problem, partly caused by wrong choice of materials. Subsequent performance on the packaging line is also important and depends very much on substrate properties.

2.11 THE NEED FOR COMMUNICATION

Almost all printing requires good communication between the parties involved. The best way to achieve this is to use as many standard specifications as possible. The ones that are especially relevant are:

- BS5750 or ISO9000 – standards defining an approach to quality management that requires all parties in the production chain to comply. It has been widely adopted in the printing of packaging but its use is spreading to all forms of print;
- BS950 Parts 1 and 2 or ISO 2664 – specify viewing conditions for the graphic arts. These standards specify the conditions under which original artwork (including transparencies), proofs and production prints should be viewed;
- BS4666 – a specification of standard ink colours for four-colour process lithographic printing. This is the only real standard for ink colours that now exists, and consequently even some non-lithographic inks are sometimes described as meeting the requirements of the standard.

There are several proofing and printing specifications aimed at different product sectors, which, while not having the authority of a BS or ISO standard, have nevertheless been accepted on a *de facto* basis. The aim of these standards is to optimize the reproduction process overall from beginning to end, and ensure that proofs are made in such a way as to simulate the printing process to be used. Such standards usually specify:

- solid densities for each process colour;
- dot gains for each process colour;
- the substrate to be used for proofing;
- the type of control elements to be used on the proof;
- screen rulings, dot shapes, screen angles, positive/negative, right or wrong reading to be used for colour separations;
- type of black printer to be used (e.g. amount of under-colour removal or grey component replacement).

Examples of this type of standard are:

- UKONS – UK Offset Newspaper Specification, for the production of colour separations and proofs of advertisements to go in newspapers.
- SNAP – the US equivalent of UKONS.
- IFRA – the European Newspaper Research Association, have also

published a similar standard which attempts to unify the standards of individual countries in Europe.
- FIPP – the European Periodical Publishers Association publishes a standard specification for heat-set web offset printing of magazines on coated and WSOP papers.
- SWOP – the US standard for web offset printing. This is actually several standards in one (including those for gravure publication printing). It is probably the most comprehensive of all the specifications published.
- FOGRA – the German Printing Research Institute publishes a standard in the form of a complete programme aimed at systematic colour reproduction.

All these standard specifications encourage a systematic controlled approach to colour reproduction which always shows benefits whenever it is implemented.

CHAPTER 3
Colour and colour matching

Colour surrounds and affects us all. This is not new. Primitive man, we may assume, was depressed by a grey sky, cheered by a blue one, and awed by the sight of a full rainbow. Certainly the colours which could be obtained from the earth and plants intrigued him to the extent that he used them for body painting and to decorate his cave dwelling. Colour in commerce followed swiftly. Cloth, glass, stones and pottery, goods that were aesthetically pleasing, could be traded for food. This commercial aspect has continued to grow through the years until almost everything in our lives is coloured. Books, films, television, packages, and now more and more newspapers, appear in colour.

In recent years people involved in marketing have realized that colour sells. Brand names are frequently associated with house colours and manufacturers have become extremely demanding in their specifications. Ink manufacturers and printers are therefore under increasing pressure. Much of the basic physics and chemistry of colour has been understood for over 100 years. Even some of the complex mathematics involved in colour measurement and match prediction was explained 50 years ago (Kubelka and Munk, 1931). We could not at that time measure colour with sufficient accuracy nor process the enormous quantity of data quickly enough to make effective use of these developments.

Colour measuring instruments which are much more sensitive than the human eye are now available. These can be directly connected or interfaced with relatively cheap computers capable of processing data at incredible speeds.

3.1 THE PHYSICAL NATURE OF COLOUR

When as children we learn to recognize the colour of everyday things, we are shown examples of objects which we are told are yellow, red or blue. Objects that are black, grey or white are not described as coloured, but the colour scientist must include them. The difference between the black, grey or white objects and the yellow, red and blue objects can be described in terms of hue. The former do not possess a hue, but they can be described in terms of their lightness. If we take a matt black object, like a piece of black velvet, we can say that it has no lightness at all, but if on the other hand we look at a piece of white chalk we can say that it has maximum lightness. The colour physicist in fact uses a block of pure barium sulphate as his 100% white object. A series of neutral grey objects can be arranged in order of their lightness, between the black and the white. If we call the black 0% light and the pure white 100% light, we can give values to our greys.

Now let us consider a similar arrangement of objects that have what we generally term colour, or more correctly hue. Newton (1730) demonstrated that what we call white light can be split up into a full range of hues, namely red, orange, yellow, green, blue and violet, using a glass prism. This he called the visible spectrum. This natural sequence of hues can be arranged in a circle since after blue we have a red tone blue, we call violet, which can lead us through purple back to red.

This circle can be divided quantitatively like the lightness scale into say 100 divisions.

There is a third dimension to colour which is called intensity, strength, purity, vividness, or better still, saturation. If we take a pure white paint and add more and more pure blue paint to it we can make a range of specimens ranging from pure white through a series of increasingly strong blues to pure blue. This could be repeated for all the different hues, and again using all the greys to produce an almost infinitive range of colours. Each step can, as with the lightness and hue, be quantified.

We therefore have a description of colour in terms of three independent variables, lightness, hue and saturation. This is why it is impossible to arrange logically a full range of shades on an ink maker's shade card, which has only two dimensions. The answer is to arrange them in three-dimensional colour space, a concept we shall return to in considerable detail later in the chapter.

With a very few exceptions, objects do not emit coloured light, but only look coloured when they are illuminated. An object that we describe as white in normal daylight, looks as it does because it reflects most of the light that falls on it. If the object absorbs all of the light it appears black, but if it absorbs some portions of the spectrum more than others it will appear coloured. For example an object that absorbs only blue light will appear yellow, and one that absorbs red light will appear green–blue, or what we call cyan.

The term reflectance needs more careful definition. A good mirror reflects most of the light falling on it, but we do not call a mirror white. The mirror reflects all the light but also the images in front of it. This type of reflectance is called specular. On the other hand a wall painted with a matt white paint again reflects most of the light, but not images. It looks like this because the white pigment particles in the paint scatter the light in all directions. This type of reflectance is called diffuse. A wall painted with a glossy white paint will exhibit both specular and diffuse reflectance.

So far only white light has been considered. If the light used to illuminate the object is deficient in some parts of the spectrum, it too will be coloured. Light from a sodium vapour street lamp is pure yellow, containing no blue. The blue vehicles therefore look almost black under these conditions.

As Newton observed, however, the rays themselves are not coloured, it is only when they react with an object or the eye that the sensation we call colour is perceived.

The colour we see therefore depends not only on the nature of the object but also the light used to illuminate it.

Light sources

If a perfectly black object like a carbon rod becomes hot, the molecules begin to vibrate and emit energy, some as light. The more intense the heat, the brighter the emitted light, but not only does the light become brighter, it changes colour. With increasing energy the colour of the light shifts from red towards the blue end of the spectrum.

It is convenient therefore to specify a light source as being equivalent to the so-called black-body radiator at various temperatures. This is known as its colour temperature. The Kelvin scale is used (°Kelvin = °Celcius + 273). A tungsten filament lamp has a colour temperature of about 2800 K and north daylight about 6700 K (Wright, 1969, pp. 4–12).

The actual light sources, however, differ from this simple idea, and some standardization is necessary. An international committee, the Commission Internationale de l'Eclairage (CIE) set up standards for illumination in 1931. They were called simply A, B and C. Illuminant A represents a tungsten filament lamp of colour temperature 2800 K, B sunlight at 4900 K and C north daylight at 6700 K. Most colour matching is carried out using north daylight, and it was found that the CIE illuminant C contains less UV than true north daylight. When a fluorescent component, for example an optical brightening agent, is present the matchings made under illuminant C are unsatisfactory. A new illuminant of colour temperature 6500 K, with a UV content near to north daylight, has been introduced. This is known as D65. The more efficient fluorescent tube lamps are now more frequently used in shops, factories and laboratories than the tungsten filament lamp.

The primary source of light in the fluorescent tube lamp originates in the collision of mercury atoms during the passage of an electric current. The light emitted from this source is made up of the yellow, green and violet line spectrum of mercury. The inside of the tube is coated with a substance which fluoresces in the mercury discharge. These substances are known as phosphors and most of the light emitted by these lamps is produced as a result of this fluorescence. A number of phosphors are blended to produce lights of different colour temperature. Typically a cool, that is blue tone, lamp, will have a colour temperature of about 4000 K and a warm, redder tone lamp about 3000 K. A new range of high efficiency lamps using rare earth phosphors, similar to those employed in colour television tubes, have recently been introduced by Philips as their Colour 80 series. The light from these, although showing a good overall colour balance, is emitted in three relatively sharp bands, rather the more normal wide band of continuous radiation.

Problems can arise when matching prints and fabrics coloured with

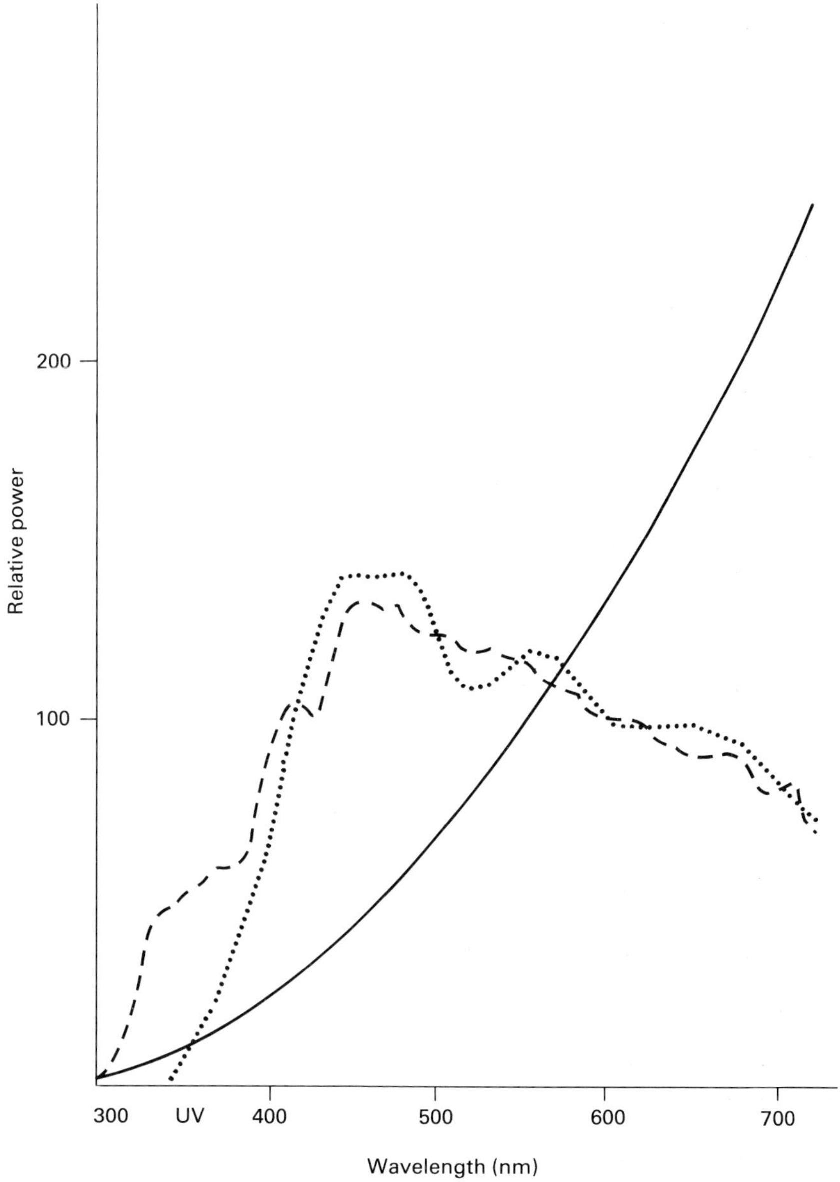

Fig. 3.1 (a) A range of illuminants: (——) a tungsten filament lamp; (----) D65 north daylight, new CIE standard; (. . . . . .) C north daylight, original standard. (Courtesy of Philips Electronics)

different pigments and dyes. Two specimens may match under daylight, but not under artificial light. This phenomenon is known as metamerism. The problem can become more acute with the high efficiency lamps.

Special fluorescent lamps are available for colour-matching areas of the laboratory and special colour-matching booths. These lamps emit light evenly over the whole of the visible spectrum (Philips Lighting). Figure 3.1(a) shows a range of illuminants and Figs 3.1(b–d) are spectra from various fluorescent lamps.

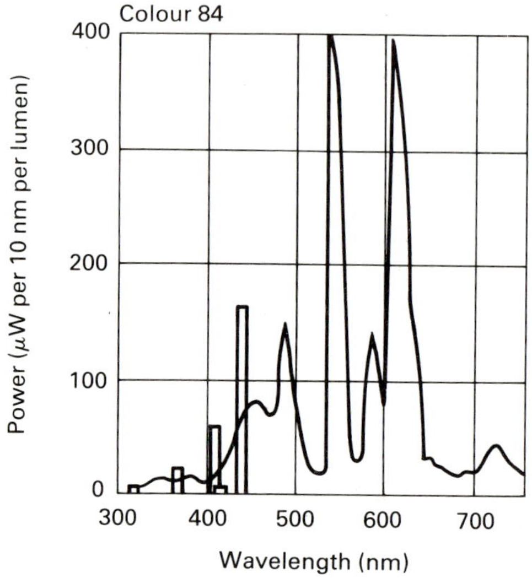

(b)

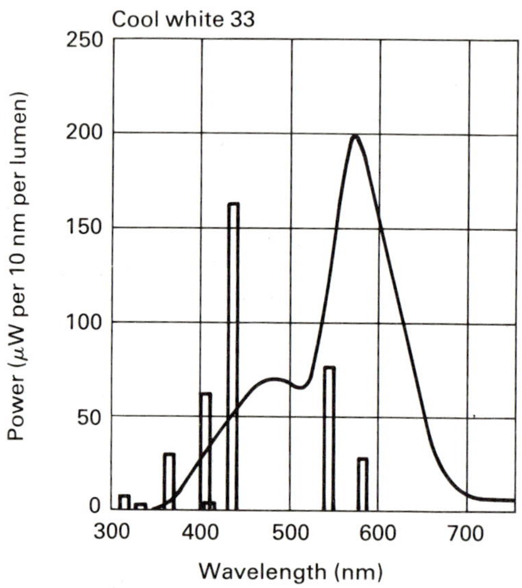

(c)

Fig. 3.1 (b) Colour 84. (Courtesy of Philips Electronics) (c) Cool White 33. (Courtesy of Philips Electronics)

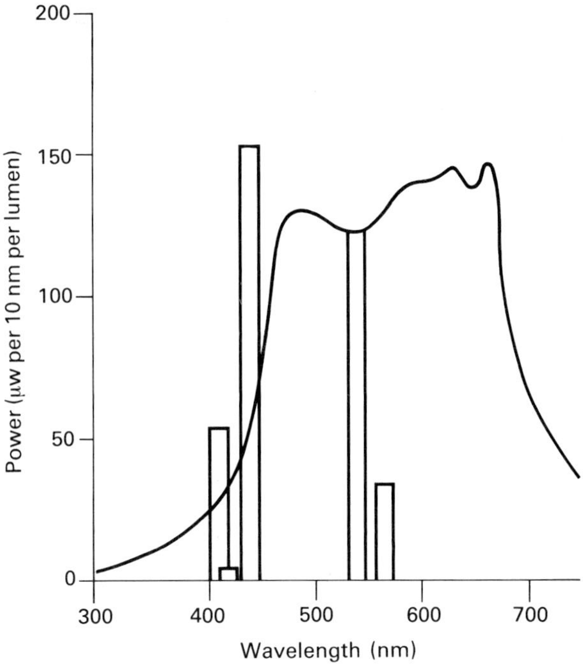

Fig. 3.1 (d) Spectral power distribution – Graphica. (Courtesy of Philips Electronics)

3.2 THE PERCEPTION OF COLOUR

The eye

A cross-section of the eye is shown in Fig. 3.2. Light enters the eye through the cornea and is brought to focus on the light-sensitive retina by the lens. The light-sensitive cells in the retina can be divided into two types, rods and cones. The rods are the most sensitive cells, but cannot detect colour. The cones are concentrated in the central area of the retina, around the fovea and are sensitive to colour. It is now generally accepted that there are three types of cone, sensitive to red, green and blue light. At low levels of illumination, when only the rods are activated, the vision is said to be scotopic. At higher levels, when colour can be perceived through the cone cells the vision is photopic.

The sensitivity of the eye varies greatly with wavelength. The rods are most sensitive at 507 nm and the cones at 555 nm (Fig. 3.3; Wright, 1969, Ch2).

Defective colour vision

The spectral sensitivity of the eye varies from one person to another Wright studied the response of 10 observers so that a standard observer could be defined.

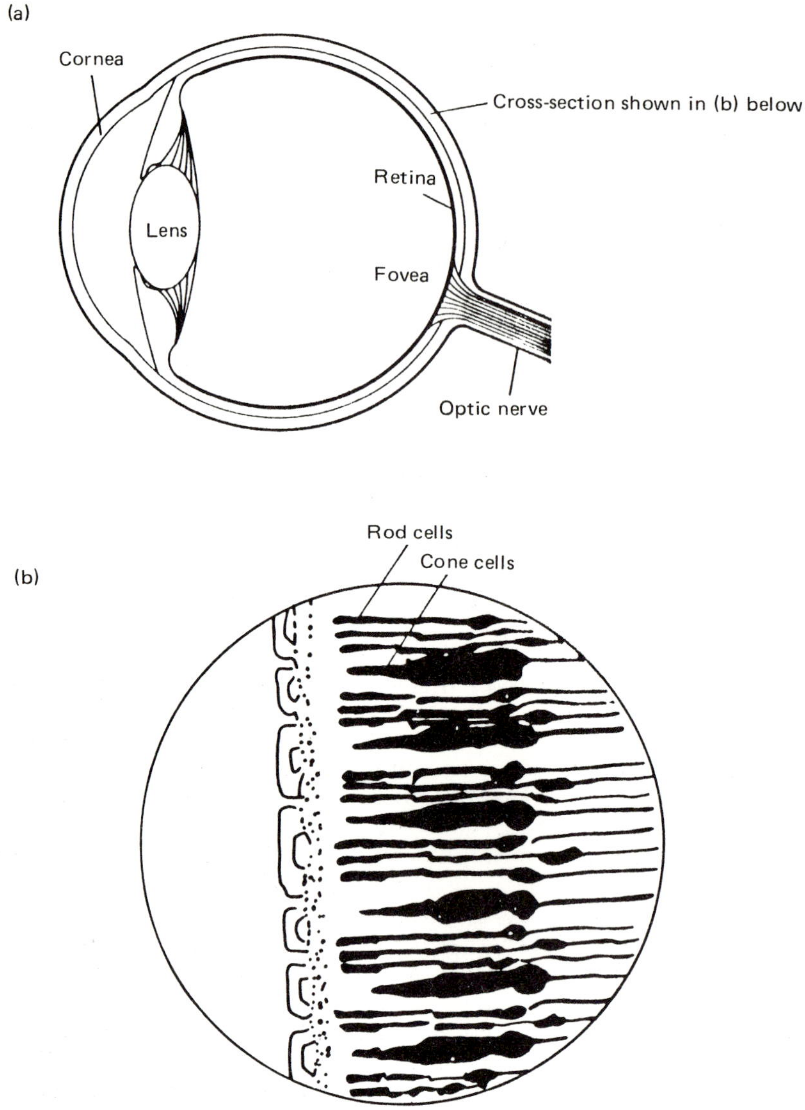

Fig. 3.2 (a) Section of the eye; (b) section of the surface of the retina.

The majority of people have vision which corresponds quite closely with this standard, but about 5–8% of all men, and a very much smaller number of women, differ markedly from this standard and are quite often termed colour blind. True colour blindness, the inability to perceive any colour, is extremely rare. The most common defect is the observer's inability to distinguish red and green at low light levels. This red–green blindness varies in degree, but it is important to be able to detect it, since

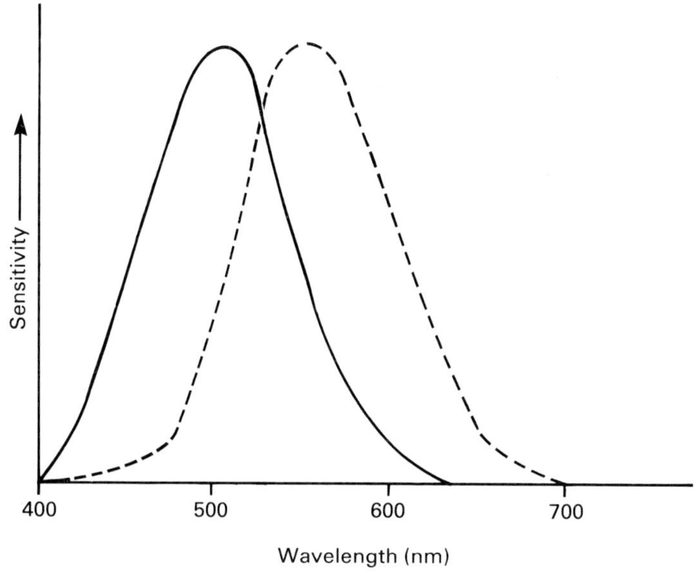

Fig. 3.3 The spectral sensitivity of the eye: (——) rods; (----) cones.

such people will generally be unable to match and assess the colour of inks and prints. The most satisfactory test was devised by Ishihara (1974). His test consists of a book of 38 coloured circles made up of a large number of coloured dots. The person under test is asked to read a number or trace a line on each plate. A reference book is provided for assessment, and if the test is carried out correctly it gives very reliable results. Even people with normal colour vision differ in their ability to match colours. Attempts have been made to assess colour matching aptitude, by selection of plastic chips made deliberately close, but not exact matches, to standards. This type of test is not generally very satisfactory, and the best colour matchers are selected by practical ink-matching tests after a period of training.

Both the cornea and the lens become yellower with increasing age and this leads to a decrease in the blue sensitivity. This is not generally a big problem, but it should be borne in mind when the inevitable arguments on the acceptability of a match arise. 'Yellowness of vision' test equipment is available (McLaren, 1983, pp. 85–88).

Chromatic adaptation and colour constancy

If a print is viewed in a light cabinet under simulated daylight and the illumination is quickly changed to tungsten light, the colours will change. The most obvious change will be that the whites will turn yellow. After a very short time, however, the eye adapts to this change and again sees the white as white. This phenomenon is called chromatic adaptation.

When the eye is adapted to the illuminant change, the colour of most common objects will be accepted as perfectly normal, although the spectral composition of the light reaching the eye will have changed appreciably. The blues for example will appear much redder, but will still be described as blue rather than purple. The objects viewed under these changing conditions are said to possess colour constancy. Some objects do not possess colour constancy and the change in illuminant will result in a change in description. Some light sources, for example certain fluorescent lamps, also change the perceived colour in such a manner that the colour description is changed.

Metamerism

If the colourants in two objects differ appreciably in their spectral distribution, they may match in one illuminant but not in another. This is known as metamerism. It becomes a problem when an ink matcher uses a series of standard bases which differ from those used in the original pattern. It is even more troublesome when an ink for a wallpaper has to match a fabric dye. The ink maker cannot use the same colourant and some degree of metamerism is inevitable.

The spectral reflectance curves of metameric matches always cross one another, sometimes several times (Fig. 3.4).

The eye sees them as a match under say D65 illuminant since the differences are averaged out in each of the red, green and blue regions. If

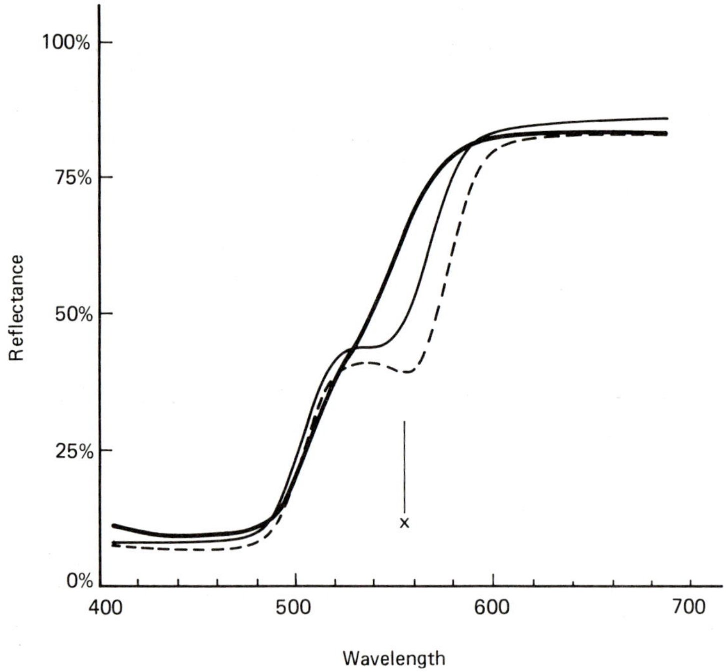

Fig. 3.4 Spectral reflectance curves of metameric matches. ——, master standard; ----, production print; ▬▬, metameric match.

the spectral profile of the illuminant changes, this may not then be the case and the shades do not match. The effect is particularly noticeable with fluorescent lamps (Fig. 3.1). If differences in the spectral curves of the specimen correspond with one or more of the sharp emission bands present in this type of illuminant marked colour differences will be seen in that region.

Matching specimens which differ in gloss are always troublesome. If the illumination is diffuse it may be possible to assess the colour differences but if the illumination is highly directional the top surface reflection, the specular component, dilutes the colour with white light, in other words, desaturates the glossy specimen. The specimens then may match when viewed at one angle but not at another. This phenomenon is termed geometric metamerism.

Dichroism

When assessing the colour of an ink, it is important to standardize the application to minimize differences due to many factors, a very important one being film thickness.

Apart from the obvious variation in apparent strength, film thickness also affects the hue. This change in hue with film thickness is called dichroism (Chamberlin and Chamberlin, 1979).

Illumination quality and levels

From the foregoing discussion it will be evident that if satisfactory matching of colours is to be made the quality and the level of the illumination used must be carefully standardized. All too often disagreements arise because of the use of non-standard lamps in the matching cabinets. The cabinets should be carefully maintained and cleaned. Objects not under test should be removed from the cabinet as these can change the illuminant effect and distract the observer.

The size of the specimens should be controlled. A very small specimen is difficult to assess against a very large standard. A minimum of 50 mm square should be used if possible.

3.3 ADDITIVE AND SUBTRACTIVE COLOUR MIXING

The additive primaries

The visible spectrum can be divided into three equal portions, the region from 700 to 600 nm being red, 600 to 500 green and 500 to 400 blue. It is impossible to produce light filters that will transmit 100% of each portion with a cut-off exactly at the boundaries, but filters can be produced to give curves as shown in Fig. 3.5.

If red, green and blue filters are placed in front of three separate projectors and the colour spots allowed to superimpose, the red and green

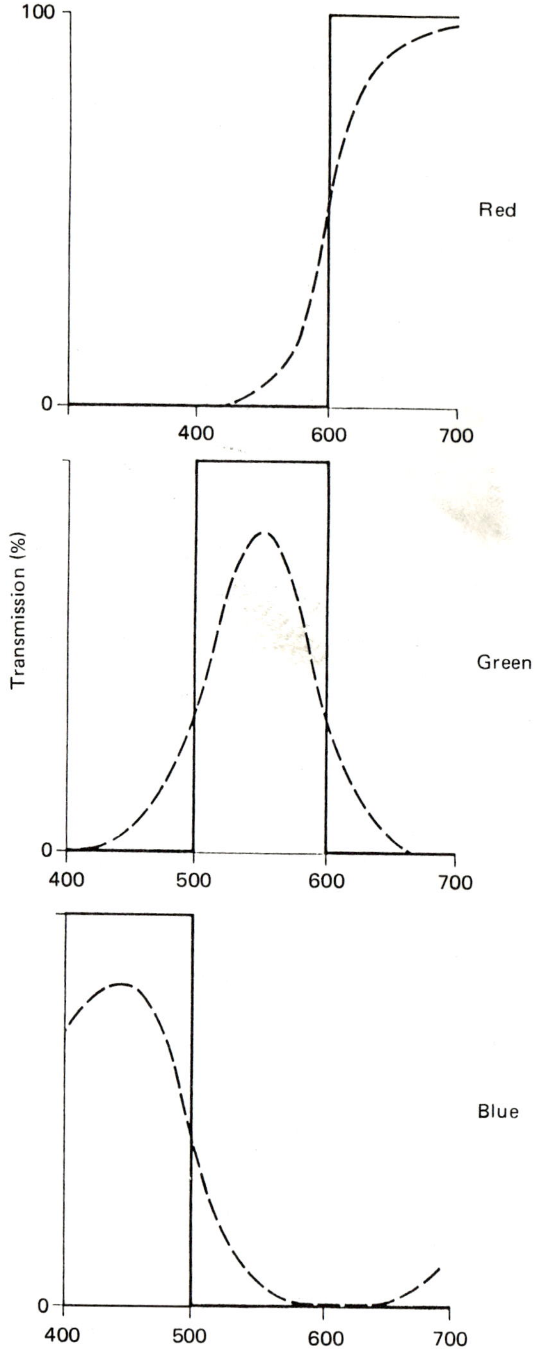

Fig. 3.5 The additive primaries: (——) spectral profile of an ideal filter passing exactly 1/3 of the total spectrum; (----) actual profile of a filter.

together will produce yellow, the green and blue, cyan and the blue and red, magenta. All three together theoretically produce white, but since in practice they differ from the ideal primaries the result will be somewhat grey. Since we are adding these lights to produce the colours, they are known as additive primaries.

The subtractive primaries

The printing of colour relies on the selective absorption of certain regions of the spectrum by dyes or pigments in the inks. Again the spectrum can be divided into three equal bands. If the blue third 400 to 500 nm is absorbed, the colour will be seen as yellow, likewise the removal by absorption of the green third 500 to 600 nm will produce magenta and removal of the red third 600 to 700 nm will give cyan.

It is theoretically possible therefore using yellow, magenta and cyan inks to produce all the required colours. Superimposition of all three producing black. This is a subtractive colour system (Fig. 3.6).

In fact the actual inks differ from the ideal subtractive primaries just as the filters differed from the ideal additive primaries. In process printing, therefore, it is necessary to correct this imperfection by the addition of a black printer and modifications of the primaries using colour masks. This procedure is dealt with under the graphic reproduction section.

The CIE system

In 1861 Clerk Maxwell (1890) studied ways of representing the additive primaries, red, green and blue, graphically so that a quantitative description of colour could be obtained. He arranged the primaries at the three points of an equilateral triangle.

If we refer to the colour diagram of the additive primaries (Fig. 3.5), it can be seen that equal parts of blue and green produce a blue–green colour called cyan. This can be represented on Maxwell's triangle at a point midway between the blue and green points. Similarly red and green produce yellow; red and blue produce magenta. Addition of all three primaries gives white which Maxwell put in the centre of the triangle (Fig. 3.7). If we start at the middle and move along a line to the blue point, we pass from white through increasingly saturated blues to the completely saturated blue primary. Similar lines can be drawn from the white point to the green and red primaries.

If now we carry out a similar exercise with the intermediate colours, cyan, yellow and magenta, we find that point C on the triangle, while being correct for hue, is not a completely saturated cyan. It is possible to produce a more saturated cyan light than that made by adding green and blue lights, since these are not ideal primaries. The saturated cyan will be at a point C_1 outside the triangle. This presents problems if we wish to quantify the colour. The desaturated cyan at point C can be described in terms of positive amounts of blue and green, say y parts of green and z parts of blue, but since the saturated cyan C_1 is outside the triangle we can only describe it by introducing a third quantity. This lies on the line from

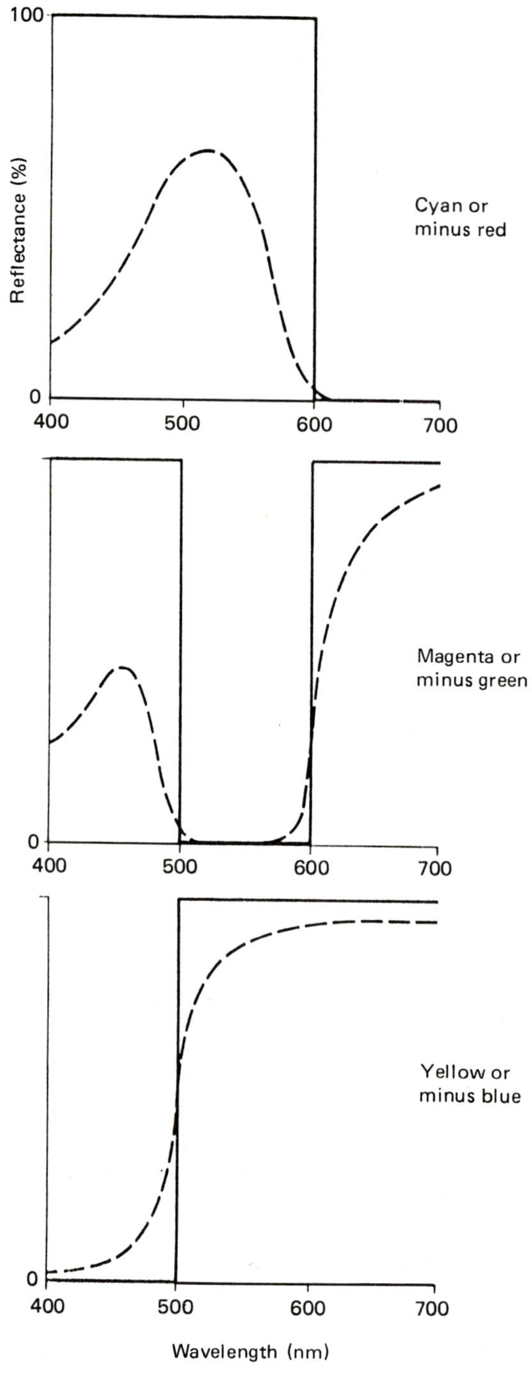

Fig. 3.6 The subtractive primaries: (——) spectral profile of ideal inks absorbing exactly 1/3 of the spectrum; (----) actual inks.

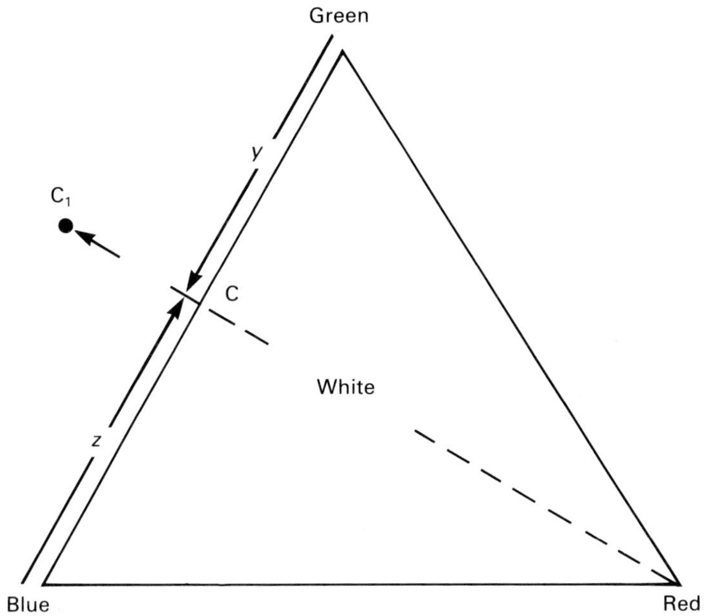

Fig. 3.7 Maxwell's triangle.

the red point through the centre, but since it is beyond the centre white point the values will be negative. We then have to describe the point C_1 on this diagram as being made up of y parts of green, z parts of blue and minus x parts of red, i.e.

$$C_1 = yG + zB - xR$$

If now all the saturated colours are plotted, they will be found to lie outside the triangle, but touching the red, green and blue points. This line is known as the spectrum locus (Fig. 3.8).

To avoid negative quantities in our description of a colour, the locus can be enclosed in a triangle X, Y, Z (Fig. 3.9).

In such a diagram the points X, Y, and Z represent colours which are more saturated than actual colours, they are therefore unobtainable with any light source and are described as imaginary primaries (McLaren, 1983, pp. 96–102). X is a reddish purple of a higher saturation than a real colour, Y has a saturation higher than any green and Z higher than any blue. Using this system we can define a colour by measuring it with an instrument and obtaining values for X, Y and Z.

These values can be represented as ratios:

$$x = \frac{X}{X + Y + Z} \quad y = \frac{Y}{X + Y + Z} \quad \text{and} \quad z = \frac{Z}{X + Y + Z}$$

then

$$x + y + z = 1.$$

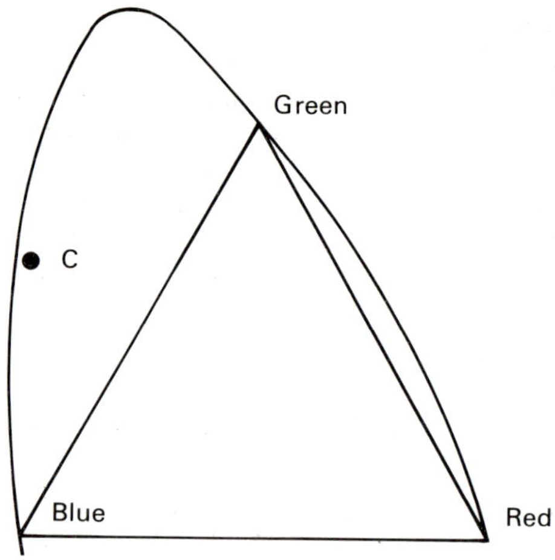

Fig. 3.8 The spectrum locus.

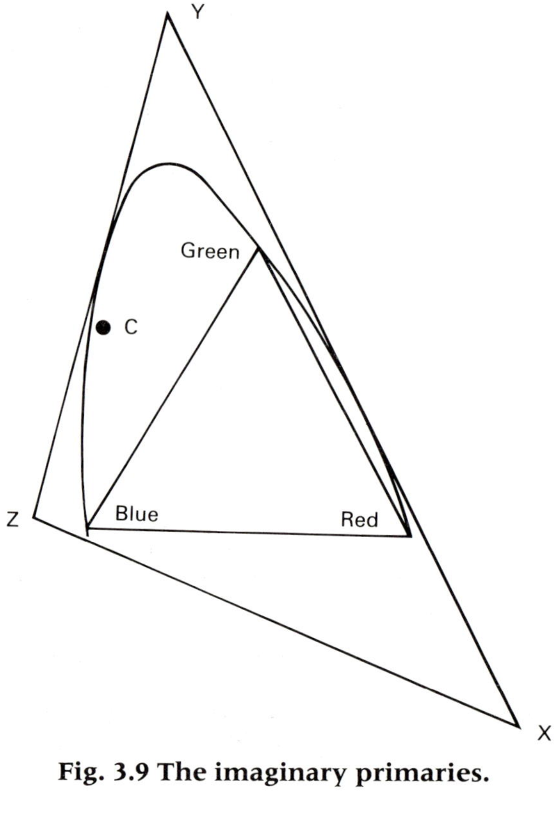

Fig. 3.9 The imaginary primaries.

It is therefore only necessary to use two of these values, known as chromaticity co-ordinates, to define the hue. The values can then be plotted on ordinary graph paper.

This description of a colour does not take luminance or brightness into account. Two colours can have the same *x* and *y* values, therefore the same hue, but differ in luminance. To quantify luminance it was decided to define the X and Z stimuli mathematically in such a way that they have zero luminance, while the value of Y, deliberately chosen to equal the response curve of the eye, gives the luminance information.

This system was adopted by the Commission International de l'Eclairage (CIE) in 1931. A full explanation is given by McLaren (1983, pp. 102–107).

3.4 ORIGINS OF COLOUR IN PRINTED MATERIAL

Nearly all printing inks contain three basic ingredients, a colourant which may be a pigment or a dye, a resin system and a thinning solvent.

Pigments

Pigments are finely divided particles relatively insoluble in the medium in which they are employed. These particles when adequately dispersed in the medium, absorb and scatter light. If the absorption is selective, the pigment will be coloured. Pigments may be entirely inorganic or organic; they may also be metallic (usually calcium or barium) salts of complex acids, or consist of organic dyes laked on to inorganic substances.

Dyes

If the colouring agent is soluble in the medium in which it is being used, it is called a dye. Dyes are normally pure organic or organo-metallic complexes. The colour of a dye is produced by selective absorption, but since no discrete particles are present, no scatter occurs and the system is transparent.

Origins of colour

The configuration of atoms in the organic pigment molecules which give rise to selective absorption of light are called chromophores (Patterson, 1967). Examples of these are:

$$-C=C-,\ -C=O,\ -C=N,\ -N=N-,\ -N=O$$

Certain other groups, while not themselves responsible for selective absorption, enhance the absorption of the chromophore and are known as auxochromes. Common auxochromes are:

$$-NH_2,\ -OH,\ -NO_2,\ -CH_3,\ -Cl,\ -Br$$

The colour of inorganic pigments is also a function of chemical composition, but is strongly influenced by the crystal form of the substance.

Transparency and opacity

Since the average particle size of pigments is close to the wavelengths of visible light, i.e. 0.4–0.7 μm, the ordinary laws of optics do not apply, and radiation is scattered by these small particles. Scatter is a complex phenomenon, primarily dependent upon the relative refractive indices of the pigment and the medium in which it is dispersed, and the particle size. Particle shape, degree of aggregation and the wavelength of the light also affect scatter.

The opacity of a pigment depends normally upon its ability to scatter incident light, but absorption also plays a part, and the overall opacity is due to a combination of both properties. In the case of a white pigment, for example titanium dioxide, little or no absorption occurs and the opacity is due almost entirely to scatter. As the refractive index of a pigment approaches that of the medium in which it is dispersed, the amount of light scattered decreases and the pigment becomes more transparent.

Transparency of a pigment therefore is partially dependent upon the medium used to disperse it. The transparency of an ink is usually expressed in terms of the ratio of reflectance of the film printed over a black background to that over a white background. A 1–5 scale is employed, 5 indicating high transparency.

Colour strength

The colour or tinting strength of a pigment or dye is a measure of its ability to impart colour to a system. Although the colour strength of a pigment is primarily related to chemical composition, particle size and distribution, the actual value in the case of a printing ink will depend also upon the film thickness and the concentration of pigment in the ink. Colour strength can be measured either by reduction with a standard white pigment in a suitable medium or by measuring the pigmentation level required to give a standard-depth print at a specified film thickness.

Substrate effects

Since most printing inks are more or less transparent, the substrate on to which the ink is printed influences the final colour of the print. The colour, absorbency and surface roughness of the substrate, together with the ink film thickness will all affect the final colour of the print.

Substrate effect is the source of most of the problems encountered when using computer-assisted match prediction systems.

Colour Index classification

The most useful classification of pigments and dyes is the *Colour Index*, prepared by the Society of Dyers and Colourists. The index is prepared in five volumes; the first three list the colourants according to type and usage and give technical information under CI generic names. For example, Lake Red C is listed as CI Pigment Red 53. Volume 4, sometimes referred to as Part 2 reference lists the colourants according to their chemical constitution. The above example is listed as CI reference 15585. The final volume lists commercial names and manufacturers' code letters.

3.5 GRAPHIC REPRODUCTION

It was shown in section 3.4 that it is theoretically possible to produce colour prints having all the possible colours from three inks, cyan, magenta and yellow, each absorbing one-third of the visible spectrum (Fig. 3.6).

The first step in graphic reproduction is to analyse the original object in terms of the absorption of light in each of these wavebands, then print the amount of each ink necessary to produce that effect (Chambers, E.).

Three-colour printing

Originally the object was always photographed separately through three primary colour filters to produce negatives from which the printing plates were made. This colour separation process and many subsequent stages in graphic reproduction is now often done by very sophisticated electronic scanning equipment. The basic principles,

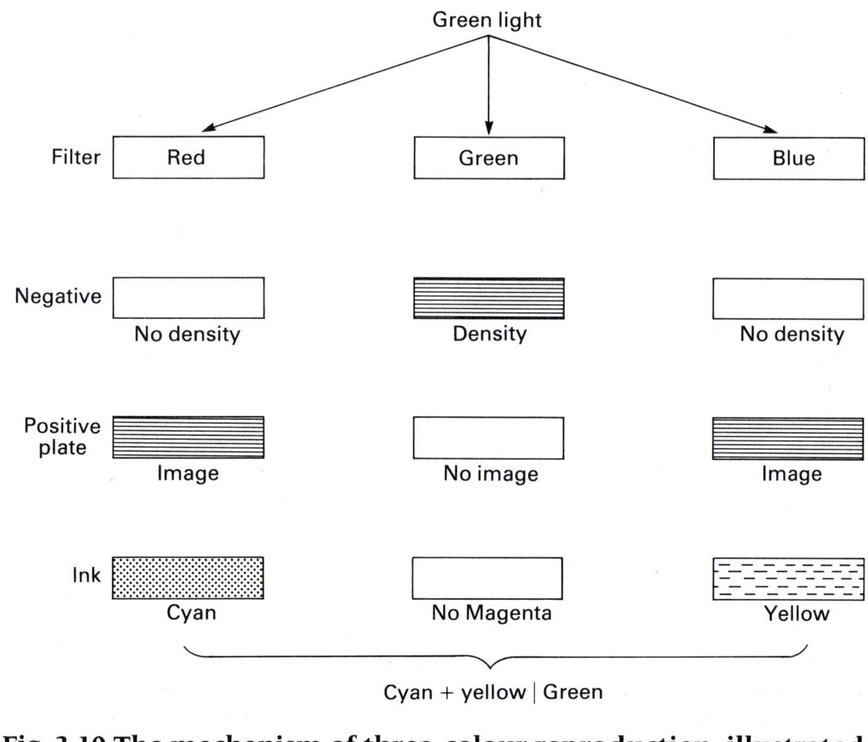

Fig. 3.10 The mechanism of three-colour reproduction, illustrated by the reproduction of a green subject.

however, are better explained by reference to the camera procedures. When the object is photographed through a red filter, the negative will have density in the red areas only. This negative is then used to produce a positive printing plate, on which the image area corresponds to the red, absorbing, that is the blue-green, region of the spectrum. This will be the cyan printing plate. Similarly the green filter will be used to produce the green absorbing, that is magenta printing plate and the blue filter the blue absorbing, that is the yellow printing plate (Fig. 3.10).

Four-colour printing

Since neither the separation filters nor the printing inks match the ideal primaries, the combination of all three inks will not produce a completely satisfactory black. It will appear as a very dark brown. To remedy this, a fourth impression, the black printer is introduced to strengthen shadows and ensure that the darkest tones are neutral black.

Under-colour removal

Rather than just adding black in this way, it is possible to analyse the separation and substitute black for the proportion of the colours producing a neutral tone. This portion is sometimes referred to as the achromatic component; the process is known as under-colour removal.

Figure 3.11 shows an example of the relative concentrations of the three process colours required to produce a particular colour. It can be seen from this diagram that much of the colour is being used to produce black, sometimes known as the achromatic value. All or part of this portion can be substituted with black thereby reducing ink cost, and sometimes improving drying.

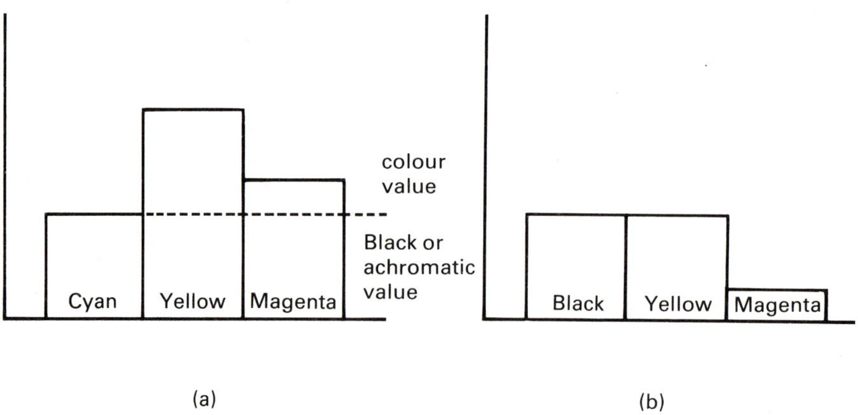

Fig. 3.11 (a) Conventional separation; (b) achromatic separation.

Masking

Sometimes a picture will have a bluish, greenish or reddish cast, which is due to an imbalance in the colours. The grey balance may be corrected by masking. The magenta ink absorbs too much blue light and the cyan too much green light. Only the yellow is satisfactory. It is necessary therefore to compensate for these deficiencies. The magenta ink cannot be made to reflect more blue light, therefore the yellow, which absorbs blue, is reduced. Similarly, the cyan cannot be made to reflect more green light therefore the magenta is reduced. Originally these alterations were made by hand retouching, but subsequently these reductions were made by masking the original negatives with low density positives produced from the other negatives, and more recently electronically by the scanner.

Halftone dots

To produce a full range of shades it is necessary to vary the relative proportions of the three inks in any one area.

In the case of conventional gravure printing, it is possible to vary the thickness of the ink film by changing the cell depth but in the lithographic process a uniform ink film is applied. The variation in colour density in this case is achieved by breaking the image up into a screened dot pattern and altering the size of the dot. Superimposition of the dots varying in size and colour would not produce the required colour effect, they are therefore juxtaposed. Since they are so minute the eye cannot discern the individual dots and a single composite colour is seen.

Dot gain

Throughout the process the size of the dot will change to some extent. The greatest effect is the increase in dot size which occurs during printing. This is partly physical as the fluid ink dots are squashed, and partly optical due to the light scattering effect of the paper. Control of dot gain is important since differences from ink to ink will change the colour balance. Dot size can be measured indirectly with a densitometer using the Murray–Davies equation; the calculation being carried out in some cases with a built-in computer.

Dot generation

The dots are produced in the photographic process by placing a ruled screen in front of the negative. It is now possible to generate these dots electronically. Very accurate control of dot size and shape is possible with this type of equipment.

3.6 THE MEASUREMENT OF COLOUR

Colour is a subjective sensation, therefore in strictly physical terms we cannot measure it (Wright, 1969). The amount of light reflected from the surface of an opaque object, or transmitted by transparent material can be measured, but the relationship between this instrumental value and some quantitative description of colour is complex. This section will deal with the measurement of the colour of opaque materials, in particular ink applied to a substrate, by instruments known as reflectance colorimeters and spectrophotometers. Those designed for the measurement of transparent materials, are known as transmittance instruments.

All colour-measuring instruments have three main components. A light source to illuminate the specimen, a system to select a specific region of the spectrum either by dispersing or filtering the light and a photodetector.

Colorimeters

The light source in these instruments is normally a tungsten filament lamp, sometimes halogen filled. The electrical supply is stabilized to ensure a constant light output.

Selection of the region of the spectrum required is by use of three coloured filters designed to correspond as nearly as possible to the three visual stimuli of the standard observer as defined by Wright (Fig. 3.12). The X stimulus has both a blue and a red component, the latter designated $\bar{X}$. It is more convenient to use separate X and $\bar{X}$ filters. The

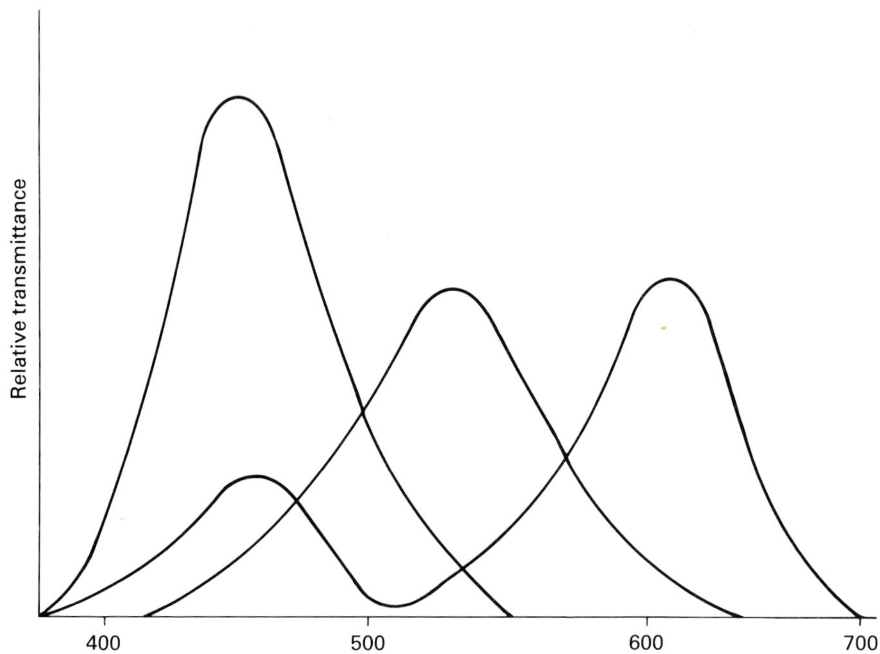

Fig. 3.12 Spectral distribution of the tristimulus filter.

so-called tristimulus colorimeter therefore quite often has four filters, X, Y, Z and $\bar{X}$ corresponding to red, green, blue (Z) and blue ($\bar{X}$). Alternatively, a proportion of the Z reading may be added to the X value to give a true X value.

Reflected light is measured using a photodetector, usually a photodiode and the value displayed by a galvanometer or printed out (Fig. 3.13). The read-out is usually given as a percentage of reflectance, the zero being set with a matt black substance, for example black velvet and the 100% by a pure matt white block of barium sulphate. In practice these standards are difficult to maintain and secondary standards, usually black and white ceramic tiles are used.

Densitometers

The reflectance densitometer is a special form of colorimeter, used to measure the four process ink colours, black, cyan, magenta and yellow on colour prints. The filters are designed to select light of the wavelength mostly absorbed by these primary colours. A red filter is used to measure the cyan print, a green filter for the magenta print and a blue one for the yellow print. The black is measured using light filtered to correspond to the average response of the eye across the whole spectrum.

High-quality filters are required for densitometers. Typical spectral profiles of the three colour filters are shown in Fig. 3.14. Our assessment of the strength of a colour is not related to the reflectance in a linear manner, but is very nearly a logarithmic relationship. The value used is called density D and is related to reflectance as follows:

$$D = \log_{10} \frac{1}{R}$$

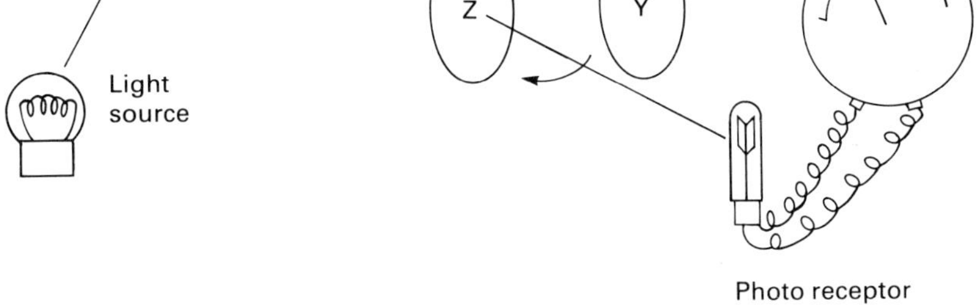

Fig. 3.13 Layout of a simple colorimeter.

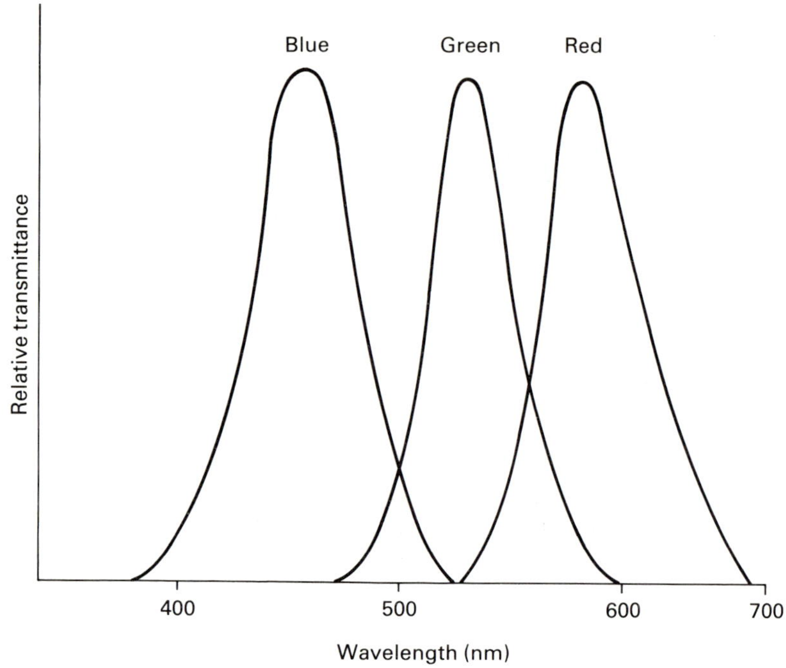

Fig. 3.14 Spectral profiles of typical densitometer filters.

where

$$R = \frac{\text{reflected light intensity } (R_1)}{\text{intensity of light reflected by the white paper } (R_w)}$$

to give

$$D = \log_{10} \frac{R_w}{R_1}$$

The sample is illuminated by a light source. The light beam first penetrates the ink layer and is thereby reduced in its intensity. The remaining light is then partly absorbed and partly scattered by the substrate. The scattered portion passes back through the ink film being further reduced in intensity before reaching the detector (Fig. 3.15).

In order to reduce directional effects, particularly with non-homogeneous samples, the light can be collected by a ring mirror before reaching the detector (Fig. 3.16).

Spectrophotometers

The light source in a spectrophotometer may be a tungsten filament lamp or a pulsed xenon discharge tube, often filtered to match D65 (diffuse north daylight).

Selection of the required wavelength of light is usually by dispersion

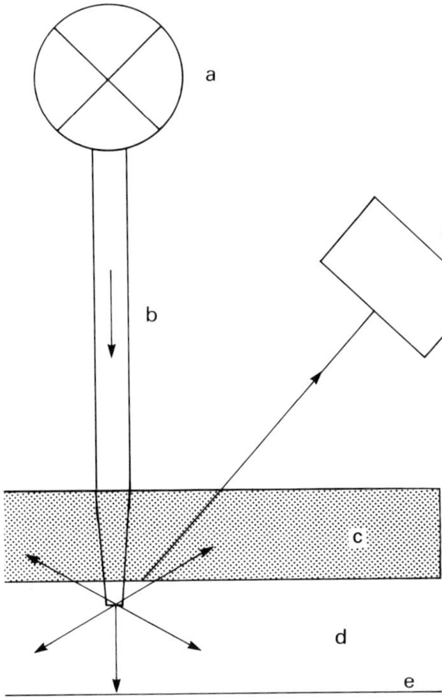

Fig. 3.15 Reflection densitometer: (a) light source; (b) illuminating beam; (c) ink layer; (d) substitute (paper); (e) substrate (filter); (f) receiver.

with a prism or diffraction grating. By scanning the spectrum it is possible to construct a continuous plot of percentage reflected light versus wavelength.

Some instruments used for colour match prediction work employ an array of 16 photodiodes which measure the output from the diffraction grating at 20 nm intervals simultaneously. A complete spectral reflectance measurement can be made in 4 seconds with this equipment (Fig. 3.17). Instruments using a series of narrow band pass filters, typically 16 with peak transmittance values at 20 nm intervals, are known as abridged spectrophotometers.

Optical geometry

The value required for colour measurement is diffuse reflectance but some of the specular component will also be included even if we use the special light traps designed to absorb this. The optical geometry of the instruments vary, and values obtained on one instrument can rarely be compared directly with those obtained on another of different design. Typical light paths are shown in Fig. 3.18.

Colour-measuring instruments are commonly interfaced, that is

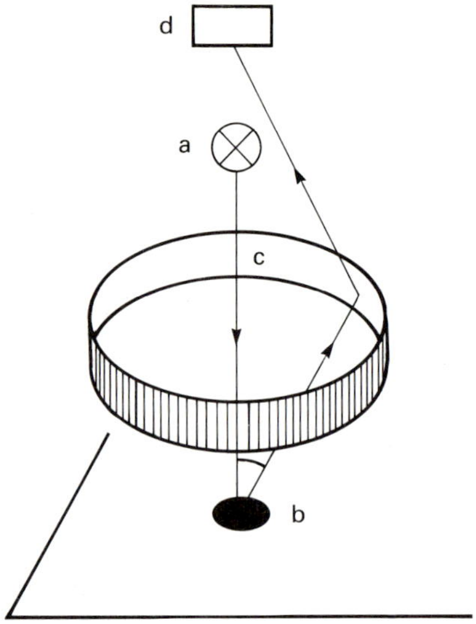

Fig. 3.16 Reflection densitometer: (a) light source; (b) measuring spot; (c) ring mirror; (d) receiver. (Courtesy of Gretag AG)

directly connected with computers, so that the often complex calculations required for calibration, colour difference and match prediction can be carried out more quickly and accurately.

3.7 THE RECORDING OF COLOUR DATA AND THE SPECIFICATION OF COLOUR

All printing ink manufacturers keep a large library of prints and standard ink samples against which production materials are compared. Prints must be carefully stored and be accompanied by data showing how the specimen was prepared, the weight of ink applied, the type of substrate employed and other details relating to special properties.

In addition to the manufacturer's own library, a number of national and international standard colour charts exist. One of the best known is the Pantone system in which eight basic colours, plus black and white are used to create a range of nearly 500 shades.

Formulations and prints on both coated and uncoated paper are included in the colour guide. Process colours are also included. The British Standards Institute issued a specification *Inks for Offset Three- or Four-Colour Printing*, BS 4666:1971, which conforms to the European Standard CEI 13–67. Colour, colour strength, transparency, lightfastness and solvent resistance are specified. The master standard inks are held by PIRA, who have also published a *Guide to Standardised Lithographic Colour Printing* which includes a densitometer calibration procedure.

One of the earliest colour atlases was produced by Munsell and is

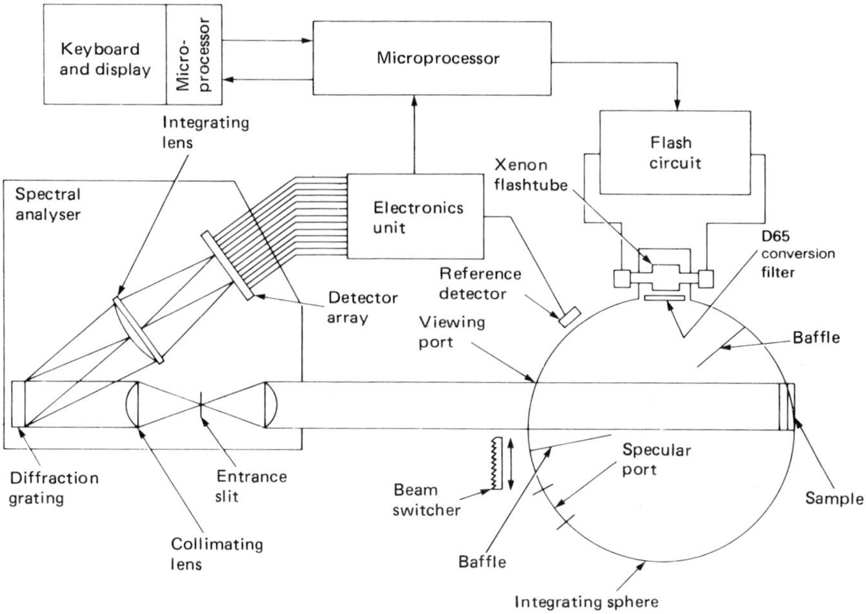

Fig. 3.17 A modern reflectance spectrophotometer – the Macbeth MS 2020. An operator-initiated command is issued at the keyboard which directs the command to the spectrophotometer microprocessor. This triggers the flash circuit which causes the xenon flashtube to flash. The light then illuminates the interior of the integrating sphere and illuminates the sample. Baffles prevent the light from directly striking the sample or the specular port. The specular port allows the inclusion or exclusion of the specular component. The beam switcher is a prism which automatically moves into the viewing port causing reference measurements of the sphere wall to be made. Light entering the spectral analyser via the fibre optics bundle is directed at the entrance slit which determines the wavelength resolution of the system. The beam is then transmitted through a collimating lens which directs the beam to the diffraction grating. The diffraction grating disperses the light into a spectrum and reflects the spectrum through an integrating lens, which directs the spectrum to the detector array. The detector array is a solid-state silicon photodiode array of 16 channels spaced at 20 nm intervals. The light is sensed by the electronics unit for processing. A reference detector is used to monitor the intensity of the flashtube, and permits compensation for pulse-to-pulse varaitions in flashtube output. (Courtesy of Macbeth Division, Kollmorgen (UK) Ltd)

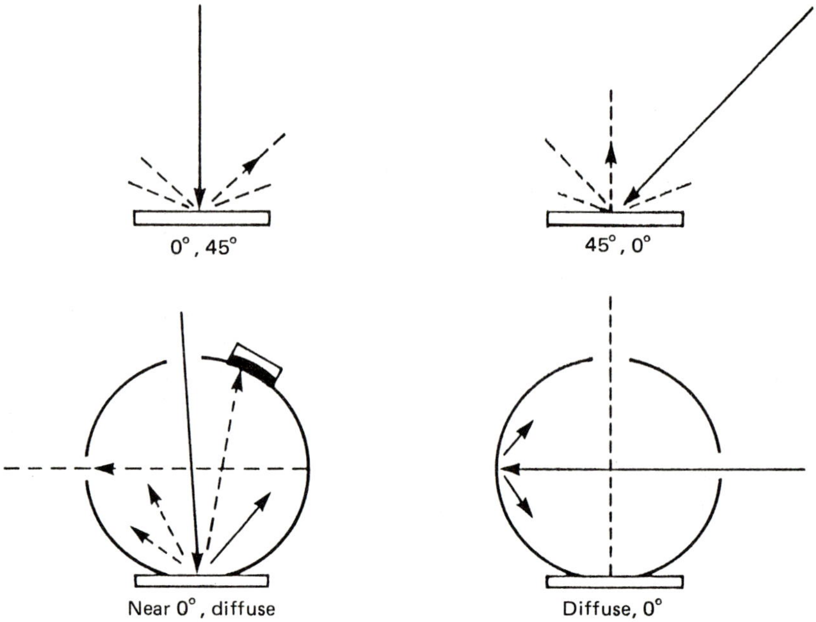

Fig. 3.18 Common illuminating and viewing conditions.

particularly valuable since the colours are arranged in such a way that there are equal visual differences in three dimensions which he called hue, value and chroma (Fig. 3.19; Munsell, 1905; Nickeson, 1969; Chamberlin and Chamberlin, 1979 pp. 22–25.)

The hue circle which follows the colour circle principle is divided into ten regions: red (R), yellow–red (YR), yellow (Y), green–yellow (GY), green (G), blue–green (BG), blue (B), purple–blue (PB), purple (P) and red–purple (RP), each region again being divided into ten sections. Value, which we now generally call lightness, is again divided into ten sections and chroma, or saturation, similarly divided on another axis, thereby producing a three-dimensional colour plot.

To define a colour by the Munsell system the hue, value and chroma are specified in that order. For example 3BG6/7 signifies a hue of 3 BG, a value of 6 and a chroma of 7.

Munsell therefore introduced the concept of colour space, a three-dimensional system in which all colours can be defined. Munsell's equal visual intervals, however, were approximate and could not form the basis of a quantitative description of colour.

If we now return to the X, Y and Z primaries introduced in section 3.4, we can construct a colour space using Y as the vertical 'lightness' axis and the x and y chromaticity co-ordinates as the other axes. This has the same configuration as the Munsell diagram, with a central vertical achromatic axis. While this is a quantitative description of colour the intervals in this colour space are not equal visual intervals (Fig. 3.20).

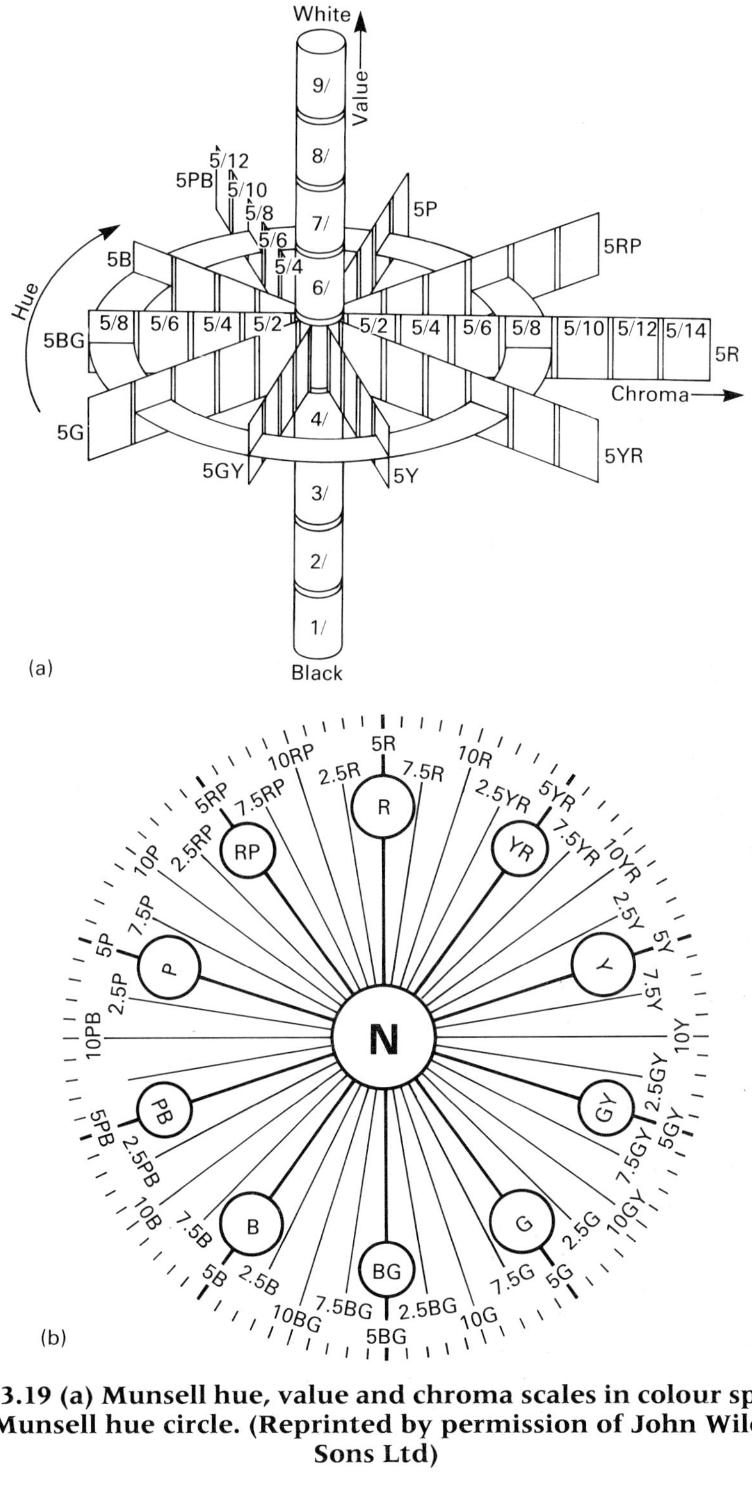

Fig. 3.19 (a) Munsell hue, value and chroma scales in colour space; (b) Munsell hue circle. (Reprinted by permission of John Wiley & Sons Ltd)

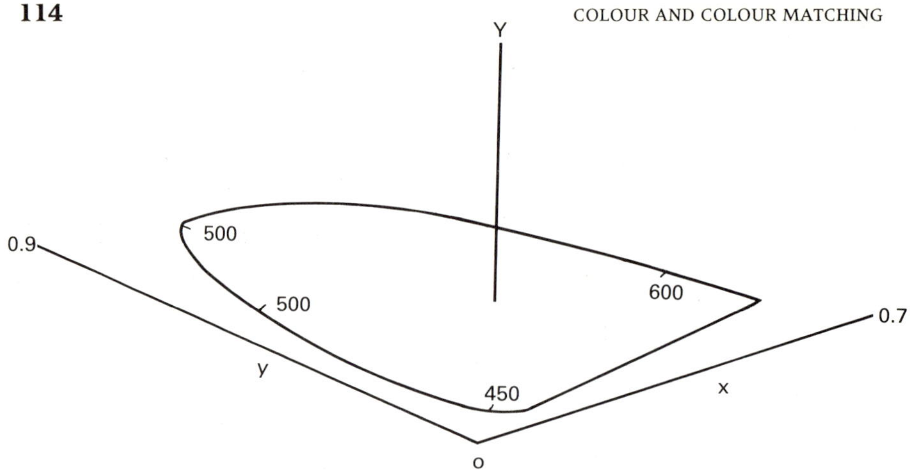

Fig. 3.20 Y, *x y* colour space.

Colour difference

The difference in colour between two specimens can be represented by their distance apart in colour space. Mathematically this can be expressed

$$\triangle E = \sqrt{(\triangle H^2 + \triangle S^2 + \triangle L^2)}$$

where $\triangle$ in this case signifies 'difference in', E the total colour difference (from the German *Empfindung*, meaning sensation), H, the hue, S the saturation and L the lightness (McLaren, 1983, Ch 10).

However, since the values derived from the colour-measuring instruments do not give an equal visual Munsell-type space it is necessary to attempt to convert these values so that $\triangle E$ is the same visually for all parts of colour space. The equations used are numerous, but in spite of many years of research, no one system satisfies all applications.

Ideally we should aim at a system where $\triangle E = 1$ is a good commercial match. A 'good commercial match' is difficult to define since it will depend upon the product and a customer's particular requirements. An acceptable match is still usually a matter of personal judgement and it is difficult to know where to draw the pass/fail boundary. In erring on the safe side we probably spend a lot of time getting matches closer than they need to be. Very frequently a fairly wide variation in strength is tolerated. On the other hand the hue is usually quite critical. A customer will often submit a standard pattern, sometimes including specimens illustrating colour tolerance boundaries. These, however, are again mainly to illustrate variations in strength.

The CIE have produced a number of equations to calculate $\triangle E$; the latest 'CIELab' equation was introduced in 1976. The L, a and b functions are derived from X, Y and Z. The chroma (C) is given by

$$C = (a^2 + b^2)^{0.5}$$

and hue (H), expressed as an angle from

$$H = \arctan (b/a)$$

with the $a+$ axis being at 0° and the $b+$ axis at 90°.

Adams and Nickerson's equation (ANLab) is also extensively used, but in 1976 McDonald used a different approach. Instead of attempting to derive a complex equation for all shades he constructed a range of ellipsoids in colour space defining the pass/fail boundaries of 55 colours, using a total of 640 matches. The axes of these microspaces are made linear with regard to visual response. Over a two-year period, Messrs J & P Coats, sewing thread manufactures, assessed 8454 matchings of 599 colours using 8 observers and these data were used to modify the original equations derived by McDonald. The equation is known as JPC 79. A similar approach was adopted by Taylor and Smart (1977) at Instrumental Colour Systems working with Marks and Spencer.

A similar approach is recommended by the Colour Measurement Committee of the Society of Dyers and Colourists using their equation CMC (l:c) (Clarke, McDonald and Rigg 1984). The ellipsoid semi-axes are calculated from the CIELab, L, C and H values, but weighting factors for lightness l and chroma c have been introduced into the equations. If these factors are set at unity the value will correspond to the perceptible colour difference. However, if wider tolerances for l and c are acceptable for a particular purpose, other values can be used. So far optimum values of l and c, for quantifying the acceptability of a colour match have only been determined for textiles. These are $l = 2$, $c = 1$. CMC (l:c) is the equation being considered by the British Standards Institution for a BS method for calculating colour difference for quantifying the perceptability and acceptability of small colour differences.

Total colour difference $\triangle E$ is normally calculated for illuminant D65. If the value changes with a change of illuminant the specimens are metametric.

Examples of a typical set of results produced on an ICS instrument are shown in Fig. 3.21.

3.8 COLOUR MATCHING

Selection of raw materials

The exact procedure used in colour matching of printing inks will depend upon the nature of the work and upon the skill and experience of the colour matcher. The pattern supplied may consist of a print, not necessarily produced by the process to be used or on the required substrate. It may be in the form of an artists colour wash, in which case the match will almost inevitably be metameric, since the pigmentation of artists materials differs from that acceptable for ink formulation.

A 'wet' sample is sometimes supplied, but although this can be very helpful it should be accompanied by a print. Before attempting a colour match, the exact requirements of the printer must be ascertained. This is

```
COLOUR DIFFERENCE PROGRAM      DATE 08/09/1986
Standard SF345 NEW STD R Code  3324
Batch     SF345 PH08328 1ST SUB U/F R Code  5000
      Equation  CMC(2:1)    10 Degree Observer
Standard
(CIELAB)L                A          B          C       H
D65            82.35    -3.47     -10.27    10.84   251.32
Delta    DE      DL       DC         DH
D65     0.67    0.33     0.15       0.56
               DARKER    MORE      BLUER
                        CHROMA   (REDDER)
A       0.52    0.33     0.09       0.39
               DARKER    MORE      BLUER
                        CHROMA   (REDDER)
TL84    0.66    0.34     0.16       0.54
               DARKER    MORE      BLUER
                        CHROMA   (REDDER)
CWF     0.58    0.33     0.20       0.42
               DARKER    MORE      BLUER
                        CHROMA   (REDDER)
```

```
COLOUR DIFFERENCE PROGRAM      DATE 08/09/1986
Standard SF345 NEW STD R Code  3324
Batch     SF345 PH08328 1ST SUB U/F R Code  5000
M&S 83A  10 Degree Observer  Pass/Fail Tolerance  0.9
PASS   D65    DE   0.7   0.6 Bluer   0.3      0.2
                         (Redder)    Darker   Stronger
PASS   A      DE   0.7   0.6 Bluer   0.3      0.1
                         (Redder)    Darker   Stronger
PASS   TL84   DE   0.8   0.8 Bluer   0.3      0.2
                         (Redder)    Darker   Stronger
PASS   CWF    DE   0.7   0.6 Bluer   0.3      0.3
                         (Redder)    Darker   Stronger
```

Fig. 3.21 Two examples of colour differences of pale-blue specimens calculated with CMC (l:c) and M & S 83A for four illuminants.

probably best achieved by preparing a formal check list.

The general type of ink, the process, initial and anticipated order quantities are the vital pieces of information required by the laboratory manager before handing the project to the colour matcher. This should be followed by details of the press and the substrate. A sample of the substrate should be obtained. General ink specification details should include finish required, setting speed and colour sequence. Fastness properties must then be considered, special attention being paid to light and product fastness.

If the ink is required for food packages, pigments meeting the required restrictions must be selected and properties such as odour, rub resistance and suitability for deep freezing have to be taken into consideration. If after printing, further processes are involved, their effect will have to be taken into account. Over-printing, waxing, heat-sealing, laminating and embossing can have a deleterious effect if unsuitable materials are chosen in formulation.

Matching techniques

When all these details have been assembled, the colour matcher can then start to examine the colour and decide the pigments most likely required. The very experienced matcher will probably be able to determine this fairly readily without reference to laboratory records and suppliers literature. In many cases, however, it will be necessary to examine full strength and reduced prints of single pigments against the pattern under various illuminants.

The colour circle

It is useful to consider the base colours in terms of a colour circle on which the purest and brightest colours lie. The arrangement is similar to that outlined in the discussion of the Munsell system. All the intermediate colours lie inside the circle with black at the centre (Fig. 3.22).

All the greens obtainable with a mixture of say a green tone blue and a green tone yellow will lie on a line joining these on the circle. Now consider the use of the same blue with a redder tone yellow. In this case the line passes closer to the centre black position indicating that the result will be 'dirtier'. In general, therefore, mixings represented by longer chords on the colour circle will be dirtier and those by short chords, cleaner. The colour circle representation does not take the third colour

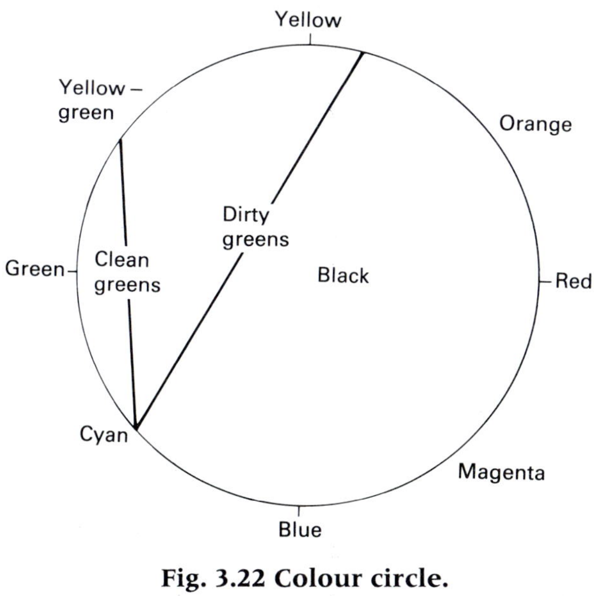

Fig. 3.22 Colour circle.

dimension, lightness, into account. Addition of white will increase this dimension but may also result in a hue shift.

Procedures

Inks are normally matched by blending pigments already dispersed in a suitable varnish in the form of finished inks, or pastes with high pigment loadings. The ink factory will normally manufacture pastes with the commonly used pigments dispersed in a range of vehicles. Quantities of these should be obtained in the laboratory and where possible matchings made with those available. This will avoid unnecessarily increasing the range of pre-dispersed pigments. Pigments with special properties, for example very high light and weather-fast materials for poster work, may be too expensive to keep in paste form. The laboratory may, however, have small quantities dispersed in suitable varnishes for matching purposes. Very small quantities of black and violet may be used in final shading. In these cases special pastes containing 5–10% of the concentrated paste may be used to avoid weighing errors. The use of dry colour in the matching process should be avoided if at all possible, because differences in degree of dispersion will add a further variable to the system.

It is sometimes convenient to carry out the initial matching in a conventional varnish even when a more unconventional system is dictated by the printer's requirements. This must then be translated into dry colour weights and checked using an automatic muller or small triple roll mill.

Oil inks

Between 10 and 15 g of the selected pigment pastes or inks is put on the greaseproof paper circles (sometimes called 'skins') and weighed to three decimal places. These are then arranged around a clean glass plate. Initially, small quantities of pastes are blended with a palette knife until the bulk, which should be about 5 g, closely matches the pattern. Using the index finger a little is 'tapped out' on to the chosen substrate. This process requires a lot of experience, but with practice a weight similar to that of each main printing process can be 'tapped out' and the masstone, undertone and strength assessed. Corrections for both shade and strength are now made, again followed by tapping out and close examination under various illuminants.

In some cases a 'drawdown', described later, is done at this stage. The bulk should now be increased to about 20–25 g, shade and strength being adjusted as before. It is advisable to work on the strong side, as it is comparatively easy to reduce strength with a clear 'tinting medium' or varnish. A weak result on the other hand will necessitate a complete rematch in order to be sure that the tint balance is maintained. The ink must now be test printed, again on the substrate to be used by the printer, using for example an IGT or Prüfbau machine set to print the desired weight. The first print will usually be some way out, more frequently for strength rather than hue. Corrections are again made.

The pastes on the greaseproof 'skins' are reweighed and the quantity of each one used is noted. A provisional formula can now be calculated and adjusted to a suitable bulk, ideally 100 units. At this stage the formula should be carefully examined and any apparently unusual features checked. The experienced matcher will have a good idea of the sort of combination likely to be required for a given shade. If it varies significantly form this, a rematch should be made.

If a very large bulk is to be manufactured, a small trial batch is made which may be printed on a suitable proofing press. Adjustments for properties such as tack can be made at this stage. When the material is put into production, it is often more convenient to use a drawdown for quality control purposes. Since the 'standard' and 'batch' are of similar composition, this method yields an adequate control procedure. If the materials differ in composition in any way, the results of a drawdown can be misleading and tapping out or test printing should be used.

The drawdowns are carried out by placing about a 0.25 ml of the inks to be compared about 2 cm apart on a suitable paper clipped on to a smooth board, or preferably a block of hard rubber. The inks are then drawn down the paper with a 7–8 cm wide knife blade. Towards the end of the drawdown, the pressure is relaxed to give a thicker film convenient for assessing masstone. It is advisable to use a paper with a black band printed across it. The drawdowns which should just touch, can then be examined over black and white substrates to assess opacity.

Liquid inks

The basic ideas discussed under oil ink procedures still apply, but the handling of these very rapid drying products presents special problems. The base inks used for the matching should be near to print viscosity, and a bulk of about 100 g should be used. It is most convenient to work in paper cups. Solvent loss should be kept to a minimum by storing the bases in tins with tight-fitting lids. Solvent lost while working must be replaced so that the viscosity of both standard and sample are the same and close to printing viscosity before application to a substrate.

The drawdown procedure using the broad knife is still useful, but on very absorbent substrates it is better to use a hand anilox roller. The anilox roller will ensure that a controlled quantity of ink is applied in a manner closer to the printing process than the drawdown.

A development of the drawdown technique involves the use of wire-wound bars, generally known as K bars. The gauge of the wire used to wind the bar determines the quantity of ink carried. The bars are numbered 1 to 8, No. 1 applying a wet film approximately 6 μm thick up to No. 8 which applies 100 μm, No. 1, colour coded yellow, is normally used for flexo inks and No. 2, colour coded red for gravure. A rubber impression bed of controlled hardness is used to support the substrate. The inks being compared must be of the same viscosity and the drawdown carried out smoothly and quickly without allowing the bar to roll.

In order to standardize the weight applied to the bar and the speed of the drawdown an electrically or pneumatically driven unit has been developed.

Proofing of liquid inks is carried out using small laboratory units of various designs, but a very convenient machine is a small flat-bed gravure type.

3.9 INSTRUMENTAL COLOUR MATCH PREDICTION

The selection of colourants and the tinting to shade of a printing ink is time consuming. The delays in printing ink manufacture and the holding of printing machines while corrections for colour are made is extremely costly. It is not surprising, therefore, that newer computer-assisted methods of colour matching, extensively used in the paint industry have been developed for printing inks. In addition to time savings, more economical use of colourants and redundant inks can be made.

Basic theory

In section 3.2 it was shown that light falling on a print is partly absorbed K and partly scattered S. The scattered light can be measured as diffuse reflectance R. The relationship between these values was worked out by Kubelka and Munk, (1931):

$$\frac{K}{S} = \frac{(1 - R)^2}{2R}$$

The model used was an ideal surface coating. The film being of even thickness, opaque, with a pigment of very small particle size compared with the film thickness, evenly dispersed throughout (Fig. 3.23). While paints are often quite close to this model, printing inks are not. The films are relatively thin and transparent, the pigment particles therefore are large in relation to the film thickness. Perhaps of more significance is the often uneven thickness of the ink film (Fig. 3.24).

At present the original Kubelka and Munk equation forms the basis of

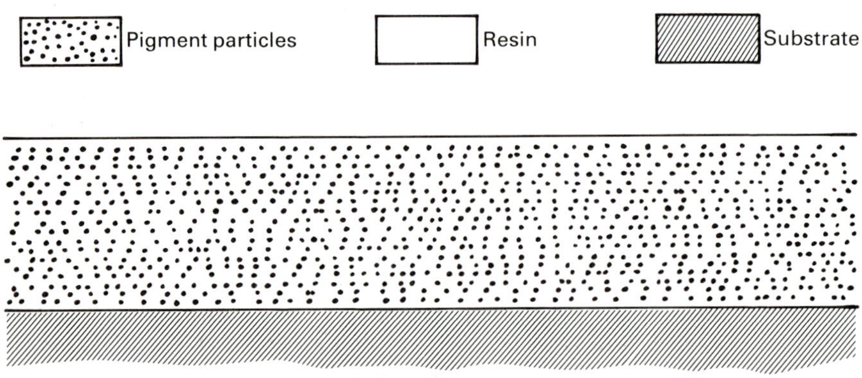

Fig. 3.23 Ideal surface coating used in Kubelka–Munk analysis.

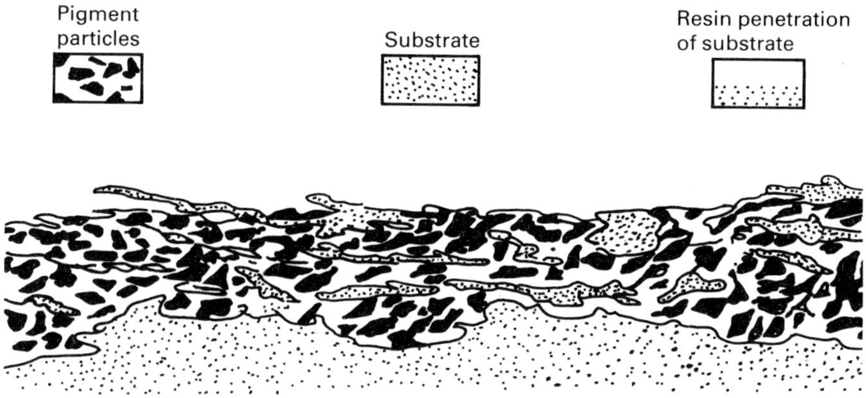

Fig. 3.24 Cross-section of an ink on an absorbent substrate.

most programmes used for colour match prediction of printing ink, but in spite of the numerous correction factors applied, results are frequently unsatisfactory. Nevertheless the procedures have proved valuable in some areas and research is continuing into more satisfactory relationships between measured reflectance and colourant composition.

In the 1940s Duncan, working at the Paint Research Station, showed that the K/S function is additive. This means that at any wavelength the K/S calculated from the reflectance is the sum of the K/S values of the constituents. Additionally the values are linear, that is, for example, by doubling the concentration of a constituent the K/S contribution of that constituent is also doubled.

The reflectance curves and the derived K/S curves for a magenta and yellow ink are shown in Fig. 3.25. It can be seen that the resulting red made from a mixture of these primaries has a reflectance curve which cannot readily be seen as a combination of the magenta and the yellow. On the other hand, the K/S curve of the red is clearly the K/S curves of the magenta and yellow added together.

The linearity of the K/S function is illustrated by a family of K/S curves for a rubine red ink, with increasing concentration of pigment (Fig. 3.26). Note that at maximum absorption, ≈560 nm, the K/S values are more nearly linear with respect to concentration, while this is not so in the case of the reflectance curves.

It is now possible to write an equation relating the reflectance at each wavelength to the concentrations of the constituent coloured pigments:

$$\frac{(1-R)^2}{2R} = \frac{K}{S} = a\frac{K}{S_A} = b\frac{K}{S_B} = c\frac{K}{S_C}$$

where a, b and c are the concentrations of pigments A, B and C. In the case of printed materials, the K/S of the substrate must also be considered, since the film of ink is relatively transparent. However deducting the K/S values of the substrate is frequently unsatisfactory, since the observed colour is not simply the colour of the ink film plus the colour of the substrate.

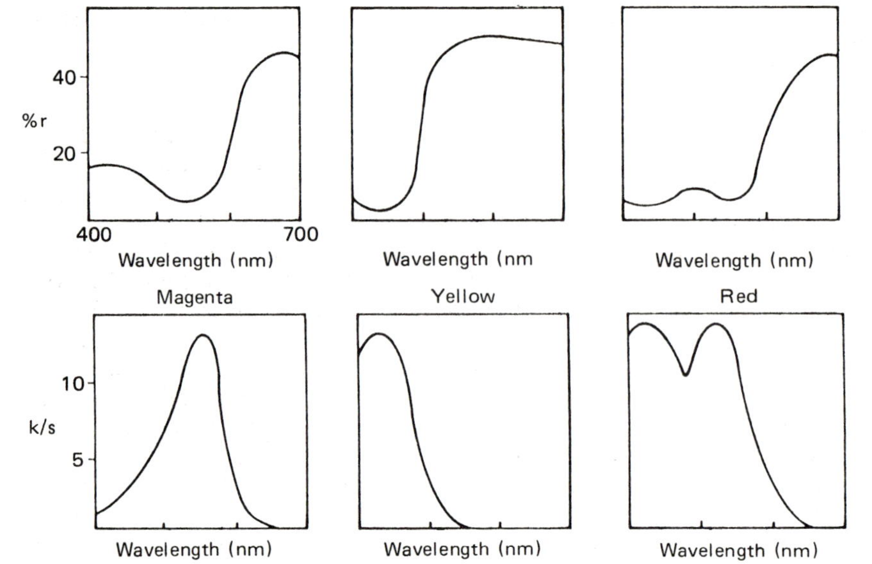

Fig. 3.25 Addition of reflectance and *K/S* curves.

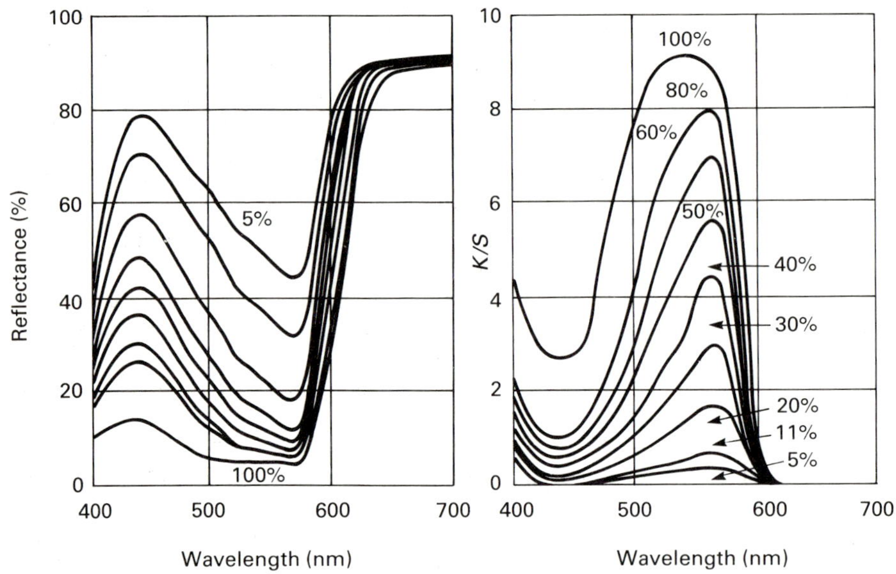

Fig. 3.26 Reflectance and derived *K/S* curves of a range of concentrations of a red ink.

To determine the concentrations of the pigments the data is weighted by the way we see the colour through the CIE *x*, *y* and *z* coefficients, combined with a specific illuminant. It is then possible to write three simultaneous equations and solve them for the concentrations of the colourants.

Match-prediction procedures

Match-prediction procedures are described in Taylor and Smart (1977). The *K/S* values for all the pigments at a range of concentrations must first be prepared. This is normally done by taking a base ink and reducing with clear medium to give a range from 5 to 100%. Usually the intervals are as shown in Fig. 3.26. Prints of these must be very carefully prepared, usually on a standard substrate using an accurate proofing press. For paste inks, a Prüfbau or Duncan Lynch press is frequently used and in the case of liquid inks an *R–K* coater. In the later case, care must be taken to ensure that the viscosity of the diluted samples is kept constant so that accurate film weights are applied. It is possible to make allowances for slight changes in substrate colour and texture, but if any significant differences occur a new set of standards must be prepared.

The reflectance curves are measured, usually by taking 16 readings at 20 nm intervals; the data are then stored on a computer disc. If significant variations occur across or down a print then preparation procedures must be modified; slight variations can be dealt with by averaging a number of curves.

The *K/S* values of all the pigments, at usually nine concentrations plus the clear medium, are then calculated and the values again stored in the computer memory, to form what is generally known as a database. Modern integrated match-prediction systems are available. The ICS system is illustrated in Fig. 3.27. In this type of system the spectrophotometer is directly connected with a computer equipped with visual display unit (VDU) and a keyboard. A set of programmes, the software, is available from the suppliers so that the storage of data and all the necessary calculations can be more easily carried out.

Equipment suppliers also offer special training facilities for staff and a technical advisory service.

The equipment should be installed close to the working area rather than in a remote laboratory. It should be kept in an environment, however, which is relatively clean, and this will often necessitate the construction of a small air-conditioned colour laboratory on the shop floor.

This system should be made the responsibility of a trained colour technician, familiar with the raw material and finished product specifications and with the production procedures.

It is possible to calculate a colour match from all possible combinations of pigments, but the ink technologist will normally wish to select a range with special properties, e.g. light or product fastness, and take into consideration cost. It is also possible to store data on redundant inks which are to be 'worked-off' in current production. A match can therefore be made from base inks plus work-off or work-off alone.

Once a shade has been matched, data can be stored on disc in a shade library. It is possible therefore to search such a library for the nearest match, before attempting a new matching. Many other programmes to determine, for example, quantitative colour differences, degree of

Fig. 3.27 A modern integrated match-prediction system.

metamerism and ink strength are available. It is now possible to interface this match-prediction equipment with dispensing and weighing systems, so that automated production of ink from bases and waste ink is possible.

Press correlation

An ink supplier has to provide a product which, when printed under practical press conditions, will meet the colour characteristics specified by the printer. In order that this may be achieved, a set of calibration samples must be prepared for each of the single-pigment bases which make up the colour range used for matching.

There are a number of small laboratory proofing presses from which a choice can be made. The following criteria should be considered when making this choice:

- the need to give a closely controlled film weight which is similar to that of the proofing machine;
- uniformity of print quality;
- repeatability of print quality;
- freedom from changes in print quality caused by wear of the press;
- ease of operation;
- quick cleaning and colour change.

Colour variation

With the proper settings once established for Press Correlation, it is important to measure the colour variation that exists within the sample print. A very slight variation is to be expected across the sample, but there may also be variations along the printing direction. If these variations cannot be removed, then the same relative position on the print must be used for all samples in making colour measurements and visual assessments.

The repeatability of the making of prints also needs to be checked.

This is done by making two successive prints and measuring the colour difference at the same relative position on each print. If this difference is equal or greater than one quarter of the pass/fail tolerance then the proofing technique should be refined.

Regular quality control of proofing reproducability is most important. The rollers or drawdown bars are subject to wear in daily use and this gradual effect can pass unnoticed. Eventually it will become significant. Therefore, it is a good practice to keep the roller or bar which is used for producing database samples as the master and to use another for routine operation. These masters should be used periodically to check the condition of the regularly utilized equipment.

Database preparation

The calibration procedure and sample preparation is broadly the same for all computer system manufacturers.

However, some systems may require database samples which are printed on specific substrates other than those being used by the printer.

It is appropriate to clarify some confusion which may arise from a misunderstanding of the terminology used by different sectors of the printing industry.

The colourant part of a formulation can be in one of four forms:

- dry pigment;
- dispersed pigment concentate in suitable medium;
- unreduced ink supplied from manufacturer;
- ink reduced to meet press requirements.

The relationship between these is given in Table 3.1.

Formulations in terms of concentrates or unreduced inks are most usual. More examples are given in Table 3.2.

Grouping of substrates

A separate database is required for each ink system, this takes into account any differences of pigment type and concentration. Also, allowance must be made for differences of flow, viscosity characteristics and suitability for the substrates which are to be used. It is common that one ink range may be suitable for different substrates.

There is thus a need to decide very carefully how the database is to be constructed to meet these variables in the most economic manner. In addition, allowance must be made for differences of substrate surface colour and absorbency.

Table 3.1 Different forms of the colourant part of a formulation

Pigment	
↓	Grind with minimum amount of solvent and varnish
Concentrate	
↓	Add technology varnish
Neat ink	
↓	Add solvent to adjust to print viscosity
Finished ink	

Table 3.2 The red, yellow and blue components of some inks

Ink	In the presence of	Acts as
Orange	Red	Yellow
	Yellow	Red
Green	Yellow	Blue
	Blue	Yellow
Violet	Red	Blue
	Blue	Red
Black	Red and Yellow	Blue

Surface colour

If the difference between two substrates is only surface colour, one being perhaps yellower than the other, then only one set of prints, made on the whiter or cleaner substrate will suffice.

In the single-constant Kubelka-Munk theory, the K/S values, the ratios of absorption coefficient to the scattering coefficient, are calculated for inks in such a manner that the colour of the substrate is removed. This is shown in the equation

$$(K/S)_{\text{sample}} = C(K/S)_{\text{ink}} + (K/S)_{\text{substrate}}$$

giving

$$(K/S)_{\text{ink}} = 1/C\ [(K/S)_{\text{sample}} - (K/S)_{\text{substrate}}]$$

where $(K/S)_{\text{ink}}$, $(K/S)_{\text{sample}}$, and $(K/S)_{\text{substrate}}$ are the K/S values of the ink, calibration sample and the substrate, and C is the concentration of the ink.

Values for $(K/S)_{\text{sample}}$ and $(K/S)_{\text{substrate}}$ are obtained from the reflectance measurements of the sample and the substrate using the Kubelka–Munk equation

$$K/S = \frac{(1-R)^2}{2R}$$

Since the *K/S* values calculated for inks are without the effect of the substrate, the *K/S* value of whatever substrate used can be added during the matching. Hence, a yellower substrate will demand a lesser amount of yellow ink in a formulation in order to produce the target colour than would be the case on a whiter paper.

In the two-constant theory, if the ink films are totally opaque, the effect of the surface colour of the substrate will be irrelevant.

For transparent colour, the colour of the substrate used in the calibration samples and that to be used in matching are handled in the following equation (Judd and Wyszecki, 1975)

$$R = \frac{1 - R_g\,[a - b\,\mathrm{ctgh}(bSX)]}{a - R_g + b\,\mathrm{ctgh}(bSX)}$$

where

a	$= 1 + K/S$
b	$= (a^2 - 1)^{1/2}$
R	= the reflectance of the ink film
R_g	= the reflectance of the substrate
K	= the absorption coefficient
$\mathrm{ctgh}(bSX)$	= the hyperbolic cotangent of bSX
S	= the scatter coefficient
X	= the thickness of the ink film.

In this case only one set of calibration prints is required. Printing on aluminium foil and light-reflective substrates, particularly if transparent inks are used, may require additional calibration samples.

Absorbency

Substrates of similar colour may also have wide differences of absorbency. This can result in significant colour differences which are sufficient to warrant separate database samples.

By examining the prints, those substrates which have similar characteristics may be rationalized into groups. In each case the substrate which is whitest and most in use should be selected.

White in ink formulations

The database requirements are different for inks which contain white, from those where the opacity due to white is absent. This depends on which theory is in use – the single-constant or the two-constant theory.

The single-constant theory is that most appropriate for inks without white, but it is still workable for inks with a very low white content. The database is always made with the addition of clear medium.

Opaque formulations are best handled with the two-constant Kubelka–Munk theory. Two samples are required for each ink: a print with ink reduced with white in the ratio 20:80, and a print with 95 parts of ink mixed with 5 parts of black ink. Both of these prints are made on black and white substrates or on suitable clear film. During colour measurements the clear film prints should be backed with black and white standard tiles.

Repeatability

Before proceeding to prepare a whole set of database samples, it is important to discover the reproducability of the sample preparation. This is done by preparing in triplicate three quantities of a light grey mix. This mix consists of 5% red, 5% blue, 5% yellow and 85% clear medium. Should the colour difference between the prints made from each mix approach one third of the acceptable pass/fail tolerance, this will indicate that sample preparation is unsatisfactory. Weighing with improved accuracy and increasing the size of the mix may improve the sample repeatability.

Quality control of raw materials

It should be emphasized that the quality of matching is always dependent on the database information. All base ingredients and mixes should be carefully checked when any changes are made. Also, fresh substrate

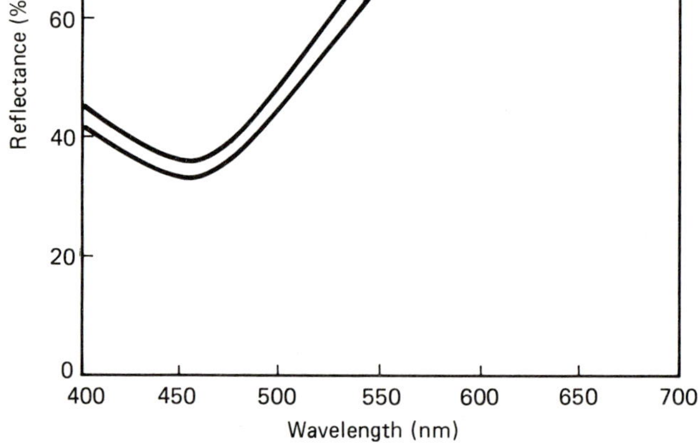

Fig. 3.28 Reflectance curves of a yellow ink printed on aluminium foil, measured vertically and horizontally.

samples should be measured for colour difference against a retained standard, other properties such as absorbency should be carefully checked.

Establishing a standard database

Database loading is one of the most important functions that determine the quality of matching.

One of the benefits of using computer-assisted colour matching is the facility of permanent storage of colour information in an easily retrievable form. In order to ensure that the measurements taken will be a true representation of the samples it is first necessary to obtain the colour variations for each of the following measurements.

- Repeated measurements of the standard white tile, taken without moving the tile, to check the spectrophotometric stability.
- Repeated measurements of a printed sample taken in one place.
- Repeated measurements of the same print, but removing and repositioning it after each reading, to check the sample alignments.
- Repeated measurements of different areas of the same print, to check the uniformity of the print.
- Repeated measurements of the same print but rotating the print through 90° after each reading. This gives an indication of directional

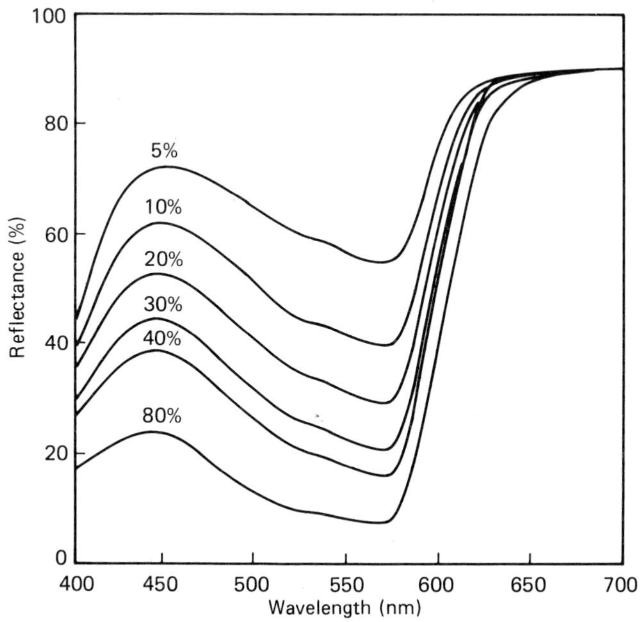

Fig. 3.29 The decrease in reflectance values with increasing concentrations of a red ink.

colour variation, aluminium foil sometimes has this property (Fig 3.28).

- Repeated measurements with both a white tile and then a black tile at the back of the print. Should there be a significant difference it indicates that light is reaching the back of the substrate and all measurements need to be made with a constant backing.

Irregularities in *K/S* values (single-constant theory)

Obvious errors in calibration sample measurement can be observed by measuring the reflectance curves. As the ink concentration increases, the reflectance values should uniformly decrease as shown in Fig. 3.29.

The higher the ink concentrations, the higher will be the *K/S* curves. The use of the *K/S* curves has the added advantage that the distance between the curves is proportional to the concentration of colour. The displacement between a 20% and a 30% sample will be similar to that between a 30% and 40% sample, as shown in Fig. 3.30.

The linearity of the calibration samples can also be checked by examining the *K/S* values against concentration at the wavelength of maximum absorption. Any obvious measurement mistakes can be seen by deviation from the theoretical straight line plot as shown in Figs. 3.31 and 3.32.

It is not uncommon that the *K/S* plot deviates from a straight line. An example is the saturation and bronzing of an ink which may give rise to

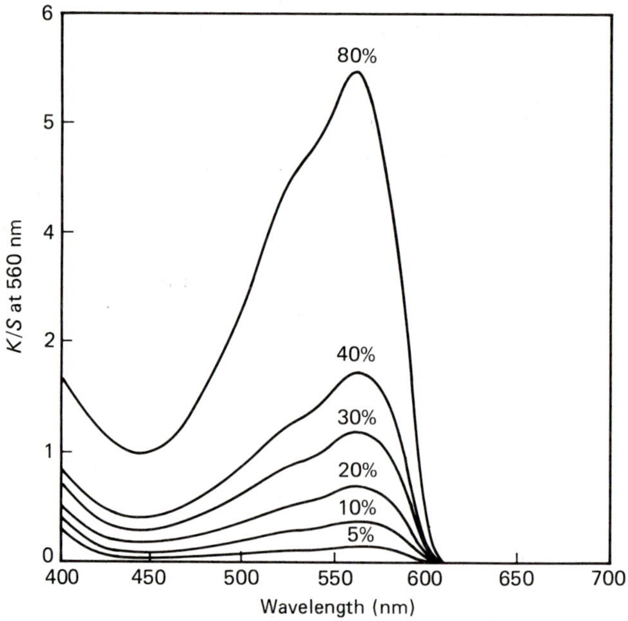

Fig. 3.30 The increase in *K/S* values with increasing concentrations of red ink.

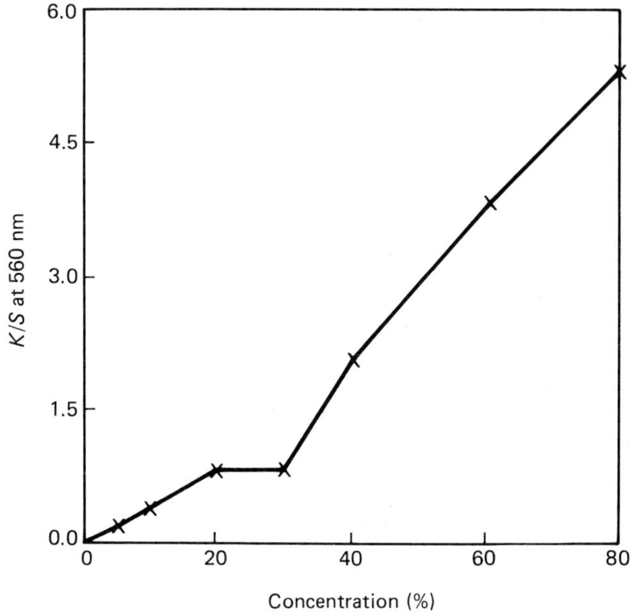

Fig. 3.31 Typical plot showing mistake of measuring a sample twice for two concentrations.

reflectance and *K/S* curves as shown in Figs. 3.33 and 3.34. This will result in *K/S* against concentration plots as shown in Fig. 3.35.

Irregularities in *K* and *S* values (two-constant theory)

The scattering coefficient *S* of a pigment depends on its particle size and the refractive indices of the pigment and dispersion medium. It also depends to a small degree on the wavelength of the light. When examining the calculated *S* values, it can be expected that TiO_2 will have the highest *S* values and the transparent organic pigments the lowest values. Since the *S* curve is a smooth one, then the absorption coefficient *K* should follow the *K/S* curve closely, as is shown in Fig. 3.36.

Use of known mixtures

The accuracy of a database can be checked by using known mixtures. This is done by taking a red, a yellow, a blue and a black, mixing three at a time with various ratios of white or clear medium. Such mixtures can be made when the calibration samples are prepared.

In general, if the predicted formulations are within an error of 5%, then when made up they should be quite close to the original samples. Once this confidence level has been reached with the four selected inks the system can be used to match unknown samples.

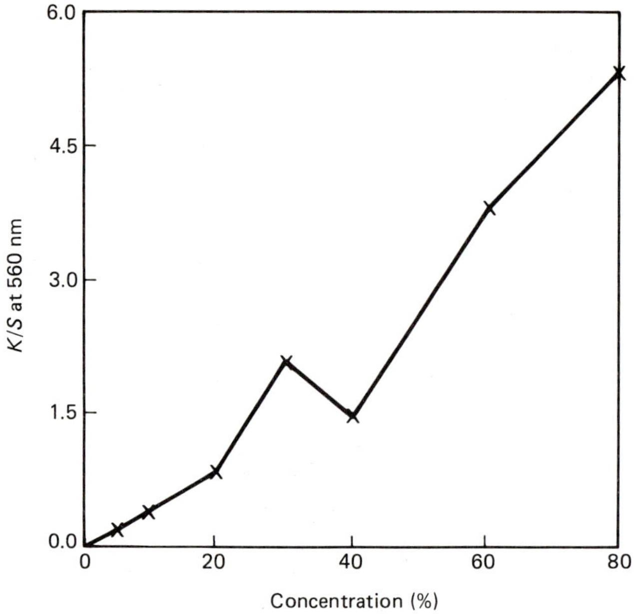

Fig. 3.32 Typical plot showing mistake of reversing the order of two concentrations in a measurement.

Fine tuning

Providing the predicted formulations of the known mixtures are within an error of between 10% and 20%, it will be possible to improve those predictions by fine tuning the database. The most commonly used process is by using the Saunderson correction and by changing the strength factors.

Saunderson correction

The Kubelka–Munk theory assumes that light travels in a continuous medium, with no change in refractive index. However, in practice the light passes first through air, then through the ink film and sometimes the substrate. Since air has an index of 1 and that of the clear medium about 1.5, there is a marked discontinuity in refractive index.

This discontinuity in refractive index can be corrected (Saunderson, 1942) by the equation

$$R = \frac{R' - k_1}{1 - k_1 - k_2 + k_2 R'}$$

where

R = the Kubelka–Munk reflectance
R' = the measured reflectance
k_1 = the top surface reflectance, sometimes called the specular reflectance

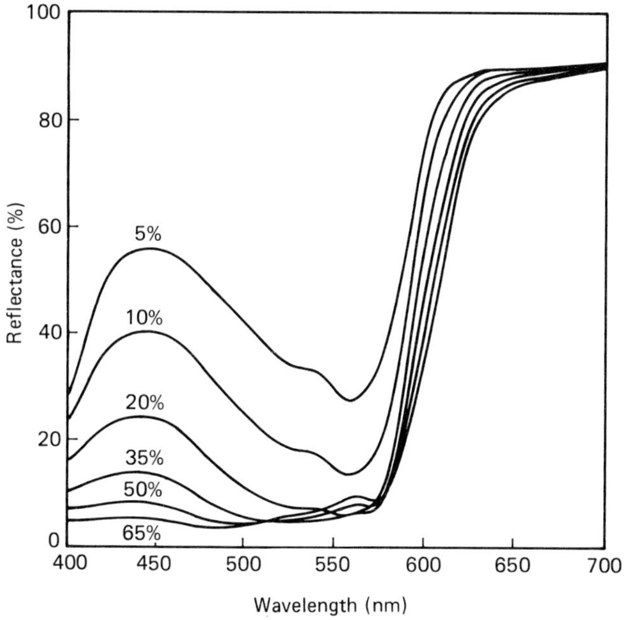

Fig. 3.33 Reflectance curves showing the saturation and bronzing of a red ink.

k_2 = the internal reflectance.

The reflectances k_1 and k_2 have certain ranges of value. It is by making changes in these values that the actual concentrations may be improved. The value of k_1 is between 0 and 4% and k_2 ranges from 0 to 60%. For single-constant databases it has been found that the Saunderson correction seldom makes any improvement, and k_1 and k_2 are set to zero. For two-constant databases a setting of 3–4% for k_1, and k_2 to either 40% or 60% works quite well. If there is a specular component then k_1 should be set to zero.

Strength factors

If the quantification of a matching is consistently less or more than that predicted by the system, then changing the strength factor may improve the accuracy of first-time computation.

However, if the prediction differs in different areas of colour space, predicting sometimes under and at other times over the true figure (and this is dependent upon the hue), then the strength factor should not be altered.

Formulations

Most systems have the capability of generating a choice of options such as:

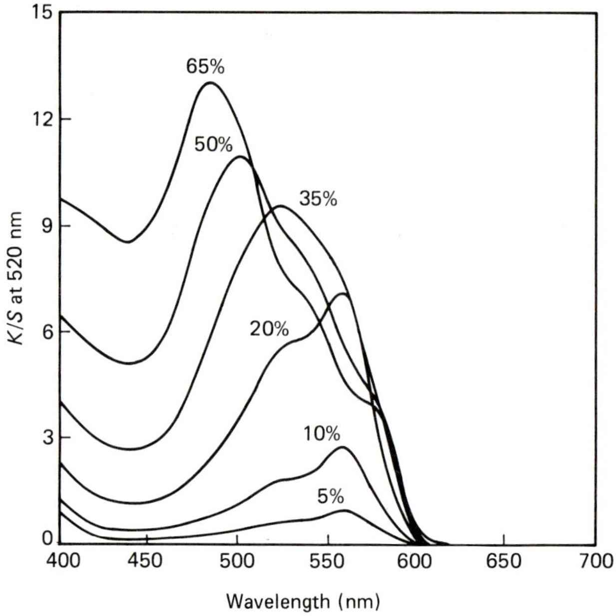

Fig. 3.34 *K/S* curves showing the saturation and bronzing of a red ink.

- combinations which meet defined acceptance limits of colour and metamerism;
- a number of formulations defined by metamerism limits;
- formulations defined by set cost limits.

While the normal choice would be those formulations having the least colour difference from standard, with low metamerism and cost, the importance of choosing a stable formulation should not be ignored.

Stable formulations

To understand the meaning of 'stable' in this context, it should be remembered that colour is three-dimensional. Colour is described by hue, lightness and chroma. This may be illustrated in the following example. By varying the concentration of black ink in a clear medium or white, a series of prints may be obtained. These will be a range of shades of grey. In this manner only the lightness has been changed.

If a fixed concentration of black ink is now chosen, then an increasing amount of red ink added, the ink will change in chroma. A series of prints with increasing chroma may be produced.

Should a third ink now be introduced, changes in hue can be made.

In order to match any colour, it is essential to match these three variables of hue, lightness and chroma. As shown above this will require three inks together with white or clear medium. Ideally, these three inks

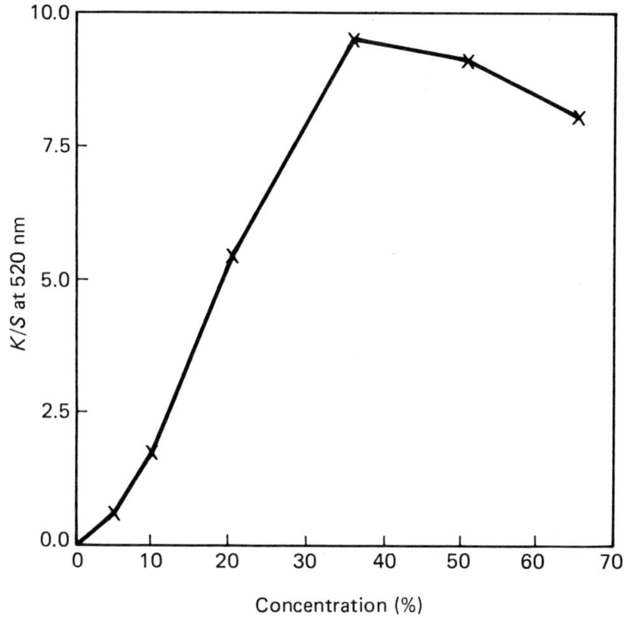

Fig. 3.35 The dropping of *K/S* values at higher ink concentrations.

should be very different from one another. The final choice will be determined by factors such as metamerism.

Formulations with less than three inks

In computer match prediction, problems may be posed to the mathematics used if the system is asked to work with less than three pigments in a formulation. Using a single ink to match a colour is trying to solve three unknowns (the tristimulus values x, y, and z) with only one equation (the single ink). It is therefore better to use three pigments even if the final print will consist of only one or two pigments. In these cases the computed formula will include traces of other inks which may safely be disregarded. These small amounts are commonly known as dummies and are present just to satisfy the computer's mathematical requirements rather than for actual necessity. Normally they need not be added to the formulation.

Formulations with more than three inks

Conversely, formulations with more than three inks may also create problems. These may arise particularly during batch correction.

For instance, a colour formulated with black, red, yellow and blue, may have a small colour difference from the target. It might reasonably be expected that a very small alteration would be needed. Instead the system may predict a surprisingly huge addition. The reason is that the computer

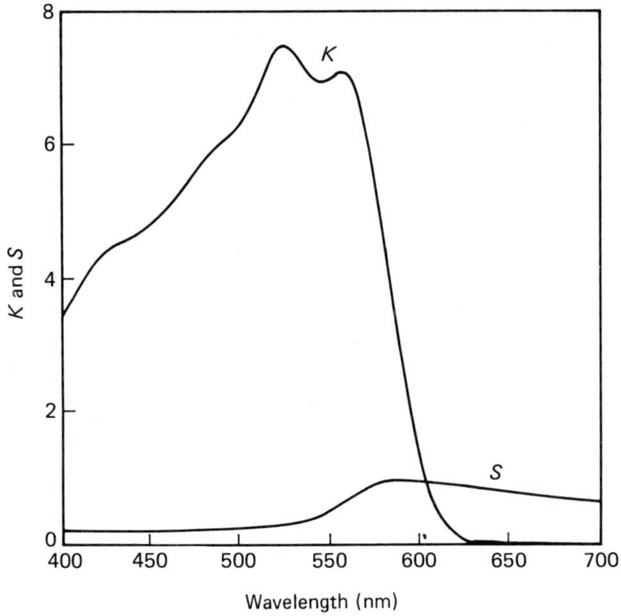

Fig. 3.36 *K* and *S* curves for a red ink.

has matched part of the black ink from yellow, red and blue. This predicts a colour grossly different from the target. Therefore, to convert the batch formula now requires a large addition.

When matching colours which do not have red, yellow and blue ink combinations, it is preferable that these components should be present in some form. A formulation with orange, red and blue would meet this criterion since the yellow component is represented in the orange ink.

There are two occasions where the use of four components is required:

- to achieve three degrees of freedom;
- to reduce metamerism.

To achieve three degrees of freedom

If a colour is formulated with green, blue and yellow, it is likely that the system will not match it satisfactorily. Only two degrees of freedom are present since green is seen as yellow plus blue. By adding the missing red component, a match should be obtained.

To reduce metamerism

When three inks are used to match a colour, the system is matching the tristimulus values x, y and z using the first selected illuminant. To obtain a metameric match, it is essential to have the colour difference under the other illuminants also as low as possible. If none of the three-ink combinations are successful, a four-ink combination may succeed. With this additional component there is an additional variable to match, the tristimulus value *x* of the second illuminant. This has been proved to be the most effective variable in minimizing the colour difference for that illuminant (Allen, 1978). When matching a colour, D65 should always be used as the prime illuminant. In this case the system will calculate a formulation with zero colour difference under illuminant D65.

Limitations in matching

Sometimes no satisfactory formulation is possible because the target shade is outside the colour scope of the ink system. If the sample is not a print but a piece of textile material, the system could possibly still match it but the formulation may be highly metameric. This is due to the different types of colourant used in textiles and inks. The failure to fulfil the defined metameric tolerance suppresses the formulation from being printed out. The metameric limits will therefore need to be widened.

Formulations containing white ink give different colouration to similar colour combinations without white. Again a satisfactory interpretation may be arrived at by widening the mis-match limits.

The substrates used are very important in determining the limitation of an ink system. The use of coloured or dirty substrates will inescapably limit the colour range that an ink can match. Any inks applied to a substrate will always lower the reflectance values (except for fluorescent ink). Brown kraft substrates cannot be printed to achieve the clean shades obtainable on white coated stock (Fig. 3.37). The addition of white may to some extent hide this substrate effect by adding opacity and may widen to some limited extent the colour range possible on a coloured substrate.

Corrections

When a batch sample requires considerable correction, it is important to discover why the colour is off shade. The following factors should be considered:

- errors due to database
- errors due to batch sample preparation
- variation in strength of newly made batches of base ink
- substrate variation
- printing conditions
- weighing errors.

Reclaiming surplus inks

One of the benefits of using colour computer systems is the efficiency with which ink can be recycled into useful material. Thus overmakes, redundant stock, and unopened and therefore uncontaminated returned inks can be used to make fresh product.

It is common to reduce the number of small amounts of variable materials by blending those of similar formulation and colour together in

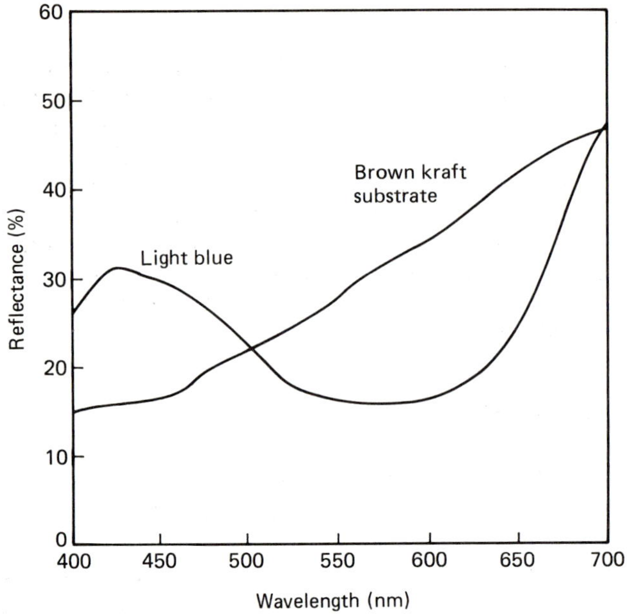

Fig. 3.37 Reflectance curves of a light blue and a brown kraft substrate.

a large storage vessel. With this method the resultant colour will become dirtier in shade as an increasing variety of shades are added to the mix. The advantage is that fewer and larger amounts are available for reclaiming.

There are several approaches to convert such bulked ink into new colours.

Database method

Reclaimed ink can be treated as normal ink. Treated as 100% concentration, a single print can be loaded into the database. It is to be expected that there will be some errors in extrapolating this one level of concentration into different concentration levels, but a correction will achieve the target colour.

Should the ink be strongly concentrated in shade, two or three reduced prints should be loaded into the database.

For a two-constant database the K and S values can be obtained by measurement of a drawdown over a black and white substrate.

Batch method

If the same three-ink combination can match the target colour and the reclaim ink, it follows that it is possible to make use of the correction routine. In this case the reclaim is treated as an offshade batch and corrected for shade to the target colour.

Conclusion

It can be seen that colour theory, as it is further developed, will to an increasing extent provide an accurate scientific measurement to the colour-matching procedure. Manual colour matching is a considerable art and takes many years to acquire. To display quickly and accurately a choice of possible ways of arriving at a colour formulation is an exciting and useful facility. The problems associated with thin, non-hiding ink films have by no means all been solved and colour prediction is not yet a panacea for the industry. However, colour measurement and the recording of formulae and shade data is virtually universal to the great benefit of the inkmaker and printer alike.

BIBLIOGRAPHY

Allen, E. (1978) Advances in Colorant Formulation and Shading, in *Color* 77 (eds, F.W. Billmeyer Jr and G. Wyszechi), Adam Hilger, Bristol, pp. 153–179.

Chamberlin, G.J. and Chamberlin, D.G. (1979) *Colour, its Measurement, Computation and Application*, Heyde.

Chambers, E. *Manual of Graphic Reproduction for Lithography*. Chapters 9–11.

Clarke, F.J.J., McDonald, R. and Rigg, P. (1984) Modification to the JPC 79 colour difference formula. *Journal of the Society of Dyers and Colourists*, **100**, 128–32.

Ishihara, S. (1974) *Tests for Colour-blindness*, 38 plate edn., H.K. Lewis.

Judd, D.B. and Wyszecki, G. (1975) *Color in Business Science and Industry*, 3rd edn., John Wiley & Sons, New York. pp. 427–8.

Kubelka, P. and Munk, F.Z. (1931) *Tech. Physik* **12**, 593.

Mclaren, K. (1983) *The Colour Science of Dyes and Pigments*, Hilger.

Munsell, A.M. (1905) *Munsell Book of Colour*, G.P. Putnam & Sons, London; Nickerson, D. (1969) History of the Munsell Colour System. *Color Engineering*, 7(5).

Newton, I. (1730) *Optics*. London Innys pp. 134–7.

Niven W.D. (ed) (1890) *The Scientific Papers of James Clerk Maxwell*, 2 vols, Cambridge University Press.

Patterson, D. (1967) *Pigments, an Introduction to their Physical Chemistry*. Elsevier.

Philips Lighting, Fluorescent Lamp Data. Handbooks 6.2.1. and 6.2.10.

Saunderson. J.L. (1942) Calculation of the color of pigmented plastics. *Journal of the Optical Society of America*, **32**, 727.

Taylor and Smart (1977) *Handbook of Industrial Colour Technology*, Instrumental Colour Systems Ltd.

Wright, W.D. (1969) *The Measurement of Colour*, Hilger.

CHAPTER 4
Raw materials

The raw materials used throughout the printing ink industry have, like those of other industries, shown a constantly changing pattern in the last 40 years.

Prior to 1950 a considerable number of raw materials used in inks were pigments, resins and solvents produced mainly for the paint and other surface-coating industries. Some were poorly adapted to meet the specific needs of what was at that time regarded by many manufacturers as a small and not very important industry.

Interest shown by raw material suppliers since then was generated by the knowledge that requirements of the industry could best be satisfied by the development of organic pigments, synthetic resins and solvents. New materials were designed to meet the stringent requirements of the more advanced inks that were being made for the increasingly sophisticated methods of printing.

More recently some suppliers have installed laboratories containing both ink-making plant and printing machines. In this way they have kept abreast of ink development and are now able to advise ink manufacturers on formulation.

Following the expansion in the number of materials in the last 15 years we have now moved into a period of consolidation. Some materials have now gone, and others have been moved to manufacture in smaller, select, and therefore more expensive batches. The other materials have been scrutinized to make sure that by-products of their manufacture are not a problem to use or dispose of (if they have, certain products have been withdrawn because of this). The remainder of available materials are now made by fewer suppliers.

A rationale is being established of the total number of materials from which it is likely we will all operate. The price, for instance, of developing a new pigment may be very high and perhaps prohibitive. The current thinking is that we will still see many suppliers with 'new' modified versions of 'old' materials instead of chemically new materials appearing.

The increasing EC and UK legislation about health and safety will demand changes, and will continue to do so. It will be for this reason that technologists will become well informed about the risks and hazards of all these materials, as well as their other properties. The dissemination of all sources of information, e.g. manufacturers data and journals, will be more important year by year because we will have to conform as well as be innovative to proceed to the future.

Section 1: Pigments

Pigments together with dyestuffs, are probably the most important items in printing ink formulation as they convey the visual identity of an ink and invariably contribute the major component.

Both pigments and soluble dyes give colour to a substrate by altering its reflective characteristics. In the case of pigments, which are multi-molecular crystalline structures, produced to an optimum particle-size distribution, insolubility is a key property. When applied in a vehicle to a substrate they either remain on the surface or have a tendency to fill the voids in paper or other irregular surfaces.

There are hundreds of different types of pigment produced. Some are formed by nature in mineral or vegetable forms, but most of them are synthetic materials produced from a variety of chemicals. A simple but not ideal classification is that of organic and inorganic pigments. Of the large number of types produced, few find their way into ink manufacture with its special demands. Many are uneconomical to produce, even in large quantities. Others do not meet the demands of tinctorial strength, particle size, cost and specific combinations of properties required for successful ink formulation.

This section is devoted to a detailed description of coloured pigments and extenders that are extensively used in all types of ink manufacture. The ones selected are those most commonly used throughout the industry.

Some pigments which are used for one specific purpose only have been omitted, together with pigments that are no longer in general use.

Each pigment is identified by names in common use, followed by the first colour index number (Chapter 3). Pigments have been grouped according to their chemical composition with the purpose of indicating their common structure.

Important physical and chemical properties have been tabulated, together with bulk value (100 divided by relative density) which indicates the volume relationship between equal weights of different pigments.

Owing to differences in flow characteristics, handleable concentrated pastes of pigment and varnish do not always follow exactly their bulk value relationships. The percentage of pigment in (twenty eight poise) 2.8 Pa s linseed stand oil required to give such a base (concentrated paste) is also listed. It is felt that the combined information gives useful knowledge of the relative consistency characteristics which may be encountered in the manufacture of paste inks.

White pigments and extenders, as well as carbon blacks, have been described separately from other pigments as special description of types and other properties is essential.

The raw materials from which pigments are prepared are stated (where known) together with the formula or structure.

4.1 YELLOW PIGMENTS

Mono Azo or Arylide Yellows

Principal CAS Nos.: Y1 2512.29.0; Y5 4106.67.6; Y4 1657.16.5; Y3 6486.233; Y74 6358.31.2; Y73 13515.40.7

Many arylide yellows are currently in production. Although they are often commercially known as Hansa, this in fact was the trade name of their originators. These yellow pigments are correctly referred to as aceto-acetanilides. Shades range from very greenish yellow to a bright warm shade as shown in Table 4.1.

They are normally fairly opaque pigments, but there are some grades that are sufficiently transparent to be suitable for superimposition. Arylide yellows are widely used for their fastness to light and tinctorial strength as well as resistance to alkali and soap. They are unsuitable for inks that require stoving because they will sublime and bleed strongly in some non-polar organic solvents, paraffin wax and plasticizers. Tinting strength is inferior to diarylides.

Arylide yellows are not normally as strong as the diarylide yellows which have replaced them in many ink applications. The dominance of diarylides is helped by the fact that arylamide yellows are bulky and have poor flow properties in certain ink vehicles.

Uses

Arylide yellows are used in both paste and liquid ink formulations. In the former they print well on letterpress and lithographic machines. They are used on soap and detergent packages. Care is necessary in photogravure inks owing to their substantial solubility in toluene.

Structural formulae

CI Yellow 1, 3, and 4, and 5.

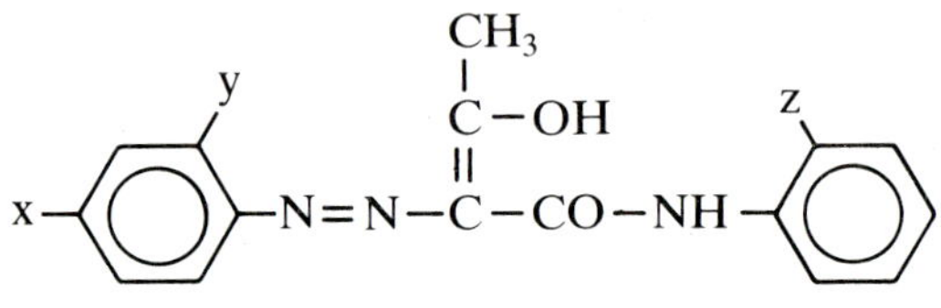

CI Yellow 65, 73, 74, 97, and 111.

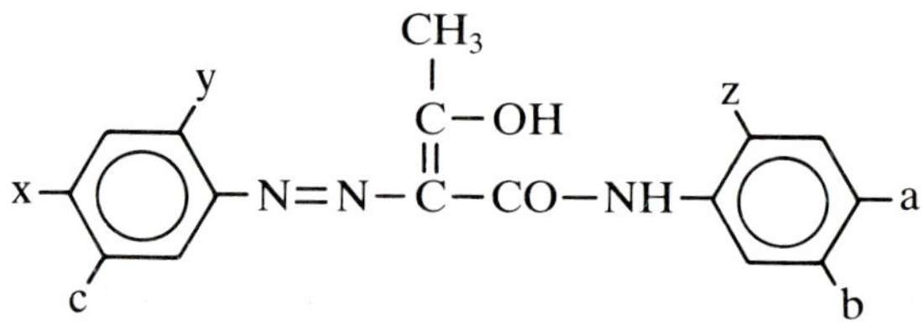

These are azo pigments formed by coupling the components given in

Table 4.2 to give a, b, c, x, y or z substituents.

Diarylide Yellows

Principal CAS Nos.: Y83 5567-15-7; Y55 6358.37.8; Y12 6358.85.6; Y13 5102.83.0; Y14 5468.75.7; Y17 4531.49.1

The diarylide yellows consist of a range of pigments based on dichlorobenzidine varying from lemon to a golden hue and from opaque to transparent. Diarylide yellows are the most widely used organic yellows. The most popular pigments in the range are given in Table 4.3.

The tinctorial strength of this important range of pigments is considerably greater than the chrome and mono arylide yellows. Lightfastness ranges from fair to excellent, resistance to heat and soap is excellent. They are readily dispersed but as their particle size is very small and oil absorption high, care is necessary in formulations to obtain satisfactory ink flow. An interesting development to improve this is the use of the easily dispersible (ED) pigment grades. The range of properties available form diarylide yellows is sufficient to satisfy most requirements of inks for most printing processes.

Uses

They are universally used in three- and four-colour process inks for printing in any sequence, but if used as the last down colour then good transparency is required. Diarylide yellows are based on diazotizing 3, 3′-dichlorobenzidine and coupling with two molecules of various acetoacetarylides.

The basic structure is:

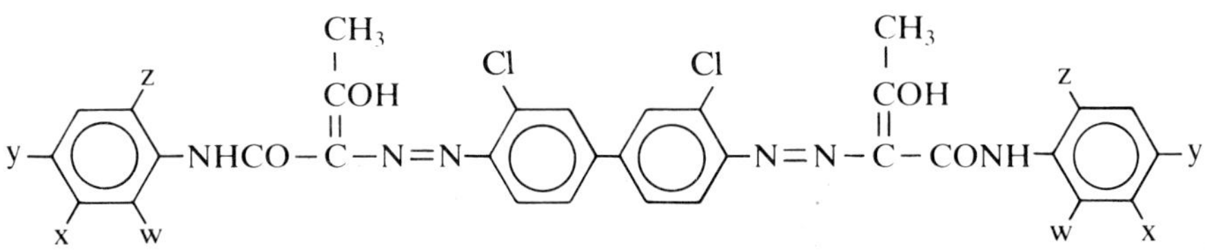

Substituent groups at w, x, y, z, yield the six diarylide yellow variants given in Table 4.4.

Iron Oxide Yellows

Pigment Yellow 42; CI No. 77492; CAS No. 20344-49-4

Synthetic iron oxide yellow is hydrated ferrous oxide produced from ferrous sulphate.

Properties
Outstanding chemical and weather resistance. Many grades are very hard to disperse. The ink maker is well advised to use the micronized grade. They are economical and opaque.

Uses
Exceptional heat resistance makes them suitable for use in printed furniture surfaces and other high light and heatfast specifications at low cost, and where low toxicity is required, e.g. 'cork tip' inks.

Tartrazine Yellow Lake

Pigment Yellow 100; CI No. 19140:1; Food and Dye and Colour Yellow No. 5; EC additive E102; CAS No. 12225-21-7

Properties
FDA (Food and Drugs Act Regulations USA) certified and typical of aluminium salts of this kind. Useful where FDA approval and few other properties are paramount.

Uses
Metal decorating inks and nitrocellulose lacquers for food packaging.

Chrome Yellows

CI Pigment Yellow 34 Nos. 77600 and 77603; CAS No. 1344-37-2

Properties
Lead chromates form a range of opaque yellow pigments varying from light primrose hue to golden orange. They have a high specific gravity; this means that the weight ratio in inks is very high, giving a low pigment volume and coverage. It is sometimes desirable to incorporate extenders to control the rheological characteristics, but, as the greener shades, being more bulky, are frequently extended with alumina hydroxide, they are sometimes prone to livering. When heated, chromes redden in colour and return to their original hue on cooling. The lightfastness is fair; they darken on exposure, especially in industrial areas; they are resistant to acids, wax, solvents, heat and soap, but decompose in the presence of strong alkali. They catalyse the oxidation of drying oils, so ink formulae must be adjusted accordingly.

Uses
The use of lead chromes has declined due to environmental factors such as effluent control and their unsuitability in inks for food wrappers or toys. They can be incorporated in all types of printing ink, especially where high hiding power is required. They were used as the first down colour in early process inks. The advent of transparent benzidines which can be printed in any sequence has superseded this. Future use will be very limited since an extremely low lead content is desirable in paper which is liable to be recycled as a secondary source of fibre.

Cadmium Yellow

CI Pigment Yellow 37 No. 77199; CAS No. N/A

Properties
Colours range from greenish to golden yellow. They are opaque, very resistant to light, heat, solvents and alkali. Cadmium sulphide pigments decompose in acids and, being sulphides, blacken when used in conjunction with lead pigments.

Uses
They are now seldom used in inks, except in applications needing vcry high resistance properties. Besides being expensive, they lack tinctorial strength and do not print well. Cadmium pigments are not permitted in inks on articles conforming to food and toy regulations. They are not recommended for outdoor poster inks where they may decompose or darken under acidic atmospheric conditions. It is doubtful whether cadmium pigments will be permitted to be used in any form, due to toxic hazards in their manufacture and use.

Other Yellows

These are becoming more interesting these days in the continued pressure on 3,3′-dichlorobenzidine (DCB) routed yellows.

CI Pigment Yellow 61; CAS No. 12286-65-6; Monoazo CI, No. 13880

Properties
Heat and lightfast; unstable to alkali and soap.

Uses
Azo metal-salt pigment weak in tint; mainly used in plastics but gaining application in inks; expensive.

Structure (as acid)

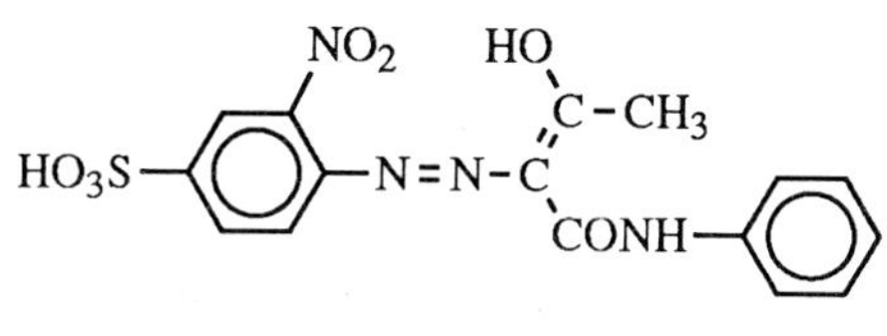

Pigment Yellow 81; CI No. 21127; CAS No. 22094-93-5

Uses
Very resistant pigment with reasonable lightfastness but excellent resistance properties. Used as green shade yellow in applications where these criteria are required; relatively expensive.

Structure

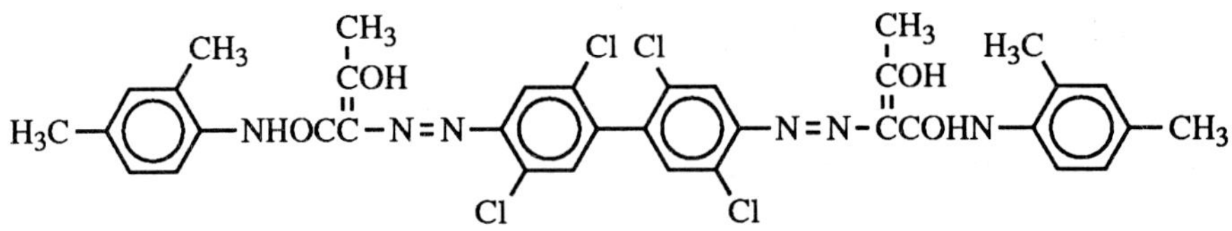

Azo Condensation Yellow; CI Yellow Pigment 95; CI No. 20034; CAS No. 5280.57.4

Properties
Lightfast, heat resistant, excellent properties, green shade yellow; expensive.

Uses
Good flowing in difficult solvents so used in liquid ink applications where cost versus properties make it worthwhile.

Structure

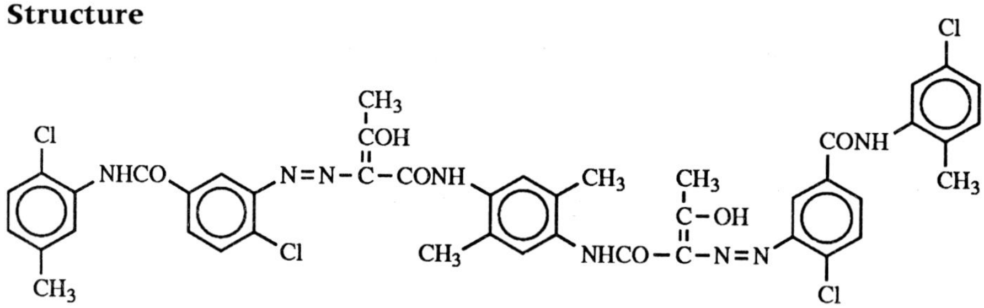

Benzimidazolone H2G; CI Pigment Yellow 120; CI No. 11783; CAS No. 29920-31-8

Properties
Very useful green shade yellow; good lightfastness; resistant to most common products.

Uses
In all applications where the cost can match the properties required.

Structure

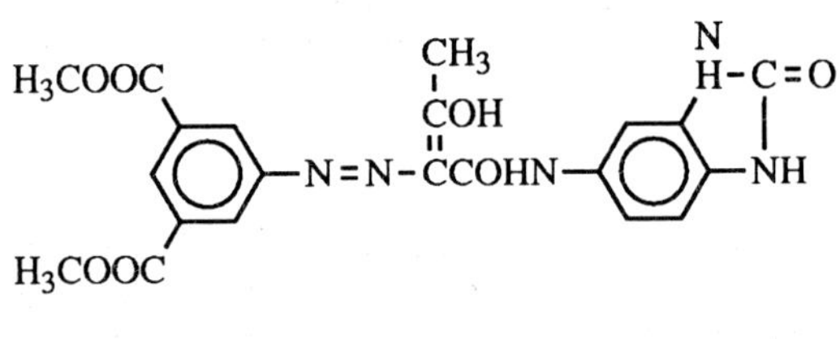

Pigment Yellow 126; CI No. not assigned; CAS No. 61815-05-0

Properties
Cheap bright greenish yellow, Similar to Pigment Yellows (PY) 12 and 13.

Uses
General yellow uses; not very lightfast.

Structure

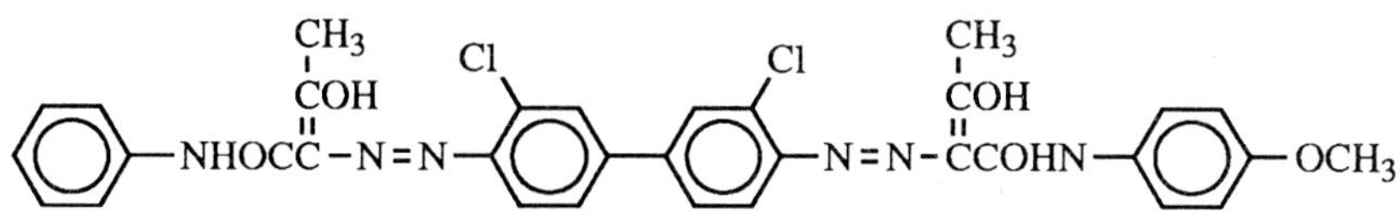

Pigment Yellow 127; CI No. 21102; CAS No. 71827-67-8

Properties
Cheap mid yellow; similar to PY13; medium lightfastness; good resistance properties.

Uses
General mid yellow, uses as alternative PY13.

Structure

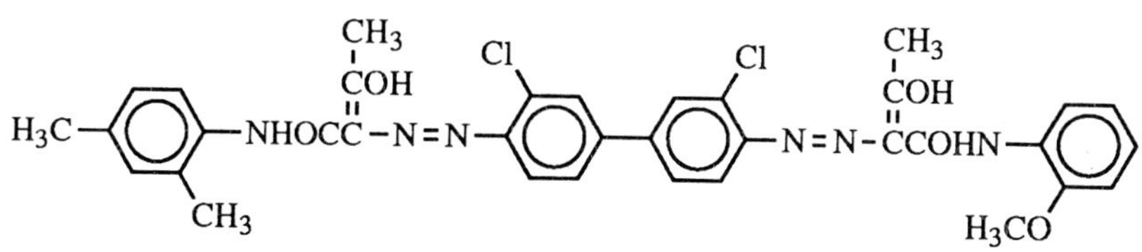

Isoindoline Yellow L1820; PY 139; CI No. 56298; CAS No. 36888-99-0

Properties
Mid red shade yellows with good fastness properties. Very heat resistant yellow; good resistance properties; expensive but has its place in the scheme of things.

Uses
Currently wallpaper and flooring inks but should be looked at as a substitute for diarylide and arylide yellows. Very useful yellow when high heat resistance is required for instance in flexo polythene inks.

Structure

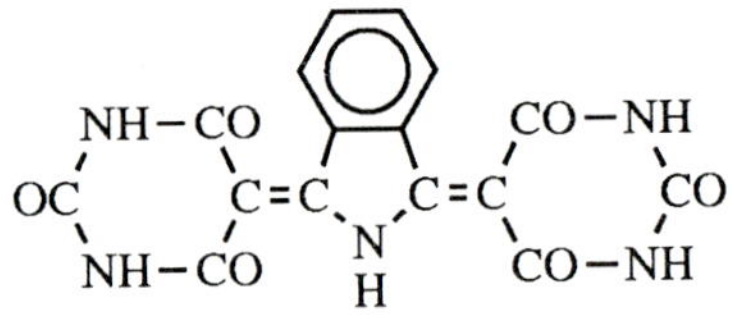

Diazo Yellow GRX; CI Pigment Yellow 176; CI No. 21103; CAS No. 90268-24-9

Properties
Good performance mid/red yellow pigment; Good resistance properties and reasonable heat resistance.

Uses
Very good flow in offset systems.

Structure

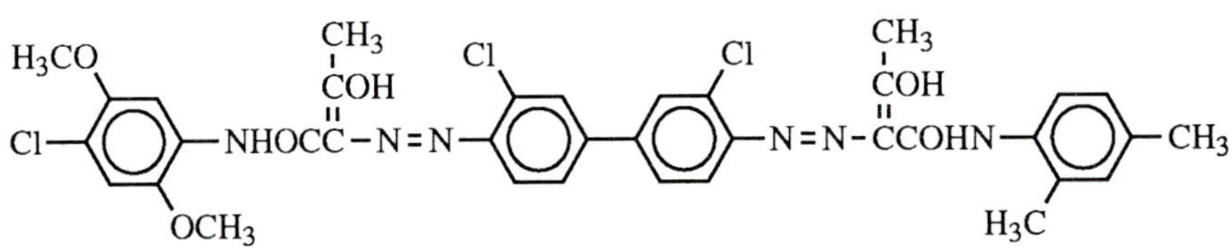

Isoindolinone Yellow; PY 110; CAS No. 5590-18-1; CI No. 56280

Properties
Very resistant green shade yellow; lightfast and very stable; expensive.

Uses
In all applications where properties are needed and cost is no object.

Structure

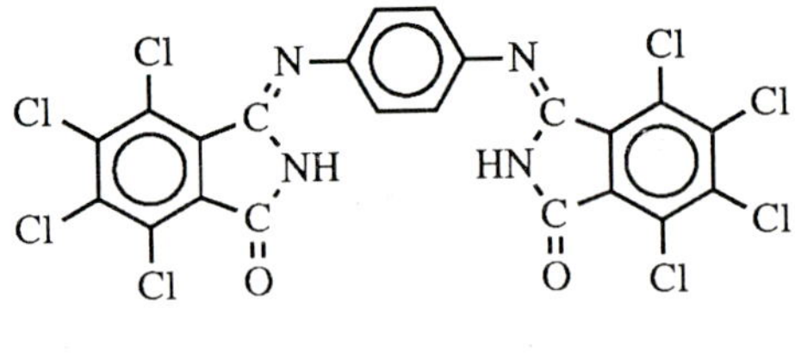

Fluorescent Yellow

Brilliant Yellow 8GF; CI Pigment Yellow 101 No. 48052; CAS No. 2387-03-3

Properties
A fluorescent green shade yellow, not very fast to light, heat or solvents, but it can be varnished or lacquered if care is taken to avoid bleeding. It is resistant to soap and detergents.

Uses
Suitable for lithographic, letterpress, gravure and screen process inks. Its main application is on cartons and display cards and it is commonly used in conjunction with other fluorescent colours on greetings cards. Brilliant yellow 8GF has also been used as a first-down process yellow to add lightness to prints and often is used as a gold on foil applications. Its price and poor lightfastness limited its use in the past. Recent developments, however, have improved brightness, and are faster to light and weathering.

Formula
2-hydroxy-1-napthaldehyde-azine

Structure

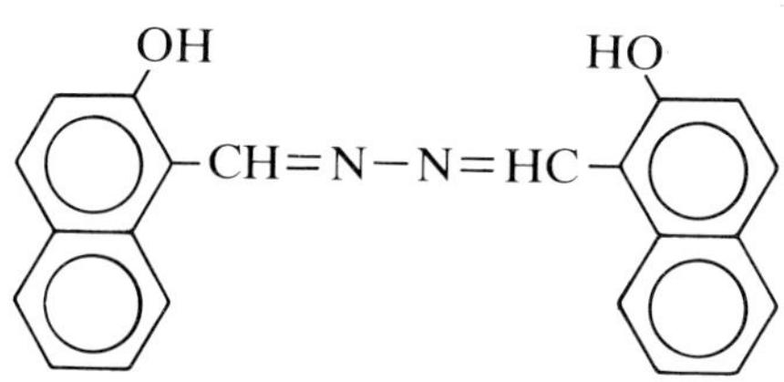

4.2 ORANGE PIGMENTS

DNA Orange

Orange 2G; Dinitroaniline Red (BS 3599/4); Permanent Red 2G; CI Pigment Orange 5 No. 12075; CAS No. 3468-63-1

Properties
Reddish orange with good lightfastness and fairly good heat resistance, with fair fastness to wax, butter, but excellent resistance to acid, alkali, and water. It has high opacity, but lacks solvent resistance.

Uses
Orange 2G has only poor flow properties in inks and finds use mainly in cheap systems for printing on kraft paper and polythene where a bright reddish shade can be combined with good covering power. There is some doubt about its toxicity at the moment and its use is under investigation.

Structure

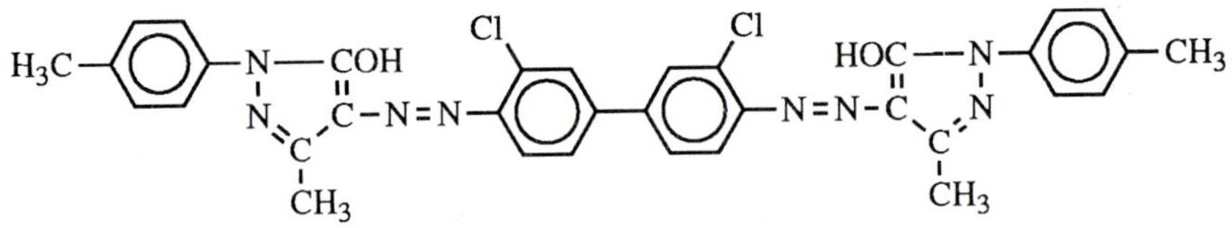

Pyrazolone Orange

Permanent Orange G; CI Pigment Orange 13; CI No. 21110; CAS No. 3520-72-7

Properties

A yellower shade orange which combines good fastness properties with high tinctorial strength. It is fast to acid, alkali, water, soap and wax. It also has good resistance to heat and to steam.

Uses

Orange G is semi-opaque and is a good working pigment which produces inks with a high finish for letterpress, litho and flexographic application. Although it does bleed in some gravure solvents, it is frequently used in liquid inks for packaging. It suffers from poor flow and printability problems.

Structure

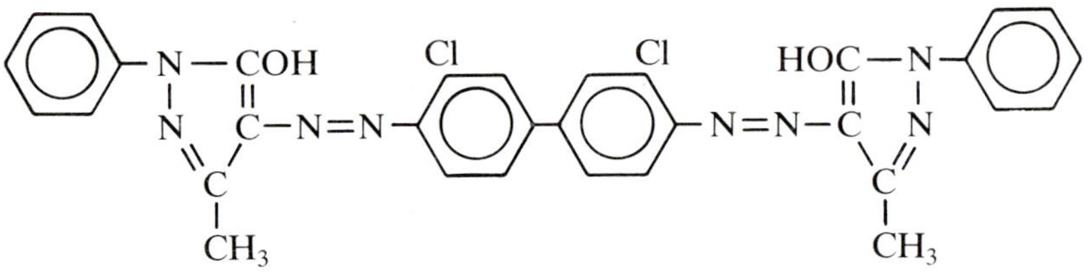

Diarylide Orange

Dianisidine Orange; CI Pigment Orange 16 No. 21160; CAS No. 6505-28-8

Properties

Typical of diarylide yellows having very high tinting strength. The hue is bright reddish orange, one of the few diarylides without a chloro group.

Uses
Often used to replace DNA orange in kraft printing because of its superior strength and brightness properties.

Structure

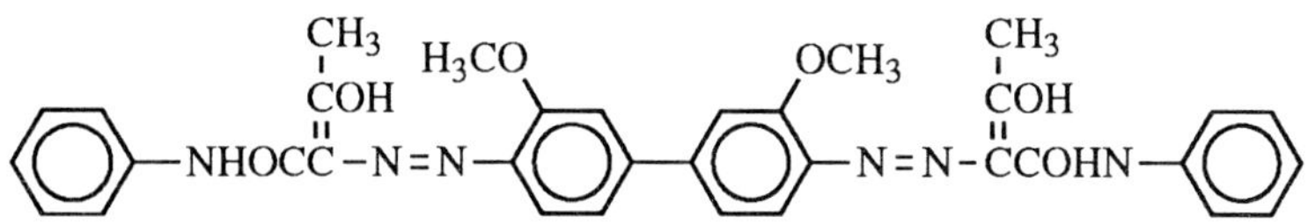

Fast Orange F2G

CI Pigment Orange 34 No. 21115; CAS No. 15793-73-4

Properties
The pigment is noted for its bright, reddish shade combined with high transparency. It has excellent lightfastness, resistance to alkali, acid soap and wax. It is fairly good for heat resistance, and steam sterilization. It is also readily lacquered or varnished. Now available for liquid ink in good flow versions.

Uses
Suitable for all systems but is recommended for oil inks where its transparency and strength enable inks with a very high gloss to be obtained. Excess transparency can cause problems in liquid inks formulae.

Structure

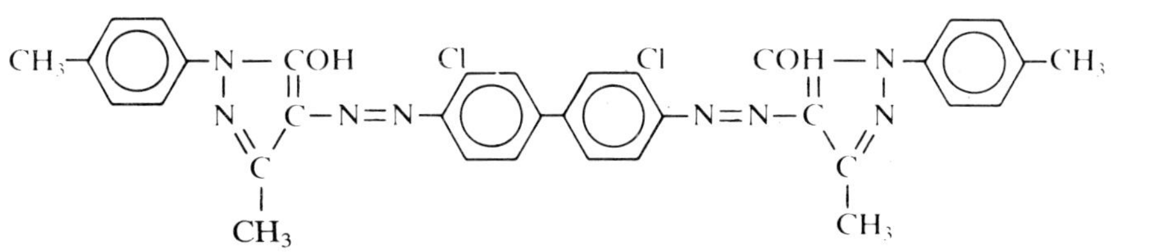

Benzimidazolone Orange HL

CI Pigment Orange 36; CI No. 11780; CAS No. 12236-62-3

Properties
Excellent light and heatfastness and solvent stability. Slightly soluble in some organic solvents, good in tints in very small amounts.

Uses
Expensive but used in products requiring these superior properties. Hue is reddish orange but dull in tint.

Structure

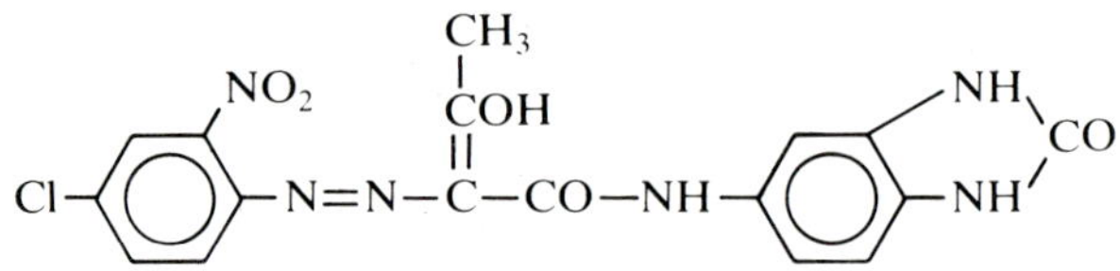

Perinone Orange

CI Pigment Orange 43 No. 71105; CAS No. 4424-06-0

Properties
Excellent pigment resistance to everything but naturally costly. Good for small quantities to tint and avoid shade fading.

Uses
Any use exposed to strong light, wall coverings, floorings and tint use for consistent colour content.

Structure

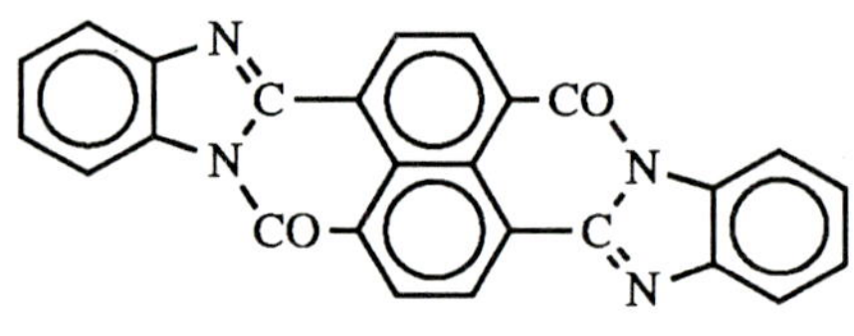

Ethyl Lake Red C

CI Pigment Orange 46; CI No. 15602; Cas No. 67801-01-8

Properties
A yellow, clean version of Lake Red C (PR 53:1). Economical pigment (like 53.1), excellent brightness and tinting strength. Poor alkali resistance, and lightfastness and can bleed (like 53.1) in strong solvents.

Uses
Useful in water-based brown kraft work.

Structure

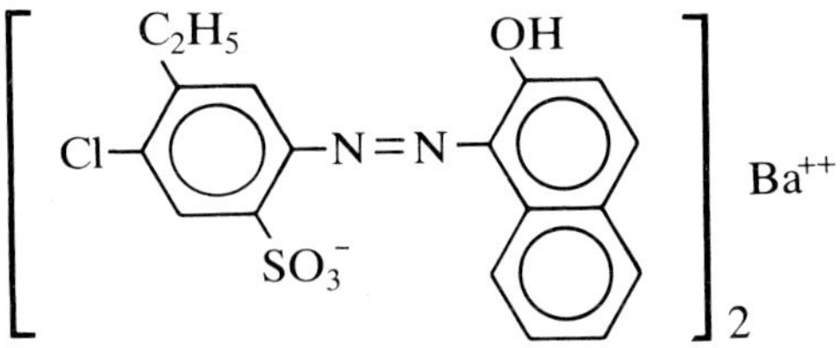

4.3 RED PIGMENTS

Naphthol Red (or permanent Red FRR)

CI Pigment Red 2; CI No. 12310; CAS No. 6041-94-7

Properties
This is a strong, bright, clean, yellow shade red, with excellent resistance to acids, alkali, soap and detergents. It is fairly lightfast.

Uses
It is used extensively in letterpress and lithographic carton inks, especially where soap resistance is important. It is also used for fertilizer sacks.

Structure

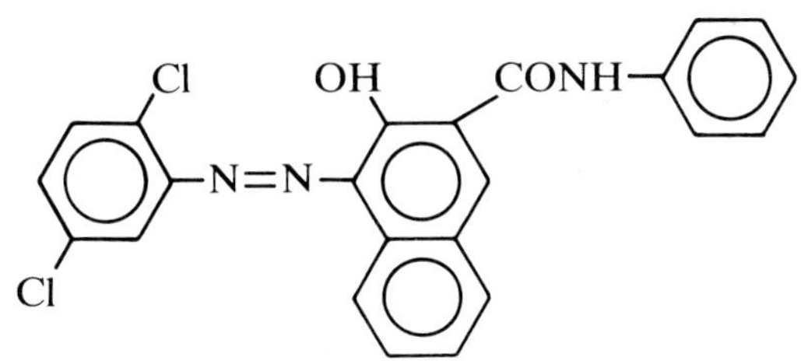

Toluidine Red

CI Pigment Red 3; CI No. 12120; CAS No. 2425-85-6

Properties
A scarlet red of medium strength but dull in undertone. Fastness to light at full strength is excellent, and resistance to alkali is good, but solvent resistance is poor and it discolours on stoving.

Uses
Used as a relatively low-cost pigment, where lightfastness is of value. It is not recommended for use at low concentration in tints.

Structure

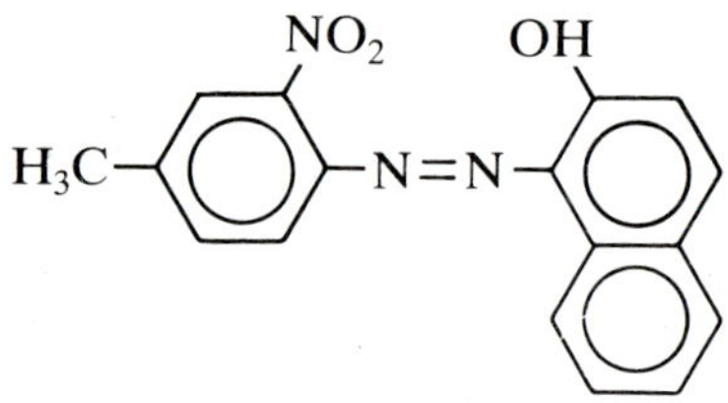

Permanent Red, 'R' (Chlorinated Para Red)

CI Pigment Red 4; CI No. 12085; CAS No. 2814-77-9

Properties
A bright orange red pigment of excellent lightfastness at full strength with good resistance to acid and alkali. Solvent resistance is poor and is not suitable for stoving. Texture is soft but dispersions lack flow.

Uses
This red pigment is used in many applications but care in formulation is required to produce inks of good printability. Used in large quantities due to its low price.

Structure

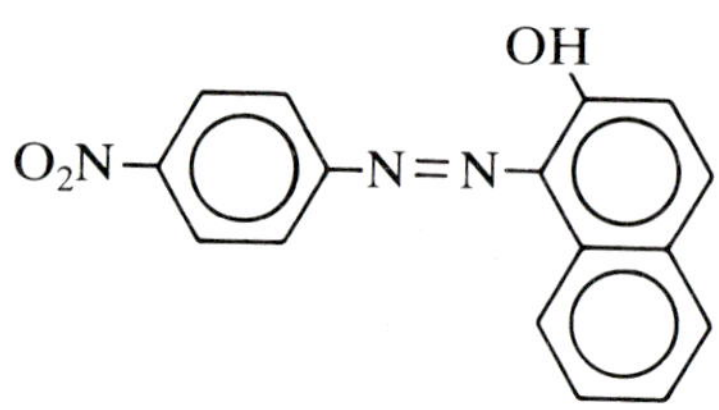

Carmine F.B.

CI Pigment Red 5; CI No. 124490; CAS No. 6410-41-9

Properties
A strong blue shade red of excellent lightfastness and resistance to water, acids, alkali, soap and detergents, possessing good stability to heat and steam sterilization. It disperses well, but lacks flow in full strength inks.

Uses
High-cost limits the use of this pigment to inks where high resistance properties are demanded.

Structure

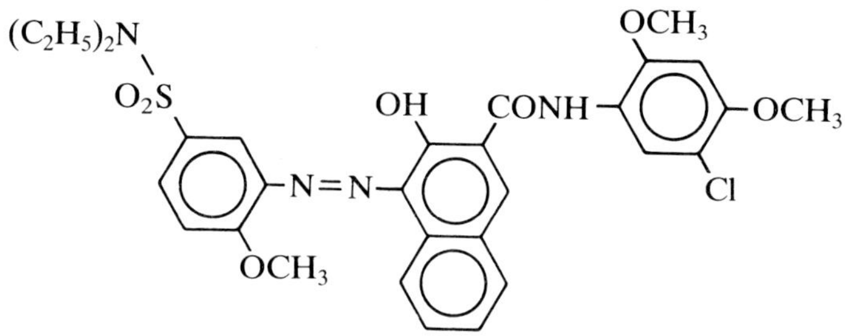

Naphthol Red F4R

CI Pigment Red 8; CI No. 12335; CAS No. 6410-30-6

Properties
A blue shade red with a clean bluish undertone, having good heat and soap resistance. Suitable for autoclave sterilization; lightfastness fair.

Uses
Used in process inks for soap cartons and for tinprinting inks. Its cost and mediocre lightfastness considerably limit its use.

Structure

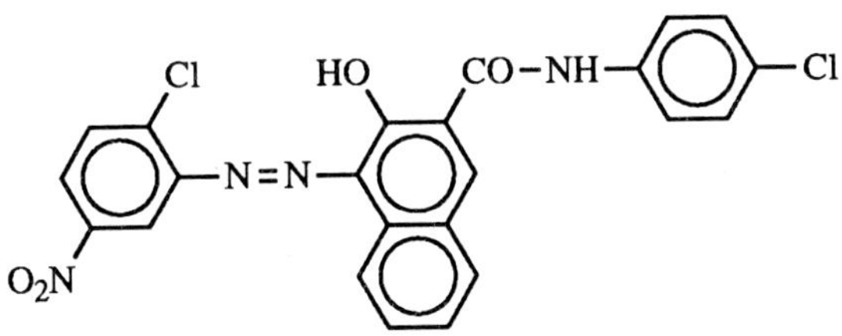

Naphthol Red L.F.

CI Pigment Red 9; CI No. 12460; CAS No. 6410-38-4

Properties
A bright, transparent, yellow shade red with clean undertone, excellent lightfastness and heat stability plus steam sterilization, alkali and detergent resistance. It lacks fastness to strong solvents.

Uses
Used in tin-printing applications, and in tints where fairly good lightfastness is needed.

Structure

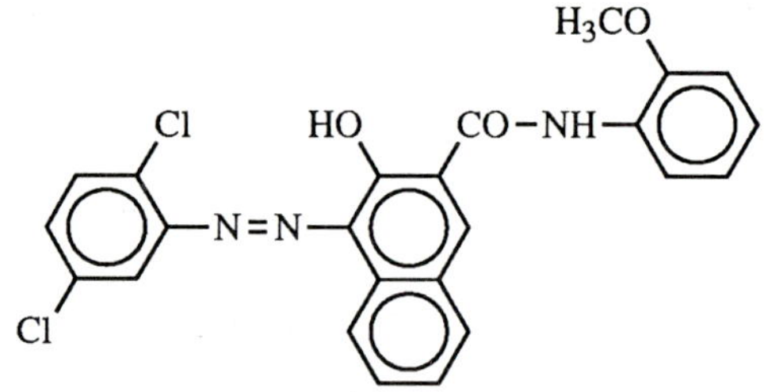

Permanent Red FRL

CI Pigment Red 10; CI No. 12440; CAS No. 6410-35-1

Properties
A bright red with orange masstone and blue undertone, similar in shade to Lake Red C but with good chemical and heat resistance as well as improved lightfastness.

Uses
Used in all types of application which warrant improved resistance properties.

Structure

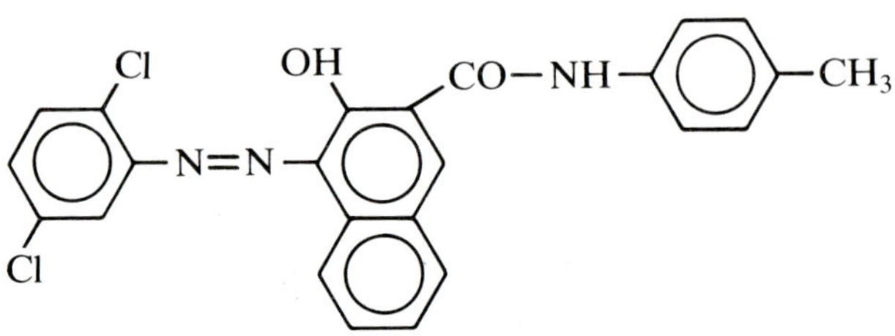

Bordeaux FRR (F4R)

CI Pigment Red 12; CI No. 12385; CAS No. 6410-32-8

Properties
Extreme blue shade red toner with high tinctorial value, the undertone and tint tone are slightly dull but extremely blue. It has good general fastness properties including high soapfastness and detergent resistance.

Uses
Used where good fastness properties are required at full and reduced strengths, and where that particular unique shade is required.

Structure

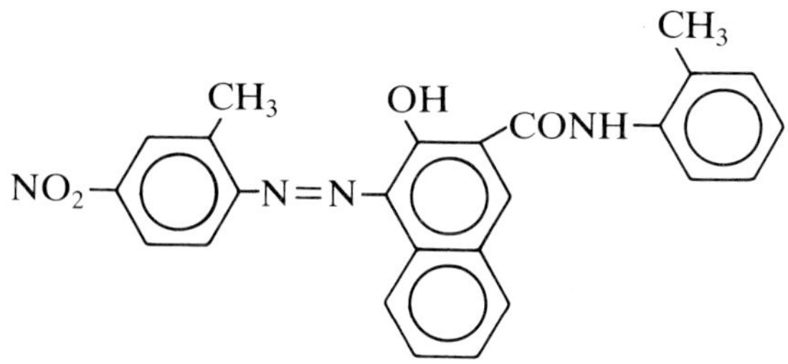

Naphthol Red Dark

CI Pigment Red 23; CI No. 12355; CAS No. 6471-49-4

Properties
A dark blue shade red with good soap and heat resistance, and fair lightfastness – useful pigment.

Uses
Used in packaging inks for shade and chemical resistance; particularly useful in nitrocellulose chips.

Structure

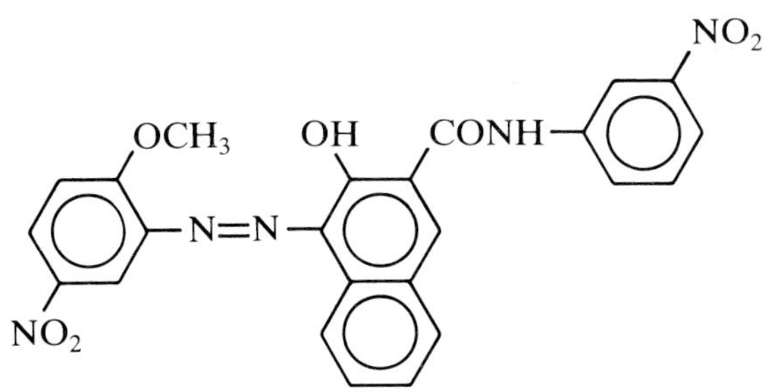

Rubine 2B, Permanent Red 2B, BON Reds

CI Pigment Red 48; CI No. 15865

CI Pigment Red 48; Rubine 2B – sodium, i.e. Na2B; CAS No. N/A
Water soluble; seldom used.

CI Pigment Red 48:1; Rubine 2B – barium, i.e. Ba2B; CAS No. 7585-41-3
Bright yellowish

CI Pigment Red 48:2; Rubine 2B – calcium, i.e. Ca2B; CAS No. 7023-61-2
Very blue shade red.

CI Pigment Red 48:3; Rubine 2B – strontium, i.e. Sr2B; CAS No. 15782-05-5
Less blue, but clean.

CI Pigment Red 48:4; Rubine 2B – Manganese, i.e. Mn2B; CAS No. 5280-66-0
Medium scarlet.

CI Pigment Red 48:5; magnesium salt, CAS No. N/A
Rarely used.

Properties
These range from a bright yellow shade red (barium) to a deep maroon (calcium). They are primarily noted for their good value for money and purity of hue. They possess adequate lightfastness, except when used as tints. They are reasonably resistant to acids and solvents, but not to alkali. They have good fastness to heat, oils and fats, but only fair fastness to heat sterilization in autoclaves.

Uses
Very widely used; the barium and calcium salts are used extensively in all types of liquid and paste ink, although the barium salt often has to be excluded due to its toxicity. Their clean shades and working properties make them invaluable for high quality inks; low to medium cost.

Structure

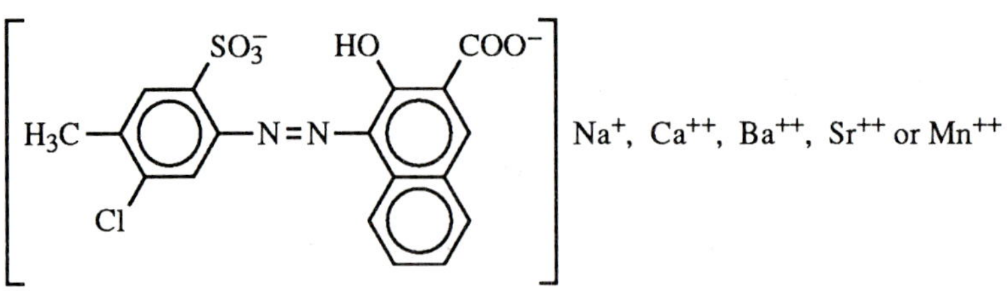

Note: strontium is less used; small amounts are sometimes added to barium or calcium salts to enhance brilliance and improve water resistance. All four (not the Na) have their uses for cost-effective mid-red application.

Lithol Reds

CI Pigment Red 49; CI No. 15630

CI Pigment Red 49, sodium salt; CAS No. 1248-18-6
Yellow shade red.

CI Pigment Red 49:1, barium salt; CAS No. 1103-38-4
Bright blue shade red.

CI Pigment Red 49:2, calcium salt; CAS No. 1103-39-5
Strong blue shade red.

Properties
The lithols range from vivid yellow to blue shades of red, dependent on the metal used. Their lightfastness and solvent fastness is poor, as is their resistance to acid and alkali. They do not have good heat resistance but are fairly resistant to oils and fats and are sometimes resinated to give transparency.

Uses
USA Publication Gravure Standard – now gradually moving over to Red 57.1. They are not used extensively in paste inks but are sometimes used for cheap jobbing colours. The barium salt gives a bright red and the calcium a deep red. These are invaluable for inexpensive liquid inks. Their poor lightfastness and solvent resistance restricts their use slightly. The very blue calcium salt gives a fine result as a liquid nitrocellulose or water-based ink.

Structure

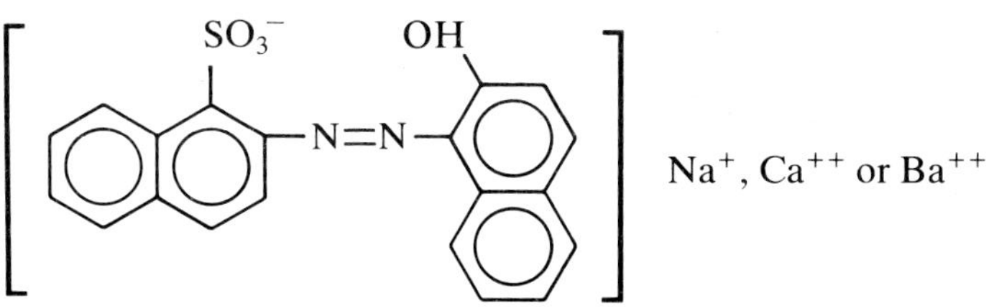

BON Red (Lake Red 'C' BON)

CI Pigment Red 52; CI No. 15860

CI Pigment Red 52, sodium salt; CAS No. N/A

CI Pigment Red 52:1, calcium salt; CAS No. 17852-99-2

CI Pigment Red 52:2, manganese salt; CAS No. 12238-31-2

Properties
A red similar to Lake Red 'C' but based upon a different coupling with the

C acid. It gives better solvent resistance, heat stability and lightfastness. It has good wax resistance. The calcium salt (the main one used in inks) is very close in shade to Rubine 4B.

Uses
A considerable amount is used in the USA in nitrocellulose and polyamide systems where good working qualities combined with bright shades are required. However, its popularity is lower in Europe. In solvent-based liquid inks, it gives good dispersion characteristics as well as a clean bright shade. Particularly well suited for nitocellulose chips.

Structure

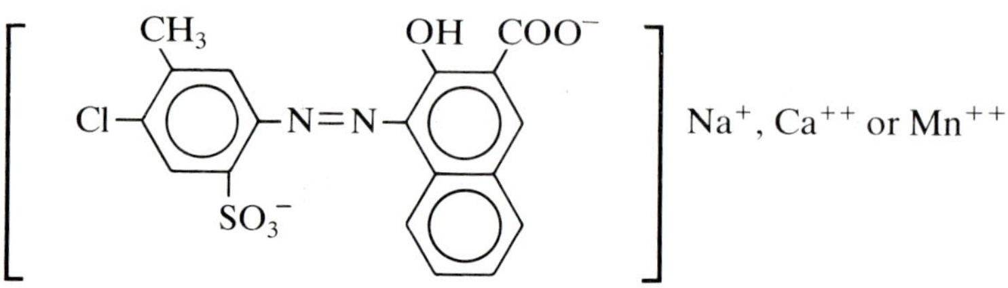

Lake Red 'C'

CI Pigment Red 53; CI No. 15585

CI Pigment Red 53, sodium salt; CAS No. 2092-56-0

CI Pigment Red 53:1, barium salt; CAS No. 5160-02-1

CI Pigment Red 53:2, calcium salt; CAS No. N/A

CI Pigment Red 53:3 strontium salt; CAS No. N/A

Properties
When referring to Lake Red 'C', the barium salt is the important one; the other three salts are seldom used. It is the most widely used of all the organic reds in printing inks. It gives a strong, bright yellow masstone combined with a blue undertone. It is a fairly transparent pigment, and this is sometimes increased in the resinated forms, which also have good dispersibility. It has good heat, fat and oil resistance as well as good acid fastness, and waxability. It is unaffected by most hydrocarbons and solvents but has only fair lightfastness.

Uses
Used in all types of paste and liquid inks, where brightness and strength are required to be combined with an economic product. It is also the basic red pigment used in matching systems. Like all other barium compounds it is toxic.

Structure

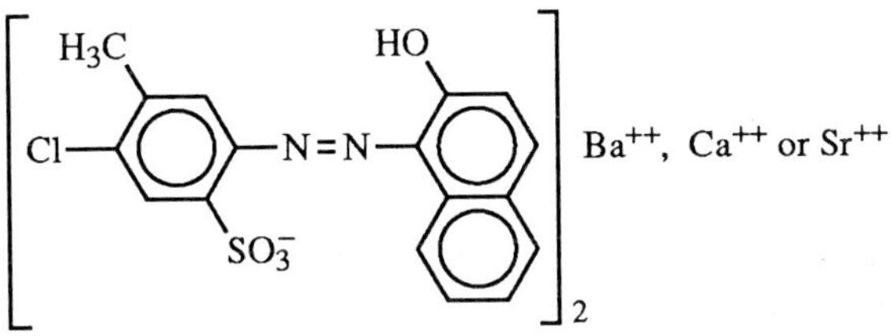

Lithol Rubine 4B

CI Pigments Red 57; CI No. 15850; CAS NO. 5858-81-1

CI Pigment Red 57:1, calcium salt; CAS No. 5281-04-9

Properties

The Rubine 4B pigment used by the printing ink industry is the calcium salt which is characterized by its brilliant deep masstone combined with a pronounced blue undertone. It has good resistance to most solvents and to oils, fats and waxes, also fairly good properties of lightfastness in dry conditions, heat resistance and steam sterilization. It has poor acid, alkali and soap resistance and only fair weatherproofness due to its sensitivity to rainwater. It disperses easily to make a good working ink. Much work has been done in recent years to achieve water stability and this has now been achieved.

Uses

The main usage of this pigment is as a process magenta in all types of inks. Earlier versions had PMTA Rhodamine additions to obtain sufficient blueness whereas the BS 4666 process sets rely upon the bluer brighter shades obtained by heat treatment and other modifications. Coated versions of this pigment are available with improved dispersibility and transparency. It is the blue shade red used in one of the base inks of colour blending systems.

Structure

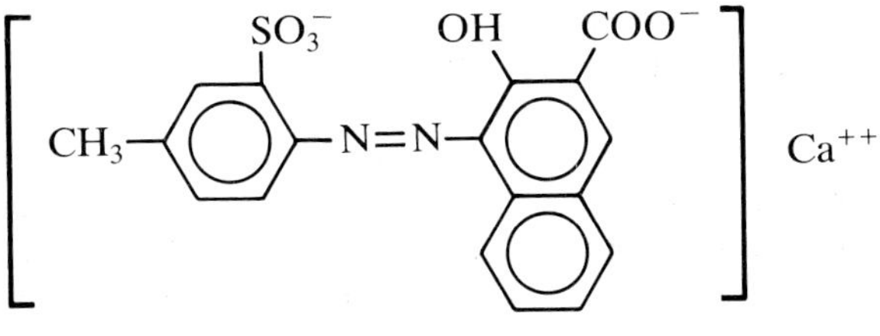

BON Maroon

CI Pigment 63, sodium salt; CI No. 15880; CAS No. N/A

CI Pigment 63:1, calcium salt; CI No. 15880:1; CAS No. 6417-83-0

CI Pigment 63:2, manganese salt; CI No. 15880:2; CAS No. 35355-77-2

Properties
Deep blue shade red; resistant to acids, moderately resistant to alkali and solvents. Resistant to fats with reasonable full tone lightfastness. Good in nitrocellulose chips.

Uses
Economical for these particular shades of Bordeaux and Maroon.

Structure

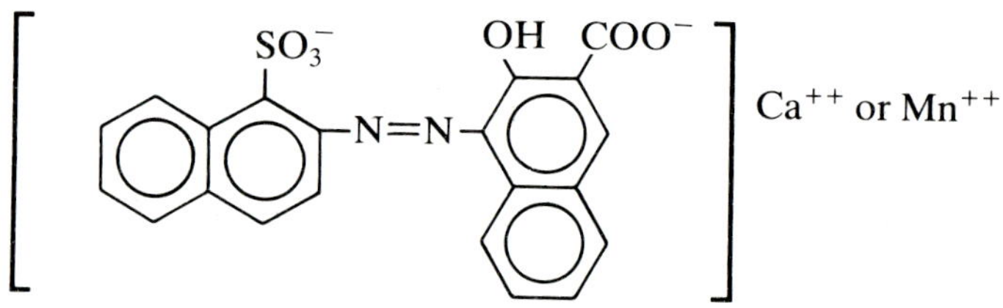

PMTA Pink Rhodamine GG

CI Pigment Red 81; CI No. 45160:1; CAS No. 12224-98-5

CI No. 45160:1, phosphomolybdotungstic salt

CI No. 45160:2, copper ferrocyanide complex

Properties and composition
Strong brilliant rose pink pigments produced by combining the basic dyestuff with phosphoric molybdic and tungstic acids known as triple salts (PMTA). It is possible to omit the molybdic or tungstic acid (PTA, PMA) or to combine with a silicomolybdic acid alone (SM). These are generally known as double salts. The original trade name of Fanal Pink is sometimes given to the whole group.

The double salts are normally stronger and cheaper but have poor lightfastness, darkening rather than fading. None are resistant to alkali or soap and they all bleed readily in alcohol. Sometimes a slight oil bleed is apparent.

Note: another pigment using the same basic dye is the copper ferrocyanide complex which has been separately classified as CI, Pigment Red 169.

Uses

Process Magenta is used in all types of ink where brilliance and tinting strength are the important factors. They print well and withstand paraffin wax but not spirit varnishing or lamination. Stabilized grades are available for liquid ink systems – unequalled in some applications for both liquid and paste ink.

Structure of CI Basic Red 1, CI No. 45160; CAS No. 989-38-8

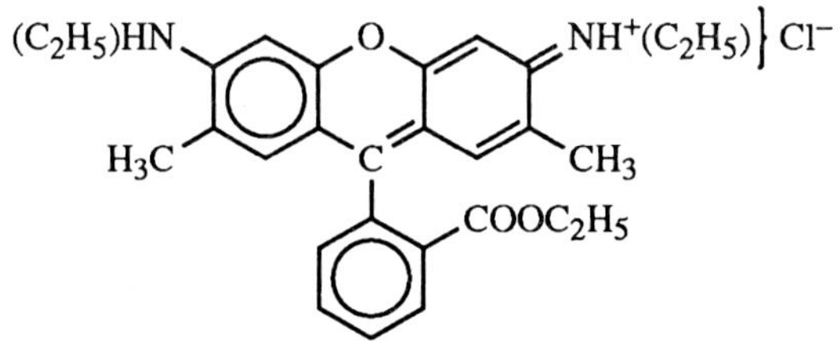

BON Arylamide Red Napthol Red FGR

CI Pigment Red 112; CI No. 12370; CAS No. 6535-46-2

Properties

This is a bright, clean, medium shade red, with bluish undertones. It has high tinctorial strength and is fairly transparent. Excellent lightfastness and resistance to acid, alkali, water and heat are characteristic. However, napthol red FGR bleeds in most solvents.

Uses

Used in all types of paste inks, especially where cheaper pigments do not have the resistance properties required. Also used for flesh-coloured tints and for making chip dispersions. Very useful soap fast scarlet.

Structure

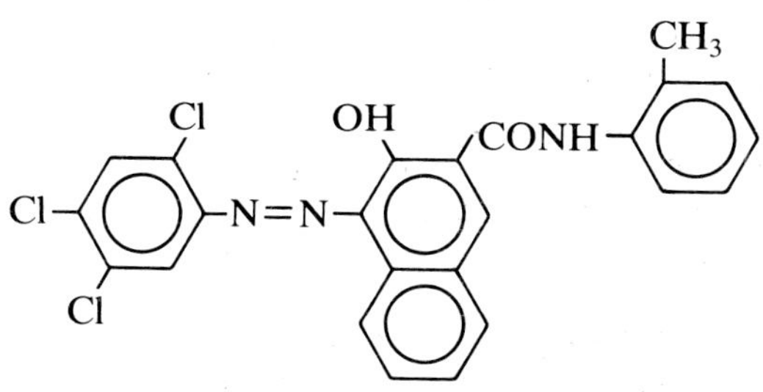

Quinacridone Magneta Y

CI Pigment Red 122; CI No. 73915; CAS No. 980-26-7

Properties
A bright blue shade red which is tinctorially weak and difficult to disperse. It is insoluble in organic solvents, unaffected by heat, light, acid, alkali, soap and wax. Very resistant pigment but usually poor flowing.

Uses
Owing to its tinctorial weakness and high price the use of this pigment is limited to the applications which cannot be met in other ways. It works well in most ink systems.

Structure

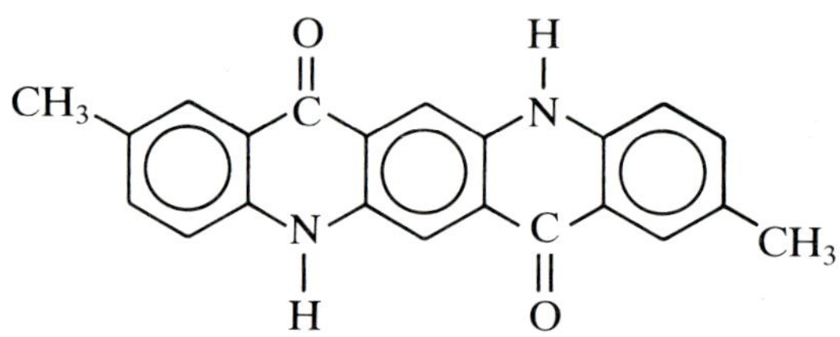

Napthol Carmine FBR

Permanent Carmine FBB; CI Pigment Red 146; CI No. 12485; CAS No. 5280-68-2

Properties
Multipurpose liquid, paste, and metal decorating, high-property blue shade red pigment. Like many other napthols needs good formulating to avoid poor flow; hue is deep carmine. It has excellent fastness properties except to some organic solvents.

Uses
Fairly expensive, but used where its superior properties are required.

Structure

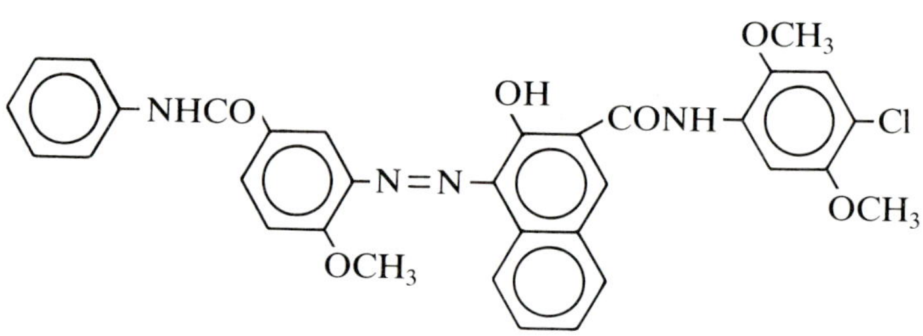

CI Pigment Red 166

Cromophtal Scarlet RN; CI No. 20730; CAS No. 12225-04-6

Properties
It is a very bright very resistant (and expensive) scarlet. Would pass any resistance tests for scarlets where 53.1, 48.1, 112 will not suffice.

Uses
Any ink where very high specification is required in that colour range.

Structure

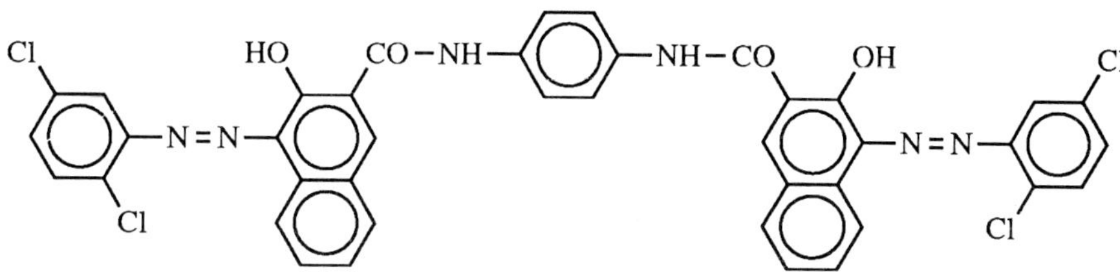

Copper Ferrocyanide Pink

CI Pigment Red 169; CI No. 45160:2; CAS No. 12224-98-5

Properties
The colour is similar to the double and triple salts (CI Pigment Red 81). Its resistance to light is lower, with less resistance to heat and some solvents. Besides being stronger that PMTA salts it has a higher oil absorption. Copper ferrocyanide pink is not soap-fast, and has poor acid and alkali resistance.

Uses
Used as a substitute for the more expensive PMTA salts in publication and water-based inks. It normally has a brighter hue and excellent working properties. Stabilized grades exist for use in some solvent combinations, but care must be taken in formulation. They flow very well and find applications where cheap bright shades are required.

Structure

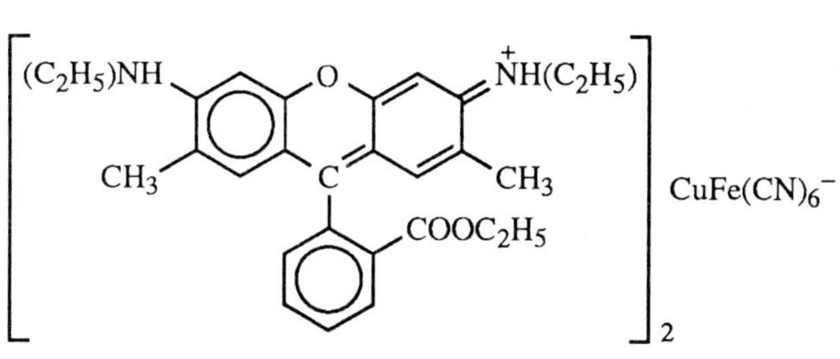

Naphthol Red F5RK

CI Pigment Red 170; CI No. 12475; CAS No. 2786-76-7

Properties
Very brilliant with a strong bright bluish red hue. Excellent lightfastness, resistant to soap, wax, grease, acid and alkali.

Uses
Suitable for all types of printing inks in virtually any system. High cost restricts its use.

Structure

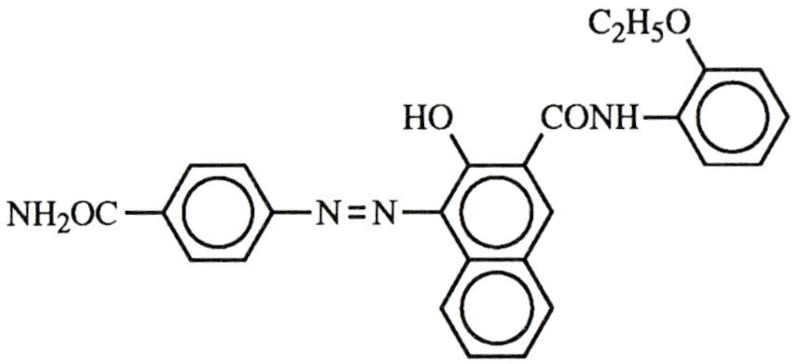

Benzimidazolone Carmine HF 3C

CI Pigment Red 176; CI No. 12515; CAS No. 12225-06-8

Properties
Very bright and transparent hue is bluish red. It is lightfast, heatstable, acid, alkali, grease and soap resistant. Widely used for packaging and metal-decorating applications. Expensive.

Uses
PVC inks, alkyd resin enamels, lacquers and many liquid packaging uses where high specification is necessary.

Structure

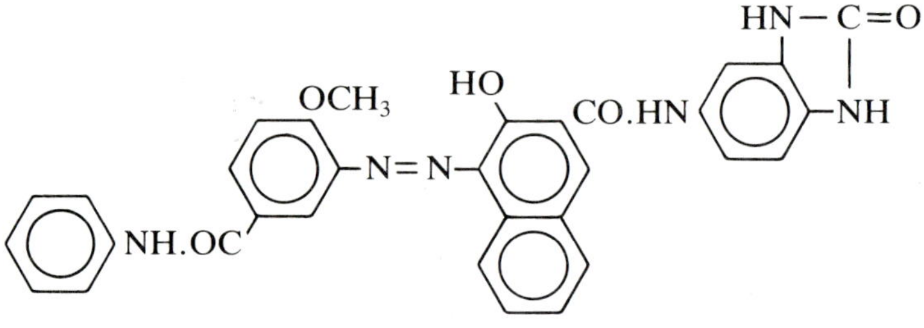

Naphthol Rubine F6B

Pigment Red CI 184; CI No. 12487; CAS No. 99402-80-9

Properties
Bluish red, very similar pigment to CI Red 146 but slightly stronger and more resistant. Good resistance properties to acid, alkali, and soap; poor lightfastness.

Uses
Packaging liquid inks. Good baking stability so suitable for metal-decorating inks.

Structure

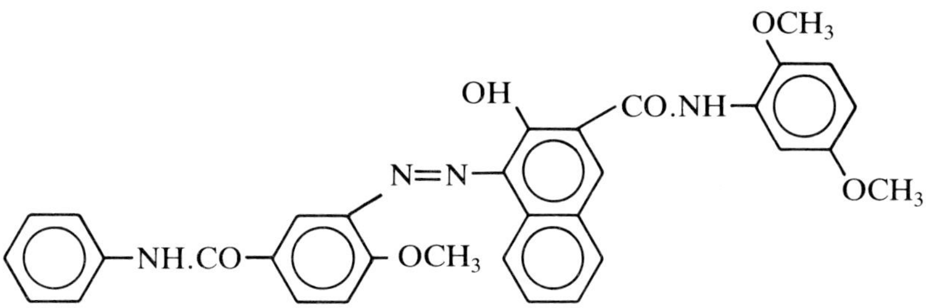

Benzimidazolone Carmine HF4C

CI Pigment Red 185; CI No. 12516, CAS No. 61951-98-2

Properties
Process magenta of very pure clean colour very resistant and superior to Red 52 and 57; rather more expensive; excellent quality pigment.

Uses
As a process magenta where stringent properties are necessary. Resistant to acid, alkali, most solvents, plasticizers, grease, wax and soap. Poor to medium lightfastness; particularly poor in tint; used widely in paste inks and metal-decorating inks.

Structure

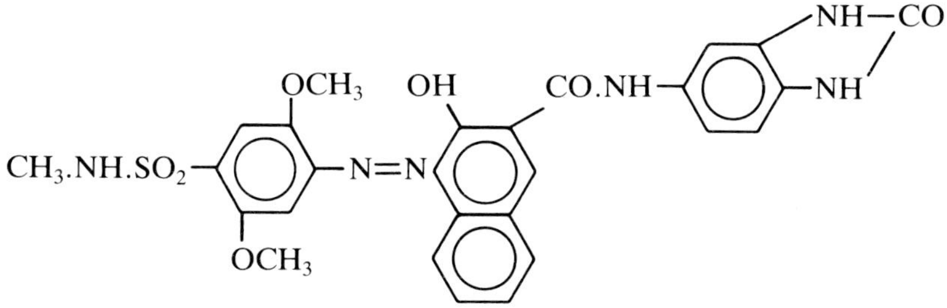

Quinacrodone Magenta B

CI Pigment Red 202; CI No. 70907; CAS No. 68859-50-7 or 3089-17-6

Properties
High-performance pigment being the most lightfast of its shade. Very similar to PR122 at full strength but much redder in tint. Resistance properties are excellent; very expensive.

Uses
Very high quality applications which require permanency of all colour properties. Usable in all systems. Care in formulation is necessary to avoid poor flow characteristics.

Structure

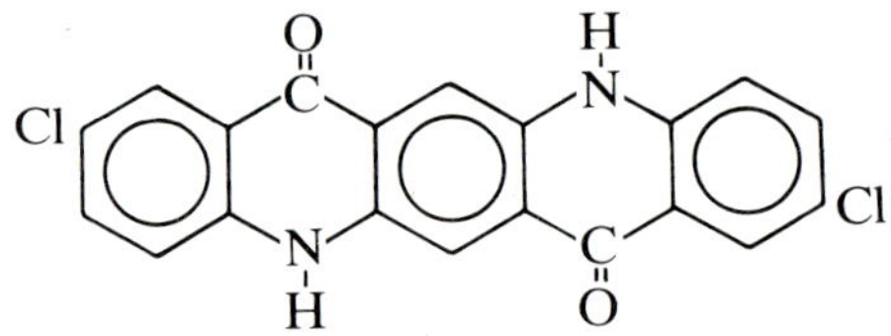

Benzimidazolone Red HF2B

CI Pigment Red 208; CI No. 12514; CAS No. 31778-10-6

Properties
A very bright medium shade red with good heat stability and light fastness. Resistant to acid, alkali, grease and soap and relatively stable to most solvents giving only slight bleed with some solvents; transparent and with high tint strength.

Uses
Liquid packaging inks and metal-decorating inks particularly to achieve strength in tint at more reasonable cost.

Structure

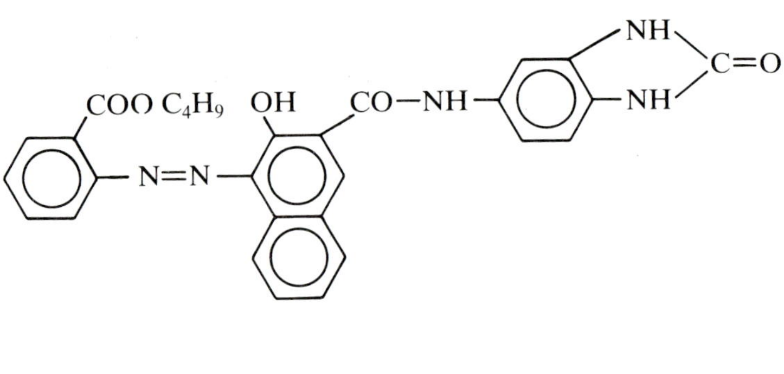

Naphthol Red F6RK

CI Pigment Red 210; CI No. 12477; CAS No. 61932-63-6

Properties
Stronger, bluer and cleaner version of PR 170. Good resistance properties resistant to acid, alkali, soap and wax. Good heat resistance but poor lightfastness, particularly in tint.

Uses
Packaging liquid inks and lacquers. Care is needed for lightfast specification inks.

Structure

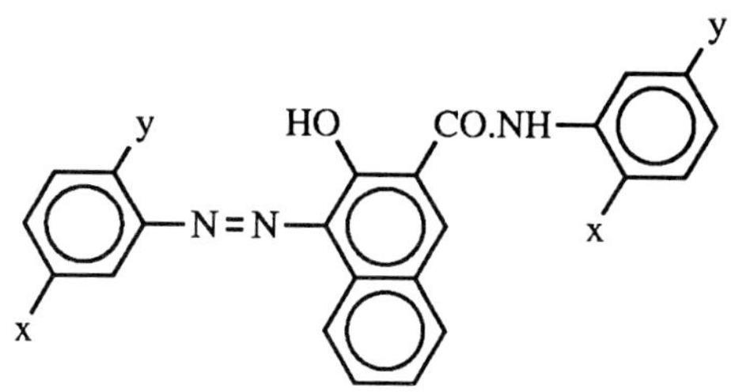

x and y not identified

Azo Magenta G

CI Pigment Red 222; CI No. N/A; CAS No. 71872-63-4

Properties
Very high resistant blueish red; excellent lightfastness and good resistance properties; strong in tint.

Uses
Liquid packaging inks in applications where high specification for this shade is required.

Structure

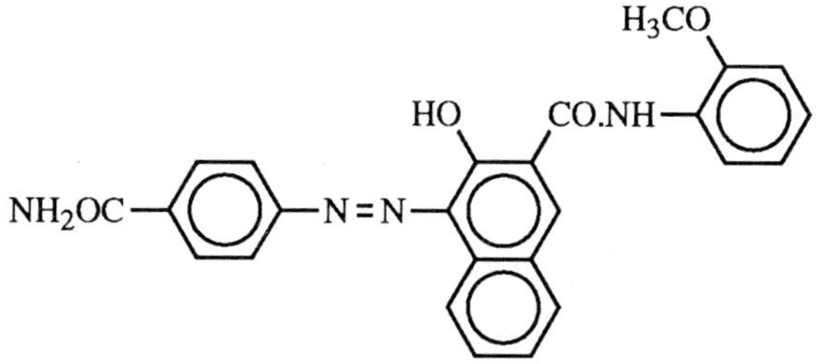

Anthraquinone Scarlet

Perylene Red Y; CI Pigment Red 224; CI No. 71127; CAS No. 128-69-8

Properties
Transparent, very yellow shade pigment with excellent tinting strength and fastness properties. Gives dull tint colours with exceptional lightfastness both at full strength and in tint.

Uses
Security printing links and other high-specification inks not requiring alkali resistance.

Structure

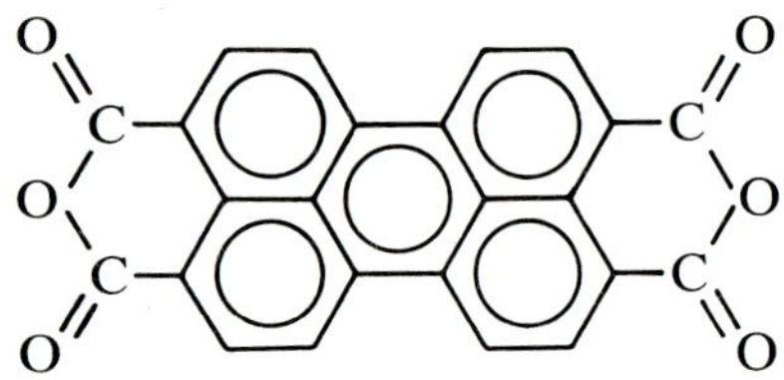

4.4 GREEN PIGMENTS

PMTA Deep Green

CI Pigment Green 1; CI No. 42040:1; CAS No. 1325-75-3

Properties
A bright bluish green of maximum strength with a clean undertone. It has fair lightfastness but tends to darken on exposure to sunlight. It has poor resistance to alkali, soap and strong solvents, but is unaffected by paraffin wax; typical of PMTA-type pigments.

Uses
Suitable for all types of printing ink, especially solvent- and water-based inks where a bright intense hue is required. Care is necessary as it is not stable in high alkaline systems and it 'burns' out; brilliant colour like a dyestuff dispersed in many types of media, very strong and bright, ideal to match a dye system to achieve better properties. Care must still be used in formulation to avoid possibilities of bleed which can always happen with PMTA colours.

Structure

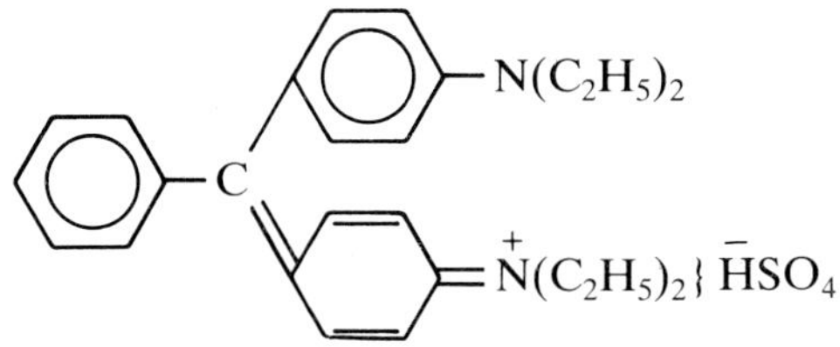

PMTA Vivid Green

CI Pigment Green 2: comprises CI No. 42040:1, CI Pigment Green 1 and CI No. 49005:1, CI Pigment Yellow 18; CAS No. 1326-11-0 and 1325-75-3

Properties
A blend of two PMTA colours (so it can bleed yellow or green). A very bright emerald green of outstanding brilliance; very strong colour but, typical of PMTA pigments, it is prone to burning out on exposure to bright light, heat or alkali.

Uses
Many printing inks exist to achieve this shade; limited in used due to lack of resistive properties. Alternative blends of pigment are often used.

Structure

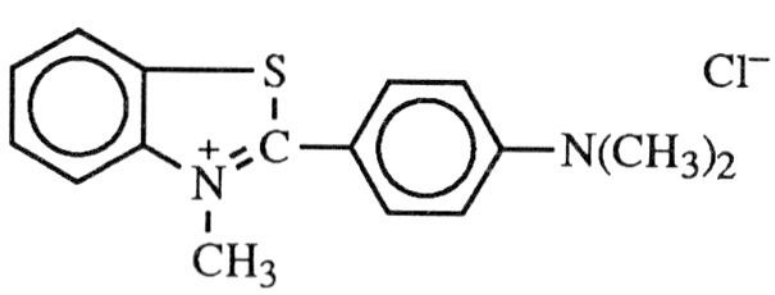

Pigment Yellow 18 is the phosphotungstomolybdiotungstic acid salt of the above.

Phthalocyanine Green

CI Pigment Green 7; CI No. 74260; CAS No. 1328-53-6

Properties
Bright green resistant pigment which is not prone to crystallization or flocculation, like its blue counterpart. Blue shade multipurpose green with superb properties. Resistant to everything. Very lightfast and heat resistant. Excellent transparency, tinting strength and colour strength.

Uses
Major green standard used in all types of printing inks. Very useful in tint as there is no lack of resistance properties. Careful formulation is necessary at high pigmentation levels in order to avoid poor flow.

Structure

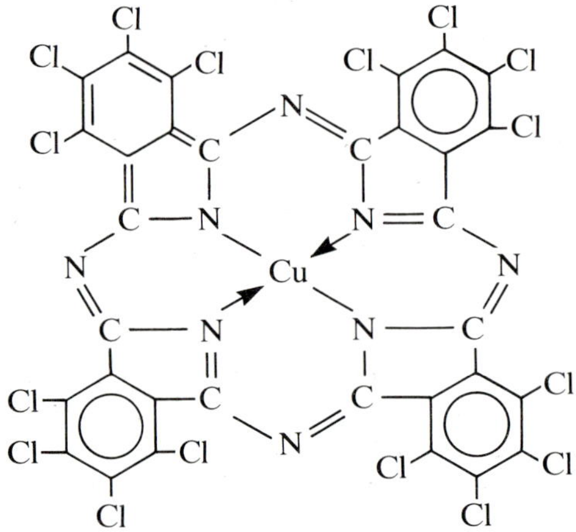

Phthalocyanine Green

CI Pigment Green 36; CI No. 74265; CAS No. 14302-13-7

Properties
Augments Green 7 to cover the colour range of the phthalocyanines from blue shade to yellow shade (the more bromine the yellower). High transparency, clean and bright, **very** resistant in all properties. More expensive and rarer than Green 7.

Uses
All types of printing inks; any use requiring resistive properties, as Green 7. It is more expensive than Green 7.

Structure
The bromine atoms vary from four to nine.

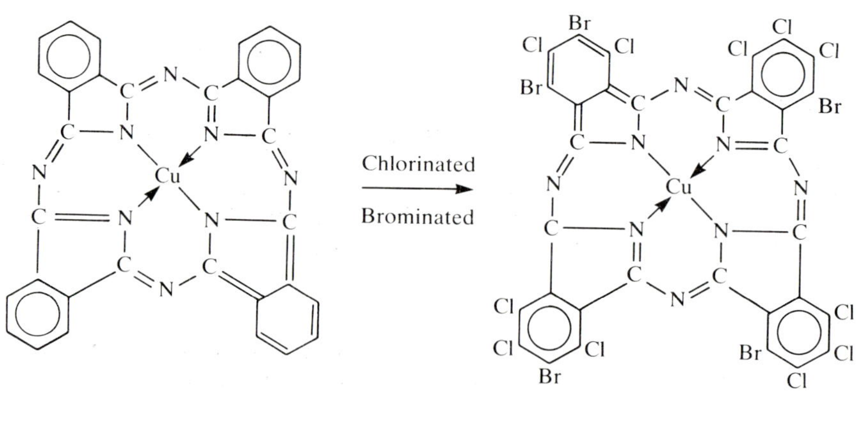

4.5 BLUE PIGMENTS

PMTA Victoria Blue PMTA Brilliant Blue

CI Pigment Blue 1; CI No. 42595:2; CAS No. 1325-87-7

Properties
Has a bright reddish blue of high tinctorial strength and purity of hue. Good lightfastness for strong colours subject to darkening on exposure and falling off sharply on reduction; readily affected by polar solvents; heat stability is encountered when varnishing or lacquering.

Uses
Used in most types of ink including water-based inks. Makes superb dispersions but rather short-flowing inks. High cost and poor resistance do not curtail its use, as the shade and strength cannot be obtained by combining other pigments. Products under this heading include the phosphotungstic molybdate salts of CI Basic Blue No 7. Like the double salts (PTA and PMA) and copper ferrocyanide complexes, they have price advantages but poor lightfastness.

Structure

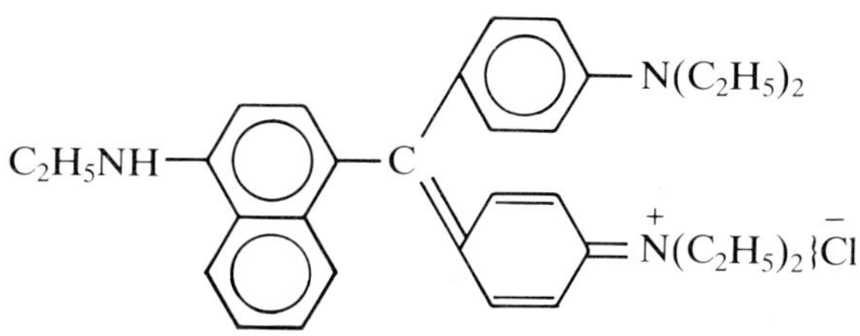

Phthalocyanine Blue Alpha Form

Pigment Blue 15; CI 74160; CAS No. all types 147-14-8

Properties
The solvent-unstable form of alpha phthalo blue; Resistive properties very lightfast, heat stable, resistant to acids, alkalis but not solvents; resistant to plasticizers, waxes, greases, and soap. Blue 15 is the reddest and strongest blue 15 type. The other pigments in this series are all modifications to correct any instability problems with this pigment.

Uses
Some usage in news-ink, but probably the least used form of the blue 15s in inks. Unsuitable for liquid inks due to solvent instability.

Structure

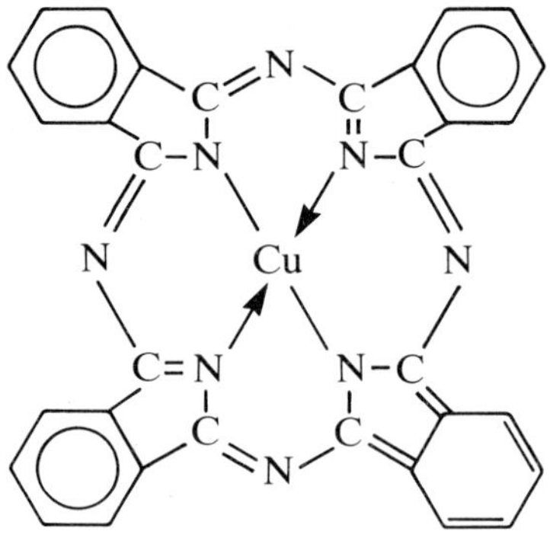

Phthalocyanine Blue Alpha solvent-stable

Phthalo Blue NC Alpha; Pigment Blue 15.1; CI No. 74160 or, CI No. 74250 (monochloride version); CAS No. 147-14-8 and 12239-87-1 for all types of this pigment

Properties
Bright reddish blue with excellent resistance properties; more heat resistant and solvent-stable than Blue 15.

Uses
The most commonly used alpha reddish shade blue, being the most cost-effective and versatile red shade blue for use in all systems. Accounts for about 25% of the B15 types usage; chosen for stability and shade.

Structure
As Blue 15

Phthalocyanine Blue NCNF

CI Pigment Blue 15:2 alpha form CI No. 74160 or 74250

Properties
Bright reddish blue more extensively treated than 15 or 15.1 giving a non-flocculating version.

Uses
Improved stability version of 15.1; higher cost restricts its use; accounts for more than 5% of Blue 15 usages.

Structure
As for Blue 15 but with more sophisticated treatment.

Phthalocyanine Blue NC Beta form

CI Pigment Blue 15:3; CI No. 74160

Properties
Bright greenish blue; completely resistant to acids, alkalis, solvents, plasticizers, greases, paraffin wax and soap; also solvent-stable, heat-resistant, and very lightfast. Transparent, strong, tinting power, very clean colour, very permanent. Slightly weaker than alpha forms and is the most-used coloured pigment (excluding black and whites).

Uses
Paste and oil ink standard; the perfect cyan for process printing; usable in every type of system and ink; accounts for over 50% of Blue 15 production.

Structure
Slightly modified form of Blue 15.

Phthalocyanine Blue B NCNF

CI Pigment Blue 15:4; CI No. 74160

Properties
Bright greenish blue, treated to prevent flocculation so a useful cyan in systems where 15:3 does not seem to work well and appears weak. All the resistant properties of the Blue 15:3.

Uses
More suitable with many liquid ink resin systems. Used more in liquid inks applications than 15:3; Slightly more expensive but with increased stability; very clean bright green shade colour; more stable in tint and strong solvent liquid ink recipes; account for about 18% of Blue 15 usage.

Structure
Slightly modified version of Blue 15 with more treatment.

Phthalocyanine Blue Epsilon form

Pigment Blue 15:6; CI No. 74160

Properties
The only very stable red shade phthalo blue. Alpha shade and will not, like 15, 15:1, and 15:2. go to Beta shade with heating. All the admirable phthalo blue resistance properties.

Uses
In cases where very clean, bright red shade blue is needed; mainly liquid ink application where shade stability can be more of a problem. Expensive; accounts for 2–3% of the total Blue 15 production.

Structure
Modified form of Blue 15.

Phthalocyanine Blue (copper free)

CI Pigment Blue 16; CI No. 74100; CAS No. 574-93-6

Properties
Heat stability better than all the Blue 15s; greenish blue shade – the greenest phthalo blue obtainable but not always very clean; excellent resistive properties like the others.

Uses
Most purposes but is more expensive than 15:3 or 15:4. Its use is restricted to materials where a metal-free product is specified, i.e. certain food wrappers.

Structure
As for Blue 15 but replacing the Cu atom with two hydrogen atoms.

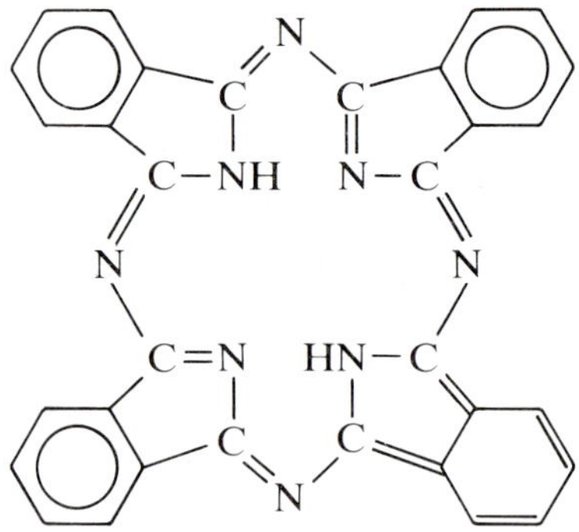

Milori Blue, Iron Blue, Bronze Blue, Prussian Blue, Chinese Blue

CI Pigment Blue 27; CI No. 77510; CAS No. 14038-43-8 and 25869-00-5

Properties
A range of strong blues from dirty-reddish tone (Prussian) to a cleaner greener tone (Chinese). A lot of surface bronzing which can vary considerably both in shade and intensity, depending on work input, drying conditions, degree of dispersion and the amount of binder in the recipe.

Iron blues possess excellent lightfastness, heat resistance, oil, fat, wax and acid resistance. They are attacked by alkalis and decompose readily. This renders them unsuitable for water-based systems, soap wrappers and fertilizer bags.

Uses
Iron blues are available in various grades to suit all types of printing ink.

Their low cost and high strength make them popular pigments. They are difficult to disperse and should not be left mixed unmilled as spontaneous combustion may occur. Easily dispersable grades are now available for certain applications.

Tints are unstable on storage as the inorganic pigment reduces to ferro–ferrocyanide, thus turning green and dirty. They need little or no extra driers as in oxidation drying ink vehicles iron, being a transition metal, catalyses the polymerization of the vegetable oils. These blues have some limitation compared with Phthalo Blue as they are dirty but in the USA they still are used as the publication gravure standards in four-colour sets.

Formula

$KFe''' [Fe'' (CN)_6] \, x \, H_2O$ or $K[Fe''Fe'''(CN)_6] \, x \, H_2O$

Note: potassium may be replaced by sodium or ammonium or mixtures of both; often referred to as potash blue or soda blue.

Ultramarine Blue

CI Pigment Blue 29; CI No. 77007; CAS No. 57455-37-5 and 1317-97-1

Properties

A red shade clean blue with low tinting strength; semi-opaque to transparent, dependent on the surface treatment and particle size distribution.

Highly resistant to the effects of light, heat, alkali, wax and solvents, but dilute acids cause fading. Uncoated grades are sensitive to water.

Uses

Mainly used in screen process, diestamping and water-based inks. As they are hydrophilic, special coated grades are available for lithography. In oil inks they are difficult to disperse, do not transfer well on three-roll mill rollers and care has to be taken in formulation to prevent piling. Suitable for food packaging but darken in the presence of copper plates or cylinders and under acid conditions on external exposure. FDA approval in certain foods and cosmetics.

Formula

$Na_{6-8}Al_6Si_6O_{24}S_{2-4}$

Alkali Blue G, Reflex Blue 2G

CI Pigment Blue 18; CI No. 42770:1; CAS No. 1324-77-2

CI Pigment Blue 56; CI No. 42800; CAS No. 6417-46-5

CI Pigment Blue 61; CI No. 42765:1; CAS No. 1324-76-1

Properties

Tinctorially the strongest blue pigments manufactured. CI Blue 56 is the strongest and most alkali resistant of this group. CI Blue 61 and 18 have poor alkali resistance. Red shade to green shade blues with a strong red–blue sheen, poor lightfastness, reasonable heat resistance, and good acid resistance. Other properties only moderate. These pigments retard oxidation drying in oil-based systems.

Uses

These pigments have traditionally been supplied as the flushed form. Either aqueous or oil based. Recently, however, powder supplies are becoming available giving the formulator more flexibility with his recipes. The powder form is very cost-effective for flexo water-based inks giving excellent strength for this type of colour. The flush form is still used in large quantities for paste inks and publication gravure inks. Mainly used as a toner in black ink (including news ink). A cheap and strong useful colour.

Structures

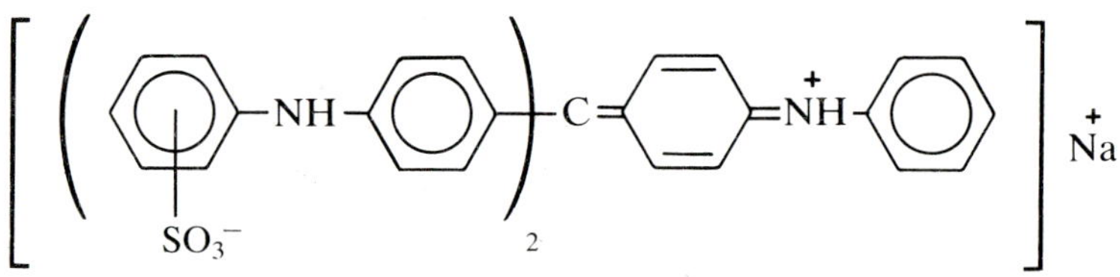

CI Blue 18 – red shade

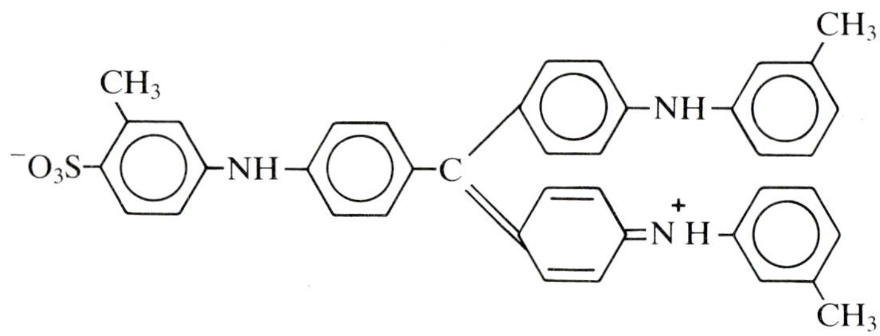

CI Blue 56 – green shade

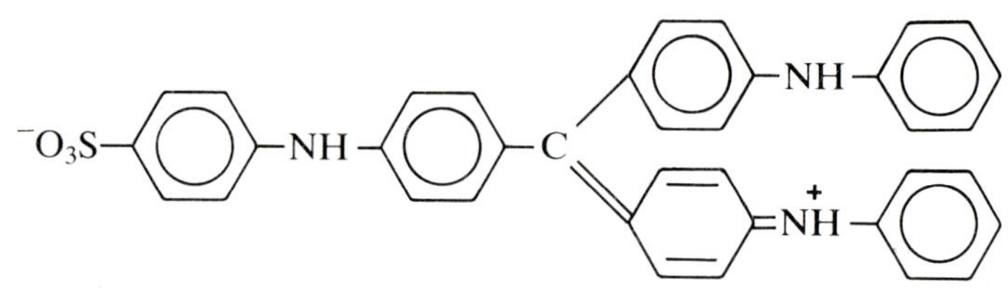

CI Blue 61 – mid shade

Indanthrene Blue, Anthraquinonoid Blue

CI Pigment Blue 60; CI No. 69800; CAS No. 81-77-6

Properties

Red shade transparent blue of good strength; extremely resistant to light, heat, steam sterilization, acid, alkali, solvents, soap and waxes; redder shade than Phthalo and very clean.

Uses

Can be used in all types of ink but is expensive; selected where outstanding fastness is the dominant requirement, as for example in gravure inks for soap or detergent packages. It disperses well but has high oil absorption which results in fairly short inks.

Structure

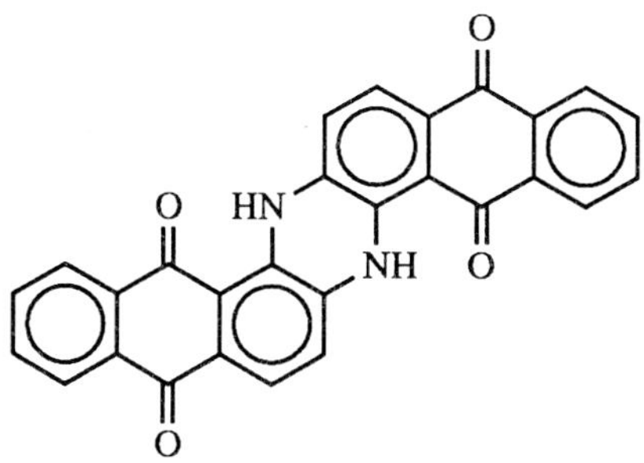

CI Pigment Blue 60

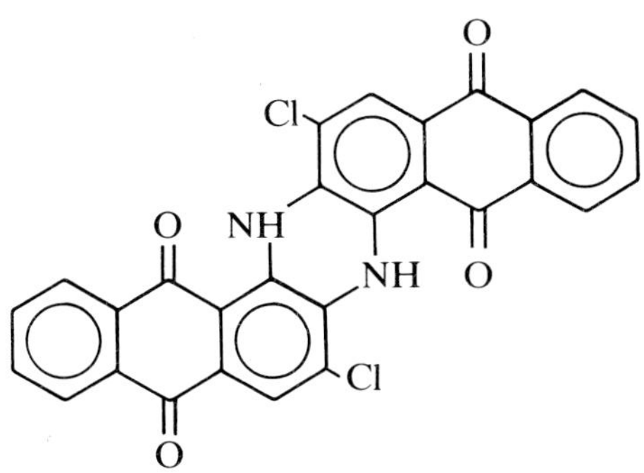

CI Pigment Blue 64

Note: the chlorinated grade is much redder and is classified as CI Pigment Blue 64 (CI Vat Blue 6, No. 69825); CAS No. N/A.

Victoria Blue CEA, Copper Ferrocyanide Blue

CI Pigment Blue 62; CI No. 42595x; CAS No. 82338-76-9

Properties
Royal blue coloured, very bright pigment not as bright as PBI (the PMTA) but a slightly more resistant and better value pigment. Useful for many applications.

Uses
Because of its shade and price is often used instead of the PMA or PMTA toners. Good for gravure and water-based flexo packaging inks.

Structure

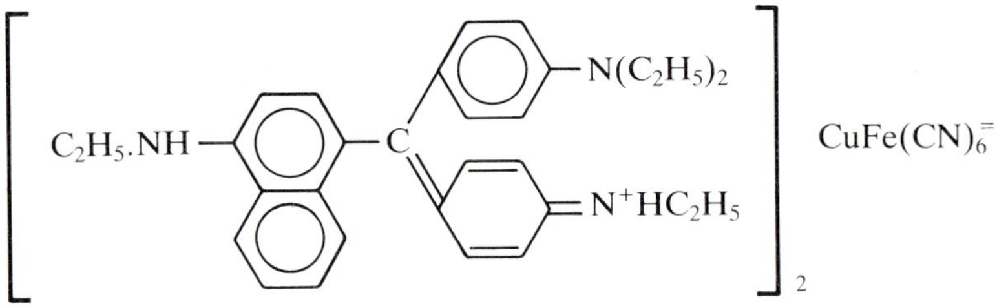

4.6 VIOLET PIGMENTS

PMTA Rhodamine

CI Pigment Violet 1; CI No. 45170:2; CAS No. 1326-03-0

Properties
Strong clean magenta shade, which is often used in inks as a blue shade red. Exhibits high purity of hue and tinctorial strength. Darkens on exposure to light and bleeds in water, solvents, oils and soap.

Uses
Declining in use. Suitable for liquid inks including water-based ones, but unsatisfactory for lithography or high-temperature steam processing. Letterpress prints can be varnished or lacquered only under special conditions. CI Pigment Violet 2 is very slightly bluer and stronger, and has similar fastness.

Structure
Pigment Violet 1 is the phosphomolybdotungstic salt of CI Basic Violet 10, CI No. 45170 (Rhodamine 'B'); CAS No. 81-88-9.
Note: double salts of this dye are made, the most popular being phosphomolybdate and silico-molybdate complexes.

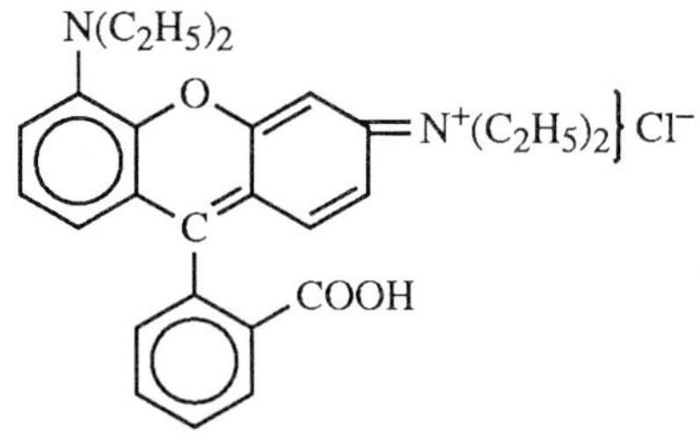

PMTA Violet, Methyl Violet

CI Pigment Violet 3; CI No. 42535:2; CAS No. 1325-82-2 and 67989-22-4

Properties
Bright bluish violet with very high strength; good resistance to light and heat; no bleed in water or aliphatic hydrocarbons but bleeds in most other solvents and in soap; suitable for steam processing. The double salts and the copper ferrocyanide complex give duller, stronger shades than the triple salt; they darken rapidly on exposure to light.

Uses
The triple salt is used in letterpress and lithographic inks. The cheaper double salt forms are used extensively in liquid inks and for toning blacks.

Composition
The phosphomolybdotungstic acid salt of CI Basic Violet 1 (Methyl Violet), CI No. 42535. The copper ferrocyanide complex is CI Pigment Violet 17, CI No. 42535:3.

Ultramarine Violet

CI Violet 15; CI No. 77007; CAS No. 12769-96-9

Properties
Brightish violet to red violet; very high colour purity similar to blue 29; good lightfastness and heat stability; quite weak.

Uses
Metal-decorating inks to tone white baking enamels to avoid yellowing.

Formula
$H_2\ Na_{4-6}\ Al_6\ Si_6\ O_{24}\ S_2$

Quinacridone Violet

CI Pigment Violet 19; CI No. 73900; CAS No. 1047–16–1

Properties
Various shades from yellowish red to bright violet; high-performance

pigment; Resistant to everything; very lightfast and heat resistant. The yellower shades are more opaque; the most lightfast organic blue shade red.

Uses
Any system where its superb properties are required. Typical of high-performance blue shade reds, it is very expensive.

Structure

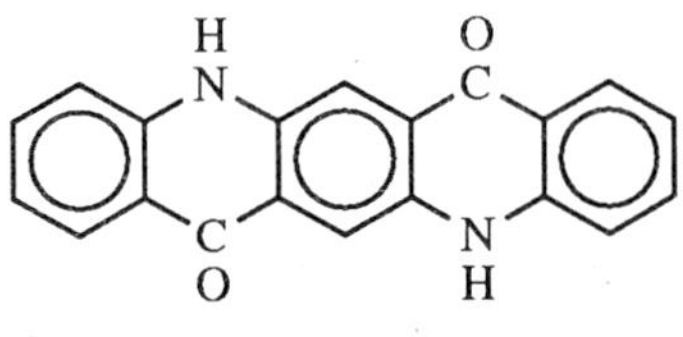

Dioxazine Violet (RL), Carbazole Violet

CI Pigment Violet 23; CI No. 51319; CAS No. 6385-30-1

Properties
A tinctorially strong, red shade purple of high transparency; superior resistance to heat, light, steam sterilization, acids, alkalis, solvents, fats and soaps.

Uses
Although an expensive pigment, it offers special advantages in formulation. It is used in all the printing processes for applications that require a stable purple, with excellent fastness properties. Also used for reddening phthalocyanines without loss of properties. Care in formulation is needed to avoid poor flow.

Structure

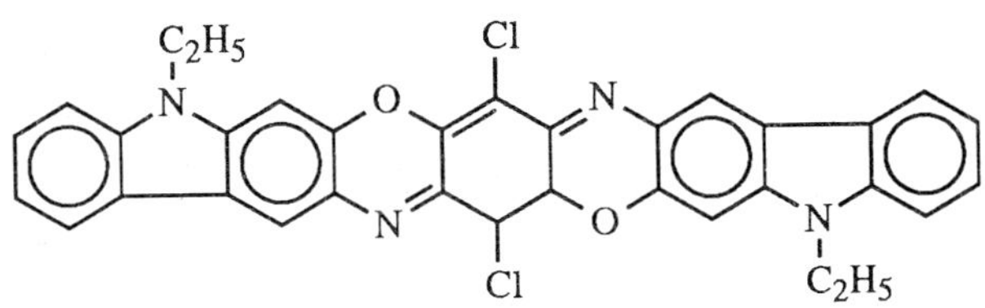

Crystal Violet CFA

CI Pigment Violet 27; CI No. 42535:3; CAS No. 12237-62-6

Properties
Very strong and bright in colour; more economical than PMTA toners (Violet 3s); not quite as clean but better properties; poor alkali and soap resistance.

Uses
Mainly water-based flexo for packaging work at right cost for bright colour, where the limitations of its properties can be accepted.

Structure
Now made by a range of suppliers – in the same range as for R169, CuFe Pink, and B62, CuFe Blue. The chloride salt is Basic Violet, the free base solvent is Violet 8 and the PTMA salt is Basic Violet 3.

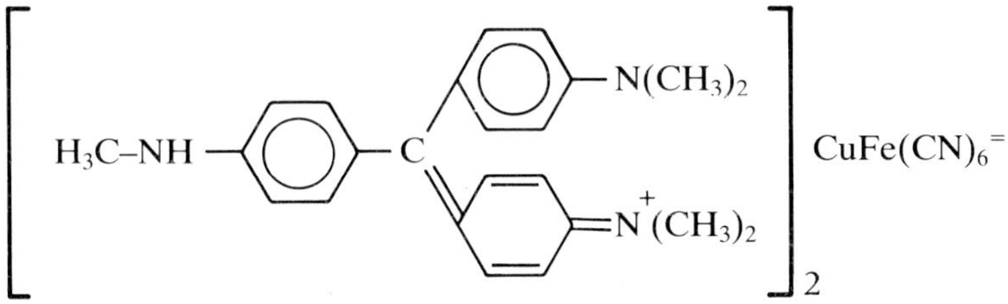

Benzimidazolone Bordeaux HF 3R

CI Pigment Violet 32; CI No. 12517; CAS No. 12225-08-0

Properties
Very bright violet red pigment; highly transparent; chemically resistant to everything except some organic solvents; very heat resistant but reasonable to good lightfastness.

Uses
Widely used as soapfast standard in this shade for both liquid packaging and metal-decorating inks. Particularly good in foil lacquers where lightfast dyes cannot be used; very transparent.

Structure

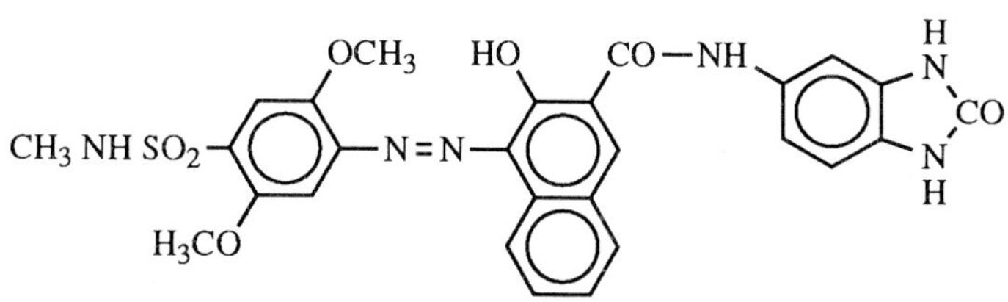

Dioxazine Violet B

CI Pigment Violet 37; CI Pigment No. 51319; CAS No. N/A

Properties
Very high performance; colour resistant to everything; lightfastness and

heat resistant; exceptional colour intensity; redder than its counterpart Violet 23.

Uses
High-specification inks only; useful for red shade work redder than Violet 23.

For very resistant Red–Violets, can work out more cost effective than Violet 23 despite being more expensive:

PV23 + B1.15	greener, very cost effective
PV23 + R202	redder, less cost effective
PV37 + B1.15	greener but same result as above; more cost effective

Certainly worth careful formulation and investigation to work out the best result. All uses are for very high-quality work in any ink system and produce an expensive ink with the best properties you can achieve with violets (both 23 and 37); choose your pigment on shade required.

Structure

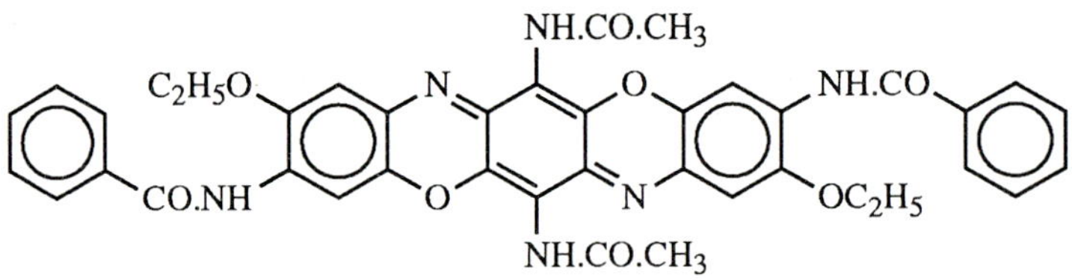

Thioindigo Red

CI Pigment Red 88; CI No. 73312; CAS No. 14295-43-3
Listed here because of violet shade.

Properties
High quality pigment; very clean colour; excellent fastness properties; slightly soluble in ketones.

Uses
Can be used for security printing; good red shade violet with high properties can prove more cost effective than other blends to achieve this shade with these properties, i.e. V37 + R202, or V23 + R202 mixes.

Structure

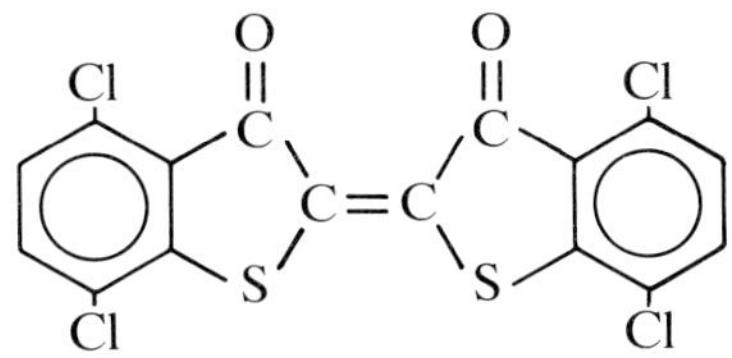

4.7 BROWN PIGMENTS

Brown Iron Oxides

CI Pigment Browns 6, 7; CI No. 77491 (77492), (77499); CAS No. various, e.g. 1309-38-2

Mars brown, synthetic brown iron oxide, natural brown iron oxide, burnt sienna.

CI Pigment Reds 101, 102; CI No. 77491; CAS No. various, e.g. 1309-37-1

Synthetic red iron oxide, mars red, mars violet, natural red iron oxide, red hematite.

Properties

Resistant to everything; exceptional lightfastness and stability; strong UV absorber; water resistant. Brown 6 and Red 101, the synthetic versions, have better colour control than the natural Brown 7 and Red 102.

Uses

Can be incorporated in a wide range of inks to reduce cost. This has become evident since the introduction of the colour-match computer. Normally they would only be used for exceptional heat resistance and wood-matching type colours. Because of their previous difficult grinding/dispersion properties they were only used specifically for mainly imitation wood printing. It is possible to use these pigments in many recipes (particularly the lighter, yellower shades) as they impart good flow and printability to ink media and help to achieve the low costs required to compete. Ideal in tinting due to their stability.

Empirical Formula

In all cases Fe_2O_3 (over 97%).

Diazo Brown 5R

CI Pigment Brown 23; CI No. 20060; CAS No. 54972-00-6

Properties

Moderately high performance pigment; acid and alkali resistance good; some bleed in soap and organic solvents; excellent grease and fat, and paraffin wax resistance; very good lightfastness and heat resistance.

Uses
Tarnish-proof bronze printing inks, lacquers, and systems like vinyl wallpaper where lightfastness is needed. More expensive than iron oxides but easier to disperse. Excellent tinting strength.

Structure

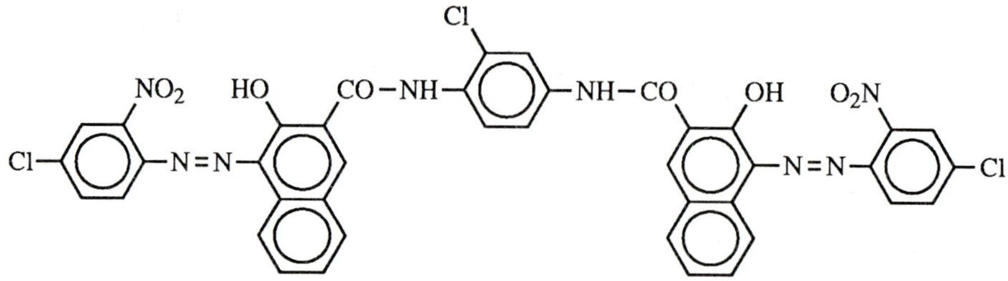

Chromium Antimony Titanium Buff Rutile

CI Pigment Brown 24; CI, No. 77310; CAS No. N/A

Properties
Buff coloured, slightly abrasive (Rutile) pigment; outstanding resistive properties; exceptional lightfastness and heat resistance (over 300°); low tinting strength slightly dirty shades.

Uses
Any high-performance ink coating, any system.

Manufacture
Crystalline mix of rutile formed by the calcination of titanium(IV) oxide, antimony(III) oxide, and chromium(III) oxide. Basic composition: TiO_2, 80–90%; Cr_2O_3, 1–6%; Sb_2O_3, 8–12%.

Benzimidazolone Brown HFR, Fast Brown HFR.

CI Pigment Brown 25; CI No. 12510; CAS No. 6992-11-6

Properties
Very transparent, very resistant chemically (some organic-solvent bleed); good lightfastness and heat resistance to 220°C. Typical of this type of pigment is Benzimidazolone (Orange 60, Red 175, Red 176, Red 185, Red 208, Yellow 120 and Violet 32), giving very good dispersability.

Uses
Where transparency is needed (rare in browns) to achieve gold effects on foil and a metal-free specification on various films (particularly when metallized). The very high heat, and other resistive properties make this

particularly suitable for laminating inks and stoving-metal decorating inks.

Structure

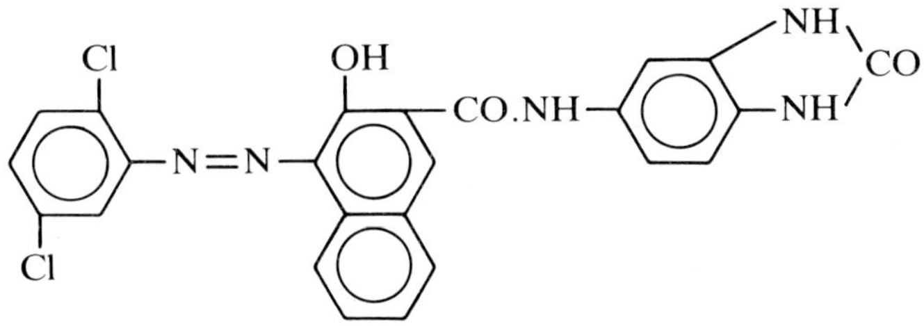

4.8 BLACK PIGMENTS

Vegetable Black (Lamp Black)

CI Pigment Black 6; CI No. 77266; ; CAS No. 1333-86-4; Nitro Pigment

Properties
A soft black with bluish undertone which is greyer and coarser than the channel blacks. It is inert but, because it has a large surface area, it tends to absorb organic colouring matter, therefore it should not be added to tints.

Uses
Oldest pigment known, vegetable blacks are used as matt blacks. They improved the strength and colour of diestamping and copperplate inks. They have now been almost entirely replaced by furnace blacks, unless cost is a factor.

Carbon Blacks (Channel and Furnace)

CI Pigment Black 7; CI No. 77266; CAS No. 1333-86-4
The most important black pigment – all types of carbon black are included under the above classification. They are pigments of extremely fine particle size, with colour varying from a grey masstone with blue undertone, to a dense jet black masstone with brown undertone, dependent on the method of manufacture. Non-rubber varieties, used in printing inks, are produced in pellet, bead, densed or fluffy form. They are chemically inert, extremely fast to heat, light, acids, alkali, solvents and soap. Most grades, however, do contain a small proportion of volatile material, the effect of which may become apparent on dispersion; accounts for over 95% black ink usage.

Furnace Black

Almost all carbon black is today produced by the furnace process, and most of this production is allocated to the rubber-compounding industry for vehicle tyre manufacture.

Some is produced by the slow vertical process in which oil is sprayed and burned under controlled conditions. The carbon is collected by cyclonic or electrostatic methods. Finer blacks are manufactured by a faster vertical process, where hot gas and oil are combined under pressure to disintegrate or crack. Carbon black produced by this process is not exposed to air and therefore absorbs little oxygen. Newer types of treated furnace black have replaced channel black in most applications. They undergo a further process which bonds them with oxygen and other surfactants to obtain required surface characteristics.

Channel Black

Now produced in only a few locations throughout the world as it requires natural gas, which is more valuable as fuel than as a source of carbon black. Natural gas consists of a mixture of gaseous hydrocarbons, predominantly methane. It is burnt in a series of small luminous flames which impinge on a water-cooled, inverted U-shaped channel. A layer of fine carbon black is deposited. The particles are continuously removed by scrapers and conveyed to beading, densing and packaging plant. They have a dense black appearance with brown undertone. Channel blacks possess good wetting characteristics despite a high oil absorption, caused by their porosity and very large surface area. This wetting is aided by the low pH that channel blacks normally have, caused by the formation of a volatile absorbed layer of between 10 and 15% by weight, including moisture. Oxygen absorbed during manufacture can be chemically bonded to produce 'long flow' channel blacks having optimum colour, density, flow, gloss and dispersion. A disadvantage of channel black in inks is the tendency to absorb both driers and alkali blue toners over a period of time. This fault is not present in the treated furnace blacks. Their use is declining.

Lamp Black

Creosote and tar residues are burned in a hooded flue, with a limited amount of air, to a smoky, sooty gas. It was formerly allowed to settle in cooling chambers, to be swept and bagged. To-day it is normally collected by more modern methods. It has a large particle size compared with channel black and it flocculates to form chain-like structures.

Formula and composition (%)

Carbon	90–99
Volatiles	1–10

Almost pure carbon.

Black Iron Oxide, Magnetic Iron Oxide, Synthetic Black Oxide

CI Pigment Black 11; CI No. 77499; CAS No. 1309-38-2 or 12227-89-3

Properties
A dense matt brown shade black of high specific gravity and low oil absorption. It is resistant to heat, light, alkali, solvents, soap and wax but it dissolves in acid. Cheap; strong UV absorber; converts to red iron oxide at 150°C.

Uses
The ordinary black iron oxide has been used in both copperplate and diestamping inks. It imparts the flow and clean wiping properties necessary for such inks, but is tinctorially weak.

Magnetic iron oxide is required in inks which print well-defined characters on cheques, for automatic cheque processing. When dry, prints using such inks will have a permanent magnetic response to reading apparatus. Better dispersability in water than Black 7. Synthetic grades are stronger and bluer than natural grades.

Composition and synthesis
Ferroso-ferric oxide. $FeO\text{-}Fe_2O_3$ (Fe_3O_4) occurs widely in nature as magnetite, the magnetic oxide of iron. The grades used in inks are synthetic products.

4.9 WHITE PIGMENTS AND EXTENDERS

Zinc white, Chinese white

CI White 4; CI No. 77947; CAS No. CAS No. 1314-13-2

Properties
Clean white of medium opacity; fine ground particle size; pure grades are FDA approved; strong UV absorber; self-fungicidal material; neutralizses acidic oxidation products. It is prone to livering in Acidic vehicles; discolours with copper compounds; yellow when hot, white when cold.

Uses
Widely used as a white extender in tin-printing and many other ink systems; it improves flow and levelling.

Formula and composition (%)

ZnO	99–99.7
Volatiles	0.3–1.0

Lithopone

CI Pigment White 5; CI No. 77115; CAS No. 1345-05-7

Properties
Chemically inert pigment, available in easily dispersable micronized forms, opaque white extender; decomposed by acids.

Uses
Gaining popularity (particularly now, due to the relentless increases in titanium dioxide) as an extender; more popular on the continent than in the UK; useful in all systems with TiO_2 usage.

Manufacture
Calcined coprecipitate of zinc sulphide and barium sulphate.

Formula and composition (%)

$BaSO_4$	60–72
ZnS	28–40

Titanium Dioxide Anatase and Rutile

Pigment White 6; CI No. 77891; CAS No. 13463-67-7

Properties
Most important white pigment currently in use; accounts for over 80% of all CI whites used (including all extenders). It has mainly replaced nearly all other whites. Resistant to everything, very lightfast and heat resistant. Inert material. Various grades available with coatings of zinc, aluminium, zirconium, or silica oxides to optimize opacity and to facilitate dispersion and durability.

Two major crystal polymorphs:

Rutile harder; more opaque (refractive index 2.71)
than anatase; durable.
Anatase less opaque (refractive index 2.55); bluer tone;
softer; less durable.

Uses
Every ink system – almost 400 grades/types available, all with different properties. Anatase is preferred in gravure ink systems for less cylinder wear. For different systems, generally the purer, less extended, non-coated grades are used for cheaper water-based flexo inks, screen-process inks, and some gravure whites. The coated grades are for higher quality, more exacting work giving higher gloss (and slightly more transparent) results. Least used in litho and letterpress as most of their substrates are already white, so used for tintwork. Used in decorating, screen and liquid ink base coats.

Composition and manufacture

Ilmenite $FeTiO_3$ natural ore is digested with sulphuric acid. The solution is hydrolized by boiling and the precipitated TiO_2 separated and calcined at approx 1000°C. It is then wet-milled to coat with hydrated oxides, washed, dried and finally ground in the presence of polyols, amines or other organics prior to packing. The mineral rutile can also be taken and reacted with chlorine to produce $TiCl_4$ which is oxidized to TiO_2. The 'chloride' process whites differ from 'sulphate' process whites, being of higher dry brightness and harder crystal, though do not differ in any chemical or pigmentary characteristic. There is, however, some reluctance by many people (a fact denied now by TiO_2 suppliers) to use chloride process because of abrasivity. Latest methods of manufacture provide easily dispersing grades of high opacity and gloss. A crystal size of about 0.2 μm is paramount to optimize the scatter of visable light.

Formula and composition (%)

TiO_2	80–99.9
(Al, Si, Zr or Zn)O	0–20 + little organic coating

Zinc Sulphide

CI Pigment White 7; CI No. 77975; CAS No. 1314-98-3

Properties

Opaque white pigment; resistant to everything but acid; heat resistant and colour stable; not as opaque as TiO_2 but less after yellowing. Excellent for abrasion resistance so a good extender for TiO_2; greater UV reflectance than TiO_2.

Uses

Printing inks of all systems except acid vehicles; metal-decorating systems – gives pure clean baking enamels; gaining favour (on cost) as an extender for TiO_2.

Formula and composition (%)

ZnS	97
ZnO	<0.3
Solubles	<0.4
Moisture	<0.2
Residue	3

Calcium Carbonate, Whiting

CI Pigment White 18; CI No. 77220; CAS No. 471-34-1

Properties

White (dull) cheap extender now used to reduce the cost of inks without affecting their printing qualities. Precipitated calcium carbonate may contain up to 3% stearate or resin coating. Both the coated and uncoated grades have superseded other extenders as tinting media. They are bulky but impart good working properties and a fairly glossy finish. Varies from opaque to semi transparent and can vary from high speed stir to need milling versions.

Uses

Coated grades are used for letterpress, lithographic and gravure inks, aqueous inks use both grades, whereas screen-process inks normally contain uncoated grades or whiting (the natural form of calcium carbonate) to obtain a matt finish. Copperplate inks also contain whiting for its wiping qualities. Selection of the correct grade aids dispersion, flow and press performance. Calcium carbonate is not recommended for use with high acid-value vehicles such as moisture-set as it reacts to liberate carbon dioxide. It also reacts with alkali-sensitive pigments like Milori Blue (B27) so care is needed in formulation. Calcium carbonate is also used extensively as a 'dry spray' powder to prevent set-off. Japanese-produced ground mineral versions once milled to a good dispersion, give results with outstanding gloss and transparency, but cost much more than the very basic forms available.

Production

Calcium carbonate ($CaCO_3$) occurs naturally in many countries as chalk or limestone which is pulverized to produce whiting. It is produced synthetically by slaking quicklime and exposing the slurry to carbon dioxide. It is also obtained as a byproduct in water purification.

Formula and composition

	Natural (%)	Synthetic (%)
$CaCO_3$	95–98	98–99
$MgCO_3$	<1–3	<1–1.5
SiO_2	<0.5	<0.1
(Al, Fe) 0	<0.5	<0.3
H_2O	<0.2	<0.2

China Clay, Bentonite, Kaolin Clay

CI Pigment White 19; CI No. 77005; CAS No. 8047-76-5

Properties

FDA-approved for coating use, white to pale grey powder; chemically inert and inexpensive. It has a higher oil absorption than most extenders except alumina.

Uses

Moisture-set and low-cost letter press inks, some gravure inks. High oil absorption grades can be used as a thickener. Useful because of its inertness and cost in a wide range of applications.

Formula and composition

Natural hydrated aluminium silicate, $Al_2O_3.2SiO_2.2H_2O$, occurs in many parts of the world. It contains impurities of calcium, magnesium, iron, quartz and mica. It is cleaned and segregated into grades by levitation in settling tanks.

	Hydrous (%)	Calcined (%)
Al_2O_3	37–45	42–45
SiO_2	44–53	51–53
(Fe, Ca, Mg)O	12–14	1

Blanc Fixe, Process White

CI Pigment White 21; CI No. 77120; CAS No. 7727-43-7

Properties

Clean and white extender with excellent resistance properties to everything; good lightfastness and heat stability. It exhibits a high diffuse continuous reflectance of light giving a pure white, which helps to retain the original brilliance and tone of colours mixed with it.

Uses

Mainly in paste ink where its tendency to matt down can be prevented by the resin solids of the ink system. Recent innovations in particle size are beginning to make available grades suitable for liquid ink (blanc fixe micro) application. It has the highest density of any extender so considerable amounts by weight can be added to an ink without loss of printing strength. It has FDA status due to its low solubility,

Formula and composition (%)

$BaSO_4$	97–99
SiO_2	0–2
Residues	0–1

Alumina Hydrate, Light Alumina Hydrate, Lake White, Transparent White

CI Pigment White 24; CI No. 77002; CAS No. 1332-73-6

Properties

White, soft-textured extender; quite transparent compared with all the others; resistant to everything except diluted HCl; heat stable and lightfast; can absorb the drier in oxidizing ink systems.

Uses

Improves flow of litho inks, and the gloss of a wide range of paste and liquid inks; reasonably priced extender easy to disperse; FDA approved.

Empirical formulae
$Al_2O_3 \cdot xH_2O$

Talc, French Chalk

CI Pigment White 26; CI No. 77718; CAS No. 8005-37-6

Properties
Pure talc is the standard 1 on the MOH scale (where diamond = 10). Lustrous white very fine powder; chemically inert; lightfast and heat stable up to 850°C. Cheap and useful but many versions can reduce gloss unless the binder system is adequate to compensate it.

Uses
In quality litho inks in particular it gives good excess tack removal and improved rub resistance to inks.

Composition
Hydrated magnesium silicate locally produced from talc, soapstone or stealite.

Formula and composition
Can vary in makeup:

	Pure (%)	Average (%)	Bad (%)
MgO	31–32	25–30	20–70
SiO_2	60–64	52–56	15–40
CaO	1	2–5	2
Fe_2O_3	1	1–3	1
Al_2O_3	1–2	1–2	5
Residues	4–5	4–10	

Silica

CI Pigment White 27; CI No. 77811; CAS No. 7631-86-9

Properties
Transparent white light fluffy matt pigment with a high oil absorption. Completely inert, unaffecting the resistance properties of inks; only the synthetic version used for ink.

Uses
Flatting, anti-settle, and bodying agent for any ink system. Perfect for reducing tack, but at the cost of a lower gloss level. It can gel when added to non polar solvents; anti-mist additive.

Formula and composition (%)

SiO_2	98
Other oxides	0.1
Ignition loss	2.0

4.10 PEARLESCENT MATERIALS

Titanium Treated Micas – Pearlescents

CI White 6 and White 20; CI No. unclassified; CAS No. mix of 13463-67-7 and 12001-26-2

Properties
Synthetic effect non-metallic shades from silver to gold; FDA approved; chemically inert; transparent with a high refractive index. Often needed between 10 and 25% pigment loading. In transparent media they produce effects due to the multiple reflection of light which match the lustre of mother of pearl; stable to 800°C.

Uses
In all cases where bronze or aluminium cannot be used and the substrate is not foil or metallized film, i.e. soap-wrappers on paper. Usually screen, offset, flexo or gravure packaging inks. Very large choice of shades with the golds achieved with oxide blends. Care must be taken to ensure the binder content of the ink is high enough to provide complete fixing of the pearlescent to avoid poor results given in underbound systems.

Composition
Blends and coprecipitates of TiO_2 rutile, mica and variations.

	Possible analysis (%)
TiO_2	40
Al_2O_3	25
K_2O	6
SiO_2	27
H_2O	2

4.11 METALLIC PIGMENTS

Aluminium

CI Pigment Metal 1; CI No. 77000; CAS No. 7429-90-5

Properties
Brilliant silvery powders (called linings) or pastes which have high hiding power. Some tend to tarnish on exposure and they will not resist alkali or acid. They are inert in hydrocarbon solvents.

Uses
Non-leafing grades are used as linings or pastes in all types of ink, either alone or with another colour to produce a 'metallic' colour effect. Many hundreds of ready-made pastes are available on the market these days, suitable for all types of manufacture. One of the latest innovations are water-stable silvers for aqueous flexo use.

Copper Powder, Bronze Powder

CI Pigment Metal 2; CI No. 77400; CAS No. mix of 7440-50-8 and 7440-22-4

Properties
Brilliant reddish gold to greenish gold; copper or copper–zinc alloys prepared for printing use. Available in many grades differing in particle size, leafing and covering power. Coarser powders generally are brighter while finer powders are matter and more opaque. Special non-tarnish grades for all types of printing are now available already pasted into different oils, solvents, or media. They are sensitive to heat, humidity, and either strong acid or alkaline media. Stabilized paste forms should be used when possible.

Uses
Decorative printing inks, particularly for packaging, mostly on paper.

Manufacture and formula
Powder, flaked by ball or stamp milling with stearic acid for leafing grades or other fatty acid lubricants for the non-leafing grades. The shade depends on the composition of the alloy.

Colour	Copper (%)	Zinc (%)
Copper	100	0
Pale gold	90	10
Rich pale gold	85	15
Rich gold	80	20
Green gold	70	30

4.12 FLUORESCENT PIGMENTS

These comprise various yellows, oranges, reds and greens and a single blue. CI Numbers are not assigned.; CAS No. N/A

Properties
Powders are created by pulverizing solutions of fluorescent basic or reacted dyes in resins. Therefore they can be unstable, not very heat resistant and able to be ground away in dispersion equipment. Dazzlingly bright colours giving a fluorescent effect in daylight. Mainly in the yellow orange and red portions of the spectrum. They appear most brilliant in the dark with a UV source of radiation to illuminate them. Fluorescent

dyes/daylight pigments powders have the property of converting short-wave visible radiation into longer wavelengths giving very brilliant colours. Blue, indigo and violets are extremely rare for daylight fluorescence.

Uses

Any type of printing advertising and packaging where the effect is strongly to catch the eye. The drawing power of fluorescent displays, e.g. price-changes, or warnings or sheer attraction (e.g. soap powder boxes) to the product can lead to requests for fluorescent printing inks. Their use was confined until the 1960s to screen inks for poster work but, more recently, all the properties have been improved, so by the late 1970s fluorescents were used in litho, letterpress, flexo and gravure systems.

Their stability has improved and they can be heat resistant up to 200°C in some cases; lightfastness has also been considerably improved. Due to the range of inks for printing purposes most reds, yellows, greens and orange fluorescents are available each one in about 8–10 different forms. All for different uses so be careful in formulation to choose the grade you need for your particular application. Useful for safety labels, greeting cards, price changes and brilliant tri-chromatic inks.

4.13 GENERAL PROPERTIES OF PIGMENTS

Tables 4.5–4.7 are based upon test results and information supplied by the main producers of pigments. Often one supplier produces several grades of pigment with a single colour index reference. These may differ in relative density, oil absorption and opacity. They may have special properties such as improved lightfastness, water resistance, and insolubility in various reagents.

The information tabulated here is intended only as a guide for selection. It must also be appreciated that the tables give estimated average test results and do not state extremes of variation. Performance in a test can depend upon a variety of factors, including vehicle composition, treatments such as rosination, metallic salt and laking complexes, surfactants, etc. as well as tinctorial strength, particle size and batch to batch variations.

It should be remembered that small differences can be significant when assessing test results; these include the percentage of pigment, the type of vehicle, hardness of drying, film thickness and possibly the substrate. Any deviation from the specified test procedure could also alter the performance and comparative test results.

With the exception of lightfastness, test results have been expressed on a 1 to 5 scale which is a cipher form of the ordered set: 1 = very poor; 2 = poor; 3 = fair; 4 = good; 5 = excellent.

Bulk value

This is equivalent to 100 divided by the relative density of the pigment. The numbers indicate relative volume per unit mass.

Concentrated paste

The maximum percentage of pigment in 2.8 Pa s stand oil or its equivalent to give a handleable printing ink base or concentrate.

Methods of test for printing inks

BS 4321:1969 (ISO 2834–2844–1974) has been taken for the basis of these tests. A précis of such tests is given for guidance.

Test method 1. Preparation of standard print: solid and 70% tone print; defined volume of ink per unit area.

Test method 2. Fastness to daylight: Blue Wool Scale ratings, 1 to 8: (1) very low; (2) low; (3) moderate; (4) fairly high; (5) high; (6) very high; (7) excellent; (8) exceptional.

Test method 3. Fastness to water: print between filter papers, 1 kg 24 h.

Test method 4. Solvents: fastness:

- 74 op meths, or
Acetone	10% by volume
Ethyl acetate	40% by volume
Ethylene glycol monoethylether	10% by volume
Meths 74 op	30% by volume
Toluene	20% by volume.

5: min immersion in test tube.

Test method 5. Alkali resistance: between filter papers soaked in 5% sodium hydroxide solution. 10 min/kg.

Test method 6. Soap resistance: between filter papers saturated with 1% soap solution for 3 h/kg.

Test method 7. Resistance to detergent: between filter papers saturated with 3% detergent solution for 1 h/kg.

Test method 8. Resistance to cheese: in contact with crust and fresh slice for 3 days at 20°C.

Test method 9. Resistance to edible oils and fats: in contact with solid fat for 24 h at 20°C plus 1 h at 4°C in refrigerator to facilitate removal of specimen. Edible oils between soaked filter papers for 24 h/kg in water-saturated atmosphere at 20°C.

Test method 10. Resistance to waxes: 5 min in melted wax, drip back through margin.

Test method 11. Results from this test, the resistance of printing inks and prints to spices, have not been included because of the variety of spices available.

Note: for more information the complete standard test methods should be studied. All tests should be carried out exactly as specified in BS 4321:1969 (Chapter 15).

Table 4.8 compares not colours in their own shades but as pigment type as this is the most recent range to be universally accepted (the Orange HL being patented in 1964) and it covers the spectrum from yellow → orange → red → violet. It is an interesting comparison by chemical type. The high cost in proofing new materials for current legislation could make this range one of the last new ranges we will have to deal with. It is more likely, with all the problems of introducing new materials, that all we will see in the future will be revamped versions of known products, particularly purer cleaner purged versions. Our job can be made easier in material selection as we are now all reducing inventories and the Zenith of the 1960s is now in decline. When additional safety data comes to hand we just eliminate suspect materials from usage. This gives us all the possibility of tidier, safer, more cost-effective ranges of materials for use in the ink industry and thus a reduction in inventories. Pressure is continually on our trade to eliminate various types of material, and, in the end, the smaller range of choice can only be for the benefit of us all, hopefully making the technicians of the future more aware in greater detail of the complete properties and effects of the materials left in a shrinking raw material inventory.

Section 2: Dyestuffs

Dyestuffs are used mainly for liquid inks, coatings and lacquers. Special disperse dyes are used in both liquid and paste inks for heat transfer printing. In this application the choice of printing process is of minor importance compared with the subsequent operation of transferring the dyestuff from paper to the textile by sublimation. Other dyestuffs used in paste inks include acid dyestuffs for double tone and invisible inks, but by far the largest use has been in cheque security inks. Basic dyestuffs in the free-base form dissolve in fatty acids to become oil-soluble and can be used as toners for black inks.

In order to distinguish between the various types of dyestuff used increasingly in printing inks it is advisable to consult the *Colour Index* published by the Society of Dyers and Colourists jointly with the American Association of Textile Chemists and Colorists. Dyes are classified for specific usage and may be included in more than one entry, for example food dyes include acid, solvent and natural dyes as well as pigments. This is followed by a classification of dyes by chemical composition and a comprehensive alphabetical index.

Usage classification as defined by the *Colour Index*:

Acid dyes
Azoic dyes
Basic dyes
Developers
Direct dyes
Disperse dyes
Fluorescent brighteners
Food and drug dyes
Ingrain dyes
Leather dyes
Mordant dyes
Natural dyes
Oxidation bases
Pigments
Reactive dyes
Solvent dyes
Sulphur dyes
Vat dyes

4.14 ACID DYES

This term applies as much to an application classification as to a chemical class. The name acid dye was originally used to indicate the presence of sulphonic acid or other acidic groups in the molecular structure. However, since acidic groups are present in other types of dye, their presence is not a sufficiently distinguishing factor. They are regarded as anionic dyes, soluble in water, mainly insoluble in organic solvents. Selected dyes are soluble in alcohol, ketones and esters. They give very bright hues, with lightfastness ranging from poor to very good.

Chemically they are azo, anthraquinone, triphenylmethane, azine, xanthene, ketonimine, nitro and nitroso compounds; an example, eosine follows.

Eosine

CI Acid Red 87; CI No. 45380; CAS No. 17372-87-1

Properties
A bright yellow–red hue, with a marked yellow–green fluorescence which provides a typical example of dichromatism. It is readily soluble in water and other reagents. Lightfastness is between 1 and 2 on the Blue Wool Scale.

Uses
It is used in water-based fugitive cheque inks, invisible inks in painting books and in dyeing paper.

Structure

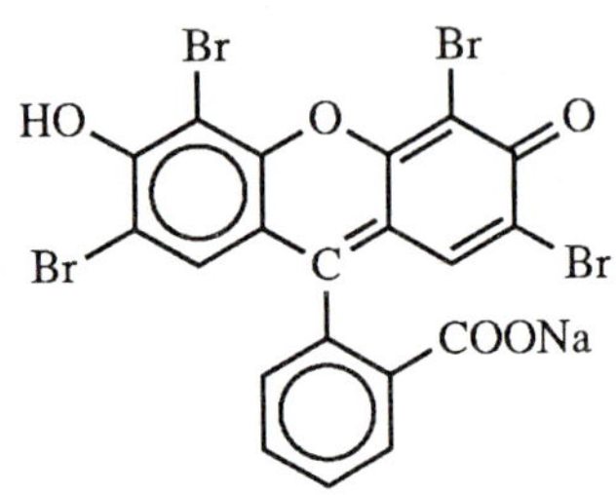

The most popular acid dyes in current use in ink manufacture are:
Acid Yellow 3, 5, 17, 23, 36, 54, 73, 121, 157, 194, 204 and 236;
Acid Orange 3, 7, 10, 142 and 144;
Acid Red 18, 52, 87, 88, 143, 221, 289, 357 and 359;
Acid Green 1, 16, 26 and 104;
Acid Blue 1, 7, 9, 15, 20, 22, 93, 129, 193, 254 and 285;
Acid Violet 9, 17, 90, 102 and 121;

Acid Brown 101, 103, 165, 266, 268, 355, 357, 365 and 384;
Acid Black 47, 52 and 194;
and some known CI and CAS Numbers are listed below.

Dye	*CI No.*	*CAS No.*
Acid Blue 9	42090	2650-18-2
Acid Orange 7	15510	633-96-5
Acid Red 87	45380	17372-87-0
Acid Yellow 3	47005	8004-92-0
Acid Yellow 73	45350	518-47-8

For detailed information reference should be made to the *Colour Index*.

4.15 BASIC DYES

The first dyes produced from coal tar were basic dyestuffs. These cationic dyes have maintained popularity with the printing ink industry due to their brilliant shades and high tinctorial strength. However, their lightfastness is poor and curtails much of their usefulness. They are soluble in water and alcohol, but have low solubility in most other organic solvents.

These dyes are often used in conjunction with a mordant or laking agent such as tannic acid, usually in amounts equivalent to the weight of the dye. This lakes the dyes *in situ* and not only fixes the dry ink, but also makes it more water resistant and helps the dye to develop its full colour and strength.

Some of the more widely used basic dyes are described below.

Auramine

CI Basic Yellow 2; CI No. 41 000; CAS No. 2465-27-2

Properties
A bright lemon yellow with poor lightfastness used in flexographic inks. The dye is dissolved in a suitable solvent in the presence of laking agents. The wax and water resistance of prints containing auramine is very much a function of the chemical formula of the laking agent. Tannic acid and dimethyl salicylic acid are extensively used in this role.

Note: the free base made from the above by addition of an aqueous alkali is CI Solvent Yellow 41, CI No. 41 000:1; CAS No. N/A.

Structure

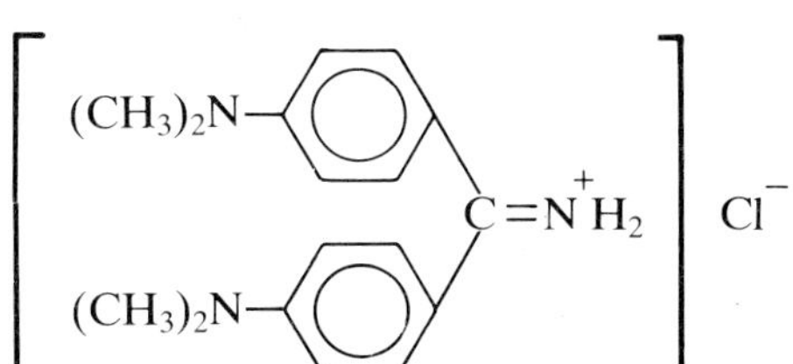

Severely declining in use in favour of Basic Yellow 37 Auramine.

Diarylmethyaniline Yellow

Basic Yellow 37; CI No. 41001; CAS No. 6358-36-7

Properties

Strong brilliant yellow dyestuff; decolourizes immediately on contact with strong alkali. Very soluble in ethyl alcohol and even much more so in the presence of a laking agent; has replaced Basic Yellow 2 in most applications.

Uses

Comic inks, cheap paper flexo printing wrappers, foil lacquers not requiring any lightfast specification. Some water-dye toners to strengthen/brighten pigmented ink systems; again dye content will fade out on prolonged exposure to daylight.

Structure

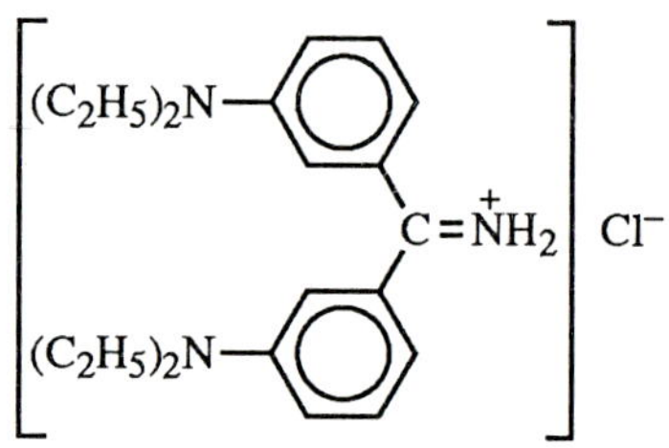

Rhodamine 6G

CI Basic Red 1; CI No. 45160; CAS No. 989-38-8

Properties

Very clean bright scarlet red, with a greenish fluorescence when dissolved in water, but with a yellow fluorescence in alcohol. It has poor lightfastness and properties similar to those described for auramine. It is used as the basis for PMTA and copper ferrocyanide salts.

Structure

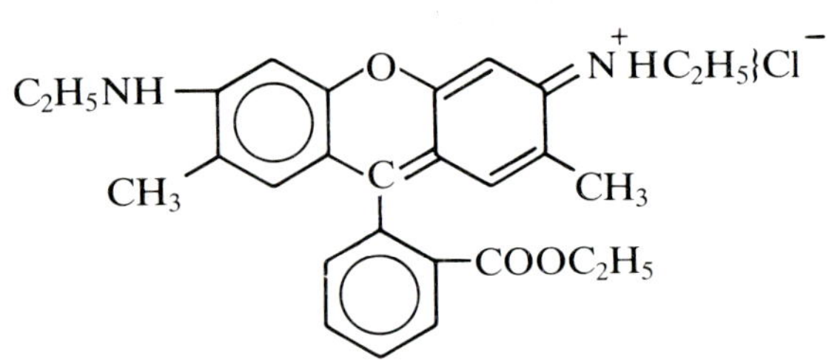

Rhodamine B

CI Basic Violet 10; CI No. 45170; CAS No. 81-88-9

Properties

Dirty and strong compared with Basic Red 1; a bright magenta red with yellow fluorescence; soluble in water, alcohol, and glycol ethers. It has poor stability to heat and light.

Note: the free base of Rhodamine B is CI Solvent Red, CI No. 45170:1. The PMTA acid salt is CI Pigment Violet 1, CI No. 45170:2.

Structure

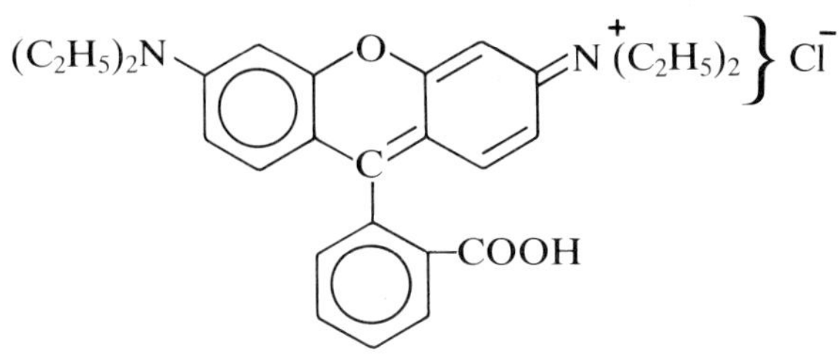

Victoria Blue

CI Basic Blue 26; CI No. 44045; CAS No. 2580-56-5

Properties

Bright royal blue, very clean, soluble in water and slightly soluble in cold water. It has poor fastness to light, tending to darken and become dull.

Uses

Used as dyestuff in flexographic inks. Further use of this dye is as the free base (CI Solvent Blue 5; CI No. 42595:1) and the PMTA salt (CI Pigment Blue 1; CI No. 42595:2) which is used extensively in all types of printing ink.

Structure

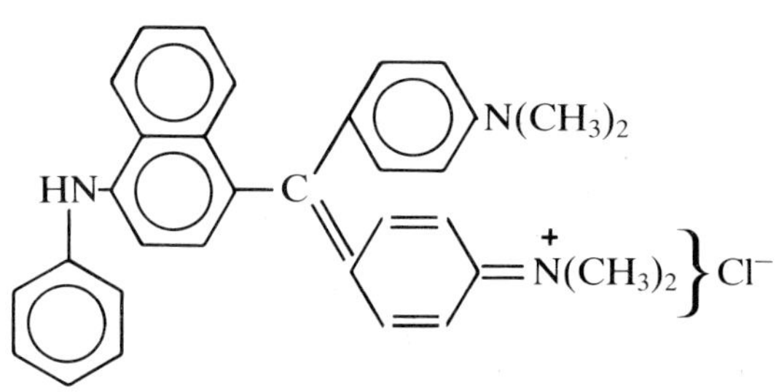

Methyl Violet

CI Basic Violet 1; CI No. 42535; CAS No. 6004-87-3 and 83968-28-9

Properties
Strong bright bluish violet, soluble in alcohol and water, but has poor lightfastness, and turns black on exposure.

Uses
Used in flexographic inks for paper bags. Widely used in the following forms, to obtain improved light stability:
CI Solvent Violet 8;CI No. 42535:1 is the free base;
CI Pigment Violet 3;CI No. 42535:2 the PMTA acid salt;
CI Pigment Violet 27;CI No. 42535:3 the copper ferrocyanide salt.

Structure

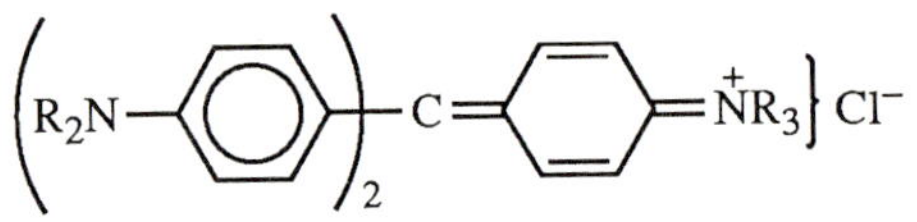

$R = CH_3$ or H

Other basic dyes commonly in use are given below but CAS numbers are not readily available.
Basic Yellow 13, 28 and 65;
Basic Orange 1, 2, and 59;
Basic Red 1:1, 2, 14, and 28;
Basic Violet 2, 3, and 11:1;
Basic Blue 1, 3, 5, 7, 8, 9, 11, 55 and 81;
Basic Green 1 and 4;
Basic Brown 1.

For detailed information reference should be made to the *Colour Index*.

4.16 SOLVENT DYES

Often known as metal-complex dyes. These are dyes in which solubility in an organic solvent or solvents is a characteristic physical property. They are drawn from a number of chemical groups and include what are sometimes described as oil dyes or fat dyes. The chemical constitutions associated with solubility in organic solvents include acid dyes of the azo chromium complex, xanthene classes and the base salt form of certain basic dyes. Sulphonic acid groups as such, are usually absent, but dyes soluble in alcohols include salts formed between sulphonated dyes (usually sulphonated azo or phthalocyanine and organic high molecular weight bases such as diaryl-guanidine, mono and di-cyclohexylamine and di-isoamylamine).

The properties of these dyes are strongly influenced by their solubility in a particular solvent, so it is usual to use mixed solvents to obtain the optimum solvency and properties. The range of extremely bright colours with fair lightfastness makes them popular in a number of applications like golds on foil and jobs where better lightfastness than a basic dye is required.

Neozapon Yellow

CI Solvent Yellow 19; CI No. 13900:1; CAS No. 10343-55-2

Properties
A green shadc yellow which has good lightfastness, it is heat stable up to 170°C and is soluble in alcohol and ester solvents, nitrocellulose, spirit and synthetic resin lacquers.

Uses
It is used for flexographic inks as well as wood staining. However, prints with this dye have poor water and alkali resistance.

Derivation
The chromium complex derived from

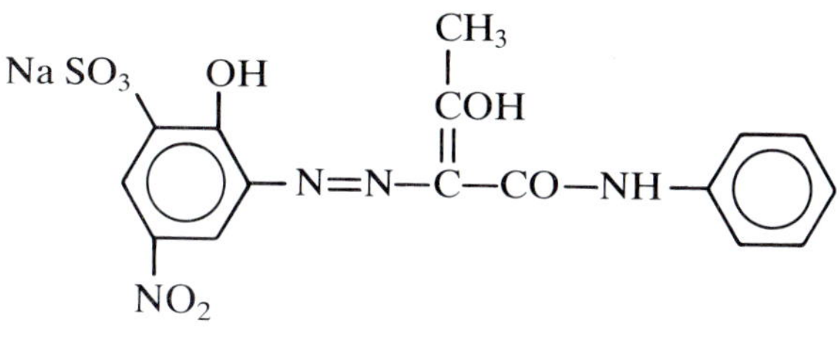

Zapon Yellow

CI Solvent Orange 45; CI No. 11700:1; CAS No. 71566-26-2

Properties
Transparent yellow shade orange, soluble in methylated spirits, diethylene glycol, esters and ketones. It has fairly good lightfastness, plus heat resistance up to 180°C.

Uses
Outlets include flexographic inks, gravure inks and lacquers based on nitrocellulose or spirit-soluble resins used to simulate gold on film and foil.

Derivation
Chemically it is a cobalt complex derived from

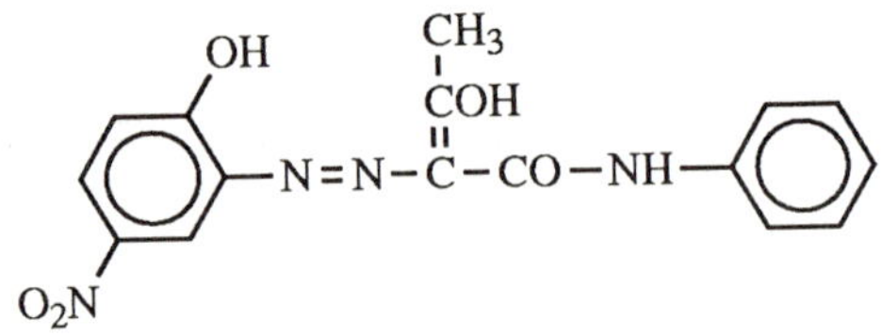

Solvent Red 8

CI No. 12715; CAS No. 33270-70-1

Properties
A blue shade red with good solubility in alcohols, ethers, esters and ketones. Heat stability up to 220°C and excellent lightfastness.

Uses
Suitable for gravure and flexographic inks and for use in stoving lacquers and two-pack polyester systems.

Derivation
Chemically a chromium complex derived from

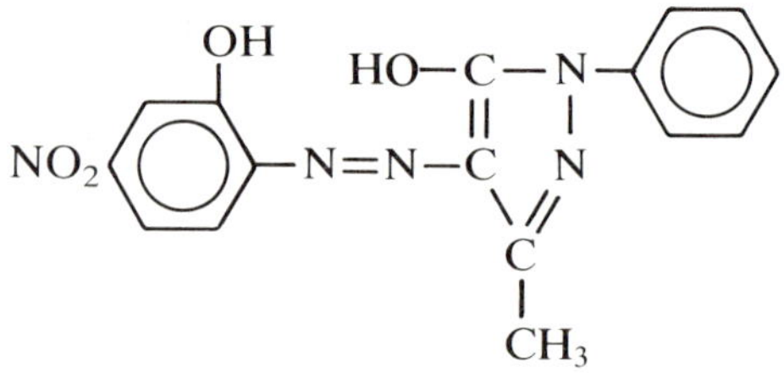

Induline (spirit soluble)

CI Solvent Blue 7; CI No. 50400; CAS No. 8004-98-6

Properties
Reddish blue to reddish navy; fair lightfastness; heat stable to 100°C; insoluble in water; soluble in ethanol and paraffin wax, oleic and stearic acids.

Uses
Recent policy in the UK is to exclude this material from printing inks because of its toxic impurities, such as 4-amino-diphenyl. Used in stamping inks. Solutions of induline base in fatty acids were used as toners in cheap letterpress, news and gravure black inks.

Nigrosine (spirit soluble)

CI Solvent Black 5; CI No. 50415; CAS No. 11099-03-9

Properties
Bluish black; heat stable to 150°C; good lightfastness; soluble in alcohol.

Uses
Used for flexographic inks and black marking inks; sometimes referred to as 'aniline black'.

Nigrosine (free base)

CI Solvent Black 7; CI No. 50415:1; CAS No. 68389-53-7

Properties
Black dye usually dissolved in a fatty acid, especially oleic or stearic acid to form the oleate or stearate. Fairly good lightfastness; heat stable to above 180°C; slightly soluble in aromatic solvents.

Uses
Used in gravure inks as well as typewriter ribbons, carbon paper and stamp-pad inks.

General
Hundreds of solvent dyes are available and they are gaining in usage. Some have very good lightfastness results giving some yellow and reds Blue Wool 7–8. All the major suppliers have ranges of metal-complex dyes, and the fact that these dyes exhibit good compatibility with a large range of resins has made them particularly suitable for flexo and gravure coloured lacquers.

4.17 DISPERSE DYES

The name disperse dyes was given in 1951 to a group of water-insoluble azo, diphenylamine and anthraquinone dyes by the Society of Dyers and Colourists because they were sold as dispersions. Modern techniques allow them to be produced as dry redispersible powders.

Nearly all disperse dyes are primary, secondary or tertiary amines of three main types: amino azo benzene, aminoanthraquinone and nitrodiaryl amines. None of these contains solubilizing sulphonic acid groups. A limited number are similar in structure to vat dyes.

Their main use in the printing industry is in heat transfer inks for printing on textiles, especially onto synthetic fibres. It is referred to either as heat-transfer printing, or sublistatic printing. The dyes are dispersed in a printing vehicle and are printed on paper by letterpress, lithography, flexography, gravure or screen printing. They are subsequently transferred to the fabric under heat and pressure. The dye sublimes and its vapour phase penetrates the fabric where it condenses. Bright, saturated colours result relative to the substantivity between dye and fibre, implying chemisorption or chemical bonding to the fibres.

There are hundreds of dyes of this type, and many of them are used in commercial heat-transfer printing. Four disperse dyes which could form the basis of a trichromatic series are described below, and are typical of other dyes of this class. It should be noted that disperse dyes do not necessarily transfer with the same strength or hue on every fibre even when the optimum temperature and pressure for the particular substrate are applied, also that mixtures of disperse dyes can, under certain conditions have, a catalytic effect which may cause premature fading. Sublistatic transfers can also be printed to transfer images to thermoplastics, metal, carpets and plastic bottles. Sometimes the dye vapour is made to cross relative large gaps (1 cm) with the assistance of vacuum.

CI Disperse Yellow 3

CI No. 11855; CAS No. 2832-40-8

Properties
A bright medium yellow hue that becomes slightly greener and more saturated when transferred to fabric. It has a very good fastness to light and is resistant to water, washing, wet and dry rubbing, acid and alkaline perspiration and dry cleaning when assessed according to the relevant BS or ISO tests.

Uses
It is used in inks for the five major processes to produce heat transfers on to polyamide, nylon, polyester and acrylic fibres, etc.

Structure

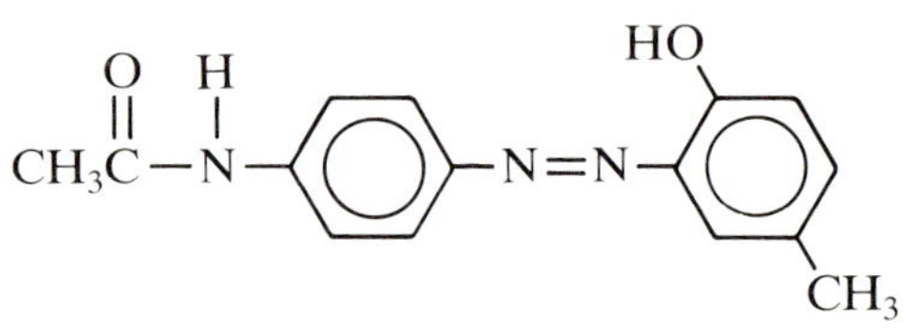

CI Disperse Red 4

CI No. 60755; CAS No. 2379-90-0

Properties
A bright scarlet anthraquinone dye which has excellent lightfastness, washing, dry cleaning and perspiration fastness.

Uses
Although expensive, it is used for heat transfer process colours and can be used in inks for any conventional printing process.

Structure

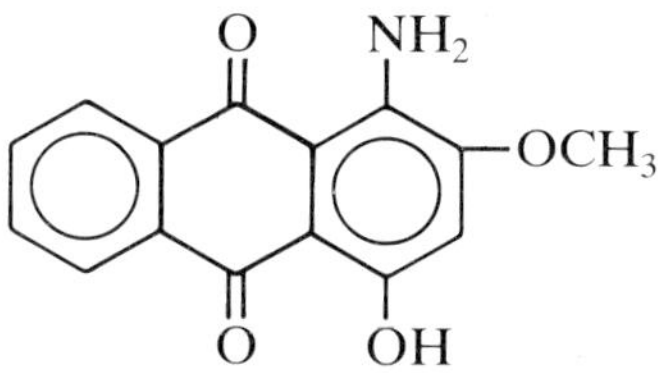

CI Disperse Blue 3

CI No. 61505; CAS No. 2475-46-9

Properties
This is a clean blue-green shade dye. When printed and transferred to fabric, it gives a shade similar to phthalocyanine blue. However its lightfastness is only fair and there are stronger blue dyes which may be used instead or blended with it. It does not transfer well on to nylon, but it does transfer to other synthetic fabrics. Its fastness to washing, rubbing and perspiration is excellent, but it must be dry cleaned with care.

Uses
Transfers using this dyestuff can be produced by all conventional methods of printing. The degree of transference will depend upon the film thickness of the print, the nature of the substrate and the transfer conditions. Excess of dye transferred will give worse fastness results than when the optimum amount is used: an upper limit to colour strength is thus set by the requirement for washing fastness.

Structure

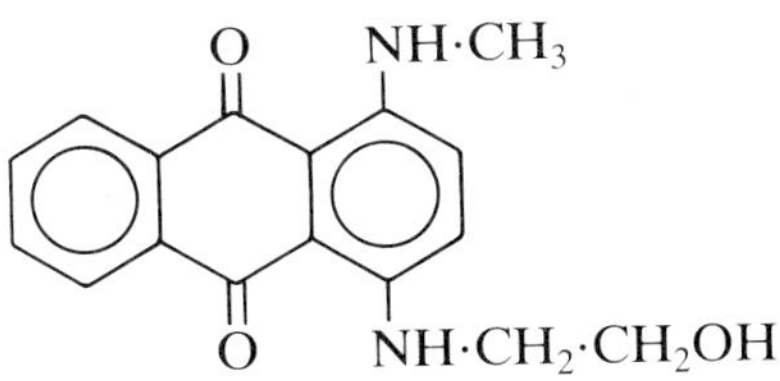

CI Disperse Red 60

CI No. 6075; CAS No. 17418-58-5

Properties
Brilliant rose-pink dye, with excellent lightfastness, excellent resistance to water, washing, acid and alkaline perspiration, wet and dry rubbing and dry cleaning.

Uses
Ideal colour for trichromatic transfer printing by the recognized major printing processes.

Structure

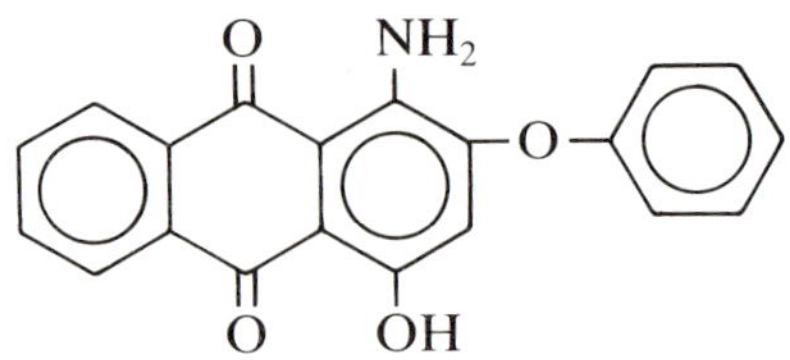

Section 3: Oils

Oils in one form or another are probably the oldest raw material in the printing ink industry, and yet they still have an important role in the formulation of printing inks and varnishes. Many oils are treated and purified or modified with other polymers to meet modern requirements so that they bear little resemblance to earlier products except that they are still derived from the same vegetable, animal or mineral sources (Table 4.9).

In the past it has been convenient to classify oils under the three headings of drying, semi-drying and non-drying, dependent on the degree of unsaturation of the raw oil, defined by its iodine value. As individual measured results differ from season to season, and from source to source, quoted iodine values are usually averaged figures. There has been some confusion in various parts of the world as to how these values should be interpreted for the purpose of classifying oils. Borderline cases are often found under different headings in different catalogues.

4.18 DRYING VEGETABLE OILS

Chemically they are glycerides or triglycerides of fatty acids, varying from those that are completely saturated (no double bonds) to those that contain three or more double bonds linking carbon-atom pairs in the fatty acid compound. This characteristic of an unsaturated compound is represented by the symbols $-CH = CH-$.

When there is more than one double bond associated with neigbouring carbon atoms $-CH = CH-CH = CH-$ it is referred to as a conjugated system. Vegetable drying oils are characterized by their power to absorb oxygen from the surrounding air and to form elastic films or skins, dependent upon whether or not complete polymerization takes place. Polymerization is defined as 'the chemical union of two or more molecules of the same compound uniting to form a single molecule'.

The amount of oxygen absorbed is in the main proportional to the iodine value of an oil, which is regarded as a measure of the extent of unsaturation of the fatty acids present, which in turn indicates the drying power of an oil. Iodine value is defined as the weight in grams of iodine that will combine with 100 g of oil. The recommended procedure is laid down in BS 684 (Chapter 16).

The nature of the fatty acids present in a drying oil, and the proportions of each, largely determine its drying performance. For convenience they are normally divided in two classes: saturated fatty acids (e.g. palmitic, stearic and oleic acid) which play little part in drying, and unsaturated fatty acids such as linoleic acid, linolenic acid and its conjugated isomer elaeostearic acid, etc. These play a vital part in the drying, colour, resistance and durability of ink vehicles subsequently manufactured from particular oils.

A high iodine value is an indication of rapid drying ability of an oil, but it is not the sole criterion for quick drying. A detailed knowledge of the actual positions of the double bonds in a molecule is essential for interpretation of iodine value because it indicates how subsequent polymerization will take place.

With the exception of linseed oil, raw oils are seldom used unmodified in ink formulations. Refined and treated linseed oil is used alone or as an adjunct to petroleum distillate in many letterpress and lithographic ink formulations. The largest outlet, however, is in the manufacture of resins, varnishes and ink vehicles (Chapter 13).

Linseed oil

Linseed oil, in the form of refined and treated oils, bodied and blown varnishes, also in conjunction with modifying resin or as an integral part of a resin system, is a prime constituent in several important classes of paste ink.

Linseed oil is obtained from the seeds of flax (*Linum*), cultivated in Argentina, USA, Canada, India, Russia, and to a limited extent in Australia, New Zealand and the UK.

Each seed contains between approximately 32 and 40% weight of oil. It is not economical to extract it all and some is left in the residual meal which is used to feed livestock.

Two methods of expelling oil from the seed are the mechanical process and the solvent extraction process. In the former, seed is first reduced to a suitable condition by crushing on a bank of five steel water-cooled rollers mounted one above the other. The crushed seed is then cooked to yield the optimum amount of oil at 80–90°C if it is to be extracted by hydraulic pressure, at 110°C if expellers are to be used. Meal from the cooker is moulded into shape, wrapped in a cloth and subjected to hydraulic pressure up to about 315 kg/cm^2. The oil is then drained into tanks. Alternatively, mechanical screw presses called expellers force the seed through a conical grating – the oil exudes from the side and the feed cake is extruded from the end.

Solvent extraction is a semi-continuous process. It gives the highest oil yield but the residue is less acceptable as cattle food. In this process the seed is crushed and is mixed with solvent (usually petroleum ether or trichlorethylene). Solvent and oil percolate from the vessel. The solvent is removed by distillation and recycled.

Raw linseed oil contains a small percentage of impurities in varying amounts. They include free fatty acids, phosphatides, carbohydrates, waxes and suspended mucilage. Unless they are removed the oil 'breaks' on heating rapidly to 300°C, i.e. it darkens and precipitates the impurities to form a gelatinous mass. On standing for long periods these impure materials precipitate by the absorption of atmospheric moisture to form 'foots'. Over a period of months or years these sink to the bottom of the tank allowing the removal of clear oil from above. Originally this was the method of preparing linseed oil for making artists' oil and varnishes, hence the name 'stand oil' which today is given to bodied linseed varnishes in the surface-coating industry.

A common method of refining linseed oil is the acid process. The vegetable oil is treated with sulphuric acid, allowed to settle overnight and the clear oil is syphoned off. This is washed thoroughly with hot water and allowed to settle once more. It is finally separated from the water and is vacuum dried. Acid refining does not remove all the fatty acids and produces refined oils of relatively high acid value. Such oils are used for their excellent dispersing qualities where the free acidity contributes to pigment wetting.

By far the most important method is alkali refining. Besides removing the break and colouring matter this process saponifies the fatty acids to form soaps, which can also be separated. Aqueous caustic soda is sprayed in and stirred until an emulsion forms. It is heated rapidly to break the emulsion and to encourage separation between oil and soapstock. The oil is thoroughly washed several times, allowed to settle and vacuum dried. There are also continuous processes in which separations are conducted by high-speed centrifugation, instead of allowing water to settle out batch by batch.

The main constituent of linseed oil is linolenic acid which has three isolated double bonds, i.e. $CH_3{-}CH_2{-}CH = CH{-}CH_2{-}CH = CH{-}CH_2{-}CH = CH{-}(CH_2)_7{-}COOH$. The other unsaturated fatty acid is linoleic acid which confers colour retention on heating. Linoleic acid has 18 carbon atoms with isolated double bonds on the 9th and 12th carbon atoms: $CH_3{-}(CH_2)_4{-}CH = CH{-}CH_2{-}CH = CH{-}(CH_2)_7{-}COOH$.

Linseed oil has always been the oil most used in ink and paint manufacture. It is also the most versatile. It is a straw-coloured oil which without driers takes almost a week to dry as a thin film on glass.

Until very recently vast quantities were bodied by heating to 305°C in open or closed pots. These were marketed in a wide range of viscosities to the ink industry as lithographic varnish.

Heat-bodied linseed oils are sometimes called lithographic varnishes, a name which is little used in Europe, but is still used in the USA. They are today made in enclosed kettles or pots, which are made of various metals or alloys, particularly stainless steel which is ideal for bodying very pale oils. Kettles are made with interchangeable configurations that include facilities to blow air through the oil, blanket the surface with oxygen or an inert gas, or to heat the oil in a vacuum. Catalysts of metal powders, gases, inorganic and organic reagents are sometimes used to accelerate the bodying process. Strict control of temperature, heating rate and time ensures that the intended oil viscosity is obtained. Control is most important in the later stages of heat bodying as viscosity increases at an accelerating rate towards gelation.

Besides using raw linseed and lithographic varnishes, the inkmaker has several other linseed oil derivatives which he can include in formulations. They include boiled oils whose purpose is to increase the drying power without unduly altering the viscosity. They are manufactured by heating the raw oil in a steam kettle to 120–140°C. The oil is stirred and occasionally air is blown in once it becomes hot, and driers in the form of oil-soluble organic salts of cobalt or manganese and lead are added.

Blown oils are produced by blowing large volumes of air through the oils at temperatures in the range of 100–140°C. They are a clear chestnut brown, faster drying with a higher viscosity and have improved wetting, dispersion and flow qualities. Inks containing blown oils must be formulated with care as they have a higher acid value and stronger odour. They are sometimes thickened by heating to 290°C for use as ink vehicles; as these are liable to body up or gel on storage, viscosity is usually kept well below 20 Pa s.

Linseed oil is very versatile. Its fatty acids are used in the manufacture of alkyd resins, and refined linseed oil is used for making varnishes. It is sometimes treated while still retaining the character of an oil, e.g. maleinized oils, where about 10% of maleic anhydride is incorporated to form an adduct. Subsequently this adduct can be neutralized to make it soluble in water.

It is also possible to produce urethane oils by reaction of the mono- and diglycerides of a drying oil with a di-isocyanate so that it becomes soluble in alcohol. This also gives faster drying times and better water resistance. However, in addition to thermal decomposition hazards, some of these can yield small quantities of isocyanates on a warm-running press or mill. This product could be toluene di-isocyanate (TDI) which can cause allergies – in particular skin sensitization or asthmatic symptoms – in confined spaces without adequate air extraction apparatus. In the UK, inkmakers are barring TDI from any formulation; therefore treated oils cannot be used without a thorough investigation.

Tung oil (China wood oil)

The source of tung oil is the seed from a nut grown on shrubs in North and South America, Africa and the Far East. The shrubs are named *Aleurites fordii* and *A. montan*. Oil from the latter, being cultivated at a higher altitude, contains less elaeostearic acid and bodies more slowly, the gel time being about 20–30 minutes at 276°C against 11–15 minutes according to the heat test of Appendix J in BS 391:1962.

The seeds are encased in hard nuts which are usually roasted to open them. They are then crushed and ground followed by either mechanical or solvent extraction similar to that for linseed oil. Tung oil has a pale straw colour with a slightly higher viscosity and refractive index than other vegetable drying oils. It dries slowly to form a soft, wrinkled, lustreless film. The raw oil has a characteristic odour, sometimes described as 'hammy' and it continues to release odorous byproducts during drying.

Like oiticica oil, it is known as a conjugated oil because about 80% of the total fatty acid component is elaeostearic acid. Three conjugated double bonds, which are very reactive sites for polymerization, occur in the molecular structure of elaeostearic acid:

$$CH_3(CH_2)_3{-}CH = CH{-}CH = CH{-}CH = CH{-}(CH_2)_7{-}COOH$$

The other 20% of the fatty acid component is made up of oleic and various saturated fatty acids in the ratio of approximately 2:1. For every three fatty acid molecules there is one molecule of glycerol in the structure of true triglyceride oils. Tung oil can be heated rapidly to a temperature of 290–310°C to ensure that it has the ability to dry to a clear glossy film. This can be a delicate operation as it starts gelling at about 275°C after seven minutes. This gelation can be retarded or prevented by adding a small quantity of linseed oil, glycerine or rosin. When properly heat-bodied it becomes the most water and alkali resistant of the drying oils.

Because tung oil gives a tough and glossy finish with rapid drying and excellent resistance properties, it is usually used as a component in oleoresinous varnishes, particularly with pure phenolic or alkyd resins where it gives a rub-proof film that is resistant to water, alkali and soap.

This makes it a good vehicle for lithographic inks for soap and detergent cartons. Other applications are in gloss overprint varnishes and metallic inks. It is normally omitted from inks to be used on wrappers for packaging food on account of odour when drying.

Oiticica oil

The only other conjugated oil normally used in ink manufacture is oiticica which is extracted from the nut of a tall tree *Licania rigida*. The tree grows wild mainly in the tropical climate of Northeast Brazil.

Shortly after extraction the oil sets to form a buttery mass, but this is stabilized to a permanent liquid by heating at 200°C for 2 hours.

Liquid oiticica oil resembles tung oil in several respects. It dries rapidly to a wrinkled matt finish and forms a gel when heated to 275°C. It has a characteristic odour and is greener than the pale straw colour of tung oil.

The principal fatty acid is *d*-licanic acid, which has three conjugated double bonds, but is unique in having a keto group in its formula:

$$CH(CH_2)_7{-}CH = CH{-}CH = CH{-}CH = CH{-}(CH_2){-}CO{-}(CH_2){-}COOH$$

Oiticica oil is used in the manufacture of oleoresinous varnishes that are quick drying and have good gloss and adhesion properties, rubproofness, water resistance and alkali resistance.

It gives a yellower more brittle film than tung oil and is sometimes combined with dehydrated castor oil, fish oil or soya bean oil to overcome this. Its usage to a large extent depends on price fluctuations relative to tung oil.

It is similar to tung oil in that it can be used to produce letterpress and lithographic inks that dry hard on non-absorbent substrates, and give scratch-proof prints on film and foil.

Dehydrated castor oil

Unmodified castor oil is a non-drying oil derived from the seed of the shrub *Ricinus communis* and as such was used in duplicator inks, typewriter ribbon inks and compounds for producing imitation water marks. It has been used as a plasticizer in liquid inks.

Castor oil mainly consists of the triglycerides of ricinoleic acid. This fatty acid contains one double bond and a hydroxyl group. The latter, together with an adjoining hydrogen atom, can be removed by dehydration. This is often carried out by heating to 280°C in the presence of a catalyst, usually sulphuric acid, using a vacuum or inert gas to prevent oxidation and to remove volatile decomposition products. The reaction produces two isomers of linoleic acid, 9:12 linoleic acid (unconjugated) and a 9:11 (conjugated) isomer of linoleic acid, usually in the ratio of 3:1.

Dehydrated castor oil (DCO) is pale and slightly more viscous than other drying oils. It has a slight smooth lingering characteristic odour. The earlier oils had a residual tackiness but this has been largely overcome.

Like other drying oils it can be bodied to produce varnishes, but these are seldom used on their own. Blown DCO is used for wrinkle finishes.

Dehydrated castor oil is mainly used in gloss varnishes to give pale coloured films that have good colour retention on air drying or stoving. It is ideal for metal-decorating inks because of its pale colour and excellent flexibility. Because of low residual odour, DCO is sometimes used in food wrapper inks. By far the largest usage is in the manufacture of air drying and stoving alkyds for white roller coatings and inks. Such fluid polyesters are referred to as DCO alkyds.

4.19 OTHER OILS

Segregated oils

These are prepared by treatment of drying oils with a solvent such as liquid propane, furfural or sometimes a solvent mixture of liquid propane and furfuraldehyde. The solvent partitions the oil in two layers.

Both layers contain oil fractions of differing iodine value, one higher and one lower than the original oil. They are then separated and the solvent is removed from each by distillation. The segregated oil with the higher iodine value gives faster air drying times than the original.

Marine oils

Fatty portions of fish and the blubber of whales are processed to yield the so-called 'fish oils'. They have a more variable composition and a wider range of fatty acid components than vegetable oils. Exploitation of these oils is limited by their high odour and colour, but they yield flexible, glossy films. To improve their drying speed and reduce odour some have been segregated, but industry has been slow in accepting them. High cost and conservation effectively prohibit further adoption of marine oils by industry.

Semi-drying oils

Almost exclusively used as components in synthetic resins, mainly alkyds. Alkyds modified with semi-drying oils are usually more flexible, are often lighter in colour and have better colour retention than drying oils.

The main semi-drying oils are tobacco seed oil, soya bean oil, safflower oil and sunflower seed oil. Of these the most important is soya bean oil which finds increasing scope in pale alkyds for tin-printing inks.

Soya bean oil

Today soya bean cultivation is of considerable importance. It produces the largest amounts of natural oil throughout the world, and is grown on all continents.

It comes from beans which are the seeds of a legume *Soya japonica* or *Glycine hispida*. These beans are a valuable source of protein. The oil is extracted by extrusion. The amount extracted is fairly small, about 20%, to allow for a proportion to be retained in the resulting foodstock.

It contains a high proportion of linoleic acid, and oleic acid with a typical odour and pale colour. It can be refined and bodied by conventional methods.

Alkyds based upon soya bean oil, either alone or with other modifications, give excellent colour retention and adhesion in tin-printing inks. They are also used in overprint varnishes and web-offset inks, despite their slower drying qualities. Their pigment wetting ability, especially with carbon blacks, is good.

Non-drying oils

Mineral oils

Mineral oil in various forms is used as a vehicle in printing inks or as a solvent or diluent in the manufacture of ink vehicles. Its use and development will increase with the exploitation of the North Sea oil fields.

A wide range of products is available. Mineral oil viscosities range from spindle oil, and oil just slightly thicker than a solvent, to heavy bodied oils. Colours range from pale straw to dark brown.

They are the products of their higher boiling fractions evaporated from petroleum at temperatures above 350°C at atmospheric pressure. They consist of varying percentages of aromatic, naphthenic and paraffinic hydrocarbons, with a small sulphur content ranging up to 4%. The pigment-wetting qualities of mineral oils depend upon the relative ratio of each hydrocarbon present. High percentages of aromatic and paraffinic hydrocarbons give poorer pigment wetting properties as well as poorer miscibility with resins. They are used alone or in conjunction with rosin, some rosin derivatives and hydrocarbon resins in newspaper inks and cheap jobbing blacks. Oils such as spindle oil are a diluent or solvent for ester-gum, resinates, isomerized rubber, phenolic and maleic resins in quick-setting inks.

News-ink oils

The essential requirement for a news-ink oil is that it should be inexpensive. Also it must show good wetting and dispersion characteristics so that it can produce an ink which has sufficient flow properties without flying or misting, together with good binding qualities, adequate penetration and speedy setting.

The oils that fulfil these requirements are dark brown mineral oils, some of which are fairly viscous, but for normal newspaper black inks a viscosity of between 0.8–1.0 Pa s at 21°C (70°F) is recommended. Such a viscosity gives a reading of about 4000 seconds on a Redwood No. 1 and 400 seconds on a Redwood No. 2 viscometer, which are the instruments normally used in the UK for measuring the viscosity of lubricating oils.

The more viscous oils are measured on similar instruments at temperatures of 60°C (140°F) and 93.3°C (200°F). The flash point for such oils ranges from about 159°C (350°F) to above 200°C (400°F).

Test methods vary with suppliers and are covered both by the Institute of Petroleum (known as IP Standards) and the American Society for Testing Materials (ASTM Standards).

Mineral oils used in inks for newspaper production have fairly high aromatic contents, which may be as low as 22% but can be double that figure. The balance is made up of naphthenes and paraffins as well as small percentages of other unsaturated polycyclic and olefinic hydrocarbons. News ink oils have to be produced at very low cost and are made in specialized refinery processes using selected feedstocks of lower petroleum distillates or residues, which give products with a relatively low aniline point.

As these are products with a high flash point they do not present handling or storage problems. Recent tests have indicated that there may be some health hazards associated with these oils, especially in the fine mists around the rollers of a fast-running press. Certain oils therefore are no longer used. Precautions should be taken to avoid prolonged skin contact with any aromatic petroleum oils.

Non-drying vegetable oils

Non-drying vegetable oils are valuable as plasticizers, lubricants and as components of synthetic resins. The most important is castor oil.

Castor oil

Castor oil is derived from beans which are seeds of the shrub *Ricinus communis* grown in tropical areas. Oil is obtained by cold pressing and solvent extraction, the finest grades being selected for pharmaceutical use. Ricinoleic acid forms from 80 to 87% of the fatty acid portion (averaged) in the castor oil triglyceride molecule. The chemical formula of ricinoleic acid is

$$CH_3(CH_2)_5CH(OH)CH_2{-}CH = CH{-}(CH_2)_7COOH$$

Castor oil is used in the manufacture of DCO by dehydration (refer to p. 214) and Turkey Red oil which is produced by sulphonation. Castor oil can also be bodied by blowing or oxidizing. Its main use is as a plasticizer for use in plasticizing alkyds, also in duplicating inks, typewriter ribbon inks, for making imitation watermarks and for calibrating viscosity, instruments. It has good pigment-wetting qualities and is sometimes used as a bridging agent, having both alcohol and oil-miscible parts in the principal fatty acid molecule.

Section 4: Resins

The term 'resin' is used to describe non-crystalline solid materials or liquids of relatively high molecular weight. Such materials often lack a precise melting point. Resins used in printing inks contribute to the properties of hardness, gloss, adhesion and flexibility in an ink. They are binders for pigments.

Resins are of two types, natural and synthetic. A few naturally occurring resins are still used in the preparation of ink vehicles and varnishes, usually after some chemical modification. Most natural resins are obtained from living vegetable matter although some are in fossilized form. Shellac is an exception, its origin being the secretion of an insect. The chemical structure of natural resins is complex and in many cases has not been entirely proven.

Synthetic resins are prepared by polymerization involving condensation or addition reactions between relatively small molecules. In many cases the composition of synthetic resins is known with considerable accuracy because they contain fewer molecular types than most natural resins. The chemical and physical properties of synthetic resins are closely related to their chemical structure and composition. Uniform synthetic products can be repeatedly manufactured, in contrast to naturally occurring resins where the feedstock composition governs the final product. A great variety of modification is available for each class of synthetic resins and for this reason such resins can only be discussed here in general terms.

Resins are subdivided into natural and synthetic products. Obviously this cannot be more than an academic distinction, which has existed during a transition from the use of purely natural products to greater use of synthetic products. In many instances synthetic resins are derived from natural products or they contain a large proportion of natural fatty acids, so that most so-called synthetic resins are dependent upon natural products.

In the past few years some natural and a few synthetic resins have been superseded by others and have not been included in this chapter. The decline was caused by the limited nature of natural resources and the changing economic climate, which made the processing and distribution of small quantities very expensive. The future outlook is that conservation, plus health and safety considerations, will encourage the use of aqueous systems as well as those derived from vegetable products. This will accelerate the decline in demand for certain materials which have served the printing-ink industry well in the past.

Chemical and physical properties of resins with usual applications in printing ink formulations are mentioned, together with a description of the origin or preparation of each resin. An indication of chemical structure is outlined where possible.

For further reference to current materials, the *Surface Coating Resin Index* published every third year by the British Resin Manufacturers Association should be consulted. This association has recently published useful guidelines for the safe handling of resins.

4.20 NATURAL RESINS

Rosin

Rosin is obtained from pine trees throughout the world. Wood rosin is obtained by crushing the stumps of pine trees after they have weathered for a number of years, then extracting rosin and rosin oil with a petroleum solvent. Gum rosin is obtained by tapping the living trees, sometimes with the aid of artificial stimulation to accelerate the flow of exuded sab.

Wood rosin and gum rosin are similar, except that gum rosin generally has a higher melting point and wood rosin tends to crystallize from alcohol solution more readily. Rosin is amber in colour with a low melting point at about 60°C by the capillary method and up to 80°C by the ring and ball method. Its saponification value is 172 and its acid value is 168–172.

Rosin is regarded as an inexpensive rosin. Its relative price has risen slightly over the last decade, but there has been little decline in its use. It is ecologically sound to produce rosin from living sources, as opposed to fossil sources and tree stumps, making gum rosin a regenerative starting material for printing ink rosins.

Rosin has been used in alcohol-based lacquers, also in combination with mineral oils in some comic publication and newspaper inks. Rosin has a strong peptizing action in breaking down gels. Otherwise it has limited use in inks in its natural state. Its main outlet is as raw material in the production of chemically modified rosin derivatives, such as dimerized and polymerized rosin, limed rosin (rosin soaps or metallic resinates), esters of rosin including ester gum, rosin/maleic adduct, modified fumaric adduct, also as a modifying resin in cresylic, phenolic and alkyd resins. Cobalt, lead and manganese salts react with rosin to give the appropriate resinate driers.

Rosin consists of approximately 90% acids and 10% neutral material. The principal acid, abietic acid, converts to laevo-pimaric acid when heated.

Abietic or laevo-pimaric acids, as can be seen from their chemical formulae; both possess two types of reactive centre, namely conjugated diene and carboxylic acid groups.

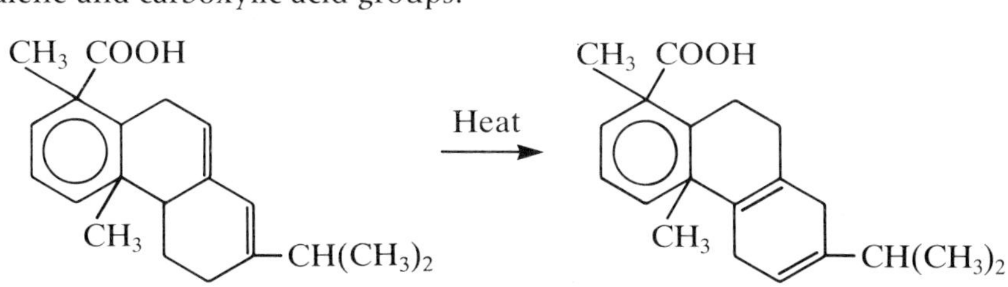

Abietic Acid Laevo-pimaric Acid

Rosin has the tendency to crystallize from solution. Varnishes made from rosin do not dry well, and possess a residual tack. These disadvantages are least apparent in rosin/tung oil varnishes.

Rosin soaps, Limed rosin, Metallic resinates

The high acid value of rosin may be lowered by reacting calcium acetate or lime or zinc oxide or a combination of them with rosin or polymerized rosin (Table 4.10). A metallic resinate with a considerably higher softening point is obtained; the increase in softening temperature is dependent on the metal content. They are inclined to give brittle films, but they have relatively rapid solvent release, and a colour similar to rosin. They are soluble in hydrocarbons but not in alcohol.

They are used in low-cost gravure inks as well as in heat-set inks. Wetting of pigments is good and they prevent some basic pigments from livering. Resinates can also be used as a peptizer in heat bodied drying oil varnishes to prevent gelation.

Rosin or polymerized rosin is heated to 235–260°C. Up to 7% lime (or calcium acetate) or up to 9% zinc oxide plus acetic acid is added slowly with constant agitation, and steam is blown through. Sometimes both calcium and zinc are added, usually in a ratio of 3:1 Zn:Ca, but still keeping the metal content of the resin below 10%.

Ester gum and other esters of rosin

Ester gum is the triglyceride derived from rosin by direct interaction between rosin and an alcohol, for example glycerol. It is an amber coloured resin with a softening point of approximately 85–90°C ball and ring (BR). It is soluble in aromatic and aliphatic hydrocarbons, drying oils and most of the usual lacquer solvents with the exception of alcohols. Varnishes made with ester gum and drying oils have good pigment wetting properties and find applications in most ink systems. Its low acidity means that it can be used in conjunction with metallic pigments. In liquid inks careful formulation is required to reduce tack.

It is prepared by heating rosin in a stainless steel kettle to approximately 260°C, then slowly adding glycerol. The paleness of colour is preserved by the stainless steel reaction vessel, also by stirring and by conducting the esterification in an inert atmosphere of carbon dioxide or nitrogen. The temperature is held at 260°C until the required drop in acid value (usually to below 10) is obtained. The theoretical requirement of glycerol is 10% but an excess of 2–5% is recommended to balance any losses that occur during the reaction.

The general formula for the glycerol ester of rosin is

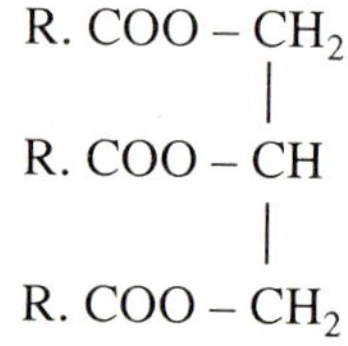

where R represents the rosin acid.

Rosin and modified rosins can also be esterified with other alcohols and esterifying agents, for example pentaerythritol.

The pentaerythritol ester of rosin is similar to traditional ester gum in appearance, but it has a higher softening point of 112°C (BR) and quicker solvent release. It also has better water and alkali resistance. For these reasons it is much more popular than the glycerol ester although it is more expensive.

Its chemical formula is

$$\begin{array}{rcl} & & CH_2-O.OCR \\ & & | \\ R.COO-CH_2 & — & C{-}CH_2-O.OCR \\ & & | \\ & & CH_2-O.OCR \end{array}$$

where R is the rosin acid moiety.

The penta-erythritol esters or dimerized and polymerizing rosin acids have even higher melting points, give harder films and have faster rates of solvent release from an ink film. They find widespread application in letterpress and lithographic inks and are gaining use in liquid ink formulae.

Maleic resins and esters

A wide range of rosin maleic and maleic-modified resins are commercially available. Maleic anhydride reacts with rosin according to the Diels–Alder synthesis. Properties vary with the molar ratio of rosin to maleic anhydride and with the manufacturing procedure. They can roughly be divided into three main classes (Table 4.11).

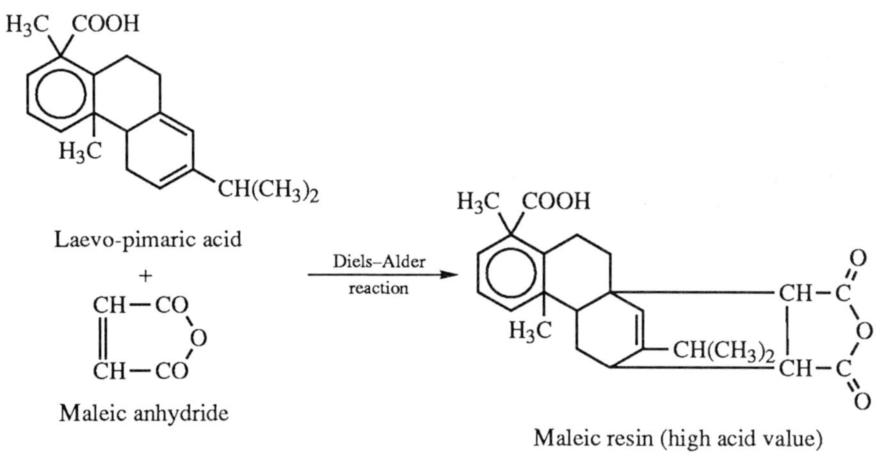

High acid value maleics are used in steam-set, moisture-set and water-reducible letterpress inks, also for water-based flexo inks, in conjunction with an amine or ammonia to secure their solubility.

To alter the solubility and reduce the acid value of the rosin maleic adduct, maleic-modified resins are prepared by esterifying carboxylic groups with a polylol such as glycerol or pentaerythritol during an *in situ* preparation.

Besides modifying the qualities of high acid value rosin maleic resins, esterified maleic modified resins find wide scope for use in inks. High acid value maleics are used in steam-set, moisture-set and water-reducible letterpress inks, also for water-based flexo inks, in conjunction with an amine or ammonia to secure their solubility.

Medium acid value maleics have applications in alcohol-based gravure and flexographic inks, in combination with polyamides, nitrocellulose or ethyl cellulose, for printing on polymer films and foil. They are also used in water-miscible letterpress inks.

Hydrocarbon-soluble maleic resins are very pale. They are used in letterpress and lithographic gloss inks, tin-printing inks, heat-set inks and overprint varnishes.

Dimerized and polymerized rosins

In order to raise the softening point of rosin it can be dimerized or polymerized by heating with a suitable catalyst. Its unsaturation makes it possible to polymerize to the dimer stage, which results in a high softening point, 150–170°C and a lower acid number of 148. Dimerized rosin has increased compatibility with other resins and varnishes: it is less susceptible to oxidation due to greatly reduced unsaturation. It can also be used at high concentration in hydrocarbons without risk of crystallization.

Further polymerization from the dimer stage gives a polymerized rosin which is very pale in colour. It has a lower melting range near 77°C by capillary method and 95°C with ring and ball. It has an acid value of 148 mg KOH/g similar to dimerized rosin.

Its main applications are in cheap photogravure inks and gloss overlacquers. It has also been used to improve the gloss, dispersion and working properties in letterpress black inks. It has also been used in heat-set inks.

The chemical constitution of polymerized rosin is not completely elucidated but reaction takes place at the olefinic centres. A dimerized rosin may have up to 80% rosin acid dimers, the other 20% being monomeric rosin acids and neutral material.

Rosin-modified fumaric resins

Fumaric acid is the isomer of maleic acid, therefore rosin-modified fumaric resins are classified together with maleic resins. In practice fumaric acid is used in place of maleic anhydride to produce rosin–fumaric resins with high acid values between 250 and 330 mg KOH/g and higher softening points, 140–150°C.

They are used in flexographic and gravure inks, in combination with other resins for non-absorbent substrates and with glycols for moisture-set inks where hardness and gloss are required.

Their chemical structure is thought to be:

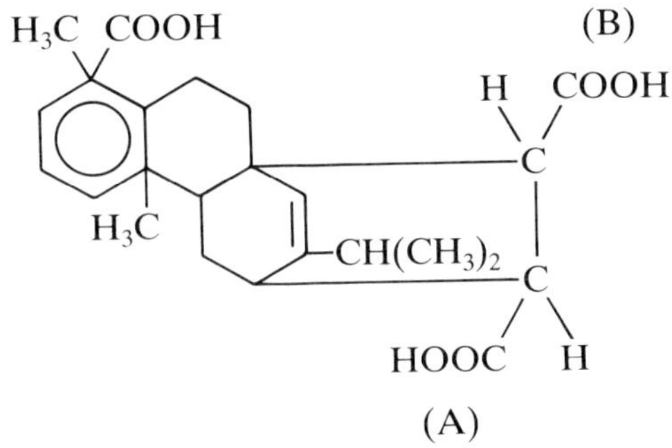

in which the carboxyl groups A and B are *trans* to one another.

Shellac

The importance of shellac in industry has rapidly declined due to shortage and high prices. At the same time the developments in acrylic emulsions and dispersions have been quickly accepted by flexographic inks formulators to fill the gap for water- and alcohol-based inks.

Shellac has an acid value of 65–80, a hydroxyl value of 260–280 and an iodine value of 15–25. Its melting range is 75–85°C. Relative density is 1.15–1.20, and its water content is between 2 and 16%. It is soluble in methylated spirits and isopropyl alcohol. It is compatible with a wide range of resins and polymers including rosin, rosin esters, ethyl cellulose, cellulose nitrate, etc. It is not compatible with polymers or copolymers or vinyl acetate.

Shellac has been widely used in the preparation of flexographic inks, and in spirit varnishes. The addition of an alkali or amine produces a sodium or amine salt which renders it water soluble so that it can be used for water-based inks.

Shellac is the flaked form of purified lac. Lac is the natural secretion of the insect *Laccifer lacca kerr*. The lac is dissolved in alcohol, filtered and allowed to dry. Bleached shellac is prepared by dissolving the resin in sodium carbonate solution, allowing the impurities to settle and bleaching the liquid with sodium hypochlorite until the solution reaches a standard colour. The resin is then precipitated with sulphuric acid, washed, dried and powdered.

Other shades of shellac include garnet lac, lemon lac and orange lac.

Shellac contains from 3 to 5% wax. It can be dewaxed by solvent extraction, or the wax may be precipitated from an alcoholic solution by the addition of water.

The formula of shelloic acid is:

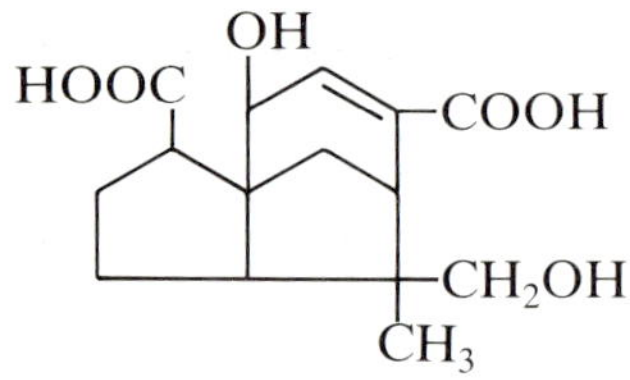

According to the Shellac Export Promotion Council for Indian Shellac, the original lac substance has the constitution:

Aleuritic acid, $HO(CH_2)_6CH(OH) - CH(OH)(CH_2)_7COOH$, m.p. 101°C	46.0
Shelloic acid, m.p. 206°C	27.0
Kerrolic acid, a tetrahydroxy palmitic acid, m.p. 132°C	5.0
Butolic acid, a monohydroxy-pentadecanoic acid, m.p. 54°C	1.0
Wax esters, derived from wax acids and wax alcohols	2.0
Neutral compounds of unknown composition	7.0
Polybasic acid esters of unknown composition	12.0
	100.0

Manila copal

Natural resin with: softening point 122–132°C (BR), $d = 1.07$; acid value 120–130; iodine value 115–130. Soluble in alcohols, esters, ketones and glycols; made oil-soluble by cooking at 310°C for 90 min. Main use was in alcohol varnishes for flexo inks, also in lacquers; used in conjunction with ethyl cellulose for hardness and water resistance. Almost entirely replaced by synthetic resins.

Obtained from the tree *Agathis Alba;* many grades were available. The softer grades were obtained by tapping and the hard ones were fossilized, copal resins consist of complex mixtures of monobasic acids, particularly those containing 8–15 carbon atoms.

Asphalts

This is a term which covers a large number of resin mixtures of complex compounds consisting of carbon, hydrogen, sulphur and nitrogen which are formed either as natural mineral compounds or as the residue from distillation of petroleum, coal tar or rosin.

They include asphaltum, pitches and bitumen which vary considerably depending on their origin or processing conditions. They range from hard brittle resins to those which are plastic or even fluid at room temperature.

Those found as minerals have such names as gilsonite, rafaelite, malthas and grahamite. These are referred to as bitumens, according to the British Standards Institute who has many specifications covering them.

They are dark solids, with fairly high softening points, and are soluble in hydrocarbons, but not always soluble in drying oils. In solution they make varnishes which vary from yellow to dark brown. Materials derived from the distillation of coal tar, rosin or petroleum are defined as 'pitches' or 'residual pitch'.

They are widely used in photogravure, letterpress and heat-set black inks. The yellow undertone of such inks is usually counteracted by addition of dye toners.

Starch and dextrin

Starch is a commonly occurring natural material, which is synthesized in plants. The main sources of starch are tapioca, maize, wheat and potato.

It consists of amylose and amylopectin, in ratios of about 1 to 4, dependent upon the source. Starch is a stable white powder at room temperature. It is insoluble in cold water, but when an aqueous suspension is heated to 60–80°C, and allowed to cool, it forms a gel. Colloidal solutions of starch react with iodine to give a characteristic blue colouration.

Starches with a large particle size are used as anti-set-off powders in sheet-fed printing; treated water-swellable starches have also been used for anti-set-off in lithographic inks. Finely ground potato starch is used to add body and to eliminate excessive tack in platen inks.

Dextrin is partially hydrolysed starch produced by heating starch alone or in the presence of hydrochloric acid. This converts it to a yellow-brown colour, which forms a viscous gummy liquid when dissolved in cold water. It replaced gum arabic in anti-set-off solutions. It is also used as an adhesive for stamps, etc. and as part of a vehicle system for water-colour inks and cheque-book inks.

Gum arabic

The main use for gum arabic was in lithographic fountain and plate 'gumming up' solutions. Gum arabic, being hydrophilic, helps to keep the non-image areas ink-repellant. (Supply shortages have shown that polyacrylamides can equally well fulfil this function.) To a lesser extent gum arabic is used in water-based inks to reproduce copies of water-colour paintings.

It is obtained from trees of the acacia species in Africa, especially in the Sudan. It consists of organic polybasic acids, related to sugars, with an estimated molecular weight of 250 000. It is readily soluble in water, but dries to give fairly brittle films.

4.21 SYNTHETIC RESINS

Pure phenolic resins

Pure phenolic resins are used in conjunction with tung oil to produce varnishes for letterpress and lithographic inks. Their pigment-wetting qualities are good and a small amount is sometimes added to modified phenolic varnishes to improve wetting and stability. If the non-reactive type is added to flushing vehicles it improves their performance. They have a fairly low softening point, 80–130°C, and are soluble in aromatic hydrocarbons, but only partially soluble in aliphatic distillates. Tung oil phenolic varnishes have excellent drying speeds and give very tough glossy films, with good alkali resistance and adhesion properties. The most serious drawback of some pure phenolic resins is their tendency to yellow on exposure to light.

Oil-soluble, oil-reactive pure phenolic resins are prepared by heating a para-substituted phenol with an aqueous formaldehyde solution under reflux in the presence of an alkaline catalyst. The para substituent must be at least a C_4 hydrocarbon chain or a phenyl group. If the substituent group is less than C_4 the resin will not be oil-soluble. After the reaction has gone to completion, the water is removed by vacuum distillation. Properties of the final product depend on the phenol used, the catalyst, molar ratio of reactants and the plant operating conditions.

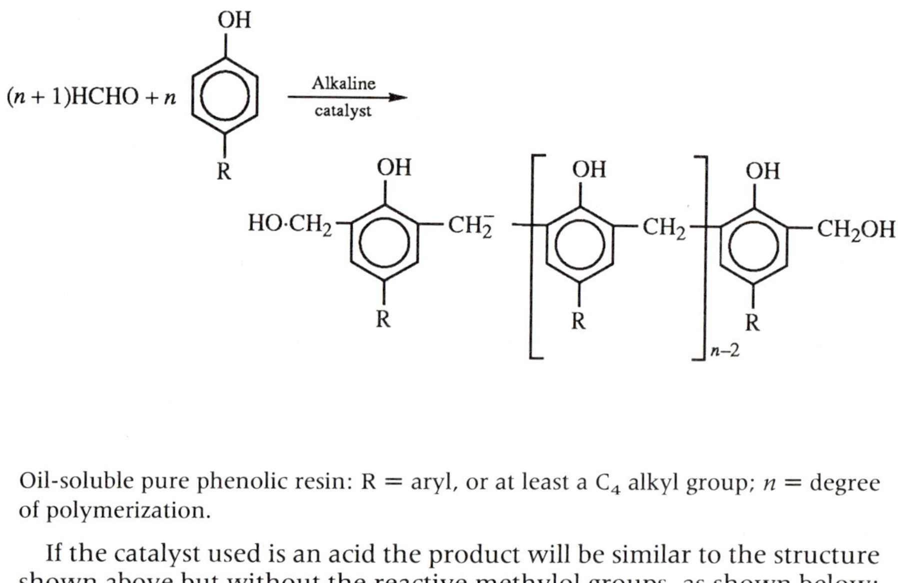

Oil-soluble pure phenolic resin: R = aryl, or at least a C_4 alkyl group; n = degree of polymerization.

If the catalyst used is an acid the product will be similar to the structure shown above but without the reactive methylol groups, as shown below:

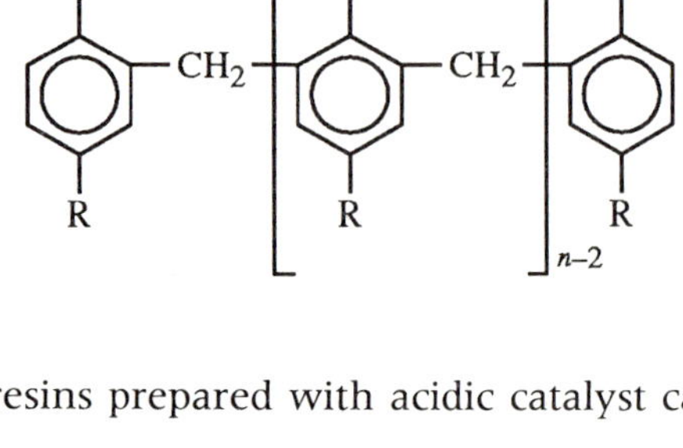

Pure phenolic resins prepared with acidic catalyst can be dissolved in drying oils, especially tung oil, without reaction occurring. Solution of the resin takes place between 200 and 250°C.

The pure phenolic resins prepared with an alkaline catalyst and which contain the reactive methylol groups can be combined with tung oil. A reaction takes place accompanied by rapid bodying of the varnish. The reaction involves the methylol groups of the pure phenolic resin and the olefinic bonds in the tung oil.

Rosin-modified phenolic resins

Rosin-modified phenolic resins have widespread use in all types of ink vehicle system and there are many grades commercially available. They have limited solubility in aliphatic distillate which makes them ideal for use in quick-setting varnishes and gels. Generally they impart tough, glossy finishes with good chemical resistance. They are compatible with alkyds and other resins to produce high-gloss inks with good rub resistance. They are also used in heat-set and web-offset inks for their solvent release properties, and in screen inks for their handling and weathering resistance. Rosin-modified phenolic resins are sometimes used in toluene-based gravure inks. It is also possible to obtain alcohol-soluble grades that have an acid value between 80 and 120 mg KOH/g, making them useful film formers in alcohol- or water-based inks.

Rosin (80–90 parts) is reacted with an oil reactive pure phenolic resin (10–20 parts) at 150°C. The temperature is then increased to 250°C in the presence of glycerol or other polyol in order to esterify the rosin acids. The functional properties of the modified phenolic resin depend on many factors including the chemical identity of the substituted phenol, the polyol used, plant operation conditions, etc. It is interesting to note that the methylol groups of the pure phenolic resin react with the unsaturated centre of the rosin acid in preference to the carboxylic acid groups in rosin.

Several chemical species are present in the final product. One possible structure for a penta-erythritol ester of a rosin-modified phenolic resin is given below

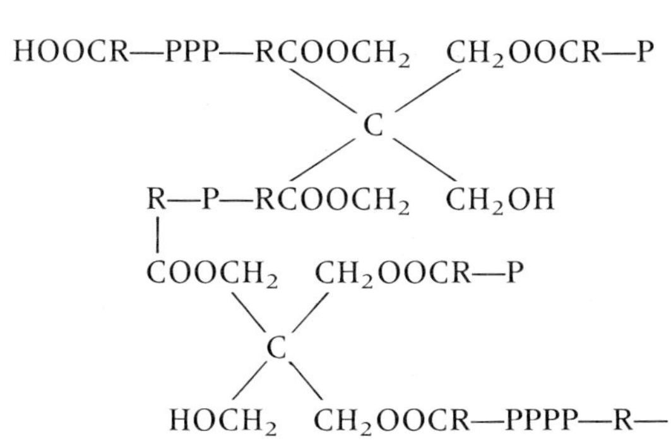

where R = rosin acid component and P = substituted phenolic grouping, e.g. where R′ = aryl or at least a C_4 alkyl group:

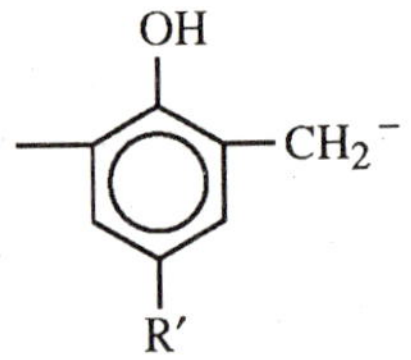

Rosin-modified phenolic resins are soluble in oils and aromatic hydrocarbons. They usually need heating to quite high temperatures to achieve solution. They also reduce the gelation rate of tung oil when the oil is heated to high temperatures.

Many resins of this class are commercially available. Their properties depend on the structure of the initial pure phenolic (or cresylic resin), the esterifying polyol, the molar ratios of reactants and the physical conditions of manufacture. The acid value and softening points of two typical rosin-modified phenolic resins are given in Table 4.12.

Alkyd resins

The use of alkyd resins in printing ink has been influenced by increasing usage in other surface coating industries and the ability of manufacturers to produce 'tailor made' resins for specific purposes. They have become a primary resin in the formulation of paste inks, especially as they are obtained at workable viscosities without the addition of solvents (Table 4.13) and are cheap and versatile in performance.

One important function in both paste and liquid inks is as a controllable resin plasticizer, where it is possible to combine control of drying rate and setting time with flexibility and adhesion. The high gloss of many alkyds ensures that they are a main film-forming constituent in quick-setting inks for letterpress and lithography. Their solubility in a variety of solvents enables them to have important applications in screen inks. When non-drying alkyds are used they make excellent plasticizers in liquid ink systems based upon nitrocellulose. Tin-printing inks rely upon alkyds as their main film former for colour retention, gloss, adhesion, steam retort processing, drawing flexibility and varnishability.

Their compatibility with other resins is excellent provided that the solvents or diluents used in combination with the other resins, or in subsequent formulation are carefully selected. A certain amount of solvent incompatibility can sometimes be used to advantage, for example in two-phase quick-setting ink vehicles.

Alkyd resins are manufactured by esterification of a polyhydric alcohol with a polybasic acid, which makes them members of the polyester family. The majority of polyesters are produced for the plastics industry. Generally alkyds include a combination of drying oils and fatty acids that modify the polyesters to allow them to cross-link and become a pliable film. These film forming properties are dependent on the quantity, type and nature of the modifying oils, fatty acids or acid anhydrides which form them. The processing methods influence the nature of the alkyd and end product.

For convenience alkyds are classed as drying or non-drying depending on the amount and the type of drying oils incorporated. These include safflower, linseed, tung, dehydrated castor, soya and tall oil. They are also classified as short-, medium- or long-oil alkyds depending on the amount of oil present. Below 45% oil is a short-oil alkyd. Normally the higher the oil length, the more soluble is an alkyd in aliphatic hydrocarbons. Most alkyds used in printing inks are long-oil alkyds, containing more than 55% of the oil or fatty acid component.

Styrenated and vinyl toluenated alkyds are prepared by reacting styrene and vinyl toluene respectively with an alkyd in the presence of hydrogen peroxide.

Uses

This brief guide to the use of alkyds in oil-based inks, is in three short sections:

- pigment interactions
- hard resin interactions
- film-forming properties.

It is possible to predict levels of performance in these areas, but there is no substitute for practical examination of materials within the framework of local formulation principles.

Pigment interactions

No alkyd exists with all its molecules having the same molecular weight; a molecular weight distribution is always observable. Frequently more than one molecular species is involved. A mean figure is derived from the molecular weight data which may have double or triple peaked distribution. The fact that the polarity or the chemical activity of individual species may differ, increases the possibility of such variations. This, together with batch blending, complicates attempts to predict performance. Surface properties of pigments, together with polarity and viscosity data for a given alkyd type, are compared to indicate potential wetting performance. The measurement of flow properties of standard pigments in a selection of alkyds augments such predictions. Pigmentation levels are chosen together with strict controls on dispersion, ageing, etc. Variation in flow properties at differing pigment volume concentrations provides much useful information. As the viscosity of an alkyd is raised, improvements are found in its versatility such that a wider range of formulation options is available.

In general, high hydroxyl value, low-viscosity phthalic alkyds show very high wetting potential irrespective of the pigment type. Isophthalic alkyds are generally found to be more specific in their wetting performance. Good wetting for carbon black is frequently quoted. Terephthalic alkyd systems are also notable in this.

Styrenated and vinyl toluenated alkyds have been found rather unsatisfactory as sole ink vehicle and are often used in conjunction with other film formers.

Hard resin interactions

The distribution of physical and chemical properties described above applies equally generally to the behaviour of alkyds in the presence of hard resins. Peptizing/solubility characteristics of a varnish prepared using an alkyd in conjunction with an oil to solubilize a hard resin vary with the amount of polar material present. Viscosity/concentration data combined with measurement of solution clarity provide a basis for ink vehicle studies which should also include measurement of gloss, tack, solvent-release and setting speed. The fact that gloss vehicles and grinding media are prepared using differing proportions of hard resin to alkyd shows that pigment carrying properties are also significant. If a pigment selectively absorbs components from the ink vehicle, it can affect the balance of hard resin to alkyd and upset tolerances in a quick-setting ink vehicle system.

Further effects arising from the loss of low molecular weight materials on to adsorbent particles should be considered, especially where four colour process inks are concerned.

Film-forming properties

Short to medium oil length alkyds that incorporate drying or semi-drying oils are used in both air drying and stoved systems. As the functionality of the polyhydric alcohol is increased (at the same oil length) drying speeds increase. As the oil length for the same functional alcohol is increased, so the flexibility of the dry film improves.

High acid value alkyds (those above 12 mg KOH/g) with a high viscosity will have a tendency to react with basic pigments, but give better wetting, good solvent release and faster setting.

In general the progression from ortho- to iso- to terephthalic alkyds, gives shorter drying times and a reduction in film flexibility, although much depends on the presence of hard resins and on relative concentrations.

Styrenated and vinyltoluenated alkyds yield tough films with short drying times and rapid setting in appropriate varnishes.

Hydrocarbon resins

Continual interest is being shown in hydrocarbon resins which are derived from the petrochemical industry. Since exploitations of the North Sea oilfields, demand and usage in them has heightened. The varying cost of oil products has sometimes slowed the expected encroachment on use of other resin systems. Hydrocarbon resin is conventionally regarded as an inexpensive resin for cheap letterpress, lithographic, and gravure inks. The advantages of chemical inertness and pale colour outweigh its poor wetting and solvent release properties in certain applications. The solubility of higher melting-point hydrocarbon resin combined with its alkali resistance makes it suitable for use in soap and detergent carton inks. The softening points range up to 140°C and acid value is less than 1.0 mg KOH/g.

Polystyrene resins and copolymers

Polystyrene is readily soluble in a wide range of solvents including aromatic hydrocarbons. It gives a water-white hard brittle film, with excellent insulation properties. However, its bad marring and poor adhesion to non-absorbent substrates limit its use in liquid inks and lacquers.

Polystyrene resins can be copolymerized with maleic anhydride to give a range of powdered resins with a variety of acid values and viscosities. These can be made water soluble and have applications in water-based gravure inks.

Polystyrene emulsions can be used in liquid inks either alone or in conjunction with shellac, etc. However, they possess the typical styrene odour.

The monomer is produced from benzene, by reaction with ethylene in the presence of a catalyst to produce ethyl benzene and by subsequent dehydration to produce styrene.

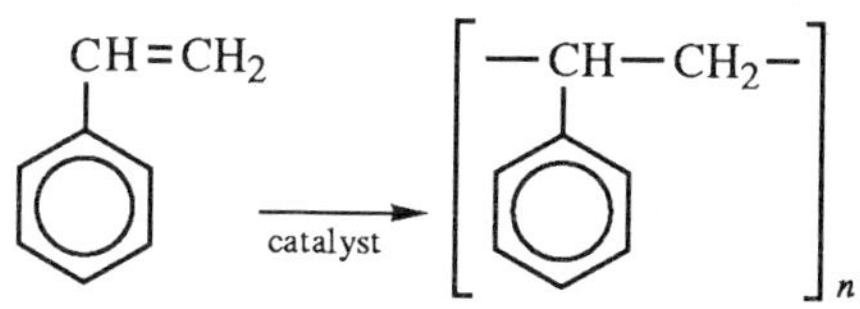

where n = the degree of polymerization.

The grades of polystyrene which are of interest for printing ink formulation have molecular weight averages from 800 to 15 000 above which the solutions become too viscous for practical purposes and 'string' badly. The melting point varies from 60 to 104°C, increasing in accordance with the molecular weight. Acid number and saponification values are below 1.0 mg KOH/g. The emulsion technique is not used for making solid resin, due to soap residue which impairs electrical resistance and optical clarity, but it does produce a valuable polystyrene latex which is gaining usage in water-based surface coatings.

Terpene resins

Clear, pale inert resins with low softening points. On economic grounds they are usually limited in their application to gloss lacquers and letterpress or lithographic overprint varnishes. They are soluble in higher ketones and alcohols, aliphatic, aromatic and chlorinated hydrocarbons to form films which are inert to dilute acid and alkali. They are also compatible with most other resins used in printing inks.

Terpene resins are manufactured by the cationic polymerization of alpha-pinene and beta-pinene monomers using Lewis acids.

Silicone resins

These have limited use in the printing ink and roller industry at present, due to high cost, but may assume a more prominent role in future. They consist of monomethyl, dimethyl, monophenyl, diphenyl, methyl phenyl, monovinyl and methylvinylchlorosilanes, and silicone tetrachloride. Resin hydrolysis is carried out batchwise or continuously, followed by the removal of acid, washing and condensation.

They are normally supplied in a solvent and can be baked to a hard film. They have been used in paints and coatings to obtain weather durability and gloss retention. Dichlorosilanes have the ability to adhere to almost any solid including polymers such as Teflon. Copolymers have also been introduced by blending and copolymerizing with alkyds and acrylic resins.

Alkylated urea formaldehyde resins

Unmodified urea formaldehyde resins are used as moulding powders in plastics. By modifying them with alcohols it is possible to produce resins that are soluble in certain solvents.

Alkylated urea formaldehyde resins are compatible with nitrocellulose, epoxy resins and short-oil alkyds. They find widespread application in two-pot high-gloss lacquers for paper, film and foil, also in flexographic and photogravure inks where chemical inertness of the finished film is essential.

They can be cured at elevated temperatures without prior addition of a curing agent and therefore find use in stoving finishes.

Preparation

The first stage of preparation involves condensation of urea with aqueous formaldehyde in carefully controlled proportions keeping the pH between 7 and 8. The reaction products are:

Monomethylol urea	Dimethylol urea
$NHCH_2OH$	$NHCH_2OH$
|	|
CO	CO
|	|
NH_2	$NHCH_2OH$

The reaction is carried out by heating the mixture under reflux. Some polymerization through the methylol groups takes place at this stage and during the following stage of alkylation of the methylol groups. This is carried out by adding an alcohol, usually butanol, to the mixture which is further heated. Finally the excess alcohol/water/formaldehyde azeotrope is distilled over, leaving the urea formaldehyde resin in the reactor:

$$-NHCH_2OH + ROH \rightarrow \underset{\text{alkylated product}}{-NHCH_2OR} + H_2O$$

R = propyl, butyl, octyl, etc.

The alkylated or etherified resin is easily soluble in aromatic hydrocarbons and is usually reduced with xylene.

Chemical and physical properties

The resins may be cold-cured by mixing with acid curing agent before application or may be cross-linked thermally when inert durable films are produced. The structure of the cross-linked resin is complex and description is beyond the scope of this book. The main reaction responsible for the cross-linking involves ether linkage with hydroxy groups which may come from another urea–resin molecule, an epoxide or an alkyd resin

$$-CONHCH_2OBu + HO{-}R \rightarrow -CONHCH_2OR + BuOH$$

The lacquers so produced are light in colour but they have poor pigment wetting properties.

Alkylated melamine formaldehyde resins

The applications are similar to those of alkylated urea formaldehyde, and when cured, melamine resins have fairly similar properties, with slightly better chemical, heat and water resistance. They form slightly more brittle films than urea coatings and are therefore formulated in conjunction with alkyd resins. Melamine formaldehyde resins are normally supplied as solutions in xylene or butanol. The cross-linking reaction is similar to that described for alkylated urea formaldehyde resins. The rate of cross-linking with an acid curing agent is faster than with urea resins.

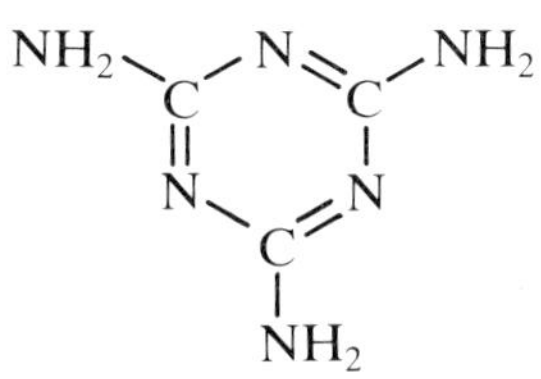

Melamine

Theoretically, six methylol groups can be introduced on each molecule of melamine, as in hexamethylol melamine. In practice a commercial product consists of mixtures of tri-, tetra-, penta- and hexamethylol derivatives. These methylol compounds are subsequently etherified with an alcohol, usually butanol.

Commercial products contain many chemical species. Butylated trimethylol melamine has the structure

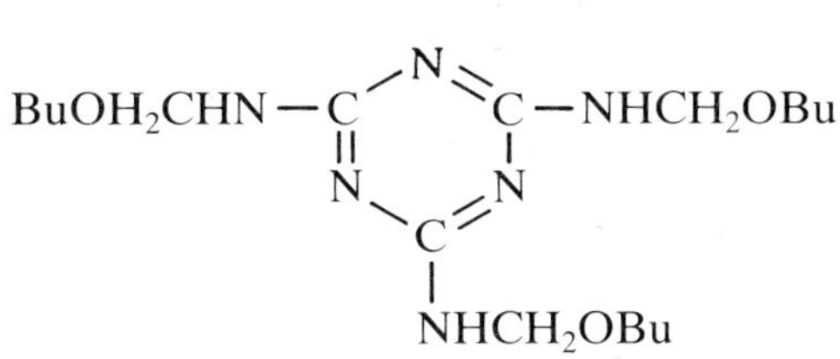

Polyamide resins

A large family of polymers that includes the amide grouping (—CO—HN—) in the main polymer chain. The largest class of polyamides is the 'nylon' group of polymers, having high molecular weight and poor solubility. They are produced for extrusion and moulding. Lower molecular weight species known as reactive polyamides are used in combination with other polymers. Thermoplastic polyamides, referred to as non-reactive, are soluble in solvents used in liquid inks. These last two general polyamide types referred to as fatty polyamides, interest the inkmaker. The reactive types are used in coatings and solventless inks. The soluble types are used in conjunction with other resins to improve film-forming properties.

The functional groups of polyamides are amino groups and the hydroxyl groups distributed in the ratio of one to one.

Chain terminators and blocking agents are commonly used to direct the condensation towards more soluble and compatible products.

The polyamides used for nylons are based upon the formation of a linear polycondensation of a diamine and a dibasic acid.

$$n[\mathrm{HOOC{-}R{-}COOH}] + n[\mathrm{H_2N{-}R'{-}NH_2}]$$
$$\rightarrow \mathrm{HO{-}[{-}CO{-}R{-}CO{-}HN{-}R'{-}NH{-}]}_n + (2n-1)\mathrm{H_2O}$$

They are manufactured either by self-condensation of amino acids or by thermal polymerization of lactams. The first process involves reaction of diamines with dicarboxylic acids, whereby interaction takes place between a carboxyl group and an amino group with the elimination of water. For example, nylon 6:6 is the polyamide formed with hexamethylene diamine $H_2N{-}(CH_2)_6{-}NH_2$ and adipic acid $HOOC{-}(CH_2)_4{-}COOH$. A preliminary step of salt formation, usually in an aqueous solution is often carried out to control the polymer chain and avoid termination.

In the case of polymerization of caprolactam or higher lactams by thermal polymerization the polymer chain formation is controlled by use of additives or adjustment of reaction time, temperature and pressures. High molecular weight polyamides are either extruded or moulded. Extruded filaments are woven or knitted into fabrics. The inkmaker is becoming more aware of these materials as they often form the substrate for transfer printing of textiles.

Reactive polyamides are obtained by the reaction of di- or polyfunctional amines with polybasic acids originating from unsaturated vegetable oils, usually in the form of dimers or trimers. Reactive polyamide resins contain reactive amino or carboxyl groups and can be prepared using an excess of the amino reagent or the carboxyl reagent. However, it is more usual for one of the reagents to be difunctional, while the other is trifunctional. When reacted in the appropriate proportions with dimer acid, they yield high molecular weight polymers containing free amino or carboxyl groups.

Example

If a dimeric fatty acid such as dilinoleic acid reacts with diethylene triamine on a mole to mole basis, one free amino group will remain for every unit of the polymer chain. This is shown in simple schematic form as follows:

$$n[\mathrm{HOOC-R-COOH}] + n[\mathrm{H_2N-\underset{\underset{NH_2}{|}}{R'}-NH_2}]$$

$$\rightarrow \mathrm{HO}-\left[-\mathrm{CO-R-CO-HN-H-\underset{\underset{NH_2}{|}}{R'}-NH-}\right]_n$$

$$+\ (2n-1)\mathrm{H_2O}$$

Alternatively, reacting a tricarboxylic acid with a diamine produces a resin containing free carboxyl groups:

$$n[\mathrm{HOOC-\underset{\underset{COOH}{|}}{R}-COOH}] + n[\mathrm{H_2N-R'-NH_2}]$$

$$\rightarrow \mathrm{HO}-\left[-\mathrm{CO-\underset{\underset{COOH}{|}}{R}-CO-NH-R'-NH-}\right]_n-\mathrm{H}$$

$$-\mathrm{O} + (2n-1)\mathrm{H_2O}$$

The amino-containing reactive polyamides have received most attention. They are capable of reacting with a number of materials, the most important being epoxy compounds. In some of these reactions using primary and secondary amino groups, the reactive portion is the hydrogen atom, not the nitrogen atom.

A combination of epoxy and reactive polyamide resins is ideal. The polyamide portion of the molecule confers toughness, adhesion and resilience, whereas the epoxy resin imparts durability, good wetting, low shrinkage and low curing temperature. Copolymers so formed are thermosetting.

Amino-containing polyamides react with acrylic resins, increasing the total unsaturation of the compound. They also react with diacrylates to bring about cross-linking between molecules, with acids to form amine salts, anhydrides to form amides and aldehydes to form methylene linkages. In conjunction with phenolic resins they produce thermosetting products having the inherent toughness and resilience of polyamide combined with the chemical and solvent resistance of phenolic resins. These polyamide resins have a low molecular weight and are usually liquid at room temperature. They form a basis for two-pot systems and other solventless inks.

The thermoplastic polyamides of higher molecular weight (2000–9000) have important roles in liquid ink technology. They are produced by condensing dimerized vegetable oil acids with a suitable polyamino compound by blending at medium temperatures and reacting until the water can be distilled off.

Thermoplastic polyamides are solid amber-coloured resins with softening points ranging from 90 to 150°C. They can roughly be divided into classes below and above molecular weight 4000. Those with a molecular weight average of 4000–7000 normally require an alcohol/hydrocarbon cosolvent blend although they are soluble in aromatic hydrocarbons and higher alcohols. They are generally compatible with nitrocellulose. Those of mean molecular weight 2000–4000 have improved alcohol solubility and readily blend with nitrocellulose. Compatibility with shellac, rosin, maleic, phenolic and ketone resins is excellent. When incorporated into flexographic and gravure inks they confer excellent adhesion to treated polyethylene, polypropylene, polystyrene and films coated with polyvinylidene chloride, as well as other difficult substrates.

High gloss, resistance to fats and grease, heat sealability, fast solvent release and clean printing are also obtained. In overprint varnishes they provide scratch resistance in addition to the above properties. Other uses for polyamides include hot-melt adhesives and heat-seal coatings for flexible substrates such as film, foil or paper. Such coatings also provide a good grease and water barrier for packaging.

Aqueous polyamide dispersions have been used in adhesives, heat-seal and barrier coatings when blended with rubber lattices, acrylic or vinyl emulsions. Reactive polyamide colloidal dispersions can be used for curing epoxy emulsions in water-thinnable coatings.

Polyimide resins

Polyimides were introduced commercially in 1961. They are polymers incorporating the imide group —CONCO— in the main chain and are produced by a two-stage reaction between an acid dianhydride and a diamine. A typical synthesis is the exothermic reaction between pyromellitic dianhydride and 4,4′-diaminodiphenylether to form a poly (amic acid).

The second stage is the cyclization of the poly (amic acid) with subsequent dehydration to form an aromatic polyimide. These resins are used mainly for high temperature and high insulation coatings.

Polymers of this class which are of interest to the inkmaker are those which are more closely related to the polyamides and are known as poly (amide–imides).

Poly(amide–imide) resins

These resins are prepared by the reaction of excess diamine with diacid dichloride to form a low molecular weight amine-capped polyamide which is then reacted with pyromellitic dianhydride in dimethylacetamide to form a poly(amide–(amide–acid)). It is possible to achieve this in three ways. The synthesis can be conducted homogeneously by adding pyromellitic anhydride to the reaction mixture of diamine and diacid dichloride in dimethylacetamide. Alternatively, the reaction of excess diamine with pyromellitic dianhydride to form a low molecular weight amine-capped poly(amide–acid), which is reacted with diacid dichloride to form poly(amide–(amide–acid)) as described above. These various reaction mixtures can each be cast into poly(amide–(amide–acid)) film which is then converted to poly(amide–imide). A third method for preparation of amide–imide polymers consists of reaction of trimellitic anhydride derivatives, such as the acid halides or acid ester, with a diamine (as illustrated). The poly(amide–(amide–acid)) is formed, which upon conversion yields a poly(amide–imide).

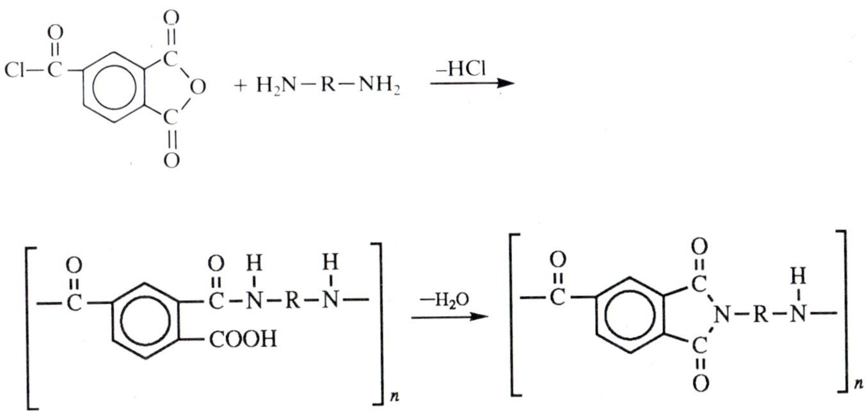

Such polymers have recently been used to give good adhesion, hardness, and resistance properties to flexographic inks, also as resins for obtaining easily dispersible pigmented chips for gravure inks.

Chlorinated rubber

Chlorinated rubber is available as a fine white powder, with up to 67% of chlorine by weight. Several grades are available, usually the higher the chlorine content the lower the viscosity of the solution obtained. It is soluble in aromatic and chlorinated hydrocarbons, esters, ethers and ketones, but insoluble in aliphatic hydrocarbons and alcohols. It is compatible with the majority of plasticizers, also hydrocarbon resins, rosin maleic, modified phenolic, acrylic, medium-oil alkyds and long-oil alkyds. It can also be dissolved in drying oil vehicles at 100–110°C and in high-boiling solvents at temperatures of up to 130°C. It disintegrates and chars at 135–140°C. Clear solutions yellow on exposure to UV light. Chlorinated rubber has a relative density of 1.6.

It is used in toluene-based gravure inks, and in publication ink formulations based on resinates to increase drying rates, gloss and rub resistance. Adhesion to plastic film and foils is good. Chlorinated rubber has good barrier properties, it lowers water permeability and aids resistance to plasticizer migration in PVC films. Other applications include use in paper coatings, heat-seal coatings, textile stencil inks and as an additive to alkyds in screen process systems to improve solvent release and film-forming properties.

Natural rubber in the form of a solid, a solution or latex is chlorinated to give an approximate formulation of $(C_{10}H_{11}Cl_7)_n$.

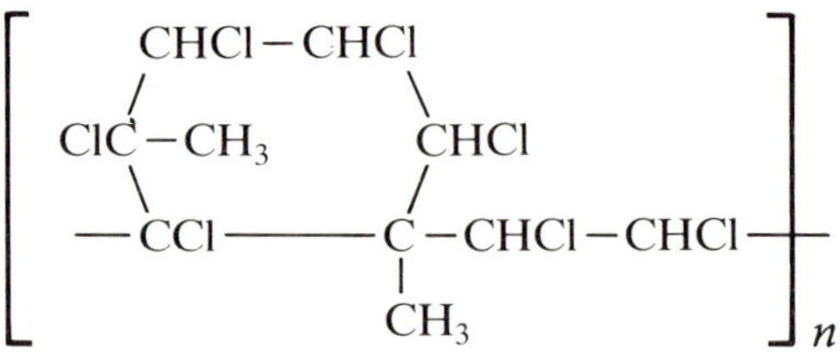

Cyclized rubber (isomerized rubber)

Several viscosity grades of the solid resin are available ranging from light to dark brown. It is also offered as a solution in white spirit or high boiling petroleum distillate. Cyclized rubber is readily soluble in aliphatic or aromatic hydrocarbons and esters. It is compatible with drying oils and minerals oils, most oil varnishes, ester gum, resinates, cellulose derivatives and hydrocarbon resins, also some maleics, modified phenolics and certain alkyds. It is not compatible with alcohols, ketones, many short-oil alkyd resins, etc. Adhesion and chemical resistance are very good and it works well in lithographic inks in conjunction with isophthalic alkyds. The softening points vary from 110 to 140°C.

It is used in letterpress and lithographic inks, for rapid setting on paper and good rub resistance when dry while remaining 'open' on the machine. Small amounts added to lithographic inks make them more hydrophobic. Their application has been limited by spraying and flying, but with recent grades this can be overcome by correct formulation. They are used in gravure and screen process inks, in combination with other resins, where the printed surface has to have good rub resistance.

Natural rubber is treated with an acid catalyst to reduce the degree of saturation. A rubber isomer forms which is regarded as a cyclic product.

The most probable structure is a polymer in which the repeating units have a structure such as:

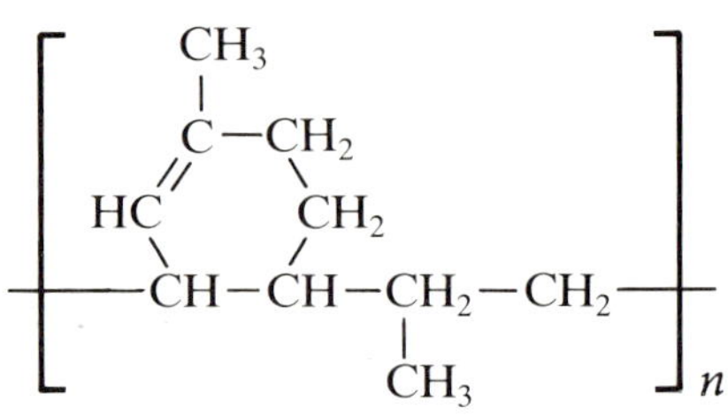

where n is the degree of polymerization, but more complex structures have been demonstrated.

Vinyl resins

Polyvinyl acetate

Vinyl acetate is produced by reacting acetylene with acetic acid using a catalyst:

$$CH{\equiv}CH + CH_3COOH \rightarrow CH_2{=}CH-O-\overset{\overset{\displaystyle O}{\|}}{C}-CH_3$$

Vinyl acetate may be polymerized by bulk, solution, suspension or emulsion techniques using peroxide catalysts, giving a head to tail structure.

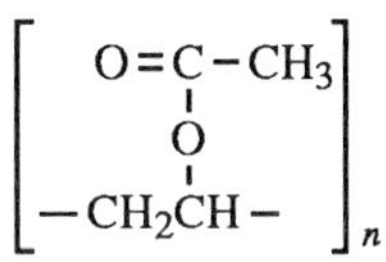

where n is the degree of polymerization.

It is frequently copolymerized with 15–20% of other monomers including alkyl acrylates, fumarates and maleates. Except where the solid resin is required, it is polymerized in water to yield lattices used in water-based paints and adhesives.

Many viscosity grades of the solid resin are commercially available. The high molecular weight and high viscosity grades give the toughest films which have more water resistance and higher heat resistance than films of lower molecular weight grades. The bulk polymer is a transparent solid with a characteristic odour of vinyl acetate monomer. The softening points vary from 32 to 86°C depending on molecular weight. The resins are soluble in most solvents, except aliphatic hydrocarbons. Compatibility with other resins is not good but they are partially compatible with nitrocellulose, phenolic resins, and chlorinated rubber. The relative density is 1.18 and acid value zero.

Used in flexographic transparent lacquers for foil and in heat sealing lacquers. Polyvinyl acetate has poor pigment wetting and is not usually pigmented.

Polyvinyl alcohol

Polyvinyl alcohol cannot be prepared from its monomer vinyl alcohol which has not been isolated in the free state. It is commercially obtained by controlled hydrolysis of polyvinyl acetate. They are cheap water-white resins.

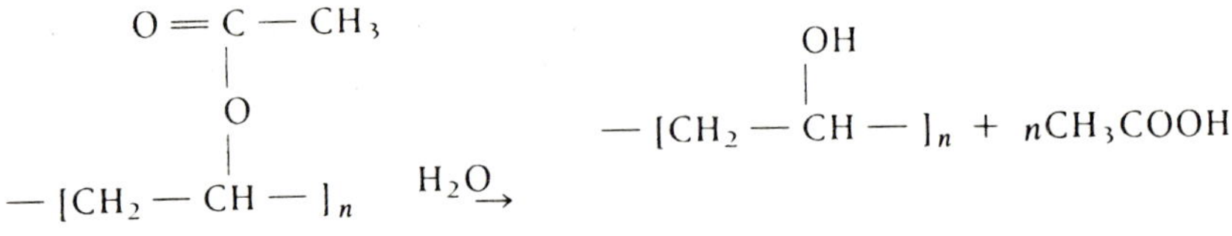

Polyvinyl acetate Polyvinyl alcohol

where n = degree of polymerization.

Several viscosity grades are available. All grades are white powders with relative densities between 1.25 and 1.35. The resins, which are essentially netural, are soluble in water but insoluble in solvents except those solvents with hydroxy groups such as glycols, amides and amines. The hydroxy groups in polyvinyl alcohol are reactive and many derivatives are possible. Some grades have a small percentage of residual acetyl groups and these are more soluble in organic solvents. The use of polyvinyl alcohol resins in inks is limited to making prints that are sensitive to water as in tamper-proof security inks. They are also used in litho plate photosensitive coatings.

Polyvinyl butyral

Prepared by the reaction of butyraldehyde with partially hydrolysed or completely hydrolysed polyvinyl acetate (polyvinyl alcohol) under carefully controlled conditions.

A simplified chemical structure is

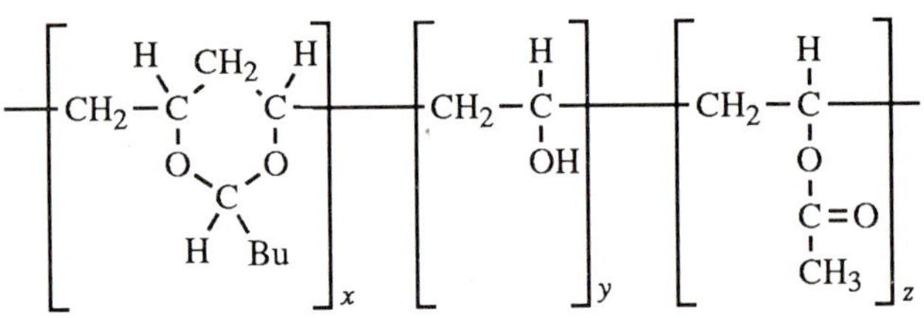

The three components are in controlled proportions and are 'randomly' distributed along the molecule.

The percentages of polyvinyl butyral, polyvinyl alcohol, and the polyvinyl acetate in most commercial products are approximately 80, 17 and 2.5.

Polyvinyl butyrals are supplied as white powders. The solubility of the resins depends to a large extent on the proportion of functional groups present. They are generally soluble in alcohols, esters, ketones, glycols and glycol esters. Films of polyvinyl butyral are very flexible. The resins are compatible with shellac, epoxides, maleic and acrylic resins, ethyl cellulose and nitrocellulose. They can be cross-linked with dialdehydes, reactive pure phenolic resins and isocyanates. Polyvinyl butyrals have a relative density of 1.1 and acid value of zero.

They have excellent adhesion to metal, glass, plastics, paper and textiles and are used in combination with other resins to modify flexographic and gravure inks. They improve the key of the ink film to the substrate, gaining use again in liquid ink applications. They are widely used in laminating systems.

Ketone resins

Very pale transparent resins ranging from almost water white to pale straw; they have excellent resistance to ultra-violet light, and are chemically inert with very low iodine and saponification values. Ketone resins are neutral (acid and hydroxyl values zero). They have good solubility in acetates, alcohols, ketones and castor oil but limited solubility in aromatic hydrocarbons. They are insoluble in aliphatic hydrocarbon solvents and glycols and water. They have wide compatibility with vinyls, cellulose, acrylics, short-oil alkyds, and in small amounts remain compatible with chlorinated rubber and polyamides. They also possess good pigment-wetting properties.

They are used in conjunction with other resins to improve adhesion, mainly in lacquers as well as high gloss flexographic and photogravure inks. They are included in nitrocellulose lacquers and white coatings to increase hardness and durability, and are very good for grease resistance.

Condensation of cyclic ketones with formaldehyde in the presence of an alkali catalyst gives rise to the formation of methylol cyclohexanones by substitution of the α-methylenes. Further alkylation of the methylol groups occurs as in urea formaldehyde resins (page 232)

Acrylic resins

The ranges of acrylic and methacrylic polymers are based upon monomers of acrylic acid, $CH_2 = CH{-}COOH$, and methacrylic acid.

$$CH_2 = \overset{\displaystyle CH_3}{\overset{|}{C}} - COOH$$

These are easily polmerized or copolymerized with other comonomers because of their highly reactive double bonds and miscibility with oil-soluble and water-soluble monomers. Polymerization can be carried out by bulk, solution, suspension or emulsion techniques using a variety of catalysts.

The general formula for an acrylic homopolymer is

$$\left(CH_2 - \underset{\displaystyle R}{\underset{|}{\overset{\displaystyle CO{\cdot}OR'}{\overset{|}{C}}}} \right)_n$$

where R represents H or CH_3, R′ represents H, alkyl, alkoxy, alkenyl or an aryl radical, and *n* is the degree of polymerization.

Various acrylic esters are available through reaction of acrylic acid with the appropriate alcohols. Acrylic esters readily polymerize under the influence of heat, light or peroxide catalysts. These and their copolymers form the bases of acrylic elastomers which can be vulcanized or cured to yield oil-resistant rubbery polymers. Synthetic rubbers based upon polyethylacrylate or polybutylacrylate copolymerized with about 10% of acrylonitrile are of interest to roller manufacturers.

Polyacrylic acids and polymethacrylic acid are often considered together because of their similarity. Both are soluble in water. They are prepared by direct polymerization of the appropriate monomer.

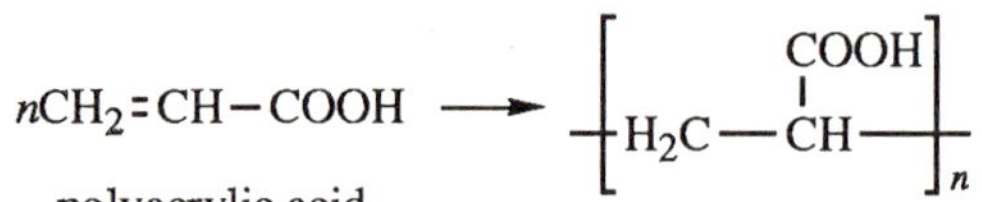

polyacrylic acid

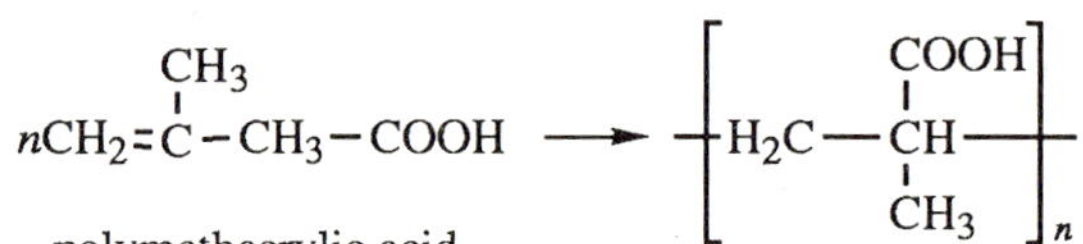

polymethacrylic acid

Acrylic resins are characterized by their clarity, chemical inertness, very good lightfastness and resistance to yellowing on stoving.

Their softening points vary widely depending on chemical formula and molecular weight.

Because the range of acrylic resins is so large examples can be found which are soluble in almost any solvent system used in modern printing inks.

Most types are soluble in gravure ink solvents but grades are available which are soluble in alcohols and both low and high boiling aliphatic distillates. Many homopolymers and copolymers of fixed composition are available in more than one viscosity grade. Acrylic resins can be formulated to be compatible with most resins used in printing inks.

Acrylic resins are used in flexographic, photogravure and tin-printing inks and high gloss lacquers. They impart good chemical resistance and improved key to the substrate. The odour of residual monomer from most acrylic resins is sufficient to exclude them from inks for food wrappers, though odourless grades can be made by purging.

The ability of acrylic monomers to polymerize with each other leads to widely differing applications using numerous combinations, for example those with thermosetting acrylic copolymers for stoving finishes. These are often terpolymers, the first for hardness and rigidity, the second for flexibility, the third providing pendant reactive groups for cross-linking with epoxy resins, amines, butylated urea and melamine resins, etc.

These polymers also undergo reactions characteristic of carboxylic acids. Recent interest has been shown in the neutralization of aqueous solutions with basic materials such as triethanolamine or ammonia which cause viscosity changes, demonstrating the presence of differing molecular conformations.

Ordinarily, aqueous solutions of these polyacids have low viscosity because the polymer is tightly coiled, being only slightly ionized. As the pH is raised, more carboxyl groups become ionized. Mutual repulsion of the charges forces the polymer chain to uncoil. In diluted solution extended coils result in higher viscosities (subsequently the viscosity would become lower if hydrochloric acid were added to lower the ionization).

The above means that solutions with high solids content can be supplied as stable fluids. When the pH is lowered they become useful vehicles for liquid inks which can then be adjusted for correct viscosity with water or water/isopropanol mixtures.

Such polymers, when converted to their alkaline salts are soluble not only in water, but alcohols and glycols. These are used in flexographic inks and in water-reducible letterpress applications.

Water-soluble homopolymers and copolymers of acrylamide are non-ionic. They are resistant to bacterial attack and have been used for gumming up litho plates.

Photopolymerization of acrylic esters can occur both with and without sensitizer. It can also be initiated by high energy radiation. They are primarily used in UV curing inks because they are stable, yet capable of rapid polymerization.

The large range of acrylic esters available, together with the numerous photo initiators used in conjunction with them, involve several mechanisms the theories of which are still not well understood.

The use of acrylic esters as the main vehicle in UV printing ink systems is well established. When used in suitable formulations they impart the ink making, printing, drying and end-use qualities needed. Pigmentation problems have been overcome satisfactorily, and special particle sizes of carbon black are selected. It is necessary to compromise between wanted tinctorial strength and unwanted absorption of the UV radiation required to photopolymerize the acrylic ester.

Epoxide resins

Epoxide resins are characterized by the presence of more than one ethylene oxide or epoxide group per molecule:

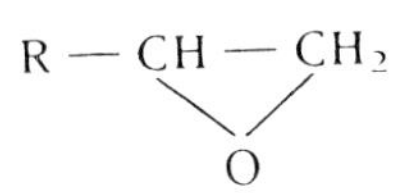

The name 'epoxy resins', currently being used for this group of resins outside Europe, is gradually becoming universal. They are linear polymers of low to medium molecular weight requiring a coreactant to obtain three-dimensional cross-linking. (Exceptions to this are the high molecular weight resins with n greater than 18.)

Prepared by the reaction of epichlorhydrin with diphenylolpropane (bisphenol A) in the presence of sodium hydroxide. The degree of polymerization of the product will depend on the molar ratio of the reactants and the physical conditions prevailing. The general formula is:

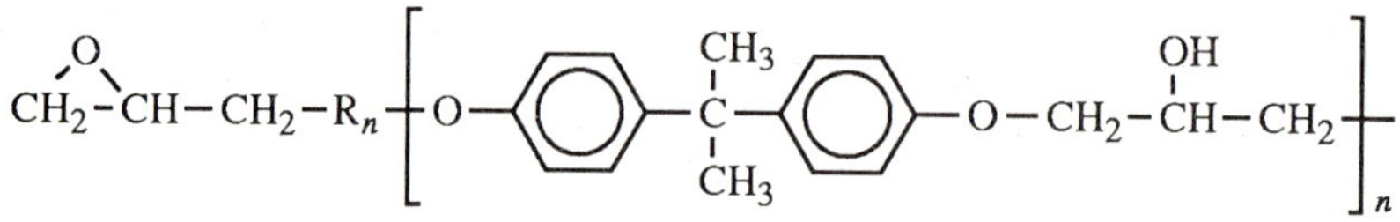

Where R is

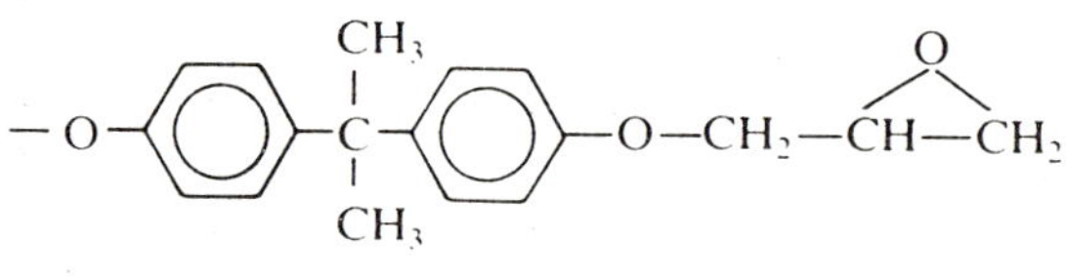

Chemical and physical properties will vary with the mean value of n, the molecular chain length, which may be obtained by using $(n + 1)$ mol bisphenol A + $(n + 2)$ mol epichlorhydrin.

The lowest molecular weights are obtained when $n = 0$, as illustrated in Table 4.14.

Epoxide resins are used extensively in the electronics, paint and lacquer industries. In printing ink manufacture they are used in two-pot lacquers and inks where extreme product resistance is essential.

Liquid epoxy resins are used in solventless coating systems and are cross-linked through the epoxide groups with aromatic and cyclo-aliphatic amines.

The solid resins are pale in colour, without appreciable odour. Their brittleness increases with increasing molecular weight. Epoxide films are fairly resistant to alkalis and promote good adhesion to many substrates. They are soluble in methyl ethyl ketone, methyl isobutyl ketone, methyl cyclohexanone, diacetone alcohol, glycol ethers and many cosolvent systems. They are compatible with certain pure phenolic, urea formaldehyde, melamine formaldehyde, polyamides and vinyl resins, but incompatible with vegetable oils, alkyd resins, and cellulose and rosin derivatives. Epoxide resins can be cross-linked to form inert flexible films with excellent adhesion by reaction with:

- amines or reactive polyamides, when reaction takes place with the epoxide group;

- organic acids, when esters are formed at the hydroxyl and epoxide groups;
- isocyanates, when reaction takes place at the hydroxyl group;
- phenol, urea and melamine formaldehyde resins.

Epoxy esters based on drying and non-drying fatty acids can be modified with styrene, vinyl toluene or rosin. Like alkyds, long-oil esters can be thinned with aliphatic hydrocarbons. Water-soluble esters are also available.

Metal-decorating inks are often protected with overprint varnish based on epoxy esters. Besides conferring good abrasion and processing resistance in such things as food and chemical containers, they protect the print from any contents spilled on the outside.

Polyisocyanates and polyurethanes

Polyurethanes are polymers obtained by the reaction of polyisocyanates with a polyhydroxy compound such as polyethers, polyesters, castor oil or glycols, etc. or with water and a stabilizer.

A great number of polyurethanes are used in the manufacture of flexible foams, rigid foams, adhesives and surface coatings as well as in roller manufacture. The most widely used systems are based upon two components. The main component is the polyisocyanate, a typical example of which is toluene 2,4-diisocyanate, obtained by the reaction of phosgene on toluene 2,4-diamine, which has been reacted with trime thylolpropane to reduce the health hazard. It will then remain stable until the curing agent (a polyol) is added. One-component systems are available. These use water as a curing agent, or are modified with drying oils or by producing a 'blocked' isocyanate which requires heat to reactivate the isocyanate groups.

Their use in inks is restricted by their effect on printing rollers, and health and safety considerations which put strict limits and controls on the amount of free isocyanates permissible. The screen process is the main outlet. Although polyurethanes obviously contain urethane groups, they also contain a multiplicity of other groups as well as aromatic rings. This results in a very complex range of possible structures. The fundamental structure of a polyurethane derived from a dihydroxy compound HO—R—OH and a diisocyanate OCN—R′—NCO can be given as:

$$\left[-R-O-\overset{\overset{\displaystyle O}{\|}}{C}-NH-R'-NH-\overset{\overset{\displaystyle O}{\|}}{C}-O- \right]_n$$

A water-diluable polyurethane can be prepared by reacting a polyisocyanate with 2,2-di(hydroxymethyl) carboxylic acid, then reacting this with ammonia or an amine.

These resins are gaining usage in liquid inks as a plasticizing resin in plasticizer-free systems.

Nitrocellulose, N/C (cellulose nitrate CN)

Nitrocellulose has wide applications in flexographic inks for paper, polymer films, foil and board, usually in combination with other resins such as maleic, polyamide or acrylic. Inks based upon nitrocellulose have excellent scratch and rub resistance. It is heat resistant and will withstand high heat-seal temperatures. Also used in overprint lacquers and in foil 'washing' formulations.

Cellulose nitrate (strictly the nitric ester of cellulose and not a nitro compound) is made by treating purified cellulose $(C_6H_{10}O_5)_n$ with excess nitric and sulphuric acids. The degree of nitration is controlled by selection of acid strength, time and temperature of nitration. When nitration is complete the product is centrifuged and washed with water. The term nitrocellulose is the standard name applied to industrial grades of cellulose nitrate.

Various viscosity grades are prepared by partially hydrolysing the nitrated product with water, under pressure, when the molecular weight of the material is reduced. Most of the nitrocellulose products used in printing ink manufacture today are 'dense grades'. These are in the form of small square platelets manufactured by replacing the water with ethanol, isopropanol, normal propanol, by washing prior to the densing operation. The final product usually contains 70% solid cellulose nitrate, with 30% of solvent or plasticizer damping.

Cellulose is composed of β-glucose anhydride units joined in a head to tail fashion. Each unit has three hydroxyl groups which can be nitrated. In practice never more than 2.25 hydroxyl groups per unit are nitrated, which corresponds to 12.2% nitrogen in solid cellulose nitrate.

Handling and storage precautions: alcohol-damped or water-damped nitrocellulose should be stored in a cool place with the lids kept firmly on containers to prevent loss of water or alcohol. If this is not done the nitrocellulose becomes dry and extremely dangerous to handle. It is advisable to ensure that half-empty containers are not left lying about.

The legal minimum of damping is 25% below which nitrocellulose is classified as an explosive within the meaning of the Explosives Act. Drums of solvent damped nitrocellulose should be opened and processed only in buildings which comply with the Highly Flammable Liquids and Liquefied Petroleum Gases Regulations 1972.

The structure of a low nitrogen content nitrocellulose product is shown below. (10.7% nitrogen roughly corresponds with two nitrate groups per unit.)

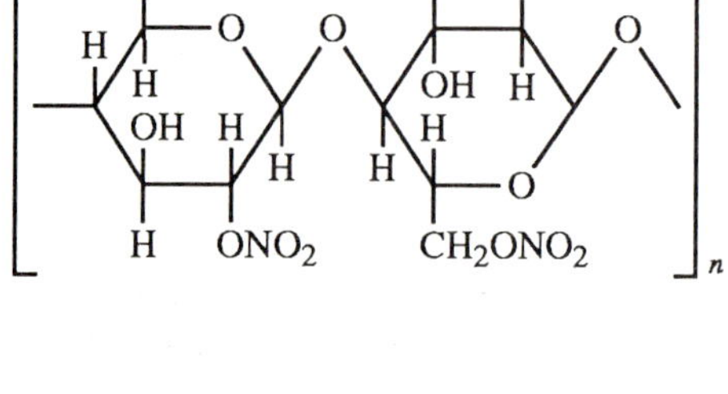

Grades of commercial nitrocellulose are characterized by their nitrogen content (Table 4.15).

For each nitrogen content there are several viscosity grades available, characterized by the molecular weight distribution after hydrolysis.

Nitrocellulose is a white fibrous material with excellent thermal stability at room temperature. At elevated temperatures it decomposes violently with the evolution of brown nitrous oxide fumes. Ignition point is 180°C. It is compatible with shellac, rosin derivatives, epoxides, short-oil alkyd resins, etc. and with most common plasticizers. The main plasticizers used are phthalates such as dioctylphthalate, and citrates. The film-forming and keying properties of nitrocellulose are outstandingly good.

Ethyl cellulose

A cellulose ether manufactured by the reaction of ethyl chloride with alkali cellulose under carefully controlled conditions whereby some hydroxyl groups are replaced by ethoxyl groups (OC_2H_5). By varying the molar ratio of reactants before etherification, grades are produced with ethoxyl content varying from 44 to 50%, that is 2.2–2.6 ethoxyl groups per anhydroglucose unit. The reaction is symbolized as

$$R{-}O{-}Na + C_2H_5Cl \rightarrow R{-}O{-}C_2H_5 + NaCl$$

where R represents the cellulose radical.

Products are divided into four types according to the degree of substitution. Most requirements for inkmakers are met by grades within the 47.5–49% ethoxyl content range which are often referred to as the N-type. A number of viscosity grades ranging from 4 centipoise to 30 poise based upon 5% by weight in an 80:20 toluene ethanol solution are marketed in powder form. Ethyl cellulose is supplied as a colourless odourless white granular powder, which is non-toxic and chemically inert. It has a relative density of 1.14 and softening points are between 150 and 170 °C depending on the degree of ethoxylation and the viscosity grade. Being chemically more inert than other chemical modifications of cellulose, it shows better alkali resistance and heat resistance and is less affected by sunlight. It also has moderate resistance to acids.

Solubility varies with the degree of ethoxylation. Grades with a lower ethoxyl content (45.5–47%) are used in plastics and films, whereas some of the high ethoxyl grades tolerate high dilution with aliphatic hydrocarbons, and have good solubility in many solvents including aromatic hydrocarbons, alcohols, chlorinated hydrocarbons, esters, glycol ethers, ethers and ketones. Cosolvent blends are used to match the solubility parameter more closely. Ethyl cellulose is compatible with rosin, glycerol esters, coumarone, gilsonite, shellac, some alkyds, including styrenated alkyds, also with maleic and phenolic resins, waxes and drying oils. It is incompatible with short-oil alkyds, acrylates, methacrylates, polyvinyl chloride, polyvinyl acetate, polystyrene and nitrocellulose.

Ethyl cellulose can be plasticized with a large variety of plasticizers. The plasticized films are not as flexible as the corresponding nitrocellulose film, nor can they tolerate such a large percentage of plasticizer.

Lacquers based on ethyl cellulose have excellent toughness and flexibility. In flexographic and gravure formulations it has been used in combination with other resins for printing on paper, aluminium foil and polymer films. Emulsions and suspensions of ethyl cellulose are also used for coating paper, leather and textiles.

Ethyl hydroxyethyl cellulose (EHEC)

Mixed ethers of cellulose, similar to ethyl cellulose, but with wider solubility characteristics. Manufactured by treating alkali cellulose with ethyl chloride and ethylene oxide water under carefully controlled conditions.

No simple formulation of its chemical constitution can be given because the structure is dependent on the molar ratio of reactants. Ethyl hydroxy ethyl cellulose is a mixed ether containing ethoxy groups as described for ethyl cellulose and β-hydroxyethoxy groups by addition of ethylene oxide:

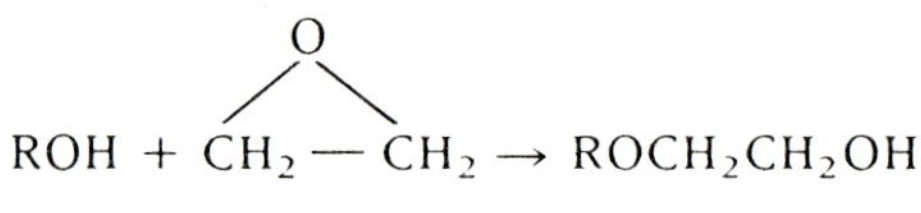

where R represents the cellulose unit. The further reaction

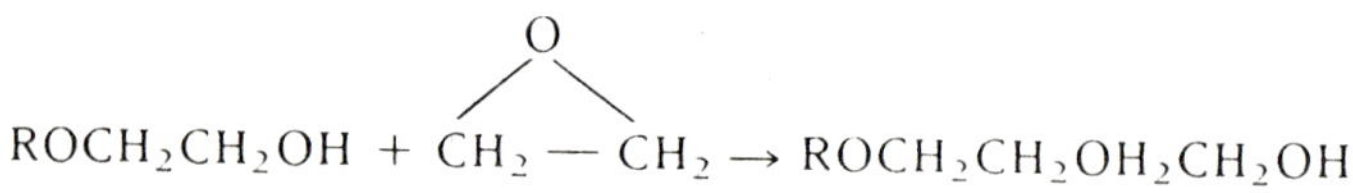

is possible.

Three different viscosity grades are produced and classified as high, low or extra low. They are marketed as white odourless powders that are insoluble in water, methanol and diacetone alcohol and only partially soluble in ethanol and isopropanol. They dissolve in a wide range of aromatic and aliphatic hydrocarbons to form hazy solutions which clear on addition of between 2 and 10% of ethanol, butanol or isopropanol. They are also soluble in chlorinated hydrocarbons, MEK, MOBK, and some esters, and compatible with zinc calcium resinate and other rosin-based resins.

Ethyl hydroxy ethyl cellulose is used mainly as a film former in screen inks and to impart toughness and scuff resistance to gravure inks. It is also used as a flatting agent in some varnishes, and as a bodying agent, especially in heat-set inks.

Cellulose acetate propionate (CAP)

Mixed esters of cellulose, very similar to cellulose acetate butyrate. The choice of procedure employed in manufacture depends upon the composition of the ester desired. It is possible to introduce amounts of propionic acid rather than butyric acid into the acetic and acetic anhydride esterification mixture to produce a uniform product containing both acyl groups. All ratios of acyl groups can be made by varying proportions and procedures, although there is far less scope than with the butyrate. Commercial grades usually contain 38–47% propionyl.

CAP is similar in appearance to cellulose acetate, with a melting range of 188–210 °C and low odour. It has excellent solubility in a wide range of solvents; some grades are soluble in alcohol/water mixtures.

Good compatibility with acrylic resins, phenolics, polyvinyl acetate and maleic resin. Because of its high hydroxyl content it can be reacted with melamine and urea formaldehyde to give long pot life, low-temperature curing inks. Easy pigment wetting of dry powders or wet presscake.

Used in flexographic, gravure and screen inks to give fast solvent release and good adhesion – useful in laminating inks. Also used as a clear overprint varnish or coating which can be applied over wet ink films. Inks formulated with cellulose acetate propionate have excellent resistance to heat, chemicals, solvents and penetration of oils or grease. Also recommended for use on paper to be recycled.

Cellulose acetate butyrate (CAB)

Cellulose ester contains more side chains than cellulose acetate, whereby some residual hydroxy groups are present.

Pure cellulose is esterified with acetic and butyric acids or more often anhydrides in the presence of a small proportion of sulphuric acid. The proportions of acetyl to butyryl may be varied from high to low. Therefore, the nature of this material will depend upon the degree of esterification of the cellulose and the molar ratio of acetyl to butyryl groups. Many grades of greatly differing composition and properties are available. They are produced as a sweet smelling white powder. The butyryl content varies between 17 and 37% with corresponding decrease in acetyl content which varies from 30 to 13%. Increase in butyryl groups increases the flexibility, moisture resistance, solubility and resin compatibility, but it lowers the softening point and hardness. The softening points of CAB vary from 120 to 240 °C. Hydroxyl contents range from less than 1% to almost 5%. Relative densities range from 1.17 to 1.26. Grades with the higher butyryl content are soluble in ketones, esters, methylene chloride, diacetone alcohol, and various cosolvents, whereas those with a lower butyryl percentage have a more limited solubility. The grades having a higher hydroxyl content are more soluble in diacetone alcohol and glycol ethers. Cellulose acetate butyrate resins are compatible with many resins particularly acrylic, phenolic, polyester, vinyl acetate, some alkyds and urethanes, and nitrocellulose. They are also compatible with a large range of plasticizers. Lacquers based on cellulose acetate butyrate have good clarity, gloss, resistance to UV light, gloss retention and solvent resistance. They are used in gravure and screen inks, as well as forming excellent media for the dispersion of pigmented chips by two-roll milling or extrusion.

Sodium carboxymethyl cellulose (CMC)

A white granular powder which is readily soluble in water or aqueous mixture of the lower alcohols and ketones. Addition of acidic materials causes precipitation. It is compatible with gum arabic, polyvinyl alcohol, starch, hydroxyethyl cellulose and certain urea formaldehyde resins. CMC can be plasticized with the usual water-soluble plasticizers, such as glycerine, glycols, ethanolamines, etc. In the food industry 'cellulose gum' is used to designate purified CMC for use in foods.

Commercially it is manufactured by treating alkali cellulose with sodium monochloracetate. The amount of sodium monochloracetate is controlled such that 0.3–1.2 (usually 0.7) sodium carboxymethyl groups are introduced per unit of β-glucose anhydride. Thus, each unit of product will usually have 2.3 hydroxyl groups. Several viscosity grades are available.

Chemical constitution

The basic formation reaction is:

$$RONa + ClCH_2COONa \rightarrow \underset{\text{Sodium carboxymethyl derivative}}{ROCH_2COONa} + NaCl$$

where R represents the cellulose radical.

Mainly used as a protective colloid and thickening agent for water-based inks and to 'gum up' lithographic plates.

It is also used as a sealer to control the porosity of paper so that it gives more gloss and consumes less ink.

Section 5: Solvents

The term solvent is widely applicable to a great number of solid, liquid and gaseous substances. The concepts of solvent and solute are interchangeable, so it is necessary to refer to a definition of solution when stating what a solvent is. A solution is a stable separation of molecules of one substance by admixture with molecules of one or more other substances. Inkmakers frequently make solutions by mixing substances (using heat if necessary) which may not spontaneously intermix on the molecular scale, but which subsequently remain in solution once mixed. Substances which spontaneously separate are non-solvents for each other: physical interactions between molecules determine the readiness with which substances intermix under given conditions of concentration and temperature.

Solvent power is the most important factor in considering the usefulness of a solvent. Solvent powder is not a general, abstract property, but a specific one relative to what is to be dissolved. The often met requirements that some solutions must have a high film-former content at a relatively low viscosity, while solvent must separate entirely from the printed film on evaporation drying, lead back to consideration of the degree of likeness of film-former and solvent. In polymer solutions of high concentration, liquids which possess the highest solvent power yield solutions with the lowest viscosity, and this property can be used as a yardstick in assessing the power of a particular solvent relative to a particular polymer. Conversely in solutions of low concentration, the solvent producing the greatest intrinsic viscosity has the highest solvent power for the polymer.

Solvents which have a high hydroxyl content are strongly polar and have a high dielectric constant, whereas hydrocarbon and other solvents have a low dielectric constant, being non-polar.

Because the equilibrium of physical interactions between molecules in a mixture of solvents is a complex mean of the component interactions, it often happens that two or more non-solvents for a polymer can be mixed in proportions such that the blend dissolves the polymer. This is the phenomenon of cosolvency. When a non-solvent for the principal resin is added to an ink it is conventionally referred to as a diluent or reducer. In fact diluents sometimes form a cosolvent blend with the ink vehicle solvents; a small change in proportions may result in greater solvent power of the blend for the film-former, but large additions of diluent cause precipitation in extreme cases. A converse of the cosolvency effect is possible. Examples exist where two solvents for a polymer form a non-solvent blend for the same polymer because a critical solvent–solvent interaction is far stronger than the same solvent–polymer interaction.

The rate of evaporation is important after solvent power when selecting solvents for inks. Slow solvents (those of low volatility) are necessary for press-stability of letterpress and lithographic processes. Controlled evaporation rates are needed in inks that dry by evaporation. The evaporation rates of solvents in a blend vary with the components, the concentrations and temperature. Physical interaction between species of molecule alters both the effective and the relative volatility of components in a blend. Volatility at a given temperature is largely determined by vapour pressure and heat of vaporization.

A solvent mixture of definite composition, having a constant boiling temperature (which may be higher or lower than that of all components in the mixture) is called an azeotrope. The term 'balance' is applied to solvent blends having a composition that remains constant as the blend evaporates. A solvent blend that is balanced for evaporation drying at room temperature usually differs from one that is balanced for heat-assisted drying (around 60°C), and both blends normally differ from the azeotrope if one exists. It is possible for mixtures to have vapour pressures such that the blend has a flash point lower than any of its constituents. Other properties that have to be seriously considered when selecting a solvent for use in printing inks include residual odour, flammability, toxicity, purity and colour. The chemical constitution of pure solvents is of less importance than that of drying oils or resins (except in consideration of solvent power), since solvents do not usually take part in chemical reactions in ink vehicle manufacture. However, the presence or absence of water is significant in certain cases.

The properties of some solvents are tabulated on page 304. They are arranged in the following chemical groups for convenience: (A) hydrocarbon (aliphatic; naphthenic; aromatic); (B) monohydric alcohol (aliphatic; alicyclic); (C) glycol; (D) glycol ether; (E) ketone; (F) ester.

The chemical formula for each solvent is given (where possible), with commercial preparation, boiling range, relative density, flash point (closed cup, unless otherwise stated), colour, the main type of resin which it usefully dissolves and finally miscellaneous properties and applications.

Legislation requires manufacturers to make available to all users details of hazards and precautions for the handling of their products, including solvents. Manufacturers have co-operated in a sensible manner and heed should be taken of their recommendations for any precautions needed. All solvents used in printing inks can be used with safety, if the precautions prescribed by factory or industrial health and safety committees are taken.

It is the user's responsibility to see that care is taken in marking and handling of materials, that sufficient ventilation is installed and where necessary protective clothing is provided and **worn**. Any risks that exist must be minimized, as nearly all solvents have an effect on the human body, serious effects being dependent on the amount of solvent and the time of exposure to it. Large doses in a short time cause acute poisoning usually by affecting the nervous system. Smaller dosage over a long period of time can cause sensitization or chronic damage to organs.

Exposure to solvents is influenced by volatility of solvents which depends upon concentration, vapour pressure and temperature, or by physical contact with the solvent itself by immersion or ingestion.

The human body reacts in many ways. Swallowing solvent or inhaling concentrated vapour can result in poisoning and causes vomiting, hallucinations, unconsciousness or death. Immersion of the hands in solvents is undesirable as toxic action may become possible by transfer to the mouth or absorption through the skin into the blood stream. Skin complaints can be caused by solvents, due to personal allergy to a particular solvent or by a degreasing of the skin which leads to irritation or infection. Barrier creams and the use of lanolin or hand cream after washing are a preventative measure in that they restore the grease level that protects the skin from bacterial attack.

Many technical grades of solvent are not pure chemicals, but contain isomers and sometimes homologues, which it may be uneconomic, and even undesirable to remove. This results in boiling ranges instead of boiling points and applies particularly to hydrocarbon solvents. In the aromatic series, terms such as 'toluol' were used to describe the less pure grades of toluene, and British Standards exist, such as BS 805:1977 for toluenes, defining three different grades. The chemical names benzene, toluene and xylene are now standard for referring to aromatic hydrocarbons. Three grades of xylene are specified in BS 458: 1977.

4.22 HYDROCARBON SOLVENTS

Low-boiling petroleum distillates – aliphatic

Various low-boiling petroleum distillates are used in gravure, flexographic and screen inks. They are produced by fractional distillation of petroleum, allowing manufacturers to sell selected brands, with specified boiling range and flash point. Distillates may be close-cut, in which the initial and final boiling points are within 20°C of each other, or they may be medium cut or wide cut solvents where this difference may be over 100°C. Products that are refined to a standard specification are known as special boiling point solvents, usually referred to as SBP with number suffixes such as 1, 2, 3, 4, 5, 6 or 11, which refer to standard grades, but not to an arrangement in sequence of solvent properties. They are non-polar, with low dielectric constants and, with the exception of SBP 6 and 11, all have flash points below 23°C (73°F).

Evaporation rates vary considerably between grades. This has become an important consideration in maintaining a working atmosphere with adequate ventilation. The aromatic content, such as small amounts of toluene, in SBP solvents ranges from 1 to 10%. A description of the naphthenic grades is given below. Legislation is moving towards aromatic free solvents.

Low-boiling aliphatic hydrocarbon distillates are good solvents for rosin, ester gum, resinate and hydrocarbon resins.

White spirit

White spirit is a higher boiling petroleum fraction than SBP6. It is cheap, evaporates fairly readily and is ideal for cleaning purposes. It is used as solvent or diluent in a few types of screen inks though its rate of evaporation is too slow for gravure and too fast for letterpress. It has an aromatic content of up to 19% and is a solvent for ester gum, isomerized rubber, hydrocarbon resins, phenolics, maleics and EHEC. Many grades of this are now available with less than 6% or even less than 1% aromatic content.

Paraffin oil (kerosene)

A water-white distillation fraction of petroleum, with a higher, wider boiling range than white spirit. It has a distinctive odour, due to its volatile 'tail', although odour-free grades are obtainable. Its aromatic content is approximately 8%. It is very cheap. Besides being used for cleaning, paraffin oil is sometimes used as a diluent in letterpress, lithographic and screen inks. The following resins are soluble in paraffin to some degree: maleics, rosin, modified phenolics, hydrocarbon resins, ester gum, calcium zinc resinates, cyclized rubber.

High-boiling petroleum distillates – aliphatic

Many close-cut high-boiling petroleum fractions are marketed, the most important being those with a 20 or 30°C gap between their initial and final boiling points. They range between 230 and 320°C, above which boiling point they are usually classified as mineral oil.

They are the most popular solvents for all types of letterpress and lithographic inks. With them it is possible to control evaporation and to obtain the press-stability needed in all types of printing machine including web-offset and heat-set machines. Many of them are branded products made to suit all types of requirement. Many distillates are now low-odour grades with an aromatic content of 1% or less: others have aromatic contents up to 24% enabling resin/solvent varnishes to be produced. They are usually water-white when supplied but deteriorate by photochemical reaction quickly to become pale or dark straw coloured.

Petroleum distillates are good solvents for rosin, ester gum, hydrocarbon resins, calcium zinc resinate, phenolics, maleics, cyclized rubber and EHEC.

Hydrocarbon solvents – naphthenic

The solvency power of petroleum distillates is largely governed by the proportion of aromatic hydrocarbons they contain. By hydrogenation of the aromatics found in hydrocarbon solvents it has been possible to produce a whole new range of solvents known as the naphthenic solvents. This range almost duplicates the various solvents in the aliphatic range above. Hydrogenation of the aromatics largely retains and sometimes increases their general solvency powers, lessens their odour and reduces aromatic, sulphur and olefin contents. This has meant that they have a more acceptable TLV (threshold limit value, referring to the vapour concentration which may adversely affect health). With changes in legislation which may be introduced within the EC, these solvents are likely to supersede many of the aliphatic and aromatic solvents at present in use.

The low-boiling fractions have improved solvency characteristics with no increase in toxicity so they can be used in liquid inks as a replacement or partial replacement for some low boiling aliphatic and aromatic solvents. They are referred to as naphthenic fractions or some identification is added to the SBP number. Higher-boiling fractions suitable for paste inks have already been accepted by the ink industry as 'low emission solvents'. The latest developments have low odour coupled with a high KB value and an aromatic content of 0.1% or less. This makes them suitable for food wrapper and packaging inks.

Aromatic hydrocarbons

Toluene (toluol); CAS No. 108-88-3

The pure solvent is derived from the fractional distillation of light coal tar oil or the catalytic cracking of petrochemicals. 'Toluene' is also the name given to commercial mixtures of which various grades are available as described in BS 805:1977. Its odour is objectionable and the vapour is narcotic.

Toluene is extensively used in publication gravure inks for which special low-odour grades are available. Care must be taken to ensure solvent release to prevent sticking and blocking of the printed reel.

It is a solvent for rubber, chlorinated rubber, melamine, urea formaldehyde, phenolics, ethyl cellulose, resinates, ester gum, polyvinylacetate and polystyrene resins.

Xylene (xylol); CAS No. 1330-20-7

Slower-evaporating solvent than toluene consisting mainly of mixtures of three isomers of dimethyl benzene. Also derived from coal tar or petroleum, its flash point is 74°F but its slow evaporation rate (equal to half that of toluene) limits it to use in gravure proofing and sheet-fed gravure inks. However, its excellent solvency power for many resins makes it a useful solvent in the removal of dried ink films.

Its solvency is similar to toluene, except that it is a poor solvent for polyvinyl acetate, but it will penetrate PVC and treated polythene.

High-boiling aromatic solvents

Mostly single-ring aromatic structures with side chains or alkyl groups attached, containing small amounts of paraffins and naphthenes. A number of proprietary brands are available, with differing boiling ranges and flash points. They are normally water-white or slightly yellow with a sweet odour characteristic of aromatic solvents.

They are used mainly in roller coating finishes and screen inks. They have KB values from 60 to 100 and are solvents for common resins such as ester gum, alkyds, maleics, modified phenolics, urea and melamine formaldehyde, chlorinated rubber and polystyrene.

4.23 ALCOHOLS

Methylated spirits (Ethanol)

CAS No. 64-17-5

Ethyl alcohol (ethanol) is not used as pure anhydrous alcohol in commercial manufacture of printing inks. Industrial methylated spirits is essentially ethanol C_2H_5OH and water, to which a denaturant such as 4% of methyl alcohol has been added.

Two grades of industrial methylated spirits (IMS) are used extensively in gravure and flexographic inks. They are 64 op and 74 op (op denotes over-proof). Proof spirit contains 57.1% alcohol by volume (49.3% by weight) and has a relative density of 0.92 at 15.6°C (60°F).

Industrial methylated spirits 64 op (density 0.821) contains 9.7% by weight of water, while 74 op IMS (density 0.797) contains 1.4% water.

Industrial methylated spirits as specified in BS 3591: 1963 consists of a mixture of 95 volumes of the ethanol/water mixture with 5 volumes of acetone-free wood naphtha approved by the Commissioners of Customs and Excise as a denaturant. The mixture must conform to the Methylated Spirits Regulations 1962.

Normal propanol (*n*-propyl alcohol)

CAS No. 71-23-8

A very pure solvent, C_3H_7OH used because it has a higher boiling point than meths or isopropanol. It is miscible with water in all proportions and with many other solvents. It has a characteristic sweet odour.

n-propanol is used in flexographic and gravure inks and overprint lacquers.

Resins that are soluble in methylated spirits are soluble in *n*-propanol, also it is a partial to good solvent for polyamides.

Isopropanol (isopropyl alcohol or secondary propyl alcohol)

CAS No. 67-63-6

A colourless alcohol C_3H_7OH with chemical and solvent properties similar to ethyl alcohol. It is supplied industrially in two grades, sometimes called the isopropyl solvents. As a first grade it contains 99.7% alcohol b.p. 82.3°C (1PS 1), the second is a water azeotrope containing a minimum of 87% isopropanol b.p. 80.3°C (1PS2). It is miscible with water, also it is used as a latent solvent in conjunction with ketones and esters in nitrocellulose lacquers, aiding their solvency and tolerance of hydrocarbon diluents.

It is a solvent for many natural and synthetic resins including ester gum, rosin, shellac, ethyl cellulose, spirit soluble phenolics, amino resins, PVAC and PVB. Like *n*-propanol it is only a partial solvent for polyamide resins, but exhibits cosolvency for polyamides in conjunction with hydrocarbon diluents.

Normal butanol (*n*-butyl alcohol)

CAS No. 71-36-3

A slowly evaporating alcohol of high purity consisting essentially of butan-1-ol, $CH_3 - CH_2 - CH_2 - CH_2 - OH$, miscible with most solvents, partially miscible with water and possessing excellent solvency power.

Used in metal coatings, some gravure and screen formulations but has considerably more residual odour than *n*-propanol or isopropanol, which precludes it from use in food wrapper inks. Used in conjunction with kauri gum to produce the 5% solution for measuring KB value. Widely used as a latent solvent in nitrocellulose and acrylic-based lacquers. Among other resins that are soluble in butanol are alkyds, urea and melamine formaldehyde, shellac, ester gum, PVAC and metallic resinates. Basic dyestuffs, oils, fats and waxes are also soluble in butanol.

Specification requirements of *n*-butanol are described in BS 508:1966.

Alicyclic alcohols

Cyclohexanol; CAS No. 108-93-0
An oily liquid with a strong but not objectionable odour. It has a fairly high flash point and boiling range. A solvent for cellulose ethers, ester gum, shellac, low viscosity silicones and polyvinyl chloride. It has a fairly limited use in screen inks. It is miscible with oils and hydrocarbon solvents.

Methyl cyclohexanol; CAS No. 583-59-5
An oily liquid with an odour similar to cyclohexanol. Its solvency power is less than that of cyclohexanol for the resins named above and its uses in ink manufacture are similarly limited.

4.24 GLYCOLS

Monoethylene glycol

CAS No. 107-21-1
Ethylene glycol is a colourless odourless somewhat viscous solvent which is very hygroscopic, miscible with water, alcohol and some ketone solvents. It is used in moisture-set inks and water-reducible letterpress inks. The relatively rapid evaporation rate leads to poor press stability.

Ethylene glycol is a solvent for gelatine, dextrin zein and maleic or fumaric resins of high acid value.

Monopropylene glycol

CAS No. 57-55-6
A colourless odourless solvent similar to ethylene glycol in most respects, but is considered as absolutely safe for use in foodstuffs and medicines. It is recommended for use in food packaging inks and formulations for marking foodstuffs.

Its solubility is similar to ethylene glycol though it is slightly more organophilic.

Hexylene glycol

A colourless liquid with no odour or taste used in the soft drink, confectionery and cosmetic industries. It is slightly slower in evaporation rate than ethylene glycol, is miscible with vegetable oils and many organic solvents. Its use is somewhat limited by its high cost. The purified grade can be used in moisture-set inks for food wrappers. Its solubility is similar to ethylene glycol though it is more organophilic.

Diethylene glycol

CAS No. 111-46-6
Colourless liquid more viscous and hygroscopic than ethylene glycol.

Miscible with water alcohols, acetone, glycol ethers and esters. Water-reducible letterpress inks formulated with diethylene glycol have improved press stability, the solvent being more organophilic and less volatile then ethylene glycol, but they have slower setting speeds. Odour level is low.

It is a solvent for high acid value maleic and fumaric resins, nitrocellulose, zein and shellac. It is currently under scrutiny for toxicity.

Dipropylene glycol

CAS No. 25265-71-8

A solvent which is very similar to diethylene glycol. It is not as safe for use with foods as propylene glycol, but has lower toxicity than ethylene or diethylene glycol. It is used in moisture-set food wrapper inks and other similar applications.

Triethylene glycol

A stable high-boiling water-white hygroscopic liquid, miscible with water and ethanol, but not soluble in hydrocarbon solvents. Its use is similar to that of diethylene glycol but inks are slower setting and possess excellent press stability.

It is a slightly better solvent than diethylene glycol especially for high acid value maleic and fumaric resins, nitrocellulose, zein and shellac.

Glycerine

A clear, colourless syrupy liquid with faint odour and sweetish taste, highly hygroscopic, being able to absorb up to 50% of its weight of water. It is miscible with water, alcohol, glycols and ketones and will dissolve oils, fats and acid dyestuffs. It is used in special water-soluble letterpress inks, e.g. cheque inks, and is a constituent of gelatine rollers. It is too hygroscopic a solvent for formulation of stable moisture-set inks. It can be used as a plasticizer, and it is used extensively in resin manufacture as an esterifying alcohol. Gelatine, sugars and gum arabic are all soluble in glycerine.

Ethoxy propanols

Propylene glycol monomethyl ether (methoxy propanol; MP) is a colourless liquid and the first of this type of solvent to be introduced into the UK in the early 1970s. It is now replacing methoxy ethanol (methyl cellosolve) in many applications as it presents less toxicological risk. It has similar solvency characteristics to methyl cellosolve, but is inferior to ethoxy propanol for solvency and wetting.

Propylene glycol monoethyl ether (ethoxy propanol) (EP) is the nearest ethylene oxide derived substitute for ethoxy ethanol. It has very similar properties without as yet any doubt regarding toxicity. Ethoxy propanol is slower evaporating than methoxy propanol but is a strong solvent with good wetting and retarding properties. It is now gaining recognition as the replacement for methoxy and ethoxy ethanol.

Propylene glycol monomethyl ether acetate (MPA)

CAS No. 108-65-6

Propylene glycol monoethyl ether acetate (EPA)

The acetate esters of methoxy and ethoxy propanol, introduced under many trade names over the past 5 years, have replaced the ethylene glycol derivatives. These are equally good solvents for many resins used in ink formulae and particularly useful as powerful solvents with low volatility.

4.25 KETONES

Acetone (dimethyl ketone)

CAS No. 67-64-1

One of the most widely used industrial solvents with relatively low cost and high solvency power. It is a colourless flammable liquid with a characteristic odour and low boiling range. Miscible with water, hydrocarbons and natural oils in all proportions. Occasionally used as a rapid-evaporating solvent in flexographic and gravure inks and in lacquers, though it is too volatile for most applications.

Powerful solvent for cellulose acetate and nitrocellulose, ethyl cellulose, CAB, CAP, PVAC, PVC, ester gum and many other natural or synthetic resins. It also penetrates polythene and methyl methacrylate.

Methyl ethyl ketone (butan-2-one)

CAS No. 78-93-3

Colourless solvent with a higher boiling point than acetone and less soluble in water, it has its own distinctive odour and is completely miscible with castor and linseed oils and with most organic solvents. Its main application is in gravure inks, the evaporation rate being about a third that of acetone at room temperature. It is often used as a substitute for ethyl acetate in nitrocellulose lacquers. Its solvency power is similar to acetone with the exception of cellulose acetate for which it is not a very good solvent.

Methyl isobutyl ketone (hexone)

CAS No. 108-10-1

Water-white solvent with a bland ketonic odour, almost immiscible with water but miscible with most solvents and oils. Used in gravure and screen inks, having an evaporation rate of only about one-eighth that of acetone. Solvent for nitrocellulose, ethyl cellulose, CAB, PVC, PVAC, vinyl copolymers, acrylics, chlorinated rubber, polyurethane, phenolic and epoxy resins.

Cyclohexanone (sextone)

CAS No. 108-94-1

Colourless liquid having a strong characteristic odour. Only slightly soluble in water, but a powerful solvent for oils and fats, miscible with hydrocarbons. It is used in special screen inks. A good solvent for most polymers, it will penetrate polythene and soften perspex.

Methyl cyclohexanone (sextone B)

This solvent resembles cyclohexanone but is slightly slower evaporating; colourless to pale yellow, with a strong odour described as similar to peppermint and acetone. It is used in special screen inks and lacquers. Its solvency action is similar to cyclohexanone.

Isophorone

Excellent ketonic solvent in the high boiling range, colourless but with a strong odour due to impurities present, only slightly soluble in water but miscible with most solvents. Solvent for oils and fats; used in metal-decorating coatings and screen inks. Regarded as a powerful solvent for nitrocellulose and vinyl resins, also for many other natural or synthetic resins.

Diacetone alcohol

CAS No. 123-42-2

Clear colourless liquid which yellows on standing. It is a ketonic alcohol, normally classified with the ketones. It has a faint odour and is miscible with water, many solvents and some oils including castor oil. It is not compatible with high-boiling aliphatic hydrocarbons. Used mainly in flexographic inks. Because it is slow evaporating it assists 'flow out', giving glossy films, especially with nitrocellulose. Diacetone alcohol is also used in some roller cleaner formulations. It is a good solvent for cellulose acetate and nitrocellulose, PVAC, shellac, zein and basic dyestuffs.

4.26 ESTERS

Ethyl acetate

CAS No. 141-78-6

One of the best low boiling solvents for nitrocellulose, water-white with strong fruity odour. Partially miscible with water (8% w/w at 20°C) but miscible with castor and linseed oil, and hydrocarbon solvents. Its solvent power is increased by adding a small quantity of alcohol. Extensively used in flexographic and photogravure inks and in lacquer formulations as a fast-drying solvent. Its evaporation rate is slightly below that of acetone at room temperature. Solvent for NC, EC, CAB, PVAC, polystyrene, ester gum and maleics; ethyl acetate penetrates polythene, perspex and PVC.

Isopropyl acetate

CAS No. 108-21-4
Properties very similar to ethyl acetate but its relative evaporation rate at room temperature is about half and its water tolerance is less.

N-butyl acetate

CAS No. 123-86-4
Colourless liquid with fruity acetate odour. Not miscible with water, but miscible with oils and many solvents including hydrocarbons. Used in small amounts in flexographic inks, gravure inks and overprint lacquers as well as some screen and metal-decorating inks. Because of its excellent solvent properties and slow solvent release it gives good gloss and reduced solution viscosities. The evaporation rate is about one-sixth that of ethyl acetate. Good solvent for nitrocellulose, hydrocarbon resins, rosin, ester gum, chlorinated rubber, vinyls, polystyrene and acrylates. In the presence of 20% butanol, *n*-butyl acetate will dissolve shellac and some alkyds.

N-propyl acetate

CAS No. 109-60-4
Water-white solvent with fruity odour. Gained recognition, particularly with the demise of cellosolve, as a retarder with good solvency and no residual odour. Now an important liquid ink solvent.

Section 6: Plasticizers

The main function of a plasticizer in an ink film is to make the dried print more flexible and pliable. Plasticizers need to be essentially non-volatile in their final ink format. Inks that dry by evaporation, especially on nonporous surfaces, tend to be brittle, and flexing can crack the film and rupture the adhesion. Plasticizers help to supply elasticity to an ink film and allow it to bend or crease without failing. They do this by acting as solvents for the film-former polymer molecules, maintaining the exact degree of mechanical freedom required.

Many chemical classes of compounds are used as plasticizers in printing ink manufacture. They are mainly viscous liquids but more solids are being used these days as they plasticize without migration problems and effect better heat resistance than liquid ones. Certain plasticizers are incorporated to promote specific properties in the final ink film such as:

- increased gloss
- deep freeze resistance
- to prevent blocking
- extra flexibility
- increased adhesion to difficult substrates
- less discolouration at high temperatures.

Very careful formulation is necessary in the initial choice of plasticizer type, and the percentage incorporated into the final ink recipe in order to achieve the optimum balance of ink properties for an efficiently safe and workable plasticized dry ink film. The major use of all these types of plasticizers is in liquid inks, paste inks having in-built plasticized resins unsuitable for liquid ink use.

The main families of plasticizers are:

abietates	polyesters
adipates	polyol esters
benzoates	ricinoleates
butyrates	sebacates
citrates	stearates
expoxidized compounds	sulphonamides
phthalates	

The major suppliers in the plasticizer market concentrate more on other industries than ink, and the most knowledgeable people tend to be in the films and plastics industries. A large amount of published information is available on the subject of plasticizers mainly because of the possible toxicity and hazard levels in contact (after migration) with foodstuffs.

Many data sources and references are available for study. These include the NPIRI *Raw Materials Data Handbook*, Volume 2, which gives a comprehensive study of 246 known plasticizers on the market.

Sucrose acetate iso-butyrate (SAIB) $C_{10}H_{16}O_9$ (CH_2OOCCH_3) $(CH_2OOCC_4H_9)$

FDA status 2520; boiling point – decomposes at 288°C; freezing point N/A; flash point 260°C; relative density 1.146.

White powder; solid plasticizer compatible with nitrocellulose, CAB ethyl cellulose, polystyrene and p.v. acetate; used as a plasticizer in inks to be laminated and when the level is carefully controlled does not affect the strength of the seal. Insoluble in water.

Triethyl citrate (TEC) $CH_2C(OH)\ CH_2[COOCH_2CH_3]_3$
Tributyl citrate (TBC) $CH_2C(OH)CH_2[COO(CH_2)_3CH_3]_3$

(Well known trade names: Citroflex 2 (TEC); Citroflex 4 (TBC)) FDA status: TBC 2520: TEC 101, 2005, 2514, 2526, 2548, 2550, 2559, 2569, 2571; boiling point: TBC 170°C: TEC 127°C; freezing point: TBC 62°C; TEC 46°C; flash point: TBC 185°C; TEC 155°C; relative density: TBC 1.042; TEC 1.136.

Clear liquid plasticizers compatible with a very wide range of resins. TEC is compatible with CAB, cellulose acetate, N/C, ethyl cellulose, PV acetate, PVB and vinyl chloride acetate. TBC is compatible with all the above and polystyrene and chlorinated rubber.

They impart excellent gloss to N/C systems but, as with all liquid plasticizers; the amount used should be carefully considered and migration tests checked with food, even though TEC is in GRAS category (generally recognized as a safe foodstuff chemical). Incompatible with water.

Epoxidized soya bean oil, $C_{57}H_{76}O_{18}$

FDA status – not considered harmful: 2005, 2514, 2520, 2526, 2531, 2548, 2550, 2559, 2571, 2572; boiling point over 400°C; freezing point +5°C; flash point 310°C; relative density 0.993.

Light yellow viscous liquid compatible with N/C, E/C, chlorinated rubber, vinyl chloride, and vinyl chloride acetate. Popular to plasticize vinyl systems and stabilize them, but useful in many applications due to its cellulose compatibility. Immiscible with water.

Phthalates

Liquid

Dimethyl phthalate (DMP) $C_6H_4(COOCH_3)_2$
Dibutyl phthalate (DMP) $C_6H_4(COOC_4H_9)_2$
Di-isobutyl phthalate (DIBP) $C_6H_4(COOCH_2CH(CH_3)_2)_2$
Di-isooctyl phthalate (DIOP) $C_6H_4(COOC_8H_{17})_2$
(Di-2-ethyl-hexyl phthalate)
Di-octyl phthalate (DOP) $C_6H_4(COOC_8H_{17})_2$

Solid

Di-cyclo-hexyl phthalate (DCHP $C_6H_4(COOCH(CH_2)_4CH\ _2)_2$)

Boiling points

DMP	282°C	DIOP	235°C
DBP	340°C	DOP	232°C
DIBP	327°C	DCHP	212°C

Freezing or melting points

DMP	–26°C	DIOP	–50°C
DBP	–45°C	DOP	–50°C
DIBP	–45°C	DCHP	+62°C

Flash points

DMP	149°C	DIOP	215°C
DBP	171°C	DOP	210°C
DIBP	174°C	DCHP	215°C

The phthalates are the largest group of plasticizer, there being 15 major phthalates used in the industry. Ease of handling and the low cost of the liquid grades account for their large usage. They are compatible with nearly every resin system:

- dimethyl phthalate (DMP) – all, but limited in chlorinated rubber and vinyl chloride;
- dibutyl phthalate (DBP) – all, but limited in CAP and cellulose acetate;
- di-isobutyl phthalate (DIBP) – all, but limited in CAP and cellulose acetate
- di-isooctyl phthalate (DIOP) – all, but cellulose acetate; limited with CAP, PVB, chlorinated rubber and acrylic;
- di-octyl phthalate (DOP) – all, but cellulose acetate, limited with CAP, PVB, chlorinated rubber and acrylic;
- dicyclohexyl phthalate (DCHP) – all.

Polyesters

A range of plasticizers, varying widely in chemical composition which can be of use in specialized formulations.

Polyol ethers

Glyceryl triacetate (Triacetin $CH_2CHCH_2(COOCH_3)_3$) is the most well known in this class and is the most acceptable. Generally regarded as safe.

It is the only member of this group with some solubility in water, a **rare** property in plasticizers.

Compatible with all forms of cellulose resins, CAP, CAB, NC, EC, CA, acrylics and poly vinyl acetate; boiling point 258°C; freezing point –50°C.

Ricinoleates

Butyl ricinoleate

$C_4H_9OOCC_{10}H_{18}CHOHC_6H_{13}$ (sold as Felxricin P3); boiling point 220°C; freezing point –23°C; flash point 207°C.

Insoluble in and immiscible with water; FDA approved 2519, 2520 and 2535. Compatible with and suitable with CAB, NC, EC, poly vinyl acetate, PVB and vinyl chloride acetate.

Glyceryl triricinoleates (castor oils)

$CH_2CHCH_2[OOCC_{10}H_{18}CHOHC_6H_{13}]_3$ (sold as castor oil std. and crystal). Melting point –15°C; no boiling point; flash point 290°C.

Insoluble in water; used in liquid and paste inks; FDA status 2514, 2526, 2531, 2535, 2548, 2550, 2553, 2557, 2559, 2562, 2571. Colour usage restricted to suitable inks.

Stearates

For example, butyl stearate $CH_3(CH_2)_{16}COO(CH_2)_3CH_3$. Melting point 19°C; boiling point 350°C; flash point 188°C.

Usage gradually dropping in favour of more easily handlable materials. One of the oldest plasticizers known in the trade, compatible with CAB, NC, EC, chlorinated robber and polystyrene; Now less used because of properties and compatibilities.

Section 7: Waxes

Waxes of many chemical types are incorporated in most letterpress, lithographic, screen, gravure and flexographic inks principally to impart mar-resistance, improved slip and water repellancy properties. They are the main constituent of carbon copying inks and other 'hot melt' inks, where printing is done at an elevated temperature and drying is by solidification.

Small additions of wax or wax paste are known to reduce tack without appreciably increasing the flow of letterpress and lithographic inks. Waxes are normally introduced into inks in one of two ways. Wax of a controlled fine particle size can be mixed or ground into the batch along with pigments or may be introduced during the final blending operations. A second way of incorporating them into formulations is to compound them into 'wax media' by dispersing or in some cases melting them into varnishes and/or solvents and adding these to the ink.

It is generally accepted that the non-rub qualities imparted by an individual wax are a function of both the particle size and the hardness as well as the melting temperature of any particular wax. It is also generally accepted that wax particles that are fine enough to show no adverse manufacturing or printing problems, but which stand just proud of the dried printed film tend to give the maximum scuff resistance. However, with the heat and movement imparted by the friction of constant rubbing under pressure, particles of the ink film can ball up and mark unprinted areas. It is therefore essential that the use of waxes be kept in perspective. With a hard-drying flexible film the quantity of wax needed to obtain excellent scuff and mar-resistance is minimal. With a poorly formulated vehicle system, more waxes are added to improve rub resistance, which in themselves will have an adverse effect on ink flow and transference, gloss, hardness and resistance characteristics.

Addition of waxes to inks almost invariably decreases their gloss and increases their drying time in both oil and liquid inks, hence the ink must be formulated in the knowledge that there is an optimum amount to give the best compromise between non-rub properties and any detrimental effect such as decreased gloss. Proofing and checking on a rub tester under controlled conditions is necessary with each change of ink formulation or substrate.

Various greases, many of which may be regarded as plasticizing wax, are used in letterpress and lithographic inks for control of rheological properties. The two most widely used, petroleum jelly and lanolin, are included in this section.

Certain wax compounds and solutions added to liquid inks, which are printed on non-porous substrates, tend to migrate to the surface of the film. This helps to prevent blocking as well as giving the printed film some slip and scratch resistance. It is possible to produce ink films from which Scotch tape can be peeled off easily, thus masking the anchorage measurement at the substrate/ink interface. This can be due to the wax floating to the top surface or to exudation of other mobile components.

Synthetic waxes, defined as those chemically produced, are now the most popular ones used in the ink industry; these are therefore given prime place, whereas natural waxes are those produced by plants or animals and include the fossil and petroleum waxes, all of which need some physical or chemical refining before they are suitable for use. The use of natural waxes has declined as they have been largely replaced by polyethylene waxes.

Since many proprietary waxes are blends of both natural and synthetic products, it is difficult accurately to describe their constituents. Some pure waxes can be identified by the type of fluorescence they give when exposed to black light from a UV lamp. This test should be carried out in both solid and liquid state, also upon a solution of the wax.

4.27 SYNTHETIC WAXES

Polyethylene waxes

These are waxy hard white solids, normally supplied in powder or pellet form. They are odourless and tasteless, and can be used in inks or coatings that have direct contact with foodstuffs. They are chemically inert, and contribute towards the chemical resistance properties of an ink film. They are soluble in a large number of common solvents and oil/resin blends at elevated temperatures, but are precipitated from solution on cooling to room temperature, when the polyethylene wax assumes the form of a very fine dispersion.

Polyethylene waxes are incorporated in all types of printing inks. They are usually added in the form of 'non-rub' or 'slip' media which are fine dispersions of the wax in the solvents, oils and resins, etc. of that particular type of ink formulation in which it is to be incorporated. For example, for use in a letterpress or lithographic ink the wax is melted in solvent, or in a mixture of solvent and resin, or in linseed stand oil. The dispersion is obtained by cooling the mixture in specialized equipment.

Today the printing ink formulator has a choice of many polyethylene waxes and it does not automatically follow that a wax which can be formulated into a medium imparting excellent rubproofness to a letterpress or lithographic ink can be formulated into a first class gravure or flexographic ink wax medium. Hence different waxes are used in differing types of ink, in order to achieve maximum rubproofness. Generally speaking the most effective type of polyethylene wax for slip properties in an ink film is the high density, high melting point isotactic polymer. This and some of the softer waxes are now available in fine powder form with a particle size of between 2 and 4μm. It is possible to incorporate these waxes both at the pre-grinding stage, or by mechanical dispersion into finished inks. These waxes are very abrasion resistant if an optimum quantity is used, between 0.25 and 2.0%.

Emulsifiable grades of polyethylene wax are also commercially availiable. These usually are waxes in which some carboxyl groupings have been introduced by partial oxidation of the hydrocarbon. An emulsion is produced by forming salts with basic compounds when the aqueous mixture is stirred vigorously at temperatures above the melting point of the wax. More recently the basic materials have been already incorporated in the commercial product, when the material is truly self-emulsified.

Polyethylene is produced by the polymerization of ethylene $H_2C = CH_2$ to form a chain structure of $(-CH_2-CH_2-)_n$. Many commercial grades are available with molecular weights between 1500 and 4000.

The lower the mean molecular weights, all factors being equal, the lower the melting point of the wax. They vary from the high-density high melting-point isotactic polymer, which is produced by the low pressure Ziegler process in which an aluminium trialkyl and a transition metal chloride is the catalyst combination, to the low-density low melting-point atactic polymer, which is produced by the high pressure (up to 1000 atmospheres) free-radical process using oxygen as the source of the propagating free radicals. High density polyethylenes have a relative density of approximately 0.94–0.96, at 24°C. Low density ones are approximately 0.90 at 24°C.

Chain branching, which is present in varying amounts in all commercial polyethylene waxes produced by the high-pressure process, as demonstrated below, has the effect of lowering the melting point of the wax.

Polymer with chain branching:

$$\begin{array}{l} -\,CH - CH_2 - CH_2 - CH_2CH_2 - \\ \quad\; | \\ \quad CH_3 \end{array}$$

Polymer without chain branching:

$$- CH_2 - CH_2 - CH_2 - CH_2 -$$

Polytetrafluoroethylene

Polytetrafluoroethylene (PTFE) is produced under a variety of trade names, i.e. Teflon (Du Pont), fluon (ICI), Hostaflon (Hoechst), etc. It exhibits useful properties over the widest temperature range of any known polymer. It is uniquely non-adhesive and antifrictional and will withstand a temperature of 320°C. As yet there is no known solvent for this material. The above properties are all ideal in theory for an anti-setoff powder for an ink which must possess a high degree of rub resistance. The coefficient of friction of solid polytetrafluoroethylene decreases with increasing load.

The relative density of such powders is 2.1 and the mean particle size 4 to 5 μm. Suitable for all types of printing inks, but especially ideal for heat-set inks, where the temperature of the drying apparatus does not permit polyethylene waxes to be used effectively. It can also be stirred into finished inks to improve their rub and scuff resistance. The relative cost of PTFE is prohibitively high for many applications. It is formed by the pyrolysis of chlorodifluoromethane at high temperature to yield tetrafluoroethylene:

$$2CHClF_2 \rightarrow CF_2 = CF_2 + 2HCl$$

This is subsequently polymerized at high pressure.

Fatty acid amides

A fatty acid amide is described as an organic acid derivative in which the OH group is replaced by an amine group NH_2 to form a long-chain amide which is a waxy material. The three fatty acid amides which are used in printing ink formulations are:

$CH_3(CH_2)_7CH = CH(CH_2)_7CONH_2$	oleamide
$CH_3(CH_2)_{16}CONH_2$	stearamide
$CH_3(CH_2)_7CH = CH(CH_2)CONH_2$	erucamide

Fatty acid amides are wax-like to the touch, vary in colour from pale yellow to dark brown, have melting points between 68 and 109°C, and are soluble to varying degrees in most solvents. They also blend well with waxes, but odour can be a problem if the refined versions are not used.

Fatty acid amides are used as rub-resistant and slip-improving additives in flexographic and gravure inks and over-lacquers. They minimize set-off and blocking, and improve the adhesion of subsequent colours. Stearamides and oleamides are used as solubilizers for oil-soluble dyestuffs. In paste inks a small amount of oleamide will improve pigment wetting. In letterpress inks stearamide will reduce stringiness without lowering viscosity and will help reduce misting problems. Often less than 1% is used.

Stearamide increases the polarity of non-polar oils and helps stabilize oil-in-water emulsions.

Fatty acid amides are chemically neutral compounds and are prepared by passing ammonia through molten fatty acid in a special reactor:

$$R.COOH + NH_3 = R.CONH_2 + H_2O$$

where R is an alkyl or alkenyl group in which the length of the carbon chain varies from C_{12} to C_{18}.

4.28 PETROLEUM WAXES

Petroleum waxes are extracted from crude petroleum. There is a variety of crude feedstocks from different world sources, which may have up to 10% wax content. Petroleum waxes are obtained by distillation of oils and subsequent refining. They are roughly classified by type, depending on the oil content and treatment.

In printing inks the petroleum waxes are used to reduce the tack of letterpress and lithographic inks, as slip agents, especially the microcrystalline waxes, and as components of carbon-copying inks.

Petroleum jelly is used as a medium to reduce tack and to improve setting or increase penetration into paper or board. It is also used to retard the speed of drying of letterpress or litho inks on the press rollers.

The petroleum waxes are composed almost entirely of complex mixtures of paraffinic hydrocarbons. Paraffin wax consists of paraffins between $C_{18}H_{38}$ and $C_{32}H_{66}$. Microcrystalline wax consists of paraffins between $C_{34}H_{70}$ and $C_{43}H_{88}$. Petrolatum is a mixture of paraffinic and olefinic hydrocarbons in which the carbon chain length of the molecules varies between C_{16} and C_{32}.

Produced by distillation of petroleum at a temperature of about 300°C a wax-oil mixture distils over. On chilling, the wax crystallizes from the wax-oil mixture and is removed by filter-pressing. The solid residue in the filter is known as slack wax and contains about 65% of paraffin wax, from which commercial paraffin wax is extracted by further refining. Petrolatum or petroleum jelly is extracted from a higher-boiling distillate of crude petroleum. The refining of the various waxes is usually by treatment with solvents of various kinds, including liquid sulphur dioxide, nitrobenzene, furfural and methyl ethyl ketone. The waxes are further refined by decolouration with Fuller's earth, charcoals, activated carbons or clay, and are then recrystallized.

Slack wax

A soft paraffin wax with up to 35% oil content and melting points between 49 and 60°C (120–140°F).

Scale wax

An intermediate paraffin wax between the slack wax and the higher refined grades with large flake-like crystals. Colour ranging from white to yellow depending on the amount of oil retained which may be up to 6%. Melting point 49–57°C (120–135°F).

Fully refined paraffin wax

Sometimes referred to as paraffins or block wax. These are hard waxes with a coarse crystalline structure. Colour translucent white to slightly yellow according to the degree of refining. Relative density (15°C) 0.916; melting point from 45–74°C (118–155°F); molecular weight 350–420.

At room temperature it is more soluble in aromatic hydrocarbons than in aliphatic solvents. At elevated temperature paraffin wax is also soluble in drying and non-drying oils, rosin, rosin esters, phenolics, maleic and alkyd resins and will not crystallize out on cooling if less than 2% is present. It is only partially soluble in alcohol.

Petrolatum or petroleum jelly

A mixture of liquid and solid hydrocarbons with a grease-like consistency. It often exhibits a green or blue fluorescence. Relative density (15°C) 0.820–0.860; melting point 45–74°C (113–165°F).

Microcrystalline waxes

High-melting waxes with a very fine crystallite particle size in contrast to the coarse crystallites of the paraffin waxes. They have an average molecular weight of 490–800 and they exhibit higher viscosities when molten. The residual oil content varies from 0.3 to 13% and the colour from yellow to white. Melting range lies between 66 and 93°C (150–200°F); relative density at 15°C is 0.930.

Ceresin wax

Originally ceresin was an alternative name for ozokerite, but today the name is applied to blended hydrocarbon waxes of variable composition and melting point. The mixture may consist of petroleum waxes only or petroleum and mineral waxes with small amounts of carnauba wax to raise the melting point. Ceresins are used for similar purposes to paraffin waxes.

Montan wax

Extracted from bituminous lignite or shale, montan wax has a colour ranging from very dark brown to yellow, which usually relates to the amount of de-resination. It also has rather an oily odour. Crude montan wax has melting point 80–83°C (176–181°F); acid value 23–28; saponification value 78–92, whereas bleached montan wax has a melting point of 72–77°C (158–170°F); acid value 93–107; saponification value 95–143. It is soluble in aromatic and chlorinated hydrocarbons when cold, but when hot it is soluble in aliphatic hydrocarbons. It contains a large amount of organic acids and can easily be emulsified. It also blends well with other waxes.

Besides finding some use in carbon paper inks and rubber manufacture it is used in the manufacture of synthetic wax esters of high molecular weight. Montan wax is obtained from brown coal by a solvent extraction process. The crude extract usually contains 70% montan wax, the remaining components being resin and bitumen. The montan wax is then purified by a further solvent extraction process. An ester is formed when the de-resinated montan wax is subjected to pyrolysis.

Montan esters

The almost pure montan wax, essentially the ester montanyl montanate $C_{27}H_{55}COOC_{28}H_{57}$, is first hydrolysed, and the alcohol is oxidized by a chromic sulphuric acid mixture to montanic acid, $C_{27}H_{55}COOH$. This acid is the raw material for most of the synthetic ester waxes, which have a higher melting point than that of montanic acid.

The most common glycols used in the esterifying reaction are ethylene and butylene glycols, when the molecular weight of the wax is greatly increased. These glycols are used in various combinations often in the presence of small amounts of saponified wax and fatty acids to give the desired properties of commercial wax blends (including the IG and Gersthofen waxes).

The ester waxes, partially saponified ester waxes, and special blends of wax, fatty acids, and other additives vary in colour from almost black, through a continuous colour range to pale yellow. Their melting points vary from 70 to 104°C (158–219°F) and their acid and saponification values from 6 to 85 and 80 to 170 respectively.

Synthetic waxes based on montan wax are used to improve the rheological properties and to control the rub resistance and water repellency of lithographic and letterpress inks.

4.29 NATURAL WAXES

Generally, natural waxes are complex mixtures of fatty acids, esters, alcohols and hydrocarbons. The acids consist mainly of mixtures of saturated fatty acids, general formula C_nH_{2n+1}, ethylenic acids C_nH_{2n-1} COOH, and many dibasic acids, etc. where n is any even number between 12 and 32. The esters are formed by reaction of the wax acids with higher alcohols from C_{12} to C_{34}. The hydrocarbons are saturated paraffins containing between 19 and 31 carbon atoms. Unsaturated hydrocarbons are rarely found in natural waxes.

Beeswax

Beeswax is a yellowish white to a yellowish brown in colour, the yellow being due to pollens carried by the bee. It has a melting point of 62–64°C (143–147°F) with a solidifying point of 61°C (142°F), acid value of 19, iodine value of 10, saponification value of 96–100, and a relative density of 0.96 at 15°C. At room temperature it is a waxy solid with a tacky surface. It is soluble in ether, chloroform, carbon tetrachloride and vegetable oils. It is insoluble in water, only partially soluble in ethyl alcohol, and insoluble in mineral oil at room temperatures. When heated it blends well with paraffin waxes.

Beeswax is sometimes used in letterpress process inks, especially in the first colour down, where it helps the trapping of superimposed colour. It is also used in mixed oil/wax compounds for reducing set-off or reducing hard drying of a film.

Beeswax is obtained from the honeycomb of the honey bee *Apis*. The wax is secreted by the worker bee and consists of (approximately): wax acid esters 71%; cholesteryl esters 1%; alcohols 1%; wax acids 14%; hydrocarbons 11%; moisture and impurities 2%. The molecular weight of pure beeswax is approximately 570.

Carnauba wax

Hard and brittle waxes that range from straw yellow to brownish grey depending on the grade. Average constants for carnauba wax are acid values 2.5–5.0; iodine value 10–13; saponification value 79–85; softening point 82–84°C (180–183°F); relative density 0.999.

Carnauba is used extensively to elevate the melting point of other waxes, especially paraffin wax. Its solubility at room temperature is low, but increases rapidly at high temperatures. The wax tends to crystallize or precipitate out of solution on prolonged heating, retaining the solvent and retarding its evaporation. In inks it is regarded as a wax which does not float to the surface after printing; as such it has been used in overprinting varnishes, gloss inks and gold media. When dispersed in drying oil in conjunction with other waxes it imparts improved slip to the dried ink film, reduces the tack and improves the setting. Small quantities are also used in carbonizing inks.

Brazil is the main country of origin of carnauba wax. It is obtained from the leaves of the South American palm (*Copernicia cerifera*). Each leaf produces about 5 g of wax which is scraped from the leaves as a powder, melted to remove coarse impurities and cast into blocks.

Carnauba wax consists mainly of esters of long-chain alcohols (C_{24} to C_{34}) with free fatty acids (C_{15} to C_{50},), and small amounts of resins, hydrocarbons, lactides, etc. An approximate composition is: wax acid esters 84%; free fatty acids 3%; hydrocarbons 2%; other materials 10%; moisture 1%.

Miscellaneous natural waxes

Of the various other natural waxes which have occasionally been used in printing inks, a large number have been superseded by the synthetic waxes. They were used for solvent and oil retention as well as surface slip. Although their use has declined in recent years, more interest is now being shown for the future of some natural products. Some waxes from this group are described under the headings below.

Shellac wax

Off-white with a tinge of pink, melting point 72–74°C (162–166°F).

Obtained from the secretion of the lac insect (*Carteria lacca*) in India by de-waxing shellac resin.

Japan wax

Hard pale ivory coloured wax, melting point 45–51°C (113–124°F). Used in letterpress and letterset (dry offset) inks to prevent 'crystallization' and to aid superimposition. Classed as a vegetable tallow extracted from the seeds of a small tree (*Toxicodendron*) cultivated in the Orient.

Candelilla wax

Brownish hard wax, easily powdered, melting point 68–70°C (154–158°F) with good slip properties. Obtained from the stems of *Pedilantus pavonis* and other related euphorbia, which are weeds found in North Mexico.

Lanolin

Anhydrous lanolin is wool grease that varies from almost white to dark brownish yellow. It is a semi-solid plastic sweet-smelling mass with a melting point of 36–42°C (97–108°F), acid value 5–15, saponification value 82–120, iodine value 15–36 and relative density 0.940.

Widely used in the ink industry as a tack reducer in letterpress inks and varnishes. It also helps setting and stops crystallization in wax compounds. A small amount is incorporated in cold carbonizing inks. Ink with pigments of high oil absorption benefit from the use of lanolin for wetting properties and for clean working. It also helps lithographic inks to take up sufficient water to improve their printing properties.

Hydrous lanolin can be produced by kneading anhydrous lanolin with water, until a soft white ointment-like material is obtained containing 25–30% of water.

Lanolin is extracted from the water-insoluble portion of wool grease produced in sheep's wool. Its extraction is an important by-product industry of the wool-producing countries.

The grease is extracted by washing or scouring the wool first with alkali and then with soap solution. The grease is centrifuged to remove dirt or left for the dirt to settle. Finally it is treated with hot water and sulphuric acid to separate the grease as anhydrous lanolin.

Wool grease differs from other natural waxes in containing a high proportion of esters of sterols together with the usual long-chain alcohol esters. An average composition is: cholesteryl and isocholesteryl estolidic esters 33%; free sterols 1%; fatty alcohol acid esters 48%; hydrocarbons 2%; free alcohols 5%; free fatty acids 3%; lactones 6%; other materials 2%.

Section 8: Driers

Catalysts used to promote oxidation of the drying oils used in letterpress and offset lithographic inks. In the presence of these, oxidation proceeds rapidly and the ink films dry hard in a few hours.

The most widely used catalysts are the inorganic salts and metallic soaps of organic acids.

Recently, more complex organic derivatives of metals have been introduced to boost the effect of conventional driers.

Metallic driers are used in two forms:

- as oil-soluble soaps, which comprise the liquid driers;
- as dispersions of inorganic salts in oils, which comprise the paste driers.

4.30 LIQUID DRIERS

Liquid driers are prepared by the conversion of suitable organic acids to their heavy-metal salts and soaps. These are soluble in oils and/or petroleum solvents to form liquids or soft pastes which readily mix with oil in inks.

The metals are almost invariably available as salts of complex fatty acids, either singly or as mixtures. The fatty acids include:

- octoic fatty acids to form octoates;
- rosin fatty acids to form resinates;
- naphthenic fatty acids to form naphthenates;
- tall oil fatty acids to form tallates;
- linseed fatty acids to form linoleates.

The chemical constituents of the organic acid component can influence dry print odour as well as the effectiveness of the drier in an ink formulation at a given percentage of metal. The metal contents range between 3 and 18% except for lead and zirconium which go up to 24%, or even 36%. An addition of between 0.5 and 4% of driers is the amount normally necessary to achieve adequate drying in printing ink films. Factors which affect the efficiency of drying of such films are the surface characteristics and the pH of the paper, stacking procedures and atmospheric conditions. It is a well known fact that inks dry more rapidly at higher temperatures, whereas a high relative humidity slows down the oxidative drying rate, especially when the pH is low.

Cobalt

The most powerful drier, also the most popular (Fig. 4.1). Violet in colour until oxidized when it becomes brown. It tends to discolour tints and whites. It is readily soluble in organic acids, so may be affected by the fountain solution on lithographic presses, by leaching out from the ink into the fountain. Conversely, cobalt acetate can be dissolved and introduced into the fount as a drying activator.

Manganese

Less vigorous catalytic action than cobalt. Brown in colour, affects whites and tints less than cobalt and is unlikely to leach out in the fountain solution. By itself only suitable for tin-printing inks, otherwise it needs

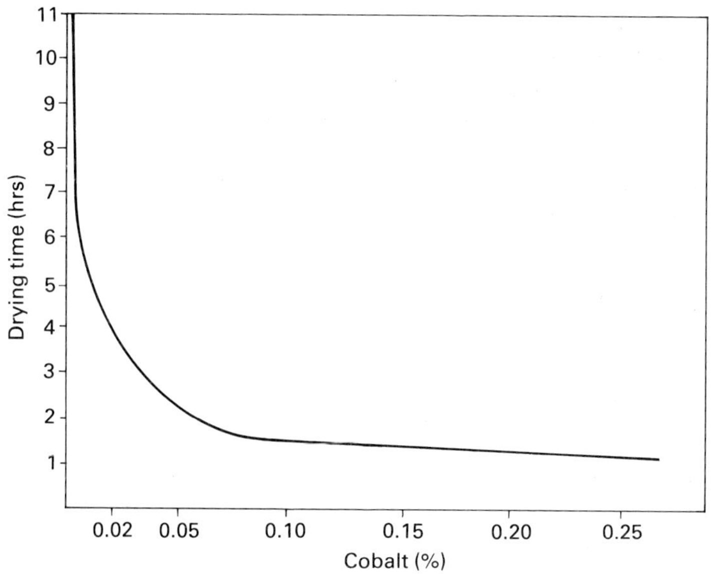

Fig. 4.1 Graph showing the relationship between cobalt content and drying time.

heat to behave as an activator. Often used in conjunction with other metal soaps.

Cerium, zirconium and lithium

Driers of medium efficiency and pale colour. They have replaced lead in many vehicles and are usually used mixed or in conjunction with cobalt or manganese soaps.

Calcium and zinc

Ineffective as catalysts, having only one valency state, and are not used much in printing inks except in some whites. Calcium is used in low-odour systems to modify the catalytic action of transition-metal driers.

Iron

Useful in tung oil varnishes.

4.31 PASTE DRIERS

These are prepared by grinding organic salts of lead and manganese in linseed oil varnishes. Lead acetate and manganese borate are commonly used. It is also possible to prepare pastes by flushing, but most formulae are confidential. The percentages vary but are of the order of 40% lead acetate, 8% manganese borate. Small amounts of soluble manganese driers may also be added, but cobalt is seldom used in paste driers.

The slow-drying characteristics of paste driers give a receptive ink surface for superimposition, therefore they are usually used in multicolour process inks. Cobalt driers are inclined to yield dried prints which are not receptive to overprinting. Paste driers, although they dry inks slowly, do dry uniformly throughout the film.

Inorganic peroxides have been successfully used as components in paste driers, but great care has to be taken in respect of **fire** during their production. These are very useful in lithographic ink, where the acidity of the fount solution, combined with the pH of the vehicle will liberate oxygen which is absorbed along the chain of the polymer.

To obtain the optimum drying effect in oil based inks, it should be noted that some pigments such as the PMTA salts require more driers than normal. Others like bronze blue of phthalocyanine require less, because they accelerate oxidative polymerization. The use of lead in driers is being phased out, being replaced by other metals such as zirconium.

Section 9: Miscellaneous additives

4.32 CHELATING AGENTS

These are chemicals which react with multivalent metal ions, to form a chelate. This is a type of atomic cocoon produced by the formation of a closed ring of atoms which are attached to a central multivalent ion, usually metallic. Normally this phenomenon is due to sharing with the central ion a lone pair of electrons from oxygen or nitrogen atoms in the chelating compound.

Chelating agents are used for locking up unwanted metal ions, to form extremely stable insoluble or soluble chelates. By chemically binding metal ions in additives, or trace metals in compounds, it is possible to improve shelf life, colour, clarity and stability. Chelating agents used in water-based systems are able to improve wetting through softening the water by locking up and rendering inactive the iron, calcium and magnesium salts.

Also by the use of chelates, it is possible to control the release of metal ions over a prolonged period or during a chemical reaction. Chelates have also proved useful in suppressing a reaction until an elevated temperature is reached. They are also used in the ink industry for cross-linking hydroxyl-containing polymers.

Some important chelating agents are ethylenediaminetetra-acetic acid (EDTA) and its sodium salts, nitrilotriacetic acid salts, sodium salts of diethylenetriamine-acetic acid, heptonates, dimethyl glyoxime and its sodium salts, as well as the alkanolamines.

A typical example is the chelation of isopropyl titanate by triethanolamine to form triethalolamine titanate, liberating isopropyl alcohol, a solution in which they are then marketed. The titanium is bonded and when liberated reacts to produce a thixotropic structure in emulsion systems.

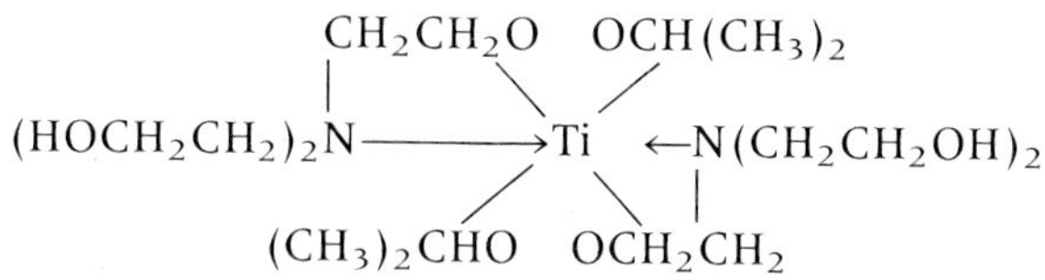

4.33 ANTI-OXIDANTS

Anti-oxidant molecules react with free radicals formed during the autoxidation process and continue to act until all the anti-oxidant molecules are exhausted. Consequently, any anti-oxidant incorporated into an ink will delay the initiation of oxidative polymerization drying.

The most common anti-oxidants are the naphthols, substituted phenols, a variety of oximes and the aromatic amines. Quantities necessary are only a fraction of 1% of the final formulation.

Press stability of lithographic tin-printing inks is usually improved by the incorporation of a small quantity of anti-oxidant which is lost during the initial part of the stoving schedule. The oxidation process then proceeds normally.

Numerous anti-oxidants are marketed for industry and many of them bear manufacturers' trade names. Some of the most effective are:

eugenol (extract of oil of cloves)
hydroquinone
pyrocatechol
guaiacol (methyl catechol)
butylated hydroxytoluene
butylated hydroxyanisole
methyl ethyl ketoxime
butylaldoxime
cyclohexanone oxime

Typical symbolic formulae are:

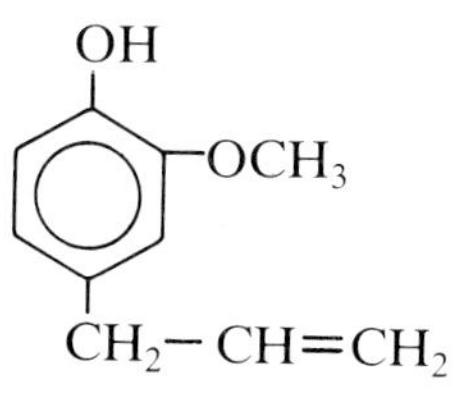

Eugenol

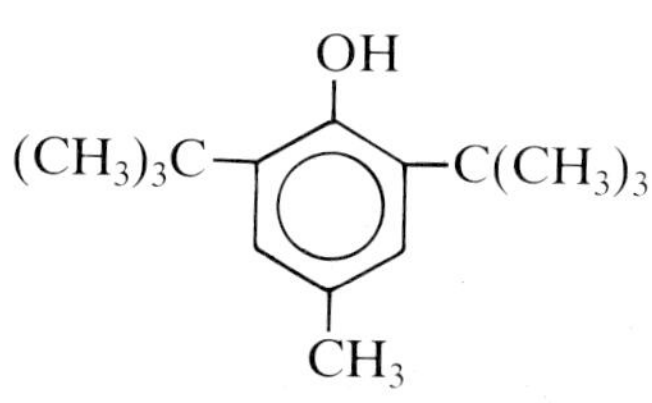

Butylated hydroxytoluene

4.34 SURFACTANTS

The name surfactant is given to substances which have the property of being adsorbed at surfaces or interfaces, thereby reducing the surface or interfacial tension. Soaps and detergents act this way in aqueous systems.

This action can be measured and demonstrated by taking the contact angle (xyz) of a drop of pure water on a non-absorbent surface (Fig. 4.2 (a)). The addition of a very minute quantity of surfactant decreases the angle of contact considerably, at the same time increasing the interfacial area (Fig. 4.2 (b)).

The same concept of wetting and spreading applies both to oleo resinous and to aqueous systems. A contact angle less than 90° enables the vehicle to wet and disperse pigments. Wetting greatly influences the physical and printing characteristics of an ink. Numerous surfactants are available, many of them with trade names. The choice is large so it is left to the ink formulator to select the best surfactant, and to include in his formulation the optimum amount to obtain the highest tinctorial strength from pigments. Incorrect use of surfactants can result in decreased adhesion to some substrates, and can lower the water resistance of a print. Surfactants can promote emulsions; therefore care must be taken that they do not also cause emulsification of lithographic inks. Selection must be made with knowledge of the properties of the surfactant, the formulation objectives and the precise amount needed to attain those objectives efficiently. This approach discourages indiscriminate use which can be harmful and costly.

Surfactants normally contain polar and non-polar groupings in the same molecule. It is the ionic or active portion of the molecule which

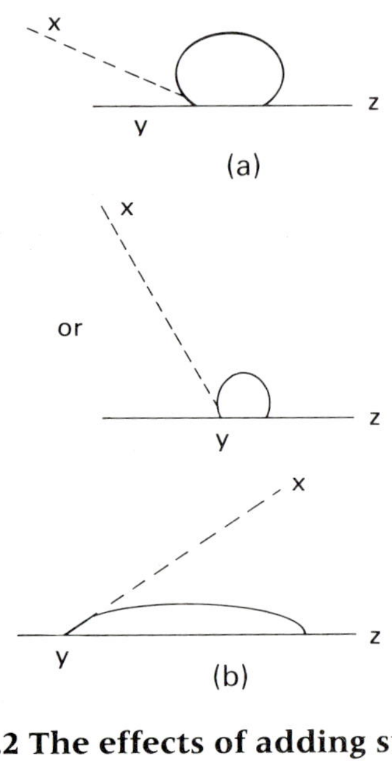

Fig. 4.2 The effects of adding surfactants.

divides surfactants into anionic (which have an active negative charge), cationic (where the positive charge predominates) and non-ionic classes. Amphoteric surfactants have both positive and negative centres of charge (Table 4.17).

Anionic surfactants

Active constituent – anions (negative charge). These attract the positive charge of other large molecules, leaving a negative charge at the surface, causing excellent wetting in aqueous systems. Anionic surfactants include common soap and are usually alkali metal, ammonium or substituted ammonium salts of long-chain fatty acids. The non-polar portion is the fatty acid hydrocarbon chain, while the carboxylate anion and its associated cation are the polar portion. Thus, when a polar pigment is dispersed in a non-polar vehicle, the polar portion becomes adsorbed on the surface of the pigment as a mono-molecular layer, giving the pigment a negative charge, and leaving the non-polar part pointing outwards into the vehicle, to which it is attracted. In this way interfacial tension is lowered, so better wetting and hence dispersion of the pigment is achieved. An electrical double layer around the pigment particle also serves to reduce flocculation tendencies.

Cationic surfactants

The reverse of the above, i.e. positive charge attracting negative charge. Ideal for dispersing a pigment in a non-agueous vehicle, the pigment assumes a positive charge and disperses well in non-aqueous systems. Cationic surfactants include quaternary fatty ammonium halides, acetates or sulphates.

Non-ionic surfactants

These have no residual electrical charge. Molecules in which part is lyophilic (solvent-liking) and part is lyophobic can behave as surfactants although they are non-ionic. To emulsify fats in water, a lyophilic (hydrophilic) polar polyethylene oxide chain is joined with a hydrocarbon chain which associates with the fat.

Amphoteric surfactants

As these contain central and balanced positive and negative charges, they exhibit behaviour of either anionic or cationic surfactants according to prevailing conditions.

The concept of hydrophile–lipophile balance (HLB) is of practical importance and HLB values may be assigned by a simple test.

Chemical formulae of some well established surfactants are given in Table 4.17.

For a more detailed description of surfactants the reader is advised to consult a standard text book on the subject.

4.35 DEODORANTS AND REODORANTS

Some of the materials used in the manufacture of printing inks have a strong odour. Some can be extremely unpleasant, causing nausea. Highly volatile solvents are probably the worst offenders for producing unwanted odour, but even high-boiling printing ink distillate odours can be very disagreeable on a printing machine which has become warm. Other materials have odours which may not be strong, but which are persistent, for example, some of the styrene and acrylic monomers. The volatile byproducts of oxidation drying or of prolonged storage, also give rise to unwanted odour if trapped within printed matter. It is required to mask these unwanted odours, reduce their intensity and convey a clean, healthy and pleasant fragrance without altering the efficient functioning of the ink or having adverse effects on the printed products.

Reducers and proprietary sundries such as blanket washes, sometimes contain specially added reodorant compounds. These are intended to make the use of the materials pleasant, to enhance sales appeal, and to disguise the ingredients, thereby adding security to the formulations.

A large range of natural and synthetic chemicals have, for many years, been used in other industries, notably food and cosmetics to mask, alter or intensify the odours of the product. Among these are included amyl and methyl salicylate, vanillin and blended derivates of essential oils.

It should be noted, however, that few deodorants or reodorants are as persistent as the odour to be masked, therefore, on ageing, it is possible for a formulation to revert to its original aroma.

Some manufacturers of products increase their sales by inclusion of a reodorant in the ink on the package in such a way that it conveys to the purchaser a desire to buy. Compounds of lavender, leather, peppermint, carnation, antiseptic, cedarwood, citron, vanillin, etc., are available.

Micro-encapsulated perfumes are sometimes used as paper coatings. When scratched these pervade the surrounding area with a fairly concentrated fragrance. They have been incorporated into inks and overprint varnishes but further development work is necessary, especially on application, so that the odours are not released prematurely.

4.36 PURE CHEMICALS

The main classes of pure chemicals used in printing ink formulations today are:

- alkalis
- organic acids and acid anhydrides
- polyols.

The most common members of each class are described below together with some relevant physical properties and their applications.

Alkalis

The major use of alkalis in printing inks is in the formulation of water-based or water-reducible letterpress, gravure, and flexographic inks. They act simply by converting the highly acidic and fairly insoluble resins such as shellac, fumaric, and maleic resins to salts which are very much more hydrophilic and soluble than the parent resin. Excellent varnishes can thus be prepared from these resin salts using water or one of the lower, more hydrophilic glycols, especially ethylene glycol, as the principal solvents.

The quantity and type of alkali used has to be calculated so that the resulting dried ink film does not remain too water soluble. The alkali used normally evaporates or penetrates into the stock leaving the original insoluble resin system on the surface as film former.

It is also possible to use alkalis to alter the complete physical structure of a vehicle. Polymers and copolymers containing acrylic and methacrylic acids are produced in high-solids emulsion form at a very low viscosity. By raising the pH and neutralizing, these resins become water soluble; at the same time the particles become ionized, and form a completely different configuration due to forces of repulsion, thereby increasing the viscosity.

These high-solids emulsions can be of great assistance to the ink formulator.

The most commonly used alkalis, together with their chemical formula, boiling point, and miscellaneous points of interest are given in Table 4.18.

Organic acids and acid anhydrides

These are used, in the main, as raw materials for resin preparation.

A list of the common acids and acid anhydrides, together with their chemical formula, melting point, miscellaneous comments and application is given in Table 4.19.

Polyols

The most common polyols used in the manufacture of synthetic vehicles for the printing ink industry (Chapter 9) are:

Glycerol

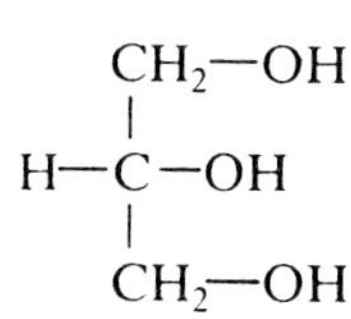

Neopentyl glycol

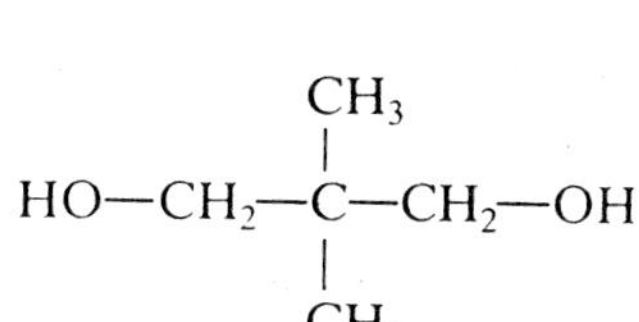

Trimethylol propane

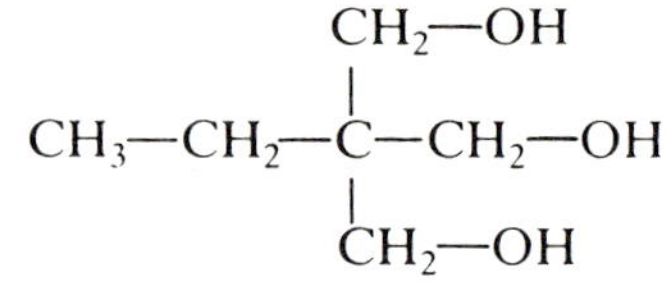

Trimethylol ethane

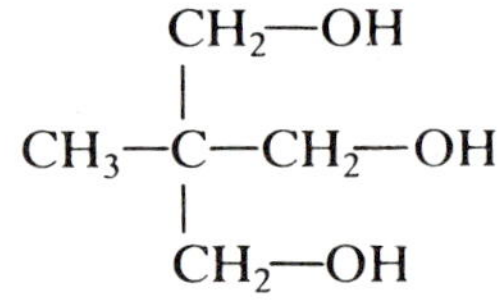

Pentaerythritol

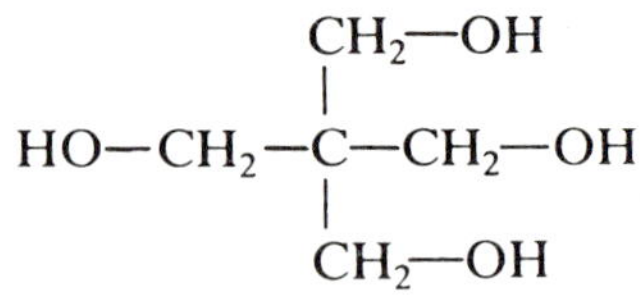

4.37 DEFOAMING AGENTS

Foaming is a condition which might happen when a monomolecular film of a surfactant which is fairly insoluble in the bulk of the liquid is present on the surface of the liquid. Elasticity of the liquid surface can cause foaming with a detrimental effect, if not controlled prior to printing. Ideally foaming should be inhibited by formulating to ensure that the surface tension is kept low. This is often difficult because of the surfactants used in the formulation of inks or in the ingredients used to make them. If foaming does occur it increases rapidly with agitation. Prevention is better than cure but where foaming does occur it should be dealt with as soon as possible.

Defoaming agents act in one of two ways.

- By being a good solvent for the surfactant – this causes the monomolecular film, so essential to foaming, to be dispersed into the bulk of the liquid. When this happens the tendency towards foaming is greatly reduced.
- By adding a small quantity of a material which is immiscible with the system and which drastically lowers the surface tension of the system – this reduction in surface tension causes existing bubbles to burst and creates a physical condition such that stable bubbles and foams cannot form.

This latter type of defoamer is nearly always added in a form which makes it readily dispersible over the surface, for example as a liquid or an emulsion.

There are many defoaming agents and they are usually classified as silicone or non-silicone, the former being the most popular type. Silicone fluids and emulsions are often used as defoaming agents. There are fluids for non-aqueous media and for emulsions ranging from 10 to 30% silicone in water-based systems.

The silicone fluids most commonly used in the printing ink industry are of the polydimethyl siloxane type having the following structure:

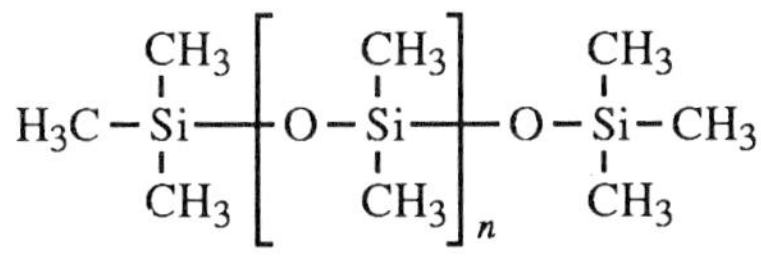

The chain length n may be from 1 to 2000 and the viscosity of the oil will be dependent upon this. When $n = 2$, viscosity will be 0.65 centistokes at 25°C, whereas when $n = 2000$ the viscosity $= 1 \times 10^6$ centistokes (i.e. 10 000 stokes) at 25°C.

Antifoam agents must be used with discretion as it is easy to overestimate the amount necessary and to initiate other printing faults such as pinholing or reticulation.

4.38 LAKING AGENTS

In both water-based and ethanol-based flexographic dye inks, laking agents are needed to stabilize the basic dye in solution.

In ethanol inks the laking agent is usually a tannic acid derivative and combination with an acidic resin like shellac or maleic can ensure a stable ink system. Properly formulated dye inks can give increased colour strength and more surface 'sit up' with good resistance to wax and water/steam bleed.

Tannic acid

The main laking agent used for water-reducible flexographic dye inks. Now becoming expensive as its production from the natural sources in remote Third World countries is gradually dropping. This scarcity has pushed ink companies to look even harder for a substitute.

Tannic acid substitutes

A number of synthetic chemicals are now available as alternatives to tannic acid. These are cheaper but not quite as effective as pure tannic acid. They are, however more easily soluble in alcohols.

They have gained wide recognition for commercial reasons of cost and availability. They tend to be used in the alcohol/alcohol:water basic dye systems for flexographic work. Mixed with acidic resins, the inks are strong, very bright and will penetrate less into the substrate. The amount of laking agent used **must** be carefully checked in each formula to ensure the resistance properties required for that formula are attained. Basic dye solutions can be made in small concentrations with acidic resins alone but this should be checked for properties carefully as small quantities of laking agents will always improve the fastness properties.

Section 10: Raw materials for radiation curing systems

One of the major growth areas of recent years, especially in the paste ink field, has been in radiation curing inks. These are inks, and clear lacquers whose components react when exposed to UV light, or when passed through an electron beam, to cure instantly to a solid polymer. This is achieved via a free-radical reaction. In UV curing inks this is started by photo-initiators, in electron beam curing by the electrons themselves. The composition of the inks is described in Chapter 11.

Radiation curing inks are basically formulated in the same way as any other ink – that is to say, they are composed of pigment, binders, diluents and additives necessary for specific applications. In these inks the binders are generally acrylates of some sort. The diluents are also acrylates and non-volatile. Pigments are as used elsewhere in printing inks, although careful selection and testing is necessary. The additives in addition to the normal waxes and surfactants, include photo-initiators and photo-activators in the case of UV inks, inhibitors and in some cases additives to improve printability. The additives, of course, especially the amines used as photoactivators, must be carefully selected to be insoluble in water for lithographic inks and varnishes.

A brief discussion of each class of component follows. For a more detailed examination of the subject there are now several excellent textbooks available (see Chapter 11).

4.39 PIGMENT SELECTION

Many of the pigments used in normal lithographic inks are suitable for use in UV systems. It is, however, essential to test each pigment very thoroughly indeed. Many pigments will react with the vehicle to give poor storage stability. Others may slow the cure unacceptably, while still others may change shade or burn out completely.

A reasonable guide to dark storage stability is an accelerated ageing test in an oven at 50 or 60°C. When compared with an ink of known stability the time taken to polymerize will allow a fair estimate of the shelf life of the ink to be made. Unfortunately it cannot be assumed that all pigments of the same chemical class are alike in stability since different manufacturers may apply different surface treatments, or even different degrees of washing. Even batches from the same supplier may vary.

Metallic inks containing bronze and aluminium powders should be avoided since gellation may occur rapidly, if not explosively, and careful formulation is necessary to avoid this.

Many manufacturers now supply lists of their pigments which have been found to be suitable in UV systems. These should be treated as a guide but full tests should still be carried out.

Cure rate can be very dependent on the pigment used. This is particularly true of black inks, and of other dark colours. Carbon black not only absorbs UV light, but also acts as a free-radical scavenger. It can therefore have an inhibiting effect out of all proportion to its concentration.

In general, to avoid generating too much heat in the dispersion stage, pigments should be selected for ease of dispersion, although those which have been surface treated to enhance this should be examined closely, in case this treatment interferes with the stability.

4.40 PREPOLYMERS

The resins currently used as binders in radiation curing inks and lacquers are predominantly acrylate esters. Those commonly available include epoxy polyester and urethane acrylates, together with oil-modified versions of these. They are used singly or in combination as the binder in inks or as the high-viscosity resin component of a lacquer or varnish. Since the undiluted form may be at too high a viscosity for convenient use many are supplied already let down in an appropriate monomer.

Epoxy acrylates

These are used widely in inks and lacquers for most applications. They are the acrylated form of bisphenol A diglycidyl ether and its derivatives. They are relatively cheap, have a good cure rate and display good gloss, chemical resistance and reasonable adhesion. They tend to brittleness, especially in lacquers, and are often combined with a more flexible component. Oil-modified versions have been supplied for pigment-wetting purposes, soya bean oil being a popular modification. This can, however, slow the cure rate, and leave ink films with an aftertack which remains permanently. Again, therefore, they tend to be used in combination with other prepolymers rather than singly.

Polyester acrylates and unsaturated polyesters

The term polyester covers a wide range of chemicals, and a correspondingly wide range of polyester acrylates are available. This variety makes it possible to select fairly accurately the desired properties. Many monomers fall into this category.

Unsaturated polyesters are typically cheaper products, and have tended to be used in applications such as wood finishing. Some more specialized products are finding limited use in both inks and varnishes, however.

Urethane acrylates

These are suitable for most applications, and are characterized by better adhesion and greater flexibility than epoxy acrylates or cheap polyesters but have a correspondingly high price. Again a wide range, both aliphatic and aromatic are available.

Recent requirements for lower odour and taint levels have led to 'stripped' versions of these monomers becoming available. These do have a much lower odour level, but as a consequence of more processing, a higher cost (Table 4.20).

4.41 REACTIVE DILUENTS

The selection of appropriate monomers is governed by the functionality of the monomer, its cutting power, and its health and safety aspects. Many of the irritation problems associated with early UV formulations were caused by the use of irritant monomers, which at the same time were all that was available. It is an unfortunate fact that many of the most efficient diluents in terms of reactivity and adhesion have turned out to be among the more irritant. The monomers dealt with below meet the SBPIM specification for a Draize figure lower than 3.5, although this can vary with the individual supplier.

Perhaps the most common monomer, certainly in those approved by the SBPIM is tripropylene glycol diacrylate (TPGDA). It is used as the major component in roller-coating lacquers, at a reduced percentage in more viscous clears, and as a tack reducer in inks. Attention should be paid in inks to its lithographic properties.

Dianol diacrylate, or its analogues (dianol is a registered trade name of Akzo Chemie), have a much higher viscosity than TPGDA. They do have a very low Draize rating, and excellent film-forming and plasticizing properties.

In those areas where health and safety are not such important considerations 1,6-hexanediol diacrylate (HDDA) is perhaps the best difunctional monomer available for adhesion and reactivity. It has a high Draize figure, however, which precludes its use in the UK in printing inks, although it, and some other monomers which are suspect for health and safety reasons, are still used in other applications.

Trimethylol propane triacrylate (TMPTA) has been the most popular trifunctional monomer until recently when it too was placed on the SBPIM exclusion list. It has good lithographic behaviour, a high reactivity and a reasonable viscosity. Recent developments in monomers have been aimed at finding a less irritant substitute with similar properties. Of these the most promising are the ethoxylated or propoxylated versions of TMPTA or glyceryl propoxy triacrylate (GPTA). Of similar viscosity to TMPTA these have lower Draize values, but reduced reactivity, or poorer lithographic behaviour, or a higher cost, but are generally acceptable substitutes. All these triacrylates convey a degree of brittleness to a lacquer film because of their high cross-linking ability.

Tetra-acrylates are rarely used, but again suffer from brittleness (see Table 4.21).

4.42 PHOTO-INITIATORS

In UV curing inks these are chemicals which become excited by UV radiations, forming free radicals, which in turn react with the vehicle of the ink or varnish to begin polymerization. The term is also used to include proton donors such as amines. The basic mechanism of photo-initiation and photochemistry in general is well documented, and can be found in many standard textbooks.

Benzophenone is one of the cheapest, most widely used photo-initiators. It requires a proton donor, such as amine, to be present in order to yield radicals easily. This synergism, two products being much more efficient than either one alone, is a common feature of photo-initiation in UV inks. Benzophenone is readily soluble and is itself a solvent, and so is useful as a viscosity modifier in both inks and lacquers. Its main drawback is its odour, which, while not unpleasant, is very strong. Not all the benzophenone content of an ink is used up in a reaction, and of what is used some readily combines to form other products with high odour levels.

Much work was done using benzophenone in combination with Michler's ketone (4,4-bis dimethylamino benzophenone). This proved a most effective system, but the health risks associated with Michler's ketone have caused it to be withdrawn from use, and as yet no effective substitute has been found.

Several other photo-initiators work by a process of hydrogen abstraction, requiring a proton donor for effective initiation. These include benzil, derivatives of benzophenone and a number of thioxanthones, of which isopropyl thioxanthone finds the widest use. These do tend to be more expensive than benzophenone to achieve the same cure rate.

Other photo-initiators operate without the need of a proton donor. These simply undergo fragmentation into reactive species under UV light. Those commercially available at the time of writing which go under a variety of trade names are mostly benzil ketals or acetophenone derivatives. Again breakdown products of some of these can cause odour problems, although these are less severe than those with benzophenone.

A wide range of amines, and acrylated amines are available as proton donors for UV systems.

The combination of photo-initiators should be carefuly chosen to provide the best cure rate at the minimum cost. It is possible by choosing the wrong combination to cause bloom to appear on lacquers, or to achieve a brittle surface cure. Often after a certain cure rate is achieved no further addition of photo-initiator will improve it. It may be convenient to make compounds of initiators as liquids for easier ink manufacture (see Tables 4.22–4.24).

4.43 ADDITIVES AND INHIBITORS

It is necessary to incorporate a small amount of inhibitor into radiation cured inks and lacquers to guard against premature gellation. Hydroquinone and its derivatives are commonly used for this purpose at very low percentages.

The shelf life of inks can be considerably extended, even when adequately inhibited, by careful processing and storage. Inks exposed to high temperatures even for a short time may start to gel, and the reaction will continue from there. Containers should have an air space left at the top since oxygen helps inhibit these reactions.

Surfactants are much the same as those used in ordinary inks and lacquers, although some specialized acrylated products are available. As with all the other components each should be checked in the system in which it is to be used to establish compatibility, and any effect on shelf life. Some will also degrade rapidly under UV light.

Tables 4.20–4.24 summarize the specialist raw materials used in radiation curing systems.

Section 11: Health and safety at work

For many years, the British printing ink industry has taken the lead in establishing standards for freedom from toxic materials in inks for food packaging use. A totally different facet of safety has, however, developed with the introduction of the Health and Safety at Work Act, etc. in 1974.

The Act places two duties on manufacturers:

- to ensure as far as is practicable that the products are safe in use;
- to provide information on how the products may be used with safety.

At the same time, an employer is required to ensure that working conditions are free from hazard and that employees are suitably trained. Finally, the employee has a responsibility to act in a responsible manner.

It may be argued that these practices have been adopted by all conscientious employers and manufacturers for many years. In common with all legislation, this Act is designed to bring the unsatisfactory few up to the standards of the majority. It is doubtful whether such an Act is necessary in the printing ink industry. However, it is now law and a good deal of time and money is being spent to prove that the manufacturer is observing the Act.

The practical effects are largely as follows.

- When any new raw material is being used in inks or coatings for the first time, its effect on the human body should be investigated. Good examples are the many new materials being introduced in the UV curing inks.
- If it is thought that any hazard may exist, special handling advice must be given.
- General advice on the recommended methods of handling printing inks should be given to all ink users so that they and any new-comers to the industry can be suitably instructed.

One important change has taken place, although not specifically prompted by this Act, the removal of lead compounds from printing inks. A programme of action has been agreed by the Society of British Printing Ink Manufacturers which resulted in the discontinuation of the use of lead driers and lead-based dry colours from July 1978.

Current legislation is covered in detail in Chapter 16.

BIBLIOGRAPHY

Society of Dyers and Colourists (1982) *Colour Index*, Society of Dyers and Colourists, Bradford.

Table 4.1 Yellow pigments

CI pigment yellow	CI formula	CAS No.	Designation	Shade
1	No 11680	2512-29-0	Arylide G	Warm yellow
3	No 11710	6486-23-3	Arylide 10G	Greenish yellow
4	No 11665	1657-16-5	Arylide 13G	Greenest yellow
5	No 21906	4106-67-6	Arylide 5G	Greenish yellow
65	No 11740	6528-34-3	Arylide RN	Reddish yellow
73	No 11738	13515-40-7	Arylide GX	Bright yellow
74	No 11741	6358-31-2	Arylide 5GX	Strong lemon yellow
97	No 11767	12225-18-2	Arylide FGL	Bright yellow
111	No 11745	69771-45-5	Arylide F4G	Mid shade G/S yellow

Table 4.2 Azo pigments formed by coupling the following components to give a,b,c,x,y, or z substituents

CI pigment yellow	Description	First component	Second component	Structural variants					
				x	y	z	a	b	c
1	G	*m*-nitro- *o*-toluidine	acetoacetanilide	CH_3	NO_2				
3	10G	*p*-chloro- 2-nitroaniline	acetoacet-*o*- chloroanilide	Cl	NO_2	Cl			
4	13G	*p*-nitroaniline	acetoacetanilide	NO_2.					
5	5G	*o*-nitroaniline	acetoacetanilide		NO_2				
65	RN	*m*-nitro- *p*-anisidene	*o*-acetoacetanisidide	OCH_3	NO_2	OCH_3			
73	GX	*p*-chloro- 2-nitroaniline	acetoacet-*o*-anisidide	Cl	NO_2	OCH_3			
74	5GX	*p*-nitro-*o*- anisidine	acetoacet-*o*-anisidide	NO_2	OCH_3	OCH_3			
97	FGL	*p*-amino-2, 5, -direthoxy- sulfonanilide	*p*-chloro-2′5′- direthoxy- acetoacetanilide	SO_2NNPh	OCH_3	OCH_3	Cl	OCH_3	OCH
111	F4G	*p*-nitro- *o*-anisidine	acetoacet-2- methoxy- 5-chloroanilidide		OCH_3	OCH_3	Cl	NO_2	

Table 4.3 Diarylide yellow pigments

CI pigment yellow	CI formula	CAS No.	Name	Shade		Transparency	Light fastness
				Masstone	Undertone		
12	No 21090	6358-85-6	Diarylide AAA	Warm–medium	Mid yellow	Transparent	Fair
13	No 21100	5102-83-0	Diarylide AAMX	Medium	Mid yellow	Transparent to very transparent	V. good
14	No 21095	5468-75-7	Diarylide AAOT	Medium	Greenish	Opaque	Good
17	No 21105	4531-49-1	Diarylide AAOA	Lemon	Green	Very transparent	V. good
55	No 21096	6358-37-8	Diarylide AAPT	Medium reddish	Warm	Opaque	V. good
83	No 21108	5567-15-7	Diarylide AADMC	Red shade	Warmest	Transparent	Excellent

Table 4.4 Diarylide yellow variants

CI pigment	Ref	Coupling component	Structural variants			
			w	x	y	z
12	AAA	Acetoacetanilide	–	–	–	–
13	AAMX	2,4-acetoacetoxylidide	CH_3	–	CH_3	–
14	AAOP	*O*-aceto-aceto toluidide	CH_3	–	–	–
17	AADA	*o*-acetoacetanisidide	OCH_3	–	–	–
55	AAPT	*p*-acetoacet-*o*-toluidide	–	–	CH_3	–
83	AADMC	4-chloro-2′5′-dimethoxy-acetoacetanilide	–	OCH_3	Cl	OCH_3

Table 4.5 Average Properties for Tests to BS 4321:1969

Colour Index Pigment No.	Name	Bulk value	Conc. paste (%)	Lightfastness Blue Wool		Water	Meths 74 op	Mixed solvent	Alkali	Soap	Detergent	Cheese	Fats and Oil	Waxes	Special comments
				Strong	Tint										
Yellow 1	Arylide G	65	45	7	5–6	5	3–4	2–3	5	4–5	5	5	3–4	3–4	
Yellow 3	Arylide 10G	65	45	7	6	5	3–4	3	5	4	5	5	3–4	2	
Yellow 4	Arylide 13G	65	45	7	3–4	5	2–3	2	5	3–4	4	5	2	2–3	
Yellow 5	Arylide 5G	65	45	7	6–7	5	2	2–3	5	3	4–5	5	3–4	2	
Yellow 12	Diarylide AAA	72	40	3–4	2	5	5	4–5	5	5	5	5	5	5	
Yellow 13	Diarylide AAMX	80	40	5–6	4–5	5	5	4–5	5	5	5	5	5	4–5	
Yellow 14	Diarylide AAOT	69	40	4–5	3	5	5	4–5	5	5	5	5	5	5	
Yellow 17	Diarylide AAOA	77	37	6	4–5	5	5	4	5	5	5	5	5	5	
Yellow 55	Diarylide AAPT	70	45	6	4	5	5	5	5	5	5	5	5	5	
Yellow 61	Diarylide AAA	63	91	7	6	5	5	4–5	2	2	5	5	5	5	
Yellow 65	Arylide Yellow RN	62	48	7	3–4	5	2	2	5	4	4	3–4	3–4	2	
Yellow 73	Arylide Yellow GX	60	50	7	4	1–5	3–4	2	5	4	4	3–4	3–4	2	
Yellow 74	Arylide 5GX	70	42	6–7	5	5	4–5	2	5	4–5	5	5	3	3–4	
Yellow 81	Diazo H1OG	71		6	5	5	4	5	5	5	5	5	5	4	
Yellow 83	Diarylide AADMC	72	40	6	3	5	5	4–5	4–5	5	5	5	5	5	
Yellow 95	Azocondensation GR	71	90	7	6	5	5	5	5	5	5	5	5	5	
Yellow 97	Arylide Yellow FGL	52	56	7	3	5	4	4	5	5	5	5	5	4	
Yellow 101	Brilliant BGF	42	55	1–2	1	5	4–5	4	4	5	5	5	2	3	
Yellow 106	Diazo 2GRL	–	–	6–7	4–5	5	5	5	5	5	5	5	5	5	
Yellow 111	Arylide Yellow F4G	40	65	7	4–5	5	4	4	5	5	5	4	4	4	
Yellow 120	Benzimidazolone H_2G	68	34	7	5	5	4	5	5	5	5	5	5	4	
Yellow 126	Diazo DGR	71.4	–	3	1	5	4	5	5	5	5	–	4	3	
Yellow 127	Diazo GRL	70	–	4	2	5	4	5	4	5	5	–	4	5	
Yellow 137	Isoindolinone GRO	–	–	8	7	5	5	5	5	5	5	5	5	5	

Yellow 138	Quinophthalone	–	–	7–8	7	5	5	5	5	5	5	5	5	5	
Yellow 139	Isoindoline 1820	–	–	7–8	6–7	5	5	5	1–4	3–4	4	5	5	5	
Yellow 151	Benzimidazolone H4G	–	–	7	5	5	4	4	3	4	4	5	5	4	
Yellow 175	Benzimidazolone H6G	–	–	8	5	5	5	–	5	5	5	5	5	5	
Yellow 185	Isoindoline 1155	–	–	7	6	5	5	5	2	3	3	5	5	5	
Yellow 189	Isoindoline 1819	–	–	7	6	5	5	5	5	5	5	5	5	5	
Orange 5	Dinitroaniline	67	45	6	4	5	4–5	3	4–5	4	4–5	4–5	4	4–5	
Orange 13	Benzidine	68	40	5	3–4	5	4–5	3–4	5	5	5	5	5	5	
Orange 16	Dianisidine	84	30	5–6	2–3	5	5	5	5	5	5	5	5	5	
Orange 34	Fast F2G	70	40	6–7	4–5	5	4–5	4–5	5	5	5	5	5	5	
Orange 36	Benzimidazolone HL	640	32	7	6–7	5	5	5	5	5	5	5	5	5	
Orange 46	Ethyl Lake Red C	55	45	3–5	2–3	3–4	2–3	3	2	5	5	5	5	3–4	Food and toy restrictions
Orange 43	Perinone	53	–	7	7	5	5	5	5	5	5	5	5	5	
Orange 48	Quinacridone Gold	75	30	7–8	7–8	5	5	5	5	5	5	5	5	5	
Orange 51	Pyranthrone	62	40	7–8	7–8	5	4	4	5	5	5	5	5	5	
Orange 60	Benzimidazolone HGL	75	30	7–8	7–8	5	5	5	5	5	5	5	5	5	
Orange 61	Tetrachloro-iso-indolinone	40	48	8	7	5	5	5	5	5	5	5	5	5	
Orange 65	Azomethine	–	–	8	7	4	2	2	5	5	5	5	5	3	
Red 1	Para	70	45	7	2–3	1–2	2	1–2	2	2	2	2	2	1–2	
Red 2	Naphthol F2R	65	45	6–7	4–5	5	4–5	4	5	4–5	5	4	2	3	
Red 3	Toluidine	68	45	7	4	5	2–3	1–2	5	3–4	5	4–5	3	2	
Red 4	Perm Red R	64	45	5–6	3–4	5	4	2	4–5	4–5	5	3	3	2–3	
Red 5	Carmine B	77	35	7	5–6	5	4	1–2	5	4–5	5	5	4–5	4–5	
Red 8	Naphthol F4R	71	40	5	3	4–5	4–5	3–4	4–5	4–5	5	5	4–5	4–5	
Red 9	Naphthol LF	67	45	6–7	6	5	5	1	4–5	3	5	5	3	2–3	
Red 10	Perm Red FRL	67	45	6–7	4–5	5	5	2–3	5	3–4	5	5	3	3–4	
Red 12	Bordeaux F4R	77	36	5–6	4–5	5	5	2–3	4–5	4	5	3–4	4–5	3–4	
Red 17	Naphthol-mid	74	40	5–6	1	5	1–2	1–2	5	3–4	3–4	1–2	1–2	1–2	Toy restrictions
Red 22	Naphthol YS	68	34	6	1	5	1–2	1–2	4	3–4	3–4	2–3	3–4	1–2	

Red 23	Naphthol Dark	83	30	5–6	3	5	1–2	1–2	3–4	5	5	5	5	5	
Red 48:1	Barium 2B	52	48	3–4	1–2	4–5	5	4–5	2	2	4	5	5	5	Food and toy restrictions
Red 48:2	Calcium 2B	62	45	5–6	4	5	5	3–4	2–3	1–2	4	5	5	5	
Red 48:3	Strontium 2B	50	45	3–4	3	4–5	5	5	2–3	2–3	3–4	4–5	4–5	5	
Red 48:4	Manganese 2B	63	45	6–7	5–6	5	5	4–5	2–3	2–3	4	5	4–5	5	Food and toy restrictions
Red 49	Sodium Lithol	66	45	2–3	2–3	4–5	3–4	2–3	3–4	2–3	3–4	3–4	4	5	
Red 49:1	Barium Lithol	67	45	2	1–2	5	4–5	3–4	4	2–3	1–2	3–4	3–4	4–5	Food and toy restrictions
Red 49:2	Calcium Lithol	64	45	2	2	5	4	4	2–3	2	3–4	3	4–5	3	
Red 52:1	BON Ca	67	40	3–4	3	3–4	4–5	4–5	2–3	1	2–3	4–5	4–5	5	Good flow
Red 53:1	Lake Red 'C'	57	45	3	2	5	5	4	2	1–2	3	3–4	4	4	Food and toy restrictions
Red 57:1	Rubine 4B	68	42	3	2	4	5	4–5	2	1–2	4	4	4	4	
Red 60	Scarlet 3B	45	55	4–5D	4D	5	5	4–5	1	1–2	4	1	5	4–5	Darkens
Red 63	BON Maroon	67	45	5–6	3	5	3	3	2	2	3	4	5	3–4	
Red 81:1	PMTA Pink	47	45	5–68D	3–4BD	4	3	2–3	4	2–3	3	4	4	4	Blues darken
Red 88	Thioindigo Red–Violet	58	40	5	4	5	5	5	5	5	5	5	5	5	
Red 90	Phloxine	40	48	2–3	1–2	3	3–4	4	1	1	2–3	3–4	4	4	Food and toy restrictions
Red 112	Bon Arylamide	66	45	6–7	6	5	4–5	3–4	5	5	5	5	4	5	
Red 122	Quinacridone	60	45	7–8	7	5	5	5	5	5	5	5	5	5	
Red 146	Carmine FBB	70	45	6	4	5	3–4	3–4	5	5	5	5	5	5	
Red 166	Azo Scarlet RN			7	7	5	5	5	5	5	5	5	5	5	
Red 169	Cu Fe Magenta	67	40	5	2–3	4	1–2	3	2	1–2	2–3	2–3	4–5	5	
Red 170	Naphthol F5RK	81	32	7	4	5	3–4	3–4	5	5	5	5	5	5	
Red 176	Carmine HF3C	70	36	6–7	5–6	5	3–4	4	5	5	5	5	5	3–4	
Red 209	Quinacridone 2Y	69	46	7	7	5	5	5	5	5	5	5	5	5	
Red 242	Monoazo Scarlet 4R	–	55	8	8	5	5	5	5	5	5	5	5	5	
Red 253	Monoazo 2GLD	–	–	8	8	5	4	4	5	5	5	5	5	5	
Red 166	Azo Scarlet RN	–	–	7	7	5	5	5	5	5	5	5	5	5	
Red 184	Naphthol F6B	85	26	6	4	5	3–4	3–4	5	5	3–4	3–4	3–4	3–4	
Red 185B	Carmine HF4C	65	40	6	5	5	3–4	5	5	5	5	5	5	5	

Red 200	Rubine 6B	40	50	4–6	4	3	1–2	1–2	2–3	1–2	1–2	4–5	5	5	
Red 202	Quinacridone	40	45	7–8	6–7	5	5	3–4	5	5	5	5	5	5	
Red 208	Benzimidazolene HF2B	65	34	6–7	5–6	5	3–4	4	5	5	4–5	5	5	5	
Red 209	Quinacridone 2Y	69	46	8	8	5	5	5	5	5	5	5	5	5	
Red 210	Naphthol F6RK	74	36	6	5	4–5	3–4	3–4	5	5	3–4	3–4	3–4	5	
Red 222	A20 Magenta G	78	33	6	5	5	3–4	4	5	5	5	5	5	5	
Red 224	Perylene Red Y	35	55	8	8	5	5	3–4	3–4	3–4	4–5	4–5	4–5	4,5	
Red 242	Monoazo Scarlet 4R	–	55	8	8	5	5	5	5	5	5	5	5	5	
Red 253	Monazo 2GLD	–	–	8	8	5	5	5	5	5	5	5	5	5	
Violet 19	Quinacridone Red	55	50	7–8	6–7	5	4	4	5	5	5	5	5	5	
Violet 32	B Bordeaux HF3R	70	40	6	5–6	5	3–4	5	5	5	5	5	5	4	
Brown 1	Azo	–	–	7	6	4	3	2	5	3	4	3	3	2	
Brown 6/7	Iron Oxide	–	–	7	7	5	5	5	5	5	4	5	5	5	
Brown 23	Diazo Brown	–	–	7	7	5	5	5	5	5	5	5	5	5	
Brown 24	Cr, Sb, Ti, Rutile Buff	24	25	7	7	5	5	5	5	5	5	5	5	5	
Brown 25	Benzimidazolone HFR	67	39	8	6	5	5	5	5	5	5	5	5	5	
Blue 1	Brilliant PMTA	59	45	5D	3D	5	2–3	2–3	4	3–4	4–5	4–5	4–5	5	Darkens on exposure
Blue 3	Sky Blue PMTA	49	45	5D	3–4D	5	4	3–4	3–4	4	4–5	5	5	5	Darkens on exposure
Blue 15	Phthalocyanine	60	45	7–8	6–7	5	5	5	5	5	4–5	5	5	5	
Blue 16	Phthalo (copper-free)	66	40	7–8	7	5	4–5	2–3	5	5	5	5	5	5	
Blue 18/56/61	Alkali	65	40	3–4	2–3	5	4–5	2–3	3	2–3	5	5	5	5	
Blue 24	Blue Lake	42	45	1–2	1	4–5	1	4–5	1	1	4	2	4	3–4	
Blue 27	Iron Blue	56	60	7–8	5–6*	5	5	5	1	2	5	5	5	5	*Decomposes
Blue 29	Ultramarine	42	75	5–6	5	4–5	5	5	4–5	4–5	5	2–3	4–5*	5	*Affected by rancid Fats
Blue 60	Anthraquinone	53	40	7–8	6–7	5	5	4–5	5	5	5	5	5	5	Good heat-fastness
Blue 62	CuFe Blue	70	35	4–5	3–4	4	3	3	2–3	2	2	2–3	2–3	5	
Blue 68	Beta Phthalocyanine	77	55	6–7	6	5	5	5	5	5	5	5	5	5	

Green 1	Deep PMTA	50	50	4D	2–3D	5	2–3	2–3	4–5	3–4	5	5	5	5	Darkens on exposure
Green 2	Vivid PMTA	46	50	4D	3D	5	3	2–3	4–5	3–4	4–5	5	5	5	Darkens on exposure
Green 7	Phthalocyanine	48	45	7–8	6–7	5	5	5	5	5	5	5	5	5	
Green 36	Phthalocyanine	35	30	8	7–8	5	5	5	5	5	5	5	5	5	
Violet 1	Rhodamine PMTA	49	50	5–6D	3D	4–5	1–2	1	2	1	2–3	3	5	4–5	Darkens on exposure
Violet 3	Violet PMTA	49	50	5D	3D	5	1	1	4	2	4–5	4	5	4–5	Darkens on exposure
Violet 15	Ultramarine Violet	40	70	5–6	5	4–5	5	5	2–3	5	3–4	2–3	3–4	5	
Violet 23	Dioxazine	70	40	7–8	6–7	5	5	4–5	5	5	5	5	5	5	Heat resistant
Violet 27	CuFe Violet	80	30	4–5	3–4	4–5	3	2–3	2–3	2	2	3–4	3–4	5	
Violet 37	Dioxazine	70	30	7–8	6–7	5	5	5	5	5	5	5	5	5	
Violet 41	Insol Monoazo	–	–	5	4–5	5	4	3–4	5	5	5	5	5	5	
Violet 44	Azo	73	93	5	4–5	5	5	2	5	3–4	3–4	4	3	3	
Violet 50	Azo	76	58	6	5	5	5	2	5	3–4	3–4	4	3	3	

* Note: 1 = poor; 5 = excellent. For lightfastness, Blue Wool scale is 1–8, D = Darkens.

Table 4.6 Black pigments

Types of carbon black	Relative prices	Toptone shade and gloss	Flow properties in ink formulation	Ultimate particle size (millimicrons)	Concentrated paste (%)	Volatile material (%)	PH of aqueous slurry	Fields of application	Special properties
Treated oil furnace	B	Blue and glossy; gloss level up to ordinary channel blacks	Excellent; vastly superior to the original furnace blacks	20–30	Approx. 53	Approx. 1.0	7.5–8.0	Letterpress and lithographic inks	
Treated oil furnace	C	Blue and matt	Very poor	25–60	43	Approx. 1.0	8.0–10.0	Gravure and flexographic inks: news inks, magazine and cheap letterpress	High oil absorption; difficult to disperse
Treated gas furnace	D	Grey/blue and matt	Low oil absorption; long flow	60–200	53	1.0	8.0–10.0	Cheap extender pigment	
Long flow channel	A	Brown and glossy	Very good; this permits higher loading of pigment	10–30	53	12–15	3.5–4.0	High-quality letterpress and lithographic inks	Glossiest type almost unavailable
Lamp black	B	Grey/blue and matt	Low oil absorption; long flow	60–200	70	1.0	8.0–8.5	Cheap extender for matt inks	Extremely low oil absorption and very weak tinctorially.

A–D represents price scale, A being most expensive.

Table 4.7 White pigments and extenders

C.I. Pigment White No.	Description	Bulk value	Concentrated paste (%)	Oil absorption	Transparency or opacity	Soap resistance	Alkali resistance	Paraffin wax bleed	Special properties
4	Zinc oxide	18	75	12	Semi-transparent	Good	Good	None	Not acid resistant
5	Lithopone	35	70	8	Opaque	Good	Good	None	Not acid resistant
6	Titanium dioxide	24	75	20	Opaque	Good	Good	None	
7	Zinc sulphide	33	70	15	Opaque	Good	Good	None	Not acid resistant
18	Calcium carbonate	38	55	30	Transparent	Good	Good	None	Not acid resistant
19	China clay	38	60	60	Semi-transparent	Good	Good	None	
21	Blanc fixe	26	80	18	Transparent	Good	Good	None	
24	Alumina hydrate	55	50	60	Transparent	Poor in alkali and soap	Poor	None	Not acid resistant
26	Talc	23	55	40	Transparent	Good	Good	None	
27	Silica	39	70	300	Transparent	Good	Good	None	

Table 4.8 Benzimidazolone pigments

	Fast Yellow H2G	Orange HL	Red HF2B	Fast Red HFT	Fast Brown HFR	Fast Bordeaux HFR	Carmine HF 4C	Bordeaux HF 3R	Carmine HF 3C	Fast Maroon HFM
Colour Index No.	Y120	036	R208	R175	B25	V32	R185	V32	R176	R171
Appearance (shade)	Yellow	Orange	Red	Reddish brown	Brown	Reddish brown	Carmine	Dark red	Carmine	Dark brown
Density (g/cm^3)	1.47	1.60	1.38	1.42	1.49	1.46	1.39	1.39	1.38	1.39
(1b/gal)	12.4	13.4	11.6	11.9	12.4	12.2	11.6	11.6	11.5	11.6
Oil absorption (lb/100 lb)	34	32	34	35	39	36.5	40	40	36	36
Optical properties										
Brightness	G	G–F	G	F	G	G–F	G	G	G	F
Tinctorial strength	G	G	G	G	G	G–F	G	G	G	G
Hiding power	G–F	G	F	F	G–F	G	F	F	F	F

Bleed resistance										
Water	E	E	E	E	E	E	E	E	E	E
Ethanol	G	E–G	E	E	E–G	E	E	E–G	E	E–G
Methyl ethyl ketone	G–F	E–G	E–G	E–G	G	E–G	G	G	G	E–G
Mineral spirits	G–F	E–G	E–G	E–G	E–G	E–G	E	E–G	G	E–G
Xylene	G	E	E	E	E–G	E	G	G	E	E
Dioctyl phthalate	E	E	E	E	E–G	E	E	E	E	E
Linseed oil	G	E	E	E	E–G	E	E	E	E	E
Chemical resistance										
Alkali	E	E	E	E	E	E	E	E	E	E
Acid	E	E	E	E	E	E	E	E	E	E
Processing										
Dispersibility	G	G	G	G	G	G	G	G	G	G
Heat stability (baking)	G	G	G	G	G	E	E	E	E	E
Lightfastness Exterior										
Full strength	E–G	G	G–F	G	E	G	G–F	G–F	G–F	G
Tint	G–F	G–F	G–F	G	E–G	G–F	G–F	G–F	G–F	G
Interior lightfastness										
Full strength	E	E	G	E	E	E	G	G	G	E
Tint	G	G	G	E	E	G	G	G	G	E–G
Durability weathering	E–G	E–G	G	E	E	G	G–F	G	G–F	G
Toxicity rating	E	E	E	E	E	E	E	E	E	E

Rating: E = excellent; G = good; F = fair.

Table 4.9 Source, uses, composition, physical and chemical constants of oils

Oil	Origin	Chemical composition of fatty acids in the triglycerides: Sat fatty acids (%)	Oleic acid (%)	Linoleic acid (%)	Other acids (%)	Relative density 15.5°C	Acid value (mg KOH/g oil)	Saponification value (mg KOH/g oil)	Iodine value (egl/g oil Wijs)	Hydroxyl value†	Application
Linseed	Seeds of flax plant: Canada, India, USA, Argentine	10	20–25	14–16	Linolenic acid 50–56	0.932–0.935	0.5–9.0	185–195	177–183	0	Oleoresinous varnishes, long oil alkyd resins
Tung	Seeds of *Aleurites fordii*: USA, Japan, Malawi	4–5	4–10	5–9	Linolenic 0–3; eleostearic 77–83	0.940–0.943	0–4.0	190–196	160–180	0	Oleoresinous varnishes, alkyd resins
Oiticica	Seeds of *Licania rigida*: Brazil	10–13	4–7	0	*d*-licanic acid 82; other acids 4	0.968	2.0–4.0	185–194	140–150	0	Oleoresinous varnishes, alkyd resins
Dehydrated castor	Synthetically from castor oil	1–4	7	62	9–11 Octadecadienoic acid 29	0.929–0.940	0–6.0	188–194	130–150	Variable, about 20	Gloss varnishes with pale colour, alkyd resins
Fish (whale and cod liver)	Whale blubber, fatty body tissues, liver, etc.	Complex mixture of acids containing 14, 16, 18, 20 or 22 carbon atoms				0.917–0.925	0–10.0	185–195	110–190	0	Oleoresinous varnishes, segregated oils, alkyd resins
Tobacco seed	Seeds of tobacco plant: USA	12	27	61	0	0.928–0.931	0–5.0	190–195	135–148	0	Alkyd resins
Soya bean	Seeds of *Glycine hispada*: Japan, USA	10–16	22–30	50–55	Linolenic 5–9	0.925–0.927	0.5–4.0	190–195	126–130	0	Alkyd resins
Sunflower seed	Seeds of sunflower: USA	7	38	55	0	0.924–0.928	0–5.0	188–195	123–135	0	Alkyd resins
Tall	Byproduct	2–10	35–50	23–35	Rosin acids 0–40	0.910–1.00	2.0–3.0	160–200	120–150	0	Alkyd resins
Safflower seed	Seeds of *Carthamus*: Europe, USA	11	25	64	0	0.925–0.928	0–2.0	186–194	140–150	0	Alkyd resins
Coconut	Fruit of *Cocus unifera*: India, Ceylon, S. Sea Islands	90	5–8	2–5	0	0.920	0–10.0	250	9	0	Alkyd resins
Castor	Seeds of *Ricinus communis*: Europe, Asia, USA	4–5	5–6	4–10	Ricinoleic acid 77–90	0.950–0.970	0.5–4.0	180–187	79–86	146–166	Plasticizer; preparation of Turkey Red oil

* The only oil listed in this table which is not a triglyceride is Tall oil. This oil is a mixture of fatty acids and rosin acids of variable constitution.

† The hydroxyl value is the number of milligrams of potassium hydroxide required to neutralize the acetic acid capable of combining by acetylation with 1 g of the oil or fat.

Table 4.10 Properties of resinates

	Percentage metal	Softening point (°C) ring and ball	Acid value (mg KOH/g)
Calcium resinate	2.4	117–122	95
Calcium resinate	3.4	140–145	85
Calcium resinate	6.2	155–165	40
Zinc resinate	4.8	130–135	100
Zinc resinate	9.0	140–150	35

Table 4.11 Properties of maleics

	Softening point (°C)	Acid value (mg KOH/g)
Glycol-soluble maleics	140–150	280–320
Spirit-soluble maleics	140–160	90–130
Hydrocarbon-soluble maleics	120–160	Less than 25

Table 4.12 Softening points of rosin-modified phenolic resins.

Esterifying polyhydric alcohol	Softening points (°C) ring and ball	Acid value (mg KOH/g)
Glycerol modified	115–145	20
Penta-erythritol modified	130–160	22

Table 4.13 Typical alkyd resins

Type	Oil length	Polyol	Dibasic acid	Percentage dibasic acid	Solvent	Percentage solids	Acid value (mg KOH/g)	Viscosity (poise; 25°C)
Linseed	72	Penta-erythritol	Isophthalic acid	20	–	100	12	150–250
Linseed	67	Glycerol	Phthalic anhydride	22	–	100	10	800
Linseed	76	Penta-erythritol	Terephthalic acid	16	–	100	12	250–350
Dehydrated castor	65	Glycerol	Phthalic anhydride	25	–	100	12	Viscous (unspecified)
Linseed/tung (styrenate)	36	Glycerol	Phthalic anhydride	20	Xylene	60	15	30–40
Soya	64	Penta-erythritol	Phthalic anhydride	23	–	100	12	450–520
Soya	52	Glycerol	Phthalic anhydride	32	White spirit	40	20	20–30
Castor	42	Glycerol	Phthalic anhydride	41	Xylene	50	50	10–18

Table 4.14 Epoxy Resins

n	Melting range (°C)	Epoxide equivalent	Molecular weight	Relative density at 20°C
0	Liquid at room temperature	175–210	380	1.167
2	64–76	450–525	900	1.206
3.7	95–105	870–1025	1400	1.156
8.8	125–132	1650–2050	2900	1.147
12	140–155	2400–4000	3750	1.190

Table 4.15 Classification of nitrocellulose

Classification	Percentage nitrogen	Suitable solvent
Low nitrogen content	10.7–11.2	74 op methylated spirit
Medium nitrogen content	11.2–11.8	Mixture of alcohols and esters
High nitrogen content	11.5–12.2	Esters

Table 4.16 Solvents

Name	Chemical formula	Preparation	Boiling range (°C)		Relative density	Flash point	
			Initial	Final		(°F)	(°C)
ALIPHATIC HYDROCARBONS							
Low-boiling petroleum distillates. Some examples are given below	No simple formula; always complex mixtures of compounds; mainly of the paraffinic type. The percentage of given molecular species is dependent on the nature of the parent feedstock	Fractional distillation of petroleum			Varies according to manufacturer's specification		
SBP 1			40	170	0.680	–50	–45
SBP 2			70	94	0.708	–40	–40
SBP 3			100	120	0.740	20	–6
SBP 4			40	160	0.710	–50	–45
SBP 5			90	105	0.728	0	–17
SBP 6			140	160	0.768	83	28
SBP 11			154	168	0.776	101	38
White spirit (the US equivalents are Stoddard Solvent and Mineral Spirit)	No simple formula; always complex and variable mixtures of paraffinic, alicyclic and aromatic hydrocarbons	Fractional distillation of petroleum	150	190	0.78	102–120	38–48
Paraffin oil (kerosene)	No simple formula; always complex and variable mixtures of paraffins with smaller quantities of alicyclic and aromatic (approx. 5%) hydrocarbons	Fractional distillation of petroleum	180	250	0.78–0.80	145–150	62–65

High-boiling petroleum distillates	Complex mixtures of higher molecular weight hydrocarbons, especially paraffinic hydrocarbons. The aromatic content is variable (1–18%)	Fractional distillation of petroleum	240	320	0.82–0.83	>150	>65
High-boiling naphthenic distillates	Low odour, low emission mixtures; very low aromatic content (0.1–1.0%)	Hydrogenation of selected petroleum fractions	210	310	0.76–0.82	>150	>65
AROMATIC HYDROCARBONS							
Toluene (Toluol, Toluole)	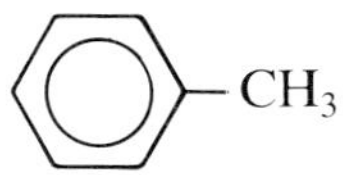	1. Distillation of coal tar 2. Petrochemical prepared by catalytic processes	105	111	0.850–0.873	45	7
Xylene (Xylol, Xylole)	Essentially a mixture of the *ortho, meta* and *para* isomers of dimethyl benzene 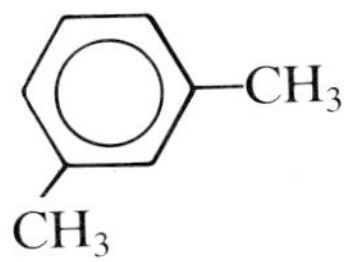	1. Distillation of coal tar 2. Petrochemical prepared by catalytic processes	136	145	0.887	78	25
High-boiling aromatic solvents	Single-ring aromatic solvent with side chain (aromatic content 95–99.5%)	Catalytic process and blending	170	270	0.900–0.990	>200	>93

MONOHYDRIC ALCOHOLS

1. Aliphatic alcohols							
Methylated spirits (ethyl alcohol, ethanol)	Essentially CH_3CH_2OH with traces of methanol and water	Catalytic hydration of ethylene and fermentation of starch	77.5	79.0	0.806	55–60	12–15
Normal propanol (*n*-propanol, propanol-l, *n*-propyl alcohol)	$CH_3(CH_2)_2OH$	Prepared by Fischer–Tropsch process	97	98	0.804	59	15
Isopropanol (isopropyl alcohol, propanol-2)	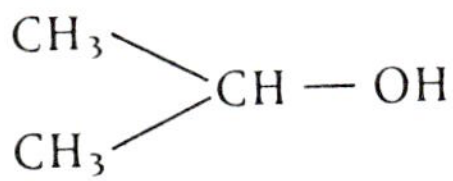	Catalytic hydration of propylene	79.5	82.5	0.784	53	11
Normal butanol (*n*-butanol, butanol-1, *n*-butyl alcohol)	$CH_3(CH_2)_3OH$	Catalytic hydrogenation of crotonaldehyde	114	118	0.813	95	35
2. Alicyclic alcohols							
Cyclohexanol (hexalin)	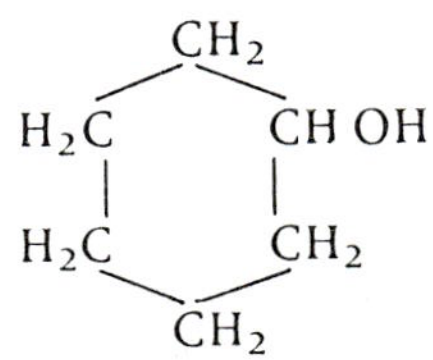	Catalytic vapour phase hydrogenation of phenol	158	166	0.963	154[a]	68

Methycyclohex-anol	Isomeric mixture of 1. 2-Methylcyclohexan-1-ol 2. 3-Methylcyclohexan-l-ol 3. 4-Methylcyclohexan-l-ol Each isomer has *cis* – and *trans*-stereoisomers	Catalytic vapour phase hydrogenation of commercial cresol	165	170	0.920–.0930	145	62
GLYCOLS							
Ethylene glycol (glycol, ethane-1, 2-diol, 1, 2-di-hydroxy-ethane)	$CH_2—OH$ | $CH_2—OH$	Catalytic hydration of ethylene oxide	196	198	1.113	230	110
Propylene glycol	CH_3CHCH_2OH | OH	Catalytic hydration of propylene oxide	187	189	1.037	225[a]	107
Hexylene glycol (2-methyl-pentane-diol)	OH OH | | $H_3C—C—CH_2—C—CH_3$ | | H CH_3	Catalytic hydrogenation of diacetone alcohol	197	199	0.924	245[b]	118
Diethylene glycol (Digol)	$O<{CH_2CH_2OH \atop CH_2CH_2OH}$	Controlled catalytic vapour phase reaction between ethylene oxide and water	244	248	1.117–1.120	290	143

Dipropylene glycol	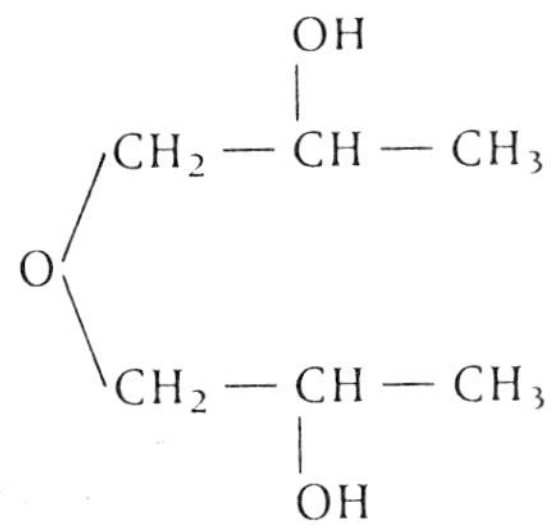	Catalytic condensation and hydration of propylene oxide	230	234	1.025	244[a]	117
Triethylene glycol	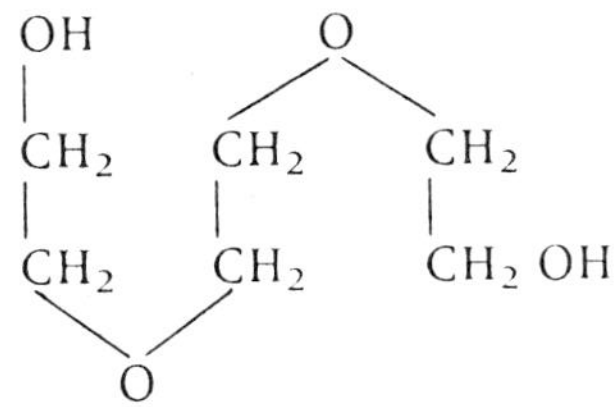	Controlled catalytic vapour phase reaction between ethylene oxide and water	278	280	1.125	330[a]	165
Glycerine	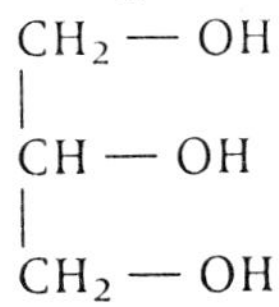	Saponification of naturally occurring triglycerides, e.g. animal fats and vegetable oils	287	293	1.26	320	160
GLYCOL ETHERS							
Propylene glycol mono methyl ether ('methoxy propanol')	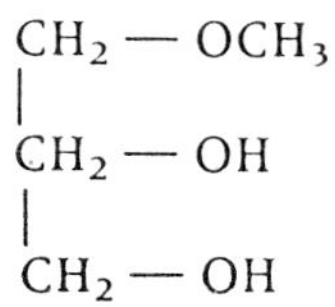	Catalysed addition of methanol to propylene oxide	120	125	0.962	95	35

Propylene glycol mono ethyl ether ('ethoxy propanol')	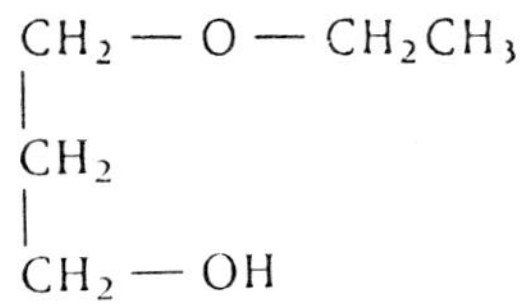	Catalysed addition of ethanol to propylene oxide	129	135	0.896	108	42
KETONES							
Acetone	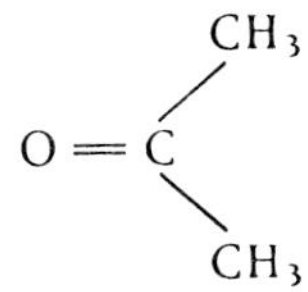	Mainly by the vapour phase catalytic dehydrogenation of isopropanol	55.5	56.5	0.79	1.0	–17
Methylethyl ketone (butan-2-one)	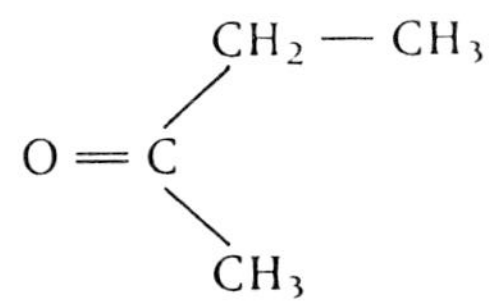	Vapour phase catalytic dehydrogenation of sec-butanol	79	80	0.805	26	–3
Methyl isobutyl ketone	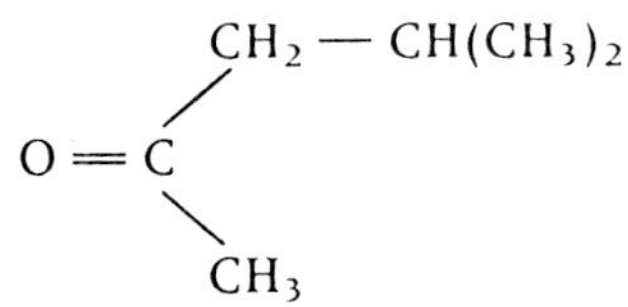	Vapour phase catalytic hydrogenation of mesityl oxide	114	116	0.800–0.801	56	13

Cyclohexanone (sextone)	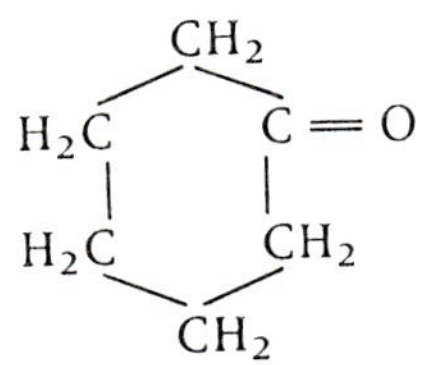	Vapour phase catalytic dehydrogenation of cyclohexanol	151	152	0.948	117	47
Methyl cyclohexanone (sextone B)	Isomeric mixture of 1-, 2-, and 3-methyl cyclohexanones	Vapour phase catalytic dehydrogenation of methyl cyclohexanol	165	175	0.914	117	47
Diacetone alcohol	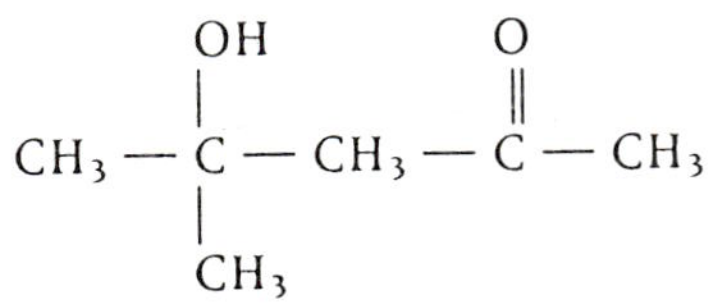	Alkaline condensation of acetone (Aldol change)	158	166	0.931	55	12
Isophorone	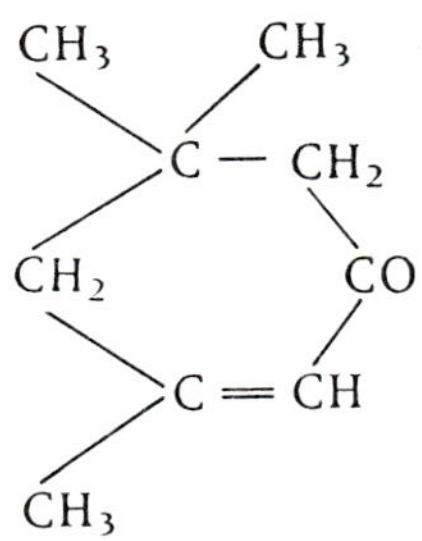	Alkaline condensation of acetone	210	216	0.923	184	84

ESTERS							
Ethyl acetate	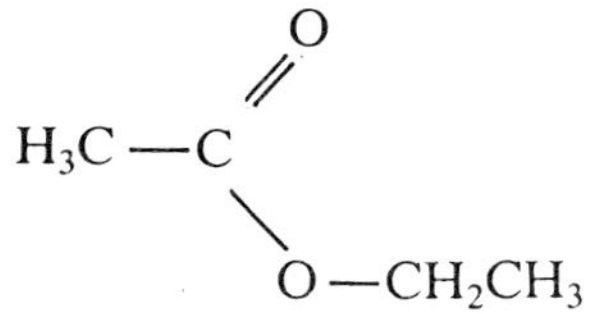	Esterification of acetic acid with ethanol	76	79	0.902–0.904	26	–3
Isopropyl acetate	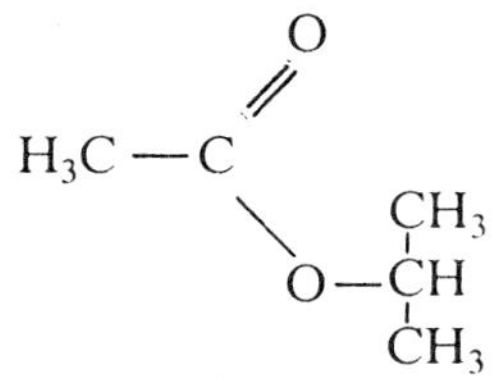	Esterification of acetic acid with isopropanol	87	90	0.872	62	16
n-Butyl acetate	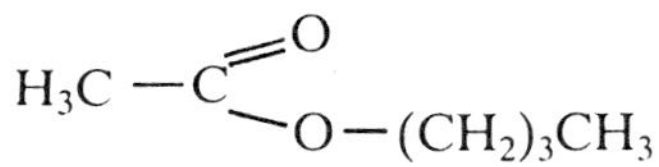	Esterification of acetic acid with *n*-butanol	124	128	0.883	99	37
n-Propyl acetate	$H_3C-COO-(CH_2)_2-CH_3$	Esterification of acetic acid with normal propanol	100	105	0.89	58	14
Ethoxy propanol acetate	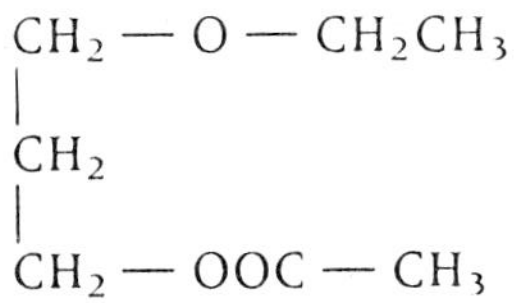	Esterification of ethoxy propanol with acetic acid	155	160	0.941	130	54

[a] Open cup.
[b] Pensky Martens.

Table 4.17 Surfactants

Anionic	$R.COO^-M^+$	Fatty acid salt R = any fatty alkyl or alkenyl group; M = NH_4 Na, K, etc.
Cationic	$R-\overset{+}{N}(CH_2)_3X^-$	Quarternary ammonium salt R = fatty alkyl radical; X = halogen, HSO_4, CH_3COO
Non-ionic	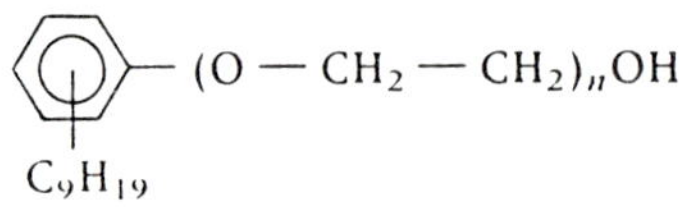	Polyethoxylated nonylphenol n = 5 to 12
Amphoteric	$R-\overset{+}{N}(CH_3)_2-CH_2COO$	Alkyldimethylbetaine

Table 4.18 Alkalis

Name of alkali	Chemical formula	Boiling point (°C) at 760 mm Hg	Miscellaneous comments
Caustic soda	$NaOH$	Not applicable	Rarely used since the inks tend to remain permanently water-soluble as alkali cannot leave ink film. Inks have excellent press stability
Ammonium hydroxide	NH_4OH	Not applicable; evolves NH_3	Industrial grade is 35% solution in water; t.d = 0.88. Its extreme volatility gives ink films with excellent water resistance but poor press stability; obnoxious odour has limited its use
Morpholine	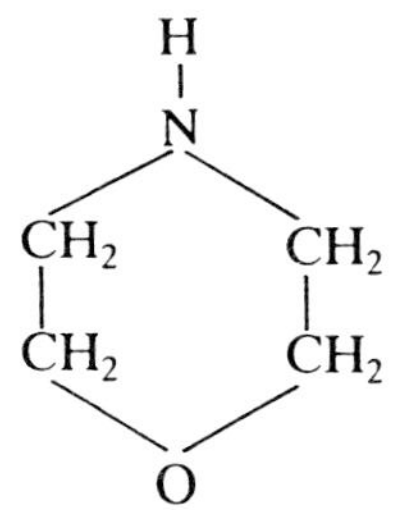	128	Most volatile of the commonly used liquid amines
Monoethanolamine	$HO{-}CH_2{-}CH_2{-}NH_2$	171	Similar to morpholine only inks more stable on press, and prints remain water-soluble for a longer period after printing

Diethanolamine	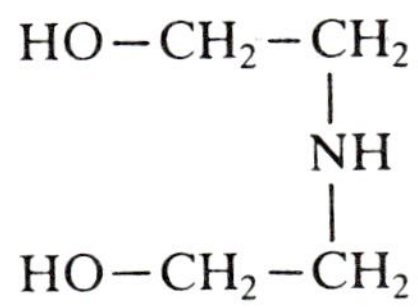	270	Functional chemical and physical properties intermediate between monoethanolamine and triethanolamine
Triethanolamine	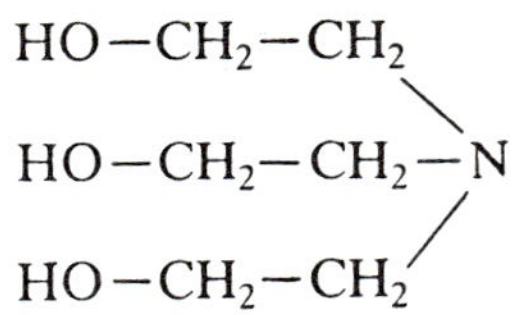	279 (at 150 mm)	The high boiling point gives inks with excellent press stability; maximum water resistance of prints is only achieved after a long period of time

Table 4.19 Acids , acid anhydrides and polyols

Name of organic acid or acid anhydride	Chemical formula	Melting point (°C)	Miscellaneous comments and application
Benzoic acid	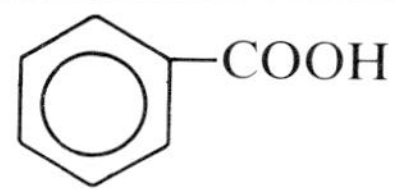	122	Finds use in medium- and short-oil alkyds as a viscosity reducer. It also gives fast drying alkyds. Alkyd resins with benzoic acid have limited application in letter press and lithographic inks.
Phthalic anhydride	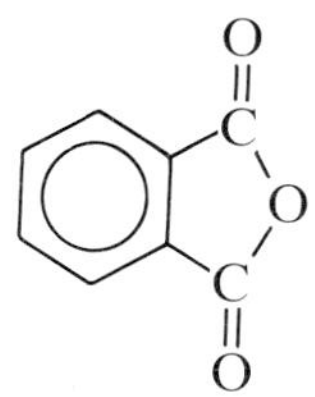	130	This is the commonest and most economic of the dibasic acidic materials used in hard resin and alkyd resin manufacture by reaction with hydroxy groups when polyesters are formed (Chapter 12). Alkyd resins with phthalic anhydride have widespread application in letterpress, lithographic, tin-printing and screen inks.
Isophthalic acid	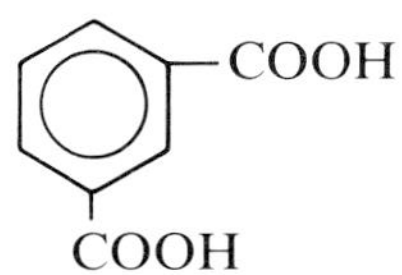	341	Application is similar to phthalic anhydride when resins with properties such as faster rate of drying, improved flexibility excellent pigment wetting properties, etc., are obtained. Its high melting point makes manufacture of the resins more difficult than with phthalic anhydrides.

Terephthalic acid	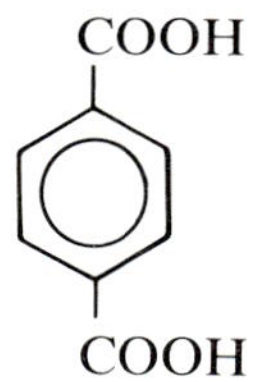	Sublimes at approx. 300	Application similar to isophthalic acid. Formulation of alkyd resins with this acid is more critical than with isophthalic alkyds. The high cost of this material limits its use.
Maleic anhydride	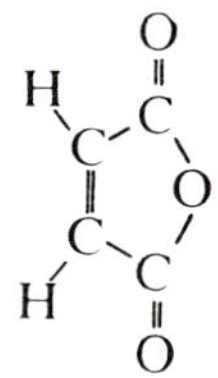	57	This material as well as being dibasic possesses an olefinic double bond which can take part in olefinic type polymerization and Diels–Alder reactions with dienes. Its main uses are:1. polyester formation with polyols which are often cross-linked with an olefinic monomer (e.g styrene) using a peroxide catalyst: 2. Diel–Alder reactions with rosin derivatives when resins of high acid values are obtained. These can be made water- or glycol-soluble by reaction with bases; 3. Diels—Alder reactions with unsaturated natural oils when the water-soluble 'maleinized oils' and alkyds are obtained on treatment with bases.
Fumaric acid	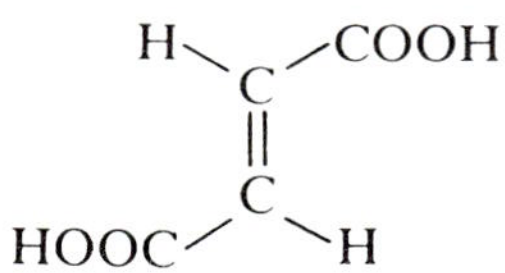	299–300 sublimes	Application and reactions similar to maleic anhydride. Fumaric acid cannot form an anhydride because the carboxylic groups are in the trans position. Under very vigorous dehydration conditions, however, it is converted to maleic anhydride.

Polyols

The most common polyols used in the manufacture of synthetic vehicles for the printing ink industry (see Chapter 9) are:

Glycerol

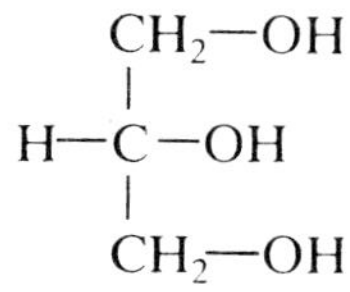

Neopentyl glycol

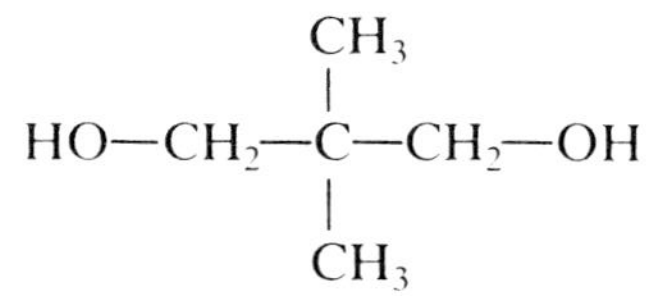

Trimethylol propane

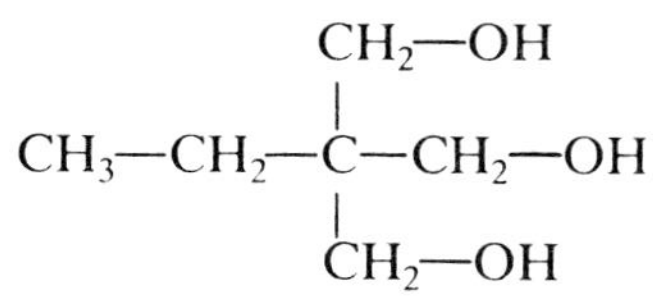

Trimethylol ethane

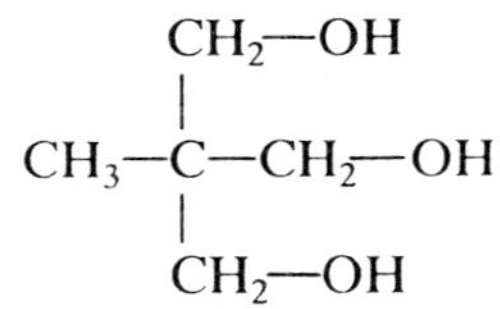

Pentaerythritol

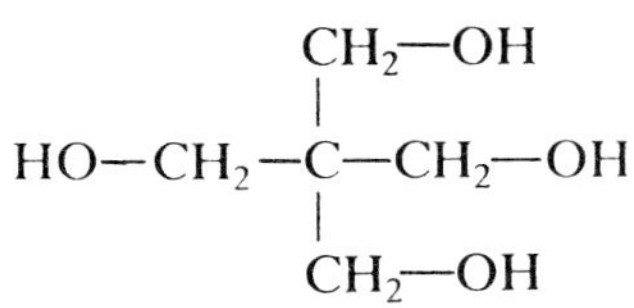

Table 4.20 Prepolymers

Bisphenol A epoxy acrylate

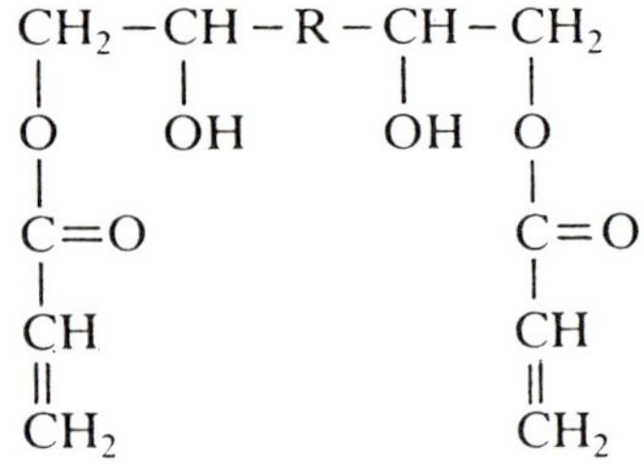

where R is

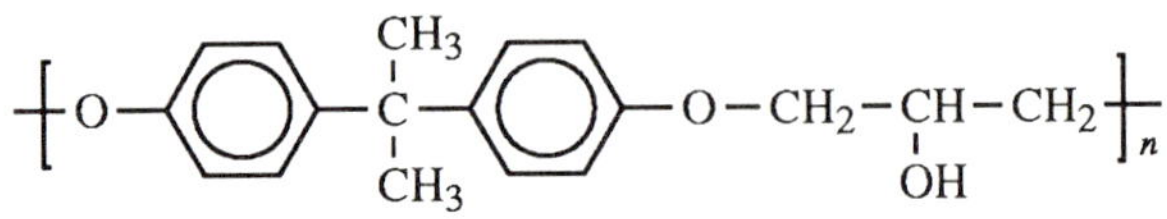

Polyester and polyether acrylates

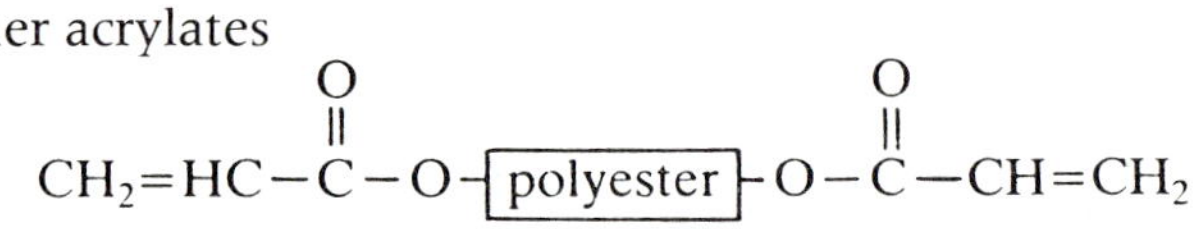

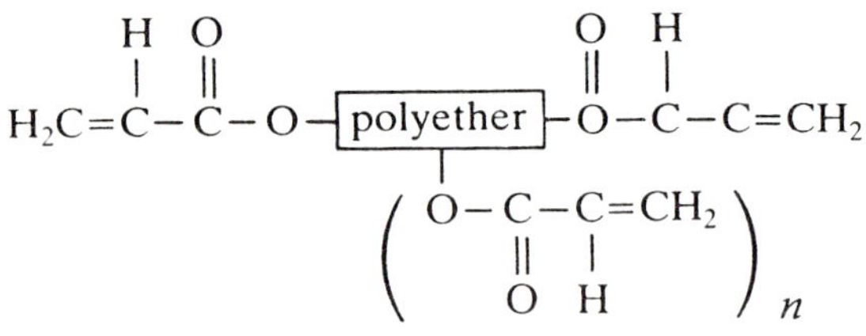

Urethane acrylates

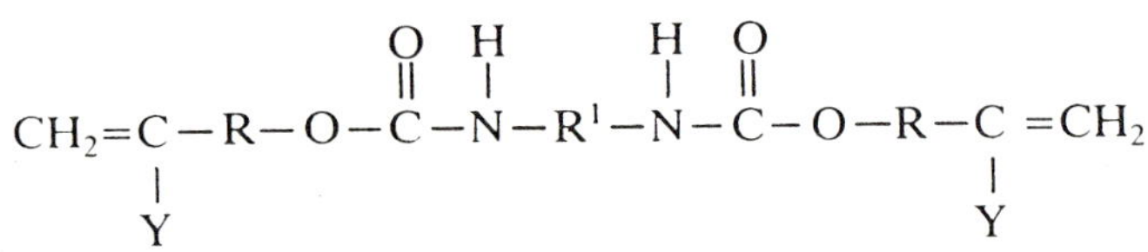

where R is $-\overset{\overset{\displaystyle O}{\|}}{C}-O-CH_2-(CH_2-OH)_n$, $n = 1$ or 2
Y is H or CH_3; R^1 is such that $OCN-R^1-NCO$
can be TDI, HMDI, IPDI, MDI

Table 4.21 Reactive diluents

Monofunctional monomer

n-Vinyl pyrrolidone

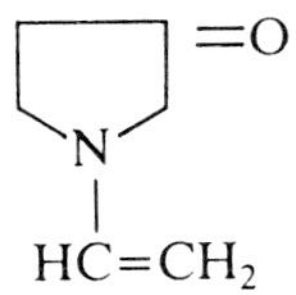

Difunctional monomers

Tripropylene glycol diacrylate (TPGDA)

Dianol* diacrylate (DDA)

*trade name

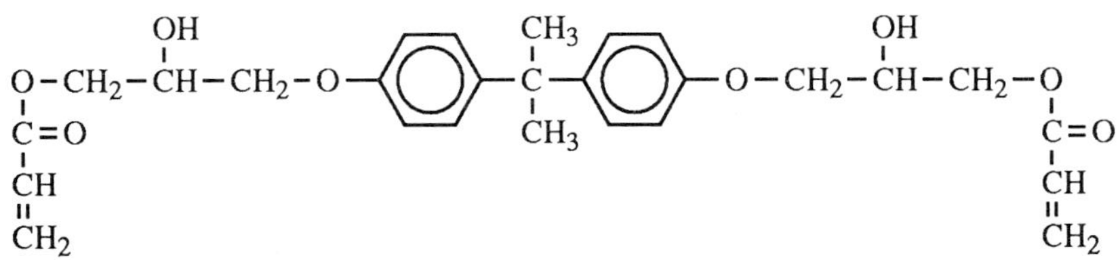

Trifunctional monomer

Trimethylol propane triacrylate
(TMPTA)

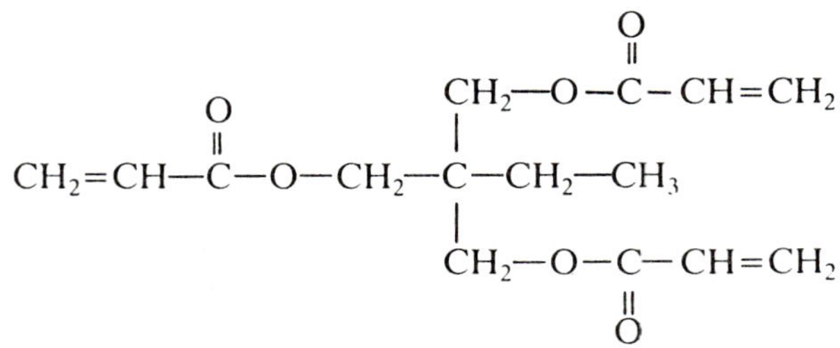

The molecular weight of some reactive diluents is increased to lower skin irritation. This is achieved by alkoxylation of the polyol with ethylene or propylene oxides prior to acrylation

Example
Alkoxylated glycerol triacrylate
e.g. OTA 480

Table 4.22 Aromatic ketone initiators used with a proton donor

Chemical name	Structure	Commercial name
Benzophenone	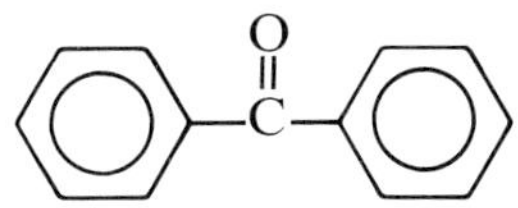	
4-Phenyl benzophenone	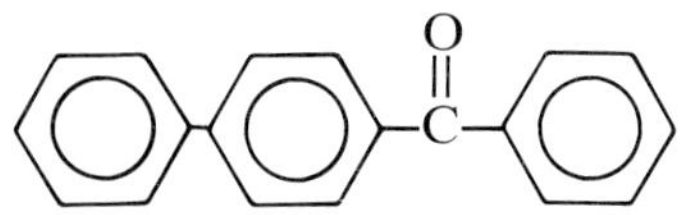	Trigonal 12
Isopropyl thioxanthone	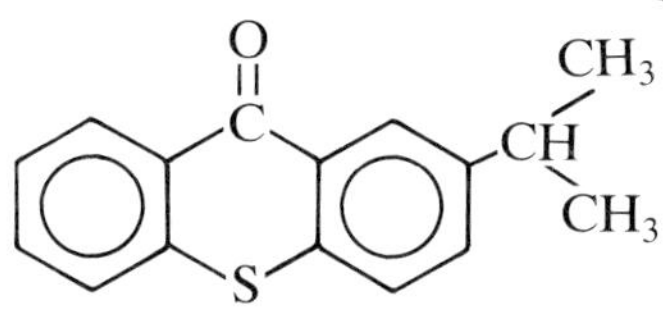	Quantacure ITX
1-Hydroxycyclohexyl acetophenone (50%) and benzophenone (50%)	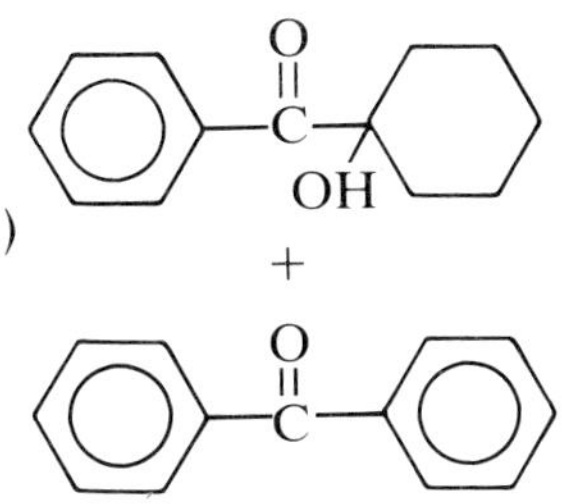	Irgacure 500

Table 4.23 Photoactivators. Proton donors used with aromatic ketone initiators

Chemical name	Structure	Commercial name
Triethanolamine	$(HO-CH_2-CH_2-)_3-N$	TEA
Methyl diethanol-amine	$(HO-CH_2-CH_2-)_2-N-CH_3$	MDEA
Ethyl-4-dimethyl aminobenzoate	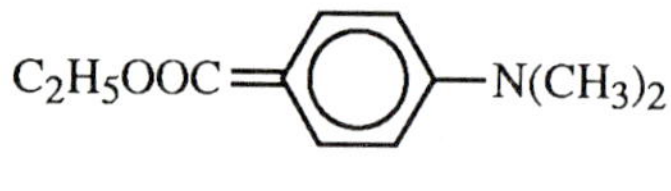	Quantacure EPD Nuvopol EMBO
2-(*n*-Butoxy) ethyl-4-dimethyl-amino benzoate	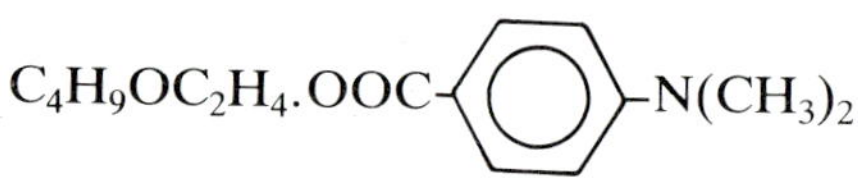	Quantacure BEA
2-Ethyl hexyl-*p*-dimethyl-aminobenzoate	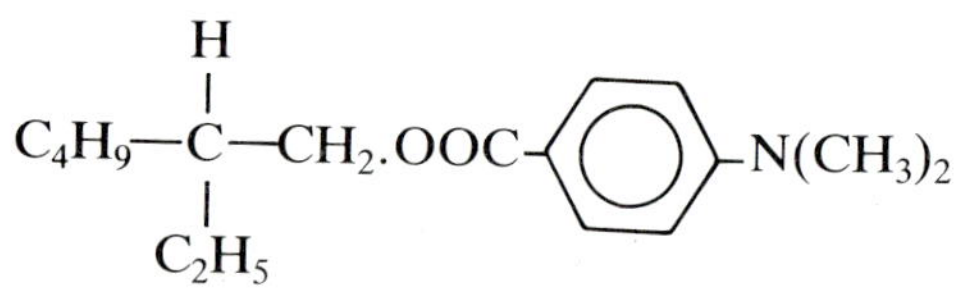	Escalol 507

Table 4.24 Acetophenone, benzoin and benzil ketal initiators used without proton donors

Chemical name	Structure	Commercial name
2,2-Diethoxy acetophenone	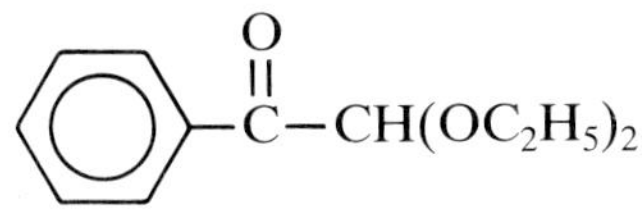	DEAP
1-Hydroxy cyclohexyl acetophenone	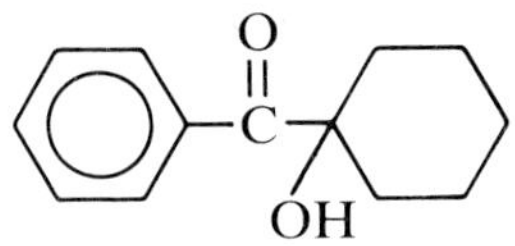	Irgacure 184
α, α-Dimethyl-α-hydroxy acetophenone	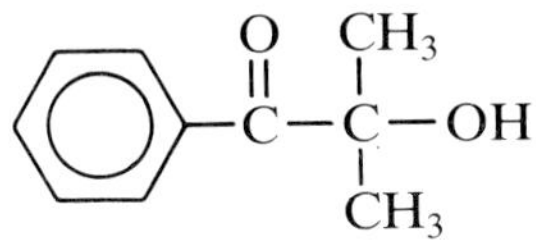	Darocur 1173
Benzoin	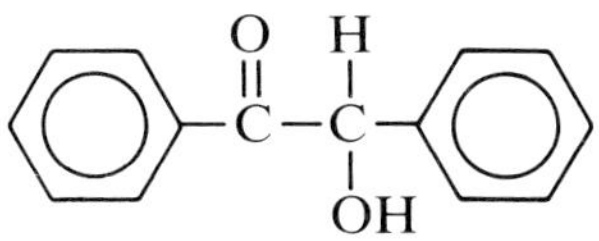	–
Benzoin ethers	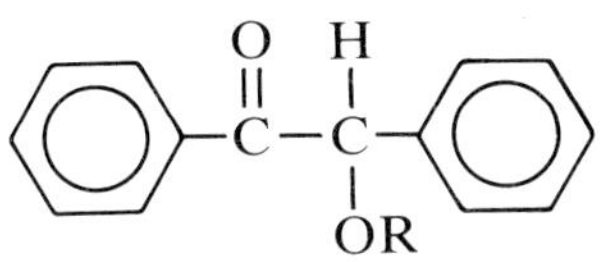 R = ethyl, butyl or isopropyl	Esacure EB3 Trigonal 14 (Butyl ethers)
Benzil ketals	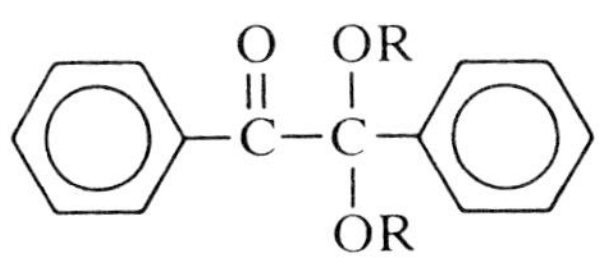 R = methyl, ethyl, butyl or isopropyl	Irgacure 651 (Methyl ether)
2-Methyl-l-[4-(methylthio) phenyl]-2-morpholino propanone-1	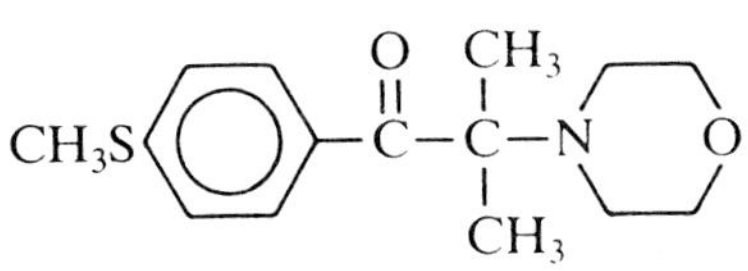	Irgacure 907

CHAPTER 5
Letterpress inks

The use of letterpress as a major printing process continues to decline, but nevertheless a significant volume of general printing throughout the world continues to be produced by this process.

Jobbing printing, continuous stationery and corrugated cases are examples where the process is still used, although on a diminished scale, since printing by alternative methods offers advantages of quality and output. A number of newspaper titles are also still printed by rotary letterpress. However, while general letterpress printing is likely to continue to have a place in the world market, it is unlikely that the process will have a long-term future for newspaper printing, in view of current trends towards the use of web-offset and flexographic presses.

5.1 NATURE OF THE PROCESS

The fundamentals of the letterpress process, together with a description of the configuration of the most widely used presses, have been dealt with in Chapter 2. The fact that the ink is applied to the paper from a raised or 'typographic' surface under considerable pressure, gives the process its characteristic properties. The printing surface can be 'hard' such as type-metal alloy, and copper or zinc blocks, or 'flexible' as rubber stereos and polymer plates. The materials used in the formulation of letterpress inks do not generally affect the various metallic surfaces of the formes, although there may be differences in their wetting characteristics. Polymer plates, however, may be affected by the materials used in some types of ink, and care has to be taken when formulating inks for these, but it is the method by which the ink is applied to the printing surface, and the printing speed which has the most influence on the formulation.

5.2 TYPES OF PRESS

Platen presses

Platen presses have relatively poor ink rolling power compared with other types of presses, and printing takes place from a rigid forme which is parallel to the paper at the moment of impression. The ink will need to be sufficiently low in viscosity to distribute over the relatively short roller train, but must be short flowing so that it will not spread down the sides of the relief characters. The ink should exhibit sufficient tack to transfer clearly under a reasonable level of impression if embossing of the type on paper is to be kept to a minimum.

The type of work printed on such presses is normally short runs of jobbing work, hand bills, invitation cards, dinner menus, etc. where a fast turn-round of the finished work is often required and therefore a fast setting and drying ink is an advantage.

Flat bed presses

These presses have increased ink rolling and distribution power, compared with platen presses, and the printing forme area can be larger. The presses are suitable for printing on a wide range of substrates, from newsprint to coated papers. The inks used, therefore, can have a wide range of rheological properties and drying rates, and compared with the ink used on a platen press will usually have a higher viscosity, increased tack and may be slightly weaker.

Rotary presses

The web of paper is printed in a continuous operation, usually at high speed. The presses are designed with complex inking systems to enable consistent inking for long periods at high speed. This necessitates the use of thinner inks which 'set' by penetration.

There are three methods of ink feed to the printing unit and each makes demands on the rheology of the ink used.

- **Duct feed** – this consists of a roller rotating in an ink fountain, the flow of the ink being controlled by a flexible blade. This system can handle inks of relatively high viscosity provided they have sufficient flow not to hold back in the duct.
- **Ink rail feed** – ink is pumped through tubes to an ink rail which is set parallel and close to the first inking drum. Separate tubes supply each 'column width' and enable the ink feed to be easily regulated across the width of the web. This system, which is fully enclosed, requires an ink which is free flowing and has a low yield value.
- **Page-Pak system** – Similar to the ink rail feed system, but each column width is supplied by an individual pump unit. Inks with low viscosity and extremely good flow are necessary.

Appearance of letterpress print

A well printed copy produced by the letterpress process should ideally have even solids, clean type and the minimum of distortion to the printed substrate caused by the impression pressure but in practice, however, this is not always the case. Where presses are incorrectly adjusted or where unsuitable papers are used, undue printing pressure is necessary in order to 'bottom' the paper to obtain a smooth print. The pressure may be such that the substrate is embossed on the reverse side of the paper and a halo appears around the edges of the solids as the ink is squashed from the print surface. These characteristics of squash and embossing are present to some degree in all typographic printing and are the usual means of identifying the process.

5.3 GENERAL CHARACTERISTICS OF LETTERPRESS INKS

To obtain the best from the letterpress process, inks should conform to the basic principles set out below.

- It is always advisable to print with the thinnest film of the strongest and stiffest ink that the paper and the printing press will tolerate. If the ink has insufficient colour strength, high film weights of ink are needed and excess squash, mottle and set off, are likely to result. At the other extreme, if the ink is too strong, it may be difficult to carry sufficient ink to obtain the required strength, while at the same time producing an even print which covers the irregularities of the paper.
- The faster an ink is required to run the lower the viscosity and the less tacky it has to be. At high speeds, stiff inks do not transfer readily on the inking rollers, while an ink with little or no tack may cause the inking rollers to skid, with the resulting loss of ink transfer.
- Printing by the letterpress process requires relatively high pressures, and inks must be formulated in a way which minimizes the unwanted effects of such pressures.
- The ink should not dry on the press during normal daily stoppages and in some instances, for much longer periods. In most press rooms, it is not economic to allow time to clean the inking rollers, either during or at the end of the run. The ink should therefore return to normal working properties, in the shortest possible time, after a break in production.

Drying mechanisms

Letterpress inks dry through various mechanisms.

In the case of inks for general or publication work, the routes will be:

- penetration – newspaper inks
- oxidation – foil inks
- combination of both – quick-set inks.

For packaging work the mechanisms are more likely to be:

- a shift in chemical balance – water-reducible inks
- UV radiation – label inks.

Since general and publication printing form by far the larger proportion of the use of the letterpress process, the drying mechanisms and the inks used will be considered in detail. Inks for packaging purposes will be considered and attention drawn to the differences involved.

Penetration

Penetration drying occurs where part of the vehicle is a low viscosity material such as mineral oil or petroleum distillate. This component penetrates into the substrate leaving a pigment/oil phase on the surface, which enables the print to be handled a short period after printing. In the case of conventional newspaper inks, the ink film does not dry further and can be easily removed from the surface of the paper by light pressure.

Cheap general-purpose inks used to print hand-outs and similar promotional work on uncoated papers can also be formulated to dry by penetration. They usually include some bodied drying oil or resinous varnish to impart some oxidation drying characteristics and to give the print a degree of rub resistance.

Oxidation

Inks drying by oxidation alone are usually used to print impervious substrates, such as foil, board or certain plastics, and print tests should be carried out on the substrate to be used to ensure that the level of adhesion is sufficient for the end-usage requirements.

The drying process is catalysed by the metallic driers described in Chapter 4. Inks drying by this process alone sit on the surface of the substrate and may take a considerable time to dry, and the prints may require an anti-setoff spray, to avoid print setoff problems in the stack.

Combination drying

The combination of penetration and oxidation/polymerization forms the most widely used drying process for letterpress inks. They are formulated to have two phases. The low viscosity component penetrates the fibres of the substrate immediately following printing, to set the ink, leaving the high viscosity vehicle/pigment phase on the surface which dries by oxidation. Depending on the rate of setting of this type of ink, the prints may require a minimum of application of spray powder to prevent setoff.

5.4 PHYSICAL PROPERTIES

Because of the large variation in types of press and printing speeds involved in the letterpress process, the physical properties of the inks used vary considerably from the long flowing, low viscosity inks used for high speed rotary newspaper presses, to short, relatively high viscosity paste inks, for use on platen machines. The properties required for each type of ink are explained with the formulation, and quality assurance methods are dealt with in Chapter 15.

5.5 RAW MATERIALS

Pigments

The range of pigments that can be used in letterpress inks is comprehensive and the process does not impose the stringent demands of 'ink fount' balance of lithography. Wherever possible, pigments and other raw materials, will have common use over the whole range of ink manufacture.

Due regard must be paid to the avoidance of certain pigments in inks designed for food wrappers and also those which are to be applied to toys or graphic instruments.

The film weights associated with letterpress printing are comparatively high and therefore the level of coloured pigmentation is lower than the offset process. It is often necessary to include transparent or semi-opaque extending pigments, such as calcium carbonate (CI Pigment White 18) or blanc fixe (CI Pigment White 21), in the formulation if the balance between colour strength, working properties and cost, is to be achieved. Reducing the colour strength by the addition of varnish alone, would tend to increase the setoff properties of the ink or reduce the viscosity to a level where it would not print properly. The addition of extender helps adjust the pigment/vehicle ratio and gives the flow, transfer and setting properties required to print evenly.

The inclusion of some opaque pigment, where the specification allows, is also recommended as it reduces the effect of the uneven solids obtained by the process.

Coloured inks for newspaper printing need to have good contrast of hue when printed on newsprint to give maximum impact and therefore brightness of colour is essential. Headlines and seals are often printed using light to mid scarlet red shades, so Lake Red C (CI Pigment Red 53:1) finds a wide usage, although other pigments with some opacity can be beneficial, e.g. chlorinated Para Red (CI Pigment Red 4), cost being the main constraint when matching other colours. However, where process colours are required Diarylide Yellows (CI Pigment Yellows 12, 13, 14 etc.), Phthalocyanine Blue (CI Pigment Blue 15.3) and Lithol Rubine (CI Pigment Red 57.1) are extensively used.

The carbon blacks used in newspaper ink formulation are usually rubber grade, intermediate particle-size furnace blacks. They may be supplied in fluffy, pelletized, or oil-beaded form. The carbon must have good tinctorial strength, have good inkmaking properties, be free from abrasive materials, but above all, must be economic. The particle size and oil absorption vary according to the manufacturer and the form in which they are supplied.

The ink properties required often cannot be achieved with the use of a single grade of carbon black and it is likely that different grades will be blended to give the optimum results.

Resins

Drying systems

The earliest drying systems used in letterpress inks were based on linseed oil varnishes, which were usually heat-polymerized to give a range of viscosities between 15 and 240 poise (1.5 Pa/s and 24 Pa/s). With the advent of presses with faster printing speeds, quicker drying systems were developed by reacting polymerized or refined linseed oil with a hard resin such as a rosin ester or rosin-modified phenolic resin, and reaching the viscosity with solvents. In current varnishes, the drying oil component has been largely replaced by the use of long-oil alkyd resins, and these, with the addition of phenolic, maleic or hydrocarbon resins, give the wetting, drying and gloss required in each particular formulation. The viscosity of this type of resin-oil varnish will usually be adjusted by the inclusion of a high boiling range petroleum distillate (see Solvents below) which will also have an influence on the setting of the ink. The inclusion of a suitable gellant may also be required if the formulation needs a structure for printing performance.

Careful formulation produces a varnish which does not skin unduly in the container, has good press stability, but rapidly sets on the paper by solvent penetration. The highly resinous phase of the ink remains on the surface and oxidizes and polymerizes to form a hard film.

For non-absorbent substrates, where the drying depends almost entirely on oxidation–polymerization, the addition of tung oil to the formulation may be required to give the speed of drying and the hardness required.

Non-drying systems

The majority of all non-drying systems especially for use on rotary presses for newspaper printing are based on mineral oil, although they may contain a minimal amount of hard resin to give a degree of rub resistance. The resins used will be drawn from the same classes as used for other types of inks, i.e. phenolic and hydrocarbon resins, etc. but the grades chosen must be soluble in mineral oil and have low cost. Since there is no requirement for fount tolerance, certain grades of zinc and calcium resinates can also be used. The mineral oils used can vary in viscosity and aromatic content depending on the source and nature of the treatment used to refine the oil.

Recent legislation has directed the suppliers of mineral oils to classify their products in accordance with the degree of treatment and refining given to the oils.

Solvent-refined oils

The oils are treated during their manufacture to remove the aromatic content. They do not normally carry any Health and Safety at Work Act warning labels.

Hydro-refined oils

This grade of mineral oil is partially treated and may require a warning label dependent on the degree of treatment.

Residual aromatic extract oils

The oils are specifically treated to remove the aromatic content whilst retaining the inkmaking properties.

Solvents/reducers

As noted above, the high boiling point petroleum distillates used will mainly have an initial boiling point of above 260°. These are not strong solvents and some quick-set varnishes require the inclusion of a 'bridging' agent to increase the solvency power of the system. High boiling point alcohols have been used for this purpose and, since they are more effective than distillates, only small additions are necessary to reduce viscosity and improve flow and press stability. Some care has to be taken to avoid the loss of these 'bridging' agents through the heat generated during the milling process, otherwise the solvent balance will be altered with consequent problems of drying on the press. In non-drying systems and those drying systems where some loss in setting can be tolerated, 0.05 Pa s 'spindle' oil makes a useful reducer and also stabilizes the system.

Additives

Additives are those materials which are present in a formulation to promote:

- drying properties
- rub resistance
- press stability
- low mist character
- dispersion and rheology of the ink.

Drying

The chemical cross-linking of a 'drying' ink system is promoted by the addition of 1–2% of certain organic soaps of metallic elements. Cobalt, manganese, cerium and zirconium compounds are the most common types of drier present in letterpress inks and may be included in liquid or paste form. They may be used either alone or as a mixed drier to achieve the end result required. The use of excess cobalt and manganese driers can lead to press drying problems without resulting in a harder print surface.

Rub resistance

The addition of waxes can improve the rub resistance of drying and semi-drying ink systems by providing extra slip. Although petroleum-based waxes (for example, paraffin and montan and natural waxes and carnauba) have been used in the past, most modern formulations will rely on polyethylene (PE) wax to improve rub and scratch resistance. Polytetrafluoroethylene (PTFE) wax will be used to reduce the friction level of the ink film surface and impart 'slip' properties.

Waxes will often be added to the formulation as 'pastes' or 'dispersions' with the ink vehicle used, but the introduction of 'micronized' grades has reduced the need for this extra processing. Blends of PE and PTFE waxes may be used and some mixtures are being marketed by suppliers.

Care has to be taken if the end use involves delayed superimposition of a second colour, UV varnish or laminate film, to avoid the use of excessive amounts of wax paste in the first colour, as this interferes with the 'lay' and keying of the subsequent film.

Press stability

Although most inks are required to dry on the substrate, unless a suitable balance is achieved, those drying by oxidation can also dry on the press. Anti-oxidants can be used to retard drying of thick films and are detailed in Chapter 4. The formulator will choose from these the material which gives him a dry print with a stable ink. Volatile anti-oxidants can cause rewetting problems in printed stacks, while non-volatile types can retard oxidation drying too far. Since these materials can be very effective in their anti-oxidant properties and only small amounts are required in the formula, they may be added in reduced form, using as an extender a high boiling solvent, oil or some of the vehicle. Other types of material which have found a use in letterpress formulations, to retard drying, have been greases such as yellow or white petroleum jelly, but these will appear mainly in non-drying or slow-drying types of ink.

Low mist additives

Many of these are available, the most common being the range of bentonite clays, which can be added as pastes or, with the newer grades, as a stir-in material. Since these also tend to reduce the flow characteristics of an ink, a balance between reducing the mist level and flow length has to be achieved.

Dispersion and flow aids

The wetting characteristics of an ink system will usually be obtained by careful choice of the resins and oils involved. The most common wetting agent for rotary news ink is an asphaltum complex solution in mineral oil. The correct addition, which may only be 2 or 3%, will improve the dispersion of the pigment and give better flow without causing excessive strike-through or ink mist. Other materials which help pigment dispersion are being marketed and several suppliers are now offering ranges of hyperdispersants applicable to individual pigments for optimum effect.

5.6 LETTERPRESS INK FORMULATION

Letterpress inks should be formulated to be as versatile as possible, to run on a wide variety of presses and substrates. Examples are given of a range of ink formulations which can be modified to suit the requirements of local conditions.

Platen ink for absorbent papers

Much of the work carried out on platen presses can involve absorbent papers and requires a quick turn-round. An ink which penetrates the paper but remains open on the press is therefore advantageous. Flat-bed newspaper inks have similar formulations but with lower viscosity and longer flow.

Carbon black (CI Pigment Black 7)	19.0
Reflex blue (CI Pigment Blue 18)	2.0
70 Pa s mineral oil	55.0
0.05 Pa s mineral oil	9.0
1.05 Pa s bodied linseed oil	10.0
Asphaltum solution	5.0
	100.0

Cylinder press ink for uncoated papers

An ink formulated for a more general range of uncoated papers will require to have some drying properties. The incorporation of some resin into the system giving the ink a degree of tack and flow, and the addition of slightly more blue toner can also be expected.

General-purpose black

Carbon black (CI Pigment Black 7)	18.0
Reflex blue (CI Pigment Blue 18)	4.0
Long-oil alkyd	16.0
0.05 Pa s mineral oil	5.0
Varnish*	55.5
10% cobalt drier	0.5
Anti-oxidant	1.0
	100.0

* Consists of a hydrocarbon resin or modified phenolic resin cooked into a blend of linseed and mineral oils.

Quick-set inks for coated paper

This type of ink will usually be based on a varnish formulated from resin, drying oil and distillates as described in the section dealing with resins. The choice of resin and ratio of resin/oil blend to distillates are determined by the quality of the substrate and the properties required from the final print.

The higher the absorbency of the substrate, the higher the viscosity of the resin/oil phase needs to be, in order that chalking of the print does not occur.

Calcium carbonate (CI Pigment White 18)	8.0
Phthalocyanine blue (CI Pigment Blue 15.3)	12.0
Gloss varnish	49.0
Setting varnish	16.0
Polyethylene wax paste	5.0
Anti-set-off paste	3.0
Cobalt/manganese drier	1.0
280°C–320°C distillate	6.0
	100.0

Letterpress ink drying by oxidation

An ink drying by oxidation for use on foil-board or other impervious substrates would be formulated on the following lines.

Hard drying black

Carbon black (CI Pigment Black 7)	20.0
Reflex blue (CI Pigment Blue 18)	6.0
Long-oil alkyd	30.0
Modified phenolic/tung oil varnish	30.0
Polycthylene wax paste	8.0
260°C–290°C distillate	4.0
Cobalt/manganese drier	2.0
	100.0

The small percentage of distillate shown can be tolerated, but higher amounts would tend to be trapped in the ink film and make it easier to remove from the surface.

Process inks

The theory of four-colour reproduction was dealt with in Chapter 3. The use of the letterpress process is now hardly known in the UK for this purpose and BS 4160:1967 (European Reference CEI 12/66) has been withdrawn by the BSI. This standard specified the colour, colour strength, transparency, lightfastness and solvent resistance of the three primary colours – yellow, magenta and cyan. It also referred to a previous British Standard for letterpress process inks (BS 3020:1959) for a suitable black ink.

Process magenta letterpress

Calcium 4B toner (CI Pigment Red 57.1)	15.0
Polyethylene wax paste	3.0
Cooked quick-set vehicle*	32.5
Gloss quick-set vehicle†	28.0
Cobalt/manganese drier	0.5
280°C–320°C distillate	20.0
Anti-oxidant	1.0
	100.0

* Based on modified phenolic resin, cooked into a long-oil alkyd and linseed oil and let down with 280°C–320°C distillate.

† Based on modified phenolic resin, cooked into a long-oil alkyd, let down with 280°C–320°C distillate and gelled.

Similar formulations for the yellow and cyan would contain 15% Diarylide Yellow (CI Pigment Yellow 13) and 14% Phthalocyanine Blue (CI Pigment Blue 15:3) respectively, to give the correct chromacity figures when printed in accordance with the standard. Because of the thicker ink films laid down, tack grading is essential, when printing colours wet-on-wet.

5.7 NEWSPAPER INKS

Newspaper colour inks

The titles and late news items are sometimes printed on very small units which have very limited rolling capacity. Due to the high operational speed of these units, considerable ink spray can be developed. Coloured inks are also widely used in provincial papers to give impact to advertising. These are usually printed on page-wide units which present the formulator with similar problems to that of the seal or headliner colours. Occasionally, if the amount of coverage of colour warrants, a complete unit will be used, but more usually, a half-deck, and the careful design of the page layout will provide the necessary colour.

A typical formulation would be

Chlorinated Para Red (CI Pigment Red 4)	10.0
Extender pigment	10.0
Resin varnish*	40.0
Mineral oils	25.0
Long-oil length alkyd resin	10.0
Wetting agent	
Anti-oxidants, etc.	5.0
	100.0

* A hydrocarbon resin or low melting point phenolic resin dissolved in mineral oil.

The amount of pigment/extender is varied depending on the properties of colour strength, etc. that are required and the ratio of resins, solvents and oils adjusted to suit the individual units being used.

Newspaper rotary black inks

The use of the letterpress process to print newspaper in the UK has dramatically declined since the lithographic process was introduced into the national newspaper printing industry. However, worldwide, considerable numbers of newspapers continue to be printed by the process. The formulations used have to meet targets which require a good tolerance for trouble-free running using a wide range of press designs and grades of newsprint paper.

The colour strength is critical in formulating newspaper inks. If the ink is over-strength, reduced film weights are used, and this may give rise to poor coverage and uneven inking. Problems of linting and transfer may also occur if an insufficient volume of ink is being replaced on the forme and ink distribution roller chain. Conversely, with low strength ink, problems of strike-through, misting, flooding the stereo and marking-off on path rollers occur, as high volumes of ink are transferred through the printing system to obtain the correct density.

The actual amounts of ink used per copy vary with the content of the edition and the absorption of the newsprint used. A guide to how much ink is used is 1.7 kg of rotary news ink per 1000 copies of a 32-page tabloid (this is based on experience rather than a calculated formula). While the amount of ink used per copy is extremely small, the total amount of ink used by a newspaper publisher can be over 3000 tonnes per annum. The cost of the ink is therefore a significant factor in newspaper production costs, and can be a limiting factor which has to be taken into account by the formulator, restricting the use of certain raw materials.

Formulation of newspaper inks

The use of refined mineral oils in newspaper ink formulation has almost completely replaced those with a high aromatic content. The oils used may be hydro-refined, solvent treated, or in combinations with sophistically treated oils which have very low aromaticity but high viscosity. The use of these oils has presented the ink formulator with a considerable challenge to achieve inks which have similar working properties to those formulated on high aromatic oils.

A general guide formulation for newspaper blacks is given below. The choice of the grade of both carbon black pigment and oil will have an influence on the formulation particularly the flow aids and surfactants.

Carbon black news ink grades	12–14%
0.9 Pa s refined mineral oil	70–75%
Surfactants, flow aid and anti-spray compounds	3–5%
Distillate	3–5%

The 9-poise oil use can be made up of combinations of viscosity grades ranging from 0.5 poise to over 100 poise. They may be solvent refined, naphthenic or high viscosity specialist mineral oils.

Keyless inking (indirect flexo)

The formulations of news inks for this type of ink-metering system are similar to those for conventional units. The viscosity of the ink needs to be low enough to fill the cells of the etched roller and the flow sufficient for the inks to be readily pumped. The level of pigmentation is determined by the depth and shape of the cells of the etched roller.

5.8 INKS FOR PACKAGING

As previously stated, these will be mainly of the water-reducible or UV-curing types, although some inks drying by pure oxidation may be used on impervious surfaces. The latter have fallen into disfavour because of the odour generated during the drying process which makes them unsuitable for most packages containing food.

Precipitation drying

Until the early 1960s 'moisture-set' inks were widely used for printing corrugated cases. These inks dry by the precipitation of the resin content of the varnish when in the presence of water. The water could come from the atmosphere, the moisture present in most papers and boards, or from an applied jet of steam.

The use of moisture-set inks declined rapidly due to the matt finish of the print and the difficulty in maintaining print quality due to press stability problems. The inks tended to dry rapidly under high humidity conditions and to lack flow properties.

To overcome these problems, water-reducible letterpress inks were introduced. These were, and still are, based on a high acid-value resin dissolved in a glycol/alkali solution. Such inks can be reduced with water and still give good flow, hence they give better press performance than moisture-set inks although they exhibit a rapid reduction in viscosity with an increase in temperature, which can lead to poor printability.

These inks dry by resin precipitation following a pH shift towards the acid side of neutral when the ink film is in contact with an acidic substrate. A key point in the formulation is to have sufficient alkali but not an excess. The latter gives prints with poor wet-and-dry rub resistance.

Much of the work which used to be printed letterpress is now produced by flexography using water-based inks; Chapter 9 gives further details. The main resin used in earlier formulations was shellac, but in view of the increased cost of this material, it has been substituted largely by maleic or fumaric resins. These differ from the resins used in oil-based varnishes in having a content of maleic or fumaric acid and in being unesterified giving the high acid value (200–300 mg KOH/g) required for solubility. Monoethanolamine, triethanolamine and urea are the main alkaline materials used to solubilize the system, while diethylene glycol and others of the same type are used to control the viscosity.

Since water-reducible inks are often used on natural kraft stocks there is a need to have some opacity to hide the colour of the substrate. Combinations of rutile titanium white (CI Pigment White 6) with a range of other pigments, are used to give reasonably bright hues. Choice of pigment is only restricted by the need for no bleed to take place in glycol, and by price.

Water-reducible red

Blanc fixe (CI Pigment White 21)	10.0
Rutile titanium white (CI Pigment White 6)	5.0
Lake Red C (CI Pigment Red 53.1)	14.0
Varnish	54.0
Diethylene glycol	8.0
Wax paste	5.0
Amine	4.0
	100.0

Varnish

High acid-value maleic resin	50.0
Glycol	40.0
Amine	10.0
	100.0

The printing involved will mainly consist of solids, type and linework, and the print usually has a semi-matt appearance. On this type of substrate, two colours can usually be superimposed, wet-on-wet without tack grading, but on more-solid boards, tack adjustments with mixtures of water and glycols will required. A low viscosity ink aids distribution on the rollers and the transfer properties from the rubber stereo to the kraft paper.

UV inks

There has been an increased demand for adhesive labels of all types, much of it generated by the requirements for product and other health and safety information.

The letterpress process, combined with the instant drying given by UV curing, has been found useful in producing what is often short-run work.

The formulation technology is similar to that used for UV curing offset inks (Chapter 11).

5.9 INK-RELATED PROBLEMS AND THEIR POSSIBLE SOLUTION

Many of the problems associated with letterpress printing are due to the relatively high film weights carried, which affect the setting and drying properties of the inks used and also contribute to the misting characteristics on rotary presses. Strike-through and filling-in problems can arise with the thin inks used on absorbent newsprints, while inks with too high a tack level can lead to linting.

Mottle

Mottle shows itself as uneven colour, caused by uneven inking of the paper. It is most likely to occur on work printed on platen presses and when the ink is too weak to obtain the correct density with normal film weights. The excess ink which has to be carried, shows up the local variations of film thickness and this is particularly noticeable when transparent blues, greens and browns are being printed. Over-reduction by the printer to overcome a tack problem is a common cause of mottle and this can be helped by recommending a gel reducer. Inks with some opacity are less prone to give this problem.

Set-off

The more an ink 'stands up' on the surface of the paper, the more it is likely to set-off on to the reverse side of the following sheet. Therefore, set-off is increased by:

- thick films of ink;
- non-absorbent substrates;
- smooth papers (which allow close contact between sheets);
- highly resinous inks which dry to give an ink film with a high tack surface;
- slow setting inks;
- faster running presses;
- large stacks of print.

Formulating the ink to have sufficient colour strength to achieve the density of colour required, at a low film weight, will therefore help reduce set-off.

The formulation of inks containing resinous varnishes must ensure that the tack level of the ink film, as it sets, does not rise sufficiently to allow it to adhere to the underneath side of the following sheet. Choice of the resin/solvent blend to give good solvent release is therefore important. Some rise in tack level will almost certainly occur and can only be prevented from becoming a problem by the use of anti-set-off sprays on the press.

Marking-off

Newspaper inks which set on top of the substrate, will tend to mark-off on to the path rollers of the press. Usually a deposit of ink and paper fibres builds up on the rollers causing a background tint of the ink, while the prints will also exhibit poor rub properties due to the heavy ink film deposited. Keeping the ink as strong as possible, with a low viscosity and reduced resin content, helps overcome this problem, but needs to be balanced against the requirements of strike-through and rub resistance.

Second impression set-off

This is a special form of set-off associated with work which is 'perfected' that is to say, printed on both sides in close succession, as on newspaper rotary machines. When the second side is being printed, the first is face downwards on the second impression cylinder, and the print is transferred under pressure.

Drying and rub-resistance

Bearing in mind the drying methods mentioned above, the recommendation of the correct ink for a particular substrate is highly important, especially since so many letterpress inks are formulated to be non-drying. Inks of the quick-setting type, which have oxidation drying capability, should be suitable for most requirements, but the use of extra wax paste and liquid driers may be necessary on harder stocks, provided this does not lead to the ink drying on the press during short stoppages.

During the production run of a newspaper, the presses will stop for various reasons. If the plates are being changed, or for web breaks, the stoppage may be prolonged. The presses are so designed that the impression between printing forme, ink rollers and paper is not released when the press stops and they remain in contact with one another. The ink must not dry on the rollers as this could cause uneven inking. Traditional mineral-oil-based inks are quite satisfactory, but the use of harder drying, more rub resistant inks, can lead to problems.

Misting

Ink mist and sling are the two components of the phenomenon which is sometimes described as 'ink fly'. The ink on a pair of fast rotating rollers is drawn out into fine filaments as the ink is transferred from one roller to the other. These filaments of ink split at each roller simultaneously and the central part is released into the atmosphere as a microscopic particle of ink. These particles are electrically charged and move away from the rollers into the atmosphere.

Long flowing inks are more prone to misting than short flow inks and the addition of 'low mist' additives is the usual cure, although a compromise with 'ease of delivery' to the press must be part of the specification.

Strike-through

This problem is always present, to a certain extent, when printing on newsprint. It can be minimized by keeping the viscosity as high as the press and set-off properties will allow. The use of refined grades of oil also helps reduce the problem because the contrast of the colour of the oil, with that of the substrate, is not so marked.

Filling-in

This occurs when the non-printing area becomes contaminated and begins to show on the printed copy. The main reason for this problem is the unsuitability of the ink for the paper. An ink with too high a tack value for the surface structure of the substrate being used, will cause the short paper fibres from the surface to be transferred to the linking system. These fibres will eventually adhere to the forme and fill-in the non-printing area until they themselves start to print. This is commonly called linting.

An alternative cause is where a low viscosity ink does not contain sufficient cohesive structure and flows down the side of the relief character, and is transferred to the substrate under the impression pressure. Poorly dispersed ink can also achieve the same result, as the unground material fills-in the non-printing area.

Hanging back in the duct

Hanging back in the duct is usually associated with short flowing inks which are liable to set-up quickly and this may be due to over-pigmentation, the inclusion of too much wax paste or very short vehicle or oil. On rotary presses, with open ducts, the lint from the fast moving webs easily contaminates the ink and unless good housekeeping is employed, can cause it to lose its flow properties.

Under normal circumstances, the ink should have sufficient flow to follow the rotating duct rollers during the press run, and in mild cases, where it 'hangs back', the addition of a bodied varnish or oil may be sufficient to overcome the problem. More severe hanging back often requires complete reformulation.

Where bulk rotary newspaper black ink is concerned, the problem will almost certainly have shown previously, because of the need to pump these inks, often for long distances. An increase in the asphaltum complex solution, used in the formulation, will cure most problems where the ink has high yield value.

5.10 SPECIAL PURPOSE APPLICATIONS

The letterpress process has been found very successful for a number of purposes using specially formulated inks.

Magnetic inks

The processing of cheques and other security documents has made possible a highly centralized clearing system.

The inks used are formulated using a synthetic iron oxide pigment (ferrous–ferric oxide) which has been dispersed in a medium with good wetting properties, such as a low viscosity alkyd resin, which may have been combined with a resin/oil varnish such as that described in the section on process inks. The ink must have sufficient flow to distribute on the press but must have sufficient body to print the characters clearly, so that the correct response is registered by the sorting equipment.

Carbonizing inks

Although carbon paper has largely been replaced by NCR (no carbon required) coatings in multipart sets, a carbon patch can be applied using a carbonizing ink. The inks need to be formulated so that the resultant film is hard enough not to smear but waxy enough to give a good copy over the life of the document. The choice of paper is important to achieve this.

The pigment is dispersed into a soft wax or lanolin-based medium and the ink must have sufficient flow from the duct, but some care is required to prevent spraying on faster running presses.

Inks with a higher wax content are still widely used where the ink duct and cylinder are heated above the melting point of the wax and the film applied then solidifies as it cools.

Densitizing inks

It is possible to prevent the duplication of part of an NCR sheet by printing a patch of densitizing ink on the coated front side of one of the parts of the set. Inks can either neutralize the chemical on the reverse of the top sheet or can form a barrier.

The film thickness of the ink must be sufficient to prevent duplication in either case, but should be kept to the minimum which is effective to prevent set-off.

Fugitive inks

As part of the countermeasures employed by the authorities against fraudulent adulteration of documents, the use of water-based 'fugitive' dye inks play an important part.

Often, security documents, cheques, passports, airline tickets, etc., include a background design which is fugitive to various chemicals and will readily show if changes have been attempted on the original.

Sympathetic inks

The term 'sympathetic' is given to inks which undergo a colour or other change when they are chemically treated or exposed to a change in external conditions, such as temperature or humidity.

While forgery detection is one use, a related colour change system is used to produce children's magic-painting books. The drawing is printed using a letterpress black containing a non-toxic water-soluble dye, which 'bleeds' when water is applied to the design. The use of a penetration drying formulation avoids sealing-in the dye.

CHAPTER 6
Lithographic inks

The theory and mechanics of the lithographic printing process have been dealt with in Chapter 2 of this manual. In this chapter the various types of lithographic ink formulations, together with their properties and the raw materials used, will be examined.

6.1 GENERAL INTRODUCTION TO LITHOGRAPHY

Lithographic printing falls into two main product groups, publication and packaging and these are produced by both the web and sheet-fed processes.

Publication products

Newspaper printing throughout the world is dominated by the use of cold-set inks which 'set' by penetration into the absorbent paper stocks used. Offset lithography is the most widely used printing process although letterpress is still used, and more recently the flexographic process has been introduced. Magazines, catalogues and brochures are all major products of web-offset. Publications printed on absorbent stock dry by absorption but heat-set drying will be required for coated substrates. The increased use of heat-set drying has resulted in the process being used to produce high quality long runs in areas which were traditionally gravure printed. The position in the market between the two processes is always shifting as the economics of the processes vary. Sheet-fed lithography may also be used where the length of the run is short.

Other major publication products printed by lithography are directories and books. Cold-set web-offset is commonly used, but some books are produced on sheet-fed perfecting presses using cold-set inks.

Packaging

Lithographic printed packaging comprises the major product groups of folding cartons, labels and metal decorating. Folding cartons are particularly diverse in their design and the requirements of the product packaged. Their design ranges from the aesthetic appeal of luxury cosmetics, spirits and tobacco packaging to functional strength for heavy mechanical components. Product-dictated performance demands a range from soap and detergent resistance on relevant cartons to low odour and no tendency to taint the contents for food and confectionery packaging. Sheet-fed lithographic printing is so successful in these areas that it is the dominant process for producing folding cartons. Gravure competes for some long run-length, fixed cut-off cartons, and a little letterpress printing remains, but otherwise sheet-fed lithography is the automatic choice. The situation is similar for paper labels, both in terms of diversity of requirement but considerable volumes are also gravure printed. However, in self-adhesive label production, flexography and UV rotary letterpress have become the major processes used on specialized narrow web presses. Attempts to formulate a wide range of wholly satisfactory inks for film, plastics and foil substrates are impeded by the limitations of the lithographic process. This is not to say that such materials cannot be printed by lithography, but achieving full ink adhesion on the stock can be more problematic than with gravure, screen or flexographic systems. UV-curable ink systems can alleviate handling and adhesion problems.

Lithography and dry offset are both used for printing metal-decorating inks, either direct on the metal surface or over a transparent size or white pigmented coating. Metal is used for packaging food, household products, etc. where its strength and wide product resistance are of great value.

Nature of the lithographic process

Lithography is a printing process that relies on a chemical distinction between the image and non-image areas in the same plane, and not on an image in relief. It is fully described in Chapter 2. With the exception of driography, or waterless plate printing, all types of lithographic plate require application of an aqueous fount solution to activate and maintain the distinction between ink-accepting image areas and ink-repelling non-image areas. To explain why films of ink and fount form in the respective areas of the plate requires a study of surface tension. Surface tension is described as the internal pressure that restrains a liquid from flowing. Polar liquids such as water have a high surface tension, non-polar liquids have a low surface tension as does a lithographic printing ink. When a drop of liquid is placed on a solid surface the relationship between the surface energy of the solid and the surface tension of the liquid determines whether the droplet will spread across the surface or form globules. If the liquid is placed on a surface of lower surface energy than itself, it will form into globules and fail to wet the surface. Conversely, if the liquid has a lower surface energy than the solid surface, the liquid will spread across, or wet, the surface. If a liquid wets a surface, it is said to have a low contact angle; if it forms globules, it is said to have a high contact angle (Fig. 6.1).

When applied to lithography, the distinction between image and non-image areas of the printing plate is sufficient to cause the following effects:

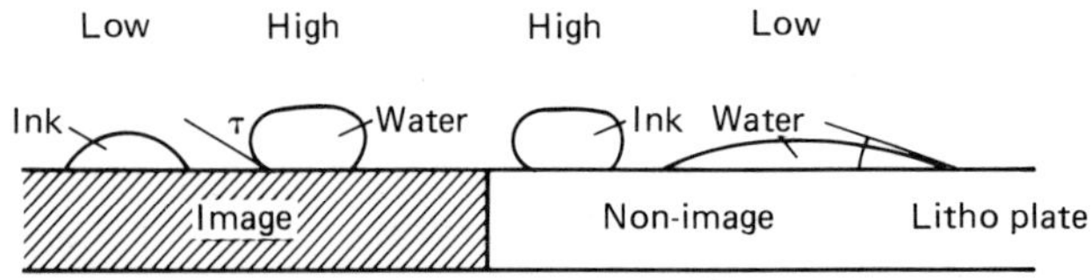

Fig. 6.1 Lithographic plate surface characteristics. The low contact angle τ of water in the non-image area allows wetting with water. Water's high contact angle in the image area causes it to form into globules.

- aqueous fount solution spreads on the non-image area because of its low contact angle;
- Aqueous fount solution in the image area is unable to form a continuous film because of its high contact angle, leaving the image free to accept ink.

Thus the image will be inked and the non-image area ink free. This highly simplified explanation of the principle of lithography is based on a static situation. In practice, the situation is complicated because ink and fount are being continuously applied from roller systems to the printing plate.

Having achieved clear and precise inking of the image on the plate, through the mechanisms described, the ink is then transferred to the stock to be printed via a resilient offset blanket. The use of a rubber blanket ensures good plate life and high quality transference even to substrates of poor smoothness. This happens in all lithographic printing with the exception of the direct, or di-litho in letterpress machines converted for the printing of newspapers and the direct printing of posters, where high film weights of ink are required.

For this reason, the terms 'lithographic' and 'offset' are virtually synonymous in the printing industry.

Lithographic plates can be prepared at relatively low unit cost as base metal preparation and the chemistry employed in coatings are sophisticated but not unduly expensive. Processing of plates through exposure, development and press preparation involves similar techniques, often akin to photographic processing, and can be partially automated to ensure high productivity. Fine-detail line and halftone work lithographic printing is generally cheaper and quicker than comparable stages in the other major printing processes.

In summary, therefore, the lithographic process is a balance between the properties of the ink fount and plate and the characteristics of each determine the success of the process.

General characteristics of lithographic inks

In order to achieve the required ink/fount balance, the degree of emulsification of ink/fount solution must be controlled. Achieving this under press operating conditions whilst retaining all the other necessary properties of the ink is the essence of lithographic ink formulation.

The mechanisms involved in the lithographic process impose both chemical and physical restrictions on the formulation of lithographic inks. In order to achieve controlled fount take-up, both the chemical affinity of the ink to fount and the rheology of the ink used have to be taken into consideration.

The wide range of printing conditions that are encountered within the industry on the various presses in use, demand inks with greatly differing physical properties. Viscosities can range from 2 to 30 PaS depending on the speed and type of printing press to be used.

Low viscosity lithographic inks can cause emulsification problems, particularly at high speed. This is because even where there is no chemical affinity between two liquids, they become more readily emulsifiable as their viscosities approach each other. A degree of emulsification is necessary to remove the unwanted fount from the image area of the plate but excess would negate the lithographic principle by destroying the distinction between fount and ink causing the non-image area to print.

For the successful application of offset lithographic inks a long ink-distribution roller train is required. This is because a thin, even ink film must be delivered to the plate and substantial rolling and distribution is essential to break down the viscous ink to this state. Inevitably, the use of such a roller system imposes its own restrictions on the ink. Evaporation of volatile materials from the ink must be severely limited as the ink is exposed for long periods in a high surface area to volume ratio on the distribution train. The necessary rolling and distribution work is achieved by using a system consisting of alternate metal- and elastomeric-surfaced rollers. The elastomers used here and for the offset blanket surface can be chosen from a wide variety of synthetic rubbers and polymers. Nevertheless, most readily available and suitably resilient materials are susceptible to attack from strong chemical solvents, particularly under the influence of the pressures and temperatures generated on an offset lithographic press. Such solvent attack leads to softening and swelling of the elastomers. The latter produces further physical stress in the form of increased pressures and softening renders the surface increasingly susceptible to physical disintegration.

Ink viscosity and distribution requirements therefore prevent the use of significant proportions of both volatile materials and strong solvents. The inability to use powerful solvents in lithographic formulations naturally produces a severe restriction on the choice of binder chemistry, since only resins soluble in weak solvents can be considered.

Chemical activity

Where there is no chemical affinity between ink and fount solution, emulsification can be difficult. The inclusions of materials with active groups in pigments and vegetable oils that can 'combine' with the fount mean that fount pick-up can be controlled. The chemical nature of the raw materials used, therefore, has a marked effect on the lithographic properties of the ink and its press performance.

Film thickness considerations

Offset lithography utilizes thinner ink films than any of the other major printing processes. This is partly because the planographic offset plate has a relatively low capacity for carrying ink, and partly due to the extra film splitting involved in transferring ink from plate to substrate via the offset blanket. This latter effect is not always as great as it might at first appear since the performance of a modern blanket, assisted by poor ink adhesion to metal and a capillary-draw effect from the paper substrate, can lead to preferential ink transfer from plate to blanket to paper. This results in greater than 50% transfer at both nips. Nevertheless, inkfilm thickness on print is normally in the 1–3 μm range.

It is therefore essential for lithographic inks that are required to print full-strength colours to be formulated with high pigmentation. Further, very careful manufacture is required to ensure satisfactory pigment dispersion for both strength development in the ink and freedom from agglomerates that will be significantly larger than the thickness of film transferred. Such large particles, if present in excessive numbers, can produce various adverse effects such as piling on the blanket reduced gloss and poor rub resistance.

Drying mechanisms

The inks used in the lithographic process dry by the 'classic' methods of penetration, oxidation/polymerization, evaporation and chemical reaction. However, because the ink film is relatively low and considerable amounts of fount are present in the ink film, it is difficult to define precisely how each type of ink dries. Inks which are described as heat-set and dry mainly by evaporation, will set by absorption of part of the ink into the paper fibres as will inks which dry by oxidation/polymerization. On the other hand, inks which dry mainly by penetration may 'react' after printing to enhance the ink film. Thus, drying is by:

- penetration
- oxidation/polymerization
- quick-set
- heat-set
- radiation.

Penetration

Penetration is the drying mechanism involved in the web-offset cold-set printing of books and newspapers on absorbent uncoated paper stocks. If drying is defined as the conversion of the fluid ink to a solid print film, then strictly speaking the ink in this case does not dry. The ink remains fluid but becomes sufficiently entrapped in the interstices between the substrate fibres for its mobility to be severely restricted. Problems can be encountered if the ink film thickness exceeds the absorptive capacity of the paper or if there is insufficient time between printing and handling to allow full absorption. Both problems can be encountered in newspaper printing but the cold-set penetration mechanism does allow the distinct advantage of high-speed production at minimum cost.

Oxidation/polymerization

Oxidation is the classical drying mechanism for lithographic inks, involving the oxygen-induced free-radical polymerization of unsaturated (drying) vegetable oils such as linseed and tung. It is a chemical process which can be catalysed further or accelerated by small amounts of appropriate metal, usually transition-metal, driers (section 6.4). Inks which dry by oxidation/polymerization are used where a degree of film strength is required to impart rub or scratch resistance. Under adverse conditions, an ink which dries on the substrate may also dry on the rollers and/or skin-over in the duct. Great care must be taken with this type of ink to ensure that the balance between runability and print properties is maintained.

Quick-set

Whilst quick-set drying is usually associated with sheet-fed lithography, both heat-set and cold-set formulations use the mechanism for part of their drying. The quick-set vehicle consists of two phases which have limited compatibility. One phase is a suspension of resin in drying or non-drying oil, the other phase is low viscosity mineral oil or high boiling-range petroleum distillates. The two components must remain stable and compatible to enable pigment dispersion and printing to take place. However, once printed on an absorbent stock, the capillary action of the stock draws in the low viscosity phase leaving the high viscosity phase on the surface. After the print has set oxidation/polymerization drying can take place within any drying oils or alkyds present. Both heat-set and cold-set formulations do have a degree of quick-set mechanism as the formulations often include varnishes and oils which have a much higher viscosity to the reducing solvents used to reduce the viscosity.

Heat-set

Heat-set inks dry by the evaporation of the solvent phase in high temperature ovens. The solid phase is usually a high viscosity colloidal suspension of resins and alkyds in low viscosity drying or semi-drying oils. These varnishes are reduced in viscosity by petroleum distillate, usually in the 260–330°C boiling range. The high temperatures (up to 50°C) that can be encountered on the roller train of high speed presses necessitate the use of non-volatile solvents. However, the lighter fractions can evaporate at these temperatures causing possible increases in ink tack and viscosity and it is necessary to formulate to prevent this happening.

It is the correct balance between press stability and fast drying in the ovens that is the key to the success of heat-set web-offset inks. The use of latent solvents with low resin solubility improves solvent release from the ink film. Other additives such as slip aids and waxes may also have a detrimental effect on drying, causing a thin layer of wax to form on the surface of the ink film which retards solvent evaporation, due to the solvent retention properties of waxes.

Radiation curing

Radiation curing mechanisms are fully described in Chapter 11. It is sufficient to note here that free-radical chain reactions are involved which produce a highly cross-linked dry film almost instantaneously.

Substantial energy input is required, either in the form of highly energetic electron particles which possess energy in the appropriate form to initiate the chain reaction directly or, more commonly in the lithographic printing industry, in the form of UV radiation. UV radiation (an electromagnetic energy form) has to be converted into usable chemical energy by processes involving special 'photo-initiator' chemicals incorporated into UV-curing inks.

Physical properties

We have seen on page 342 that the lithographic process dictates certain physical characteristics of the inks. These can be described under the headings of rheology and tack and ink/fount relationships. Chapter 14 covers in detail the theory of ink rheology and tack and the practical measurement of these properties. Consequently, these topics are only outlined in this section.

Rheology

Four main rheological properties are relevant to lithographic inks:

- viscosity
- yield value
- thixotropy
- flow

Viscosity

Viscosity is defined as the ratio of shearing stress (shearing or deformation force per unit area) to shearing rate (velocity gradient of flow order deformation). Lithographic inks are substantially non-Newtonian as this ratio is not constant but is shear-rate dependent. This is primarily due to the high content of dispersed pigment present and, indeed, unpigmented lithographic varnishes frequently approach Newtonian character. The non-Newtonian behaviour necessitates viscosity measurement over the full range of shear rates that inks can encounter in use. This range is very wide since very low shear rates are encountered in removing ink from the can and in the ink duct, whereas exceedingly high rates of shear exist in the roller nips of fast running presses. Consequently, it is not normally possible to measure viscosity under all relevant conditions on a single viscometer. A variety of techniques are employed. The two most common instrument types are the falling-rod viscometer for high shear rates, and the cone and plate viscometer for lower shear conditions.

Yield value

One important feature of the non-Newtonian behaviour of lithographic inks is that they usually demonstrate a finite yield value. That is, a distinct shear stress or force is required before any deformation or flow takes place. It is difficult to obtain a direct measurement of yield value without sophisticated and expensive controlled stress rotational viscometers. Normally an estimation of yield value is made by extrapolating stress results obtained at much higher shear rates to the zero shear rate, using a falling-rod viscometer. As with any extrapolation, validity has a degree of uncertainty.

Thixotropy

Lithographic inks are generally thixotropic as well as non-Newtonian. Their viscosity depends not only on the rate of shear at a given point in time but also on the previous shearing history of the sample under investigation. In everyday terms, inks thicken on standing, and this 'structure' can be broken down by stirring. Thixotropic behaviour is a major complication in the measurement of ink rheology. Methods employed on the commonly used viscometers are usually specifically designed to eliminate the influence of thixotropy as far as possible.

Flow

A simple study of ink rheology may be conducted by allowing an ink sample to flow under the influence of gravity down some form of inclined plane. This technique enables a quantitative determination of flow to be made by recording distance travelled by the ink sample against time. However, flow in this sense is not a fundamental rheological property but rather a complex interaction of viscosity, yield value and possibly thixotropy, the latter depending on the technique used for sample presentation to the inclined plane.

All rheological properties are greatly influenced by temperature. For this reason all measurements and assessments must either be carried out on temperature controlled equipment, preferably operating in a controlled, environment, or in a strictly comparative manner against reference samples of known performance.

Relevance of rheology

The relevance of rheology to press and print performance can be seen by considering the transport of the ink from the can or container to its final position as the printed image.

High viscosity, substantial thixotropy and high yield value can all produce problems in transferring the ink from the container to the duct of the press. This is particularly relevant where pumps are used to transport the ink from bulk delivery units or storage tanks to the press, unless the pumps are of the follower-plate type which are able to move viscous or poor-flowing ink by a basic high-pressure pushing action. Conversely, low viscosity, minimal thixotropy and low yield value can pose problems of dripping with inks that are hand-fed, usually by some form of palette knife, from container to duct. A frequent problem area is cold-set web-offset printing inks supplied in pails rather than in some form of bulk pumping system. Due to the demands of the presses and substrates used, inks are usually relatively low viscosity products and problems with retaining all the ink on the palette knife during feeding may be encountered if the ink has too low a yield value or if its thixotropy is very readily broken down. The problem may be much more than just nuisance value if ink falling off the knife lands on a newsprint web in such a way as to cause a web break.

Once in the duct, it is important that the ink flows readily into the gap between the duct blade and roller in order to begin its journey up the ink distribution train. This will not occur if the ink has a yield value above the relatively low shear stresses that exist in the duct, nor if there is too much thixotropy. Thixotropic structure may inhibit ink feed if it builds up significantly during the dwell time of the ink in the duct or if it was inadequately broken down when the ink was fed onto the press. Whether due to thixotropy or yield value, poor flow in the duct leads to inadequate or sporadic feeding of ink up the inking rollers, a condition known as 'hanging back' (Fig. 6.2). This problem may be alleviated by manual agitation of the ink in the duct or by the use of automatic duct-agitators, which rotate as they traverse through the ink and are shaped in such a way as to push the ink towards the duct-gap. Clearly, both techniques break down thixotropy and serve to raise shear stresses above the ink's yield value. Again, a converse problem can exist, most commonly on newspaper presses, where low viscosity free-flowing ink drips from the sides of the duct and possibly even from under the duct blade. The situation is aggravated where printing units are supplied with ink even when they are not being used, as can happen in cold-set web-offset production of newspapers on multiunit presses. As with knifeability difficulties, higher ink yield value and more rapid thixotropic set-up reduce the problem although duct engineering is obviously an important factor.

As discussed on page 345 a long ink distribution train is employed to roll out relatively viscous lithographic inks to an even film for application to the offset plate. While such substantial rolling power is a design feature of lithographic presses, it is still necessary for the ink to have the correct rheological characteristics to ensure even distribution. Too high a viscosity and poor flow can impede distribution, particularly where the image demands rapid ink replenishment in certain areas. From the inking rollers through to transference to the substrate, the rheological characteristics of the ink may be modified by the presence of emulsified fount solution. This must always be borne in mind when attempting to relate laboratory rheological studies with commercial press performance. The two properties are interrelated as emulsification is influenced to a large extent by viscosity (page 344). A

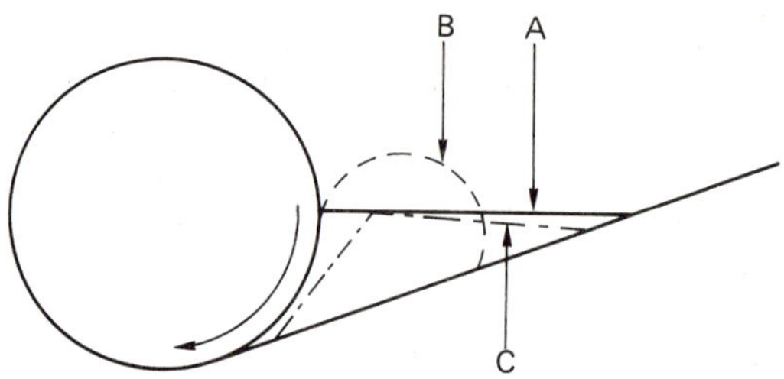

Fig. 6.2 Ink profiles in a duct: A, low yield value and low viscosity – feeds: B, high yield value and high viscosity – feeds; C, high yield value, low viscosity – hangs back.

large measure of the ink formulator's skill centres around producing inks which show rapid, but controlled, emulsification that does not inhibit distribution or transference. Chemistry, as well as rheology, has a major influence here, but generally lower viscosity inks emulsify more rapidly but may be closer to the danger point at which distribution and transference can be impaired.

Problems at the ink distribution stage may also be encountered if droplets of ink are expelled or 'fly' from the roller train. The droplets may be formed by cavitation and multiple string rupture during the film splitting that occurs at the exit side of an ink roller train nip. Cohesive, and even electrostatic, factors are believed to be involved in flying, but rheological properties also influence string formation. Low viscosity, and particularly long flow, appear to promote flying and the problem is often eliminated by incorporation of particulate structuring agents which increase viscosity and significantly restrict flow.

In the transference stages from plate to blanket to substrate, the rheology of the ink, or, more precisely, that of the fount-in-ink emulsion, affects both the amount of ink printed and the fidelity of the reproduction. Generally, low viscosity and yield value will promote transference but these factors will definitely result in a greater physical dot gain in halftone printing (page 348). The latter phenomenon arises from the physical squashing of the fluid ink that inevitably results from the pressures existing in the plate/blanket and blanket/impression transference nips. Naturally, high viscosities and yield values give reduced flow, and hence less squash, at a given pressure.

Tack

Tack is a measure of the forces required to split a single film of ink into two. Such film splitting is influenced by rheological and adhesive properties, in addition to the internal cohesion of the ink. Various instruments have been developed to measure tack but all rely on the principle of measuring the force exerted on a roller as it splits a single ink film in a nip into two films at the nip exit. A subjective, comparative assessment of tack can be made using fingertips to separate a thin film of ink on a slab into two films, one portion remaining on the slab and the other adhering to the fingertip. A skilled technician, using repetitive dabbing of two inks with adjacent fingers, can make a surprisingly good comparison in this way. However, the method suffers the disadvantages of all such approaches which depend on individual skill and which do not produce a quantitative result.

Tack does not vary as rapidly with temperature as does viscosity. Nevertheless, there is an influence and all competent tack measuring devices incorporate appropriate temperature control of the roller system. As an empirical quantity, tack can be significantly influenced by film thickness and it is important that the volume of ink applied to tack measuring devices is carefully controlled. Even with temperature and film weight control, different tack machines can produce different results on given inks. Where the machines are not of a single type, two main factors are involved.

- Since tack is an empirical quantity, there is no scientifically defined unit of tack. Various manufacturers of tack instruments have established their own arbitrary scales.
- Different designs involve different roller surfaces, geometries and speeds. These variants produce different interactions of the cohesive, rheological and adhesive components that contribute to tack. This accounts for the phenomenon where ink A is higher tack than ink B on one instrument, but B is higher than A on another. Such differences also prevent simple scale conversion factors which might otherwise be a solution to the problem of differing arbitrary scales.

Instruments of the same design from a single manufacturer would be expected to overcome these problems of inter-instrument comparability. In reality, agreement is not always perfect. It appears that this is due to the basic tack machine approach being very susceptible to influences from bearing wear, roller surface condition and other mechanical and frictional factors.

Despite all this, tack is a major influence on ink performance and currently available instrumentation, whilst not perfect, is fully satisfactory in enabling control of this parameter.

Relevance of tack

Tack is relevant to all stages where distribution or transference takes place. Too high a tack can cause some form of rupture in the substrate since this may require less work than producing the very high forces necessary to split the ink film. In the most severe cases, part or all of the paper will be pulled onto the surface of the offset blanket and the press will have to be stopped and cleaned immediately. Where the tack is only marginally too high for the strength of the substrate, a small degree of coating pick or fibre linting will occur at each impression. This will obviously impair the smoothness, and possibly the gloss, of the print and will eventually result in a deposit of fibre or coating on the blanket which begins seriously to affect print quality, necessitating a blanket wash.

From the point of view of minimizing the risk of substrate disruption, particularly with weak, low-quality papers, it is clearly advantageous to use inks with as low a tack as possible. However, very low tacks can produce a variety of problems from the ink roller train right through to final print quality.

- Lithographic press distribution trains are usually engineered such that only alternate rollers are driven by the press motor. Every other roller relies on frictional contact with a press-driven roller to ensure rotation. Inks with very low tack may have insufficient cohesion to transmit the frictional forces across the nips, resulting in roller slippage and inadequate ink feed and distribution.
- In wet-on-wet printing on multi-unit presses, it is necessary, in relevant image areas, for ink to transfer or 'trap' on top of a previously printed, wet ink film. If the initial colours printed are very low in tack they may be susceptible to an unacceptable degree of 'back-trapping' from subsequently printed, higher tack, inks. This will occur if less

force is required to split the previously printed ink film from the paper than is necessary to split the subsequent ink from its blanket. Naturally, such poor trapping completely disrupts colour reproduction and may easily result in an unsatisfactory final print.

- The influence of rheology on physical dot gain in halftone printing was referred to in the previous sub-section. Studies have shown that dot gain is also dependent on tack. It is less easy to identify a simple explanation for this effect but it seems possible that the low cohesion of dots of low-tack ink provides less resistance to the squashing action at the plate/blanket and blanket/substrate impression nips. Thus, excessively low-tack inks may produce unacceptably high dot gain.

Ink–fount relationships

During the lithographic printing process, fount is emulsified into the ink and may change its physical properties. In order that the printing characteristics of the ink remain consistent, it should retain its original properties over a wide range of emulsified fount additions. The tack of an ink will invariably drop with emulsified fount and this must be taken into consideration if ink transfer is to be maintained.

6.2 COLD-SET LITHOGRAPHIC INKS

The term 'cold-set' is used to describe a printing method in which the ink is applied by the lithographic process to an absorbant substrate and 'dries' by penetration (page 346). External heating is not normally applied. Newspaper stock is the principal material used in this process.

The inks do not normally include any raw materials which can be polymerized or cross-linked by the oxidation drying process although a small percentage of drying oils may be added to modify the rheology of the ink. Volatile solvents are not usually present in the formulation and therefore the use of external driers does not improve the film strength of the ink. Some work has been carried out with the use of IR but this is used to improve the setting of the ink. The statement that the final characteristics of the ink film on the substrate are the same as the ink when delivered oversimplifies the properties of the final printed product, but does give a good insight into the unique characteristics of the process.

The fact that 'non-drying' inks are used indicates that the process is inherently stable. The inks used do not skin over in the ink duct or dry on the ink-distributing rollers, making the process ideal for long production runs with many short press stops. However, an ink which is non-drying on the printing press is also non-drying on the paper, and this fact highlights the major disadvantage of the system. The printed ink film lacks the film strength of a polymerized or heat-set system. However, for its major use involving long production runs with many planned and forced stoppages printing on absorbent stock, this disadvantage is overshadowed by the need to have a very stable non-drying ink.

The first provincial newspaper to be printed by the lithographic process was about 25 years ago and since then the majority of national newspapers printed in the UK and abroad have followed suit. This has resulted in a print revolution in which many of the major newspapers, worldwide, have introduced colour and converted from letterpress production.

Modern presses capable of over 75 000 copies per hour have been installed. The control of ink, fount, plates and paper is highly automated, and once the parameters have been set, the presses operate with the minimum of attention. This places considerable limitations on the suppliers of all these materials to ensure that their products operate successfully over a wide range of operating conditions.

Many other types of publication are printed by this process and include telephone directories, local business guides, advertising, etc. and in fact any product where the very highest quality is not demanded, but where economy and long production runs are required. A typical press is shown in Fig 6.3.

Newspaper inks

The printing of newspapers, where production can often be measured in hundreds of thousands of copies and within the production run many stoppages can occur, as the editions are changed, is ideally suited to the cold-set concept

Web-offset newspapers are designed for high speed continuous run-

Fig. 6.3 Modern newspaper cold-set press. (Courtesy of Perschke Price Service Organisation)

ning which is necessary to meet the tight schedules of newspaper production. The inks must be formulated to give the required quality on the substrate but at the same time cause no delay to production. The print specification often calls for an ink which sets very quickly to give a dense sharp image with minimum rub off. A cold-set web-offset newspaper ink should therefore possess the characteristics described below.

Low cost

The cost of printing ink in relation to the total cost of production of newspapers is below 5% and one may be forgiven for assuming that with such low value the cost or printing inks is not a significant factor. However, the volume of newspaper ink used in production may be so large that the total expenditure on printing ink could be very considerable over a period of a year.

It is against this background of high volumes and large gross ink expenditure that the need to produce inks with low cost must be measured.

Low cost covers a number of characteristics of the ink such as:

- cost per kilo
- mileage
- wastage produced.

Each of these topics will be dealt with in detail later but is sufficient to highlight at this stage that selling price alone does not necessarily indicate good value or a more economic ink. Nevertheless, the selling price of newspaper inks does play a large part in the success of any ink project and the need to keep the raw material cost to the lowest levels consistent with all other properties is paramount in the formulation of this type of ink.

Press performance

The pressures of achieving production targets place great demands on the press crews that operate newspaper presses and the time that they are able to devote to the adjustment of ink–fount relationships is minimal, particularly during the critical start-up period. Add to this the variation in plate and paper quality and it is easy to understand why newspaper inks must have excellent lithographic properties and be press friendly.

During the start-up or commencement of the production run the plate is invariably fully inked and then excess fount is applied to clean the non-image area of the plate. It is essential that this operation takes place as quickly as possible in order that paper wastage is kept to a minimum. The ink must therefore quickly allow the fount onto the non-image area and be capable of emulsifying the excess fount applied at this stage. As soon as clean copy is produced and the quality of print is suitable, the speed of the press is increased until production levels are achieved. During this operation the ratio of ink to fount applied to the plate is controlled on the press by a small computer program, described as the ink–fount curve (Fig. 6.4).

The amount of fount applied for a given speed is set to that which will just keep the plate clean. As can be seen from the diagram, the ratio does

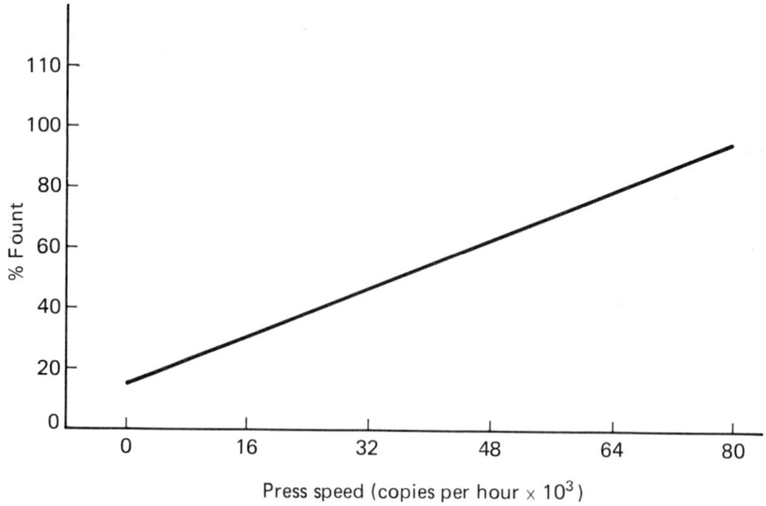

Fig. 6.4 Ratio of fount to press speed. MAN Colorman 40 press 1.5% fount concentration.

vary greatly as the speed increases and the whole operation from start to running speed demands an ink which will operate under widely differing ink–fount conditions without causing any decrease in quality.

The temperatures of both ink and fount increase as the length of time the press operates at high speed continues, and this also has an effect on the lithographic properties of the ink. As temperature rises, the viscosity of both ink and fount decreases. The surface tension characteristics of both ink and fount also change with temperature and the ink must be formulated to run under these wide variations of conditions.

Because of the high cost of running web-offset newspaper presses, trials of new inks are restricted. It is very important that parameters such as ink–fount balance are studied under laboratory conditions first, where much information can be gathered and formulations suitably adjusted.

A number of successful laboratory tests have been devised to determine lithographic properties and they usually involve measuring various rheological properties of the ink containing amounts of emulsified fount. Interpretation of the results is relative since the test procedures used relate to the energy required to emulsify fount and the static conditions of both the emulsification and the measurement of the related properties are the nearest one can get to actual press conditions.

Formulation principles

Cold-set web-offset inks have severe economic restrictions imposed on them by the nature of the work to be printed and this together with the influence that the ingredients have on the lithographic properties of the ink dictates very careful selection of the raw materials used.

Table 6.1 indicates the influence that each raw material category has on

Table 6.1 Lithographic properties

Material	Rheology	Tack	Ink–fount relationships	Rub	Cost	Setting
Pigments	4	2	5	3	5	2
Resin vehicles	5	5	4	4	3	4
Solvent oils	5	5	4	3	2	4
Additives	4–5	5	5	1	1	2
Waxes	1	2	1	5	1	2

the properties of the final ink. The influence has been graded from 1 to 5 with 5 being the strongest influence.

Pigments

The choice of pigment has an influence on many properties of the ink and the correct choice is critical for both carbon blacks and colour. The main pigment used in cold-set formulation is black, but considerable amounts of pigment yellow 12, 13, Rubine PR 57 and Cyan PBl 15:3 are used.

Smaller amounts of other organic colours are used to obtain 'spot' colours. Spot colours are areas of single colour which are printed in the publication to highlight an editorial story or are used in an advertisement. The hue and brightness and exactness of colour shade are such that it is often not possible to obtain the shade by the four-colour process and therefore bespoke inks must be used.

Cold-set web-offset formulations

Black inks

Cold-set web-offset blacks fall into the three general categories:

- low cost general-purpose inks
- low rub inks
- vegetable oil-based inks.

Inks from each category will have to meet the same broad specification for press performance and the choice of which ink to use depends under what market influences the printer is operating.

Low cost general-purpose inks

Inks are formulated under this heading when low cost is the major factor. They are used to print newspapers and other publications where the highest print quality specification is not required. Good press performance is needed but setting speed, rub resistance and dot gain do not have to reach the highest standards. High-structure rubber-grade carbon blacks are used, with mineral oil as the vehicle. Small amounts of varnish (typically gilsonite or hydrocarbon-resin based) can be added to control the lithographic properties and bentonite may be used to control the rheology.

Formulation

Carbon black (Pigment Black 7)	18.0
0.9 Pa s mineral oil	70.0
Varnish	10.0
Additives	2.0
	100.0

Low rub inks
The print specification issued by many publishers precludes the use of inks in the low cost category and requires one of much higher quality. The mineral oil can be all or partly replaced by vehicles containing high percentages of resins, depending on the degree of rub resistance required. Low structure carbon blacks are used often in conjunction with rubber grades to obtain the required colour strength. Solvents, bentonites, alkyds, etc. are again used to control the rheology and the lithographic properties.

Formulation

Carbon black (Pigment Black 7)	20.0
0.9 Pa s mineral oil	38.0
Varnish	35.0
High boiling point aliphatic solvent	5.0
Additives	2.0
	100.0

Vegetable-oil inks
There is increasing pressure for environmental reasons to replace mineral oils with vegetable oils; vegetable oils are perceived as safe renewable materials. The relatively high and variable price precludes their use from all but those customers who consider the extra cost sustainable.

Formulation

Carbon black (Pigment Black 7)	18.0
Varnish*	50.0
Refined soya oil	30.0
Additives	2.0
	100.0

* Phenolic resin cooked in gelled Alkyd resin and soya oil.

Colours
The design of many modern four-colour web-offset presses does not allow units to be used for the printing of spot colours. However, many printing houses still rely on spot colour to give impact to their publication and while national newspapers will only use them where specified by advertisers, spot colours are still widely used in all other printing areas, utilizing additional printing units.

Bright, clean colours are often required, many of which are not possible to achieve by using process inks, and a wide range of pigments can be used. The lithographic properties of each pigment chosen must be carefully checked as the usage of ink can vary greatly from run to run, thus affecting ink–water balance and transfer properties. Amongst the pigments commonly used are Lake Red C (CI Pigment Red 53), copper ferrocyanide (CI Pigment Blue 62) phosphomolybdotungstic acid (PMTA) pigments (e.g. CI Pigment Blue 1, CI Pigment Red 81) (Chapter 4).

The vehicle used can be vegetable, mineral oil or resin based according to the printer's requirements, but care must be taken to ensure that the colour of the vehicle does not detract from the colour of the ink, particularly if dark coloured resins are used.

Process inks

The use of colour, particularly in newspapers and directories has increased to the level where it can be regarded with the same importance as the black. Where colour is widely used, the total cost of the coloured ink consumed can easily exceed the cost of black ink, although of course the volume of the latter is considerably higher. The choice of pigment, vehicle and method of manufacture all play a vital role in deciding the quality (and quantity) of the ink produced.

The colour hue of pigment types used is determined by the colour separation standards incorporated in the pre-press systems used by the platemakers. UKONS recommendations (PIRA, 1990) are widely used by many printers. This means that while the hues of the colours of process sets are broadly similar, colour strengths can vary greatly to cater for the different dot gains that are achieved. Presses will vary in their ability to control ink fount balance and this in turn can influence the colour strength of the ink. The formulator, therefore, must take careful consideration of all aspects of the press, plate, paper and fount solution. The most widely used pigments are CI Pigment Yellow 12 and 13, CI Pigment Red 57:1 and CI Pigment Blue 15:3.

The pigments can be used in dry form, but the use of flushed pigments is becoming more popular. Flushed pigments are said to give better tinctorial strength and eliminate dry pigment handling, but some formulators consider that their use limits the flexibility of formulation. Nevertheless, flushed pigments are widely used in the USA particularly in conjunction with vegetable oils.

Typical formulation types

Process yellow (using flushed pigments):

Diarylide Yellow (Pigment Yellow 12) (40% flushed in varnish)	40.0
Process oil	48.0
Aliphatic 300–330°C distillate	10.0
Additives	2.0
	100.0

Process magenta (using Dry Pigments):

Lithol Rubine (Pigment Red 57:1)	15.0
Process oil	60.0
Hydrocarbon varnish	7.0
Phenolic resin varnish	10.0
Aliphatic 300–330°C distillate	6.0
Additives	2.0
	100.0

Process cyan (using Vegetable Oil)

Phthalocyanine Blue (Pigment Blue 15:3)	18.0
Extender (Pigment White 18)	5.0
Varnish*	50.0
Refined soya oil	25.0
Additives	2.0
	100.0

* Phenolic resin cooked in a gelled alkyd resin and soya oil, then gelled.

The future of cold-set web-offset inks

The technology of cold-set web-offset inks has changed considerably over the past few years, and the standard of print quality now expected is of a very high level. The widespread use of colour has also required inks which give good performance when superimposed. The original cold-set web-offset inks were based on mineral oils which gave the ink good press stability, but which produced prints with poor rub resistance, dot gain and strike-through. The initial improvements in print quality were usually obtained at the expense of press stability, but high specification inks which are easy to run can be formulated with the correct choice of resins and oils. The future technology of cold-set web-offset inks will be to enable inks to perform at ever increasing press speeds with the minimum of variation to operating conditions. The use of vegetable oils may become more widespread in colour formulations but its price precludes it being used more generally in black formulations.

6.3 WEB-OFFSET HEAT-SET INKS

Since the 1950s when heat-set web offset was introduced as a printing process it has seen remarkable growth and is now firmly established as a major printing process. It is used, primarily, for the production of magazines, catalogues and brochures and competes effectively with sheet-fed offset in terms of quality. Print run lengths as low as 20 000 are now done by heat-set which has further eroded the market traditionally held by sheet-fed printing. The process yields high quality and high gloss prints which are printed simultaneously on both sides of the paper and dried in an oven.

Most modern web presses run at speeds of 1400–1800 feet per minute but speeds of 2000–2200 feet per minute are not uncommon (Fig. 6.5). Press speeds are set to increase substantially over the next decade and Heidelberg–Harris are already building a press which will run at 3000 feet per minute. The increase in speed places severe demands on an ink's performance with only fractions of a second being available for emulsion formation and, possibly, a greater risk of misting.

All heat-set inks are expected to fulfil exacting criteria. High gloss is demanded on even the poorest quality paper, the inks need excellent press stability to avoid the risk of piling or poor transference but are expected to dry quickly in an oven. Lithographic performance must be robust enough to allow the inks to run on a wide range of presses fitted with different dampening systems which may, or may not, run with alcohol. They must be capable of forming stable microemulsions very quickly to run satisfactorily on high speed presses and to print a consistent, sharp dot. Sufficient rub resistance and slip must be formulated into the inks to ensure that no marking occurs when the print travels over the turner bars of the press, to ensure that the signatures resist sticking together when ram bundled on the end of a press and to allow them to be processed in bindery operations without marking.

Ovens and chill stacks

Heat-set inks are dried by passing the printed web of paper through an oven using high velocity hot air, sufficient to raise the temperature of the

Fig. 6.5 High speed Baker Perkins G16 web-offset heat-set press. (Courtesey of Rockwell International).

web to 100–140°C. Most modern ovens have two or three zones in which the temperature can be individually varied to ensure that the web is progressively heated up rather than being subjected to the maximum temperature at the beginning of the oven. This may cause the water vapour contained in the web explosively to boil out and rupture the paper surface, a condition known as blistering. Around 80–90% of the distillate contained in the ink is removed in the oven with the other 10–20% being retained in the ink/paper. The oven should be run at a temperature sufficient to dry the ink and prevent marking on the chill, turner bar and folder. If the oven temperature is too high the gloss of the ink will be adversely affected.

Following the oven, the web is passed over a series of chilled rollers to reduce the temperature to near ambient before it enters the folder. Most presses have between three and six chill rollers in the chill stack. The chill rollers are normally set so that the web passes over the warmest one first and then passes over progressively cooler ones. Ideally, the web should be cooled at an even rate and the reduction in temperature from the oven exit to the end of the chill stack should be such that the web loses the same amount of heat as it passes over each chill roller, i.e. the temperature reduction should be linear. Figure 6.6 shows an ideal web temperature gradient, measured at the point the web leaves each chill roller.

If the web is cooled too quickly, the surface of the ink is hardened and solvent is likely to be trapped underneath the ink surface which

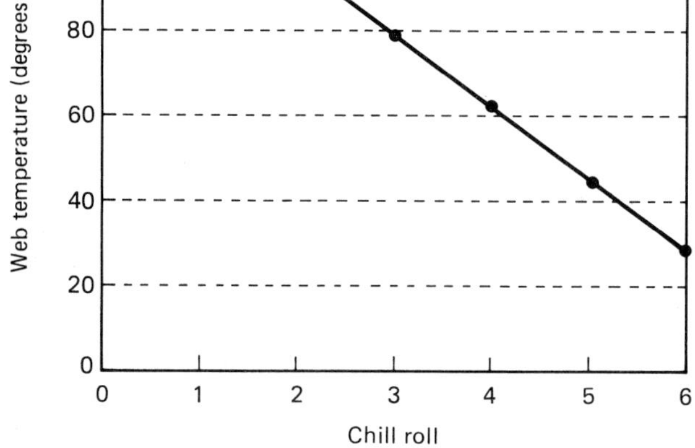

Fig. 6.6 Chill stack efficiency.

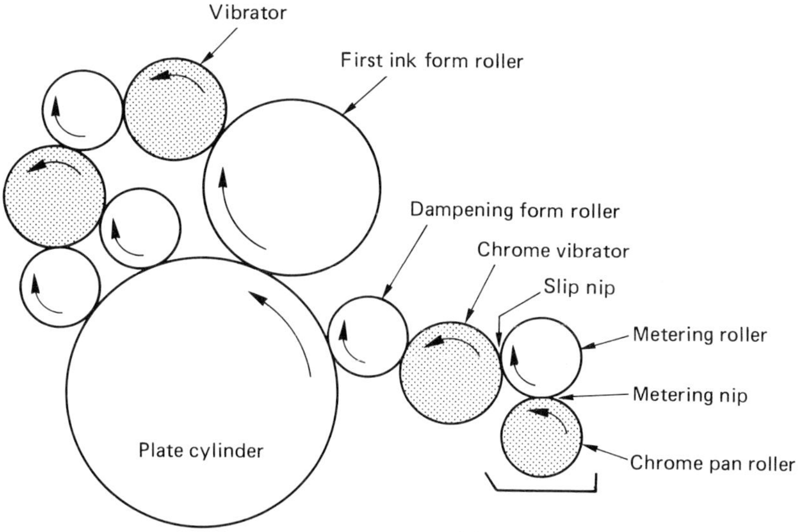

Fig. 6.7 Harris Duotrol dampening system.

may resoften the film and cause marking on the following chill rollers. If the web is insufficiently cooled, the ink will remain tacky and this may cause marking in the folder or on the turner bars. An aqueous solution of silicone is normally applied to the web after it has passed over the chill rollers to increase its slip and thereby help to prevent marking as it passes through the folder.

Dampening systems

Three common types of dampening systems are used on web presses; Duotrol, integrated or semi-integrated and contact-free systems such as brush or spray. Of these, the Duotrol type of system is by far the most common and may be represented schematically as in Fig. 6.7.

The Duotrol system offers good control of the dampening levels carried on the press and is responsive to minor changes in their settings. It was developed, initially, to run with alcohol and is probably the most common type of dampening system found on web-offset presses. As there is an accelerating movement away from alcohol, for reasons of environmental safety and cost, this type of system is now running alcohol-free. To do so successfully, it is important that the Shore hardness of the rubber dampening rollers is correct, according to the manufacturer's specification, to allow the transport of the lower viscosity dampening fluid through the rollers of the dampening system. A Shore hardness of 32 degrees is normally the maximum recommended for running alcohol-free.

Integrated dampening systems may be represented schematically as shown in Fig. 6.8.

Dampening in whole or in part via the inking rollers is the most

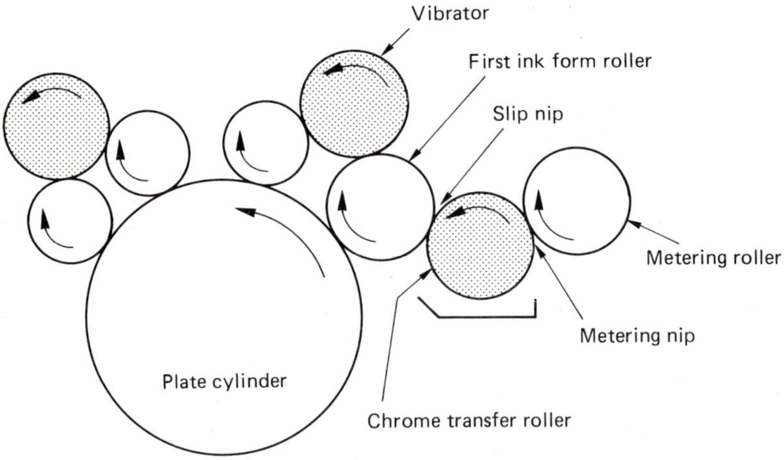

Fig. 6.8 Dahlgren integrated dampening system.

responsive of all types. Web presses fitted with this type of system normally run between 5 and 12% of alcohol. They can be run alcohol-free but the ink–water balance may be difficult to control.

Contact-free dampening systems may be depicted as in Fig. 6.9.

These types of brush dampening systems, or the similar spray-dampening systems, being contact free, avoid any problems of inks feeding back onto dampers but are, by their very nature, less responsive than other types of dampening systems.

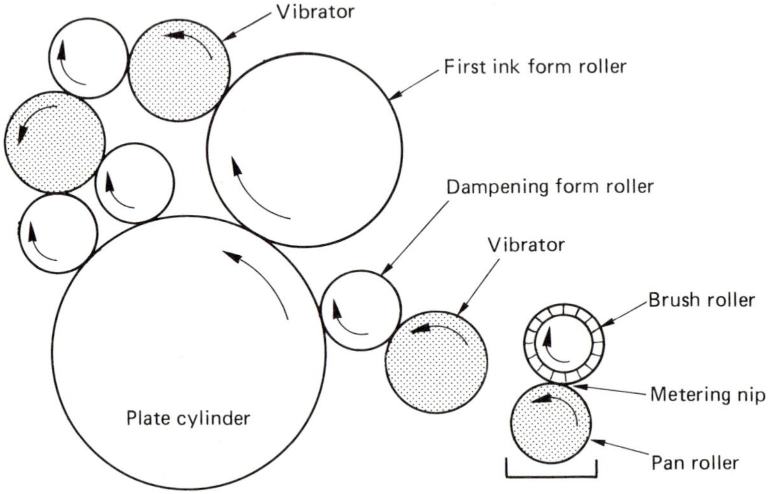

Fig. 6.9 Harris brush dampening system.

Emulsion formation

The lithographic process depends on a satisfactory ink-in-water emulsion being formed during printing and the speed of wet presses makes the choice of fountain solution vitally important as the ink and fount must react quickly to form a stable emulsion. The type of dampening system also influences the best fountain solution for a given press.

Although of fundamental importance for good lithography, the interaction between ink and fountain solution under the dynamic conditions which prevail on a printing press is difficult to assess under laboratory conditions. Various tests have found favour over the years with one of the most enduring being the water pick-up test developed by Aage Surland (1980). This test departed from the then existing practices as it sought not only to measure the **amount** of water picked up by an ink but also the **rate** at which the ink picked up the water.

The test method, or variations of it, are straightforward. Using a Duke water pick-up tester or a similar instrument, 100 g of fount is added to 100 g of ink and stirred for one minute at a fairly low speed of 90–100 rpm. The excess fount is decanted and measured and then put back in the ink and stirred for a further minute. This procedure is repeated for ten minutes. As the figure for fount absorbed by the ink in each one-minute test is then known, the cumulative total of fount absorbed over the duration of the test may readily be calculated and the results represented graphically as water pick-up against time.

Surland maintained that the shape of the graph was critical and used it to predict an ink's performance, as in Fig. 6.10.

The interpretation Surland placed on these types of curve was as follows.

- Curve A – unsatisfactory lithographic performance as the ink continues to absorb water without ever reaching equilibrium. Inks of this type have no water balance and are prone to scumming and emulsification piling.
- Curve B – unsatisfactory because, although approaching it, the ink never reaches equilibrium. Such inks are likely to print with a reduced density which leads the printer to increase the amount of ink carried which, in turn, requires more damp to be carried which leads to reduced density. The printer therefore is constantly adjusting ink and water settings to maintain print quality.
- Curve C – Surland considered this to be the ideal ink as it reaches equilibrium about halfway through the test. Inks of this type have a wide water-balance on press, print with maximum density and sharpness and require little adjustment of ink or damp settings throughout the run.
- Curve D – unsatisfactory as the ink reaches equilibrium too quickly and is therefore likely to become waterlogged during running, leading to reduced print density and a tight water balance.
- Curve E – unsatisfactory as the ink fails to absorb enough water to print satisfactorily and has no water balance at all. Inks such as these fail to transfer through the press satisfactorily, are prone to causing

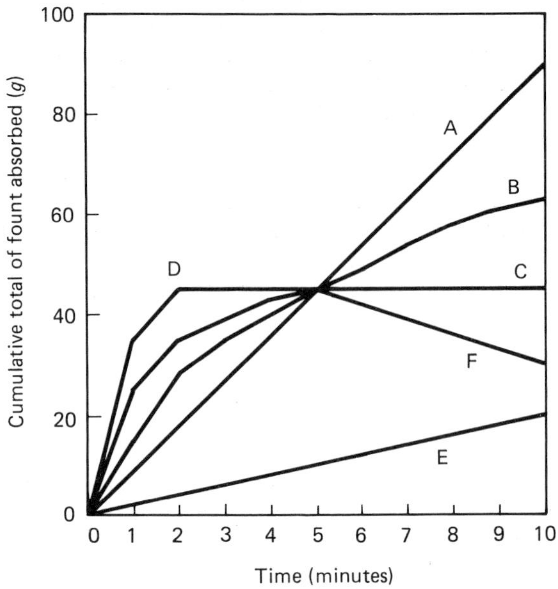

Fig. 6.10 Surland curves.

roller stripping and to print with very much reduced density.

- Curve F – unsatisfactory because, although these inks appear to reach equilibrium in the normal way, the emulsion is unstable and breaks after a period. Inks of this type generally have very poor lay, print with water-marks and may also cause roller stripping.

The major drawback of the Surland test is that it measures water pick-up under conditions of low shear. Since its inception press speeds have increased markedly with a consequent increase in the shear rates to which inks are subjected and a decrease in the time available to form an ink/water emulsion. Inkmakers now place more emphasis on the gradient of the graph produced by a Surland test, rather than the absolute amount of water absorbed, as this is an indication of the **rate** of water pick-up which must increase as presses run faster. Inks of type B, and even A, are now more common than inks of type C, as experience has shown that they tend to have a wide water-balance and the rate of fresh ink replenishment due to speed prevents over-emulsification.

In addition to the rate of water pick-up, good lithography requires that the emulsion be of the correct type. The internal (water) phase of the emulsion should be as finely divided as possible as a microemulsion has little adverse effect on the rheology of the ink which will retain the viscosity, yield value, tack and flow characteristics necessary to transfer properly through the printing operation. Fine emulsions also have greater stability than do coarse ones. Two factors primarily dictate the formation of a good emulsion, the chemical affinity of the ink and fountain solution and the mechanical energy available for emulsion formation.

The ink's vehicle system must have the ability to form a stable microemulsion when in contact with the fount solution. Vehicles of similar types or rheology may have widely differing characteristics in this respect. The type and amount of wetting agent used in fountain solutions can have a significant effect on the ability of the ink to form the correct type of emulsion.

Mechanical energy is provided by the press and depends, in part, on the type of dampening system used. It should not be surprising, therefore, to find that while a given ink–fountain solution combination works well on one type of press it may be totally unsatisfactory on another. Dampening systems may be classified as low, medium and high energy types.

Dampening system	*Energy input*
Dahlgren or semi-integrated	Low
Duotrol or similar	Medium
Brush, spray or similar	High

The lower the energy input provided by the dampening system, the more optimized the ink/fountain solution relationship needs to be which is why the Dahlgren or integrated dampening systems present more problems in use to an inkmaker than do the other types.

The perceived crudities of the Surland test have led inkmakers to explore alternative methods of measuring water pick-up. Generally these have been methods which utilize higher shear rates to form the emulsion as this leads to a closer approximation of what happens on press. The use of high speed mixers with a saw-tooth blade often provides a result which more closely mirrors the realities of running on a high speed web press. In addition, inkmakers now normally seek to assess the quality of emulsion formed. Methods for doing this range from simply examining the emulsified ink under a microscope and measuring the size of the water droplets to checking the rheology of unemulsified and emulsified ink on viscometers to quantify the change in rheology upon emulsification. Sophisticated (and expensive) instruments such as the Grapho Metronic are also used to attempt to predict lithographic behaviour on press.

Bassemir and Krishnan (1987) have shown the effect of differing types of emulsions on the water balance of inks on press. Three inks with different emulsion characteristics, ranging from coarse to fine, were run on a press and the water balance was checked from just above catch-up to just below washout with the following results:

Ink	Maximum water droplet size (micron)	Size distribution	Water feed (%)
A	2	Very good	65–80
B	5	Good	75–85
C	20	Poor	85–90

The results illustrate the importance of promoting a stable microemulsion and the beneficial effect this has on the ink's water balance.

To ensure rapid start-up on a press and therefore to reduce wastage to a minimum, it is necessary to have a low interfacial tension between the ink and fountain solution. The use of isopropyl alcohol, still common in the UK and Europe but less so in the USA, has a two-fold effect. First, it reduces the interfacial tension between the ink and dampening fluid and secondly it increases the viscosity of the dampening fluid which allows better control of the damp being fed to the plate. In situations where the use of alcohol has been eliminated, the fount concentrate contains surfactants to lower the interfacial tension. These are normally less effective than alcohol and tend to build up in the fountain reservoir tanks. For these reasons switching a press to running alcohol–free very often leads to problems.

The design of the press dampening system will have a major influence on the thickness of the film of dampening solution applied to the plate. Brush, spray or conventional dampening systems carry a thicker film of damp than do the bareback type of dampening systems. With brush dampening systems, a fountain solution of relatively high surface tension may be run, and even if not optimized, the sheer volume of damp will tend to minimize or eliminate any unfavourable surface energetics. Print quality with this type of system is not, however, normally as high as that achieved with the more controllable bareback types. The following table forms a useful guide to the optimum surface tension of fountain solution required by each type of dampening system:

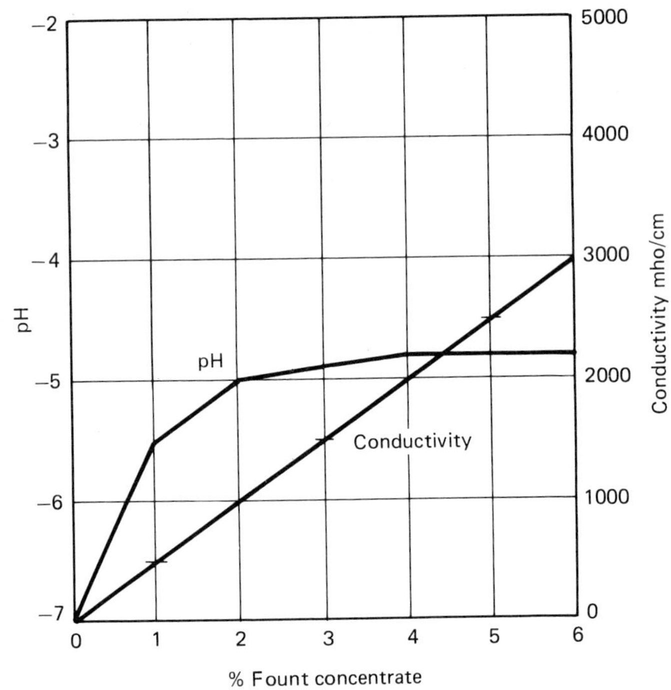

Fig. 6.11 Conductivity and pH versus fount concentration.

Dampening system	Surface tension of fountain solution
Brush or similar	40–50
Duotrol or similar	30–35
Dahlgren or similar	28–33

pH and conductivity

Most fountain solutions are buffered to maintain the required pH over a range of operating conditions at the correct percentage addition of concentrate. This is normally 2–4%. However, the very fact that they are buffered makes control of the correct dosing rate difficult to maintain by measuring the pH alone and it is therefore normal to measure, and control, the dosing level by checking the conductivity. It is readily appreciated from Fig. 6.11 that the fountain concentrate illustrated will yield similar pH figures for dosing rates of 2–6% while measurement of the conductivity will accurately indicate the actual concentration. Most founts are dosed at their optimum level when the conductivity of the fountain solution is about 700–900 mho above the reading for the plain water being used. Alcohol lowers the apparent conductivity of a fountain solution and care must be taken to make allowances for this when calculating the dosing rate of concentrate. It is a simple matter to plot calibration graphs of the concentrate without alcohol and then to remeasure the conductivity with different percentages of alcohol added.

Conductivity offers far more precision in the control of fountain solutions than pH as it measures the concentration of all the charged ions present rather than merely the hydrogen/hydroxide ions measured by pH. Precise control of the dosing rate is important because fount solutions are present in the final dampening fluid at low concentrations but contain a number of powerful surfactants which, if overdosed, are likely to have a very adverse effect on the ink–water relationship.

The vehicle system

The backbone of any ink is the vehicle system as it has a major influence on properties such as lithographic performance, transference, gloss, drying speed and stability, viscosity, tack and flow.

While there is a multitude of raw materials available to an inkmaker, most heat-set vehicles are formulated from hard resins, alkyds, distillates and gelling agents.

Hard resins

The restrictions and requirements of lithographic printing means that the resin chemistry has to meet the fundamental criteria of:

- solubility in weak solvents;
- controlled emulsification characteristics;
- low surface energy;
- good cohesive properties.

The range of basic resin types which meet these criteria are limited and, in the case of heat-set inks, the resins used are normally rosin-modified phenolics, maleics, rosin esters and hydrocarbon types. Although the range of basic resin types may appear limited, there is, in fact, an almost bewildering number of different grades of resins available and even small changes in a resin's formulation can lead to significantly different properties.

Classical rosin-modified phenolics are fairly easy to categorize. High melting-point resins are high molecular weight and tend to produce vehicles of limited solubility which have high viscosity, low tack, relatively poor gloss, fast setting, good solvent release and poor pigment wetting. Low melting-point resins, on the other hand, are low molecular weight and tend to produce vehicles with high solubility which have low viscosity, high tack, good gloss, slow setting, poor solvent release and good pigment wetting. It is common to use a mixture of hard resins within a vehicle formulation to optimize the required properties.

The latest generation of high molecular weight rosin-modified phenolics seek to optimize the properties seen as beneficial by inkmakers. Vehicles formulated from these new types of resin exhibit high viscosity coupled with high solubility and low tack which allows the inkmaker to formulate vehicles with good rheology. Such resins are often referred to as self-gelling due to their inherently high structure/viscosity. In fact it is rare to use these products without any gelling agents at all as these help to counter any batch to batch variation in the resin. However, vehicles may be produced with lower levels of gelling agents than was possible in the past. The high solubility of these resins aids in transfer on high speed presses and they tend to produce stable microemulsions quickly.

Hydrocarbon resins were first introduced into lithographic inks to cut costs as they tended to be cheap. The increase in oil prices since the 1970s have eroded their commercial advantage but they are still used as they have some useful properties in modifying other hard resins. They tend to be extremely soluble in distillates and varnishes with up to 60% resin may easily be produced. However, the tack of such varnishes is very high. Hydrocarbons are used to increase the solubility of less soluble resins, to reduce the viscosity of inks and to reduce the water pick-up of an ink as they tend to be water repellent. They may aid the transfer of inks on high speed presses. There are now a number of phenolic-modified hydrocarbons on the market which have some interesting properties.

Maleics and rosin esters may be used in place of phenolics in lower cost formulations. They tend to have inferior lithographic performance but have good gloss and fast solvent release.

Alkyds

Alkyds are used in heat-set inks primarily as pigment-wetting vehicles although they also improve the stability of the ink and may affect the water pick-up. They may also help to improve the gloss. Because heat-set inks dry by evaporation rather than by oxidation, the amount of alkyd used is normally kept to a minimum to avoid drying-related problems. It is unusual to use more than 10–12%. Long-oil alkyds are generally favoured although medium-oil lengths are also used because of their higher viscosity. Linseed or soya types are the most common oil modifications.

Distillates

High boiling petroleum distillates are used as the diluent/solvent in heat-set formulations. Three boiling ranges are commonly used: 240–270°C, 260–290°C and 280–320°C.

The choice of the boiling range to be used is dictated by the balance required between roller stability and drying speed in the oven. Figure 6.12 shows the influence of the distillate boiling range on the tack stability of a vehicle. Both vehicles were formulated with the same resin and alkyd, the sole difference being the boiling range of the distillate.

This balance may also be influenced by the aromatic content of the distillate used. Three types of distillate are available within each boiling range, each having a different aromatic content:

Grade	*Aromatic content (%)*
Regular	Above 16
Low odour	5–6
Aromatic free	Less than 1

Figure 6.13 shows the results of changing the aromatic content of the distillate. The first vehicle was made with a regular distillate containing

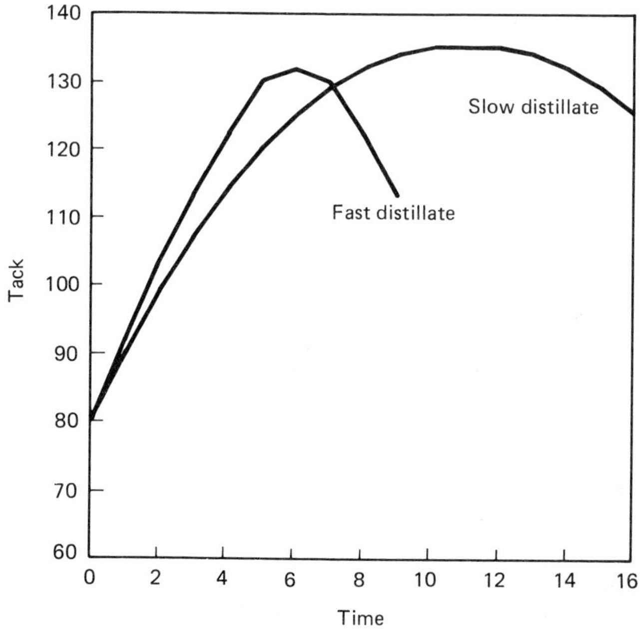

Fig. 6.12 Influence of boiling range on stability.

about 18% aromatics, whereas the second was made with an aromatic-free distillate.

Stability, therefore, may be influenced by changing either the boiling range or the aromatic content of the distillate used in an individual ink.

A further consideration is the feedstock used to produce the distillate which may be paraffinic or, by hydrogenation of the aromatics in hydrocarbon oils, naphthenic. Naphthenic grades have better solvent power than the paraffinic grades and a correspondingly lower aniline point.

Gelling agents

Most heat-set varnishes are gelled to some extent to increase the viscosity and yield value, while maintaining or reducing the tack. Gelling a varnish also reduces the gloss and water pick-up slightly. The most common gelling agents used are the aluminium isopropoxide types or polymeric organic aluminium compounds.

These types of product react with the hydroxyl and carboxyl groups on the resin and alkyd to form micelles, which increase the viscosity and yield value of the varnish.

The level of gelling agent used depends on the strength of gellation required and the reactivity of the other materials present but is normally between 1 and 3%.

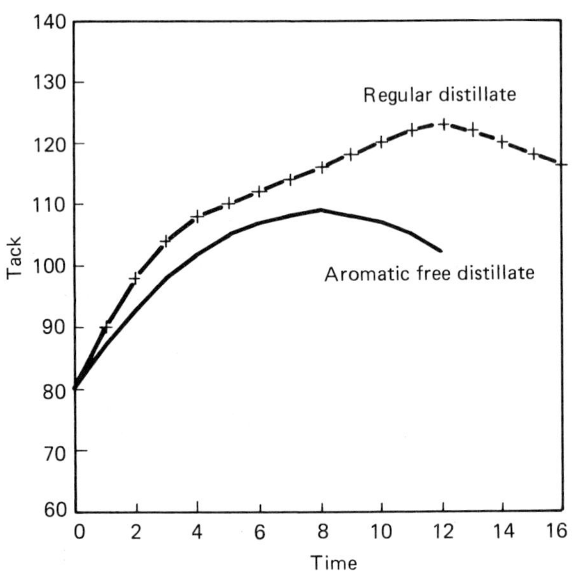

Fig. 6.13 Influence of aromatics on stability.

Pigments

Heat-set printing, for the most part, consists of four-colour process printing. The pigments used are CI Pigment Yellow 12 or 13, CI Pigment Red 57:1, CI Pigment Blue 15:3 and CI Pigment Black 7.

Black inks are very often toned with one of the reflex (alkali) blue pigments such as CI Pigment Blue 18 (red shade), Blue 56 (mid shade) or Blue 61 (green shade) and/or with Milori Blue, CI Pigment Blue 27. The reader is referred to Chapter 4 for a full description of the chemistry of these products.

Heat-set inks may be formulated using either dry colour or flushed pigments. In the USA, virtually all heat-set inks are made from flushed pigments and in European markets the use of flush-based inks is increasing. Flushed pigment is manufactured by taking the presscake and placing it under vacuum in a powerful, heated mixer. The water which is evaporated from the presscake is replaced by varnish which preferentially wets the pigment. The resultant pigment paste is a highly concentrated pigment dispersion in lithographic varnish. Because the pigment is never dried, flush-based inks have some properties which cannot be achieved by dry colour manufactured inks. The dispersion of pigment in a flush is finer than can be realized with dry colour and inks made from flush have, generally, better gloss, better transparency and better colour strength per unit of pigment than equivalent dry colour inks. Inks manufactured from flush are quicker and easier to make than inks made from dry colour as they require no energy-intensive grinding stage. They also tend to have better batch to batch consistency in terms of rheology than dry colour inks as the pigment does not continue to wet out after manufacture.

Why, then, does their use not dominate the European market as it does that of the USA? The reasons are straightforward and rarely technical. Some European inkmakers claim that the use of flushed pigment introduces a significant amount of material (the flushing vehicle) into the formulation which is of unknown composition and not necessarily the material which the inkmaker would choose. It has been said that flushes reduce the flexibility of a factory as no one flush is suitable for different ink types. In addition they claim that the use of flush reduces inkmaking to a 'mix and pot' operation and diminishes the unique technology offered by individual inkmakers. None of these reasons is really valid. Pigment manufacturers will frequently disclose exactly which varnish has been used in the flushing process and will, if the volume requirement is sufficient, flush into vehicles chosen by the inkmaker. Heat-set inks are now often made in high volume dedicated plants and the requirement for the flexibility of dry colour does not exist. Flushes do allow mix and pot manufacturing techniques to be used which produces very consistent products from batch to batch. They do not remove the unique technology of individual inkmakers as the ink's vehicle system is of paramount importance to an ink's performance.

The reason for some European inkmakers choosing dry colour is simple: it has been a matter of availability. Pigment manufacture in

Europe has concentrated on dry colour production and flushes have been unavailable unless imported from the USA. A great deal of development effort has been expended to improve the dispersive characteristics of pigments to allow dry colour production to be maximized and most pigments are now surface treated to aid dispersion. High quality inks can be, and are, produced from either dry colour or flushed pigments.

Other additives

A number of additives are used to control, or modify, particular properties. In the case of heat-set these are waxes, rheology modifiers and miscellaneous additives.

Waxes

Waxes are used in heat-set inks to provide sufficient slip to get the print over the chill stack and turner bars, through the folder, through binding operations and carriage to the point of sale, without marking.

Rub resistance and slip are a function of the particle size of the wax as well as its hardness (which is generally related to molecular weight) and the best results are normally obtained with a particle size just big enough to stand proud of the printed film but not large enough to cause printing problems such as piling. The most common waxes used in heat-set are polyethylene and PTFE (polytetrafluoroethylene), the former to provide rub resistance and the latter to provide good slip.

Waxes may be added as a dry powder or as a compound. Dry waxes are available in micronized form to ease dispersion into the final ink but they can be difficult to wet properly. Compounds are easier to add to the ink but have the disadvantage of introducing the vehicle system in which the wax has been dispersed into the ink. Compounds may be manufactured either by heating the wax above its melting point in combination with a suitable vehicle, distillate or oil and crash cooling the mixture on a three-roll mill to prevent re-crystallization into large particles or by passing the molten wax mixture through a heat exchanger to provide controlled cooling and therefore controlled particle size. Most ink manufacturers who make their own compound favour the former method, whereas most industrial wax compounders prefer the latter because of its better productivity. The wax content of compounds can vary between 25 and 40% in the case of polyethylene, and 45 to 55% in the case of PTFE.

The processing temperature of inks should not exceed the melting point of any waxes included as they may re-crystallize into large particle-size agglomerates when the ink cools and this can lead to printing problems such as blanket or plate piling as well as reducing the gloss of the ink.

Inks which are destined for UV varnishing should not contain significant quantities of PTFE as this will prevent satisfactory adhesion of the UV varnish to the ink. A maximum of 0.5% is recommended.

Rheology modifiers

The use of organic aluminium compounds to produce particular rheological characteristics in heat-set vehicles was referred to earlier on page 371.

Ink rheology can also be manipulated by the incorporation of additives directly at the ink manufacturing stage. Ideally, it would only be necessary to have a series of materials in which each was capable of producing a significant change to a single rheological parameter when added in small proportions. This would provide the formulator with the maximum flexibility to achieve the exact balance of rheological properties being sought. In reality, the additives available all produce more complex rheological effects and considerable care must be exercized to ensure that the benefit in one area is not achieved at the expense of unsatisfactory effect on another property. Such potential pitfalls can extend beyond rheological considerations into other properties such as gloss, stability, water balance, etc. With these thoughts in mind, the major rheological modifiers may be summarized as follows:

Montmorillomite clays (bentones)
These are now available as stir-in grades for direct incorporation into inks, thus avoiding the polar solvent activation and shear requirements that previously existed. Viscosity, yield value and thixotropy are all increased generally with little influence on tack beyond that expected as a result of the higher viscosity. They can help to reduce misting or flying of inks on press and may also be used to increase the thixotropy of an ink to prevent it dripping from a duct. Levels in excess of 3% may have an adverse effect on gloss.

Fumed silicas
These are very fine particle-size powders which produce substantial increases in thixotropy and yield value but do not increase viscosity to any great extent. They have little influence on tack but can have a marked adverse effect on gloss. These materials are available as hydrophobic and hydrophilic grades with the former being more commonly used in lithographic inks. Their usage is similar to that of bentone, detailed above.

Polyamides, aliminium chelates and amines
Liquid polyamides and aluminium chelates may be added directly to an ink in addition to their more widespread use in the production of gelled vehicles. The effect, however, is similar with a fairly pronounced effect on yield value and thixotropy rather than a simple rise in viscosity. Polyamides, in particular, produce a structure which is readily broken down even under conditions of low shear. Amines, such as triethanolamine, may be added to lithographic inks and will increase the yield value and viscosity without making the ink unduly thixotropic. Additions of 1–2% are normally sufficient. Perhaps surprisingly for an amine, this type of material has no adverse effect on the lithographic performance of the ink. A drawback with all these additives when they are incorporated at ambient temperature is that the structure may continue to develop over an extended period of time. Gloss and tack are relatively unaffected by these materials.

Micronized hydrocarbon resins
Low molecular weight and melting point grades are available as stir-in additives to increase the tack of inks. Some increase in viscosity, and perhaps a reduction in yield value, usually also occurs. Large additions of these materials can slow down the drying of heat-set inks and leave the print sticky.

Supersolvents
Inks may be modified by using so-called supersolvents such as tridecyl alcohol (TDA), texanol isobutyrate (TXIB) or other high alcohols. As these materials make the resin system more soluble and cut down the gell structure of vehicles they have the effect of reducing viscosity and yield value and increasing the tack and flow of the ink to which they are added.

The main usage of any rheology modifier should be in quality control where they may be used, in small quantities, to bring an ink into specification. It is not good practice to use them as a basic formulating tool.

Miscellaneous additives

Additives are sometimes used to overcome deficiencies in the lithographic performance of an ink. Oleates or amine derivatives such as oleamides, being effective emulsifying agents, may be used to increase water pick-up, and proprietary products such as Aquapel from Diamond Shamrock may be used to decrease it. Again, it should be stressed that the use of such materials does not compensate for an inadequate formulation and their use as a basic formulating tool is questionable.

Ethylenediamine tetra-acetic acid (EDTA), its sodium salt, the sodium salt of diethylenetriamine acetic acid or similar sequestering agents are sometimes incorporated to complex any free calcium ions, to prevent them from taking place in any unwanted reactions. Calcium ions may be available from certain grades of Pigment Red 57.1 (Rubine), paper coatings or hard water. They will react with some buffering agents used in fountain solutions, most notably citric acid, to form water-insoluble salts which precipitate out on the rollers of a press and cause roller stripping.

Typical vehicles and inks

Heat-set inks may be formulated using a number of alternative approaches depending on the individual formulator's particular experience and bias. He may be influenced by his own company's formulating philosophy, commercial contracts obtained from different raw material suppliers, previous good or bad experiences with specific raw materials and the processing equipment available to produce his formulations. For example, inks may be formulated from dry colour or flushed pigments, from vehicles produced in house or proprietary vehicles purchased from outside sources, from different hard resins and alkyds and from dry or compounded wax. There is, therefore, no 'right' formula although, despite widely differing formulating principles, all successful inks will share some common key features. The viscosity, yield value and tack of all heat-set inks fall into a fairly narrow band because of the physical requirements of the printing process and substrates. The overriding feature of all heat-set inks should be trouble-free press performance at high speed, under conditions of high shear and on a variety of substrates.

The formulation of heat-set inks requires experience and expertise from the chemist but, to be truly successful, it is essential to add one more consideration to the list of specific properties being sought and raw materials to be used – that of cost. Heat-set inks are almost a commodity item and many otherwise competent formulators neglect this aspect of their work. A small reduction in the raw material cost of the ink can make a significant difference to their company's profit and loss account.

Vehicle formulations

Maleic resin-based heat-set varnish

High melting point maleic copal resin	36.00
High viscosity soya alkyd	16.00
Aromatic-free distillate 260–290°C	46.80
Aluminium chelate	1.20
	100.00

Phenolic resin-based heat-set varnish

High molecular weight soluble phenolic resin	45.00
High viscosity linseed alkyd	12.00
Aromatic-free distillate 260–290°C	24.00
Aromatic-free distillate 240–270°C	18.50
Aluminium chelate	0.50
	100.00

Hydrocarbon resin-based heat-set varnish

Hydrocarbon resin, melting point 120–140°C	60.00
Regular distillate 260–290°C	25.00
Regular distillate 240–270°C	15.00
	100.00

Ink formulations

Flush-based heat-set inks

	Yellow	Magenta	Cyan
Flushed pigment	30.00	35.00	30.00
Medium gell heat-set varnish	15.00	10.00	15.00
High gell heat-set varnish	44.00	45.00	44.00
Polyethylene wax compound	4.00	4.00	4.00
PTFE wax compound	1.00	1.00	1.00
Aromatic-free distillate 240–270°C	6.00	5.00	6.00
	100.00	100.00	100.00

Dry colour heat-set inks

	Yellow	Magenta	Cyan	Black
Dry pigment	12.00	16.00	14.00	18.00
Medium gell heat-set varnish	63.50	57.00	62.50	53.50
High gell heat-set varnish	10.00	12.00	10.00	10.00
Low viscosity linseed alkyd	5.00	7.00	5.00	6.00
Micronized polyethylene wax	2.00	2.00	2.00	2.00
Micronized PTFE wax	0.50	0.50	0.50	0.50
Alkali blue heat-set flush	–	–	–	4.00
Aromatic-free distillate 240–270°C	7.00	5.00	6.00	6.00
Sequestering agent	–	0.50	–	–
	100.00	100.00	100.00	100.00

Ink-related problems and their possible solutions

The number of variables in the printing process may lead to a variety of problems being evident either on the final print or with the running of the press. The cause of the problems may be ink, paper, press conditions or fountain solution. The inkmaker is very often called upon to offer advice as to the cause of a particular problem as the ink offers a visible manifestation of printing problems even when it is not the cause. Any particular problem may have a number of different solutions although some solutions may be more practical than others.

There are five main areas where difficulties can arise with heat-set printing:

- lithographic problems
- rheology
- tack and tack stability
- drying
- reproduction and print appearance.

Lithographic problems

Lithographic problems occur when the ink–water balance breaks down and leads to the following types of situation.

Tinting or feedback

Tinting or feedback occurs when pigment, with or without other ink ingredient, become solubilized or emulsified into the fountain solution. The fount then takes on a weak colouration from the ink which may be transferred as a weak wash of colour to the non-image areas of the print. The problem may be caused by pigments which have been incompletely washed free of soluble components during their manufacture, or which contain materials which are reactive with certain vehicle components and form water-emulsifiable soaps. Changing to an alternative pigment grade will cure this problem.

The more common cause is the formation of an inverse emulsion, i.e. when the emulsion being formed is ink-in-water rather than the necessary water-in-ink. The formation of an inverse emulsion will cause the ink to transfer back through the dampening system where it will either contaminate the fountain trough or pile on the dampening rollers. Where it does the latter, it may prevent the transference of fount through the dampening system to the plate which may lead to the plate becoming insufficiently damped and therefore picking up ink. The most common solution is to increase the emulsification capacity (water pick-up) of the ink to prevent inverse emulsions being formed.

Scumming

Scumming occurs when non-image areas of the plate accept and transfer ink to the blanket and from there to the print. Minute flecks of scummed ink are present on virtually all lithographic prints and their presence only becomes a problem when they are visible to the naked eye or when they affect tonal rendition by increasing the apparent density of halftone areas. It may be caused by a variety of factors such as poor plate development, insufficient gum, incorrect plate exposure, dirty or poorly set dampers, water-soluble materials in the paper coating, insufficient acid in the fountain solution or inks which are too low in viscosity and tack and are, therefore, greasy.

Scumming may be cured by resetting dampers or cleaning dirty ones, by making a new plate, by running a more acidic fountain solution or by changing the ink to one with a higher viscosity and tack.

Scumming is often confused with tinting – see above – and the two terms are sometimes used interchangeably.

Stripping

Stripping occurs when ink fails to transfer along all areas of the ink roller train surface because they start accepting damp rather than ink. The rollers may strip completely or in patches. This adversely affects ink feed to the image area in line with the affected roller area. It may have various causes including badly set rollers or rollers in poor condition. Copper rollers may be sensitized and require cleaning with a mildly abrasive polish. So far as the ink is concerned, it may be caused by the precipitation of hydrophilic materials formed from the reaction of ink and fount ingredients onto the rollers. For example, free calcium, if present in the magenta pigment (or paper coatings or hard water for that matter), may react with buffering agents in the fountain solution such as citric acid to precipitate calcium citrate. This may be cured by the addition of sequestering agents, such as EDTA, to the ink.

Occasionally, the duct roller may refuse to accept ink. This form of stripping is usually caused by water travelling back up the roller train to the duct and may be cured by increasing the water pick-up of the ink so that it is capable of emulsifying the fountain solution before it can travel back to the duct.

Piling

Piling is the term used to describe the general condition of an excessive build-up of ink on the rollers, plate or blanket which fails to transfer to the substrate. The problem is visually apparent on a press and manifests itself on the print as a lack of continuous and even coverage of all image areas. This is normally first observed along the trailing edge of solids or type. If the press is not washed either by hand or with the auto blanketwash system fitted to many web presses, the problem usually gets worse to the point where the print becomes unacceptable and physical damage may occur to either the blanket or plate because of the increased pressure in the affected area. There are a number of potential causes of piling and these may be interconnected in complex ways.

Piling may occur on the rollers if the ink becomes over emulsified and loses the rheology and tack required to continue to transfer satisfactorily down the roller train. This may be cured in a number of ways. The ink may be changed for one which is capable of emulsifying the level of damp being carried on press and retaining the rheology and tack required for good transference, i.e. the water pick-up should be increased. Conversely, the water pick-up of the ink may be reduced to prevent the ink becoming over emulsified. This route may lead to the type of duct roller stripping discussed above occurring if the unemulsified fount travels back up the roller train. The fount solution may be changed for one that is less emulsifying with the ink being used. Again, if the amount of damp being carried on the press is not reduced, any unemulsified fount may travel back up the roller train and cause stripping of the duct roller.

Lack of tack stability may cause the ink to become too tacky or viscous to transfer properly. This may sometimes lead to picking or linting of the substrate. The common solution on press is to carry more fountain solution and this is sometimes successful if the ink can emulsify the increased level of damp. The inkmaker can cure the problem by using distillates of a higher boiling range or higher aromaticity, or by incorporating some supersolvent into the ink and thereby increasing the solubility of the resin system to gain greater stability. The formulator must ensure that the solution to this problem does not lead to poor drying in the oven. The balance between press stability and drying speed is always a compromise.

Plate and blanket piling can be caused by a filtration process whereby most of the liquid components of the ink transfer but some hard, particulate materials remain behind, bound by a small proportion of retained vehicle. Commonly, such materials may be poorly dispersed pigment or poorly wetted micronized waxes. Emulsification may make the situation worse by promoting the dewetting of particles which are not completely coated in vehicle. The only cure for this type of problem is to ensure that well dispersed and thoroughly wetted inks are supplied.

Finally, piling may involve significant proportions of components which originate from the substrate and which have been removed by either the fount or the ink. These may be fibres, coating or fillers. Again, the on-press solution is to increase the amount of damp carried, providing the ink can emulsify the additional damp, as this should effectively lower the printing tack of the ink. The formulator should reduce the tack of the ink to avoid linting problems.

All the above refer to situations where ink piles on its own unit but sometimes ink piles on subsequent units of the press, e.g. the black ink may pile on the magenta or yellow blanket. This is called subsequent-impression piling or carry-over piling. Normally, ink is transferred from the substrate to the blankets of subsequent units and reaches an equilibrium where as much ink is transferred back to the substrate as is being transferred from the substrate to the blanket. If the ink sets too quickly on a particular substrate it may reach a tack and viscosity by the time the web is passing through subsequent units such that it only transfers from the substrate to the blanket but not back again. Carrying more damp on the unit of the press which is experiencing the piling may reduce the problem by providing a film of water on the blanket through which the ink cannot transfer. From a formulator's point of view, the problem can be cured by slowing the setting rate of the inks which are printed on the first units of the press. Setting rate should not be confused with drying rate and the solution is not to use higher boiling distillates but rather to decrease the resin to oil ratio of the vehicle or increase its solubility be the use of some supersolvent.

Rheology problems

Incorrect rheology can be either the root cause of other problems such as emulsification and the related difficulties of tinting, scumming or piling, or the lack of transference of ink to the press and down the roller train.

Viscosity is related to temperature and at the operating temperature of a web press, usually 30–40°C but occasionally much higher, the ink has a significantly lower viscosity than it does at 25°C which is the temperature normally used for laboratory testing. An ink which is over 10 Pa s at 25°C may be as low as 3 Pa s at 40°C. In addition to increased temperature under operating conditions, the ink is also subjected to very high rates of mechanical shear by the roller train, which breaks down any structure. Normal operating conditions therefore lead to low viscosity and low structure inks being run and such inks are more prone to lithographic problems than inks of the same formula with higher viscosity and structure. The low viscosity and structure may lead to serious lithographic problems. The only 'cure' for this type of serious problem may involve a complete reformulation to obtain an ink with better rheology under the operating conditions to which it will be subjected. If the problems being experienced are borderline, the formulator should increase the viscosity and structure of the ink by the use of high gelled vehicles, bentone or aerosil.

Most heat-set inks are now sold in 1-tonne containers and the ink must have sufficient flow at ambient temperature to flow out of the silo and into the ink pumps which force the ink down pipelines to the press. If the structure of the ink is too high or if the product is unduly thixotropic, it may fail to pump out of the container. Highly structured or thixotropic inks may fail to follow the duct roller, resulting in a problem known as hang-back, or may fail to transfer down the roller train satisfactorily. The solution is to reduce the structure of the vehicles used, to limit the amount of bentone or aerosil types of compound in the formulation, or to increase the solubility of the system.

A satisfactory rheology is a compromise between the requirements of being able to pump the ink from containers and maintaining a rheology to produce satisfactory lithographic performance under the operating conditions on a press.

Tack and tack stability problems

An ink's tack may be controlled through the use of gelled vehicles and the use of distillates or gelled distillates. Additives, such as micronized hydrocarbon resins, are available as stir in grades to enable small upward adjustments to be made to the tack of a finished ink which otherwise would have to be rejected and replaced with a batch of a modified formula.

Inks of too high a tack may cause picking and linting problems or even web breaks if the tack is very high or the paper quality very poor. Inks of too low a tack tend to be greasy and fail to transfer well, which may cause lithographic problems as well as poor dot definition.

Drying problems

The drying problems associated with heat-set printing centre around marking on the chill rollers, marking on turner bars or in folders, signatures sticking together when ram-bundled and marking in binderies or during transportation of the final print. The importance of the correct chill stack temperature gradient has been discussed previously (p. 361). Badly set chills can cause the signatures to mark on turner bars, in the folder and to stick together when ram-bundled as well as causing marking on the chill rollers themselves. If the cause of the problem is the ink itself being too slow drying, this needs to be corrected by the use of a lower boiling distillate. If this fails to cure the problem then further measures such as using distillates of low aromaticity, the use of higher molecular weight resins or a reduction of the oil to resin ratio, all of which will yield a harder and less plastic film, need to be considered. It is sometimes useful to increase the content of waxes with excellent surface slip characteristics, both to reduce marking on the press as well as in the bindery and to provide protection to the print during transportation and handling.

Reproduction and print appearance

Dot gain, the situation where the dot on the paper is larger than the dot on the original film, is an inevitable consequence of lithographic printing. It occurs during platemaking for a variety of reasons and during printing because of the mechanical squash to the ink as it is transported from plate to blanket to substrate. It should, however, occur in a sufficiently controlled and reproducible manner so that appropriate allowances may be made at the colour separation stage. It becomes a problem when it is outside the tolerances specified.

Dot gain is made up of two components, optical and physical. Optical dot gain is caused by light scattering under the printed dots in the surface layers of the substrate. Consequently, it is almost entirely controlled by the characteristics of the substrate. Physical gain is influenced by the press, blankets, substrate and ink. It may be caused by an ink which is over emulsified or one which is too weak in colour strength and therefore needs a heavy filmweight to be carried to achieve the required density. It can also be caused by inks of low viscosity and structure as well as low tack. Techniques for controlling all these features have been discussed elsewhere in this section.

Dot loss is the opposite of dot gain and relates to the situation where the dot on the substrate is smaller than the dot on the original film. In serious examples the fine halftone dots may be lost altogether. It may be caused in platemaking by the overexposure of positive plates or, on the press, by fountain solutions which are too acidic or inks which are too water resistant. The last two result in incorrect emulsion formation and may be corrected either by changing the fount to a more neutral one or by increasing the emulsification capacity of the ink.

Heat-set web-offset prints are renowned for their high gloss which is almost a feature of the process. The gloss can be adversely affected by running the oven at too high a temperature (page 361) but it can also be affected by other factors. The gloss of a printed film is dependent on the formation of a smooth, continuous film. It is strongly influenced by the absorptivity and surface smoothness of the substrate as well as by the chemical and physical properties of the ink. Poor ink gloss may be caused by poor pigment wetting, poor transference and flow or the presence of large particle-size material. Poor pigment wetting will result in light scattering within the ink film which reduces gloss. It may also cause over emulsification of the ink which will inhibit the transference and flow properties, vital in the formation of a smooth ink film. Particulate matter which protrudes through the ink film also acts as light scattering sites which reduce gloss.

The use of flushed pigments is particularly advantageous in improving gloss levels because of their excellent dispersion and wetting. The gloss of an ink manufactured from dry colour can rarely equal the gloss of flush-based inks but if they have to be used, the pigments chosen should be of an easily wetted nature and easily dispersed on manufacturing equipment such as three-roll mills or shot mills. Within appropriate limits, a reduction in the pigment content of the ink will help increase gloss due to the heavier film which will be necessary to achieve the required density.

Waxes, too, should be chosen for their ease of dispersion and wetting.

The vehicle system may be modified to improve its solubility and hence its flow and transference by either using more soluble resins or by incorporating some supersolvent, although this may result in a lower viscosity which, in turn, may lead to the types of lithographic problems already discussed. High molecular weight resins, particularly the latest generation of high viscosity/high solubility ones, tend to have superior gloss to low molecular weight or insoluble resins.

Recent and future trends

Heat-set printing is now a fully established, mature printing process and changes, both to the process and the ink technology which serves it, tend to be evolutionary rather than revolutionary. Relevant evolutionary trends can be identified in three main areas:

- the lithographic printing process

- print finishing and added value
- heat-set ink raw materials and manufacturing techniques.

Lithographic printing

The major trend in heat-set web-offset is to higher productivity either through the use of faster running presses, the introduction of short grain presses and in-line print finishing for added value.

Both the press manufacturer's maximum mechanical rated speed and the average actual production printing speeds have increased significantly over the last decade. No web press running at less than 2000 feet per minute is now considered worthy of special mention and presses are already being built to run at 3000 feet per minute. Although only an evolutionary change, the increase in speed places severe demands on an ink because of the very high shearing forces present which break down an ink body, the very short time available for emulsion formation and the increased risk of misting due to the high peripheral speed of the rollers. The inkmaker is therefore required to undertake a critical review of such aspects of his formulations as rheology, chemistry and dispersion, to ensure continued satisfactory press performance. He must constantly strive to achieve a chemical and rheological balance that will not break down at high shear rates nor produce unsatisfactory emulsions in the time available. Better dispersion has been found to be essential in avoiding piling as pigment agglomerates seem to render an ink particularly susceptible to coarse over-emulsification with the subsequent loss of transference.

Much about emulsification and emulsion formation has been said in this section as it is this, the most fundamental aspect of a heat-set ink, which dictates the press performance. Considerable effort must still be made fully to understand the driving forces of lithography before any inkmaker will be in a position to predict and control his products performance over a wide range of operating conditions.

Waterless lithography

The 1980s have seen a resurgence of interest in waterless lithography and Toray Industries of Japan have both positive and negative plates available. The image area of a waterless plate is a photopolymer similar to that used on conventional plates but the non-image area is made from a siliconized rubber which is ink repellent. The technology has a number of advantages and disadvantages.

The major advantage of the system is the elimination of water from lithographic printing. Emulsion formation is one of the biggest variables in lithography and its removal reduces or eliminates most of the problems associated with the process. Press start-up is achieved faster without water and dramatic savings of paper through reduced wastage offer considerable cost benefits. The waterless-plate prints a slightly heavier film of ink than a conventional plate and this, coupled with the fact that the ink has not been diluted with fount, leads to an increase in the density of colour on the print. The lack of fountain solution also improves the dot-gain of inks printed by the waterless process with figures of less than

10% in a 40% screen not being unusual. Bright, bold solids in conjunction with excellent halftone reproduction are the hallmark of waterless lithography. An unforeseen advantage in heat-set is that the oven temperature can often be reduced when the requirement to dry the water contained in conventional inks is removed.

The system has some inherent disadvantages with the coating on the non-image area which is fairly easy to damage by scratching if handled carelessly. Hard surfaced substrates or those with high levels of lint and debris tend to damage the surface of the plate and therefore limit its life. The absence of water during printing leads to the press running hotter which has an adverse effect on the rheology of the ink which must be kept as high as possible to avoid problems of ink being picked up in the non-image areas. Presses need to be fitted with a system of temperature control to hold the inkers and plate at around 32°C. Roller settings are more critical than with a conventional system if problems of background toning are to be avoided. Despite these drawbacks, waterless lithography seems to offer a route toward better productivity and quality for the future. Most of the inks which are available today are formulated with traditional ink making materials but with the removal of the demand for ink–water balance, this need not be the case. Future developments seem certain to evolve around new and novel materials which will be chosen for their refusal to ink up the non-image areas.

Print finishing and added value.

Heat-set printing is a competitive business and printers are constantly seeking ways of improving their margins by producing added-value products. A number of magazine covers, on papers of up to 150 gsm, are now printed web-offset and UV-varnished either in line or off press. This places a requirement to provide UV-varnishable inks on the inkmaker and means that the use of materials such as PTFE must be limited. As it is not unusual for the same silo of ink to feed a number of presses, not all of which will be producing work which will be varnished, the ink must retain good rub resistance even when not varnished. Most inline coating is done via a dedicated coating unit fitted to the press after the oven.

The printing of documents on which additional information may be subsequently laser printed by a computer is now common. Most laser printers operate at high temperatures and the ink used on the pre-printed form should, ideally, not be thermoplastic. Heat-set inks are, by their nature, thermoplastic and for satisfactory results need to be based on very high melting point resins coupled with a higher than normal amount of drying oil.

Raw materials and manufacturing

Although the raw materials used for the formulation of heat-set inks are also subject to evolutionary rather than revolutionary change, the raw material suppliers have devoted considerable effort toward improving their performance.

Pigments

The major developments in dry pigments has been to make them easier to disperse, and work in this area is likely to continue. The use of hyperdispersants and special pigment coatings is now widespread. Pigments are available which reach an almost adequate level of dispersion during the pre-mix stage and thereafter require only a light grinding run on dispersion equipment such as three-roll mills or shot mills.

The advantages of flushed pigments have already been discussed elsewhere in this section and it only remains to be said that work to produce flushes of higher strength to maximize formulating flexibility is continuing. The availability of these products in Europe is improving and it seems likely that the benefits of better gloss and dispersion will lead to their increased usage, particularly for high speed presses.

Resins

Resin formulation has changed in a fairly radical manner in the last five years with the introduction of the new generation of highly structured, high viscosity and highly soluble resin-modified phenolics becoming available. These resins offer benefits in performance when compared with the more traditional types available hitherto and it seems that the technology can be pushed even further to produce yet higher viscosities for a given solubility. These types of resins offer superior drying, gloss, lithographic performance and viscosity stability under conditions of high shear and temperature, and are ideally suited to the development of inks for tomorrow's high speed presses.

Manufacture

Heat-set ink manufacture, in Europe at least, tends to be a high volume/low margin business. To succeed, ink manufacturing costs should be kept as low as possible, while meeting the quality demands of the market.

Batch sizes have increased with the major manufacturers producing batch sizes of at least 4 tonnes. The large batch size gives production economies of scale while improving the consistency of the product from batch to batch.

Three-roll mills are rarely used due to their slow production rates and the use of shot mills as the dispersion equipment of choice is now widespread giving products of high gloss and consistency.

Inks based on flushed pigments do not require any high energy dispersion techniques to be employed during production, and this undoubtedly is an advantage in terms of production economics. Large batches can be manufactured quickly, economically and consistently by the mix and filter method.

Summary

Heat-set web-offset printing is now a firmly established process and grows at the expense of other printing processes on an annual basis. It is competitive, both technically and commercially and produces high quality print. It is unlikely that it will undergo any revolutionary changes but is certain to continue developing to exploit the maximum benefit from the process in terms of increased productivity. If the commercial and technical advantages of waterless lithography are fully realized the process may be destined to enter a new era. The role of the inkmaker will continue to be a significant factor in meeting the expectations of the heat-set printer.

6.4 SHEET-FED INKS FOR PAPER AND BOARD

Sheet-fed printing is used for the production of publication and packaging work on paper and board. Publication work consists of magazines, brochures, catalogues, calenders and books whilst packaging work consists, primarily, of folding box board and label printing on paper.

Sheet-fed publication work has declined over the last few years because of fierce competition from heat-set web-offset (section 6.3) and is now restricted, in the main, to short run work. One notable exception to this trend has been that of book production which is still undertaken on sheet-fed Perfecta presses (Fig. 6.14).

Many folding cartons of diverse design and with differing requirements imposed upon the printed product by the contents are printed by sheet-fed lithography although the traditional choice of oil-based inks is under attack from UV inks. Product-dictated performance requirements

Fig. 6.14 High speed sheet-fed Speedmaster press. (Courtesy of Heidelberg UK).

range from soap and detergent resistance to low odour and taint properties for food and confectionery packaging.

The situation is similar for paper label printing in terms of the diversity of product dictated requirements placed upon the finished print. Sheet-fed is the dominant process by which paper labels are printed. However, for self-adhesive label production, flexography and UV rotary letterpress, utilizing special narrow width web presses, have become the dominant processes.

Lithographic considerations

The lithographic process demands the formation of a satisfactory oil-in-water emulsion. This topic has been fully covered in section 6.3 and elsewhere and the reader is recommended to read these sections as the subjects covered are applicable to all types of lithographic inks. Similarly, pH and conductivity have been covered in section 6.3 and elsewhere.

Dampening systems

Three types of dampening systems are commonly employed on sheet-fed presses: conventional, bareback and integrated. The major difference between the three is their responsiveness to changes in the damp settings.

Conventional systems

Despite their name, these types of dampening systems are now only used on older, slow running presses because they are considered difficult to control and unresponsive. The Molleton cover on the ductor roller acts as a reservoir for dampening fluid which results in changes to the damp settings, taking some time to feed through to the plate. They also tend to carry a lot of water and this tends to make them prone to causing over emulsification. The damper sleeves pick up significant quantities of ink over a period of time and have to be regularly taken off the press for cleaning or replacement. This type of dampening system may be shown schematically as in Fig. 6.15.

Bareback systems

These types of dampening systems are fitted to most modern presses and, as the name implies, they run with a system of uncovered rollers. They generally run with a fount /alcohol mixture in which isopropyl alcohol is dosed into the fountain solution mixture at levels between 5 and 15%. The alcohol lowers the surface tension of the dampening fluid and allows it to wet the plate more effectively, so enabling a lower filmweight of damp to be used on the plate. As less damp is carried on the plate the risk of over emulsification is greatly minimized and because there is no reservoir of fount being carried in a damper cover these systems are very responsive to changes made in the damper settings.

However, as the alcohol lowers the surface tension of the fount there is a risk that inks may fail to print satisfactorily if they are not correctly formulated.

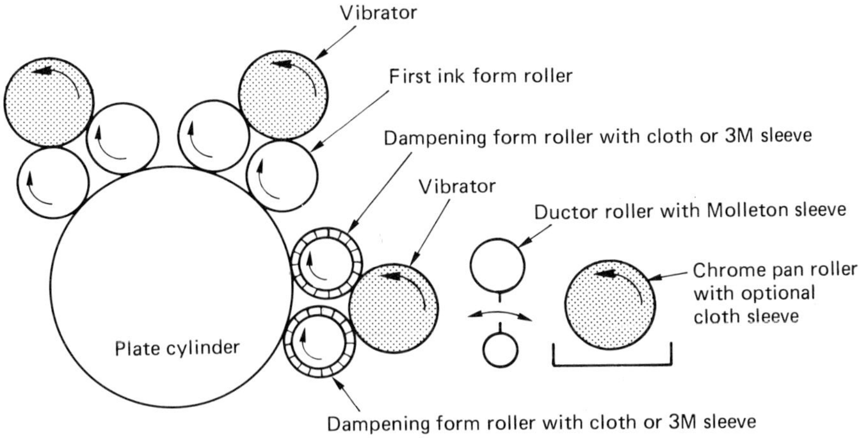

Fig. 6.15 Conventional dampening system.

The Roland–Matic dampening system is typical of bareback configurations, shown schematically in Fig. 6.16.

Integrated dampening systems

Integrated systems, dampening in part via the inking rollers, are the most responsive of all dampening systems. They usually require the use of alcohol at levels of between 5 and 12% to ensure adequate reduction of the surface tension of the dampening fluid and to encourage the formation of a satisfactory emulsion on the inking rollers. They are extremely responsive to only minor changes in the damp setting and allow minimum films of damp to be carried which helps to ensure good quality print. However, as they damp, at least in

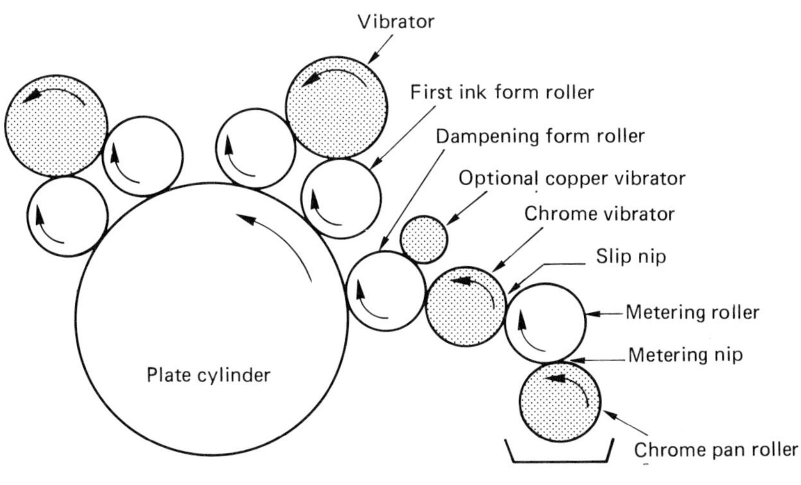

Fig. 6.16 Roland–Matic dampening system.

part, via the inkers, care must be taken to formulate inks which resist over emulsification and which retain good transference when emulsified.

The Heidelberg Alcolor dampening system fitted to Speedmaster presses amongst others typifies these systems and is shown schematically in Fig. 6.17.

Drying mechanisms

Sheet-fed lithographic inks dry by a mixture of processes which may be categorized as penetration quick-setting/phase separation and oxidation/polymerization.

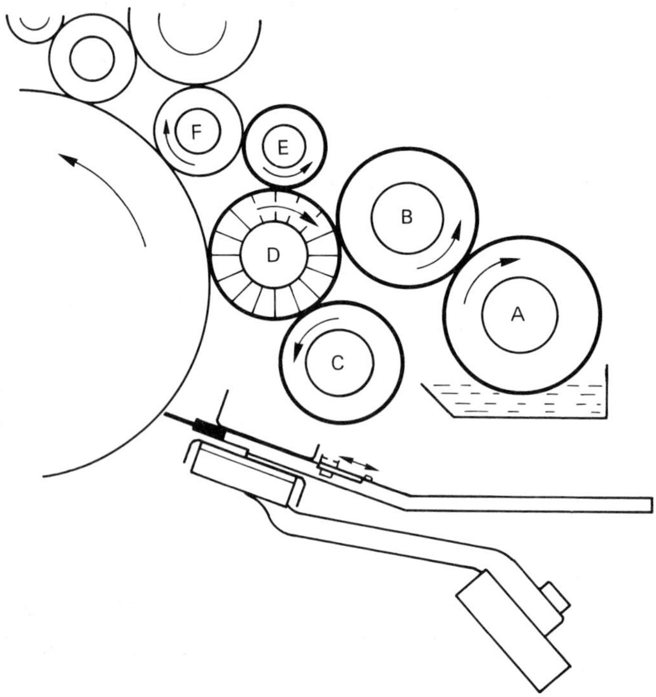

Fig. 6.17 Heidelberg Alcolor dampening system. All rollers operate without cloth covers. The roller surfaces have been selected according to individual function. A water pan roller (rubber); B metering roller (high-polish chromed); C distributor (mat chromed); D plate damper (rubber); E intermediate roller (Rilsan); F first plate inker (rubber). Operation: A/B premetering and initial formation of dampening film through splitting; B/D formation of an extremely thin dampening film through slippage; C/D final stage of dampening preparation by working into the ink; D/E/F connection between inking and dampening units.

Penetration

Penetration drying is the first phase of the drying cycle which occurs on uncoated papers and boards. The mass of the printed ink film penetrates the interstices between the fibres of the substrate and becomes immobile. Providing that the ink film thickness does not exceed the absorptive capacity of the substrate and enough time is allowed for full absorption to take place, the printed substrate may be handled with care relatively soon after printing. Final drying to a hard, rub resistant film then proceeds via the mechanism of oxidation which is discussed below.

Quick-setting/phase separation

Quick-setting is the first phase of the drying cycle which occurs on coated papers and boards. It occurs as the distillate contained in the ink is drawn into the paper coating leaving the resin and higher viscosity materials, such as alkyds, on the surface of the paper which rise rapidly in viscosity to the point where the printed film becomes immobile. At this stage the ink is said to be set and ceases to transfer readily which prevents it marking the reverse of subsequent sheets in a sheet-fed stack of printed work. Setting may take from as little as two minutes to over half an hour depending on the ink formulation, the filmweight printed, the substrate and ambient conditions, particularly of temperature. Following the setting process, oxidation drying within the drying oil or alkyd leads to polymerization and the formation of a three-dimensional cross-linked network of chemical bonds. This process, considered in more detail in the following section, is usually well advanced in 6–16 hours although it may take some days for complete reaction to occur and, therefore, for the ink to reach its maximum level of rub and scuff resistance. This cycle is shown in Fig. 6.18.

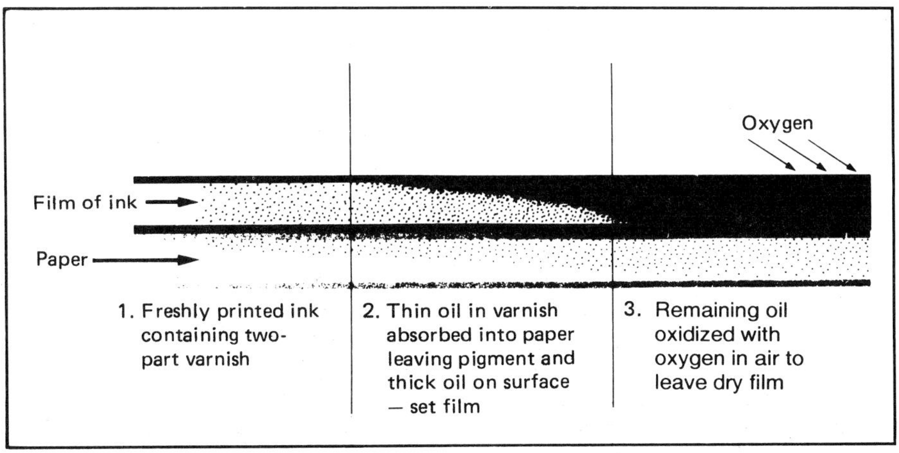

Fig. 6.18 Drying process of quick-setting inks – the principal method used by lithographic inks. It relies on rapid separation of thin mineral oil from the ink film followed by oxidation/polymerization of the drying oil.

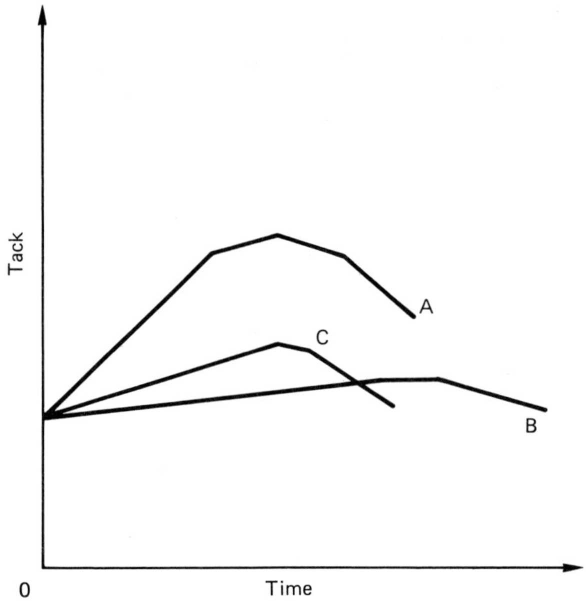

Fig. 6.19 Tack peak versus time.

Quick-setting systems are employed extensively for printing paper and board and may be formulated using a variety of approaches. The 'classical' route was to use two resins with different solubilities in distillates. An insoluble resin yielded vehicles which were fast setting and had low tack peaks but had poor gloss whilst soluble resins produced glossy but slow setting vehicles with a high tack peak. By blending the two resins in a single vehicle or by using two vehicles made from each of the single resins, the formulator was able to optimize the properties of gloss, tack peak and setting speed in any particular ink. The setting rate and the tack peak which an ink reaches during the setting phase is important as they determine the likelihood of set-off occurring in the printed stack. If the tack peak is high, the ink is likely to stick to the underside of the following sheet in the stack and if the setting rate is too slow the ink will remain wet and be capable of transferring to the subsequent sheet. To avoid set-off, inks need to be quick-setting and to have a low tack peak during the setting phase. As setting may be considered as the point at which an ink becomes immobile and ceases to transfer, the setting speed can be approximated by a tack–time curve where the effect of distillate loss is shown by an increase in tack to the point where it becomes difficult to split the film and the apparent tack starts to drop.

Figure 6.19 shows the effect of tack peak on set-off. Curve A shows an ink with a high tack peak which is likely to cause set-off. Curve B shows an ink with a low tack peak but slow setting speed. This type of ink is also likely to cause set-off. Curve C shows the ideal quick-setting ink – rapid setting but with a low tack peak.

The new generations of hard resins offer the possibility of making quick-setting inks from a single resin which is at the same time fast setting, glossy and has a low tack peak. Obviously the choice of hard resin becomes critically important in these cases and the formulator needs methods of categorizing the properties of different hard resins. This topic is discussed more fully on page 400 which deals with hard resins.

Oxidation/polymerization

Oxidation is the classical drying mechanism for lithographic sheet-fed inks and involves the oxygen-induced free-radical polymerization of unsaturated (drying) oils such as linseed and tung. It is a chemical process, the primary reaction of which is thought to be that of autoxidative polymerization. This proceeds by several stages:

- peroxide/hydroperoxide formation
- decomposition to form free radicals
- polymerization
- termination.

Peroxide/hydroperoxide formation

This stage involves the attack by atmospheric oxygen at activated sites on the drying oil fatty acid chains. One such site is the methylene group adjacent to one of the carbon–carbon double bonds present in all drying oil molecules:

$$- \overset{*}{C}H_2 - CH = CH -$$

* active methylene group

If we consider linseed oil as an example of a typical drying oil, which has linoleic and linolenic acids as principal fatty acid components, the following reaction scheme has been proposed.

Taking linoleic acid as a specific example, which occurs naturally in the *cis* configuration:

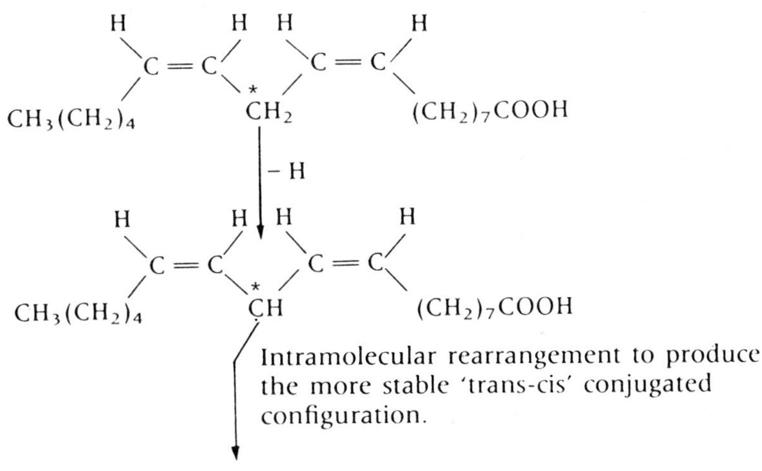

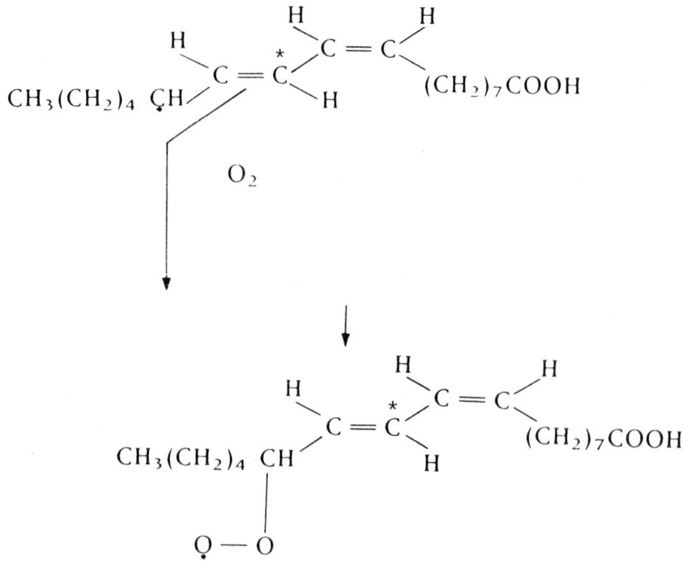

This peroxy free radical can abstract a proton from an active methylene group on another fatty acid chain producing the hydroperoxide:

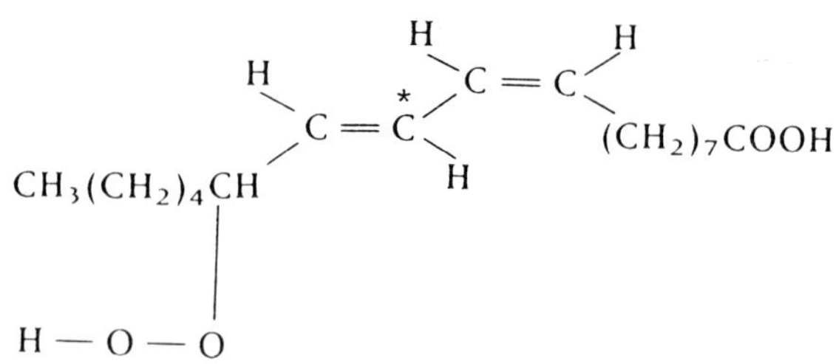

Note that the – CH_2 – group marked with an asterik, one of three methylene groups adjacent to carbon–carbon double bonds in the molecule, is more reactive than the others because it is adjacent to two double bonds and is therefore the major reaction site.

Where conjugated structures are present, as in tung oil, direct oxygen addition to the conjugated system to form a 1,4-cyclic peroxide is possible.

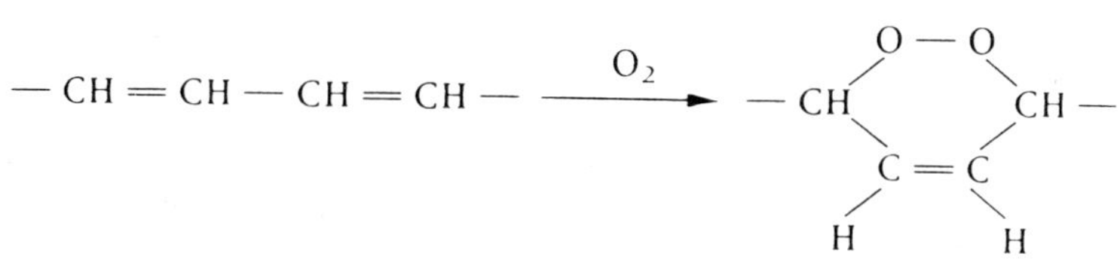

It follows, therefore, that any further oxygen addition to the conjugated hydroperoxide formed by the oxidation of linoleic acid will also give a 1, 4-cyclic peroxide.

Decomposition to form free radicals

Once hydroperoxides or cyclic peroxides have been formed, they can decompose into free radicals:

$$ROOH \rightarrow RO^{\bullet} + {}^{\bullet}OH \qquad (1)$$
$$2ROOH \rightarrow RO^{\bullet} + ROO^{\bullet} + H_2O \qquad (2)$$

The following reactions are then possible:

$$RO^{\bullet} + RH \rightarrow ROH + R^{\bullet} \qquad (3)$$
$${}^{\bullet}OH + RH \rightarrow R^{\bullet} + H_2O \qquad (4)$$

The $R^{\bullet}$ radicals formed in reactions (3) and (4) can react with atmospheric oxygen to form peroxides and hydroperoxides, thus propagating the chain reaction.

Polymerization

The radicals formed in the above reactions can add on to another molecule of drying oil, and these addition reactions can continue, increasing the molecular weight until termination occurs. There is also more than one site for the growing radicals to attack and this produces cross-linked molecules. The increasing molecular weight and the cross-linking reactions cause the ink vehicle to become a solid material, encapsulating the pigment.

Termination

The $R^{\bullet}$ radicals formed in reactions (3) and (4) can also react with each other leading to the termination reaction:

$$R^{\bullet} + R^{\bullet} \rightarrow R-R \qquad (5)$$

Other, minor reactions involving the other free radicals formed include:

$$R^{\bullet} + RO^{\bullet} \rightarrow R-O-R \qquad (6)$$

and

$$RO^{\bullet} + RO^{\bullet} \rightarrow R-O-O-R \qquad (7)$$

Nowadays, oxidation as the sole drying mechanism is normally only encountered in inks for printing impervious substrates such as foils and plastics.

The oxygen-induced polymerization of drying oils and drying oil modified alkyds is accelerated by incorporation of small quantities of certain metal compounds. These drier catalysts are mainly based on transition metals that are able to exist in stable forms in more than one oxidation state. This provides reaction pathways which speed the oxidation reaction and enable a highly cross-linked film to form in a matter of hours rather than days. Metal soaps of long-chain carboxylic acids are usually employed because of their solubility in the oleoresinous vehicles found in oxidation drying and sheet-fed quick-set inks.

With the demise in the use of lead for toxicological and environmental reasons, cobalt and manganese are the principal metals used today. Their soaps are normally used in the form of relatively dilute solutions in hydrocarbon solvent in order to assist handling and make weighing of small metal concentrations less critical.

Cobalt is probably the most widely used drier catalyst in oxidation drying systems at present due to its outstanding ability to accelerate the drying reaction. The catalytic activity of cobalt relies on repeated transitions from the Co^{2+}oxidation state to Co^{3+}.

We have seen earlier that a key step in the oxidation drying mechanism is the formation, and subsequent decomposition, of hydroperoxides. These hydroperoxides decompose at elevated temeratures to give free radicals:

$$\mathrm{ROOH} \rightarrow \mathrm{RO}^{\bullet} + {}^{\bullet}\mathrm{OH} \qquad (8)$$

However, in the presence of small quantities of metals such as cobalt, this decomposition reaction proceeds readily at ambient temperatures. The reaction can be represented as follows:

$$\mathrm{Co}^{2+} + \mathrm{ROOH} \rightarrow \mathrm{Co}^{3+} + \mathrm{RO}^{\bullet} + \mathrm{OH}^{-}$$
$$\mathrm{Co}^{3+} + \mathrm{OH}^{-} \rightarrow \mathrm{Co}^{2+} + {}^{\bullet}\mathrm{OH}$$

The free radicals generated during both the above oxidation and reduction reactions accounts for the increase in the rate of polymerization seen when cobalt is used as a drier.

Generally, drying efficiency, in terms of speeding cross-linking and producing a harder final film, increases with drier content up to a limiting value. Additions above this level may give more rapid drying but can also result in soft films. This phenomenon is believed to be associated with the formation of large numbers of relatively low molecular weight polymers by extensive cross-linking. Over-addition of driers can also result in press stability problems due to skinning, particularly on the roller ends where there is little take-off and so ink dwell time as a thin film is maximized. Both quantity and choice of driers can affect the formation of odorous oxidation byproducts during the drying reaction and hence can influence the odour and tainting potential of the final print. Thus, drier combinations for inks to be used on food and confectionery packaging must be carefully formulated to minimize the formation and retention of oxidation byproducts, such as volatile aldehydes, ketones and carboxylic acids, which are odorous and mobile.

Volatile aldehydes, ketones and carboxylic acids occur as byproducts of oxidation drying of drying oils through decomposition of the peroxides and hydroperoxides formed as intermediates in these reactions. The following general scheme for production of these materials is:

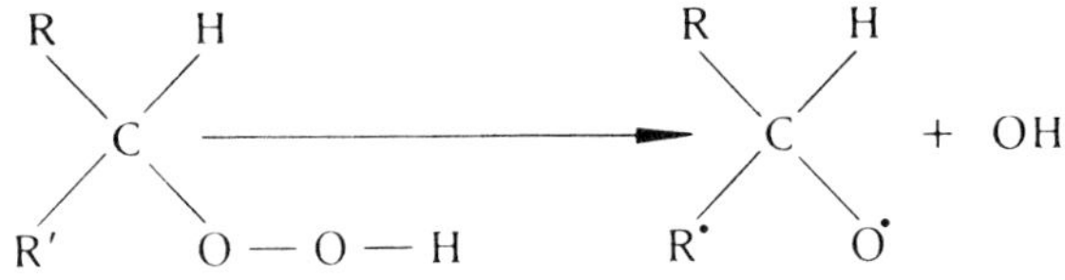

(9)

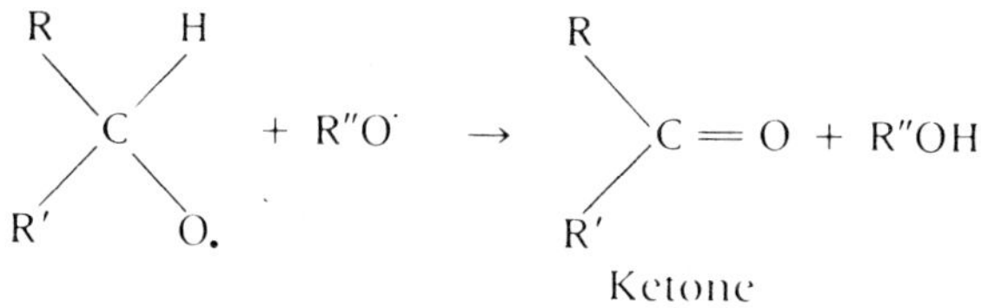

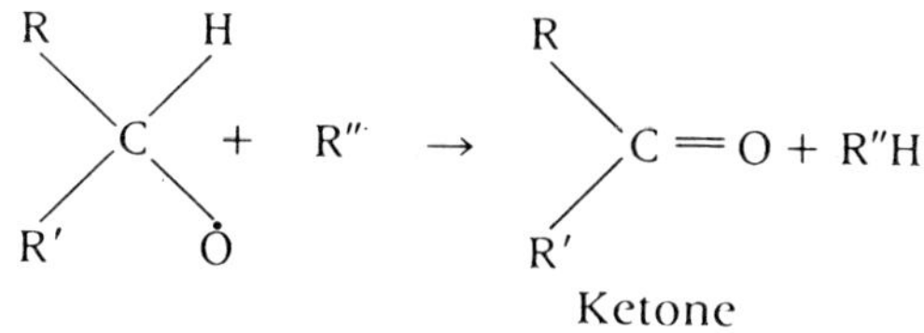

(10)

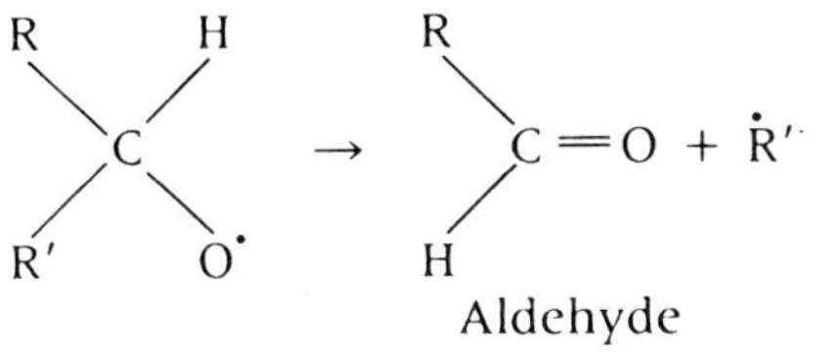

(11)

The aldehyde produced in reaction (11) can then oxidize further to produce carboxylic acids:

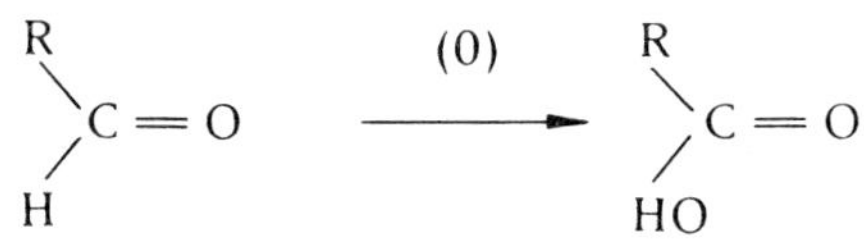

To illustrate how such low molecular weight materials can be generated, the following reaction scheme, starting with the hydroperoxide of linoleic acid is proposed. The scheme follows reactions (8) and (9) above:

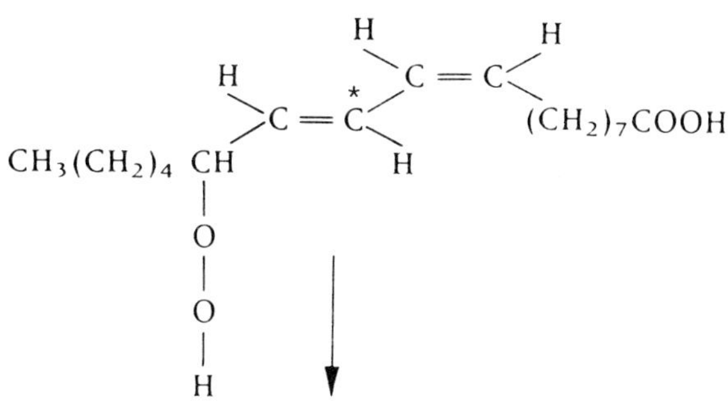

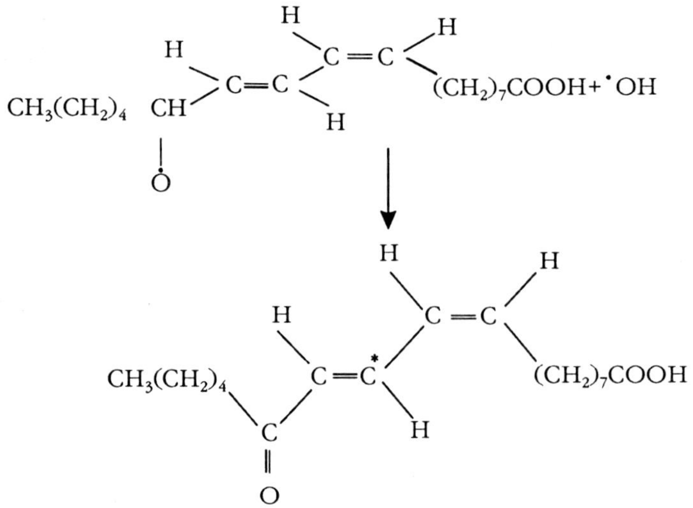

Ketone

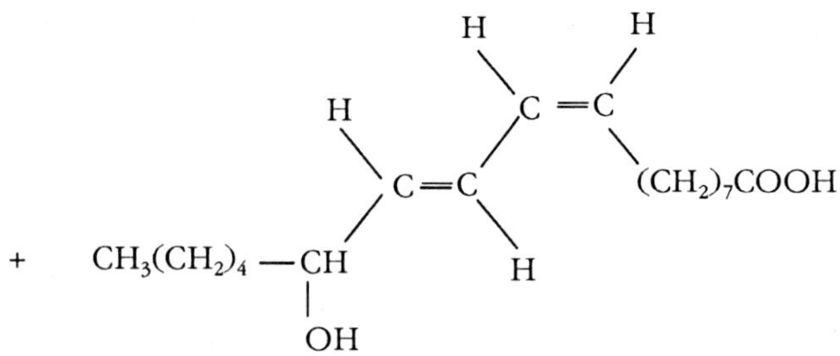

Alcohol

The unsaturated ketone (molecular weight 246) produced by this mechanism is capable of further oxidation to give

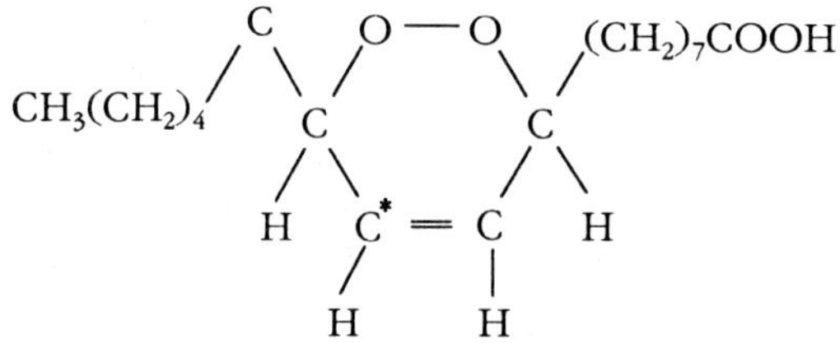

The cyclic peroxide then decomposes in the following way to give a diketone of molecular weight 84, and aldehyde of molecular weight 100, an acid of molecular weight 144 and a keto-acid of molecular weight 242.

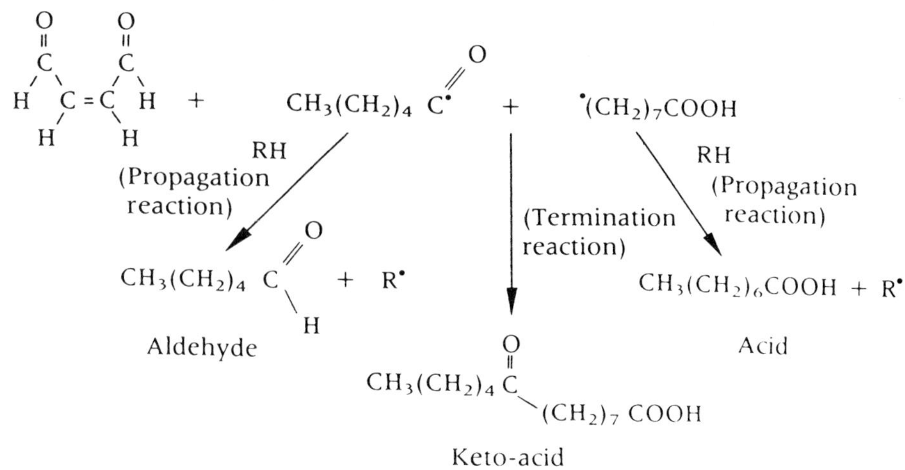

The low molecular weight diketone, aldehyde and acid formed are potential sources of off-odours and taints; the diketone produced in this scheme is noted in chemical references as being a lachrymatory liquid which will obviously have significant potential to cause odour and possible taint problems.

Drying and setting tests and measurement

Before leaving the fundamentally important areas of drying and setting it is worth considering the assessment of these properties in the laboratory.

All setting checks involve taking a print in a controlled manner (with respect to substrate and film thickness) and subjecting it to contact under pressure against a virgin piece of stock at appropriate intervals throughout the likely setting period. Unless a temperature and humidity controlled environment is available, it is usual to conduct all tests directly in parallel against a standard ink of known performance.

Set-off in the stack may be assessed by the controlled building of a weighted stack on top of a print in contact with virgin stock over an appropriate period of time. As with setting, comparison against a known standard ink tested in direct parallel is usually required. This type of testing assesses the set-off produced both during initial setting and during the subsequent drying stages, which may sometimes involve periodic ink softening.

Two types of ink drying test are used for oxidation drying and quick-setting formulations. The first involves printing the ink on a relevant substrate and assessing rub resistance at appropriate intervals. The second involves producing ink films on impervious surfaces and periodically assessing them until they are touch dry. This latter technique isolates the the oxidation drying mechanism from any penetration or setting effects. However, if the film is freely open to the atmosphere there may be substantially greater availability of oxygen than exists in a stack of printed work. For this reason tests are sometimes employed where the ink film is stacked or sandwiched in some way to limit severely atmospheric contact. Oxidation drying is greatly influenced by ambient conditions of temperature and humidity, therefore all tests should either be done in a suitably controlled environment or against a standard ink of known performance. The latter approach is not wholly reliable since ink-drying characteristics may change with ink ageing and it is therefore difficult to ensure that standards are truly of known performance in this respect.

Formulating principles

We have seen in section 6.1 how the nature of the lithographic process imposes certain fundamental requirements on the chemical, physical and strength characteristics of a lithographic ink. The achievement of those required characteristics and those dictated by the typical end uses to which the inks are put will now be considered in terms of raw material selection and formulating principles.

Sheet-fed lithographic inks are formulated from vehicle systems, solvents/diluents, pigments and additives, each of which is discussed in more detail in the sub-sections below.

The vehicle system

The vehicle system of a sheetfed ink is the component which has the greatest effect on the physical properties of the ink, such as setting speed, gloss, viscosity, yield value, flow, tack, hardness of drying and emulsion formation. It will dictate whether the ink will perform satisfactorily or not. Correct vehicle formulation, necessary to achieve the desired properties in an ink, is therefore the single most important aspect of the formulator's work.

Lithographic ink vehicles may be divided into two major classes. These are the oleoresinous (hard resin and drying oil alkyd) systems used to produce quick-set, heat-set and oxidation drying inks and the acrylate systems used in radiation curing inks. The latter are described in detail in Chapter 11. However, it is worth noting here the relevance to radiation curing vehicles of all comments in section 6.1 and below, which relate to the emulsification, cohesion, surface energy and pigment-wetting dictates of the lithographic process, and the requirements that they pose in both chemical and physical characteristics. In other respects this sub-section is restricted to a consideration of oleoresinous vehicles, whether drying exclusively by oxidation or by the quickset mechanism:

Oxidation drying vehicle	*Quick-set mechanism vehicle*	
Hard resin(s)	Hard resin(s) }	A
Drying oil/alkyd	Drying oil/alkyd }	
	Petroleum distillate }	B

The vehicle is formulated such that there is limited compatibility between A and B. We have already seen (page 365) that the lithographic process places stringent demands on the chemical nature of the binder resins and the physical form of the vehicles based on these materials.

Chemistry

The resin chemistry used has to meet a number of fundamental requirements:

- solubility in weak solvents;
- controlled water tolerance – neither fully miscible nor totally water repellent;
- cohesive
- low surface energy

The last two are requirements when used in appropriate weak solvent-based vehicle. Hardly surprisingly, the range of basic resin types which comply with this demanding set of requirements is very limited. In fact, in the major area of oleoresinous systems the vast majority of the hard resins used fall into only two classes. These are the modified rosin ester and modified hydrocarbon groups of resins. The drying oil and alkyd components used, which can essentially be regarded as liquid resins, also represent a narrow range of chemistry. Linseed and soya are the dominant vegetable oils, either for direct incorporation into litho ink vehicles or for alkyd modification. The majority of alkyds employed are long oil modified isophthalic esters of either glycerol or pentaerythritol. This is not to say that other classes of resin and oils or alkyds cannot be found in litho ink formulations. Some, such as cyclized rubber resins for certain small offset applications and tung oil for very hard drying requirements, may be essential choices in certain formulating situations. However, it is a distinctive feature of litho ink vehicles that the range of basic chemistry available to the formulator is severely limited. Attempts have been made to broaden this base and, for example, oil-compatible acrylic resins suitable for litho applications have been developed within the last few years. These resins have not made a major impact and it seems unlikely that future developments will achieve major departures in terms of novel lithographic vehicle chemistry at acceptable costs.

On this basis, the formulator of oleoresinous lithographic vehicles would seem to be provided with only a handful of raw materials with which to perform the task of producing a variety of vehicles for sheet-fed applications. While the pure chemist might be tempted to agree, the ink technologist will recognize that resin manufacturers have produced a bewildering range of resin grades, albeit on the basis of the restricted fundamental chemistry that we have already identified. Both chemical modifications, as in the phenolic or maleic modification of rosin esters, and molecular weight variants exist, and even minor changes have a marked influence on printability and print characteristics. One of the prime properties that can be altered is the solubility or compatibility of the resin with petroleum distillate.

The correct choice of hard resin to obtain optimum properties in a quick-setting system was mentioned on page 391 with an indication that the solubility of a resin was an important characteristic. The solubility of a resin may be characterized by its precipitation temperature which varies with the concentration of resin in the vehicle, rising as the resin concentration increases. The resin concentration of an ink increases dramatically at the moment of printing as distillate is lost from the ink to the paper. The solubility of the resin in the ink on the paper surface increases equally dramatically and the solvent release of the resin slows down at an equal rate. The net result may be a very high tack peak and a slowing down of the setting rate which is a feature of soluble systems and leads to an increased likelihood of set-off in the stack.

The precipitation temperature of a resin may be measured by simply weighing a known amount of resin into distillate, heating the mixture to a point where a clear solution is obtained and measuring the temperature at which the resin precipitates out of solution. The higher the precipitation temperature, the less soluble is the resin. A common test uses 40% of resin in distillate which may be heated in a test tube. The formulator has the choice of either using a standard distillate to categorize all resins tested or of using the distillate blend intended for use in a given ink system. The data yielded by a simple measurement of the precipitation temperature is limited, particularly with the new generation of soluble but fast setting resins now available, and it is useful to consider either an alternative type of test or a full scale evaluation in a trial ink.

One useful test which has been developed over the last few years is the V-graph used by Hercules Incorporated to categorize resins by solubility and viscosity as a function of concentration, and from which conclusions about the resin's performance may be drawn.

Examination of the graphs in Figs. 6.20 and 6.21 will aid in understanding the principles of the test procedure. The left-hand side of the graph has a vertical axis showing the precipitation temperature in degrees centigrade and the horizontal axis in the middle of the graph shows the resin concentration. The precipitation temperature is checked at three different resin concentrations in either distillate or a mixture of distillate/vegetable and oil/alkyd, and the results plotted on the graph. Inset into the top right-hand quadrant of the graph is a Laray chart on which are plotted the results of viscosity measurements on varnishes made at three different resin concentrations in the same distillate or distillate/oil blend as that used for the precipitation temperature measurements. Because the same solvent blend is used for the generation of both sets of data, the two areas of the graph may be interrelated. By choosing a standard viscosity for comparison (Hercules use 20 Pa s) the resin solids required to obtain that viscosity can readily be calculated and different resins compared. If a vertical line is dropped from the standard viscosity and the line showing precipitation temperature is extrapolated to intercept it, the precipitation temperature of the resin at that viscosity may be calculated. This has more meaning than comparing precipitation temperatures of different resins at the same resin concentration and ignoring the viscosities produced by different resins under conditions of normal usage. A number of predictions about the resin's performance may be made by interpretation of the results. The higher the precipitation temperature of a resin dispersion at a given viscosity, the faster will be the setting speed. As the precipitation temperature of a resin dispersion depends on the concentration of the resin, then it follows that the setting speed depends on the resin concentration. The slope of the precipitation temperature/resin concentration line on a V-graph, indicated the solvent release of the resin. The steeper the line the slower setting will be the resin and the higher will be the tack peak reached during setting. Resins which yield more horizontal lines are faster setting and have lower tack peaks during setting.

Tack is influenced by a resin's solubility and the lower the precipitation temperature at a given viscosity, the higher will be the tack.

Similarly, the lower the precipitation temperature, the higher will be the gloss.

In an ideal world a resin would have a high resin-solid's content for a

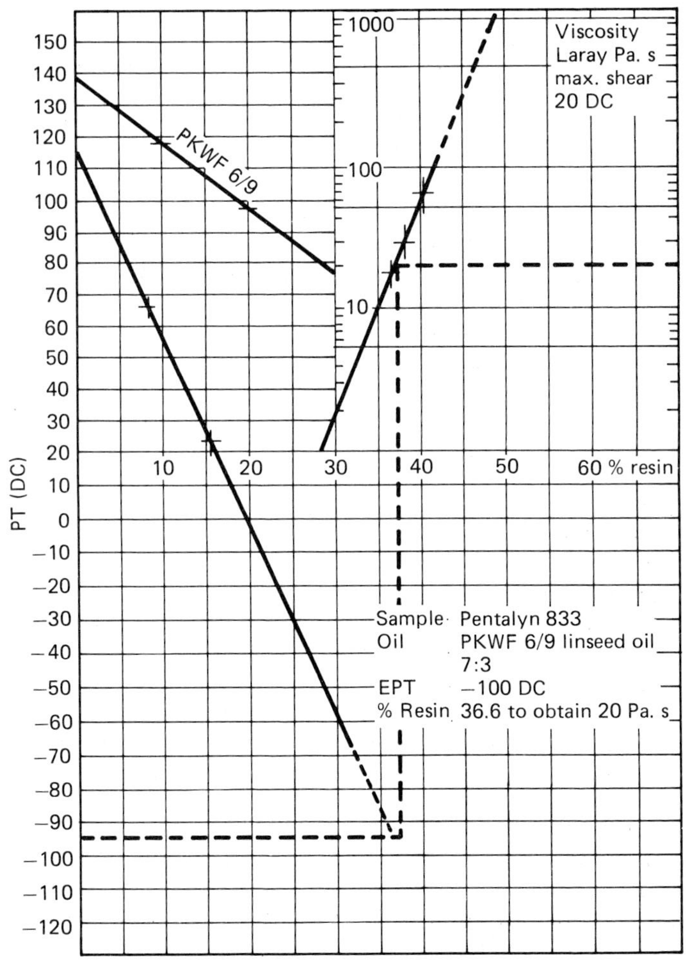

Fig. 6.20 Examining resins by V-graph (pentalyn 802).

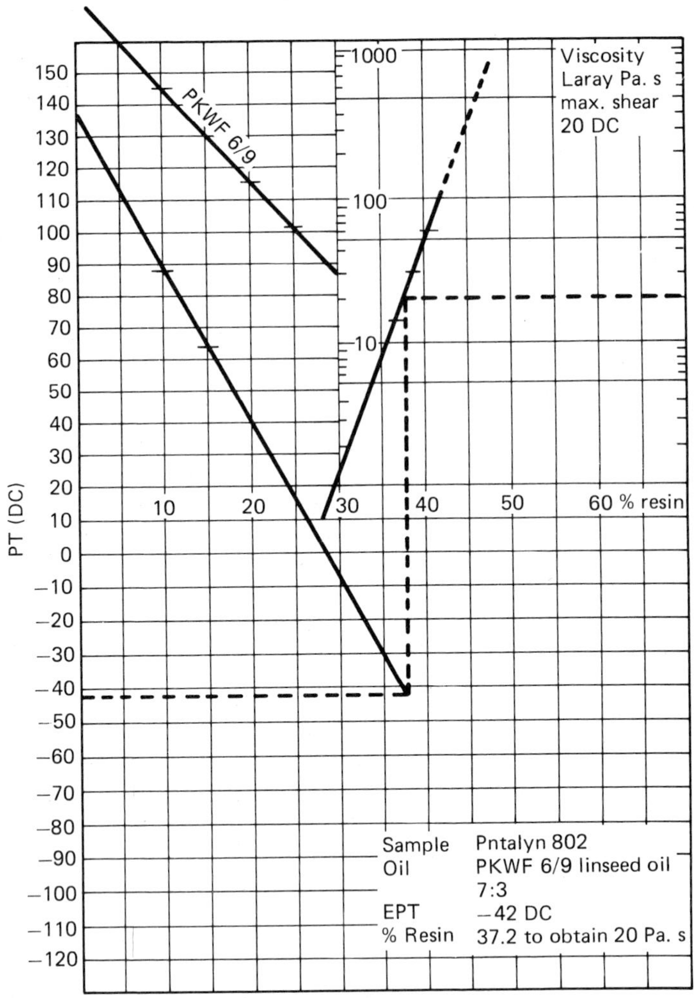

Fig. 6.21 Examining resins by V-graph (pentalyn 833).

given viscosity coupled with a high precipitation temperature which remained almost unaltered by changes in resin concentration.

Producing resins that have good solubility, for maximum stability, coupled with rapid viscosity rise on solvent release, in order to give the fastest possible setting, has been a major goal of the resin industry. The compatibility of a vehicle is not controlled by resin choice alone but is also influenced by oil and distillate types and proportions. The role of distillate will be considered in more detail later.

Vegetable oils and alkyds have a major solubilizing influence on the hard resins used in oleoresinous lithographic vehicles. This effect is usually proportional to the alkyd or oil content and it is, therefore, customary to consider quick-set and heat-set vehicles in the context of their resin-to-oil (and/or alkyd) ratio. It must always be borne in mind that resin and distillate choice influence vehicle compatibility and hence the stability/setting balance of the formulation. Nevertheless, when these factors are controlled, the resin-to-oil ratio relates directly to the compromise achieved. Thus, a high resin-to-oil ratio minimizes compatibility and promotes setting at the expense of stability. The converse applies for low resin-to-oil. Other properties are influenced by compatibility and hence by resin-to-oil ratio. Gloss is a prime example, being generally improved by high vegetable oil or alkyd contents. Within the quick-set mechanism, resin-to-oil ratios of vehicles may vary from 7:1 to 0.5:1, dependent on the choice of resin, oil and distillate and the end application of the vehicle. The latter may range from high-speed heat-setting to high-gloss, low-speed sheet-fed.

Vehicle compatibility can be influenced by manufacture as well as by raw material choice and proportions. This effect is exploited in the production of so-called 'cooked' varnishes. In these the hard resin and drying oil components are heated together at elevated temperatures, usually in the 220–240°C range. Here, appropriately modified rosin ester grades will undergo chemical reaction, largely involving ester interchange, with vegetable oils. The reaction modifies the distillate tolerance of the resin/oil phase. In this way a resin which is incompatible in a given distillate grade may be cooked to a tightly controlled tolerance before crash-cooling and completion of vehicle manufacture by incorporation of the distillate.

Manipulation of vehicle components is also carried out in order to achieve the optimum chemical balance for the particular lithographic performance required. We have already seen in this subsection that hard resin and alkyd chemistry is severely constrained by the requirement for controlled emulsification. Even so, within the materials that are potentially appropriate, there are subtle variations in water tolerance that can make a critical difference to machine performance under adverse conditions such as high press speeds, poorly controlled dampening, and plates of poor image/non-image distinction. Within the limited range of chemistry that is available, combinations of materials sometimes enable improved lithographic compromises to be achieved. One of the classic examples of this is the combination of modified rosin ester and modified hydrocarbon resins in precise proportions to achieve particular water tolerance characteristics. The simple guiding principle in this area of vehicle formulating is that the hydrocarbon is more water repellent than rosin ester. In practice, much empirical work is normally necessary before the exact proportions of specific grades are defined such as to yield rapid, controlled emulsification with minimum impact on transference and distribution.

Physical form

The required viscosity for good lithographic performance can generally be achieved in oleoresinous vehicles by manipulation of the resin-solids content. Oxidation drying vehicle viscosity can be adjusted by altering the drying oil content and similarly quick-set vehicles can be adjusted by varying the amount of distillate. Nevertheless, certain limitations apply since too low a resin-solids content may produce insufficient hardness when the vehicle is acting as the binder on the dried print. Even more important, too much distillate in a quick-set vehicle can result in poor pigment wetting. This impedes dispersion and is likely to produce ink of unacceptable rheology – high viscosity and yield value and excessive thixotropy – even though the vehicle appears to be of appropriate viscosity. This is a particular problem in high resin-to-oil ratio systems for heat-set and very rapid setting sheet-fed inks. Therefore, resin grades must be selected with due consideration of the vehicle viscosity that they will produce. In a given formulation where the alkyd and distillate types and contents are defined, the viscosity produced by a particular resin grade will be dependent on its inherent viscosity (as measured in strong solvent solution) and its compatibility, or distillate tolerance. This latter parameter influences vehicle viscosity in quick-set systems since the vehicle is not a true solution. The resin is best regarded as being partially dissolved and partially in colloidal dispersion. Forces of association between the dissolved and dispersed resin components ensure that the vehicle is a stable fluid exhibiting bulk homogeneity. However, these forces also impair vehicle flow and thereby raise viscosity. Consequently, compatibility, or distillate tolerance, and vehicle viscosity are generally inversely proportional.

Careful raw material selection and formulation can yield vehicles which produce inks of the required viscosity for lithographic printing while maintaining resin-solids that ensure good pigment wetting and adequate hardness on print. However, it is often difficult to achieve the optimum balance between viscosity and tack. In simple terms, lithographic ink vehicles based solely on appropriate hard resins, oils, alkyds and, possibly, distillates often yield inks which are rather too high in tack at optimum viscosity. Lowering the tack by distillate or vegetable oil addition produces low viscosity and may lead to emulsification and inhibition of distribution or transference. To maintain viscosity and reduce tack, the resin-solids need to be lowered.

A frequently used technique for achieving this is to produce a structured gel varnish at reduced resin-solids. The gel structure is created by introducing a material, most commonly some form of aluminium complex, which will increase intermolecular bonding, hence inhibiting flow and raising viscosity. Gel varnish production, despite having been established in the industry for decades, can be problematic since the reactivity of both the resin and gellant can vary. The result is not a simple increase in viscosity but rather the introduction of a yield value and thixotropy into the vehicle. This further complicates the rheological behaviour of inks based on these gels. Despite these drawbacks, gel varnishes are frequently used in lithographic ink formulation for fine balancing of tack and rheology.

Solvents/diluents

The volatility and solvency dictates of the process (page 344) restrict the lithographic ink formulator to the use of very weak solvent power, high boiling petroleum fractions. Indeed, few people outside the lithographic industry would regard these materials as solvents and in many ways they might be considered as diluents. The very low solvent power is sometimes boosted in a controlled manner by the use of small proportions of stronger solvents which may have alcohol or ester functionality. Tridecanol is an example of one such 'bridging' or 'co-solvent'. High molecular weight materials are chosen since volatility restrictions apply. While these are strong solvents to the lithographic ink chemist, they are still weak solvents in absolute terms as a result of their molecular weight and low functionality. Because of their higher solvent compared with distillates, a small addition of bridging solvent will give a significant increase in compatibility. Thus, they can be useful for reducing viscosity with little effect on tack. They can also produce an improve balance between compatibility and phase separation, particularly in heat-set web-offset inks and may, therefore, be used to promote stability and gloss with minimum sacrifice on setting.

Typical bridging solvent contents in lithographic inks may be of the order of 2%, whereas distillate contents can be in the range 20–50%. Petroleum distillates are therefore major components of inks which dry by the quick-set mechanism and produce subtle effects beyond their simple role as viscosity and tack adjusters. The two prime properties of distillates are their boiling range and aromatic content:

Boiling range (°C)	
240–260	Heat-set
260–290	Slow heat-set/ultra-fast quick-set
280–320	Quick-set

	Aromatic content (%)	
Conventional distillate	Above 15	Highest solvent power
Low aromatic distillate	5–6	Medium solvent power
Zero aromatic distillate	0–1	Low solvent power

Volatility restrictions necessitate a high boiling range. However, slightly more volatile grades are acceptable in high-speed shhet-fed or web-offset printing where the ink dwell-time on press rollers is minimized. Boiling ranges are narrow or 'close-out' in order to achieve the optimum balance between press stability and phase separation. A broad boiling range can result in both poor stability, due to evaporation of the more volatile, lower-boiling fractions from the ink distribution rollers, and slow setting, caused by retention of poorly mobile high-boiling fractions.

Modern resin development has made it increasingly acceptable to use low or zero aromatic distillates to give adequate resin-solids levels which would only have been possible with 'conventional' relatively high solvent power distillates a few years ago. There are additional advantages in using these weaker solvent power distillates. They often give a better balance of rapid setting with adequate stability because of the inherent solubility of the resin. In low odour and taint systems for food packaging, and where heat-set oven emissions are subject to controls, the inks supplied must be formulated on the lowest aromatic content distillates available.

Before leaving this consideration of the solvent power of the distillates used in lithographic inks for drying by the quick-set mechanism, it should be mentioned that aromatic content is not the sole controlling factor. Two different distillate grades of zero aromatic content may differ substantially in the proportions of naphthenic and olefinic unsaturates present. A grade which contains a high proportion of these materials will be a stronger solvent than one which is mainly or exclusively paraffinic in nature.

Pigments

As previously stated off-set lithography prints ink films of lower thickness than any of the other major printing processes. Since the final print is generally required to match the range of colour strengths available by other printing techniques, it follows that lithographic inks must be produced at the highest colour strengths. Hence pigments selected must be inherently strong and able to develop their strength when dispersed in lithographic ink vehicles. Also, they must not produce unacceptable rheology, in terms of high viscosity, high yield or excess thixotropy, even when incorporated at high loadings. Dispersibility to fine particle size is necessary, not only for strength development but also to ensure that the presence of particles substantially bigger than lithographic ink film thickness is limited. Overriding all these requirements, it is essential that the pigments chosen must be fundamentally insoluble in, and unreactive with, the aqueous or aqueous/alcohol fount solution employed.

The nature of the printed work produced by lithography can impose further restrictions on pigment choice. Transparency is required where ink films or halftone dot patterns are to be superimposed, either by wet-on-wet or wet-on-dry multicolour printing. However, there are instances, particularly when printing on stocks of poor quality and whiteness, where a degree of opacity can be beneficial since it helps 'cover' or hide the less than perfect substrate. Publication printing rarely places any particular resistance demands on pigment selection, except for the excellent lightfastness demanded in poster printing, but packaging, both cartons and labels, can have stringent requirements in terms of chemical resistance, lightfastness and toxicity.

Cartons and labels generally require at least moderate lightfastness, in order to cover the occasions when merchandise is displayed in shop windows, and may have more stringent lightfastness specifications when greater exposure is anticipated. Where such performance is required, poor lightfastness pigments are either avoided altogether, or are used in a careful balance with pigments of superior performance such that light exposure produces only a slight lightening or darkening rather than an unacceptable fade or shade-shift. Chemical resistance requirements are as diverse as the products packed or labelled with lithographic print. Major areas are soap and detergent resistance on relevant cartons, and solvent resistance on cosmetic labels. Solvent resistance may also be dictated, both in packaging and publication work, if the print is to be further processed by spirit varnish or lamination.

The major pigment types which have the right performance to comply with the fundamental demands of the lithographic printing process and which have established bulk use in lithographic ink formulations are:

- The process colour pigments (for four-colour halftone printing):
  CI Pigment Yellows 12 and 13
  CI Pigment Red 57:1
  CI Pigment Blue 15:3
  CI Pigment Black 7 (usually toned with CI Pigment Blue 18);
- Other yellows:
  CI Pigment Yellow 17
  The Hansa yellows (CI Pigment Yellows 1, 3 and 5) are restricted in use by their poor strength, rheology and transparency. However, their soap resistance and lightfastness characteristics lead to some use in relevant applications;
- Oranges:
  CI Pigment Orange 13
  CI Pigment Orange 34;
- Other reds:
  CI Pigment Red 2
  CI Pigment Red 4
  CI Pigment Red 48:2
  CI Pigment Red 53:1
  CI Pigment Red 81;
- Other blues:
  CI Pigment Blue 1
  CI Pigment Blue 15 and 15:1
  Some CI Pigment Blue 27 (iron blue) is used but it is generally difficult to disperse adequately for modern lithographic requirements;
- Violets:
  CI Pigment Violet 1 (PMTA Rhodamine)
  CI Pigment Violet 3
  CI Pigment Violet 23;
- Greens:
  CI Pigment Green 1
  CI Pigment Green 2
  CI Pigment Green 7;
- White:
  CI Pigment White 6 (Rutile grades);
- Extenders:
  CI Pigment White 18.

Additives

Lithographic inks, in common with all other inks, are usually modified with small percentages of various additives to improve the printability or print performance of the basic pigment/resin/solvent/oil combinations from which they are formulated. The main classes of additives used in sheet-fed inks are:

- driers
- anti-oxidants
- waxes
- anti-set-off compounds

- lithographic additives
- rheology modifiers.

Driers

The oxygen induced polymerization of drying oils and drying-oil modified alkyd is accelerated by the incorporation of small quantities of certain metal compounds. These drier catalysts are normally based on transition metals which are able to exist in stable forms in more than one oxidation state. This provides reaction pathways which speed the rate of oxidation reactions and enable a highly cross-linked film to form in a matter of hours acids are usually employed because of their solubility in the oleoresinous vehicles found in oxidation drying and quick-set sheet-fed inks. They are used as relatively dilute solutions in hydrocarbon of small metal concentrations easier.

Various types of driers, having different reactions or reaction rates, are used for sheet-fed inks.

Primary liquid driers are usually organic salts of cobalt or manganese which are often used in combination with each other as they appear to have slightly different, but complementary, effects on the oxidation process. As long ago as 1953, Berry and Mueller (Berry D.A. and Mueller, C.R., American Chemical Society, Paints Plastics and Printing Ink Division, 1953) suggested that cobalt, which is capable of repeated transitions from the Co^{2+} oxidation state to Co^{3+}, is a more effective catalyst for the formation of hydroperoxide, whereas manganese is more effective as a catalyst for the succeeding reactions. The use of both these driers in a single ink therefore gives a fast start to the oxidation reactions with the cobalt and fast subsequent reactions with the manganese.

Secondary driers, sometimes called auxiliary driers, are the organic salts of barium, zinc and calcium which, although they have no direct catalytic effect on oxidation when used on their own, act as synergists when used in combination with primary driers and greatly speed up oxygen uptake by the drying oils. Calcium is the most effective while barium is now rarely used because of its high toxicity.

Coordination driers are usually compounds of aluminium or zirconium. Zirconium is the most widely used and, although it plays no part in the oxidation/reduction cycle, once electron-donating groups are present, it promotes coordination polymerization with a consequential increase in the drying rate.

Water-activated driers are sometimes incorporated into sheet-fed inks to assist with drying on difficult substrates. They are normally manganese or cobalt perborate compounds which release oxygen into the ink when they emulsify fountain solution during printing. They are not particularly effective and are never used as the sole drier but they do provide an extra boost to the drying of inks under difficult conditions.

Fountain driers are water-soluble compounds of cobalt, such as cobalt acetate, which are added to the fountain solution as a means of adding a small amount of extra drying potential to the ink as it emulsifies the fountain solution during printing. Again, they are not particularly effective but can be useful when difficult substrates or drying conditions are experienced.

Anti-oxidants

An increasing number of inks are required to be non-skinning in containers or ducts while retaining rapid drying once printed. It is not possible to obtain this balance of properties by simply reducing the amount of driers used in a formulation and anti-oxidants are used to control the oxidation drying potential of lithographic printing inks. Anti-oxidants function by reacting with the free radicals formed during oxidation, thereby preventing polymerization occurring. Their use always involves a compromise between drying speed on print and the length of time for which the ink remains free from skin. They should be used with care and only the minimum amount needed to obtain the non-skinning time required should be used as even a slight excess can retard the drying on the substrate.

The main types used in sheet-fed inks are given below.

- Oximes, e.g. methyl ethyl ketoxime. This material is an excellent anti-oxidant for preventing skin formation in containers and ducts and, because it is volatile and evaporates from the roller train during printing, has only a minimal effect on the drying of the print. The fact that it is volatile makes it unsuitable for use in inks which are formulated to be non-drying on the rollers of a press overnight. It has a fairly strong odour and this has limited its use. Other oximes such as butyraldoxime and cyclohexanone oxime are less efficient and find only limited use.
- Substituted phenols, e.g. butylated hydroxy toluene (BHT). These materials are very powerful anti-oxidants and even a slight excess can have a marked effect on an ink's drying properties on paper. They are non-volatile and so remain in the ink film.
- Quinones e.g. hydroquinone. These are now the most widely used types of anti-oxidant as they provide a reasonable balance of properties in terms of skin prevention and effect on final drying. Hydroquinone has been found to leach out of the ink and into the fountain solution during printing leaving as little as 30% of the initial amount in the ink on paper.

Waxes

As with inks for other printing processes, waxes are incorporated into lithographic inks to produce some or all of: slip, for in-line handling; scratch resistance, and rub resistance, for end-use requirements.

The major chemicals types of wax used in modern lithographic ink formulations are polyethylene (PE) for rub and scratch resistance, and polytetrafluoroethylene (PTFE), for good surface slip. Numerous different grades are available but generally these are molecular weight (and hence melting point) or particle size variants of these two chemicals. Some materials are now available which combine PE and PTFE, either by chemical reaction or by simple physical blending. These are designed to enable an optimum balance of surface slip and rub/scratch resistance to be achieved by incorporation of a single additive.

Low melting point grades promote surface slip but are usually inferior for rub and scratch resistance. Problems may also be encountered in ink manufacture if the wax is processed at any stage at a temperature near or above its melting point. Recrystallization of the wax on subsequent cooling can result in the formation of wax particles of uncontrolled size which can adversely influence ink rheology, printability and final print characteristics.

Large particle-size wax grades are particularly effective for both slip and rub/scratch resistance performance. This is because the particles protrude significantly from the ink film and thus the print surface is essentially wax rather than ink vehicle binder. However, the presence of such large particulate matter in the ink can produce printability problems, such as blanket piling, and print deficiencies, such as poor gloss.

Certain substrate requirements can make the choice of wax combination absolutely critical to the performance of the final print. A case in point has been the introduction of matt-coated papers and boards. The optically matt surface has generally been achieved by the incorporation of significant proportions of calcium carbonate extenders into the coating formulations. Unfortunately, this has tended to produce harsh and abrasive surfaces which give severe ink scuffing problems when print is subjected to face-to-face rubbing. Ink formulation changes cannot always completely overcome this substrate problem, although the utilization of large additions of highly effective PE and PTFE grades to tough binder vehicles can alleviate the situation.

Waxes may be incorporated into lithographic inks either directly, as finely micronized powders, or via an intermediate wax paste dispersion. The latter route involves easier blending but destroys some formulating flexibility by dictating incorporation of the vehicle present in the wax paste. Micronized grades have improved substantially over recent years and now rarely produce blanket piling problems involving separation of the wax from the ink vehicle, unless too large a particle size grade is chosen or unless the wax is improperly incorporated into the ink.

Anti-set-off compounds

The quick-set mechanism (page 390) is a major boost to achieving set-off-free stacks of sheet-fed print. However, even the fastest quick-setting ink is likely to take a couple of minutes to set. In this time on a fast running press a significant weight of stack can build up on top of the sheet. Set-off during this period is avoided by the use of anti-set-off spray powder on the press and by careful adjustment of the delivery to ensure minimum frictional contact between the sheets. Anti-set-off spray is a nuisance in the press room and the printer is always looking for ways to minimize its use. Infra-red drying units have been a major help in this area, but the inkmaker is also able to play a part by the incorporation of anti-set-off compounds. These compounds are based on particulate materials such as silica or starch which have a particle size slightly greater than the printed ink film thickness. Thus, the particles protrude through the film and separate adjacent sheets, functioning as it were as 'internal spray powder'. Naturally, the exact size of the particulate material employed and its content in the ink must be carefully controlled to minimize adverse effects such as blanket piling, poor print gloss and poor rub resistance. Due to these limitations, anti-set-off compounds can never be total solutions but merely another tool in the reduction of set-off.

Lithographic additives

We have already seen that inks will only perform satisfactorily in the lithographic process if their physical properties and fundamental chemical make-up are correct. Additives cannot be used to produce adequate lithographic performance from an ink that is based on chemistry inappropriate to meet the emulsification, cohesion and surface energy requirements of the process. However, additives may be beneficial in controlling undesirable reactions between fount and ink that might, in certain circumstances, jeopardize the otherwise good machine performance of a basically sound formulation. One example of this is the use of soluble salts of tartaric acid and ethylenediamine tetra-acetic acid (EDTA), which are able to form complexes with soluble calcium ions which may be present with certain pigment grades, or which may originate from the coatings of paper and board substrates. If they are not removed, these calcium ions can react with other materials and create problems of roller stripping and ink-in-water emulsification. The former is caused by the formation of ink repellent deposits of insoluble calcium salts produced by reaction between calcium ions and certain acids used in fount solutions. Ink-in-water emulsification can result if calcium ions are free to form soaps with fatty acids present in some vegetable oil based vehicles.

There are a number of proprietary products available which will either increase or decrease the emulsification tendencies of an ink. Those which increase the water pick-up are normally oleates or amine derivatives such as oleomides. They are effective emulsifying agents which, if used with care, can improve lithographic performance.

Rheology modifiers

The purpose of rheology modifiers in sheet-fed ink formulations is similar to that in heat-set inks and is covered fully on page 375.

Typical inks and varnishes

The formulations of lithographic inks and vehicles are dependent on the intended use of the product and a variety of approaches may be used. There is no 'right' formula for a given application but all inks intended for that application will share a number of key characteristics such as viscosity, yield value, tack, strength, setting speed, drying and rub resistance. Many of the physical properties are dictated by the press design, whereas others are influenced by market forces and competitive activity. The following sub-sections cover the requirements and formulations of inks and varnishes for the major product areas.

Inks and varnishes for paper printing

Inks for paper printing use the quick-set mechanism followed by oxidation as the standard drying process. The formulating approaches which follow can be used to produce process colours, spot colours and colour blending system inks. As discussed on page 400, properties such as setting speed and gloss may be optimized either by using two vehicles with different properties, or by using a hard resin which has the required properties in its own right. The former approach uses vehicles of the following type.

General-purpose quicksetting varnish:

Insoluble rosin-modified phenolic resin	30.00
Hydrocarbon resin	5.00
High viscosity linseed alkyd	25.00
280–320° C aromatic free distillate	20.00
260–290°C regular distillate	19.50
Aluminium-based gelling agent	0.50
	100.00

General-purpose gloss varnish:

Soluble rosin-modified phenolic resin	42.00
Low viscosity linseed alkyd	22.00
280–320°C regular distillate	35.25
Aluminium-based gelling agent	0.75
	100.00

The exact proportions of these vehicles can be varied to optimize the balance between gloss and setting speed as desired, in formulations of the type:

Organic pigment	18.00
Gloss vehicle } Quick-set vehicle }	67.00
Polyethylene wax compound	5.00
Anti-set-off compound	3.00
Cobalt driers	0.50
Manganese driers	1.00
Anti-oxidant compound	0.50
280–310°C distillate	5.00
	100.00

One of the advantages of using two vehicles is that individual inks within a four-colour set can be optimized in terms of performance. For example, it is normal to make the first one or even two colours slower setting than the subsequent colours, to avoid problems of carry-over piling on four-colour presses. The last colour may be made either very fast setting to minimize set-off or glossier to enhance the gloss of the whole sheet.

The use of single vehicle formulations, although lacking flexibility for fine tuning of this type, offer a good compromise of properties which, with the correct resin choice, should satisfy most requirements. The vehicle may be formulated using one of the new generation high viscosity, quick-setting, soluble types of rosin modified phenolic which also exhibit good gloss.

Quick-setting vehicle with good gloss:

Soluble, fast-setting rosin-modified phenolic	40.00
High viscosity linseed alkyd	20.00
280–310°C aromatic free distillate	20.00
260–290°C aromatic free distillate	19.25
Aluminium-based gelling agent	0.75
	100.00

To obtain the maximum rate of setting, quick-set systems are often formulated to be slightly stronger than general purpose inks so that the minimal possible filmweight is needed to achieve the correct colour density. Conversely, gloss inks tend to be slightly weaker than normal to ensure that relatively high filmweights need to be run to obtain the correct colour density which helps to maximize gloss.

Black inks of the highest density require the incorporation of specific wetting vehicles, generally based on low viscosity resins and alkyds to give high solids in the vehicle at moderate viscosity. This enables the vehicle to wet fully the substantial surface area associated with high loadings of fine particle-size carbon blacks.

Good pigment wetting is essential for satisfactory ink rheology, particularly flow and transference, to the promotion of gloss and a true black jetness.

Quick-set overprint varnishes for sheet-fed lithographic application to paper substrates are based on rather different formulations for anything usually encountered in vehicles for inks. A typical varnish would consist of the following:

Maleic-modified rosin ester (pale colour)	30.0
Tung oil	15.0
Long-oil alkyd	15.0
PE wax paste	7.5
Cobalt driers	2.5
260–290°C distillate	30.0
	100.0

The pale colour of the resin is necessary to minimize yellowing on top of non-ink-image areas and to avoid shade distortion of underlying colours. Tung oil is introduced to give a touch dry film for maximum rub and scuff resistance. As setting of overprint varnishes is often partially restricted by the ink films already present, oxidation drying is speeded by the use of high drier contents, and solvent separation is encouraged by the use of more mobile and slightly more volatile, lower boiling range distillate fractions. These latter generally give acceptable roller stability since the throughput of varnish on the press is usually high due to the relatively high film weights and coverage which are commonly applied.

Specific ink formulating approaches are utilized where there are particular substrate requirements, such as with label papers, or particular pigment characteristics, as in metallic inks, or where there is a distinct variation in the characteristics of the printing process, e.g. small-offset.

Sheet-fed label inks

Plain paper labels for the can, jar and bottle industries are generally printed on large format sheet-fed machines with a single sheet carrying many individual labels. The substrates used are normally low in base weight and coated on the front side only. Large sheet size, high ink coverage (in order to meet the visual impact demands placed on labels) and lightweight paper produce severe sheet curl problems unless specially formulated low tack inks are used. Adequate viscosity is preserved in such inks by the use of gelled vehicles and gelled reducer. Typical vehicle and ink formulations are now given.

Vehicle for label inks:

High melting point rosin modified phenolic	35.00
Medium viscosity linseed alkyd	22.00
280–310°C aromatic free distillate	19.80
260–290°C aromatic free distillate	22.00
Aluminium based gelling agent	1.20
	100.00

Sheet-fed label ink formulation

Pigment Orange 13	20.00
Label vehicle	57.30
High viscosity alkyd	7.00
Polyethylene wax compound	5.00
Anti-set-off compound	3.00
Gelled reducer	5.00
Cobalt drier	1.00
Manganese drier	1.00
Anti-oxidant compound	0.70
	100.00

More flexibility in the tack/viscosity/gloss relationship may be gained by using two vehicles with resins of widely differing solubilities as shown in the previous sub-section.

Metallic inks

Gold and silver inks are based on bronze and aluminium linings respectively. These linings are powders of a specific physical form which consist of extremely small, flat particle. In order to achieve a satisfactory metallic lustre at the low film weights printed by offset lithography, very high loadings of the finest particle size grades are utilized. Even then, optimum brilliance and lustre can only result if the particles are able to 'leaf' in the ink vehicle, i.e. orientate themselves such that the flat surfaces are overlapping and parallel to the substrate surface. Care is required in formulating metallic lithographic inks in order to ensure that the vehicle both permits leafing and retains adequate tack and rheology characteristics when very high proportions of the metal linings are present. Rheological stability and freedom from tarnishing are also major considerations because reactions can take place between the metal powder and ink vehicle components.

The most commonly used silver and gold metallic inks are now one-pack systems. The most demanding feature of such products is that they retain the lustre of the metallic pigment over an extended period of time. This is achieved by vacuum packing the product at the point of manufacture and by using resins of extremely low, preferably zero, acid value to prevent reactions occurring between the pigment and the vehicle system which diminish the metallic lustre. Hydrocarbon resins are commonly employed for this purpose as they have the required low acid value and can maintain the necessary tack with high pigment loadings. The use of cobalt driers has also been found to have an adverse effect on the shelf life of one-pack systems and its use is normally avoided. Silver inks are less likely to tarnish on storage than gold inks.

Two-pack systems in which the gold or silver lining is supplied in the form of a pre-wet paste and mixed with the vehicle system immediately prior to use are less convenient to use but do not suffer from any problems of tarnishing on storage. The following formulations are suitable for these one- and two-pack systems.

Single-pack gold varnish:

Hydrocarbon resin	55.00
Soya or dehydrated castor oil modified alkyd	15.00
280–310°C aromatic free distillate, low sulphur	30.00
	100.00

Single-pack gold ink:

Bronze lining paste (90% bronze lining in AF distillate)	50.00
Single pack gold varnish	40.00
Polyethylene wax compound	7.50
Manganese drier	2.00
Anti-oxidant	0.50
	100.00

Two-pack gold varnish:

Soluble rosin modified phenolic resin	38.00
Low viscosity linseed alkyd	15.00
280–310 aromatic free distillate, low sulphur	47.00
	100.00

Two-pack gold ink:

Bronze lining paste (90% bronze lining in AF distillate)	50.00
Two-pack gold varnish	39.30
Polyethylene wax compound	7.50
Manganese drier	1.50
Cobalt drier	1.00
Anti-oxidant	0.70
	100.00

Silver inks utilize a lower lining content than golds, due to the lower specific gravity and superior lustre and opacity of aluminium linings. A typical level of incorporation would be 20%.

In order to preserve the leafing potential of the linings, all metallic inks must be manufactured by processes involving low shear only. Thus, slow mixing and blending are appropriate but high-speed stirring and three-roll milling are not.

Small-offset

The presses used in small-offset are designed for ease of operation, since they are often run by staff who are not fully trained lithographic printers, and for economic construction, in order to provide a low-cost facility for the duplication of type matter. Inking and dampening systems are frequently rather different from those employed on larger, sheet-fed presses and are often extensively integrated. The ink distribution system is restricted in rolling power. Additionally, a wide range of plates is used, extending down to very low cost grades which rely on a tough paper as the base support material. Such plates may have only a limited distinction between image and non-image areas.

These factors impose a unique set of requirements on the physical and chemical characteristics of small-offset inks. Primarily, the inks must show good distribution and transference, controlled emulsification even on integrally damped presses (Fig. 6.22) with their associated founts (which may contain glycols or glycerol), and high cohesion in order to prevent ink transfer to poorly protected non-image areas. The inks are usually required to remain non-skinning over extended periods because the presses are not routinely washed up, even at the end of the day or week, particularly where they are only utilized for printing black. This does not pose too many drying problems when the inks can be formulated to dry by penetration into uncoated paper stocks. Here rubber resin-based inks can be used as well as inks based on oxidation drying vehicles with appropriately high additions of anti-oxidant. Where the role of small-offset presses is extended to include the printing of coated papers, oxidation drying systems with a very careful balance of driers and anti-oxidants must be utilized, usually to give extended non-skinning with relatively slow drying.

Typical formulations for oil-based and rubber-based small-offset blacks are as follows:

Carbon black (CI Pigment Black 7)	20.0
Reflex blue (CI Pigment Blue 18)	2.0
Oxidation drying vehicle*	70.0
Anti-oxidant paste†	2.0
Alkali-refined linseed oil	6.0
	100.0

* Based on modified rosin ester in linseed oil.

† Butylated hydroxy toluene dissolved in oleoresinous vehicle.

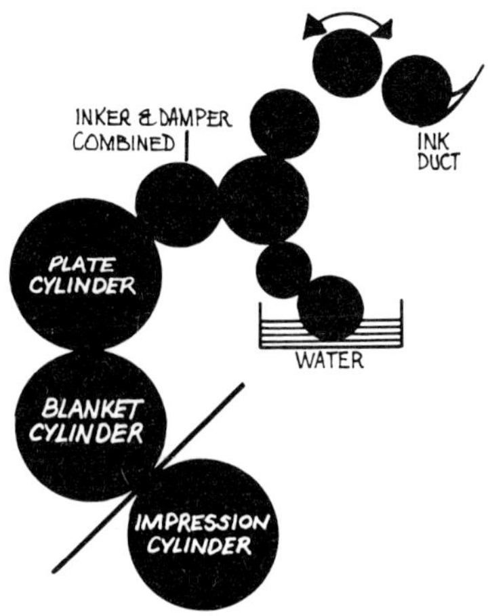

Fig. 6.22 Integrated inking system – small-offset.

Carbon black (CI Pigment Black 7)	20.0
Reflex blue (CI Pigment Blue 18)	2.0
Micronized talc*	10.0
Cyclized rubber vehicle†	65.0
280–320°C distillate	3.0
	100.0

* Incorporated to control the severe flying tendency of cyclized rubber-based system.
† Cyclized rubber resin dissolved in relatively high aromatic content distillate, at low resin-solids.

Inks and varnishes for sheet-fed carton board printing

A significant proportion of sheet-fed offset carton printing is now carried out using UV curing systems. Ink and varnish formulations for this process are fully detailed in Chapter 11. This subsection will consider those formulations for board substrates which dry by the quick-set mechanism. Although inks and varnishes produced for paper printing can be applied to board stocks, optimum performance to meet the stringent and unique demands of the folding carton packaging industry is usually achieved with specialized formulations. The major requirements are:

- hard drying with maximum scuff resistance in order to avoid marking on filling, distribution or retail of the packaged goods;
- very rapid setting to minimize set-off problems and the need for anti-set-off spray, aggravated by the substrate weight and its rapid development of pressure in the stack;
- minimum odour and minimum tendency to impart off-flavours (taint) when used for food and confectionery packaging;

- Suitability of varnishing or lamination in relevant instances.

Maximum scuff resistance is achieved through the use of appropriate wax compounds and the inclusion of hard drying vehicles in the ink. The hardest-drying vehicles used contain little or no distillate and a large amount of drying oil. Such vehicles have formulations similar to those given below.

Hard drying vehicle:

Insoluble rosin-modified, high melting point, phenolic	40.00
Alkali-refined linseed oil	22.00
Tung oil	33.00
260–290°C regular distillate	5.00
	100.00

General-purpose carton vehicle:

Insoluble rosin-modified, high melting point, phenolic	35.00
Medium viscosity linseed alkyd	20.00
Tung oil	5.00
280–310°C regular distillate	39.50
Aluminium-based gelling agent	0.50
	100.00

A combination of the two vehicles above can be used to produce inks with maximum scuff and rub resistance combined with good gloss in the following type of formulation.

Rub resistant carton ink:

Naphthol red (C.I. Pigment Red 2)*	20.00
General-purpose carton vehicle	41.70
Hard drying vehicle	25.00
Micronized polyethylene wax	3.00
PTFE wax (low particle-size)	1.50
Cobalt drier	1.00
Manganese drier	2.00
Anti-oxidant	0.80
280–310°C regular distillate	5.00
	100.00

* Resistant red pigment for soap and detergent carton requirements.

Inks of the above types are relatively slow setting and require the use of adequate amounts of suitable spray powders to prevent set-off. Where a less hard drying ink can be tolerated, a reduction in the amount of hard drying vehicle combined with an increase in the general purpose vehicle will improve the setting rate.

Low odour carton inks are traditionally formulated from materials having intrinsically low odour and taint characteristics including aromatic free distillates, maleic resins and semi-drying oils. The use of semi-drying oils is questionable in the reduction of odour created by the formation of volatile aldehydes and ketones as by-products of the oxidation process.

Semi-drying oils, by definition, do not have a high potential for oxidation and are therefore slow drying. This may result in a low level of odour initially, but the background odour will remain constant as the oxidation process continues over a considerable period of time. Fully conjugated drying oils such as tung oil, on the other hand, yield high levels of odour as they dry but this reduces to a very low level once the oxidation reaction is complete. Provided that inks based on this approach are allowed to dry fully, and the stack is aired thoroughly to dissipate any residual odour, the final print is surprisingly odour free. If the distillates in the above systems are changed to aromatic free, the resulting ink will fulfil all but the most demanding criteria for odour and taint. Where odour is of critical importance the following types of formulations have been found useful.

Low odour carton varnish:

Soluble rosin-modified low odour phenolic resin	20.00
Soluble rosin-modified maleic resin	20.00
Low viscosity soya alkyd	25.00
280–310°C aromatic free distillate	34.00
Aluminium-based gelling agent	1.00
	100.00

Low odour carton ink:

Rubine 4B (C.I. Pigment Red 57:1)	20.00
Low odour carton vehicle	67.50
Micronized polyethylene wax	3.00
PTFE wax (low particle size)	1.00
Cobalt driers	1.00
Manganese driers	2.00
Anti-oxidant	0.50
280–310°C aromatic free distillate	5.00
	100.00

The removal of the PTFE and a reduction of the polyethylene wax to 2.00% from either of the types of ink shown above will render them suitable for vanishing with UV products or lamination.

Overprint varnishes for sheet-fed lithographic application to food and confectionery packaging have to be formulated with due consideration to the points about odour and taint discussed earlier. An additional consideration is to minimize any yellowing or distortion of underlying colours by using resins with a dark colour. A typical low odour formulation is given below.

Low odour overprint varnish:

Rosin modified maleic resin (pale colour)	30.00
Soya alkyd	25.00
Polyethylene wax compound	7.50
Cobalt drier	1.50
Manganese drier	1.00
260–290 aromatic free distillate	35.00
	100.00

Ink for sheet-fed impervious substrate printing

Offset lithography is heavily orientated towards the printing of paper and board substrates which possess a significant degree of absorbency. This enables penetration and setting to play major roles in drying and adhesion in most sheet-fed work. However, a range of foil and plastic stocks are printed lithographically, either with oxidation drying or UV curing formulations. The latter are detailed in Chapter 11.

Oxidation drying inks for impervious substrates represent a difficult compromise between press stability and hard drying. They must be non-skinning in the duct and on rollers, plate and blanket, but free from set-off and must give good key and adhesion to the substrate. In practice, stability times are trimmed to the bare minimum in order to promote rapid, hard oxidation drying. Even so, anti-set-off spray powder and small stack heights are essential to avoid severe set-off and blocking. Bearing in mind the restrictions that apply to the chemistry of lithographic vehicles as discussed in section 6.1, it is not possible to achieve key to plastic substrates by incorporating resins of similar chemistry to the plastic in the binder vehicle of the ink. Key and adhesion have to be achieved through oxidation drying producing a hard, yet flexible, polymerized film. Because of this lack of formulating flexibility lithographic inks cannot readily cope with the wide range of impervious substrates that are printed by gravure, screen and flexography.

A further complication arises in the area of ink–water balance. Since the substrates are impervious they are unable to transport fount solution away from the impression area of the press in the way that paper and board substrates can. This increases the likelihood of over emulsification of the ink. As well as being a potential source of poor reproduction quality, such emulsification can also retard oxidation drying and provide another obstacle to satisfactory ink adhesion on print.

Despite all these potential pitfalls, correctly formulated oxidation drying inks in the hands of skilled lithographic printers regularly produce fully satisfactory work on foil boards and on plastic sheets. A typical formulation is given below.

Oxidation drying vehicle:

Insoluble rosin-modified phenolic resin, high melting point	40.00
Alkali-refined linseed oil	25.00
Tung oil	35.00
	100.00

Ink for impervious substrate:

Phthalocyanine green (C.I. Pigment Green 7)	20.00
Oxidation drying vehicle (above)	70.00
Micronized polyethylene wax	3.00
PTFE wax (low particle size)	1.00
Cobalt drier	3.00
Manganese drier	1.00
Alkali-refined linseed oil	2.00
	100.00

Ink-related problems and their possible solutions

There are five main areas where difficulties can arise with lithographic printing inks:

- lithographic
- rheology
- tack
- setting and drying
- appearance.

Lithographic problems

Lithography runs into problems when ink transfers to the fount solution or to non-image areas of the plate; when it fails to retain a continuous film on the inking rollers, or when it builds up on rollers, plate or blanket. The names given to these various difficulties are, respectively, tinting, scumming, stripping and piling. All can occur in inks whose chemical and physical properties meet the demands of the lithographic process identified in section 6.1. However, the exact detail of these characteristics may lead to lithographic problems under particular circumstances, as will be seen from the following descriptions.

Tinting

Tinting occurs when pigments, with or without other ink ingredients, become solubilized or emulsified into the fount solution. The fount takes on a weak colouration from the ink which transfers as a wash of colour on to the non-image areas of the print. This problem is usually caused by pigments which have been incompletely washed free of soluble components during their manufacture, or which contain materials that are reactive with certain vehicle components to form water emulsifiable soaps. It can therefore be avoided by using an alternative pigment grade. Tinting caused by ink-in-water emulsification resulting from calcium soap formation may be overcome by adding a complexing lithographic additive, as described on page 376.

Scumming

Scumming occurs when non-image areas of the plate accept and transfer ink such that the print does not remain completely clear in these areas. Minute flecks of scummed ink are present on virtually all lithographic print. In fact, their presence represents the only simple way to distinguish lithographic print from high quality letterpress work. Obviously, scumming becomes a problem when ink transfer in the non-image areas becomes apparent to the naked eye or when it affects tonal rendition by increasing apparent density in halftone areas. It may be caused by a variety of factors associated with plates and their processing, with fount solutions, or ink characteristics, both chemical and physical.

When ink chemistry imposes a narrow range of water tolerance the chances of scumming are increased. Thus, if ink is too water repellent, dampening levels will have to be kept very low to avoid the fount from inhibiting receptivity and transfer of ink to sections of the image area (so-called 'watermarking'). At these low settings there may be insufficient fount present to protect fully all non-image areas from accepting ink. Such problems are aggravated by irregular demands for fount and ink due to image layout across the width of the press. While there is control of ink feed to different lateral zones on all presses, any such control of dampening feed is very rare. Thus, it is usual for a standard fount application to apply across the plate regardless of image/non-image relationships. Variations in the latter have to be accommodated by ink–water balance tolerance if problems such as watermarking and scumming are to be avoided.

The physical characteristics, of the emulsified ink can have a major impact on scumming. Problems may arise because the ink has too low a viscosity possibly caused by over-reduction or elevated temperature. Ink may then transfer over the top of the fount film on the non-image areas (page 346). Excessive emulsification may also lead to a greater attraction between emulsified ink and fount, leading to the same effect.

Most ink-related scumming problems are due to a complex mixture of the particular physical and chemical properties of the ink, often linked through emulsification. Sometimes it is possible to produce a solution by a simple increase in viscosity and tack. This increased inherent cohesion and reduces the degree of emulsification. Equally, there are occasions when a change in formulation is required to reduce water repellency and increase water tolerance of the ink, or to give a more cohesive emulsion.

Stripping

Stripping occurs when the ink ceases to transfer along all areas of the roller train surface. This can adversely affect ink feed to the image areas of the plate and can result in a print that is weak or of uneven density. The problem can be caused by poor setting or condition of the rollers, but when related to ink rather than machine, it is most often caused by a reaction between ink ingredients and fount components which can produce hydrophilic deposits on the rollers. This reaction and its control were discussed in the paragraph on lithographic additives on page 409.

Piling

Piling is the name given to the build-up of excessive ink film thickness on rollers, plate or blanket. The immediate cause is a lack of distribution or transference in the relevant areas. The problem is visually apparent on the press and manifests itself on the print as a lack of continuous and even coverage of all image areas. This is normally first observed along the trailing edge of solids when the sheet is examined in the direction that it was printed. If the press is not stopped and relevant areas cleaned, the problem usually increases to the point where print quality is seriously impaired and physical damage is caused to the plate and blanket by the increased pressures in the areas of piling.

There are a number of potential causes of piling and these may sometimes be interconnected in a complex manner. First, piling on the rollers, plate or blanket may occur due to emulsification having an adverse influence on rheology and tack properties and hence hindering distribution and transference. Alternatively, plate and blanket piling in particular may be caused by loss of distillate, with the resulting poor press stability causing the ink to dry up and lose transference. Both these causes essentially involve the ink as a whole piling. There are also a number of instances where the build-up involves a type of filtration process whereby most of the fluid components of the ink transfer, but hard, particulate material remains behind on the plate or blanket, bound by a small proportion of retained vehicle. Common examples of this latter phenomenon are piling of poorly wetted micronized waxes and piling of coarse or poorly dispersed pigment particles. Emulsification can make things worse by promoting de-wetting of particles that are not completely coated in vehicle. Finally, piling may involve significant proportions of components which originate from the substrate, and which have been removed by the influence of fount and ink.

With this diverse set of causes there is clearly no single solution which will overcome all instances of piling. Changes in the basic chemical and physical characteristics of the ink may be necessary if emulsification inhibits distribution and transference. However, piling associated with poor press stability may be overcome by addition of bridging solvent or by the use of a distillate grade of higher boiling range or stronger solvent power. Filtration type piling clearly requires improved ink manufacture, or a change in the grades of the relevant particulate materials used to ensure that satisfactory wetting of the particles with vehicle is retained, even when thin films are sheared and emulsified with fount. If materials originating from the substrate are a major component of piling, then ink is not the sole cause of the problem. However, it may still be possible to avoid the problem by ink modification, for example by reducing tack in order to reduce the force pulling fibres or coating from the stock at the point of impression.

Rheology problems

One of the major complications in lithographic ink technology is the interaction of rheology and emulsification, with each property being a major, but not sole, influence on the other. Thus, as has been discussed, rheology may be involved in complex ways in lithographic problems such as scumming and piling. With this background it is easy to understand why a considerable array of techniques has been developed to enable ink rheology to be manipulated. These range from straightforward viscosity reduction with distillate, through techniques for producing various degrees of structure in vehicles and inks, to flow promotion by incorporation of low viscosity alkyds or other wetting resins. With these tools the formulator is able to tailor lithographic ink rheology to overcome problems, whether these be directly caused by the physical properties of the ink or associated with emulsification.

Tack problems

Tack is controlled by the use of gelled vehicles and by incorporation of distillate or gelled distillate into the ink. Additives are also available which enable minor upward adjustment of the tack of a finished batch of ink that would otherwise have to be rejected and replaced with a new make of a modified formulation. This can be helpful where trapping problems are encountered. Consequently, inks can be manipulated through a reasonable tack range by additives or, if this is insufficient to overcome particular tack problems, they can be reformulated using different balances of gelled and ungelled vehicles and reducers.

Setting and drying problems

Lithographic inks generally represent a major compromise between the conflicting requirements of press stability and drying. The former demands that thin ink films must retain fluidity throughout the distribution and transference processes on the press, whereas the latter necessitates rapid formation of a hard, solid film at only slightly lower film thicknesses. Press stability considerations normally result in penetration, quick-setting and oxidation drying inks being retarded in some way to maintain printability. Thus, setting and drying characteristics may be jeopardized, particularly under adverse press or print conditions. In this context, high press speeds, high ink film thickness and substrates which are heavy or which restrict penetration or oxidation can all produce problems. Among the specific difficulties encountered are set-off, carry-over piling, slow drying and poor rub/slip/scratch.

Set-off

Set-off occurs in sheet-fed printing when ink transfers from the print to the reverse of the adjacent sheet in the stack. It can be influenced by a whole range of substrate and machine factors, such as:

- absorbency of the stock – lack of absorbency inhibits setting;
- smoothness of the stock – smoothness promotes contact and hence set-off;
- weight of the stock – heavier stock produces more pressure and hence more set-off in the stack;
- speed of the machine – faster running gives a more rapid rise in stack pressure;
- adjustment of the delivery – any lateral sheet movement increases marking;
- spray-powder setting – insufficient spray-powder allows sheet contact;
- maximum stack height – greater pressures exist in higher stacks.

However, ink performance may also be a prime cause of set-off. Problems can arise if the ink is too slow setting, or if it passes through a very sticky phase as it sets. Either characteristic can make the printed ink film prone to set-off as the stack is being built in the press delivery. Even an ink which sets rapidly and without passing through an excessively tacky stage can cause set-off in the stack if it is susceptible to 'sweat-back'. As oxidation drying gets under way in a stack of work heat is generated by the chemical reactions involved. Thus, temperatures are raised and the affinity of the set ink film for distillate fractions which have penetrated into the substrate may be increased. If this occurs, such solvent components may migrate back into the warm ink film, softening it and rendering it prone to set-off or 'sweat-back'.

All ink-related set-off problems are associated with the compatibility balance of the quick-set vehicle. Slow setting speeds are a result of too great an affinity between the resin and the distillate in the same way that 'sweat-back' is associated with too much residual compatibility between the binder resin and the separated distillate. Where the ink has set satisfactorily initially, 'sweat-back' is most often due to the presence of a small proportion of stronger solvent components in the distillate. This problem may be avoided by the use of cleaner, tighter-cut distillates in the ink and ink vehicle. Slow setting problems may be reduced by the use of weaker solvents or an increase in resin-to-oil ratio to decrease compatibility. If the ink passes through a sticky stage this may necessitate a change in resin to a grade which rapidly builds viscosity, but without increase in tack, as distillate is lost in the setting mechanism (Fig. 6.23).

Anti-set-off additives (page 412) help to separate sheets in the stack and can therefore reduce the tendency of inks to set-off. They will also assist in fully oxidation drying systems where no actual phase separation setting can occur. However, these latter inks generally require severe limitations on stack heights and appropriate use of spray powder to avoid excessive set-off.

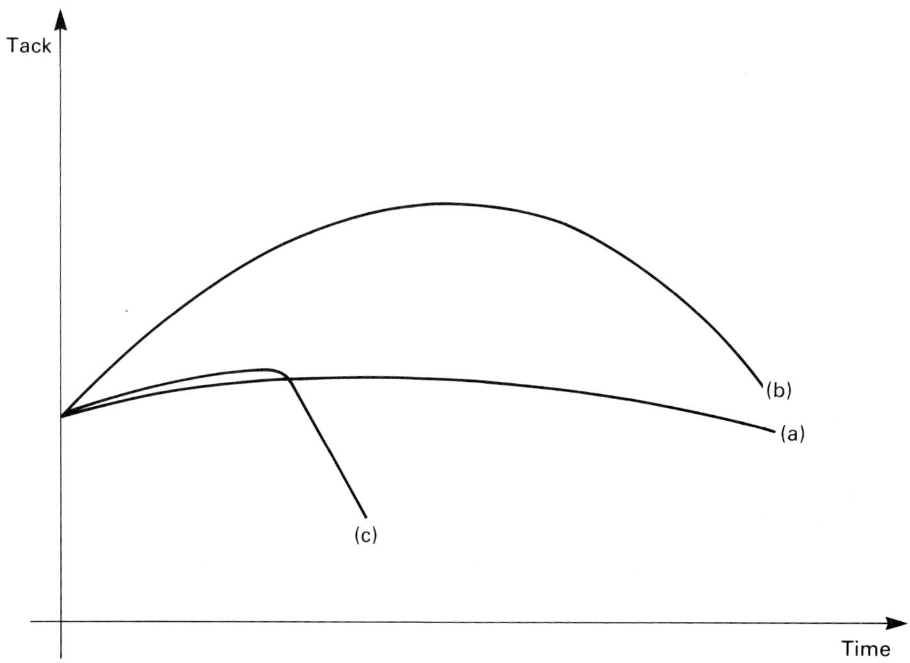

Fig. 6.23 Set-off and the influence of tack characteristics of an ink after printing: (a) slow setting, low tack rise – some set-off; (b) slow setting, high tack rise – severe set-off; (c) fast setting, low tack rise – ideal situation, minimal set-off.

Carry-over piling

So far all problems associated with setting that have been described have been to some extent involved with a lack of setting speed. Carry-over piling is a problem that can occur on multi-unit presses if the ink sets too rapidly on the particular substrate being printed. Even an ink that has a sufficiently suppressed tack rise on setting to avoid set-off will, nevertheless, go through some increase in tack as distillate separation commences. It is therefore possible for an ink printed on the first printing unit of a multicolour press to be at a sufficiently high tack level as it passes through subsequent units for it to transfer back from the print to the blanket. Thus, print quality is impaired and a build-up of ink occurs on a later colour blanket. The problem may be overcome by slowing the setting speed of inks to be printed on early stations of a multicolour press. This may be achieved by decreasing the resin-to-oil ratio of the vehicle or by some other technique to increase compatibility, such as addition of bridging solvent.

Slow drying

As with set-off, slow drying may be due to a range of factors.

- Low temperatures – oxidation drying, like other chemical reactions, is retarded by reduced temperatures.
- Acid stocks – at high humidity (which is almost always present in lithographic printed stacks) acidity can severely retard oxidation drying.
- Lack of oxygen – this can be particularly problematic in large stacks, especially if these are wrapped for dust protection.

The main ink-related causes of slow drying are inadequate oxidation potential in the vehicle, or lack of available drier catalysts. The former may be rectified by incorporating more oil or alkyd into the vehicle or by the use of fully drying materials such as linseed or tung oil in place of semi-drying oils such as soya. Ineffective driers may be due to inappropriate quantities being included in the formulation, or to an effect known as drier absorption. This involves pigments with a high surface area, notably carbon black, and prevents the drier being available for reaction with the vehicle since it is absorbed into the pigment surface. Drier absorption is linked to the age of an ink as a given ink's drying potential can deteriorate with time. Such drying problems can clearly be rectified by addition of further catalyst to the ink. When anti-oxidant has been added to extend skinning times, such a simple approach may be more difficult to implement. Careful balancing of drier and anti-oxidant types and proportions is essential to give adequate drying without jeopardizing stay-open requirements.

Ink-formulating approaches can also assist in instances where poor drying is not solely an ink problem. Poor drying due to lack of oxygen availability can be boosted by the use of calcium perborate-based driers which liberate oxygen when mixed with water, as occurs when fount becomes emulsified into the ink. Drying on highly absorbent substrates, such as cast-coated papers, may be impaired by vehicle penetration into the substrate or its coating. This then leaves insufficient binder on the surface and the pigment is prone to 'chalk' from the print. This gives the appearance of poor drying. The problem may be alleviated by increasing the resin and oil binder of the ink vehicle.

Poor rub/slip/scratch

Lack of rub and scratch resistance or inadequate surface slip characteristics may be associated with poor drying. However it is also possible to have problems in these areas even when all possible oxidation drying has taken place. This is because dry lithographic ink films have a very high pigment-to-binder ratio and it may not be possible for the polymerized binder on its own to produce all the desired surface protection. Thus, various waxes and slip aids are used in lithographic ink formulations.

Consequently, it is possible to encounter rub, scratch and slip problems if the incorrect type and quantity of these additives is present. Rub and scratch resistance problems may be alleviated by using greater quantities, coarser grades or harder grades of PE and other waxes. Surface slip characteristics may be improved by appropriate use of PTFE wax or silicone slip aids.

Appearance problems

A lithographic print may fail to produce the desired appearance through lack of fidelity in the printing process or because of unevenness or poor gloss in the printed ink films.

Dot reproduction

Reproduction is most critical in halftone dot areas since variations affect both tonal balance and colour rendition. Dot gain, due to both optical and physical effects, is an inherent feature of offset-litho printing on paper and board substrates. It is not, as such, a fault, provided that it occurs in a sufficiently controlled and predictable manner to enable appropriate allowances to be made at the colour separation stages (Chapter 3). Optical dot gain is caused by light scattering under the printed dots in the surface layers of the substrate. Consequently, it is almost entirely controlled by the characteristics of the stock. Physical dot gain is a result of the squash or flow-out that occurs when the fluid ink dots are transferred from plate to blanket to substrate under impression pressures. It is, therefore, influenced by press, blanket and substrate characteristics as well as ink factors.

Inks must be formulated to minimize the degree of sideways spread during the transference stages. Rheology, in terms of viscosity and flow, and cohesion (tack), require appropriate control to maintain dot gain within identified tolerances. Further, the tinctorial strength of the ink is also important since variations in this property will be compensated for by adjustment of ink film thickness to maintain colour density. Thicker dots will naturally be more prone to dot gain through edge squash under printing pressures. Techniques for controlling the rheology, tack and colour strength of litho inks have already been covered in detail in preceding sub-sections of this chapter.

Poor lay

Lay is the term given in the printing trade to the smoothness and evenness of a printed ink or varnish film. Poor lay means that the solid areas of ink or varnish are not of a completely uniform film thickness. This may be caused by variations in the receptivity of the substrate surface or to inadequacies in transference and flow characteristics of the ink or varnish.

Receptivity problems may be an inherent deficiency of the given stock or, particularly in the case of over-varnishing, may be related to the surface characteristics of underlying ink films. An ink film which has too low a surface energy, either due to the nature of the polymerized binder or from the presence of low surface energy rub and slip additives, will cause the varnish to reticulate since its surface energy is greater than that of the underlying ink. This is a severe case of poor lay where the substrate surface is non-receptive to the varnish film. Sometimes the problem may be overcome by additions of surfactant materials to lower the surface tension of the varnish. Alternatively, it may be necessary to reprint the job using a different ink and substrate combination to ensure that the surface is more receptive to the varnish.

Borderline transference characteristics can lead to poor lay since some image areas may not transfer as well as others. Transference is dependent on both rheology and emulsification characteristics at the point of impression and other properties, such as press stability, have an impact on these. Therefore, there is clearly no single solution to overcoming poor lay due to borderline transference. However, factors which improve the compatibility of the vehicle, such as minor additions of oil, alkyd or bridging solvent, frequently produce sufficient change to eliminate this marginal deficiency.

Poor flow characteristics of the ink film initially transferred to the substrate can result in print showing poor lay. This is because completely even transfer is not always achieved and a certain amount of flow is required to enable a film of completely even thickness to form. Again, a variety of factors influence the flow characteristics of the printed, emulsified ink. A small addition of a flow promoting oil or alkyd may be sufficient to improve lay. However, if emulsification is severely restricting flow then it may not be possible to effect improvements to the lay without a significant adjustment of ink chemistry or rheology.

Poor gloss

Gloss, or the mirror-like reflection of the ink film, is a much demanded property for high quality print, particularly where it is needed to make a marketing impact. Since gloss is dependent on the formation of a smooth, continuous film it is strongly influenced by the absorptivity and surface smoothness of the substrate and the chemical and physical properties of the ink. Poor ink gloss may be caused by poor pigment wetting, poor transference and flow, or the presence of large particle size material. Incomplete wetting of the pigment particles by the ink vehicle will result in light scattering within the ink film which reduces gloss. It may also promote over-emulsification of ink with fount which will inhibit the transference and flow properties that are so important to ensure the formation of a smooth ink film on print. Particulate material which protrudes through the thin ink film will act as light scattering sites and will impair gloss.

A number of options are open to improve gloss by ink modification. Increasing the oil or alkyd content of the ink is beneficial in a number of ways. It improves pigment wetting and boosts compatibility. The more compatible vehicle has improved press stability and hence superior transference and flow on print. Use of a lower viscosity resin in the ink vehicle enables solids to be increased and thus promotes hold-out of a smooth ink film on the substrate surface. This gives superior gloss compared with an ink film that has penetrated extensively into the stock and which therefore allows light to be scattered from protruding paper fibres and coating particles. Choice of pigment and wax grades which are readily wetted and which break down to fine particle sizes in the ink manufacturing stages are also clearly beneficial. Within appropriate limits, gloss can also be promoted by a reduction in pigment content, and hence tinctorial strength, which results in a thicker ink film that boosts hold-out and flow. The disadvantage of this approach, and indeed of the use of more compatible vehicles, is that set-off becomes more problematic. If this is not controlled by appropriate stack handling techniques advantages in gloss may be lost, either due to marking in the stack or through the light scattering produced by excessive application of coarse anti-set-off spray powders. In this respect the only techniques that can improve gloss without detriment to setting are in the areas of pigment choice and manufacturing method to improve wetting and reduce the presence of over-size particles. Pigment flushes can be particularly advantageous for gloss inks because of their final level of dispersion and excellent degree of wetting.

6.5 THREE-PIECE TIN-PRINTING INKS

Metal is used to produce a wide variety of decorated products such as cans for foods, paint, oil, beer and beverages, aerosol cans, tins for biscuits, caps and closures for bottle and jars, drums, trays, advertising signs and many other products (Fig 6.24). Although there are certain products which are printed on pre-formed metal, notably the two-piece can for beers and beverages, the majority of products are produced from sheets of tinplate, tin-free steel (TFS) or aluminium which are decorated as flat sheets and subsequently formed into the products with which we are so familiar.

The decoration of sheet metal may involve the use of a whole series of size, lacquer, enamel, ink and varnishing processes. However, we will concentrate on the nature and role of the printing inks used, the other coatings being a study in themselves which falls outside the scope of this chapter.

Although some printing is carried out by the dry-offset process, the majority of work is printed lithographically, even for relatively simple solid designs, due to the ease of production, and greater flexibility of design lithographic plates afford. In practice, rather than carrying two

Fig. 6.24 Examples of sheet-fed metal printing.

ranges of inks, printers generally use lithographic inks throughout and these are simply referred to as 'tin-printing inks', designating their suitability for use in the decoration of metal products (not just tinplate as the name might suggest).

There is nothing intrinsically different about the process of lithographic printing on sheet metal than printing on paper or board, and as such the printing process has been covered in the preceding sections. However, there are extra requirements for tin-printing inks which relate to the following three factors:

- substrate properties
- flexibility and adhesion
- heatfastness.

Substrate properties

Metal is heavy, non-absorbent and non-compressible under the pressures associated with printing. This hard, non-compressible nature requires heavier pressure between the printing blanket and the sheets to be printed to obtain consistent, even print across the sheets. Due to the weight of the sheets, it is essential that the inks are well dried before they reach the stack if set-off is to be avoided and the non-absorbent nature of metal requires the ink to be dried by the application of heat or some other form of energy.

Printing may be on the metal surface or over a transparent size or pigmented enamel previously applied by roller coating. For each print pass the ink must be dry when it reaches the stack.

The traditional technique of using a long, travelling oven is still widely used. The printed sheets are picked up on an endless chain of racks (known as 'wickets') which carry them through the oven in a vertical position. Hot air is blasted at the sheets as they pass through the oven and the total stoving time is in the region of 10–15 minutes. With some exceptions, the inks are protected by a suitable finishing varnish.

The curing process is the slowest stage of the printing operation and the development of inks places strong emphasis on faster curing rates which will allow shorter stoving times and higher line speeds. Ultra-violet curing has secured a niche in this market and several printers have set-stack machines which no longer require thermal curing ovens. However, very few metal printers are totally dependent on UV inks and the majority have opted for a combination, i.e. a UV set-stack machine followed by a conventional press which can apply an overvarnish wet-on-wet and where one thermal stoving cures both inks and varnishes.

Flexibility and adhesion

One of the most severe tests for a tin-printing ink occurs when the sheets are being stamped and formed into shape.

The manufacture of crown closures presents a severe test of coating materials. The stamping process requires toughness and good adhesion to avoid damage and the finish must be very hard and abrasion resistant to withstand packing, transport and hoppering in the filling plant.

Roll-on closures, the threaded caps used on bottles, present their own particular coating requirements. The inks must be able to withstand the severe distortion as the metal is drawn into shape from a flat sheet. This is followed by the thread forming and knurling operations and the ink must not suffer from lack of adhesion or surface damage.

Drawn and redrawn, or DRD, cans are finding increasing favour for food packaging, particularly for patés and pastes. Flat sheets are precoated and sometimes distortion printed, then formed into can bodies by drawing in several stages to the full height. The ink performance must be of a high order to withstand the difficult tooling requirement without loss of appearance or adhesion.

Heatfastness

In addition to its own stoving, and those of subsequent print passes, in the region of 130–170°C, the ink is likely to undergo other processes involving higher temperatures.

For most applications, tin-printing inks are overvarnished and stoved at around 180°C for approximately 10 minutes. Subsequently, containers for food will be given an internal coating on the reverse of the sheets after printing, typically an epoxy phenolic coating stoved at 200 or 210°C for 10 minutes.

When cans are produced the side seam may be soldered at temperatures reaching up to 370°C for a few seconds. Lug caps and crown closures have a liquid gasket applied and stoved at up to 260°C for 2 minutes and food products are generally sterilized in steam and water for over an hour at elevated pressure. In all these cases, there must be no tendency for the ink to fade, discolour, bleed, blister or scorch.

Throughout the rest of this section it will be these extra requirements necessary for printing on metal, which will be given most attention, rather than the general principles of lithographic printing which have already been dealt with in previous sections.

Formulating principles

The correct properties for a tin-printing ink are achieved by the use of suitable raw materials and the correct formulating principles. As with any ink, the formulations comprise:

- pigments
- resin/vehicles
- solvents/diluents
- additives.

Pigments

It is generally desired to decorate metal products with designs as strong, clean and bright as those produced on other substrates, notably paper or white board. Indeed, graphics for most designs will initially be produced on white board and then converted to the first proof on metal on a proofing press. Therefore, although ink may be printed on a clear size or even base metal occasionally, for the majority of printed work, the dark grey of metal would not enable the achievement of the correct colours.

Generally, therefore, a white coating is applied to the metal prior to printing to allow colours to resemble more closely those printed on paper or board.

However, although white-coated aluminium comes close to the whiteness of paper, this is only a small proportion of the total metal decorating market (mainly caps and closures) and white-coated tin plate or TFS produce a much dirtier white than paper or board, because it is not possible completely to hide the blue/grey nature of the substrate without multiple passes of white enamel which is not normally practicable. Thus, as well as the normal requirement for pigments with high colour strength for the lithographic process, it is also important to use the cleanest, brightest pigments possible to try and counter the dirtiness of the substrate. For the same reason, with the exception of the process colours, which are required to be transparent for the four-colour process, it is normally preferred to use more opaque pigments for inks to be printed on white-coated metal. As with all lithographic inks the pigments must not produce unacceptable rheology, such as very high yield values, at the high pigment loadings required and must not be soluble in nor react with the aqueous or aqueous/alcohol fount solution employed.

In addition to these standard requirements the pigments may well have to withstand high temperatures after their own print stovings, as described in the introduction under 'heatfastness', due to the application of further coatings and various post-print processes associated with the production and preservation of the final product. The pigments must therefore have better than average resistance to heat under both wet and dry conditions.

Since colour change or fade becomes more noticeable at low colour strength, it is normal to have a standard range of pigments, with moderately good heat resistance, for use in strong colours at a reasonable price, while having certain fast colours for use in special applications or where small amounts of colour are required, e.g. in a cream or buff where even slight colour fade would be very noticeable.

Given these requirements, the following pigments have established bulk use for metal decorating lithographic ink formulations.

- Process colour pigments (for four-colour halftone printing):
  - CI Pigment Yellow 12 and 13
  - CI Pigment Red 57:1
  - CI Pigment Blue 15:3
  - CI Pigment Black 7 (often toned with CI Pigment Blue 15:3)
- Other yellows:
  - CI Pigment Yellow 17
  - CI Pigment Yellow 83 (superior heatfastness)
- Oranges:
  - CI Pigment Orange 34
  - CI Pigment Orange 36
- Other reds:
  - CI Pigment Red 48:2
  - CI Pigment Red 48:3 (not suitable for steam sterilization)
  - CI Pigment Red 112
  - CI Pigment Red 185 (superior heatfastness)
- Other blues:
  - CI Pigment Blue 15:1
- Violets:
  - CI Pigment Violet 23
- Greens:
  - CI Pigment Green 7
- White:
  - CI Pigment White 6 (Rutile Grades)
- Extenders:
  - CI Pigment White 18
  - CI Pigment White 24.

Resin/vehicles

Vehicles for tin-printing inks may be divided into two major classes: oleoresinous systems for high temperature oxidation drying inks, and acrylate systems used in radiation cured systems. The nature of the raw materials within these two classes will be dealt with separately, but let us first consider the general requirements for a tin-printing ink vehicle.

The basic requirements for a lithographic ink have already been considered in previous sections of this chapter and so will only be dealt with briefly here.

In addition to providing the medium for effective wetting and dispersion of the pigment during manufacture, and binding the ink together when cured, the ink vehicle is the main source of the physical and chemical properties of the ink. It must therefore have the correct balance of water tolerance to ensure that ink is repelled by water on the non-image areas of the lithographic plate while readily emulsifying with inevitable droplets of water on the image areas. The resin must have good solubility in the solvents used to reduce ink viscosity since these are by necessity of low solvency power (to prevent swelling of the printing blankets and rubber rollers in the inking train). If the resin has insufficient solubility the ink is likely to have unacceptable rheology in terms of not only viscosity, but excessive yield value and tack. High yield value causes problems in handling the ink and may lead to ink starvation due to the ink hanging back in the duct, i.e. not flowing fast enough to replenish ink as it is removed by the ink duct roller. Tack is generally less critical for tin printing than other lithographic processes, but if it is very high it can result in sheet lifting, which can potentially damage equipment. High tack can also result in back trapping of a wet ink from a previous printing unit in tandem printing or even in extreme cases, stripping of a cured ink back onto the printing blanket.

Insufficient solubility of the resin will also adversely affect transference of the ink through the printing press and onto the printed sheet, thus affecting gloss and lay. Gloss is generally not a problem since tin-printing inks are normally overvarnished. However, lay can be critical since poor lay can result in a speckled or striated appearance to solid printed areas.

Compared with other lithographic processes, tin printing uses higher pressures between the printing blanket and substrate to ensure even printing. This requires higher viscosity inks, typically 25–35 Pa s, to reduce the degree of ink squash which would result in excessive dot gain and loss of clean edges to solids. Since the easiest way to increase viscosity is to decrease the amount of reducer present this increases the need for good resin solubility.

The ink vehicle is also responsible for the hardness of the cured ink and adhesion to the substrate. This is particularly important for metal decorating due to the heavy weight of stacks. Soft ink can set-off onto the backs of other sheets and, if adhesion is poor, portions of ink may completely transfer to the back of another sheet in the stack.

In addition to the basic lithographic ink requirements, with their greater or lesser importance for tin printing, there are other properties which tin-printing ink vehicles must exhibit, relating to the various post print processes which the inks may have to undergo.

In many applications the last ink to be printed is wet-varnished. Whereas this ink is most likely to be black or dark blue, variations in design mean that practically any colour could be wet-varnished at some time. During this process there must be no tendency for colour to bleed back into the varnish or a coloured wash will be seen across the white sheet due to the contamination of the varnish. Neither must the varnish bleed from the edges of print areas between varnish application and cure. The resin must, therefore, show minimal solubility in the varnish solvent system.

In practice, due to the constraints already placed on the ink formulator to obtain sufficient solubility in the ink reducer, it is the responsibility of the varnish formulator to choose a suitable solvent system. However, it is a matter of which the ink formulator should be aware.

The varnish will be stoved, typically at 180°C for 10 minutes, and subsequently the decorated sheets may undergo further stovings or processes involving high temperatures. As well as the requirement for heatfast pigments, discussed earlier, it is important for the ink resin to have good colour retention. Both the oils used to manufacture alkyds, used in oxidation drying systems, and certain oligomers, in radiation-cured systems, show a tendency to yellow at high temperatures. This effect is controlled by the choice of driers/initiators used. Since these components are necessary parts of the formulation they cannot simply be removed and as such a compromise may be required between colour retention and other properties, such as drying speed or adhesion.

In practice, yellowing is seldom a major problem except in white inks and so it is common to have a basic ink range based on an ink vehicle with moderate colour retention properties and a special white for applications where improved colour retention is necessary due to further high temperature processes.

Another additional requirement for tin-printing ink vehicles is flexibility with good adhesion to cope with the various tooling operations the decorated sheets may be subjected to in order to be converted into the finished product. The cured resin must be flexible, to survive being stretched or compressed without pulling apart or losing adhesion to the underlying enamel or size coating on the metal, during tooling. The cured ink may be considered on a molecular scale as a mass of small tangled polymer chains (the original resin) held together by various numbers of cross-links (produced when the ink cures). When the ink is stretched it is the polymer chains which are able to be unravelled and provide flexibility. The cross-links hold the stretched chains together and provide adhesion to the substrate. If the ink has too low a cross-link density the chains will both pull away from the substrate, causing loss of adhesion, and pull apart from each other, visible as short cracks in the ink wherever it fails.

Conversely, if the cross-link density is too high, i.e. the length of chain between cross-links is short, then the chains will snap. This type of failure tends to run and so gives rise to long visible cracks in the ink. It is the aim of the ink formulator to produce an ink from a vehicle which will cure to give the desired cross-link density for good tooling.

Although this is achievable for oxidation drying systems, where cure occurs over several minutes at high temperature, radiation-curing inks need very high cross-linking vehicles to achieve cure in a few seconds under the UV drier and additional stoving is not normally required. UV inks are therefore generally not suitable for applications requiring a high degree of flexibility.

A final process, which should be considered in conjunction with flexibility, is processing in water and steam, often called retorting, generally used to sterilize the contents of food containers once they are filled and sealed. A typical cycle would be $1^1/_2$ hours at 1.05 kg/cm^2 and 121°C, but higher times, pressures and temperatures can be used. The cured ink must show no tendency to pick up water or steam, since this might lead to blushing (slight colour change in areas that have taken up most water) or loss of hardness or adhesion.

Tooling failure, resulting in micro cracks, invisible to the naked eye and thus unobserved, may develop during retorting into much larger, visible cracks, and total loss of adhesion in heavily deformed areas of print. This occurs because the relatively long time at elevated temperature provides the energy required to allow stresses built up in the ink during deformation to relax.

Unfortunately, in the case of microcracks, they achieve this by running and joining up to produce larger cracks and/or loss of adhesion.

Because of this relaxation effect upon reheating tooled inks, it is common practice when testing for flexibility, to reheat laboratory tooled items, typically wedge bends or screw caps, for a few minutes at 200°C as a quick alternative to lengthy retortings and then re-examining for breakdown of the ink.

Having considered the requirements of a lithographic vehicle suitable for a tin-printing ink let us briefly consider which materials meet these requirements.

Oxidation drying systems

The physical and chemical requirements of tin-printing inks, over and above those of other lithographic inks, severely limit the choice of suitable ink vehicle. The hard resins used in other lithographic inks cannot cope with the flexibility, temperature and processing requirements of tin-printing inks and, in practice, the ink formulator finds himself limited almost exclusively to long-oil alkyd resins. Drying oils, and inert resins, such as saturated polyesters and plasticizers, may be used to modify ink properties but, where such modifications are required, they are generally better achieved by modifications to the alkyd resin used. Resin manufacturers are generally very helpful in this area and their co-operation in the past has led to a very wide range of alkyds with differing chemical and physical properties being available to the ink formulator, albeit that they are all based on similar chemistry.

Suitable alkyds are normally based on the polyesters formed between phthalic anhydride or isophthalic acid and either glycerol or pentaerythritol, modified with coconut, dehydrated castor, soya bean or linseed oils or their fatty acids. Long-oil lengths are used since they confer greater solubility in the aliphatic hydrocarbons used to reduce the inks, and also improve drying. Hydroxyl values are generally high to aid pigment wetting but acid values need to be low to avoid potential problems with water balance on press and water uptake during steam sterilization of the finished product.

It is becoming more common for the alkyd manufacturer to supply alkyd with anti-oxidant incorporated, to improve shelf life and press stability, and reduced to a reasonable viscosity for handling with a suitable reducer such as 260–290°C petroleum distillate.

For more information on alkyds refer to Chapter 4.

Radiation curing systems

The chemistry of UV-cured inks is dealt with in detail in Chapter 11 and so will only be dealt with briefly here.

The entire liquid phase of a UV-cured ink can be considered as the vehicle since even the reducer is fully convertible to solid upon cure. The vehicle will consist mainly of acrylates, present as oligomers, which dictate the main physical and chemical properties to the ink, and monomers, used to reduce the ink and improve ink transfer, wetting and adhesion. There may also be a high molecular weight, inert thermo-softening resin present to increase viscosity and aid adhesion.

Unfortunately there is always a compromise between hardness and cure speed with flexibility and adhesion and in practice it is not possible to produce fast curing, highly flexible UV inks with current technology.

It is worth noting as a final point, that although the inks are dried by UV light, they harden up considerably when stoved and, since practically all work is overvarnished with a conventional varnish, the inks harden during the varnish stoving to give their final properties.

Solvents/diluents

As with all lithographic inks, tin-printing inks are restricted to the use of solvents of low solvency power. 260–290°C petroleum distillate is the main reducer used, since it is sufficiently high boiling to remain in the ink on the press, yet readily evaporates in the oven. Bridging solvents such as tridecanol or other high molecular weight alcohols may be used to improve the lay of the ink. They achieve this by their relatively high solvency power which enables the ink to flow out better as it heats up in the oven by keeping the resin in solution for a longer period as the reducer evaporates.

Bridging solvents can be added at up to 5% where necessary, whereas the petroleum distillate will be present at levels typically in the range 5–25%, including any distillate which may be present in the alkyd, to improve handling properties.

Additives

The additives used in tin-printing inks to modify the print performance of the basic pigment, resin and solvent combination are generally the same as those used in other lithographic inks and so have been dealt with in more detail in the earlier sections of this chapter. We will therefore confine ourselves here, to the particular applications relevant to tin-printing inks. The main classes of additives used are:

- Driers
- Waxes
- Anti-oxidants
- Lithographic additives
- Rheology modifiers.

Driers

The main drier combination used in tin-printing inks is a 3:1 mixture of manganese and cobalt driers. The manganese driers give good through cure and the cobalt driers excellent surface cure. However, the cure of an ink is pigment-dependent and this combination, while good for most inks, is not the best for blacks or whites.

For black ink the cobalt drier can be too efficient, effectively skinning the ink, slowing down the rate of oxygen penetration to the rest of the ink and so resulting in a soft ink. For this reason a mixture of manganese and zirconium driers is more favoured.

With white inks the problem is colour. With both manganese and cobalt driers being dark in colour it is therefore prefered to use a mix of cerium and zirconium driers which have a much lighter colour and so aid colour retention.

Few other driers have found any bulk usage in tin-printing inks.

Waxes

Tin-printing inks are normally overvarnished and so while there is occasionally a need for a non-varnish ink, normally wax is only required to give sufficient slip for protection during handling. Where a non-varnish ink is required, relatively high levels of polyethylene (PE) and polytetrafluorethylene (PTFE) are used; 1–3% micronized wax would be typical for this application.

For a standard tin-printing ink PE is the preferred wax since it gives the best scuff resistance and hence protection from sheets rubbing over one another during handling. However, wax levels are kept to a minimum, typically 0.1–0.3%, to avoid varnishing problems due to the very low surface energy of these waxes which can give rise to pinholing, i.e. de-wetting of the varnish around wax particles in the surface of the ink.

As with all lithographic inks, relatively small particle-size waxes of typically 4–5 microns must be used, since, although larger particles give better slip, they can cause press problems such as piling.

The waxes may be added to the ink as a powder in micronized form, but it is more common to add them as a pre-dispersed paste, typically 30–40% wax in a mixture of resin and reducer.

Anti-oxidants

Since tin-printing inks are dried in an oven at high temperature, typically 130–170°C, it is easier to achieve a compromise between press stability and drying by the use of an anti-oxidant and it is not necessary to use weak anti-oxidants for fear of drying problems. However, very strong anti-oxidants, such as hydroquinone, can sufficiently retard drying to give potential set-off problems and so it is normal to use a moderate anti-oxidant such as butylated hydroxy toluene (BHT) to obtain fairly long open times on press with good drying in the oven.

It is becoming more common for resin manufacturers to supply alkyd resins with anti-oxidant already added at a level suitable for the final ink formulation. This has the advantage of improving the shelf life of the resin and makes it unnecessary for the ink formulator to add extra anti-oxidant unless there are specific problems experienced on press.

Lithographic additives

It is essential that tin-printing inks are formulated to have the necessary physical and chemical properties for the lithographic process. However, although each printer will control his fount between strict limits, there are a great many different founts currently in use which vary quite considerably in both pH (typically between 4 and 8) and alcohol levels (0–20% isopropyl alcohol) to suit the particular press and water type.

Since it is important that the ink has the correct balance between affinity and repulsion for the fount solution, it is not always possible to formulate an ink which will perform perfectly with all the founts it encounters. To adjust for this the ink supplier may have lithographic additives which either increase the water resistance of the ink, or much more rarely, increase water affinity, in order to obtain the best balance of properties for a particular press/fount combination.

Rheology modifiers

For a given resin/solvent system, different pigments will wet-out and interact differently with the ink vehicle to give substantially different rheologies for each pigment used. As well as adjusting the resin contents of the ink to allow for this the ink formulator has a choice of additives which modify rheology by both physical and chemical means.

High tack can be alleviated by reducing the viscosity of the ink. However, this can give rise to other printing problems and so it is generally desirable to maintain some body to the ink as it is reduced. This may be achieved chemically by the use of polyamides or aluminium chelates which increase the amount of intramolecular binding within the resin, reducing flow and thus increasing yield value and thixotropy.

Such additives are generally added to the resin system prior to addition of pigment, to form a gel varnish which is then used to produce the finished ink. A drawback of these additives is that gellation may continue to develop over an extended period after manufacture.

As an alternative to gel formation, the yield value of an ink may be increased by the incorporation of fumed silicas or china clays which physically restrict flow of the system and thus increase yield and viscosity.

China clays and fumed silicas can also be dispersed in reducer, or a low viscosity gel varnish can be produced, to form a gelled reducer which enables the printer to reduce tack on line, while maintaining some body in the ink. Gelled reducers are also easier to mix in the duct than straight reducers.

The only other rheology modifier of note for tin-printing inks is talc which is useful for introducing slight thixotropy into the ink at small additions, typically 1–2%, without affecting other properties greatly. The introduction of thixotrophy is important since it will greatly reduce any tendency for the ink to mist or fly.

Typical ink formulations

Since tin-printing inks are generally expected to be suitable for whatever end use they are put to, formulations tend to be similar. Those variations that do exist can best be considered under the headings:

- inks for matching, spot and four-colour work
- metallics
- non-varnish applications
- UV inks.

Inks for matching, spot and four-colour work

The majority of inks produced for tin printing are either part of a base range of typically 15 to 20 inks, which enable the customer to match any colour he requires, or spot colours, which are matched by the manufacturer from his own internal base range and supplied in bulk. Such base ranges may either be based upon single pigments, in which case the pigments dictate the colour range, or they may be matched to a specific colour range preferred by certain customers. In either case the range will be chosen to include a standard four-colour set which can then be used either for four-colour work or for matchings.

A typical formula for a base range ink would be:

Pigment	25.0
Long-oil (dehydrated castor oil) alkyd	60.0
Anti-oxidant	0.5
PE wax paste	0.5
Talc (anti-fly)	2.0
260–290°C distillate	10.0
Cobalt/manganese driers	2.0
	100.0

Whereas the above recipe is suitable for the majority of base range colours, it is normal to have a special white and black.

The white ink will differ from the other inks in the range in that it will contain a much higher pigment level, due to the opacity required. Also, anti-oxidant will be omitted and the alkyd and driers used will be chosen for good colour retention, to ensure suitability for high temperature stovings and processing. A typical formula would be:

Rutile titanium dioxide pigment	50.0
Long-oil (soya bean oil) alkyd	43.5
Talc (anti-fly)	2.0
260–290°C distillate	3.0
High boiling alcohol	1.0
Cerium zirconium driers	0.5
	100.0

Most pigments become harder to grind as they are broken up into smaller particles. However, carbon black pigments will continue to grind down to particle sizes well below those of other pigments, exposing more surface area each time a pigment particle is broken up. Thus black inks will differ from others due to the grinding properties of carbon black pigment which make it very hard completely to wet out the pigment and leads to poor yield values. This can cause problems with wet varnishing and so is overcome by the use of a pigment-wetting agent and incorporation of a filler to increase yield value. Carbon black pigment also affects drying and so a less potent drier combination is used to prevent the surface skinning too rapidly in the oven.

A typical wet varnishable black ink would be:

Carbon black pigment	25.0
Long-oil (dehydrated castor oil) alkyd	58.5
Pigment-wetting aid	1.0
Anti-oxidant	0.5
Talc	4.0
Fumed silica	1.0
260–290°C distillate	8.0
Zirconium/manganese driers	2.0
	100.0

In addition to the colours in a base range it is necessary to have a transparent media to reduce colour strength in certain matchings. The ink vehicle itself could be used but it is more normal to produce a filled media which exhibits improved fly and yield properties.

The following is an example of a filler medium:

Calcium carbonate extender	15.0
Long-oil (dehyrdated castor oil) alkyd	77.5
Anti-oxidant	0.5
260–290°C distillate	5.0
Cobalt/manganese driers	2.0
	100.0

Metallics

To produce a silver ink for tin printing it is normal to incorporate a 'silver' paste based on aluminium powder wetted out with a suitable solvent. Generally, it is necessary to use surface-treated aluminium grades since, although these have inferior brightness, they are less prone to oxidation during stovings and so more readily maintain their colour once printed. Such pastes tend to produce inks of poor yield value and so a suitable filler may be incorporated to boost yield value.

A suitable formulation for a silver tin-printing ink would be:

'Silver' paste	30.0
Long-oil (dehydrated castor oil) alkyd	58.0
Anti-oxidant	0.5
Reodorant	0.5
Fumed silica	4.0
260–290°C distillate	5.0
Cobalt manganese driers	2.0
	100.0

Gold inks may be produced using a similar formulation to that of a silver ink but using a suitable 'gold' paste based on a copper alloy. These pastes are generally supplied as pale, rich/pale, or rich, in order of ascending redness.

Alternatively, if a specific gold is required, it may be matched by adding the appropriate mix of red and yellow inks to the silver ink. In this case it is more likely that the silver ink would be supplied to the printer as an extension to his base range and he would mix his own golds. The second technique can also be used for any other metallic colours.

It is worth noting that, due to the appearance of metal itself, where printing is carried out on clear sized plate, as opposed to white enamelled plate, golds and other metallics can be matched without the use of gold or silver pastes, relying on the substrate to provide the metallic appearance required.

Non-varnish applications

In the majority of applications, tin-printing inks are overvarnished and so receive protection from damage on the final product. However, it is occasionally required that a non-varnished ink is used. In this case it is necessary to use the hardest-drying alkyd possible and provide as much slip and scuff resistance as possible by incorporation of suitable waxes. An example of a non-varnish black ink is as follows:

Carbon black pigment	25.0
Long-oil (linseed oil) alkyd	63.5
Anti-oxidant	0.5
Talc (anti-fly)	2.0
PE wax	1.5
PTFE wax	0.5
260–290°C distillate	5.0
Zirconium/maganese driers	2.0
	100.0

UV inks

Radiation-cured systems are dealt with in detail in Chapter 11 and so at this point it will be sufficient to give an example of a typical tin-printing UV ink. Formulations vary greatly in exact composition but in general will contain a lower pigment loading than oxidation drying inks since pigments absorb UV light, to varying degrees, and hence interfere with cure response.

A typical formulation would be:

Pigment	20.0
Inert resin	10.0
Epoxy acrylate	25.0
Urethane acrylate	26.5
GPTA† (reducer)	8.0
PE wax	0.5
Talc (anti-fly)	2.0
DMPA* (photo-initiator)	4.0
ITX‡ (photo-initiator)	2.0
Amino benzoate synergist	2.0
	100.0

* 2,2-dimethoxy-2-phenylacetophenone
† glycerylpropoxytriacrylate
‡ isopropyl thioxanthone

For more detailed information refer to Chapter 11.

Ink-related problems and their possible solutions

As with all lithographic inks, printing problems can be considered under the five main headings:

- lithographic
- rheology
- tack
- drying
- appearance.

Lithographic problems

Tin-printing inks suffer from the same problems as other lithographic inks and these are covered in Chapter 6 for paper and board printing. However, while scumming and piling are still commonly encountered, stripping is uncommon and tinting is extremely rare, since the pigments used for tin printing are totally insoluble in water. In each case, where problems arise, the solutions are the same as for other lithographic inks.

Rheology problems

Due to the limitations on choice of pigments suitable for tin-printing inks and the high pigment loadings required to obtain the required colour strengths, it is seldom possible to have the optimum relationship between viscosity and yield value (ideally 30–35 Pa s and 1500–2000 dynes). Yellow inks, for instance, tend to have high yield values and, although this is compensated for by reducing viscosity slightly, it is still possible for the ink to 'hang back' in the duct which can lead to ink starvation (Fig. 6.25). Duct agitators fitted to the press help to overcome this problem. Once a knife is run along the duct two or three times the body disappears and the ink flows, but it is a source of inconvenience for the printer since after a short while the ink may body-up again. Reducing with standard reducer will alleviate the problem.

Rheology will affect other factors such as scumming, piling and dot gain, this being common to all lithographic inks. It is also a major factor in wet-varnishing problems since if the ink has too little yield value the pressure of wet varnishing will 'wash' the dots in halftone areas in the direction of varnishing, increasing dot size and thus strengthening colour. Where this occurs it is usually due to over thinning of the ink on press. It can therefore be solved by replacing the ink in the duct with fresh ink.

If it is necessary to reduce the ink then it is preferable to use tack gel reducer if wet varnishing is to be carried out.

Tack problems

Tack is less critical for tin-printing inks than most other lithographic printing processes.

Provided tack is not so high that it becomes hard to separate the sheets from the printing blanket, or so low that the ink does not transfer properly, down the inking roller chain, then it is not likely to cause problems. Where tack is too high it can easily be lowered with tack gel reducer.

Drying problems

Since tin-printing inks are stoved at high temperatures, to accelerate drying, there is less of a compromise between press stability and drying. However, the heavy weight of the stacks necessitates that the inks be fully

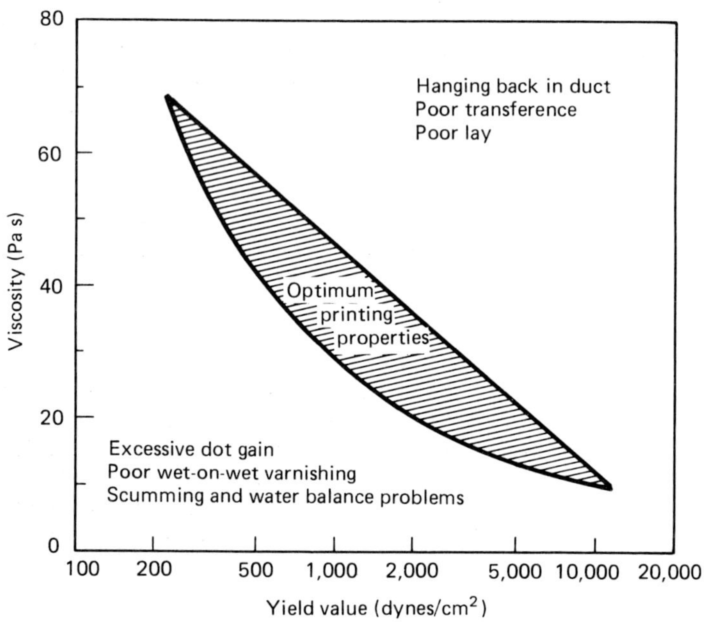

Fig. 6.25 Relationship between viscosity, yield value and printing properties of tin-printing inks.

dried by the end of their stoving and so conditions such as high press speeds and high ink film weights can still produce problems.

Specific difficulties may be considered under the headings of:

- set-off
- carry-over piling
- non-drying
- poor rub/scratch resistance
- UV-drying problems.

Set-off

Set-off occurs when ink transfers from the print to the reverse of the adjacent sheet in the stack.

Since metal is impervious, drying is entirely from the surface of the ink down. If there is insufficient drier in the ink there may be too little cure at the substrate surface to give good adhesion and set-off will occur. This can be made much worse if the ink is printed onto an enamel or another ink which has been over cured, since a proportion of ink adhesion is due to chemical bonding, or association, with unreactive functional groups in the underlying coating.

Printing onto bare aluminium or TFS may also give set-off since it is harder to adhere to bare metal than to an enamel or coating.

Such problems are generally best overcome by increasing the oven temperature up to around 160–170°C to ensure good cure right through the ink. Alternatively, more drier can be added or wax paste incorporated to aid sheet separation.

Black inks may surface dry to an extent that they skin, effectively stopping through-cure completely. In such cases the ink is dry to the touch but rubs off with firm pressure and has a high chance of setting-off. In this case the oven temperature should be reduced to give the bulk of the ink more time to dry before the surface skins. Addition of driers in this instance may make the problem worse.

Carry-over piling

On a two-colour press, the ink on the first deck may lose sufficient reducer on press to tack up slightly and transfer back from the sheet onto the second printing blanket. This is not common with tin-printing inks but can be solved by reducing the ink on the first deck.

Non-drying

When an ink is still tacky after curing it generally indicates a lack of surface driers which can be remedied by addition of a small amount of cobalt driers. However, with certain enamels, containing for instance anionic surfactants or high levels of acid catalysts, drying of the over-printed ink can be severely retarded. In such cases addition of driers makes little difference and it is necessary to increase the stoving temperature to obtain maximum drying speed.

Poor rub/scratch resistance

Tin-printing inks are generally overvarnished and so protected on the final product. However, they must withstand handling prior to overvarnishing.

Lack of rub and scratch resistance may be due to insufficient driers but can also be caused by too much drier since a high drier level causes the ink to dry with a higher number of lower molecular weight polymer chains which results in a softer ink. High pigment to binder ratios, necessary for ink strength, can also lead to rub/scratch problems. Rub resistance is generally imparted by addition of PE wax paste but can only be added in a limited amount to avoid causing overvarnishing problems. PTFE may also be added in combination with, or in place of PE wax.

Silicone slip aids are not used in tin-printing inks due to potential wetting problems with subsequent coatings, i.e. inks and varnishes applied directly and internal coatings applied subsequently to the reverse of sheets where back contamination may have occurred in the stack.

UV-drying problems

The curing of acrylate based inks is inhibited by oxygen and hence good rub/scratch resistance requires a hard fast-reacting system. Unfortunately, for good adhesion, it is necessary to use softer, slower curing oligomers. Thus there is a compromise between adhesion and surface hardness.

In practice, since the inks are overvarnished they only need to survive handling prior to varnishing and so the ink formulator aims to have the minimum surface cure necessary for handling rub resistance, in order to obtain maximum adhesion and hence minimize risk of set-off.

As with oxidative tin-printing inks, PE and PTFE waxes can be used to improve rub resistance.

Appearance problems

Although tin-printing inks are nearly always varnished and thus rely upon the varnish for their final gloss and surface lay, there may still be appearance defects due to problems in the printing process or unevenness in the printed film.

Dot reproduction

Reproduction is most critical in halftone-dot areas since variations affect both tonal balance and colour rendition. Dot gain due to physical effects, is an inherent feature of tin printing. It is not, as such, a fault, provided that it occurs in a sufficiently controlled and predictable manner to enable appropriate allowances to be made at the colour separation stages. Physical dot gain is a result of the squash or flow-out that occurs when the fluid ink dots are transferred from plate to blanket to substrate under impression pressures. It is therefore influenced by press, blanket and substrate characteristics as well as ink factors.

Inks must be formulated to minimize the degree of sideways spread during the transference stages. Rheology, in terms of viscosity and flow, and cohesion (tack), require appropriate control to maintain dot gain within identified tolerances. Further, the tinctorial strength of the ink is also important since variations in this property will be compensated for by adjustment of ink film thickness to maintain colour density. Thicker dots will naturally be more prone to dot gain through edge squash under printing pressures. Techniques for controlling the rheology, tack and colour strength of lithographic inks have already been covered in detail in previous sub-sections of this chapter.

Poor gloss

Gloss is not generally a problem in tin printing, since this is controlled normally not by the ink but by the overvarnish. Occasionally, however, non-varnish inks are used and in these cases there is a compromise between gloss and rub resistance since PE and PTFE waxes, incorporated to help protect the ink from damage, also have a slight matting effect. When these inks are used it is generally accepted that a silk finish is sufficient rather than a full gloss.

Poor lay

Lay is the term used to describe the evenness of a print. Poor lay indicates that the ink in solid areas has printed with a non-uniform film thickness.

Poor lay can cause a loss of gloss due to non-uniform reflection from an uneven surface. This is not a major problem for tin printing due to overvarnishing. However, when the lay is sufficiently uneven to cause variations in colour it becomes unacceptable.

Poor lay is usually the result of transference problems. These may arise due to the ink having too high a viscosity, failing to emulsify properly on the image areas of the plate, or a build up of particles in the ink which compete in size with the print film thickness. Such particles could be a result of large particles of wax in the ink or simply poor grind.

Poor flow characteristics of the ink, due to high yield value, will also add to poor lay, since any unevenness of print will fail to flow out.

Generally, all these problems can be eased by reducing the ink, especially if a bridging solvent is used. However, this will only help in marginal cases and severe problems may require the ink to be reformulated and/or the press conditions to be investigated. It should also be noted that overthinning of the ink will give rise to dot gain and wet-varnishing problems.

The nature of the stoved ink can also affect the lay of the overvarnish. Under cured ink may have a sufficiently low surface tension to resist good wetting of the varnish. High levels of PE or PTFE wax can also cause de-wetting of the varnish. This may be overcome by the use of a surfactant to lower the surface tension of the varnish. Alternatively, it may be necessary to use a different varnish or reprint the job with a different ink to ensure a more receptive surface.

Metal marking

White pigment is ceramic and very hard. If a piece of metal is rubbed against a white ink it will leave a mark where the metal has been rubbed off onto the ink. This mark may look like damage to the ink but can be rubbed off with a finger showing undamaged ink underneath.

This is not normally a problem with white itself since white ink is seldom printed over a large area of the sheet. But it can be a problem for large areas of creams, buffs and pastel shades.

When metal marking is observed it is necessary to find the part of the equipment where contact is being made and mask the area with tape. A more permanent solution is achieved by re-matching the colour with a proportion of the white replaced by transparent media to make the ink less abrasive.

Recent and future trends

Since tin printing is a well established process changes tend to be evolutionary rather than revolutionary and generally involve improving rheology and dispersion properties to improve press performance with the ever increasing range of controlled dampening systems.

Improvements are also being made in the heat resistance and tooling properties of inks, and equipment changes are being introduced to reduce down time and make the printing process easier to control.

A completely new area of development, however, is likely to become of ever increasing importance. Tin-printing inks, containing relatively low levels of reducer, have been considered as high-solids systems in the past. However, with new legislation aimed at reducing total levels of volatile organic components (VOC's) in the atmosphere and new varnishes and enamels being developed with very low solvent levels, inks will also need to reduce their levels of volatiles even further. This could be achieved by developing further UV-curing inks and water-reducible systems and coatings. The above points which relate to lithographic printing in general have already been dealt with in previous sections but two broad areas which are directly relevant to tin printing will be considered: resin chemistry and, processes and equipment.

Resin chemistry

As well as general improvements in the lithographic properties of resins for tin-printing inks, more revolutionary changes are being made to try and improve colour retention of whites and pastel shade inks, and to deal with the increasing pressure to reduce all sources of VOCs.

Polyester resins

Conventional tin-printing inks are currently based almost exclusively on alkyd technology. Unfortunately alkyds are prone to yellowing during high temperature stovings, such as those used to cure an internal lacquer on food containers, due to reactions involving residual unsaturation in the cured ink.

Polyesters are basically oil-free alkyds and as such have no unsaturation to cause yellowing during subsequent stovings. Unfortunately it is the oil portion of alkyds which is primarily responsible both for cure and lithographic properties requiring water repellance. Cure can be effected by the use of an acid catalyst to cross-link the residual acid and hydroxyl groups in the polyester but is again very sensitive to fount conditions which can lead to leaching of the catalyst.

Although cross-linked systems have been available for many years, polyester inks are now becoming available for certain applications where non-yellowing is the predominant factor.

UV curing resins

UV inks are more expensive than conventional inks but offer cost savings in other areas where heavy tooling is not a requirement. Present technology is almost entirely based upon acrylate, free-radical curing systems since all alternatives incur a heavy cost penalty.

However, UV offers an obvious way to achieve VOC reductions if the poor tooling characteristics can be overcome. Such inks can be produced from other technologies, the most dominant being cationic curing using cyclo-aliphatic epoxides, which cure by the use of photo-initiators which cleave under ultra-violet radiation to produce strong acid catalysts. The technology is restricted by the limited choice of suitable pigments.

While these systems are more expensive than acrylate technology, bulk usage may well reduce costs, and legislation against VOCs may make this approach favourable.

More details on cationic systems are given in Chapter 11.

Water-reducible resins

Another way to reduce VOCs is to use water-reducible resins and thus replace the ink volatiles with water. Unfortunately this poses major problems for lithographic printing due to problems controlling interactions with the fount solution. This may make water-reducible resins appear unsuitable for tin printing. However, the recently renewed interest in driography which eliminates the need for fount makes such systems feasible.

Processes and equipment

The printers' desire continuously to improve quality and reduce waste is being met by advances in press technology.

Driography

The concept of driography is to produce printing plates that are unreceptive to ink in the non-image areas and so do not require the use of a fount solution. This greatly simplifies press operation since half the printing problems associated with lithographic printing are related to controlling the dampening system.

Since it is the plates which make this system work, existing equipment can be used simply by disconnecting the dampening system.

In the past, driographic plates have only found limited use since they are physically more delicate than lithographic plates and it has proved very difficult to eliminate ink receptivity completely in the non-image areas on long runs, leading to a background tint.

Improvements in the plates combined with certain modifications to the printing press and ink rheology have now made driography viable for tin printing. This system also offers a solution to the problems of running polyester inks and even water-reducible inks by eliminating the dampening system.

Driography is not completely without problems, especially where UV inks are involved, but is still likely to become more predominant in the future.

Control systems

Another way in which lithographic printing problems can be reduced is by obtaining more control over the operation of the press. To this end presses are becoming more automated and enable the printer to set specific pressures and dampening levels which are then maintained by microchip control as conditions change during a run. This has a further advantage that, once the optimum conditions for a particular print run have been found, they can be recorded and if the same job is carried out in the future it can be started with the same conditions as before.

Summary

Offset lithography is now well established for the decoration of sheet metal. This process is both technically and economically competitive and unlikely to undergo any major changes, although the use of driography is likely to find its place alongside lithography with time.

Ink formulation, however, is likely to show greater change as legislation concerning VOC emissions encourages the use of alternative chemistries.

BIBLIOGRAPHY

Bassemir, R. and Krishnan, R. (1987) *A Study of the Lithographic Performance; Mechnical versus Thermodynamic Considerations*,

PIRA (1990) *United Kingdom Offset Newspaper Specification*, Newspaper Society and PIRA, Leatherhead.

Surland, A. (1980) *Laboratory Test for Predicting Lithographic Performance*, TAGA, London.

CHAPTER 7
Dry offset inks

This chapter is divided into two sections, one covering the printing of plastics and the other the printing of two-piece cans. Each section will cover formulation principles and the problems encountered during ink application.

Section 1: Dry offset inks for plastic

Dry offset, also known as letterset, uses features found in two other printing processes; the raised surface plate of letterpress and the rubber blanket of offset lithography. Typically, up to six colours are transferred consecutively from raised plates onto a common impression blanket and then printed on the container in one pass. A decorative or protective

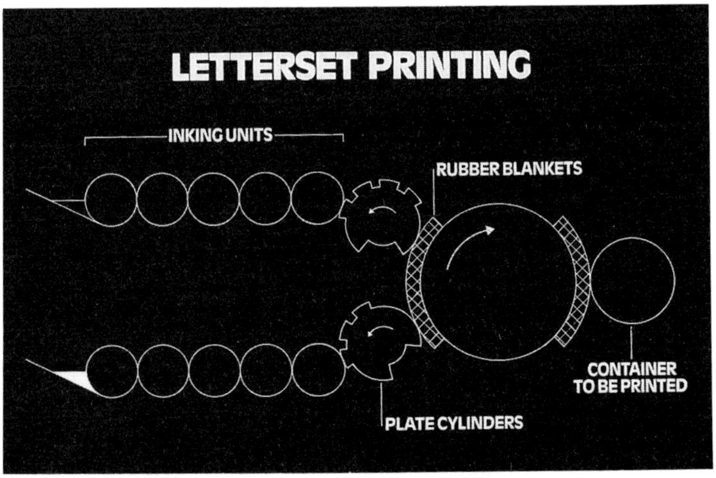

Fig. 7.1 Principle of dry offset printing.

overvarnish may be applied by a satellite roller-coating unit dependent on the type of packaging being produced. Ink is fed to the printing plate from an ink duct down a chain of distribution rollers, the chain is short, usually of seven or eight alternating metal and rubber rollers which produce a thin, even ink film on the plate. After subsequent transfer to the blanket and then the container a 2–4 μm film thickness is obtained (Fig. 7.1). Dry offset printing is used to decorate a diverse range of plastic containers, the main groups being:

- dairy products tubs, cups and lids containing margarine, yogurt and cream, etc. (Fig. 7.2);
- building trade products such as emulsion paint cans, rigid tubes for mastic and sealant, buckets and pails for adhesive;
- cosmetic and toiletry tubes;
- pharmaceutical containers and syringes;
- polylaminate (a metal and plastic laminate) tubes for toothpaste;
- detergent, bleach and motor oil plastic bottles.

The print speeds of the modern machines are relatively high, up to:

40 000 per hour double line lid printer
24 000 per hour cup printer (dairy products)
4500–8000 per hour pail printer (paint cans)

Fig. 7.2 Examples of printed tubs.

7.1 DRYING MECHANISMS AND THE INFLUENCE OF THE SUBSTRATE

There are four main mechanisms by which dry offset inks dry:

- evaporation of solvent
- absorption of solvent and plasticizer
- catalytic cross-linking
- radiation UV curing.

This has led to the development of four distinct types of dry offset ink formulation, each utilizing one of these drying methods.

The most appropriate drying mechanism for any dry offset ink is partly dictated by the type of plastic substrate it is to be printed on. For the purposes of drying, substrates can simply be split into two groups determined by their resistance to the ester solvent contained in conventional dry offset inks. The two groups are non-absorbent substrates – the polyOlefins, polyethylene and polypropylene – and absorbent substrates – polystyrene and acrylonitrile styrenated butadienes (ABS), and polyvinyl chloride (PVC).

Drying methods

The actual drying method used depends on the type and size of the container being printed. Infra-red predominates in food packagaing, while hot air is used for non-food containers. Both use conventional inks, but UV-curing inks have found increasing use in all areas.

Infra-red

The printed tub is rotated while being carried on a conveyor belt past a row of electrically heated ceramic panels which emit medium to long wave IR radiation (3–10 μm wavelength). The drying line is long, up to 10 metres, and the tubs will take 15–20 seconds at 80–100°C to dry before being nested together and packed.

Hot air

The printed container will circulate in a basket or on its mandrel for 2–5 minutes at 80–100°C in a fan-assisted hot-air oven. The drying speed of conventional inks can be increased in three ways:

- using a higher drying temperature although this will be limited by the melting point of the plastic substrate;
- more highly pigmented inks which can be printed at a lower film-weight. However, with some pigments poor flow in the ink duct may result or even ink piling on the rollers, plate and blanket;
- Using faster evaporating solvents in the ink but poor press stability is then possible.

Ultra-violet curing

Standard medium pressure mercury arc lamps are used arranged in one of two configurations: curing on mandrel (COM), or off-line tunnel oven.

With COM the UV lamps, usually two, are fitted around the mandrel immediately after the print station as the lamps are housed inside the printing machine – a very compact, space-saving design results. The other arrangement feeds containers on a conveyor belt through a tunnel oven of 2–3 metres in length containing two diagonally opposed UV lamps. The containers are not rotated as uniform exposure is achieved by the highly reflective, polished aluminium interior. For further information see Chapter 11.

Pretreatment of substrate

Polyethylene and polypropylene are non-polar polyolefins with an inert, low energy surface. It is necessary to alter their surface chemistry and create a thin, oxidized, microroughened surface layer if ink adhesion is to be obtained. This is done by either electric corona discharge or gas flame treatment. Corona is usually used on thin walled food containers, gas flame on heavier containers like bottles and paint cans. Pretreatment improves ink adhesion as it leads to increased wetting of the substrate, mechanical keying and electrostatic bonding between polar oxidized surface groups and the reactive/polar groups on the resin molecules in the ink. Polyethylene and polypropylene always need pretreatment before being printed with any type of dry offset ink. Polystyrene is not pretreated when printed with conventional inks because adhesion is easily obtained by attack of the ester solvents which dissolve the surface. However, polystyrene is a non-polar material and, as UV inks contain no solvents, adhesion on the untreated substrate is only moderate, but by using one of the two treatment techniques described, good adhesion and improved print strength and density are obtained as the UV ink will wet the polystyrene surface more readily. To obtain adequate adhesion on these plastics their surface tension, after treatment, should be at least 42–44 milli Newtons per metre (dynes/centimetre).

7.2 FORMULATING PRINCIPLES

Food packaging application

It is accepted by dry offset ink producers in the UK that these inks should be formulated according to the Society of British Printing Ink Manufacturers (SBPIM) recommendation regarding inks for immediate food wrapping in view of the widespread use of dry offset printing to decorate food containers and lids, particularly for the dairy product market. Other countries have separate recommendations. The moulded container forms a functional barrier between the food inside and the printed surface outside. The containers should be printed and dried in such a manner that set-off in the printing process is avoide. This ensures that the inside surface of the container in contact with the food is free of printing ink. As a further precaution, dry offset inks should not be formulated using materials in the SBPIM exclusion list or with materials known to be toxic.

While this principle guides the choice of all materials used, the selection of pigments and plasticizers is especially strict.

Pigments

When selecting pigments for dry offset inks the following points are considered:

Chemical resistance

A diverse range of container types are dry offset printed and the pigments selected must be capable of withstanding accidental contact with the product being packed. Inks for food containers require resistance to water, fats like margarine and dairy products such as yoghurt, cream and soft cheeses. Inks used on non-food containers may have to resist detergents, washing up liquids and motor oils.

Lightfastness

A very high level of lightfastness is not normally required for container printing inks as they are not recommended for outdoor use or exposure to high levels of direct sunlight. In normal practice they only have to withstand artificial lighting and some indirect daylight when printed containers are displayed in shops. Pigments with a Blue Wool rating of at least 5 at full strength are considered acceptable.

The following is a list of typical pigments used in a standard dry offset range of nine colours plus black, white, and an extender base.

		CI No.
Yellow	Arylide yellows	PY1,3
	Diarylide yellow	PY13
Orange	Pyrazolone orange	PO13
Red	Naphthol red	PR2 and 8
Magenta	Lithol rubine 4B (calcium salt)	PR57:1
Violet	Dioxazine violet	PV23
Blue	Phthalocyanine blue (β form)	PBl15:3
Green	Phthalocyanine green	PG7
Black	Carbon black	PB7
White	Titanium dioxide (rutile form)	PW6
Extender	Calcium carbonate	PW18
	Talc	PW24
	Silica	PW27
Metallics	Aluminium (leafing and non-leafing grades)	PM1

Resins

The restricted solvent and plasticizer system used in conventional dry offset inks means that the choice of resins is limited. Cellulosics like ethyl cellulose or nitrocellulose are used as the main film former and are modified with a hard resin component whose function is to improve adhesion and gloss, and also to raise the application solids. Suitable classes of hard resin are cyclic ketones, isophthalic polyesters, rosin-modified maleics, fumarics and phenolics. A general purpose dry offset ink for use on polyolefin and styrenic containers would be based on a high nitrogen-content extra low viscosity grade of nitrocellulose which would be ester soluble and give an acceptable varnish viscosity at the required solids, together with a low acid value rosin-modified maleic resin (for adhesion and product resistance, particularly to water). Catalytic dry offset inks have specialized applications on polylefin containers used, for example, to pack emulsion paints or detergents. Here, the cross-linked epoxy resin film has excellent adhesion, scratch hardness and is highly chemical resistant. A solid grade bisphenol A epoxy resin (n=2) dissolved in a glycol ether, such as diethylene glycol monobutyl ether (boiling range 210–231°C) would form the ink vehicle, the ink drying initially by solvent evaporation followed by an epoxy–polyamine cross-linking reaction.

Solvent and plasticizers

The nature of the dry offset process requires an ink based on a system with low volatility but strong solvent power. This gives inks with the necessary press stability on the ink distribution rollers and plate, while having rapid drying by absorption/evaporation on the plastic substrate. In practice, this need has led to the almost exclusive use of a high boiling point glycol ether ester solvent combined with a phthalate plasticizer. Typically, diethylene glycol monobutyl ether acetate (boiling range 235–250°C) and diethyl phthalate (boiling range 292–299°C) are examples.

It is important that the roller, blanket and plate are made from materials that are resistant to ester solvents, otherwise swelling and stickiness will occur as the ink solvents attack and dissolve them. Plates are made from either plastic using nylon polymers, or a metal alloy of zinc and magnesium, whereas rollers and blankets are made of resistant rubbers like butyl or EPDM (ethylenepropylenediene monomer). To avoid problems with press stability due to solvent absorbtion, it is usual practice to prime new rollers and blankets by washing them over thoroughly with the appropriate ink thinner before they are used in production.

Ink additives

Anti-flying agent

To prevent ink fly or mist from the ink distribution rollers, micronized talc or silica is used at up to 2%. The level in each colour is optimized to minimize flying while retaining good flow properties.

Wax

Waxes impart slip and rub resistance to the dry ink film, properties that are particularly important in a dry offset ink used on food packages, where tubs, cups and lids are nested one inside another for packing and no ink set-off must occur. The most common types of wax used are polyethylene or PTFE which either migrate to the surface or sit proud and so provide a high-slip surface.

Plate 1
The Pantone Matching System Color Formula Guide

(a) Yellow

(b) Magenta

(c) Cyan

(d) Black

(e) Yellow + magenta

(f) Magenta + cyan

Plate 2
Four colour printing showing colour separations and superimpositions. Assembled on Crossfield Studio 9500 version 4.35. Images scanned using a Crossfield Magnascan 636

(g) Yellow + cyan

(h) Yellow + magenta + cyan

(i) Yellow + magenta + cyan + black

(a) Cyan + magenta + yellow + black

(b) Black

(c) Cyan + magenta + yellow

0% u.c.r.

(d) Cyan + magenta + yellow + black

(e) Black

(f) Cyan + magenta + yellow

50% u.c.r.

Plate 3
Showing under colour removal at 0%, 50% and 100% of black. Assembled on Crosfield Studio 9500 Version 4.35. Images scanned using a Crosfield Magnascan 636

(g) Cyan + magenta + yellow + black

(h) Black

(i) Cyan + magenta + yellow

100% u.c.r.

Plate 4
Munsell Colour Tree

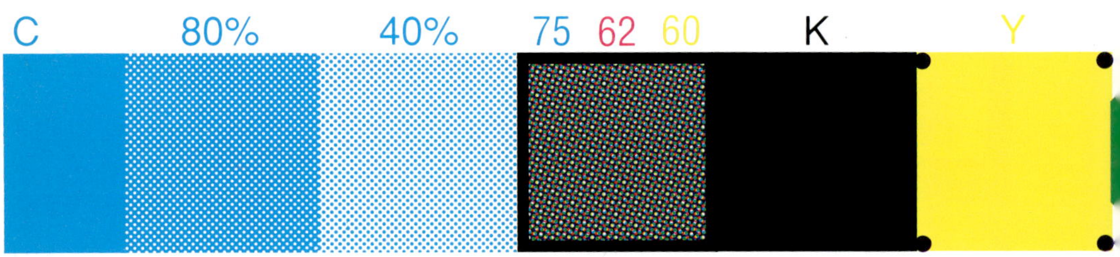

Gretag colour control bar
Plate 6a) Expanded to show register

Plate 6b) Normal size

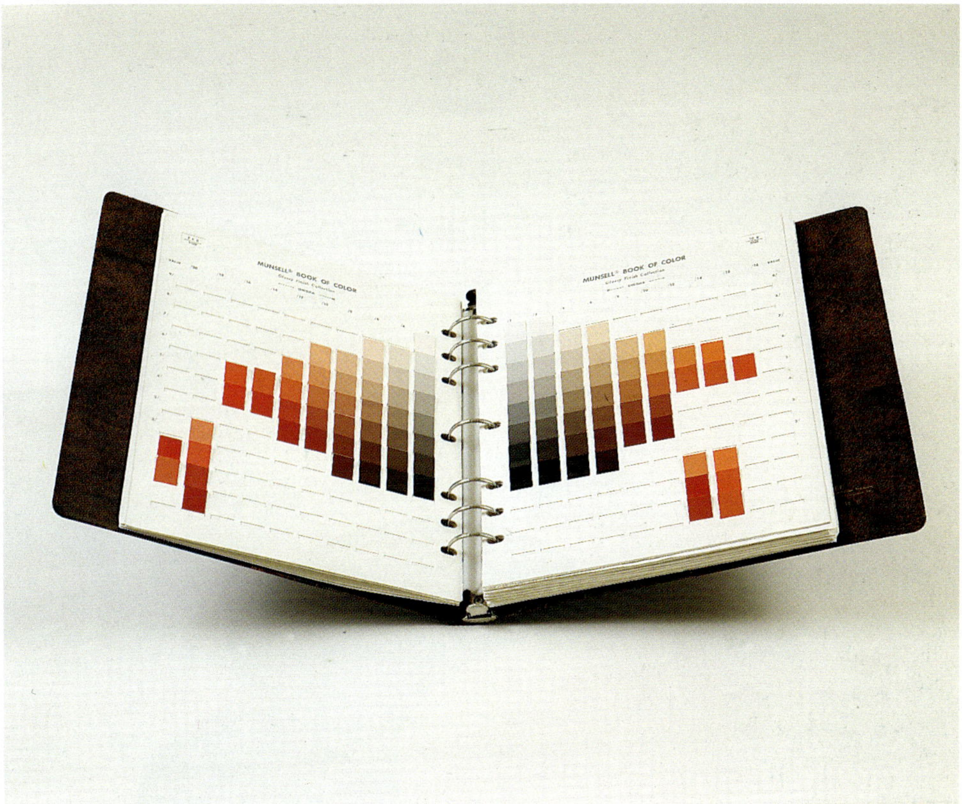

Plate 5
Munsell Book of Colour

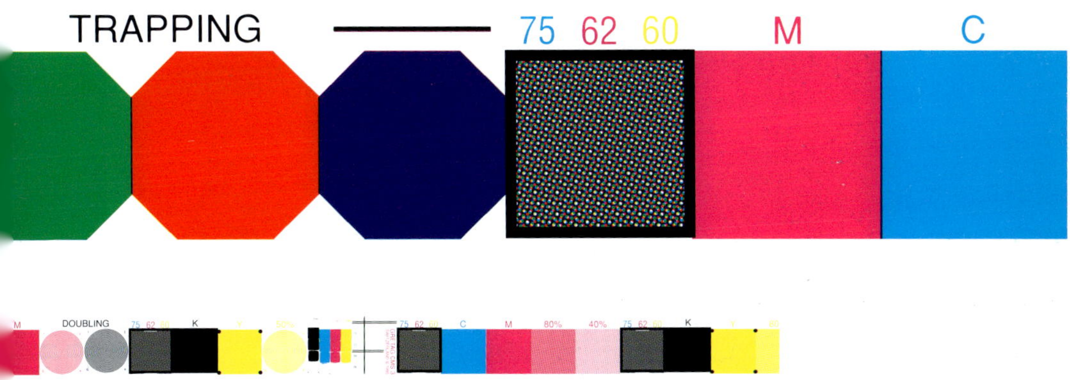

Plate 7
Additive colour theory, where red, green and blue combine to make cyan, magenta and yellow. Where all three primaries meet, white is produced.

Silicone
A silicone fluid of the polydimethyl siloxane type is used at a low level of 0.5–1.0%. This not only adds extra slip, but gives good ink wetting without pinholing or reticulation on substrates with a low surface tension.

7.3 GENERAL CHARACTERISTICS OF DRY OFFSET INKS

Viscosity

The viscosity of dry offset inks is more easily measured on a cone-and-plate rheometer than on a 'falling-bar' type, although either can be used. Conventional inks will be in the range of 15–25 Pa s, while UV inks will have a higher viscosity range of 25–35 Pa s. The inking units on a dry offset machine may experience a modest temperature rise to around 30–35°C when running. This presents no problem to a correctly formulated conventional ink but the viscosity and tack of a UV ink are more sensitive and will drop rapidly as temperature rises, thus affecting transference. For this reason UV inks are supplied at higher viscosities.

Press stability and tack measurement

The press stability of a dry offset ink can be assessed by plotting its tack versus time curve, at 25°C using a tackmeter, and comparing it with the curve of an ink of known performance. As the tack of an ink influences its distribution and transfer properties, tack change against time is used when formulating an ink as a reliable indicator of the expected press stability. The relationship between tack and viscosity with an established dry offset ink formulation is such that, in practice, tack–time curves are not needed as a general quality assurance test. The measurement of viscosity alone is sufficient.

Dispersion

The degree of dispersion is tested using a standard twin channel fineness-of-grind gauge. As dry offset is a thin-film printing process using highly pigmented inks, it is desirable that the maximum colour strength is developed from the pigment. This means that an excellent level of dispersion is necessary and grind specifications require low readings.

Product resistance

As a container-decorating system, dry offset inks must possess adequate resistance to the product packed in the container and in certain cases to specific processing and end uses to which it may be put.

Inks may have to resist water, dairy products such as yoghurts and cream, fats like margarine, washing up detergents, bleach and motor oil. Retortable laminated polypropylene containers are used now in niche market areas like pet foods. The pre-printed containers are filled with the foodstuff, sealed, then retorted in a steam pressure atmosphere at around 120°C to cook the contents. Pharmaceutical containers and syringes are commonly gamma-radiation sterilized after being printed. To resist these processing conditions two-pack catalytic or UV curing dry offset inks are used due to their higher level of resistance and a protective roll coat lacquer may also be applied.

Pressure curing

Under extreme conditions a UV dry offset ink can prematurely polymerize in the ink distribution unit, resulting in the formation of hard, dried ink bits which contaminate the ink causing film weight control problems due to their piling up on inking rollers, plate or blanket. The machine will have to be stopped, the affected areas cleaned and the duct reloaded with fresh ink. Pressure curing is primarily caused by too tight a setting of the doctor blade, against the metal duct roller, generating pressure and heat leading to localized ink polymerization. This occurrence can be minimized by having a low rotation speed of the duct roller and by maintaining ink temperature in the duct below 35°C. Inks are also formulated to contain suitable thermal and light stabilizers which inhibit premature UV curing.

Halftone printing

Dry offset printed designs are normally built up using single colours placed beside and sometimes on top of another on a common blanket, all of which are then transferred to the container in one pass. The classic wet-on-wet four-colour process technique, as found in offset lithography, is also used in dry offset although two points need particular attention, namely colour sequence and ink tack. The lightest colour should be transferred to the blanket first, and the darkest colour last, to minimize the contamination of subsequent colours which may be caused by the back-trapping of ink up the inking rollers. An offset four-colour build up would therefore be deposited from the printing plates onto the blanket in the order:

Yellow	1st colour
Red	2nd colour
Blue	3rd colour
Black	4th colour.

To ensure that the inks trap properly over each other the tack of subsequent colours can be lowered slightly by a small addition of solvent or monomer of between 2 and 5% in order to produce a tack graded set of inks. Some typical formulations are given below.

Conventional solvent-based dry offset ink

A general purpose ink suitable for all types of plastic and drying mainly by evaporation to give a matt finish would have as a typical formulation:

Pigment	20.0
Ketone resin (food grade type)	27.0
Ethyl cellulose resin	7.5
Butyl carbitol acetate solvent	44.0
Polyethylene wax	1.0
Silicone fluid	0.5
	100.0

UV curing dry offset ink

A gloss finish, general purpose UV cured ink for printing on all substrates could be formulated as follows:

Pigment	18.0
Acrylate prepolymer	30.0
Modifying hard resin	25.0
Acrylate monomer	16.0
Photo-initiator and amine photosynergist package	9.0
Polyethylene wax	1.0
Silicone fluid	1.0
	100.0

7.4 THE FUTURE

Viewed from a printing ink manufacturer's standpoint, the progress of dry offset printing has been along the following routes; first, faster printing speeds requiring quicker-drying inks and, secondly, a wider range of both substrates and container types being decorated. Historically, this has lead to oil-based oxidation drying inks being superseded by conventional solvent and plasticizer-based inks and then to acrylate-based UV curing inks. The adoption of UV gives several advantages, as they can usually be used on all absorbent and non-absorbent plastics, and they have infinite press stability and dry instantly to a hard chemically resistant film that rarely needs a protective overvarnish. New dry offset presses fitted with UV drying can print at high speeds using the rapid cure of modern UV inks and have a space and energy saving design in comparison to older machines using IR or hot-air ovens. So, although it is a mature and firmly established process, all those concerned with dry offset printing would identify a gradual tendency to use UV as the preferred drying method.

Section 2: Two-piece can decoration

The two-piece beer and beverage can is a major force in the drinks packaging market in many parts of the world with sales during 1992 of approximately 150 billion cans world-wide. All these cans are externally decorated, the vast majority either with inks printed on top of a basecoat as is most usual in Europe (non-varnish decoration), or by printing directly onto the metal substrate and subsequently applying a varnish to the wet inks.

The cans are produced in highly automated plants at very high production speeds (currently 850 to 1500 cans per minute for each production line) in a process which is continuous from the coiled flat sheet to the final, palletized, empty can bodies. It is designated a two-piece can because the finished product comprises two components: a combined base and body produced from a single piece of metal (either tinplate or aluminium) to which the second component (the lid) is seamed once the can has been filled. The body is manufactured by the draw and wall ironing process whereby a cup is stamped, or drawn, from strip metal and is then rammed through a series of dies of reducing bore. This has the effect of thinning or ironing the walls of the cup and causes them to grow taller. During this operation the base shape or 'profile' is also formed. The cans are then thoroughly washed and dried before the external decoration and internal lacquer(s) are applied.

If a basecoat is to be utilized this is applied first and cured in an oven before the inks or inks plus varnish, which are also passed through an oven. On tinplate cans particularly, the application of a white basecoat is very common as it provides a bright, attractive base colour for the design. Where a clear basecoat is utilized (more commonly on aluminium cans) its function is more protective than decorative and is an alternative to varnishing.

The other major operation is to form the top profile – necking and flanging – where the wall at the top of the can is reduced in diameter or necked-in and then flanged outwards ready to accept the end. This is undertaken either before or after the internal lacquer is sprayed on to the inside (and external bottom surfaces) followed again by curing in an oven.

7.5 METHOD OF PRINT APPLICATION

The inks are applied to the can by the rotary offset process using purpose designed 'decorators' (Fig. 7.3). These comprise a large drum made up of individual offset printing blankets around which are set between 4 and 6 printing heads. Each printing head or 'inker' has an ink duct, a set of inking rollers and a printing plate with a raised image area which carries one part of the printed design. As the drum rotates each printing plate adds its image to the blanket and the full design is then transferred to the can in one operation as it rotates in contact with the blanket. This method restricts the type of designs that can be printed, as any areas of the design where inks touch or overlay one another do so while the inks are still wet on the blanket (wet-on-wet printing). As already mentioned the wet inks on the can are sometimes varnished at this stage using an application unit attached to the decorator. The final part of the printing process involves passing the cans through an oven to cure the inks and varnish.

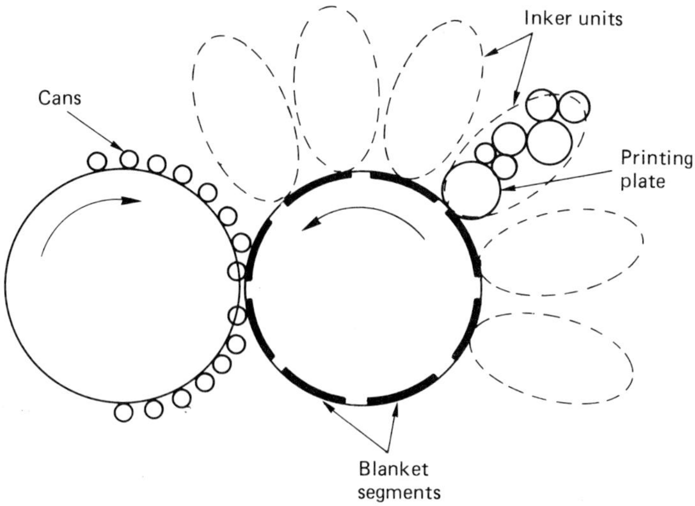

Fig. 7.3 Two-piece can decorator.

Fig. 7.4 Two-piece DWI cans.

7.6 INK FORMULATION

The resin type by far the most commonly used is a reactive polyester/alkyd combined with a high boiling polar solvent, such as an alcohol (typical boiling range 200–300°C), anti-fly additives, and pigment. The nature of the resin used is crucial to the final ink's performance, and also each pigment type and grade must be matched to the resin for optimum results over a wide range of properties. The pigment grade will affect for instance ink lay, gloss, flow, cure and slip as well as the more obvious properties of shade, strength, light and heatfastness, pasteurization resistance, etc. In order to obtain high pigment strength with resin systems such as these, that are not particularly good pigment wetters, dispersion must be as effective as possible and dispersion aids may also be incorporated.

Whether or not the can is to be varnished will determine whether any other ingredients are required. For varnished cans the cure of the ink is often achieved through reaction with the melamine formaldehyde resin and catalyst in the varnish applied on top of the inks while they are still wet. Additionally, the varnish acts to protect the ink and also provides mobility. Consequently it is frequently possible to produce satisfactory varnishable inks which contain no other additives.

Where no varnish is applied the inks need to contain their own reactive ingredients and slip aids. Such non-varnish inks therefore additionally contain melamine formaldehyde resin, acid catalyst, lubricant and possibly a stabilizer to prevent premature cure during storage.

The acid catalyst causes the melamine formaldehyde resin to react with the free hydroxyl and carboxyl groups on the polyester/alkyd resin. The curing process is complex but a typical condensation reaction mechanism between the hydroxyl groups on the polyester/alkyd resin and the melamine formaldehyde resin is:

$$R'N(CH_2OH)_2 + 2\,R''OH = R'N(CH_2OR'')_2 + 2\,H_2O$$

where R′ is the melamine formaldehyde resin and R″ is the polyester/alkyd resin. Some typical recipes are now given.

Varnishable ink:

Pigment	25
Polyester/alkyd resin	50
Anti-fly paste	3
Solvent	22
	100

Non-varnish ink

Pigment	20
Polyester/alkyd resin	40
Anti-fly paste	3
Solvent	14
Melamine formaldehyde resin	12
Catalyst	2
Lubricant	7
Stabilizer	2
	100

A typical range of pigments used in two-piece can inks would consist of:

Colour	*Pigment Type*	*Colour Index No.*
Yellow	Azo Yellow	PY13
Yellow	Azo Yellow	PY138
Orange	Azo Orange	PO34
Red	Lake Red C	CPR53:1
Red	Azo Red	PR214
Magenta	Permanent Red 2B	2BPR48:2
Violet	Dioxazine Violet	PV23
Blue	Phthalocyanine Blue (β form)	PB15:3
Green	Phthalocynanine Green	PG7
Black	Carbon black	PB7
White	Titanium dioxide (rutile form)	PW6
Extender	Calcium carbonate	PW18
	Talc	PW24
	Silica	PW27

7.7 INK PROPERTIES REQUIRED

The properties required of inks for two-piece can printing can be considered in two parts';those required to enable the ink to be applied satisfactorily and those required of the ink after its application to the can.

Application properties

The high printing speeds and pressures involved result in conditions of high temperature and shear with the inks on the decorator usually reaching temperatures between 40 and 65°C. The rheological properties of the ink must be carefully controlled in order to minimize problems of ink fly and filling-in while ensuring that the ink still has adequate flow both to transfer efficiently from the rollers to the plate, then to the blanket and finally the can. It must also quickly flow out on the can in the short time before the ink has been cured in the oven.

The optimum balance of these requirements are obtained from high strength inks with good transference, at viscosities between 7 and 20 Pa s and yield values from 0 to 6000 dynes. In addition the inks must not dry on the rollers too quickly, nor body up in the ink duct where they may remain for several hours at these high temperatures. They must not hang back in the duct and thus fail to transfer down the roller chain due to too high a yield. Finally in this category, the inks must remain stable after manufacture for at least one year even though they need to possess high reactivity to ensure extremely rapid cure in the oven.

Post-application properties

As already mentioned the inks must cure satisfactorily in the oven (under stoving conditions ranging from oven air temperatures of 400°C for 6 seconds to 185°C for 60 seconds). After stoving the cans are transported at high speed to the next stage of the manufacturing process and the external decoration needs to be sufficiently hard and abrasion resistant to withstand the harsh mechanical handling involved as well as having a high degree of slip. Even higher speeds are encountered in filling lines where 2000 cans per minute are now not unusual. The cans must move along conveyors and rotate against each other with the minimum of friction to ensure a smooth passage with no delays in their progression.

The necking process involves stretching of the can wall and therefore the inks need to be both flexible and adhere satisfactorily to the basecoat (or metal substrate in the case of varnished inks). The nature of the equipment used to carry out the necking operation will affect the flexibility and abrasion resistance required with the spin/spin necking of tinplate cans being the most demanding in both these respects.

The ink stoving itself and the subsequent stoving(s) for the internal spray lacquer require the ink to have considerable resistance to the affects of heat. The internal laquer stoving conditions vary from 230°C for 90 seconds to 210°C for 4 minutes. However, there is a distinct possibility that whenever the production line stops some cans will remain in an oven for longer periods and resistance to degradation is required for at least double the normal stoving time.

Even though a can is not expected to be exposed to long periods of direct sunlight during its useful life, some degree of resistance to fading is required. The lightfastness is specified as a minimum on the Blue Wool scale and this minimum normally ranges between 4 and 6.

Many products in two-piece cans are pasteurized after filling to improve their storage life, this particularly (but not exclusively) applies to beer. Typical pasteurization conditions involve a multistage process whereby the temperature of the can and its contents is gradually raised to 65–70°C over a period of up to 30 minutes using a hot water spray. It is often necessary to include additives in the water to prevent growth of bacteria and/or algae and some of these additives can be aggressive to a number of commonly used pigments. In addition, the conveying lines and the cans are sprayed with a dilute solution of soap-based additives to aid mobility and again this can attack certain pigments if the solution is allowed to dry on the cans or finds its way in significant quantities into the pasteurizing water. Pigments used must at least be resistant to pasteurization in plain water; whether pigments with greater resistance are required will then depend on the conditions in the particular pasteurizer (particularly the concentration of additives in the various zones).

Non-varnish inks are also required to have high gloss and the latest generation frequently match the gloss of varnished cans.

7.8 PRINTING PROBLEMS

The problems encountered with two-piece can inks can conveniently be considered under the headings of those associated with the application of the ink to the can and to the subsequent treatment of the can.

Ink application

Problems most commonly encountered in this category are:

- poor flow out or 'lay'
- ink sling and misting
- filling-in
- ghosting
- repeating
- hanging back in the duct
- piling on the rollers, plate or blanket
- drying up.

Post application

Problems encountered after application include:

- low gloss
- inadequate adhesion to the basecoat or substrate
- inferior abrasion resistance
- insufficient slip or mobility
- pasteurization failure
- poor lightfastness
- ink rub out
- necking problems.

Varnishable inks

In addition, varnishable inks must interact beneficially with the varnish. The most common problems encountered in this area include:

- low varnish gloss
- poor varnish flow
- ink reticulation
- poor intercoat adhesion between the ink and varnish
- pigment bleed into the varnish
- insufficient cure
- insufficient hardness/mar resistance of the ink plus varnish system.

7.9 PROBLEM SOLVING

Application problems

Poor lay

This can be caused by inadequate printing pressures – especially the pressure between the can and the blanket, by a basecoat which is not receptive to the inks, by the basic lack of flow of the ink formulation, by too high a pigment to binder ratio or by too high ink viscosity. In addition some pigments produce better flowing inks than others. The darker shades tend to give most problems exacerbated by the fact that poor lay is more noticeable with dark colours when, as is often the case, they are printed over a white or at least light-coloured basecoat. Given that the machine adjustments have been eliminated as the source of the problem it unfortunately is not always possible simply to reduce ink viscosity to improve lay as this will lead to a greater tendency for ink fly and filling-in. Neither can the pigment to binder ratio always be reduced for the same reasons.

Sling/misting

Although these are being considered together they are not exactly the same. Here, sling is the tendency for the ink to be flung out from the rollers in relatively large droplets which find their way onto other parts of the decorator in close proximity to the point of origin, whereas misting relates to extremely fine droplets of ink which can hang in the air and float for considerable distances away from the point of origin. The filmweight of ink is a very important factor in ink sling and should be kept to a minimum by using full strength inks whenever possible. Increasing ink viscosity is also effective in reducing sling while anti-fly additives plus the resin system chosen are most important in controlling misting.

Filling-in

Common causes include too high printing pressures – especially forme rollers to plate and most importantly, plate to blanket pressure – too high an ink film weight, basecoat powdering on the blanket, and too low ink viscosity – which could be exacerbated by too high running temperatures. The most important factors to bear in mind are to ensure that the printing pressures are not too high and that the ink is as strong as possible. If the lay on the can is extremely good it may be possible to increase the ink's viscosity/tack. Finally it should be realized that the individual ingredients in the ink can play a very important role in the tendency for filling-in, both in their selection and their degree of dispersion.

Ghosting

This phenomenon is exhibited as an area of higher ink filmweight appearing on a large solid print in the shape or 'ghost' of a different part of the design. It is caused by the inability of the inking rollers fully to replenish the ink on the printing plate between each impression. The most common cause is inadequate forme roller pressure although it must be noted that the majority of decorators in current use have only two forme rollers and a minimum number of distribution rollers so that ghosting problems occur quite frequently. Too low tack/viscosity will lead to increased ghosting. Also, the more transparent is the ink the more noticeable will be the different filmweights which are the effect of ghosting. Thus the most effective remedies lie with the decorator itself, and some can manufacturers have added an extra 'dummy' forme roller to the inkers. However, this has only limited effectiveness unless the inker is modified to ensure the extra roller is fed with ink (a relatively large engineering exercise).

Repeating

Superficially this can resemble ghosting but closer examination will show that the outline of a different part of the design seen on the solid print is caused by a deposit of one of the other inks onto it. In this case the solid plate has touched the edge of an adjacent ink print and picked up a thin line of ink which has been redeposited onto it by a forme roller. It is most frequently seen where a solid background print in a light colour has a dark adjacent print. The means of solving the problem mainly lies with the decorator settings (pressures and fit of the individual parts of the design) so that the two ink prints do not touch. This is affected, however, by the strength and tendency of the inks to spread (as in filling-in).

Hanging back in the duct

This occurs when the ink yield value is too high relative to its viscosity and is governed by the nature of the ink vehicle, the pigment to binder ratio and the pigment itself. Some pigments are much more prone to hanging back in the duct and in particular the diarylide yellows are notorious for this problem. Care needs to be exercised in assessing an ink's performance in this respect as a freshly produced ink may well be much worse after it has been stored for a period of time. The immediate remedy is significantly to increase the proportion of binder to pigment in the ink but if this is not effective reformulation will be required.

Piling

This is caused by inefficient transfer of ink between the decorator component where the piling is occurring and the next part of the transfer process (rollers, plates, blankets). It is mainly a feature of the pigments in the ink and is most commonly seen with aluminium inks. It occasionally also occurs when a chiller unit is used to cool the rollers and is an indication that the temperature of the rollers is too low. Assuming this is not the case, the solution lies in reducing the level of the offending pigment in the formulation.

Drying up

This can be a significant problem for designs where one element/colour is present at a low level. In this situation, and bearing in mind that the decorator can reach 70–80°C, whenever the machine stops there is a danger that the thin film of ink on the rollers and the blanket will dry up and stop transferring. The choice of resin and how quickly it releases solvent, the extent to which the resin is still mobile once the solvent has evaporated and the rate of evaporation of the solvent all affect this problem.

Post-application problems

Gloss

Very careful selection of all the ink ingredients is required to attain the high gloss levels required of current two-piece can inks. Resin and pigment selection are particularly important with an extremely close interdependency existing between the two. Within a given formulation the most significant factors affecting gloss are the pigment to binder ratio which should not be too high, the tack/viscosity which again will assist gloss if it is reduced and the degree of dispersion of the pigment.

Adhesion

Resin and melamine selection are the most important factors which affect adhesion. The other elements are the cure level of the ink, which should not be too low and the nature and cure of the basecoat, which should not be too high at the print application stage.

Abrasion resistance

In addition to the basic resin and melamine formaldehyde used it is also essential to ensure that cure takes place at a sufficient rate at least partially to cross-ink the system after the first (sometimes extremely short) stoving. The rate of cure will be significantly affected by the pigment combination used and also by the basecoat underneath (where utilized). Assuming that the ovens have been eliminated as the source of the poor cure, an addition of catalyst and possibly melamine will assist. An effective lubricant is also essential to ensure good abrasion resistance.

Slip/mobility

Lubricants, used to impart slip and hard waxes, such as PTFE, are commonly used. While soft lubricants have been used, silicones in particular are no longer favoured because of their tendency to cause de-wetting of the internal spray lacquer if ink should find its way onto the inside of the can (due to ink fly for instance). Cans are transported at these high speeds both empty and when filled. The filling process can also involve a pasteurization stage which places additional resistance requirements on the inks.

Pasteurization

It is essential to utilize a resin/melamine formaldehyde system which will withstand the pasteurization process without softening or water take-up. Additionally the pigments must also be capable of pasteurization in plain water. However, as previously mentioned, aggressive additives are sometimes introduced into the water and where this could occur extra care must be taken. As with many types of resistance properties the level of pasteurization resistance is dependent not only on the type of pigment used but also the quantity present in the ink film. Thus pigments such as Lake Red C which are already marginal in this respect will be very much worse at low percentage levels.

Lightfastness

Many of the pigments used exceed the minimum standard. However, as mentioned above, in selecting pigment combinations for a particular shade of ink it is necessary to take into account both the inherent lightfastness of the pigments used and their concentration, as low levels of pigment will have significantly poorer lightfastness.

Ink rub-out

This problem is seen as colour being easily removed by rubbing with the finger or as a deposit on the side rails on can conveyors. Assuming that the ink film is fully cured, the problem is likely to be caused by pigment on the surface of the print which is not fully bound by the resin. This in turn is often caused by pigment degradation during stoving or an incompatibility between the pigment and the resin system (the latter is frequently seen with blue/purple pigments). Careful pigment selection is therefore required and testing under the severest stoving conditions which are expected to be encountered.

Heatfastness

Pigments need a degree of heat fastness to withstand the stovings involved in the can manufacturing process and again the pigment's resistance to degradation will be affected by its concentration. For light-coloured inks, and especially white, the ink vehicle must also have adequate resistance to yellowing. When a problem occurs on the production line the actual temperatures being attained in the oven should be checked as these do occasionally rise above the recommended level.

Necking problems

These fall into the two main categories of flexibility and abrasion problems.

Flexibility

The flexibility of the external decoration will depend on the resin/melamine formaldehyde system, the degree of cure, the level and type of pigment and the thickness of the film. The proportion of melamine formaldehyde in the recipe is extremely important with a higher level giving a harder ink but a less flexible one. If a flexibility problem occurs the degree of cure and the thickness of the film are the two areas which should be checked first. If there is scope to increase the strength of the ink this will enable a thinner film to be applied and therefore aid flexibility. While if the cure is excessively high a reduction in pigment concentration will be beneficial.

Abrasion

Again the resistance to abrasion will depend on the degree of cure, the resin/melamine formaldehyde system, the lubricant type and level and, to some extent, the pigment type. An increase in cure will reduce abrasion although this will be far less effective if cure is satisfactory to start with. Extreme cases of poor adhesion to the substrate can also result in abrasion problems.

Varnishable inks

Careful choice of the resin and solvent system is required to ensure that the problems listed previously do not occur with varnishable inks.

The exception to this is the problem of pigment bleed which depends on the choice of pigment itself and requires pigments which are not affected by the solvents in the varnish (it should be noted that most varnishes in use today are described as water-based but still contain significant quantities of solvent).

Modern non-varnish inks often do not exhibit good lay when wet and rely on their ability to flow out in the oven to achieve their lay and gloss. When varnish is applied (to the wet inks) the opportunity to flow out further is lost and consequently such inks need to have better initial lay. This is achieved by careful choice of the resin/solvent combination combined with a lower viscosity/tack than their non-varnish counterparts. Fortunately, the inks can be formulated with higher strength because large quantities of additives are not required to achieve cure, hardness and slip so that many of the problems associated with low viscosity do not arise.

If insufficient cure/hardness does occur and the oven conditions have been shown to be satisfactory, an addition of melamine formaldehyde, and possibly catalyst, will be beneficial.

7.10 FUTURE TRENDS

The future of the two-piece can itself seems relatively secure. It has proved to be an extremely successful package, dominating the take home drinks trade in many parts of the world. As with many forms of packaging the most serious threat at the present time comes from environmental issues. The canmakers have continued progressively to reduce the 'energy cost' of the package through reductions in the thickness of the metal used and reduction in the diameter of the end. In addition, recycling of cans is an already established process with the proportion of cans recycled growing each year. Solvent emissions are another area requiring increasing attention due to concerns over their effect on the environment. The major contribution to emissions from two-piece can-plants arise from the basecoats rather than the inks and these are increasingly being changed to water-based products. The challenge for the ink supplier will be to ensure that the inks perform satisfactorily with these products and that the solids of the inks are as high as possible to minimize even the low level of emissions that do occur.

Increases in production efficiency require continuous increases in line speeds which in turn place greater demands on the inks both in terms of resistance to damage in the can handling arena and the performance on the printing press. In this respect higher speeds cause a worsening of ink lay which the ink technologist must overcome, while at the same time also improving the ink's resistance to fly, which will also worsen at higher speeds. This will require increasingly sophisticated formulation techniques as these two requirements normally demand opposite remedies.

To date the decorator manufacturers have not made any radical changes in the basic design or printing technique used for two-piece can printing and none are on the horizon. Improvements in the quality of printing have been obtained through general improvements in the expertise of the printers, platemakers, blanket and ink suppliers. This can be expected to continue with, for instance, the use of superior printing plates made from harder materials such as metal instead of the normal photopolymer materials. Such plates are capable of holding finer type and halftone dots and are already becoming established in the industry.

Although one can manufacturer in the USA has developed a system utilizing UV curing inks and overvarnish, this has not been taken up elsewhere as the demanding properties required of two-piece can inks have not been sufficiently met with UV curing versions.

CHAPTER 8
Gravure inks

The gravure process has become established as a major part of the printing industry. At the time of its invention, at the beginning of the twentieth century, it offered the means of high quality reproduction on paper coupled with high press speeds, which in due course gave it an advantage over the established litho and letterpress processes, especially for publication printing. In later years the process was to prove ideally suited for printing the rapidly developing non-porous substrates, principally coated and uncoated cellulosic films and aluminium foils. As the era of the plastic films emerged, flexographic printing also developed and combined with gravure to become the major sources of printed flexible packaging. Gravure printing has many other areas of application, including paper and board packaging materials for confectionery, foodstuffs and the tobacco industry. It gained favour in the printing of cigarette cartons because of its lower production costs. It also offered superior metallic golds when required. The lower costs can be attributed to the use of the in-line cutting and creasing operation which produces finished blanks from the reel. This is made possible by the drying characteristics of gravure inks. There are many non-packaging applications where the gravure process has been adopted. Postage stamps, football coupons and catalogue printing are typical examples. In the industrial and home areas, plastic laminates, wallcoverings, decorative wall panels, PVC cladding for buildings and PVC floor coverings, all represent products that are dependent on the ability of the gravure process to print consistent, fine-tone reproduction.

From the description of the gravure process in Chapter 2, it is apparent that the system requires inks to be mobile, low in viscosity and rapid drying. In order to print, the ink must fill the cells of the gravure cylinder and after removal of the excess by the doctor blade it must be sufficiently fluid to transfer to the substrate at the point of impression, flowing out to give a smooth print.

For inks to have the fundamental fast drying properties necessary for the gravure process, the use of highly volatile solvents is required. These solvents are invariably highly inflammable and sometimes harmful. The harmful effects of the solvents have been reduced over recent years by improved press extraction and ventilation. In addition certain harmful solvents have been replaced in gravure ink formulations by more acceptable alternatives.

8.1 GENERAL CHARACTERISTICS

Viscosity

Unlike the offset and letterpress processes discussed in previous chapters, the gravure process requires low viscosity inks. However, they are similar to offset and letterpress inks in that they contain resinous binders and pigment colourants. Gravure inks dry by the evaporation of highly volatile solvents and normally there is no oxidation or chemical change once the ink film is dry. It is this need for low viscosity, combined with rapid solvent evaporation, that determines the essential characteristics and performance of gravure inks.

The range of ink viscosities met with in gravure printing is generally between 15 and 25 s Zahn Cup No. 2 at 25°C, dependent on the conditions which influence ink viscosity on the press. The main factors are the:

- rheological characteristics of the ink;
- speed of printing;
- rate of evaporation of the solvent system;
- shape and range of cell depths;
- wiping characteristics of the ink by the doctor blade;
- print design parameters;
- nature of the substrate.

Printing at high speeds demands faster drying inks and hence lower viscosities, to enable the ink to flow rapidly out of the cells at the point of impression. Absorbent substrates require lower viscosities than non-porous surfaces, in fact some substrates require the addition of retarding solvents to enable an acceptable print to be achieved. It is not until all these conditions have been taken into account that the correct viscosity can be established.

Ink viscosity must be adjusted to meet several critical factors. Too high a viscosity results in an inadequate flow of ink from the cells, giving a pattern to the print in some areas that follows the cell walls, commonly known as 'screening'. Too low a viscosity results in a 'slur-out' or 'halo' occuring on the trailing edge of the print, appearing as a thin film of ink printing beyond the limits of the design. It is important to note that any final strength and shade adjustments to an ink should only be made when the correct viscosity and printing speed have been determined.

Flow and rheology

As indicated previously, a gravure ink has to change from a low viscosity liquid to a solid in a matter of a few seconds. The ability to do this lies primarily in the rapid evaporation of the solvent content of the ink, followed by the resin attaining a tack-free state as soon as the solvent has left the ink film.

The solvent and resin components are the carrier or ink vehicle which carries the colourant, while the resin binds the colourant onto the substrate. Pigments and dyes provide the colour, additives being used for a variety of purposes, particularly to improve press performance and enhance the characteristics of the printed ink.

Prior to considering the rheology of gravure inks it is necessary to understand the conditions prevailing on the gravure press.

The gravure process consists of a design engraved or etched in the form of minute cells into the surface of a cylinder or occasionally a plate (Fig. 8.1). The cylinder A rotates in an ink duct or reservoir which is supplied with ink pumped from the ink box. Excess ink is wiped from the cylinder with a reciprocating doctor blade C, leaving the unetched area free of ink and a deposit of ink in each of the cells. The substrate F is passed over the cylinder with a rubber impression roller D forcing the substrate on to the surface of the cylinder at the nip B, thereby transferring the ink from the cells on to the substrate.

It is important that the correct machine settings are maintained, the doctor blade must be set at the correct angle and pressure. Too much pressure and incorrect angling of the blade can lead to similar print defects to those described for incorrect viscosity adjustment. The impression roller must have the correct Shore hardness for the type of substrate being printed, with only sufficient pressure used to ensure proper transfer of ink to substrate.

It is when the ink is transferred to the substrate that the rheological

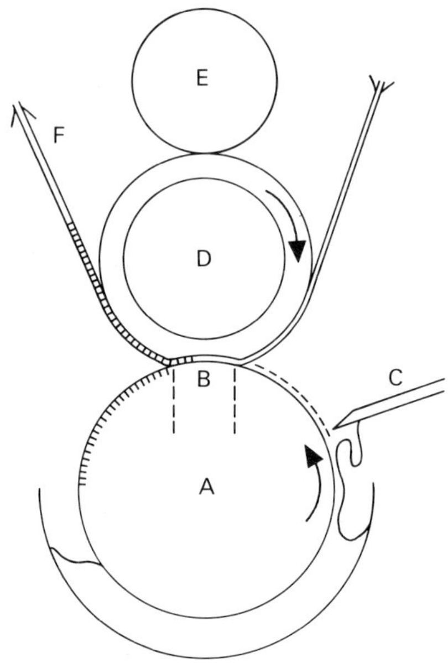

Fig. 8.1 The gravure printing system.

properties of the ink influence the print quality and smoothness, i.e. lay characteristics – the ink must flow out of the cells to give as smooth and unblemished a print surface as possible.

On occasions this is not the case, and close examination of the print shows that bubbling has occurred which subsequently leaves minute surface craters. Excessive pigment flocculation can occur, manifesting itself as irregular lay, characterized by 'crows' feet'-like patterns in the print. Frequently the overall appearance of poor lay is a 'hills and valleys' configuration which is attributed to the ink splitting unevenly and failing to flow out before it dries.

Clearly, factors other than ink performance affect lay; substrate, machine speed and particularly the type of engraving, all contribute to the print quality achieved. Two traditional types of cylinder engraving are conventional (Fig. 8.2), and direct or hard dot (Fig. 8.3). In recent years mechanical methods, e.g. Helioklischograph, have been used to engrave cylinders, resulting in solids with smoother lay and fewer imperfections, when compared with similar designs using conventional engraving. This has proved helpful when printing difficult colours such as greens and browns (Fig. 8.4).

The printability and lay characteristics of a gravure ink can often mean the success or failure of a specific formulation. Unfortunately there is no exact way of predetermining if a combination of pigment, resin, solvent or additives, dispersed by a chosen method will produce an ink with acceptable lay. This is due to a number of parameters which cannot be reproduced in the laboratory, e.g. press speed, doctor blade angle and pressure.

The influence of components on characteristics

Gravure inks are composed of resins, pigments, dyes, plasticizers, waxes, additives and solvents, each playing an important role in the characteristics and properties of a particular ink.

Pigment and dyes

The prime purpose of pigments and dyes is to confer colour. Basic dyes are rarely used in gravure packaging inks due to their reactivity and poor lightfastness. Occasionally they are used in publication inks to brighten and strengthen the blue and red. Lightfast dyes are primarily used on aluminium foil where their transparency is complemented by the metallic reflectance of the foil. However, they are expensive in comparison with transparent pigmented chips which are more widely used.

The choice of pigments most suited to the gravure process can be complex and their effect on the behaviour of an ink can be critical. Ideally a pigment should be resistant to the solvents in the ink, and should not give rise to adverse flow properties. Pigments which produce inks with poor flow characteristics generally give poor lay due to the lack of ink flow into and out of the gravure cells. A suitable gravure pigment must therefore exhibit good printability characteristics in the chosen resin/solvent system.

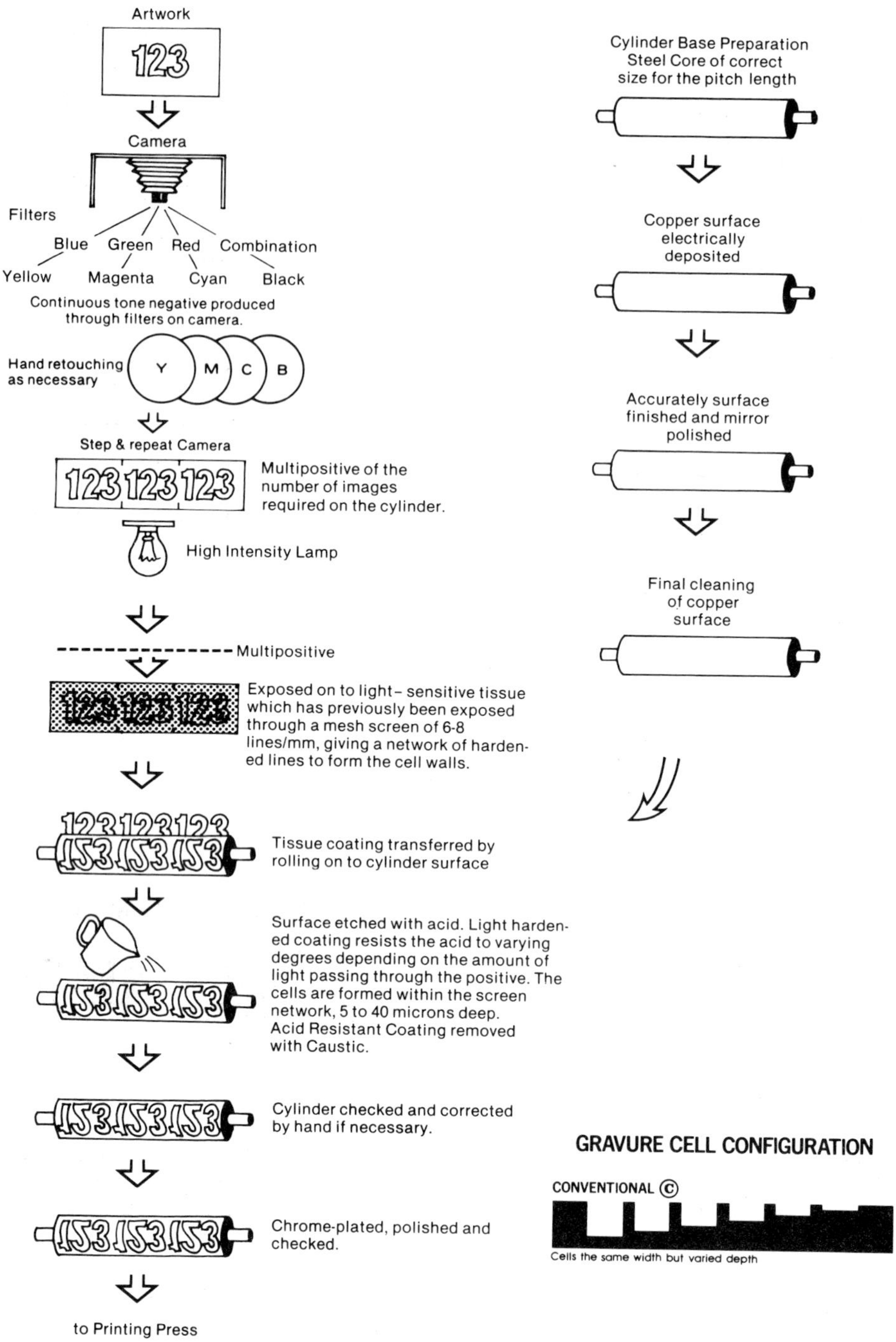

Fig. 8.2 A conventional gravure engraving. (Courtesy of Gravure Cylinders Ltd, Derby)

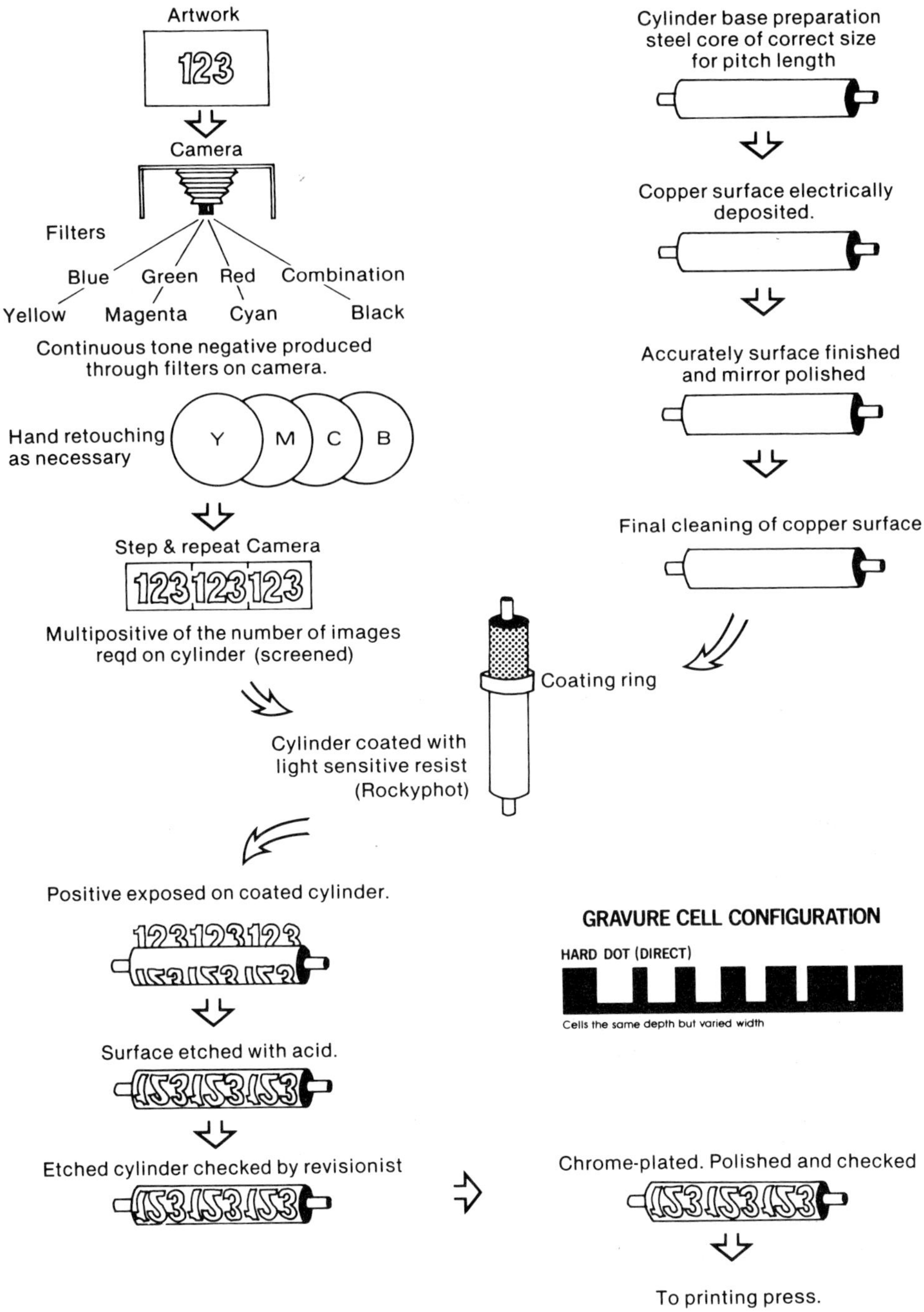

Fig. 8.3 Direct engraving. (Courtesy of Gravure Cylinders Ltd, Derby)

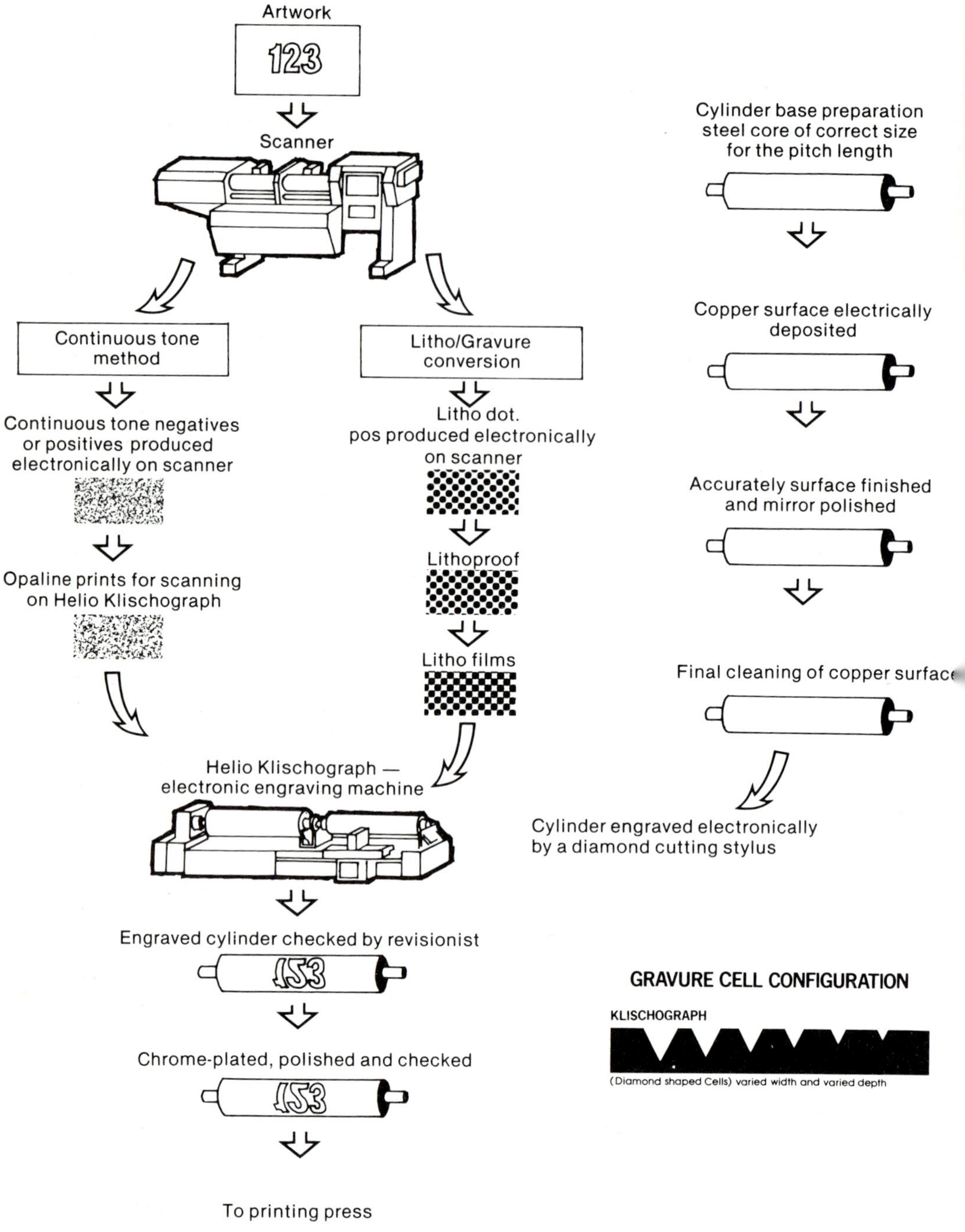

Fig. 8.4 Electronic engraving. (Courtesy of Gravure Cylinders Ltd, Derby)

As the pigment in most coloured gravure inks is the most expensive part of the formulation, the economics of pigment selection is of vital importance. The normal practice is for inkmakers to have a preferred standard range of pigments against which alternatives can be assessed.

In addition to the basic requirements outlined above, consideration should be given to their tinctorial strength, ease of dispersion and gloss characteristics. Where it is envisaged that a pigment is to be used in a concentrated base scheme it must have good flow properties at high pigmentation.

In comparison with flexographic, offset and letterpress printing the gravure process can, when required, deposit fairly heavy film weights of ink, which are governed by depth of engraving and hence the volume of ink held by the cells. At press viscosity the maximum pigmentation for organic pigments is unlikely to exceed 15%, most shades being matched with 5–10% of pigment (by weight).

With inorganic pigments, of which titanium dioxide is the most common, higher pigment levels of 25–35% are common, particularly for backing whites on transparent films and aluminium foil.

Extenders such as china clay and precipitated calcium carbonate are used in many gravure inks to control the gloss level and improve the lay characteristics in formulations with an otherwise high binder/pigment ratio. In addition, because of their low cost they can effectively be used to cheapen gravure inks. While extenders are used in publication inks to assist tack reduction, they are primarily used in this type of product to control the colour strength of the ink. Occasionally silica-based pigments are used as matting agents, although only limited amounts can be employed before the flow properties of the ink are substantially impaired.

In many areas of gravure printing, pigments need specific properties to meet demands made on them during the processing of the print and in the end use of the printed material, e.g. printing of PVC wallcoverings, deep-freeze polyethylene bags.

Resins

It is unusual to find an individual resin that will impart all the desired properties to a gravure ink and therefore the formulator will usually select two or possibly three resins to achieve a combination giving the required characteristics. Nevertheless, it is important to ensure that the principal resin in the formulation has most of the essential properties. These properties should include:

- adequate adhesion to the substrate;
- good solubility in the preferred solvent system;
- combination of good solvent release properties with the ability to dry to give tack-free ink films;
- capability of providing the nccessary level of gloss;
- good pigment-wetting properties and acceptable printability characteristics;
- toughness, to ensure adequate rub resistance of the print;
- flexibility, to ensure non-cracking of the finished printed product.

Secondary properties of the resin system will be largely determined by the end use of the ink. These might include heat resistance, product resistance, e.g. to soap and alkali, and low odour.

Solvents

Choice of solvent type is largely determined by the material to be printed and its end use. With the demand for low odour packaging by many end users, severe restrictions are placed on the solvents that can be used. For example typical solvents for food packaging would be chosen from ethanol, isopropanol, ethyl acetate, isopropyl acetate, acetone, methyl ethyl ketone, selected aliphatic hydrocarbons and as retarder certain glycol ethers.

As the toxicity of solvents is increasingly taken into account the use of those that are considered harmful is in decline. However, there are areas where solvents such as toluene and xylene are still used; the former often as a co-solvent with ketones in PVC inks and with alcohols for polyethylene inks. Football coupons, postage stamps and bottle labels are currently printed with toluol-based inks, advantage being taken of their superior printability in the fine tones. Inks for gravure publication printing can contain both toluol and aliphatic hydrocarbons, being formulated to give excellent tonal printing at high press speeds with the lowest possible cost. It is particularly important that the solvents used in inks for this process have excellent lubrication properties to reduce wear on gravure cylinders during long print runs.

With increasing pressure being made on printers to reduce solvent emissions, attention has recently focused on waterborne gravure systems. The substitution of solvent based inks by water based products is not straightforward and consideration of the press drying capacity, air flow and speed must be made, in addition to the specification of the design being printed. However an increasing number of printed specifications are now being produced with water-based materials and it is likely that this trend will continue. To print designs where rewetting of a previously printed ink occurs it is often possible to replace one or more solvent inks or varnish with water-based products, so overcoming the problem.

To summarize, solvents, albeit only a temporary part of gravure inks, play a vital role in their printability characteristics; their contribution must be to have good solvency for the chosen resin system and have an evaporation rate commensurate with the speed of printing. With increasing attention being paid to health and safety in industry, solvents should be utilized that are the least hazardous in their nature under the particular press conditions.

Additives

The overall performance of most gravure inks can be improved by the use of additives. They perform a wide variety of functions, some of which are essential to the success of the ink, while others enhance its characteristics.

Some additives are primarily functional during the manufacturing stage, e.g. pigment-wetting aids, others perform as stabilizers and bactericides ensuring storage stability. Many resin systems require the addition of chemical plasticizers or plasticizing resins, in order to prevent ink drying in the cells on the press and to ensure satisfactory adhesion and flexibility on the substrate.

Water-based systems must contain selected alkalis, such that the resin can be saponified, but they must be sufficiently volatile to leave a water-insoluble dried ink film. Water systems will also contain additives to reduce surface tension properties and so enhance lay characteristics.

The surface characteristics can be considerably modified by additives which promote rub resistance and slip. The most common and probably best additives for these purposes are polyolefin, paraffin and amide waxes. Care should be taken when selecting additives that such inclusion whilst imparting desired characteristics do not denigrate other properties, e.g. loss of interlayer adhesion due to excessive wax content, or fish-eyeing due to the presence of unsuitable silicone compounds.

8.2 PHYSICAL PROPERTIES OF INKS AND THEIR MEASUREMENT

Dispersion

The process of dispersion of the pigment into the vehicle to produce a gravure ink can be achieved by several quite different manufacturing techniques. The subject of ink manufacture is considered in detail in Chapter 13, but the relationship of the dispersion process to the end properties of the ink is of great importance and will therefore be discussed further.

The options available to the ink manufacturer for dispersion of the colourant, range from the use of chips and pre-dispersed pigments to conventional dispersion equipment such as bead mills, ball mills, rod mills or even high-speed stirring. The choice of dispersion technique will depend on both the desired end use properties and the batch size requirements.

The use of pigment chips and to a lesser extent pre-dispersed pigments, tends to give short thixotropic inks with very high gloss and transparency. Dissolving is usually carried out on a high shear mixer and this is normally sufficient to produce an ink with a dispersion of well below 10 microns as measured on the Hegmann gauge (Chapter 15). In the event of poor dispersion a mill pass may be necessary but this is usually indicative of a poor quality chip either through an inadequate predispersion technique or degradation/oxidation of the resin carrier.

There are other advantages in using chips, particularly with pigments that are difficult to grind, where advantage can be taken of the increased brightness and tinctorial strength the process achieves, e.g. PMTA colours (PR 81, PBL 62, PV 3, PG 2) and lithol rubine 4B toners (PR 57). The high cost of the pre-dispersion process, together with decreased flow characteristics which places a limitation on the attainable ink strength, means that such techniques are, however, only normally used when no other option is feasible.

Bead mill dispersion is now the most common process for the manufacture of gravure inks. Various modifications are available which enable different diameter bead sizes to be utilized. In general terms the smaller the bead size the better the dispersion but the shorter the flow and hence a compromise has to be reached. If glass beads are used then special care must be taken to ensure total separation of the beads from the ink to prevent the risk of cylinder wear.

Ball mills are becoming less popular due to their lack of batch size versatility and the relatively long times required to achieve dispersion. Ink flow, however, is excellent and high pigment loadings can be used. High speed stirring is not an efficient method of dispersion for most organic pigments. For titanium dioxide, however, it is possible to achieve a reasonable dispersion by this technique. The final ink lacks the gloss of a bead mill dispersion and is dependent on consistent quality pigment to ensure that no oversized particles are present. The presence of oversized pigment particles can lead to printing defects, e.g. scumming, doctor blade lines.

Stirring techniques are used for dyestuffs, but this is a solubilizing process as opposed to a dispersion technique.

Strength, hue, transparency and gloss

The transparency requirements for an ink vary with the substrate to be printed. For printing on foil or other metallized substrates and the 'pearlized' films, inks with a high transparency and gloss enhance the appearance of the substrate, whereas on paper a glossy transparent ink will appear weak in comparison to a matt opaque ink. In film printing, gloss is usually paramount for surface printing but of no value in reverse printing where the design is viewed through the substrate. Transparency is not of value where coloured inks are used in conjunction with backing whites.

Pigment strength is normally assessed by use of the drawdown technique and by 'bleaching' (Chapter 15). The drawdown will also give an indication of the transparency of the ink if a black bar is used (Chapter 15) and the hue can be compared on the undertone and the mass tone. A wide variation in hue should be avoided in any gravure ink colour match since this will lead to a change in hue at different ink film thicknesses. In any dispersion technique, a range of variable sized pigment particles will be produced. In general the lower the dispersion energy input, the wider the particle size distribution. For maximum colour strength a narrow distribution of small sized particles is required since the colour strength will be proportional to the surface area of the pigment present. Care must therefore be taken when carrying out pigment strength comparisons using the bleach technique that interpreted results reflect pigment comparison and not dispersion technique comparison. In general it is advisable to assess relative pigment strength using products manufactured by the same dispersion process.

Relative strengths of inks produced by differing manufacturing processes must still be assessed since the pigment is normally the most expensive raw material in a gravure ink formulation. Reliance is therefore placed on the standard drawdown method with variations in gloss and bronzing produced by the different dispersion processes eliminated by overdrawing the inks with a matt varnish.

The instrumental measurement of gloss is discussed in Chapter 15 but although the use of a gloss meter will provide comparative prints with a numerical correlation, the printing industry still relies heavily on a visual assessment.

The gloss level achieved on a print is governed by the choice of raw materials, the dispersion process and the drying mechanism on the printing press. For example, an ink which dries rapidly on a hot drum to give a high gloss finish may not achieve the same gloss level when dried by hot air. This usually occurs when the last solvent removed is a non-solvent for the resin system causing a precipitation of the binder, and flocculation of the ink. In the case of drum drying, all solvents are evaporated together and hence this situation does not arise, but this drying technique has limitations due to the effect on the substrate.

High gloss levels, while normally demanded for surface printing for visual appeal, are not always desirable from a print quality viewpoint and in certain cases, e.g. wallpaper printing, from an appearance aspect. In process printing a high gloss first colour (usually yellow) will give inferior trapping of the superimposed colours to a matt or even semi-gloss colour. Poor trapping will detract from the overall visual appearance of the print and hence the first colour down must be carefully formulated to give a balance in between the gloss of the solid areas of this shade and the print quality of superimposed areas of colour.

Wallpaper printing will be discussed in greater detail later in this chapter but matt inks enhance the appearance of the print, especially after embossing since they reduce surface reflection effects. The use of silica matting agents to produce such finishes can give rise to packing in the etch if the raw materials and the dispersion process are not carefully controlled.

Viscosity

The importance of obtaining the correct printing viscosity to achieve uniform lay has already been discussed in the previous section. In

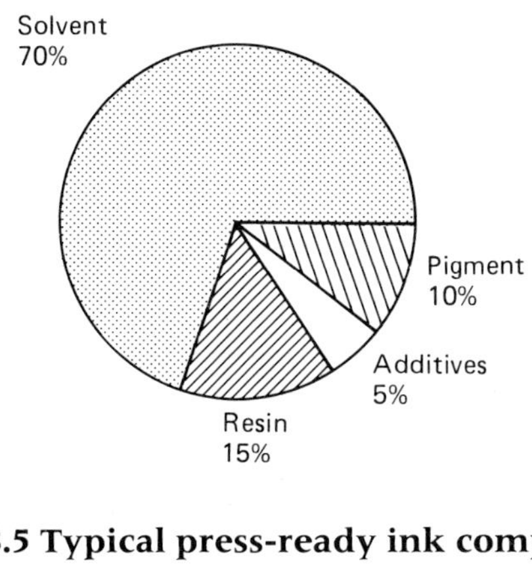

Fig. 8.5 Typical press-ready ink composition.

addition the solvent plays a role in acting as a lubricant between the cylinder and doctor blade thus reducing wear on these components. On high speed presses the cylinder is moving at speeds of up to 500 m/min past the doctor blade and the force which has to be applied to the blade to wipe the cylinder clean is highly dependent on the ink viscosity. In general terms the lower the ink viscosity the lower the force which has to be applied to the blade and consequently the wear rate is diminished

However, in practice there is a limit to lowering the ink viscosity since when a critical point is reached the wear rate actual begins to increase. The reason for this phenomenon lies in the belief that a microscopic layer of ink, which is not visible to the naked eye, actually passes beneath the doctor blade even in the non-etched areas of the cylinder, acting as a lubricant between them. When the viscosity of the ink approaches that of pure solvent there is insufficient binder present to act as lubricant and in such cases the formulation requires to be adjusted to include additives to take on this lubricating role.

Wear on doctor blades and cylinders will in turn result in print defects and while blade wear can be easily corrected by fitting a replacement, wear on a cylinder can be costly and may result not only in press downtime, but, in the extreme, rechroming.

Since the viscosity of the ink is a prime factor in the successful operation of the gravure press, an accurate, quick and reproducible method of measurement is required. Traditionally, printers have used flow cups for viscosity measurement at the press side (Fig. 8.6). While these cups are fast and reliable for liquids with Newtonian flow properties they give false results with thixotropic liquids. Pigmented inks exhibit complex rheological behaviour and print defects due to overreduction of the thixotropic product are commonplace.

Small differences in the viscosity of the ink can produce major variations in print quality and therefore the accuracy of the flow cup must be considered. Table 8.1 shows the inaccuracy values obtained for a Newtonian fluid with various cups, for both a single operator and multioperator repeat measurements. Pigmented inks are likely to have errors greater than these values.

As an alternative to efflux viscometers, Brookfield or Rotothinner viscometers can be used for all ink types since these pieces of apparatus record a viscosity at a particular set rate of shear.

Once the viscosity of the ink in the duct has been set it must be constant during the print run. This can be achieved by regular manual readings

Table 8.1 Precision of flow cups (NPIRI)

Cup	Test method	Single operator inaccuracy (%)	Multi Operator inaccuracy (%)
FORD	ASTM D1200	8	20
ZAHN	ASTM D4212	11	33
SHELL	ASTM D4212	9	18
ISO	ISO 2431	5	10

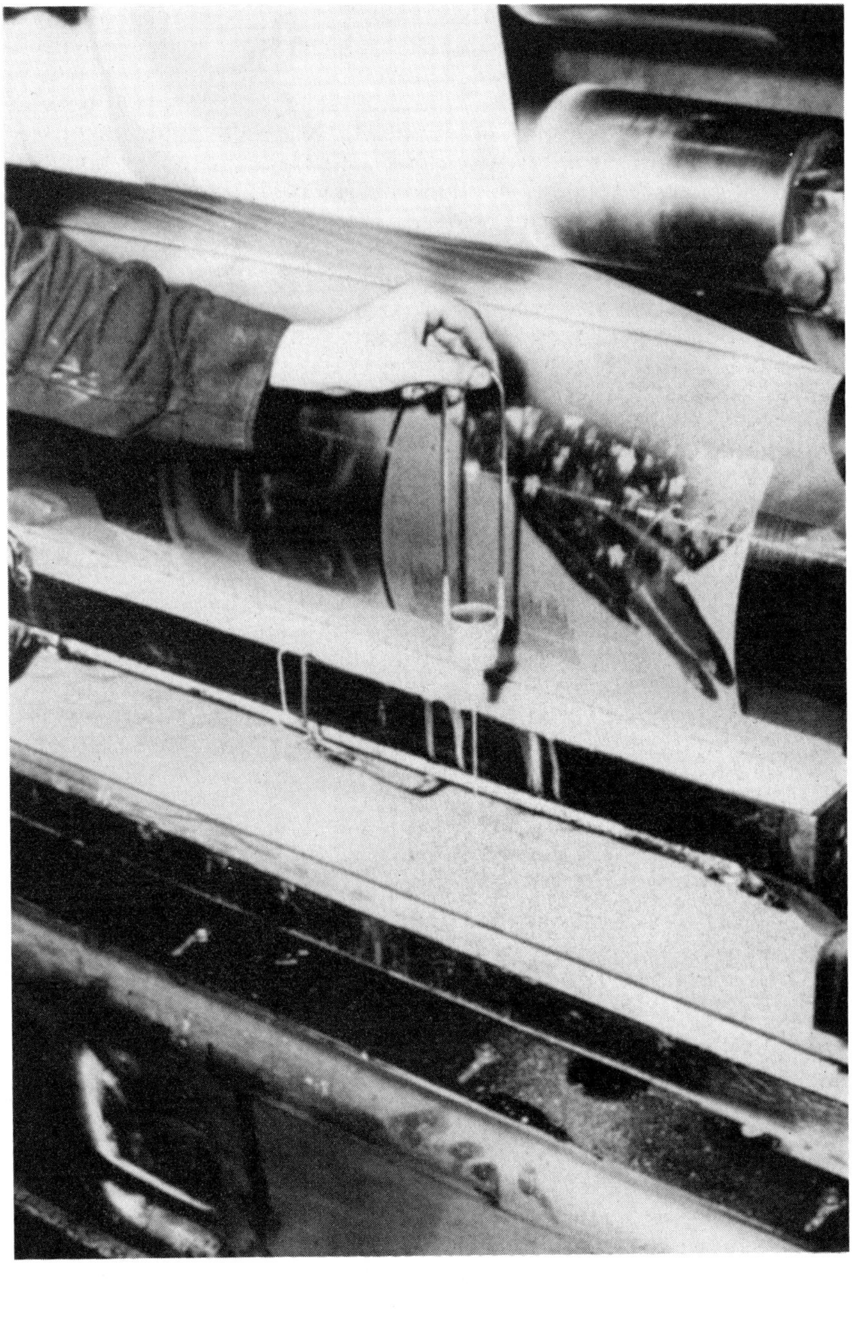

Fig. 8.6 A gravure unit with the duct open to allow a check on viscosity.

and corrections or by using an automatic viscosity controller (Fig. 8.7). Such equipment allows for solvent from a header tank to be metered into the duct ink to replace evaporated solvent loss. The equipment will also normally compensate for temperature differences, fluctuations in which will cause viscosity variations. Frequently the viscosity of the duct ink will drop during the print run, as the localized air temperature rises. It must be emphasized then that in any set of measurements the temperature should not only be kept constant but recorded along with the viscosity reading.

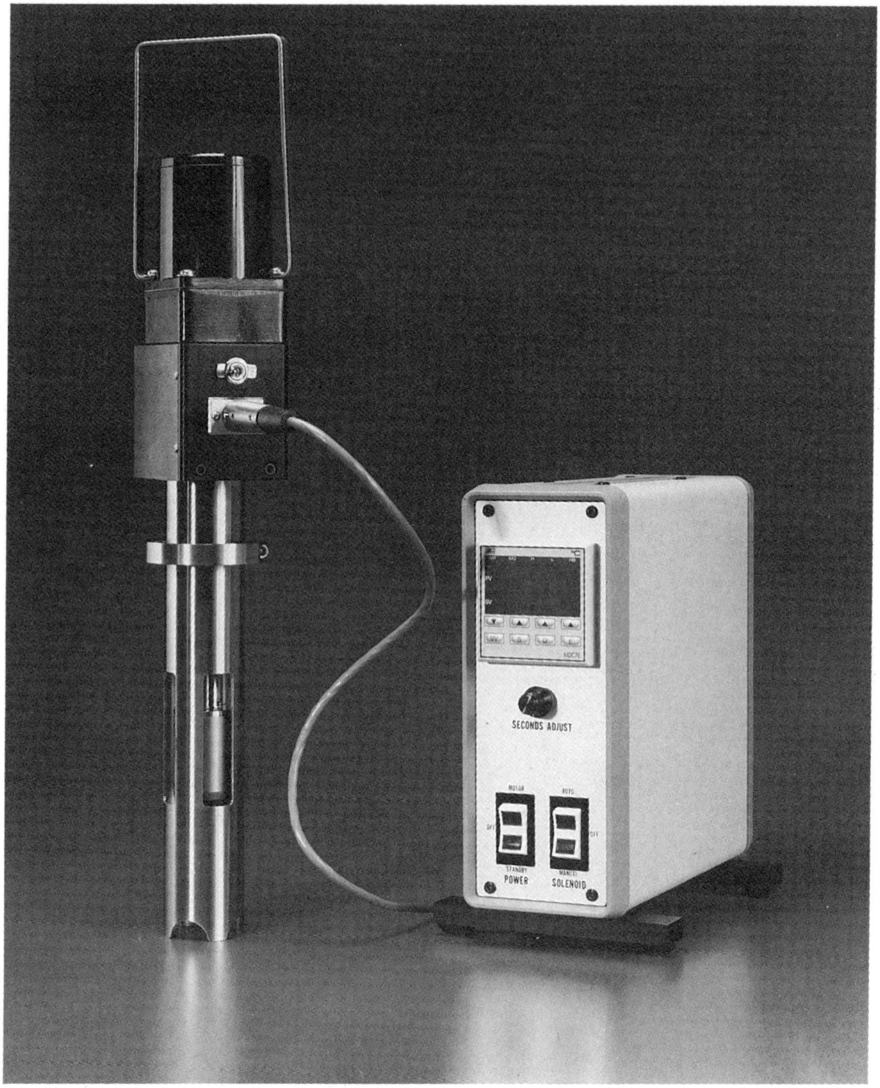

Fig. 8.7 An automatic viscosity controller. (Courtesy of Fullbrook Systems Ltd)

Gravure printing inks are normally supplied at viscosities higher than those used for printing. This then enables the printer to adjust the viscosity and solvent balance to suit his press and application, e.g. the addition of retarder may be necessary for slow press speeds. Additionally, the low viscosities required for gravure printing can give rise to pigment settling on storage and therefore higher viscosity supply will minimize this property.

Drying speed

The drying speed of an ink is of extreme importance. An ink which is too slow drying may give:

- trapping and pick off problems where two inks are superimposed on one another;
- marking on the turner bars;
- retained solvent in the final print;
- set-off or block in the reel or stack.

In the alternative situation where the ink is too fast drying, printability problems can occur due to screening (especially in the fine tones) and perhaps surprisingly it may lead to retained solvent. This is because certain resin systems tend to skin rapidly and if this occurs prior to the solvent being released then the excess solvent becomes trapped within the ink film.

The drying speed of an ink is dependent on the choice of solvents, the resin system and the substrate. The evaporation rate of the solvent is an obvious factor, but equally important is the solubility of the resin in the solvent and the ease with which the binder will release this solvent. A resin which is soluble in a mixture of solvents but not in the individual solvents themselves, will release the solvents on the point of drying, faster than resin which is soluble in one solvent alone. The porosity of the substrate and the affinity of any substrate coating for the ink solvents will also affect the drying speed of the ink system.

Measurement of the drying speed is difficult and can only be truly carried out on the actual production press. The inkmaker will, therefore, normally resort to the Hegmann gauge to compare the drying rate of a test ink against that of a known standard. This test, which is more fully described in Chapter 15, has the disadvantage of excluding the role of the substrate and, to overcome this problem, a comparison with a known ink can be made by using drawdowns at a number of film thicknesses on the actual substrate to be printed. These are prepared using metal rods wound with steel wire of various diameters.

Tack

The gravure printing process requires that wet ink is transferred from the gravure cells to a dry surface, i.e. wet-on-dry printing. This is unlike the lithographic process described in Chapter 6 where the process requires wet ink transferring onto a non-dry surface, i.e. wet-on-wet printing. The term 'tack' is, however, used for both processes and an understanding of the distinction must be made.

The correct adjustment of tack or 'tack grading' is essential for lithographic inks so as to allow one ink to trap over another wet ink. In gravure printing the tack of the wet ink is of little importance to the printing process but the residual tack of the dried ink film may have major consequences on the finished properties of the printed material, e.g. ink pick-off and/or set-off within the reel.

In gravure printing the ink formulator is solely concerned with the tack of the dried ink, especially in areas of two or more colour superimposition. Factors influencing tack include:

- choice of resin
- retained solvent
- heat retention.

Selection of the correct resin is important and this will depend not only on the stickiness or dryness of the resin at room temperature, but also on its solvent release characteristics and on its actual softening point, since retained heat in a web or in a stack may contribute to an apparent abnormal situation. This situation may be particularly prevalent in high-speed publication gravure printing where a high solids, relatively high tack resin system is used for maximum gloss. Build-up of tack can then occur, especially in areas of superimposition. Such a problem can be overcome by inclusion of a low tack binder, e.g. ethyl cellulose (EC) or ethyl hydroxyethyl cellulose (EHEC). Alternatively, a pigmented extender medium can be incorporated, but this will normally detract seriously from the gloss level of the print.

8.3 FORMULATING PRINCIPLES

The composition of gravure inks varies widely dependent on the choice of substrate, press parameters and end use of the finished print. A typical ink will, however, contain the following ingredients:

Pigment (or dyestuff)	4–12%
Extender pigment	0–8%
Resin	10–30%
Solvents	40–60%
Plasticizer/wax/additives	2–10%

Pigments

The initial factor in pigment selection is that it should be chemically suitable for the end-use specification, e.g. an acidic pigment should not be adopted for a design requiring alkali resistance. Secondly, the pigment should disperse readily in the selected vehicle system giving an ink with near Newtonian flow properties, which when reduced to press viscosity will give good gravure printability. Thirdly, the pigment must exhibit good dispersion stability, both as the ink is supplied and also at press viscosity, since press returns may be stored for lengthy periods prior to reuse.

All pigments are treated by the pigment manufacturer to render them more easily dispersible in particular resin systems. Hence, even within the same chemical class (as defined by its *Colour Index* reference) the ink formulator will require several pigments to obtain inks with adequate dispersion and flow properties in the wide variety of resin systems at his disposal. For example it is rare for the same phthalocyanine blue pigment to be used in publication gravure ink as well as a nitrocellulose packaging ink.

The gravure process will frequently require the same ink to be printed as a solid and tonal colour and it is therefore imperative that where pigment blends are adopted they should be of a similar particle size. This prevents the problem of the finer particle size pigment being preferentially printed out in the shallow tones with a consequent variation in hue from the same ink in solid areas.

Below is an abridged list of pigments which are particularly suitable for gravure inks. Their properties are discussed fully in Chapter 4 and only points pertinent to the gravure process will be highlighted.

Yellows

Monoarylide (PY 3, PY 74, PY 98) and diarylide (PY 12, PY 13, PY 14, PY 17, PY 83) yellows are almost exclusively used; although caution must be exercised with the monoarylide grades due to their solubility in toluene, poor fat and wax resistance and poor heat resistance particularly at tint strength.

Oranges

These fall into similar chemical classes to yellows and hence the same comments apply. In general oranges of *Colour Index* reference Pigment Orange 5, 13, 16, 34 are used.

Reds

Lake Red C (PR 53:1) Permanent Red 2B (PR 48) and the Lithol Red 4B (PR 57) toners find wide application due to their high tinctorial strength at economic cost. Care should be taken with the 4B toners if a water-based overprint varnish is to be applied, due to their partial solubility in water. With the exception of the manganese 2B toner (PR 48.4), all the above pigments have poor to moderate lightfastness and if an improvement in this property is required then the para and toluidine reds (PR 2, PR 3) can be utilized. These pigments, however, are highly variable in their inkmaking and chemical resistance properties and hence the permanent scarlet (PR 166), permanent rubine (PR 184) and permanent carmine (PR 185) grades are sometimes adopted. These grades bear a relatively higher price.

Finally, for brightness and cleanliness of shade PMTA grades also find specific applications.

Blues

Phthalocyanine (PBL 15), bronze (PBL 27) and PMTA (PBL 62) blues are the principal grades used for gravure printing. Care should be exercised with the bronze blue as it is sometimes difficult to disperse and can give rise to streaking and/or scumming if a temporary lift of the doctor blade is caused by oversized particles.

Copper-free phthalocyanine blue (PBL 16) is sometimes used in applications where there is a possibility of trace copper turning the food product rancid, e.g. some ice-cream applications.

Violets

PMTA (PV 2, PV 3) and indanthrene (PV 23, PV 37) violets are almost exclusively used, although the latter gives inks with poor flow characteristics and its use is avoided unless high fastness properties are required.

Greens

Phthalocyanine (PG 7) and PMTA (PG 2) greens are the most common green pigments used. Care should be taken with phthalocyanine greens, since despite improvements to their non-abrasive characteristics they can cause excessive wear on chrome cylinders on long print runs.

Whites

Although both rutile and anatase grades of titanium dioxide (PW 6) can be used, the rutile is generally adopted as it has greater opacity. Grades produced by the sulphate process are preferred as the chloride process grades tend to give greater cylinder wear problems. For certain applications, even the sulphate grade rutile pigments can cause excessive wear and then either the anatase grade or preferably zinc sulphide is utilized with a decrease in opacity. When using zinc sulphide, care must be taken to ensure that the binder is non-acidic or hydrogen sulphide gas may be generated.

Extenders

Alumina (PW 24), calcium carbonate (PW 18) and china clay (PW 19) are the most commonly adopted extenders. Their addition to the ink tends to aid printability and improve smoothness of lay; however, they do have a detrimental effect on gloss. Since extender pigment particles are not readily coated by the resin in the dispersion process, they also tend to settle easily and any settlement is generally very difficult to restir into the ink system. The choice of type of extender depends on the binder system but calcium carbonate must not be used in the presence of an acidic binder or decomposition will occur.

Blacks

Furnace blacks are the most common though on occasions the more expensive channel blacks are used for their better flow and gloss characteristics.

Metallics
Bronze and aluminium powders are used but the vehicle system must be chosen carefully to provide a stable system.

Pigments	*Colour index*
Monoazo Yellows	PY 3, 74, 98
Diaryl Yellow	PY 12, 13, 14, 17, 83
Pyrazolone Orange	PO 13, 34
Monoazo Orange	PO 5
Monoazo Reds	PR 2, 3, 166, 184, 185
Monoazo Reds – Laked	PR 48, 53, 57
Phthalocyanine Blue (β form)	PBl 15, 16
Bronze Blue	PBl 27
Phthalocyanine Green	PG 7
PMTA pink, green, blue, violet	PR 81, PG 2, PBI 2, PV 2, 3
Permanent violet	PV 23, 37
Carbon black (furnace channel)	PBk 7
Whites	PW 6, 7
Metallics	PM 1, PM 2
Extenders	
Alumina	PW 24
Calcium carbonate	PW 18
China clay	PW 19

Dyestuffs

In packaging inks solvent dyes are sometimes used in preference to pigments for optimum transparency on foil and metallized substrates. Their solubility depends to an extent on the chemical structure but they all tend to prefer polar solvents and in particular the lower alcohols. In contrast to the basic dyestuffs used in flexographic printing, most of these dyes exhibit excellent lightfastness.

Dyestuffs are also available which are soluble in toluene and these find some application in toning publication gravure inks though their use is limited by their poor lightfastness.

Resins

The general requirements of the resin will be apparent from the tasks it has to perform during the printing operation. The resin should be film-forming, hard, amorphous, low odour, usually colourless or pale, transparent, and soluble in the solvents to be used in the ink. It should also assist the wetting of the pigment to ensure good pigment dispersion and stability. The viscosity of the varnish prepared from resin in solvent should increase as the resin solids increase and a thin film of varnish when exposed to air and heat should release its solvent quickly to give a tack-free homogeneous resin film. Clearly, the resin must impart adhesion of the ink to the substrate to be printed. On paper, resins can be chosen to give penetration into the substrate and thus a low gloss level, or less penetration giving higher gloss levels.

The melting ranges of resins vary considerably and in many instances softening point is less critical than adhesion, rub resistance or lightfastness. Resins should not interact with the other constituents of the ink otherwise gelation usually occurs. Thus, basic pigments in the presence of an acidic resin will more often than not give problems. Acidic resins do help to prevent pigment flocculation in low viscosity inks at low resin solids. Under certain conditions, the resin can react with another specific constituent of the ink in a controlled manner. These are cross-linkable resin systems, which may also be catalysed, and will be discussed later.

Resins are complex organic compounds of high molecular weight. Originally, the choice of resin was restricted to a limited number of processed natural products, some of which are still used, e.g. shellac, gilsonite and modified rosin derivatives. Synthetic resins have much more to offer the ink technician because they can be tailor-made to suit individual requirement. In general, they are higher melting and lower in odour and colour than the natural resins.

The major classes of resin are examined below in relation to the gravure print process but for more general data the reader should consult Chapter 4.

Cellulose resins

Undoubtedly the most important resin in this class is nitrocellulose because of its excellent inkmaking characteristics, wide compatibility with modifying resins and additives, excellent solvent release, low odour, good colour retention and heat resistance. Nitrocellulose is available not only in a range of viscosity grades but also with varying nitrogen content. This enables the solubility characteristics to be varied from alcohol–ester to alcohol–toluene blends or even to straight ester solvents. Compatibility with other resins is also affected by variation in the nitrogen level.

Nitrocellulose therefore finds wide application in the gravure packaging field and has only a few limitations which the experienced inkmaker cannot overcome. For example, in multicolour gravure printing, trapping of one nitrocellulose ink over another can be problematical and in extreme cases the use of an alternative resin system for one or two of the colours may be necessary. Another drawback is in biscuit roll-wrap applications where a water-based PVdC overlacquer is applied for maximum seal integrity at low pressures. Inks based on nitrocellulose cannot be utilized because of poor interlayer adhesion between the ink and PVdC overcoat. A further limitation is in metallic inks where the use of bronze and to a lesser extent aluminium pigments, can cause decomposition of the resin, resulting in the liberation of nitrogen dioxide gas.

Ethyl cellulose, in contrast, is more expensive and has relatively poor pigment-wetting characteristics and a tighter compatibility spectrum than nitrocellulose. It gives very tough, flexible ink films and will dissolve in both toluene and alcohols depending on the grade used.

Toluene gravure inks based on ethyl cellulose are noted for their fine halftone printability and often there is an advantage in using at least one ink based on this resin to improve the tonal reproduction on films.

Both ethyl and ethyl-hydroxyethyl cellulose (EHEC) are useful resins for mixing with other resinous systems in order to remove the tackiness which is sometimes present during the drying of the ink film. This is particularly helpful on multicolour superimposition when picking can be a problem. In publication gravure inks EHEC is often used because of its solubility in aliphatic solvents.

Cellulose acetate propionate has found increasing use in the packaging field especially in conjunction with acrylic resins for printing on PVdC and acrylic-coated films where excellent low temperature adhesion is given. This resin blend can also be successfully used for PVdC overcoating where nitrocellulose cannot be utilized. Two grades of resin are currently available. The first, half-second propionate (HSP), has excellent heat resistance but limited application in colours due to its poor wetting properties with organic pigments; the second, alcohol soluble propionate (ASP), contains a higher hydroxyl content giving improved solubility in ethanol–ester (or even ethanol–water) blends. This latter grade shows superior wetting characteristics to the half second propionate but requires greater selectivity of pigments in comparison to nitrocellulose.

Polyamides

The use of polyamide resins for printing on polyethylene is well established. Historically the cosolvent grades, which are soluble in blends of alcohols and hydrocarbons, were used due to their excellent pigment-wetting properties, high gloss, rapid solvent release and good deep-freeze resistance. The trend towards hydrocarbon-free inks for food wrappers has resulted in an increased use of the alcohol soluble grades. The slower solvent release of these resins means that for non-blocking properties to be achieved it is necessary to use the alcohol grades in conjunction with nitrocellulose.

The emergence of cold-seal packaging has given a new lease of life to polyamide resins. Since the cold-seal adhesive has to be applied in line with the printing stations, the ink–lacquer system is in direct contact with the adhesive during storage of the printed reel. Polyamide resins have been found to be practically unique in providing release to the cold seal adhesive while not impairing (poisoning) the strength of the bond formed when the adhesive is sealed to itself during the packaging operation.

Polyamides also find value in lacquers because of their high gloss and resistance to products such as soap and margarine. Typical end uses are on detergent cartons, confectionery and margarine wrappers.

Further applications are limited because of the low softening point of the resins and unsuitability for use in inks for adhesive lamination.

Acrylics

There exists a wide range of acrylic resins available to the ink formulator. The monomers used in the resin polymerization, acrylic and methacrylic acids and esters, impart widely differing resin properties. However, apart from areas already discussed in blends with cellulosic resins, acrylics have little scope because of their poor pigment wetting properties and low softening points.

This latter property can, however, be used to advantage in formulating gravure heat-seal lacquers, e.g. foil to polystyrene. Acrylic resins also have some applications as vehicles for metallic inks.

Solutions of carboxylated acrylic resins in aqueous alkaline media are the basis of water-based inks and lacquers and these will be discussed fully later in this chapter.

Vinyls

There are two types of resins that are grouped under the heading of vinyls, namely copolymers of vinyl chloride and vinyl acetate, and polymers of vinyl acetals. The vinyl chloride copolymers generally give inks with poor flow, gloss, heat resistance and solvent release. Their applications have therefore been limited to printing vinyl wallpapers, polyester films for lamination, vinyl films or vinyl washed foil. Pigment selection is highly important to give inks with reasonable flow and good gravure print characteristics.

Inks with good resistance properties, e.g. perfume, acid and alcohol, can be produced using selected resins. As with acrylic resins the low softening point of the vinyl allows some use in heat-seal lacquers. There is an increasing concern over the use of chlorinated resins (PVC, PVdC, chlorinated rubbers) due to the evolution of HCl and other hazardous gases during the combustion of the printed waste. However, to date these resins have unique resistance and adhesion properties. In recent years vinyl acetal polymers have found increasing use as resin binders. Polyvinylbutyral has been used for many years as a modifying resin for N/C foil inks, imparting increased flexibility to the film former. Its wider use was restricted by its low heat resistance and relatively high cost. However, it has now found application as a plasticizer-free film former in conjunction with suitable release lacquers for cold seal specifications on polypropylene films.

Hydrocarbon-soluble resins

The major application for such resins is in the publication gravure field. Calcium and zinc–calcium resinates are soluble in both aromatic and aliphatic hydrocarbons giving high solid but low viscosity varnishes with excellent solvent release. In addition their low cost makes them an attractive proposition for publication gravure inks, but they do suffer from the drawback of forming powdery films which tend to give poor rub resistance.

Rosin-modified phenol formaldehyde resins have high gloss, good solvent release and better rub resistance in comparison to the resinates. However, they are more expensive than the resinates and their use is mainly confined to toluene-based inks because of their limited tolerance of aliphatic hydrocarbon solvents.

Hydrocarbon resins, while giving high solid low viscosity varnishes, are inferior to resinates for solvent release and are therefore only of limited value.

Gilsonite is a natural resin which is deep brown in colour and hence its use is confined to black/brown inks. Pigment-wetting properties are excellent and black/brown inks with excellent gloss and good solvent release can be obtained. Varnishes on storage do have a tendency to gel and deposit insoluble matter unless a small quantity of alcohol is included in the formulation. Gilsonite is incompatible with cellulose resins and its use should be avoided if these resins are present in the ink or reducing medium.

In packaging inks, chlorinated rubber is used for high quality gloss inks on paper. As well as being soluble in aromatic hydrocarbons chlorinated rubber is soluble in esters, but totally insoluble in alcohols. The system has excellent chemical and abrasion resistance and its non-curl properties on paper make it an ideal basis for label work.

Modifying resins

A wide variety of chemical types of resin are used to modify the main vehicle system usually as a means to increase gloss or reduce cost. Typical of resins in this category are rosin-modified maleics which can be produced with or without acidic functionality to vary the solubility into polar or non-polar solvents respectively. Typical use of such resins are in combination with nitrocellulose or chlorinated rubber (low acid value grades only), and since flexibility on film and foil is limited the main application is on paper and board. The resins have a typical rosin odour and care should be exercised in food packaging applications.

Ketone, aldehyde and styrene–allyl alcohol resins are all used to promote gloss improvement of nitrocellulose inks. Again all suffer limited flexibility on film and especially on foil and will detract from the heat resistance and solvent release of a nitrocellulose ink.

Shellac, despite its wide application in flexographic inks, is only rarely used in gravure inks because of its poor resolubility characteristics. This leads to resin packing in the gravure cell and a progressive loss in colour strength of the print.

Solvents

The solvents in gravure inks are temporary ingredients which are present purely as a means of applying the vehicle solids to the substrate by way of the printing unit. In theory the solvent is then eliminated, mainly by evaporation and/or absorption, and takes no further part in the properties of the printed film. However, in practice this can be an over-simplification and the consequences of trapped solvent will be discussed in more detail later.

The choice of solvents in a gravure ink is governed by:

- the resin system to be employed;
- the press speed;
- direct or offset gravure process;
- nature of the design;
- the substrate;
- end-use properties of the print;
- health and safety consideration.

The chemical nature of the resin system will often severely restrict the choice of solvents available to the inkmaker, and also with most resins a blend of solvents will give lower viscosity than a single solvent. Selection of solvent can be predicted by consideration of the solubility parameter, and the possible hydrogen bonding associated with a solution of the resin in the solvent. While this may appear complex, details of solubility parameters of individual resins can normally be obtained from the resin supplier and the ideal solvent balance can easily be predicted without the need for theoretical knowledge. Information obtained by this technique is particularly useful since it can predict where a non-solvent for a resin can be used in conjunction with a true solvent to give a lower viscosity solution.

Solvent selection will also be dependent on the pigment type within the ink. Where the selection is incorrect it will produce inks with poor flow characteristics.

Press speeds may vary from about 10 m/min on an in-line extruder and printer to several hundred metres per minute on a publication gravure press. The ink must remain open during the printing stage while the ink flows from the gravure cells onto the substrate, but then the solvent must be rapidly removed within the drying cycle. Hence, the evaporation rate of the solvent must be geared to the length of time in between excess ink being removed by the doctor blade and the removal point of the solvent. This time span is clearly governed by the press speed and hence solvent selection will also be determined by press speed. The time span will also be related to whether a direct or offset gravure process is utilized.

The nature of the design to be printed also plays a major role in the speed of evaporation required in the solvent system. Solvent loss from ink in a shallow cell will be liable to cause drying-in problems more quickly than the same solvent loss from ink in a deep cell, simply due to the quantity of ink and therefore solvent present in the particular cell. Hence, the amount of tone work on a design will govern the choice of solvents required in the individual inks.

Many print formats will involve inks with a high coverage and others with very low usage. The low usage inks may require little or no replenishing even on long print runs and any loss of solvent by evaporation in the duct must be replaced in order to retain a constant viscosity. While this is a simple task for single-solvent inks it is a rare situation and in the more common situation of inks containing blends of solvents, solvent replenishment must be made after considering the evaporation rates of the individual solvents or printing problems may occur, e.g. resin throw-out, causing blocking of the cells.

A practical example of this problem would be an ink based on 9:1 mixture of toluene and ethanol. As the ink loses solvent by evaporation the solvent vapour consists of a 54:46 ratio of toluene to ethanol. Hence to maintain the solvent balance of the ink at the 9:1 mixture, the replenishing blend must be 54:46 rather than the original ink solvent balance.

Substrate can have two main effects on the selection of the solvent system. First, with permeable substrates the rate of absorption of individual solvents can affect the flow-out properties of the ink in between the point of impression and the ink losing its solvent and this can give rise to a variation in the smoothness of lay. Secondly, some solvents can attack or be absorbed into certain films and substrate coatings. Such solvent absorption can have an influence on the anchorage of the coating or even on the properties of the substrate itself.

Food packaging in particular must not be tainted by the effects of residual solvent. Therefore, special attention must be given to the release characteristics of the solvents employed from the chosen resin system so as to give the minimum retained solvent levels in the finished print.

The new EC labelling regulations introduced at the beginning of 1986 specify the hazard code label which must be attached to all ink systems. While this is not restricted to just the ink solvents, these in the main are the constituents which dictate the label required on a particular gravure ink.

In recent years changes have been made to the Occupational Exposure Limits (OEL) of certain solvents. These changes have reduced the maximum concentration of the solvent vapour allowable in the working atmosphere. Such changes have resulted in the virtual elimination of some traditionally used solvents, e.g. 2-ethoxyethanol and its corresponding acetate. They have been replaced with glycol ethers and esters whose OEL values are higher, and which are usually those based on propylene glycol.

The most common solvents found in gravure packaging ink formulations for paper, film and foil are ethyl, isopropyl and *n*-propyl alcohols, acetone, ethyl, isopropyl and *n*-propyl acetates, methoxy and ethoxy propanols, aliphatic hydrocarbons and water. Publication inks are almost totally restricted to toluene and aliphatic hydrocarbon solvents. Some properties associated with these solvents, along with some other usable solvents are listed in Table 8.2.

Ink additives

A wide variety of different chemical compounds are incorporated into gravure inks for a host of reasons. Generally, levels of addition are small, usually below 5%, but even at 0.1% the effect on printability must be carefully examined.

Slip agents must be incorporated with great care as silicone materials usually give lay defects and micronized waxes if not finely and uniformly dispersed can give packing in the etch or even streaking problems. Over print properties can also be affected by the choice of wax or slip aid. Care must also be exercised in choosing slip aids which will not migrate within the printed structure.

Dispersion and surface-active agents can be included for easier pigment wetting and also to prevent long-term pigment (or extender) settlement by imparting a weak gel structure to the ink.

Other additives such as epoxides, amines, phenolics, organic acids are incorporated as stabilizers, anti-oxidants, etc. to prevent ongoing side reactions which may occur on storage of the ink or finished print.

Table 8.2 Properties of gravure solvents

Solvent	Density at 4°C	Boiling point (°C)	Evap. rate (Bu Ac = 10)	Flash point (°C)	Hazard classification	General nature of risk	Risk phase Nos. R††	Safety phase Nos. S#	Long term exposure limit 8 hour TWA ref. period (ppm)	Short term exposure period 10 minute ref. period (ppm)
Acetone	0.79	56	115	–15	–	Highly flammable	11	9,16 23,33	750	1500
Ethyl acetate	0.90	77	62	–5	–	Highly flammable	11	16,23 29,33	400	–
Methyl ethyl ketone	0.80	80	57	–7	–	Highly flammable	11	9,16 23,33	200	300
Hydrocarbon SBP 2	0.71	70–95	54	–40	Class IIa	Highly flammable, harmful	11 20,21 40	9,16 29,33	600*	750*
Isopropyl acetate	0.87	88	50	4	–	Highly flammable	11	16,23 29,33	–	200
Hydrocarbon SBP5	0.73	90–105	35	–22	–	Highly flammable	11	9,16 29,33	600*	750*
Ethanol (74 op spirits)	0.79	75–81	33	13.9	–	Highly flammable	11	7,16	1000	–
n-propyl acetate	0.89	102	27.6	14	–	Highly flammable	11	16,23 29,33	200	250
Isopropanol	0.80	81	23	12	–	Highly flammable	11	7,16	400	500
Hydrocarbon SBP 3	0.74	110–120	22	–9.5	–	Highly flammable	11	9,16 29,33	600	750
Toluene	0.87	111	21	4	Class IIc	Highly flammable	11,20	16,29 33	50	150
Water	1.00	100	18	–	–	–	–	–	–	–

Methyl iso-butyl ketone	0.80	118	16	14	–	Highly flammable	11	9,16 23,33	50	75
n-propanol	0.80	98	11	22	–	Highly flammable	11	7,16	200	500
Butyl acetate	0.88	126	10	21	–	Highly flammable	11	16,23 29,33	150	200
Propylene glycol methyl ether	0.92	121	8.3	36	–	Flammable	10	24	100	300
Xylene	0.86	137	6	25	Class IIc	Harmful, flammable	10,20	24,25	100	150
Propylene glycol ethyl ether	0.90	132	4.9	49	–	Flammable	10	–	–	–
Propylene glycol methyl ether acetate	0.96	140–50	3.4	47	–	Flammable	10	–	–	–
Cyclohexanone	0.95	157	2.3	44	Class IIc	Harmful, flammable	10,20	25	25	100
Methyl cyclohex-anone	0.92	175	1.8	48	Class IIc	Harmful, flammable	10,20	25	50	75
Diacetone alcohol	0.94	166	1.4	52	–	Irritant, flammable	11,36	7,16 24,25	50	75
Dipropylene glycol methyl ether	0.95	188	0.2	79	–	–	–	–	–	–

* Values based on pentane isomers

Risk phrases: †R 10 flammable; R 11 highly flammable; R 20 harmful by inhalation; R 21 harmful in contact with the skin; R 22 harmful if swallowed; R 36 irritating to eyes; R 40 possible risk of irreversible effects.

Safety phrases: #S 2 keep out of reach of children; S 7 keep container tightly closed; S 9 keep container in a well ventilated place; S 16 keep away from sources of ignition; S 23 do not breathe gas/vapour/fumes/spray; S 24 avoid contact with skin; S 25 avoid contact with eyes; S 29 do not empty into drains; S 33 take precautionary measures against static discharges.

Anti-pinhole agents are used in inks for printing on the nitrocellulose-wax coated cellulose films (e.g. MS film) in order to lower the surface tension of the ink and allow a smooth pinhole-free coating to be obtained. Other films such as the corona discharge-treated polyolefines will often require additives like titanium acetyl acetonate to ensure good adhesion of a nitrocellulose ink system.

A comprehensive list of additives cannot be attempted within the scope of this section but further details of specific materials will be given in the section on ink formulations.

8.4 INKS AND VARNISHES FOR SPECIFIC END-USE APPLICATIONS

Before discussing individual ink systems it is important to understand the basic ink requirements for the different substrate types, the end-use properties, and details of certain after processing techniques such as lamination, PVdC coating and cold seal application.

Gravure printing of paper and board is relatively straightforward using in the main opaque coloured inks directly on the substrate. In contrast inks for foil and metallized substrates utilize transparent coloured inks to emphasize the metallic nature, but if a white is used then it should be formulated with maximum opacity to prevent the greyness of the foil showing through. In cases where a gloss opaque white is required on foil a primer or wash coat under the white ink may be necessary.

As a means of enhancing the brightness and strength of the colours, multicolour designs on film are usually printed with a backing white. In surface printing the white is printed first and provides the vital property of base adhesion to the film. The colours may solely be printed over white, but occasionally an overdraw can mean that adhesion of the colours to base film is also required. In reverse printing the colours are printed first, followed by a backing white and hence all inks must have adhesion to the base film. The gloss level of inks for reverse printing is of no importance and in some instances is detrimental, since the design will be viewed through film.

The end-use property requirements of gravure inks will vary widely but all inks must show good basic adhesion to the substrate, good flexibility or resistance to cracking and good resistance to scuffing.

Inks for packaging applications will have to withstand tests designed to reproduce packaging and storage of the finished product. Tests for packing line simulation will involve resting for slip and in certain cases heat seal resistance. Increases in packing line speeds mean that slip control is vital and on heat sealing, not only must the ink not be plucked from the surface, but it must also rapidly release from the sealing jaws. To this end inks now contain additives specifically aimed at providing hot jaw release.

Product storage tests will be varied ranging from odour and taint tests to resistance tests for soap, alkali, oils, fats, waxes, light and deep-freeze to name but a few. For details on all the above test methods consult Chapter 15

In addition, the Society of British Printing Ink Manufacturers has published lists of raw materials which they recommend should be excluded from use on immediate food wrappers. This list primarily covers pigment, plasticizer and solvent selection. Furthermore, print for a specific end use, e.g. bundling tissue for tobacco wraps, may be subject to even more stringent safety regulations.

End-property testing on inks for non-packaging applications can also be very stringent. For example inks for wallpaper and melamine laminates will have to withstand a whole variety of tests to simulate marking, staining or rub by a wide range of different chemicals.

Lamination

Lamination is a means of bringing two or more substrates together by either the use of adhesive, wax, heat or an extrusion process. In adhesive lamination (Fig. 8.8) the design is normally reverse-printed on film and the adhesive is coated over the ink system. Any solvent is then removed prior to bringing the second web into contact, often by the use of a hot nip.

The adhesive can be solvent- or water-based or even a 100% solids systems which can be applied via a special applicator. Whilst the solvent-based still predominate, water-based adhesives are increasing in usage due to proposed solvent emission legislation. Adhesives fall into two main categories.

- Polyurethane adhesives which are available as single-pack moisture-curing types or two-pack reactive systems, the latter providing the best

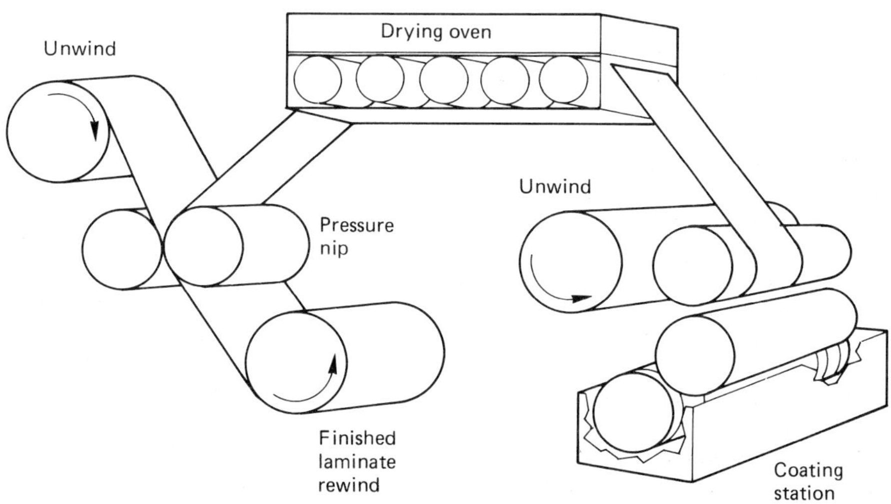

Fig. 8.8 The process of dry-bonding adhesive lamination.

results for high specification demand, e.g. boil-in-the-bag and retortable pouches. The chemical nature of these products means that only water-free ester and ketone solvents can be used as true solvents, though limited dilution can be made with aliphatic hydrocarbons.

- Polyether type adhesives cured with an epoxy resin which can be used in alcohol solutions have found increasing application since they are isocyanate free and give less rewetting when applied over most common laminating inks. This is solely due to the use of alcohol solvents as opposed to the more aggressive ester and ketone solvents in polyurethane adhesives, and is especially applicable to in-line print and lamination when the underlying inks have not had time to harden off before application of the adhesive.

Few guidelines can be given in formulating inks for adhesive lamination. Clearly the ink must show a good level of adhesion to the base film but even then this does not guarantee that good bond will be given on adhesive lamination. Polyamide resins should be avoided as poor bonds will result and PMTA pigments can, under certain circumstances, give a bleed in contact with the adhesive.

In extrusion lamination (Fig. 8.9) two substrate plies are bonded together by extruding a thin layer of polyethylene (or similar material) on to one surface and nipping in a second layer of substrate while the polyethylene (or extrudant) is still molten. The process can be applied to two identical or quite dissimilar substrates.

The best results are obtained when the resin system of the ink has an affinity for the extrudant. Hence, nitrocellulose inks modified with polyamide or shellac usually suffice for polyethylene extrusion laminations.

In general terms bond strengths are inferior to those in adhesive lamination, but can be further improved by application of a primer over the inks prior to extrusion. Volatile components such as solvents or water in the printed web will reduce the strength of the finished bond due to a bubbling effect on application of the hot extrudant.

Extrusion coating is a similar process to that of extrusion lamination

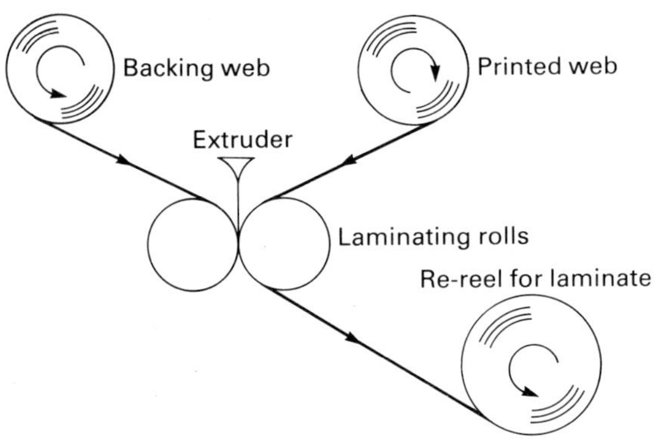

Fig. 8.9 The process of extrusion lamination.

but instead of a second web being brought into contact, the extrudant is simply passed over a chill roller.

Wax lamination is basically similar to extrusion lamination but in this case the extrudant is molten wax, usually paraffin wax. The vast majority of this work is in foil–paper laminates for the confectionery and snack food markets. While in adhesive and extrusion lamination, printing is carried out before lamination (to reduce wastage) the printing of foil–paper wax laminates is often done after the webs have been combined to prevent the problems of printing unsupported foil. However, to guard against poisoning of the foil surface by chemicals in the paper the foil surface must be printed or at least primed within hours of the lamination process taking place.

Heat lamination is now only rarely used and it utilizes a web printed with a thermoplastic heat-sealable ink or varnish which is brought into contact with a second web before passing through a hot nip under pressure. The process is shown in diagrammatic form in Fig. 8.10.

PVdC coating

Aqueous PVdC coatings are applied primarily to obtain better seal integrity of the package and also to promote gloss levels. The extra barrier properties given by the layer of PVdC are normally superfluous especially when a PVdC-coated film is used.

The coating can be applied in-line with the gravure printing operation. Drying of the PVdC coating is an important factor since if excess moisture is present in the final film then poor adhesion to the underlying inks results. Hot air driers are the most effective with the facility to increase the web temperature gradually to prevent surface skinning of the coating which can give rise to trapped water or blistering of the surface layer.

Nitrocellulose-based inks have been shown to be unsuitable for PVdC

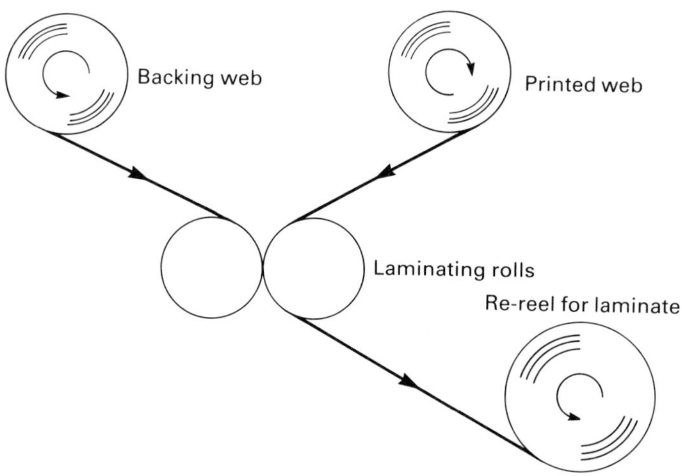

Fig. 8.10 The process of heat lamination.

coating giving poor interfilm adhesion on long-term storage of the final print. Pigment selection is also an important factor since interlayer adhesion can be affected (e.g. with bronze blue pigments) and those giving a water bleed (e.g. some 4B and 6B toners) will discolour the PVdC during the coating operation.

Cold-seal applications

Cold-seal packaging has increased rapidly during the 1980s, offering the advantages of higher line speeds, less wastage and no side effects on the end product due to the heat sealing process. In this process the design is surface printed on the substrate in the normal manner and then the web is reversed and a cold seal adhesive applied to the back of the substrate in pattern form. The adhesive must be printed in line for registration purposes, and is formulated to seal to itself at the packaging stage.

Cold seal adhesives are all based on natural rubber latex, modified with varying amounts of synthetic latices to adjust their properties to suit the end-use requirements. After application the adhesive is in direct contact with the inks within the reel and hence no interaction between the two must take place during the storage period. Typical problems which can occur are:

- poisoning of the adhesive by set-off of the inks on to the adhesive or vice-versa;
- migration of additives, e.g. waxes or slip agents from the ink or varnish into the adhesive again causing a poisoning effect;
- chemical interaction by components in the inks with those in the adhesive – for example metallic pigments will react with the ammonia in the lattices;
- bleed from pigments or dyes into the adhesive and hence risk of contamination of the product being packaged.

Publication inks

The gravure process was once a major source of publication print. However, in recent years it has been faced with the competition of the heat-set web-offset process (Chapter 6). The quality of the competing process is not as good but the economics of shorter runs, e.g. below 500 000 copies, make it very advantageous and consequently the gravure process has lost market share.

Publication gravure inks are formulated to achieve the lowest possible cost; the main demands are on their press performance. Once printed their prime function is to provide text and illustrations, merely requiring adhesion to the substrate and a moderate degree of scuff resistance and lightfastness.

Paper in publication printing

In publication printing the nature of the paper surface plays a critical role in the print quality achieved. The gravure printing process is dependent on direct contact between ink and paper at the point of impression.

Because the surface of a gravure cylinder is smooth, the ink cannot be easily forced into any indentations in the paper. In line printing where the engraving is relatively deep, lack of contact with very small areas of paper are not necessarily too critical, as there is sufficient ink in the adjacent cells to flow out into these areas. However, in the tonal areas of publication printing there is insufficient ink in the adjacent cells for this to happen. Therefore, if the paper has an uneven surface with a predominance of indentations then undesirable speckle occurs (Fig. 8.11).

In efforts to improve the quality of paper produced for gravure publication printing, paper manufacturers have resorted to a number of tests designed to indicate the degree of speckle that might be expected from a particular quality or batch of paper. For the test to be relevant all other possible variables have to be eliminated so that at the point of testing any change in the degree of speckle is due to the surface of the paper only.

This can be achieved by using a mechanically driven reel fed laboratory proofing press which must have a pneumatically operated impression roller, a reciprocating doctor blade, with a rewind sufficiently distanced from the point of impression to provide an adequate web carriage. Additionally, the machine should have an effective speed control and an inking system which will enable consistent ink viscosity control to be maintained during the tests.

The test inks should be of proven quality consistent with a typical publication ink formulation. Black ink is normally chosen as it provides the greatest contrast between ink and paper. The cylinder used for the test is engraved to a depth which represents the tonal values where speckle is the most critical.

A standard paper is selected which has a predetermined and acceptable

Fig. 8.11 The effect known as dot skip or speckle.

level of speckle. The standard paper and test paper are attached adjacent to each other on the web and run through the press at the same time. The machine is stopped before the papers reach the rewind reel. By visually counting the number of speckles on the test paper, over a given area of print, its suitability for gravure printing can be assessed against the number of speckles showing in the same area of the standard print. The test should be repeated a number of times, to assess different areas of the test paper, i.e. both sides, centre and edges. By logging the results, numerical data can be accumulated to give both development and batch records.

Gloss and absorbency

Given inks of acceptable gloss characteristics the gloss level of a print will depend on the quality of paper selected. Papers ranging from absorbent newsprint, highly calendered mechanical paper through to good quality blade coated paper are used in different areas of publication printing, with each type of paper having a different absorbency characteristic.

Only the first colour down, usually the yellow, prints directly on to the paper surface, subsequent colours print not only on the paper but also on a dry ink film of varying thickness and coverage.

Lower ink viscosities are required for absorbent papers to enable the interstices between paper fibre and filler to be penetrated, thereby covering the whole of the surface of the substrate before ink flow is prevented by solvent evaporation.

With highly absorbent papers, the penetration of the ink into the paper results in poor gloss and dullness of colour, particularly with the red and the blue. On calendered mechanical papers the gloss and colour brightness is greatly improved, as the paper fibres and fillers have been compressed to form a less absorbent surface on which the ink penetrates to a lesser degree.

Publication gravure printing with the highest gloss and colour brightness, though not necessarily speckle free, is achieved on blade-coated papers. This is due to the coating of titanium dioxide and method of application which produces a glossy white surface with a low degree of absorbency, thereby enhancing the gloss and brightness of the inks.

Papers that are excessively absorbent allow penetration of the ink vehicle which can cause undesirable 'strike through'. A similar problem occurs when the ink vehicle has a refractive index close to that of cellulose, which renders the print area more transparent.

Formulation requirements

As indicated earlier the greatest demands on publication inks are made during the actual printing operation. Prior to use, the ink should have viscosity stability during storage. Due to the large quantities involved, inks are often kept in tanks ready to be pumped to the machine room. On the press they should have good lubricity and wiping properties to enhance cylinder life. They should exhibit good pigment suspension at low viscosities and be tolerant to solvent shock. When printed there should be immediate and rapid solvent release, followed by the formation of a tack-free ink film, thus avoiding set-off on the turner bars. The ink should print with a sharp definition and provide a gloss level consistent with the type of substrate being painted.

A number of alternative resin systems can be used for publication printing. The majority of inks are based on zinc–calcium resinates with ethyl hydroxy ethyl cellulose (EHEC) used to improve hold out and reduce surface tack. Occasionally cyclized rubber is also used, detracting less from the gloss than EHEC. However, both these resins are expensive and can only be used to a very limited degree. To assist in reducing tack it is normal to use either china clay or precipitated calcium carbonate as extenders; care in selection is required to ensure that colour brightness and gloss levels are not adversely affected. Resinous plasticizers with extremely good wetting properties, can be employed to ensure pigment suspension at machine viscosities.

Other approaches to publication ink formulation have been made with the use of phenolic resin systems, which because of their more complex molecular structure are capable of exhibiting better hold out and gloss. However, their inferior solvent release and drying characteristics have meant that they must be modified to give good performance on the modern publication presses, where machine speeds of up to 500 m/min are common. Well dispersed polyethylene wax is usually used to achieve the necessary rub resistance.

Choice of solvent systems is extremely limited in publication printing, normally a combination of toluene and aliphatic hydrocarbons with a very small percentage of xylene. In some cases printing houses prefer 100% toluene, especially where solvent recovery is installed and the printing units can be effectively sealed to prevent solvent loss to the surrounding atmosphere. Where a solvent system with a low aromatic content is preferred, the overall solvent power can be enhanced by the introduction of a small percentage of *n*-propanol. This is particularly helpful when the formulation contains EHEC.

The choice of pigments for the four process colours normally employed in gravure publication printing is dependent on them fulfilling the pigment-orientated properties outlined above. The yellow pigment should have a degree of opacity especially when used on less than white absorbent papers. The pigment chosen should be a neutral yellow to ensure bright greens, but not too light or greenish to produce weak scarlets and reds. The rubine or magenta should be as bright as possible particularly in the tones, thereby enabling reasonable mauves and violets to be printed. The blue should be midshade and again as clean as possible. The choice of black depends principally on the preferred tonal hue; most publication houses choose a neutral or slightly blue black.

Over the years developments in pigment technology have taken place to a high degree. Pigments are coated with surface active agents, hyperdispersants, resins and other chemical compounds during their manufacture resulting in products which are ideally suited for the high speed production of publication inks.

Techniques of production using these newer pigments vary since some are fine powders which disperse easily by triturating with solvent, e.g. toluol, while granulated dust-free types require mixing at high pigment loadings to produce shear. They have strong deflocculating properties and the mixes produced then require only the addition of varnishes and additives to produce the finished ink. Typical formulations are given below.

Conventional:

Diarylide Yellow (CI Pigment Yellow 12)	6.0
Calcium carbonate	6.0
EHEC resin	2.0
Zinc–calcium resinate	37.0
Polyethylene wax	1.0
Resinous plasticizer	3.0
Aliphatic hydrocarbon solvent	20.0
Toluene	22.5
Xylene	2.5
	100.0

Low aromatic:

Rubine Red (CI Pigment Red 57)	7.5
Calcium carbonate	5.0
EHEC resin	2.0
Zinc–calcium resinate	37.0
Polyethylene wax	1.0
Resinous plasticizer	3.0
Aliphatic hydrocarbon solvent	39.0
n-propanol	2.0
Toluene	2.5
Xylene	1.0
	100.0

Inks for catalogue printing

The ink requirements for catalogue printing are somewhat different from those outlined above. The coated paper normally used is highly calendered with a rather more brittle surface than magazine papers. This means there is less ink penetration into the surface and a reduction in mechanical bonding.

In addition to high gloss, inks are required to have tough flexible films with good rub resistance and printability. As there is little assistance from the substrate, rapid solvent release and hard film forming properties are essential to ensure adequate drying. Because of the nature of the paper, press speeds are usually lower than those used for magazine printing.

Pigment selection for catalogue inks is to some extent dependent on the requirements of individual printers, but in general the pigments are similar to those used in other areas of publication printing. A suitable set of inks can be formulated using diarylide yellow, phthalocyanine blue, calcium 4B and a blue-toned carbon black pigment. As high gloss level is essential only low concentrations of extender are possible.

Unlike weekly magazines, catalogues have to withstand considerable handling over varying lengths of time. This sometimes calls for resins that have tougher film-forming properties than can normally be achieved with resinate/EHEC or resinate/phenolic combinations. A particularly suitable resin system is chlorinated rubber modified with a phenolic resin, which gives the toughness and flexibility required. Rub resistance can be achieved by using carefully selected well dispersed polyethylene waxes. Careful plasticization is required to ensure a correct balance between flexibility and tack.

The solvent system is usually toluene or possibly toluene/ester.

Typical formulation of a catalogue ink would be:

Phthalocyanine Blue (CI Pigment Blue 15:3)	8.0
Chlorinated rubber	15.0
Phenolic modified resin	10.0
Resinous plasticizer	5.0
Polyethylene wax	1.0
Toluene	51.0
Ethyl acetate	10.0
	100.0

Packaging inks for paper and board

Labels

The use of labels in packaging is widespread and varied, normally the preprinted label is attached to the product container to identify its contents. Unlike printed wrappers, labels do not have close contact with the product, which is normally packed in glass, plastic, metal or some other impervious material. This allows, within certain limitations, a fairly wide choice of solvent and resin combinations to be utilized for label-printing inks. When printed labels are produced for returnable bottles, the ink film should be penetrable during the washing process, which utilizes a dilute solution of caustic soda, thus allowing easy removal of the label.

Pigment selection will be dependent on colour requirements. A reasonable degree of lightfastness is obviously desirable and some labels may require product resistance. Where high levels of gloss are important those pigments specifically developed for high gloss in liquid inks should be utilized. Bead milling is the usual method of manufacture.

Advantage can be taken of the low viscosity grades of chlorinated rubber to help promote gloss.

Typical fast-drying formula for labels:

Rubine 4B (CI Pigment Red 57)	5.0
Lake Red C (CI Pigment Red 53)	7.0
Chlorinated rubber	12.5
Phenolic resin	20.0
Resinous plasticizer	4.0

Dioctyl phthalate	3.5
Polyethylene wax	1.0
Toluene	27.0
Ethyl acetate	10.0
Aliphatic hydrocarbon	10.0
	100.0

Some label printers have a preference for alcohol/ester-based inks choosing to avoid toluene for environmental reasons. This reduces the choice of resin systems considerably and normally means the use of nitrocellulose, maleic and ketone resins.

Non-curl properties and easy wash-off can be achieved by correct plasticization and the choice of a suitable maleic resin. The main problem is to formulate suitably glossy inks that are sufficiently hard drying, especially when the label is not varnished. The use of specially developed, easily dispersed pigments give reasonable results. But as these pigments are a limited range, it is often found to be expedient to use nitrocellulose chips as a means of equalling the gloss of the chlorinated rubber-based inks.

Typical formula of a chip-based non-curl label ink:

50% Rubine 4B nitrocellulose chip	10.0
50% Lake Red C nitrocellulose chip	14.0
Maleic resin	10.0
Dioctyl phthalate	3.5
Wax dispersion	5.0
Ethanol	40.0
Ethyl acetate	14.0
Selected glycol ether	3.5
	100.0

A third option for labels are cosolvent polyamide-based inks, where advantage is taken of their natural non-curl properties and potentially high gloss levels. There are some limitations to their use due to the high degree of resistance to weak alkalis, which prohibits them being used on wash off labels. Also ink films based on polyamide tend to weld when ram punched or guillotined, therefore they cannot be used on the edges of the labels.

Over recent years there has been a small but significant increase in the use of metallized paper for labels. There have been problems associated with ink adhesion on this substrate, mainly due to the oxidation of the aluminium surface. Lacquering immediately after the vacuum depositing process has been completed has alleviated the problem to great a extent. However, as the surface is basically non-absorbent, problems can occur with ink trapping as is the case with aluminium foil and certain films. In these circumstances it is useful to use a cosolvent polyamide backing ink.

Typical cosolvent polyamide ink:

Phthalocyanine blue (CI Pigment Blue 15:3)	10.0
Cosolvent polyamide	22.0
Nitrocellulose	3.0
Polyethylene wax	1.0
Isopropanol	12.0
Toluene	37.0
Isopropylacetate	8.0
Ethanol	7.0
	100.0

Metallic label inks

Many labels require the use of metallic inks as part of the overall design. The various shades of bronze inks are particularly popular, usually they are the last colour to be printed in a multicolour design and often overprint one or more of the preceding inks.

Aluminium-based inks are used less frequently than bronze inks on labels. Primarily they are used as background to give a metallic effect to overlying colours. Care is required in the selection of a suitable grade of aluminium, if this is the case, to ensure interlayer adhesion. A typical formula for a bronze medium into which bronze powder would be added at 30–40% by weight is:

Chlorinated rubber (medium viscosity)	30.0
Dioctyl phthalate	10.0
Wax dispersion	7.0
Stabilizer	1.0
Toluene	52.0
	100.0

and a bronze ink:

Bronze powder (gravure ink lining)	40.0
Acrylic resin	20.0
Wax dispersion	5.0
Stabilizer	1.0
Toluene	27.0
Isopropyl acetate	7.0
	100.0

Although both chlorinated rubber and acrylic systems are excellent binders for bronze powders there has been an increasing requirement to move away from toluene as a solvent. Consequently many metallic label inks are now based on ester soluble acrylic resins:

Bronze powder (gravure lining grade)	40
Acrylic resin	16
Wax dispersion	3

Stabilizer	1
Ethyl acetate	15
Isopropyl acetate	25
	100

Paper wrapper inks

With paper wrappers being used predominantly for food, confectionery and toiletry packaging, the choice of raw materials in gravure inks is largely determined by their suitability for these end uses. It is of paramount importance that the dried ink film is free from odour or potential taint that might be derived from retained solvent or odorous resins. Although inks are not intended to come into direct contact with foodstuffs, they should not contain pigments or dyes which are likely to bleed into food constituents, for example animal fats, vegetable oils and in the case of chocolate products, cocoa fat. Migratory materials should be avoided where possible and harmful pigments, chemicals and additives should not be used.

For guidance on this subject the Society of British Printing Ink Manufacturers has issued an advisory booklet *Printing Inks for Use on Food Wrappers and Packages* and a list of materials which are recommended to be excluded from inks for food wrappers.

After the exclusion of harmful pigments and certain dyes, the possible choice of colourants for wrappers is wide and largely dependent on their end use. As indicated earlier specific product resistance is often a factor, with standard tests available to predetermine a pigment's suitability. Soap wrappers for example require inks that are resistant to 2% sodium hydroxide solution and also to a 25% toilet soap gel.

Because of its low odour characteristics, coupled with its fast solvent release properties, nitrocellulose has been generally accepted as the ideal film former for paper wrappers. It can be used with plasticizer only or modified with carefully selected maleic and ketone resins.

Ethanol with small quantities of esters and glycol ethers are the preferred solvent combinations. When the label is not varnished, nitrocellulose chips may be required to achieve an acceptable gloss level. Otherwise bead milled inks can be used, with the overprint varnish providing gloss. As with labels, inks for paper wrappers should be adequately plasticized to prevent curling during the packaging process. One of the inherent problems with varnishing paper wrappers for foodstuffs is the choice of solvents for the overprint varnish. Having formulated the inks on low odour nitrocellulose and ethanol, there is little point in introducing odorous solvents such as toluene and ester solvents into the overprint varnish. These would not rewet the inks and would provide an acceptable gloss level, but there would always be a danger of retained solvent leading to possible taint. Alcohol-based varnishes can cause rewetting and glossy versions are often prone to tackiness. If they are used to overprint inks containing PMTA pigments or dyes they can quickly become discoloured and tint the unprinted areas of the paper.

It was the introduction of water-based overprint varnishes in the late 1960s that largely overcame this problem. Initially vinyl chloride vinylidene copolymer emulsions were used. They satisfied all the main demands of gloss and low odour, and because of their high water and low alcohol content did not rewet the inks. The main drawback was their high temperature of coalescence, which meant that only machines fitted with extended drying capable of reaching a temperature of 120°C could use this type of varnish successfully.

Following these came the acrylic-based varnishes which require a much lower coalescing temperature. With a wide range of acrylic solutions, dispersions and emulsions becoming available, the formulator is now able to provide water-based varnishes for paper wrappers with a wide range of properties.

A typical ink for a food wrapper would be:

Calcium Rubine 4B (CI Pigment Red 57:1)	10.0
Nitrocellulose	15.0
Maleic resin (low odour)	5.0
Wax dispersion	2.5
Dioctyl phthalate	5.0
Ethanol	54.0
Ethyl acetate	5.0
Glycol ether	3.5
	100.0

Water-based overprint varnish, using specially selected low odour materials:

Acrylic resin dispersion	40.0
Acrylic resin solution	28.0
Wax emulsion	12.0
Wax dispersion	5.0
Anti-foam	1.0
Water	6.0
Iso propanol	8.0
	100.0

Inks for cork tipping

The first requirement for cork-tipping inks is that their constituents should be fully FDA approved. They also should be manufactured under very carefully controlled conditions to avoid possible contamination.

Pigmentation is achieved with specially selected iron oxides and titanium dioxide, all chosen for their inert properties.

Cork-tipping inks are normally based on nitrocellulose combined with a specially selected plasticizer, which when pigmented is tack free and insoluble in human saliva.

Carton inks

This is one of the areas in which gravure printing competes with offset litho and comparisons in print quality are often made. For this reason the lay characteristics of a carton gravure ink is of critical importance. Also many of the cartons for well known brands of chocolate confectionery and cigarettes have considerable coverage of a colour with which their name is associated. Therefore particular attention is given to the print quality to ensure that it is smooth and unblemished.

Before discussing inks it is worth considering the role played by the engraving on the gravure cylinder and its influence on lay. With some shades, dark greens, browns and deep blues, it is an advantage when printing on board to have an engraving of greater depth than normal, this ensures good ink coverage and flow-out reducing the tendency for mottle to occur. It also enables the inkmaker to include a reasonable amount of titanium dioxide in the formulation to give increased opacity and improved lay.

Recent developments have shown that Helioklischograph engraving can also reduce the lay deformities typical of certain colours.

In common with other packaging inks for food and tobacco, low odour is of prime importance for cartons printed for these purposes. The inks are similar in characteristics to those discussed for paper wrappers where comparable requirements apply.

If inks with acceptable lay are to be formulated then careful attention must be given to pigment selection; not only individually, but in the combinations chosen to match a particular shade. For example browns are often best matched from an orange or red which when darkened with black has the nearest hue to the shade required. Bronze blue and yellow (PY 83) are a useful combination with black for matching dark greens. Sometimes, despite their cost, PMTA pigments can be used to advantage to achieve good lay.

Almost invariably, cartons are varnished and as with paper wrappers water-based overprint varnishes have become increasingly important. All cigarette cartons and many chocolate boxes are overwrapped with transparent film which is heat-sealed directly over the printed surface. Obviously this requires the inks to be heat resistant and the overprint varnish to have heat-seal release properties against the particular film employed.

In addition to heat resistance properties, the water-based overprint varnish should be formulated to give an acceptable gloss level and also be based on hard drying acrylic resins, which have the ability to release water from the drying varnish film as rapidly as possible. In this respect the amine content should be critically evaluated and, if tolerable, ammonium hydroxide should be considered. Sometimes the best overall results can be realized by a carefully balanced combination of two amines.

Heat-resistant carton ink:

Rubine 4B (CI Pigment Red 57.1)	10.0
Orange 5Y (CI Pigment Orange 5)	2.5
Titanium dioxide (PW 6)	7.5
Nitrocellulose	13.5
Maleic resin	5.0

Wax dispersion	5.0
Dioctyl phthalate	5.0
Ethanol	42.0
Isopropylacetate	7.0
Glycol ether	2.5
	100.0

Heat-seal resistant varnish:

Hard acrylic resin	15.0
Isopropanol	20.0
Water	15.0
Amine or ammonium hydroxide	2.0
Wax emulsion	5.0
Acrylic emulsion	35.5
Wax dispersion	5.0
Release agent	1.5
Anti-foam	1.0
	100.0

When metallic inks are required for carton printing the overall design has to be carefully considered. If the area of metallic print is small and if the ink has to overprint other colours, it is likely that the best printability will be achieved with a toluene-based acrylic resin type of ink described previously. If the ink is required to print directly on to the board then an ester-based acrylic metallic ink should perform satisfactorily.

Problems can occur when a large area of gold is required to overprint a nitrocellulose based colour and, because of the low odour requirements, a toluene system cannot be used. The best approach is to use a vinyl-modified acrylic with a carefully balanced ester solvent system, suited to the machine speed and its drying capacity.

The choice of materials for bronze inks is discussed in detail in the section dealing with metallic inks.

Metallic bronze gravure ink for cartons:

Bronze powder (Superfine lining)	40.0
Ester-soluble vinyl resin	2.5
Acrylic resin	17.5
Wax dispersion	3.0
Stabilizer	1.0
Ethyl acetate	10.0
Isopropyl acetate	16.0
Glycol ether	10.0
	100.0

Foil inks

Aluminium foil is widely used in packaging both unsupported and as a laminate to paper and board. Having been used for a wide variety of packaging outlets over many years, aluminium foil is now being challenged by metallized papers and films, the former because of lower cost, the latter on product appeal.

Virgin foil

When unsupported foil is printed it is normally untreated, the inks being required to key directly to the virgin aluminium surface. During the annealing process any rolling oil left on the surface is removed. When exposed to the atmosphere aluminium oxidizes, rendering ink adhesion more difficult. In severe instances a suitable primer must be applied although normally the foil is sufficiently tightly wound to prevent oxidation occurring too rapidly.

One of the problems when printing virgin foil, especially the very thinnest gauges, is to achieve adhesion and flexibility of the ink film without leading to subsequent blocking. This can be particularly hazardous where colours overdraw or where the nature of the design creates a build-up of ink in certain areas. This can lead to excessive pressures in those areas in the rewind reel thus increasing the danger of set-off or blocking.

To take advantage of the metallic lustre of the aluminium many colours printed on foil are transparent. The required transparency is usually obtained by the use of pigmented nitrocellulose chips and occasionally lightfast dyestuffs, or a combination of both. As with paper wrappers the choice of pigments and dyes is fairly extensive. The most common use of lightweight foil is for chocolate and confectionery wrappers where the colourant will need resistance to vegetable oil, cocoa butter and possibly lactic acid. The colourant will also be selected for its low odour characteristics in the vehicle system.

Providing the foil is free of oil and oxidation, adhesion to the surface can be achieved with plasticized nitrocellulose modified with polyvinyl butyral. The plasticizer content should be balanced carefully to ensure sufficient flexibility without incurring a soft or sticky ink film that could subsequently set-off or block. Adhesion on foil is usually assessed by the self-adhesive tape test and crinkling the printed foil between finger and thumb.

Solvent choice is limited to those that are suitable for food packaging as discussed previously. Particular attention should be paid to the use of any retarding solvents, as these can remain in the ink film giving a false degree of flexibility, which is not enduring. Also their presence increases the danger of blocking and residual odour.

Formulation for lightweight aluminium foil:

50% pigmented nitrocellulose chip	25.0
Polyvinyl butyral	4.0
Dioctyl phthalate	4.0
Wax dispersion	3.0
Ethanol	54.0
Isopropyl acetate	10.0
	100.0

Not all chip-based inks key to foil satisfactorily: certain pigments can be difficult if the overall chip content exceeds a certain level. Typical examples are Bronze Blue and Lake Red C: it may well be the dispersing aids used during the manufacture of the chips that are responsible for the problem.

Harder grades of foil laminated to polyethylene are used to package medical tablets individually. Hard foil is produced by varying the annealing process and is of a heavier gauge than that used for confectionery wrappers. A thin film of polyethylene is coated on what is to be the reverse side of the foil by hot melt application and is used to heat-seal two strips of foil with the tablet sandwiched inside. The encapsulation is done at very high speeds using crimp heat-sealing and to achieve the required throughput the sealing temperature is between 180 and 200°C.

The heavier grades of aluminium foil are usually referred to as capsule foil and are primarily used for milk bottle tops and polystyrene cream and yoghurt tub closures. Bottle tops are printed by both the gravure and flexographic process. The main requirement is for wet rub and water resistance. When printed gravure either vinyl based or plasticized nitrocellulose based inks can be used.

Vinyl-based ink for milk bottle top:

Vinyl chloride/vinyl acetate copolymer	15.0
Phthalocyanine Blue (CI Pigment Blue 15:3)	8.0
Wax dispersion	2.0
Methyl ethyl ketone	60.0
Ethyl acetate	15.0
	100.0

Yoghurt lids are normally printed with plasticized nitrocellulose often modified with melamine resin to improve mechanical flexibility. As previously indicated base white inks will often include zinc sulphide in place of titanium dioxide to reduce the abrasion on the cutting tools. Printed materials are heat-sealed to the tub by means of a heat-seal varnish which is applied to the reverse side of the foil. A modified vinyl chloride/vinyl acetate resin combined with a suitable soft acrylic is normally used, so as to give a sufficiently firm bond when sealed, but allow the closure to be easily peeled from the tub.

Typical heat-seal coating:

Vinyl chloride/vinyl acetate copolymer	15.0
Acrylic resin	10.0
Methyl ethyl ketone	60.0
Ethyl acetate	15.0
	100.0

Foil laminates

Foil to paper laminates are to be found in the food confectionery, tobacco and toiletry industries. To prevent poisoning and oxidation of the foil after lamination it is coated with a wash of nitrocellulose, shellac and, in some cases, vinyl. Therefore the ink no longer has to key to aluminium but to a resinous base which enables a wider range of inks to be employed.

When heat resistance is not essential, advantage can be taken of employing polyamide or acrylic backing whites overprinted with nitrocellulose colours, thus ensuring good trapping. Some difficulty can be met with the rewetting of nitrocellulose on foil, both as the basecoat and when inks of this type overprint each other. With inks the solvent sensitivity of the nitrocellulose can be reduced by modification with a hard acrylic; the solvent system however will have to be predominantly ester based. Inks of this type plus plasticized nitrocellulose can be used where heat resistance is required and also for soap wrappers.

Shellac-washed foil rarely presents a problem with rewetting due to its tendency to insolubilize on ageing. Because of this alcohol and ester based inks can be printed without difficulty.

Wax-laminated wrappers are commonly used for chocolate biscuits. Preprinting on virgin foil, then lamination with hot melt wax, requires inks with a reasonable degree of heat resistance to prevent picking off immediately after lamination. Prelaminated material must be printed within a few hours of lamination to prevent poisoning of the foil surface, thus making ink adhesion difficult. Heat resistant polyamides ($>$120°C) can be used in conjunction with nitrocellulose-based inks. Care is required in pigment and dye selection to ensure there is no bleed into the wax.

Foil-board laminates

Printing foil-board laminates was at one time almost exclusively done by sheet-fed litho. The foil was usually coated with a gold tinted nitrocellulose lacquer. After printing by offset the sheets were then varnished, often as a separate operation. A similar approach was followed when the gravure process was originally adopted to print cigarette cartons on foil-board laminate. Using a sheet-fed machine the gold lacquered foil-board was printed in one operation. Problems were encountered with printability and adhesion on the gold nitrocellulose base lacquer.

It is now more customary to print on foil-board laminates which have been coated with shellac. Often the darker colours are printed directly on the substrate with the light colour, usually a gold, printed last also playing the part of an overprint varnish.

Modern reel-fed machines with in-line cutting and creasing demand inks that are hard drying and flexible. The adhesion and flexibility should be sufficient to withstand a 180° fold followed by ten firm finger rubs on the ink surface. All cigarette cartons are overwrapped with film, requiring the inks to be heatseal resistant. The surface slip of the final overprinting colour should be carefully adjusted to meet the requirements of the high speed cigarette packing machines.

To achieve the adhesion and heat seal resistance requirements, nitrocellulose modified with melamine resin are suitable combinations. On occasions where there is superimposition of the darker colours the base colour can be hardened with an acrylic resin to reduce rewetting. Mainly nitrocellulose chips and occasionally lightfast dyes are used as colourants.

For example.

50% Yellow 83 nitrocellulose chip	3.0
50% Orange 5 nitrocellulose chip	2.0
Nitrocellulose	20.0
Modifying resin	10.0
Wax dispersion	5.0
Ethanol	46.5
Isopropyl acetate	10.0
Glycol ether	3.5
	100.0

Inks for polyethylene film

The vast majority of polyethylene film is converted by the flexographic process because of either the simplicity of the design or the length of run which makes the gravure process commercially unviable. However, designs of a more complex nature still utilize the gravure method or alternatively the offset gravure process which can allow a blend of flexo and gravure printing stations.

Conventionally cosolvent polyamide resins have been used in gravure inks for polyethylene because of their excellent pigment-wetting properties good solvent release characteristics, good water resistance and enable a high gloss print to be obtained. The trend to remove hydrocarbon solvents from packaging inks has meant a change to the use of alcohol-soluble polyamide resins. Unlike their cosolvent counterparts the alcohol-soluble grades have inferior solvent release and water resistance properties and modification with nitrocellulose is essential to provide inks of the desired quality.

For heat-resistant applications polyamide resins cannot be employed and inks based on nitrocellulose modified with polyurethanes are used. These systems will be discussed more fully under the section 'Inks for treated polypropylene films'.

Regardless of the resin system employed, pigment selection plays a vital part in the determination of the water resistance properties of the final ink film and the particular pigment selected will not necessarily be the same for the above resin systems.

In polyamide inks PMTA pigments (especially PBI. 1) should be used with caution so as to ensure there is no bleed into the base film on storage. Certain pigments based on transition metals may react with polyamide resin to give severe odour problems. For example manganese 2B toners and copper bronze are both liable to give odours, whereas copper phthalocyanines are satisfactory since the copper has already formed a highly stable chemical complex.

Polyamide inks frequently gel at low temperatures and the ability to recover to a fluid system on warming varies widely from one resin to another. In cosolvent resin systems a small quantity of water greatly facilities the case of gel recovery despite having no effect on the original temperature at which gelation first occurs. Water has no effect on alcohol-soluble inks, gel recovery being adjusted by resin selection.

Ink based on cosolvent polyamide resin:

Phthalocyanine Blue (CI Pigment Blue 15.3)	12.0
Cosolvent polyamide resin	25.0
Anti-oxidant	0.5
Isopropanol	32.0
Toluene	14.0
Aliphatic hydrocarbon	12.0
Water	2.0
Amide wax	1.0
Polyethylene wax	1.5
	100.0

Ink based on alcohol-soluble polyamide resin:

Phthalocyanine Blue (PB1. 15.3)	12.0
Alcohol-soluble polyamide resin	13.0
Nitrocellulose	18.0
Dioctyl phthalate	2.0
Anti-oxidant	0.5
Ethanol	40.0
Ethyl acetate	7.5
n-propyl acetate	5.0
Amide wax	1.0
Polyethylene wax	1.0
	100.0

Inks for treated polypropylene films

Treated polypropylene films have continued to find increasing use in gravure packaging outlets. The biaxially orientated homopolymer film (BOPP) is unsuitable for heat seal operations and the corona discharge treatment loses its effectiveness with time. Its main usage is a second ply in laminate structures. In recent years an opaque grade has been introduced which is corona discharge-treated on the print surface and flame treated on the reverse, making it suitable for cold-seal outlets. The grade tends to contain negligible slip additives and therefore compensating wax levels must be included in the ink formulations. Care in pigment selection must be made with this type of vehicle system since some colourants are unstable in alkaline media, e.g. Pigment Red 57, Pigment Violet 3.

White ink for printing treated opaque BOPP:

Titanium dioxide	32.0
Polyvinyl butyral	12.0
Ethanol	50.0
Ethyl acetate	3.5

Micronized PE wax	1.5
Polyethylene imine adhesion promoter	1.0
	100.0

Treated coextruded polypropylene films are mainly used for heat resistant applications while the 'pearlized' grades have attained tremendous growth in cold-seal outlets. The pearlescent effect in these films is produced by air entrapped in the film rather than by use of mica pigment. The insulating properties of these air bubbles makes heat sealing difficult within the dwell time on a packaging line.

For coextruded films, inks based on nitrocellulose modified with polyurethanes and titanium acetyl-acetonate (TAA) are used for both surface and reverse printing as well as for lamination. Titanium acetyl-acetonate acts as the adhesion promoter to the treated side of the film but has a drawback in that it may also react with the pigment to give an unstable ink. The same system can also be employed for printing on the treated side of homopolymer polypropylenes though higher levels of the adhesion promoter may prove necessary with a consequent risk of ink instability and high residual odour.

White ink for treated side of co-extruded polypropylene:

Titanium dioxide (CI Pigment White 6)	30.0
Nitrocellulose resin	11.0
Dioctyl phthalate	3.0
Polyurethane	8.0
Polyethylene wax	1.0
Erucamide	1.0
Ethanol	27.0
Ethyl acetate	16.0
Titanium acetyl-acetonate (TAA)	3.0
	100.0

While the above ink system is widely adopted both in the UK and in other countries, it has the drawback of relatively poor gloss levels for surface printing. Special two-pack reactive white ink systems for use with conventional overprinting colours have therefore been developed to overcome such a deficiency. The white can be based on epoxy–amine, polyl–isocyanate or alkyd–melamine chemistry, though the latter is little used because of the evolution of formaldehyde in the heat seal process during the packaging operation. Isocyanate systems are no longer used in the UK for health and safety reasons but they are still frequently adopted in some countries. The epoxy–amine system is therefore preferred in the UK and has the additional advantage that if used in conjunction with cellulose acetate proprionate–acrylic colours it can also be suitable for PVdC overcoating.

Cold-seal packages will normally use an overprint varnish to give release to the adhesive within the printed reel though in certain cases, dependent upon the design, the release properties can be built into the inks themselves. Polyamide varnishes frequently contain anti-oxidants either as an additive in the varnish or in the resin itself to reduce the risk of unwanted odours and the anti-oxidant, if phenolic in its chemical nature, will react with transition metals to form coloured complexes. Hence, if a TAA-based ink system is used in conjunction with a polyamide varnish then extreme care must be taken to ensure no discouloration occurs on storage of the printed reel.

Since in cold-seal work there is no heat resistance requirement, polyamide inks can be employed with polyamide varnishes as an alternative to TAA-based ink system. While this might appear to be an attractive solution to the problem, other difficulties still have to be overcome. The use of alcohol-soluble polyamide resins throughout can give rise to excessive tack levels due to poor solvent release and an all cosolvent resin situation can lead to undesirable retention of hydrocarbon-based solvents. A combination between the two resin grades for the inks and varnish provides a workable system, however TAA-based inks are still widely used.

It is important to ensure that the chosen ink–lacquer combination gives the required slip characteristics and since the amide waxes normally used with polyamide resins can migrate into and poison the cold-seal adhesive, such additives must be avoided. For this reason micronized polyethylene or polypropylene waxes are used to give the correct coefficient of friction levels to suit the packaging requirements.

Cold seal release lacquer for pearlized OPP:

Cosolvent polyamide resin	30.0
Ethanol	34.0
Aliphatic hydrocarbon	34.0
Micronized PE wax	2.0
	100.0

Red ink for pearlized OPP:

Calcium 2B toner	10.0
Alcohol-soluble polyamide resin	12.0
Nitrocellulose	10.0
Ethanol	52.0
Ethyl acetate	14.5
Micronized PE wax	1.5
	100.0

Recently, there has arisen a concern over plasticizer migration within the printed reel to the reverse side of the pearlized substrate, (i.e. there is a possibility of direct food contact). Whilst the degree of this potential migration would appear to be negligible, a plasticizer-free system has been introduced. This system is based on polyvinylbutyral resin.

Plasticizer-free ink for pearlized OPP:

Phthalocyanine Blue (CI Pigment Blue 15:3)	8.0
Polyvinyl butyral	20.0
Ethanol	65.0
Ethyl acetate	5.0
Micronized PE wax	1.0
Polyethylene imine adhesion promoter	1.0
	100.0

Coated polypropylene films

The PVdC-coated polypropylene films are heat sealable and have excellent gas and moisture barrier properties. Modification of the PVdC latex with a methacrylate can broaden the sealing range of the films with only minimal effect on the barrier properties. In recent years pure acrylic coated films have become available which have the advantage of lower cost but inferior barrier properties and this has been followed by combinations of one side acrylic and one side PVdC and pearlized grades.

The coatings on all the above films are affected by hydrocarbon solvents and high ester content inks which can destroy adhesion of the coating to the base film. In addition the acrylic coatings are sensitive to glycol ether solvents and the slower ester solvents (e.g. *n*-propyl acetate) leading to very high solvent retention and ink set-off.

Coated polypropylenes are primarily used where the cheaper uncoated grades have insufficient barrier properties or where the seal format of the latter is unsatisfactory. In addition the ability to overcoat the inks with PVdC emulsion enables very high bond strengths to be obtained on front/front (i.e. A/A) and back/back (i.e. B/B) seals. Here A and B refer to the front surface and reverse surface respectively. Such seal compatibility can only be realized on the treated films by coating both sides of the film with PVdC and this is usually impracticable. In or out of line PVdC coating of the inks is used particularly on biscuit roll wraps where the bunch seals and the undulations of the pack on the side seam enable only low sealing pressures to be used.

For the majority of end applications cellulose acetate propionate (CAP) in combination with acrylic resins are used as these offer excellent print adhesion at low web temperatures and hence minimize press registration problems. Such systems give low retained solvents and are suitable for both surface and reverse printing, PVdC overcoating, lamination and with a polyamide overlacquer, cold-seal applications. The major problems with these inks are on fine typework and tone areas where the rapid solvent release can cause premature drying. Retarding solvents are not a straightforward solution due to the coating sensitivity described above and in case of extreme difficulty, and where the print format allows, a nitrocellulose system may be preferred.

Yellow ink for printing on coated polypropylene film:

Diarylide Yellow (CI Pigment Yellow 12)	8.0
Alcohol soluble propionate	10.0
Acrylic resin	6.0
Ethanol	45.0
Ethyl acetate	30.0
Polyethylene wax	1.0
	100.0

Cellulose films

The use of cellulose films has declined in recent years due to competition from biaxially orientated polypropylene. The latter films are both cheaper to produce from an energy requirement and also from a raw material basis. Polypropylene films also enable greater yields per kilo to be achieved due to their lower density and now occupy the major market share enjoyed previously by cellulosic films. Cellulosics are still utilized for specific end uses, e.g. twist wraps and moisture containing product packaging, where their characteristic properties are superior to polypropylene. For convenience the films can be divided into three categories:

- Single side coated films, using either a PVdC or nitrocellulose coating. The uncoated side (designated PT) must be printed and then laminated to another film to protect it from moisture. Inks are relatively simple, being based on plasticized nitrocellulose but high in ester solvents to prevent excessive retention of alcohol-based solvents.

Green ink for reverse printing on PT film lamination:

Diarylide Yellow (CI Pigment Yellow 13)	7.0
Phthalocyanine Blue (CI Pigment Blue 15.3)	2.5
Nitrocellulose (high nitrogen grade)	14.0
Dioctyl phthalate	5.5
Polyethylene wax	1.0
Ethyl acetate	70.0
	100.0

- Two side coated nitrocellulose films are available varying in the content of wax and plasticizer in the coating to alter the properties of the film. While plasticized nitrocellulose inks offer excellent adhesion to these films great care must be exercised in the resin to plasticizer ratio to prevent blocking. The high wax content in the film coating also causes the ink to pinhole and the surface tension of the ink must be adjusted, usually by the addition of a phenolic resin.

White ink for surface printing on N/C coated films:

Titanium dioxide (CI Pigment White 6)	30.0
Nitrocellulose	12.0

Dioctyl phthalate	7.0
Phenolic resin	1.0
Polyethylene wax	1.0
Ethanol	30.0
Ethyl acetate	19.0
	100.0

- Two side PVdC-coated cellulose films form the third grouping with the PVdC being applied via an aqueous emulsion or solvent-based varnish. Ink systems are similar to those described for PVdC-coated polypropylene films though since the coatings have better anchorage to the cellulose base retarding solvents can be used in moderation.

Polyester films

The high tear and impact strength of polyester film makes it particularly suitable for packaging both heavy and abrasive goods and its high heat resistance lends the film to boil-in-the-bag and retortable pouches.

For reverse printing and lamination on uncoated polyester a vinyl ink must be used since other resin systems give poor bond strengths either initially or after a period of time. It is especially important to select the pigments carefully for the vinyl carrier in order to obtain maximum flow and aid printability characteristics. Solvent retention can also be a problem especially with ketone/ester-based laminating adhesives and alcohol-soluble grades and therefore preferred as virtually no rewetting of the ink is then given.

Red ink for reverse print and lamination on uncoated polyester film:

Barium 2B red toner (CI Pigment Red 48.1)	9.0
Vinyl chloride/vinyl acetate/vinyl alcohol copolymer	10.0
Ethyl acetate	40.5
n-propyl acetate	40.0
Polyethylene wax	0.5
	100.0

For surface printing of the uncoated grade it is possible to use the vinyl ink systems but it is preferable to use systems with improved printing characteristics. Polyamide inks can be used for non-heat resistance applications and where heat resistance is required a polyurethane-modified nitrocellulose resin system with TAA as adhesion promoter is utilized.

To make printing easier a number of coated polyester grades have been introduced. CAP acrylic systems, as described in the section of PVdC coated polypropylene, can be used for both surface and reverse printing of PVdC-coated polyester. For printing acrylic and special coated polyester films there exists a number of unique systems specific to each ink manufacturer. These systems are tailor-made to meet specific end uses and vary from CAP acrylic to plasticized nitrocellulose.

Nylon films

Nylon is primarily used for its thermoforming properties for example in wrapping cheese and in bacon pouches. The nature of the surface of nylon film appears to vary with the source of manufacture and therefore unlike most films a simple ink recommendation cannot be made and ink systems differing widely in choice of resin systems have been adopted.

Vinyl films

Ink systems for printing on vinyl films differ not only in choice of resin system but also in pigmentation types dependent on the plasticizer content. For unplasticized grades (UPVC) either nitrocellulose–acrylic or CAP–acrylic systems can be adopted without any special pigment limitations. However, for plasticized vinyl films not only does the inkmaker need to take special care on the adhesion and non-blocking properties of the resin system, but also pigments which do not cause bleed into plasticizer have to be chosen. For further details the reader should consult the later section on inks for vinyl wallpaper.

Metallized films

Metallized films have been included as a separate section because of the special problems associated with the printing of such films. To date metallized PVdC-coated cellulose, coextruded polypropylene and polyester films have been made commercially available. Difficulties have arisen since the adhesion properties of inks vary not only with the age of the films after metallizing but also vary from batch to batch of film. Hence, it is virtually impossible for the inkmaker to guarantee that an ink system will adhere satisfactorily to any batch of film. The phenomenon which causes such variable ink receptivity is still not readily understood and attempts to classify the behaviour as due to surface oxidation or migration have not been substantiated by chemical evidence.

Polyamide-based inks do show a greater tolerance to film variation than most other resin systems and hence for cold-seal applications such inks can be recommended provided reasonable precautions are taken by the printer. For heat-resistant applications individual ink manufacturers have formulated their own specialist systems which cope in variable degrees with the film inconsistency, and for extremely difficult cases primers have also had to be utilized.

Wallcoverings

A variety of processes are used to print wallcovering, with gravure, flexo and ink emboss predominating where surface printing once reigned supreme. Many manufactuers use all three processes and it is not uncommon to see combinations of them used to print the same reel of wallcovering.

A modern trend is for printers to carry the same range of inks for printing vinyl-coated papers and plain paper by both gravure and flexo. This is done to reduce the range of inks they have to stock. Normally the inkmaker supplies a pallette of colours along with reducing mediums, pearlescent bases and specials such as non-tarnish golds. Colour matching is carried out on site by the printer.

Inks for wallpaper

Gravure inks for wallpaper are normally expected to withstand the spongeability test without varnishing after embossing and the washability test after varnishing. As the base paper plays an important role in resistance to these tests, it is sometimes difficult to judge whether a failure is due to paper or ink.

The pigments used for wallpapers must have a minimum lightfastness of 5 on the Blue Wool scale at tint strength. The resin system should be flexible, but tough enough to withstand cold embossing and when modified with suitable wax additives, provide an ink film with excellent rub and mar resistance. As tonal printing is common on wallpapers an essential feature is that inks should have excellent printability. Toluene is usually the preferred solvent with chlorinated rubber and phenolic resin being used to give properties described above.

It is important that the ink should print to give a matt finish, even where there are overlying colours. This is to prevent distorted light reflections, especially when the wallpaper is viewed from an oblique angle. The mattness of the inks is normally controlled by the use of transparent white extenders.

Typical formula for a gravure wallpaper ink:

Phthalocyanine Blue (CI Pigment Blue 15.3)	10.0
Calcium carbonate (CI Pigment White 18)	12.5
Chlorinated rubber	12.0
Modified phenolic resin	17.5
Resinous plasticizer	2.5
Wax dispersion	7.5
Toluene	38.0
	100.0

Vinyl-based varnishes can be applied by air-knife coatings or in-line gravure application when a washable standard of paper is required.

Vinyl-coated wallcoverings

Two types of inks are used on vinyl-coated papers for gravure printing, those which are for gravure printing only, with a solvent base of methyl ethyl ketone (MEK) and toluene, also the dual purpose, but more expensive type, which can be printed by both gravure and flexo and has an ester solvent base.

As the processing and end use are identical, the choice of pigments for both types of inks is comparable. Their selection is governed by the specific requirements of the production process and the overall specifications demanded of vinyl wallcoverings. All pigments must have a lightfastness of 5 on the Blue Wool scale at tint strength. They must be fully resistant to bleeding in plasticized PVC and be heat resistant at 200°C during the embossing process. As pigments with these properties tend to be very expensive, particularly oranges, reds and violets, care should be exercized in pigment selection to achieve the greatest economy of use.

The resin systems for the gravure and dual purpose inks are not identical but have much in common, being based on vinyl chloride/vinyl acetate copolymer with acrylic for the former and modified vinyl resin plus acrylic for the latter.

As indicated above, most vinyl wallcoverings are heat embossed at about 200°C. In addition to enhancing the appearance of the wallcovering, the heat and pressure integrate the ink into the surface of the vinyl. This process is usually sufficient to ensure that correctly formulated inks will pass the scrubbability test.

When printing on vinyl, it is an advantage if the solvent system softens the substrate slightly, thereby giving the ink 'bite' and ensuring good adhesion. With straight gravure inks this is achieved by a correct balance of MEK (the strong solvent) and toluene (the diluent). If the formulation has excess ketone there is a danger of softening the vinyl too much leading to screening. With the dual purpose inks the solvent power of esters for vinyl is less than ketones and therefore they can be used without diluents.

Gold printing is normally done using simulated golds, where aluminium-based inks are tinted with transparent yellows, oranges and scarlets. These are used to avoid the tarnishing which eventually occurs with bronze powder-based inks.

Background mica finishes are based on gravure printed inks based on pearlescent pigments. These can be plain silver pearl or tinted pastel shades, giving a pleasing lustre to superimposed colours.

Typical formula of a vinyl wallcovering gravure ink:

Organic pigment (heat and bleed resistant)	10.0
Extender	7.0
Vinyl chloride/vinyl acetate copolymer	6.0
Acrylic resin	12.5
Wax medium	7.5
Methyl ethyl ketone (MEK)	23.5
Toluene	28.5
Butyl acetate	5.0
	100.0

Typical gravure/flexo wallcovering ink:

Organic pigment (heat and bleed resistant)	12.0
Extender	5.0
Modified vinyl resin	8.0
Acrylic resin	10.5
Wax medium	7.5
n-propyl acetate	57.0
	100.0

Expanded polyethylene

Expanded polyethylene is a third substrate type used for wallcoverings. Problems with long-term adhesion are associated with polyamide-based inks. Specially formulated inks have been developed to overcome this using an ester-based resin system. The expanded polyethylene base gives a wallcovering that is distinctly warmer and softer to the touch than either vinyl- or paper-based products.

Decorative wall panels

This is one of the few processes where offset gravure printing is exclusively used. It is principally designed to print sheets of hardboard or plywood, either to produce a woodgrain effect or simulate bathroom or kitchen tiles.

The sheets of hardboard are fed along a roller train where initially a base coat is applied and stoved dry. The offset gravure unit is situated above the roller or blanket which then transfers the ink on to the hardboard as it passes underneath. Unprinted ink is wiped from the offset roller after printing and when the system is idling. The inks dry by solvent evaporation, assisted by heat absorbed by the board during the earlier stoving operation. Finally an acid-cured precatalysed varnish is applied by curtain coat to protect the print.

Pigments should exhibit a high degree of lightfastness, iron oxides were originally used for woodgrain designs. Nowadays organic pigments have gained favour due to their less abrasive nature.

Nitrocellulose, ketone resins, ethyl cellulose and polypale resin can be employed as ink binders.

The solvent system must be sufficiently slow drying to ensure that the ink remains fluid when it is transferred to the rubber roller and subsequently to the substrate. Solvents should be selected with care as they must not swell or distort the rubber roller unduly.

Melamine laminate inks

Gravure-printed laminated plastics are used to produce decorative panels and sheeting for household furnishings, particularly in kitchen and bathrooms where exposed surfaces have to be highly resistant to water and heat. They are produced by printing paper which has been impregnated with either melamine formaldehyde or urea thermosetting resin.

The laminate is produced by laying the thoroughly dried print over layers of unprinted impregnated paper to a thickness slightly greater than the laminate required. A sheet of pure thermosetting resin is overlayed on the print. The stack is sandwiched between highly polished metal plates and subjected to a pressure of 600–1000 lb/in^2 for 45–90 min at a temperature of 150–200°C. The temperature and pressure compress the stack inducing the thermosetting resin to flow and cure to produce a solid plastic. Finished laminates should resist boiling water for two hours without delamination.

Pigments should have a high degree of lightfastness and have an acceptable resistance to colour change and development during the lamination process. There is a major difficulty in producing consistent colour matching to standard patterns, especially with multicolour tone work. A final assessment against the standard can only be made after a test laminate of the print has been produced. As each test takes nearly an hour the overall colour-matching time can be lengthy. Normally pigment selection is form phthalocyanine blues and green, naphthol reds and certain diarylide yellows, e.g. PY 83.

The resin system used for laminate inks should be compatible with the type of thermosetting resin used in the laminating process. Polyvinyl butyral resins are used for solvent-based inks with styrenated acrylic systems being preferred for water-based products.

The use of additives such as waxes and silicones should be avoided if possible as they are a source of blistering and delamination.

Low pressure laminates

The majority of wood finishes used on modern furnishing are some form of gravure printed laminate. Low pressure laminates are printed on paper with a woodgrain design then coated with an acid catalysed lacquer to give protection. The print is adhesive laminated to block or chipboard. Apart from reducing costs, one of the main advantages of this method is that the grain matching required with natural wood finishes is eliminated. This is particularly important if part of the furnishing is damaged during manufacture, when a printed matching replacement panel can be used. With natural wood grains this is not always possible.

For home use heavy gauge plasticized PVC is surface printed with either woodgrain or decorative designs, given a light emboss then reverse coated with self-adhesive and protected with a release film.

Stamps

The paper used for stamp printing is a smooth highly calendered coated paper with low absorbency. Designs may call for single colour or multicolour with overlaying tone and line work involving very fine detail.

Pigments should be carefully selected for their printability properties in the chosen resin system. They should have a lightfastness of 5 on the Blue Wool scale and be resistant to household cleansing agents and bleaches. They should be well dispersed during manufacture to ensure problem-free printing. Filtration of the finished ink requires care to ensure there are no large particles present which might give rise to streaking.

Care is required in both the formulations and manufacture of gravure inks for stamps to ensure they have excellent printability, rub and mar resistance. Choice of resin system is of great importance, as it should be tough and flexible and have excellent resolubility in the solvent system to maintain consistent tonal reproduction. Both phenolic resins modified with low viscosity ethyl hydroxy, ethyl cellulose and ethyl cellulose/maleic resin systems are used successfully with toluene as the main solvent. Any waxes should be finely dispersed and used sparingly.

Speciality systems

Metallic inks

For many years ready-mixed metallic gravure inks were of minor interest to ink manufacturers, as gold and silver inks were mainly blended on site by the printer.

It was the development of frame-board for cigarette cartons, using metallic bronze inks to match the existing gold lacquered foil-board laminates, that introduced the need for premixed inks. Printers lacked the facilities to mix the large quantities of ink involved.

Inks for this purpose require careful formulation and must:

- be viscosity stable – although a slight increase in viscosity as the bronze powder is wetted out is normal, long-term stability should be achieved;
- have good solvent release coupled with a low level of odour after printing – the use of toluene should be avoided;
- exhibit a high degree of brightness even after prolonged storage as the frame-board is often part of a foil-board laminate pack;
- print with maximum coverage and minimum striation, for which all the components are important including the correct solvent balance;
- the ink film should be tough, flexible and have good rub resistance – as complete rub resistance is difficult to achieve, frame-board is normally given a light coat of varnish.

The choice of resin systems for frame-board golds can only be established after stability tests have been carefully evaluated. Resins should be neutral in nature and non-reactive with the copper content in bronze powders. An increase in viscosity, dulling of the bronze when printed, and in extreme cases a green or blue tint in the medium are all indications of instability. Suitable resins are certain grades of acrylics with or without selected vinyl resins as modifiers.

With toluene being excluded and the adverse effect on storage stability of some alcohols, the normal choice of solvents for frame-board golds are esters with selected propylene glycols as retarders.

An example of a frame-board gold would be:

Bronze powder	40.0
Acrylic resin	16.0
Wax dispersion	3.0
Stabilizer	1.0
Ethyl acetate	10.0
Isopropyl acetate	25.0
Glycol ether	5.0
	100.0

Bronze inks for film

When printed on transparent film, bronze inks are normally backed with a colour to enhance the overall effect. Reds are particularly pleasing with the paler shades of gold, with greens and blues complementing the rich shades. The choice of resin system will largely depend on the type of film to be printed. If the resin selected to meet end use requirements has limited stability with bronze powders, then it is possible to premix them just prior to printing, disposing of the overs at the end of the run.

When polyamide resins are used as binders, care should be taken to ensure that sufficient anti-oxidant is incorporated into the formulation to prevent reaction between the resin and the bronze powder. Incorrectly balanced formulae can lead to objectionable odour rendering the printed material worthless.

Mention should also be made of the necessity to overlacquer a gold ink where contact with cold seal adhesive is like to occur. The absence of such a protective lacquer may result in discolouration of the gold and poisoning of the adhesive.

Aluminium-based inks

These are used in two different ways. On the one hand as a self-colour providing a metallic silver backing for superimposed inks. On the other hand as a means of providing coloured metallic effects when blended with transparent organic pigments.

To achieve the best results for both uses, careful selection of the aluminium base is needed. It is necessary that an aluminium gravure ink film should provide a surface on which superimposed inks can achieve acceptable adhesion.

A very fine particle size aluminium is required to enable the formulator to produce metallic colours with pleasing decorative appeal. These inks are particularly suitable for printing gift wrappers, labels and cosmetic cartons. Poor results are obtainable on absorbent stock, due to the preferential absorbance of the smaller pigment particles leaving the large aluminium particles on the surface. Ideally only suitably coated papers and boards should be used.

Transparent yellows and orange/reds are of particular interest, as from these and aluminium a wide range of simulated golds can be produced. Correctly formulated inks of this type can compete for visual effect with bronze-based inks. Because they have a lower specific gravity they also have a greater mileage and when formulated correctly can be used to coat board on which it is possible to print and obtain good adhesion with selected offset inks.

Advantage is taken of simulated golds to produce copper-free inks which may be an essential requirement for printing wrappers for certain types of foods, e.g. prawns.

When bronze-based metallic inks are used on wallcoverings, there can be a reaction between sulphur-based compounds and the copper content in the powder resulting in the formation of black copper sulphide and subsequent print spoilage. Additionally, because of the presence of sulphur oxides in the atmosphere, particularly in heavy industrial areas, bronze-based golds used on wallcoverings tend to tarnish quickly. Simulated golds are free from this problem maintaining their brightness over the natural lifespan of the wallcovering.

Typical formula for vinyl wallcovering simulated gold:

Vinyl Yellow chip base (CI Pigment Yellow 83)	10.0
Vinyl Orange (CI Pigment Orange 34)	5.0
Aluminium base	15.0
Vinyl resin	7.0
Acrylic resin	10.0
Wax dispersion	5.0
MEK	20.0
Toluene	23.0
Methyl isobutyl ketone	5.0
	100.0

Pearlescent inks

The use of pearlescent pigments to replace pure mica has been a growing practice in the gravure printing of wallcoverings. Their advantage is that they are more flexible and provide a considerably more cohesive surface than the old dextrin bound micas, thus ensuring adequate adhesion of superimposed colours.

Pearlescent pigments are titanium dioxide coated with mica and are dispersed into the gravure media by stirring. There are different shades available in different particle sizes, the most popular being the silver or pearl white grades. Pleasing effects can be obtained by small additions of tinting colours to produce, pinks, light blues and greens.

Also of interest, although somewhat expensive, are interference pearl pigments which reflect different colours from different light sources. They also exhibit a change of colour when printed over a black or dark background.

Cosmetic and soap packaging designs take advantage of pearlescent inks to enhance customer appeal. As with wallcoverings the pearlescent print often forms a background to the whole design. Overprinted colours are usually transparent thus allowing the pearl effect to be transmitted.

Typical formula for pearlescent ink for wallcovering:

Pearlescent pigment	20.0
Vinyl resin	10.0
Acrylic resin	15.0
Wax dispersion	5.0
MEK	20.0
Toluene	25.0
MIBK	5.0
	100.0

Fluorescent inks

Fluorescent gravure inks are not widely used; often their main role is to emphasize some point of sale feature. However, there has been a consistent use of this type of ink on detergent packs. Occasionally they are used to good effect on wrapping paper and they can be effectively utilized on wrappers for children's sweets.

Being resin based, fluorescent pigments are adversely affected by strong solvents which swell the resin giving inks with poor flow. Aliphatic hydrocarbons are suitable solvents but they limit the choice of resins considerably; cyclized rubber, vinyl toluene alkyds and certain specially formulated acrylics can be used.

Achieving a satisfactory dispersion with fluorescent pigments is not easy and unless the pigment is adequately wetted-out, hard settlement can occur. As an alternative to pigments, soluble dye toners offer a means of formulating inks on a more conventional and wider range of resins such as nitrocellulose, polyamide and cellulose esters using alcohols, esters, toluene and MEK solvents. This enables films, foils and papers to be printed with resin systems that have the requisite properties.

Where possible, fluorescent pigments should be used as single colours to ensure the maximum brightness. Additions of small amounts of conventional pigments can be used to adjust shade, but this is inevitably accompanied with a corresponding loss of fluorescence.

It should be noted that the lightfastness and fat resistance of these types of pigment is generally poor. However, it is possible to improve the light fastness by backing the fluorescent ink with an ink based on conventional pigment of similar shade, thereby minimizing any colour change. Also a degree of protection can be achieved by overprinting with a suitable lacquer.

8.5 PRINTING INK FAULTS

The identification and resolution of many gravure printing faults, in common with other printing processes, can be difficult, time consuming and expensive. Most faults manifest themselves as some form of irregularity in the print appearance or as a blemish on the unprinted area of the substrate. Many faults may be due to incorrect adjustment of the machine or the ink and will be overcome when the correct conditions are restored. Other faults are more persistent and require further investigation.

Streaking

Gravure printing is unique in that on occasions it suffers from this particular fault. As the term implies it occurs when the doctor blade fails to wipe the cylinder cleanly allowing a narrow streak of ink to pass under the blade. If the machine is printing at the time, the streak is deposited on the unprinted area of the substrate. The reason for the streak can often be determined by its position and its nature. A heavy streak near the ends of the cylinder is often caused by ink skin which can form in the corners of the unit, eventually becoming dislodged and finding its way under the doctor blade.

If the resulting degree of print spoilage is unacceptable, the machine should be stopped, the skin removed from the duct and the ink carefully filtered. Any draughts impinging onto the ink should be excluded. Should the skinning and subsequent streaking persist, reformulation of the ink may be required to eliminate the problem.

During the printing of paper and board, fibres and particles of coating find their way into the ink. Where hydrocarbon-based inks are concerned the fibres and coating remain finely dispersed in the ink system and rarely give rise to streaking problems.

With alcohol-based inks there is a tendency for the fibres to form nibs which can lodge under the doctor blade resulting in streaking. Fine filtering of the ink can reduce the problem but if streaking persists then either a change of substrate or ink may be the only solution. Streaking may be aggravated by an ink that has a tendency to scum and therefore to remove loosely bound fibres or coating from the substrate.

The most difficult type of streaking to eliminate appears as a very fine streak, often only a few inches in length, originating from the print area. The streak is too short to be caused by a solid particle lodged behind the doctor blade. Normally, filtering of the ink has little lasting effect on the phenomena. The most likely causes are from pigment dispersion, especially hard pigments, e.g. bronze blue, agglomerates of wax particles, small particles of gelled ink or even static. On occasions the addition of a reducing medium with a high pigment extender content can eliminate the problem.

Non-polar hydrocarbon solvent based inks are occasionally affected by static originating from the friction of the substrate as it passes through the machine. Its effect is to produce spiky protrusions around the edge of the printed area. Anti-static additives, highly polar in nature, normally overcome the problem.

Scumming

Scumming is characterized as a thin, hazy deposit of ink on the non-printing surface of the cylinder, which subsequently transfers to the substrate. The problem can originate from a number of causes, all of which have in common the failure of the doctor blade to wipe cleanly. Although appearing perfectly smooth to the naked eye, it is possible for the surface of a polished chrome cylinder to be sufficiently porous as to retain a thin film of ink which cannot be removed by the doctor blade. Often the addition of a fast drying solvent may allow this faint tint to dry between the doctor blade and the impression cylinder. After a few hours the grainy chrome should be sufficiently polished to resume normal solvent use.

Often inks containing highly dispersed pigments are prone to scum. The addition of a small amount of aromatic hydrocarbon or the addition of extender medium often enable the doctor blade to wipe the cylinder more cleanly due to increased lubricity of the ink system. Such additions are not always practical due to design requirements, e.g. hydrocarbon-free or high colour strength.

Scumming can also be exaggerated by a print design which bears large oblong window areas aligned in the web direction. Scumming can be reduced by reducing press speeds or adjusting the hardness of the impression roller. However, an ink change is often necessary.

Screening

This fault appears as a screen pattern in what should be a solid print area and often shows at the leading edge of the print. It is caused by non uniform or insufficient ink flowout between neighbouring cells. The fault can normally be overcome by reducing the ink viscosity or by retarding the drying rate of the ink or by generally improving the flow characteristics of the ink.

If during the print run there is a persistent recurrence of what appears to be screening, especially if it is repeated in the same area of the print, then either drying in or packing in has occurred.

Drying in

Drying in is basically due to the resin system failing to rewet rapidly enough in the solvents, so that the ink retained in the engraving after printing gradually forms an inert layer at the bottom of the cells that is too viscous to combine with the ink in the duct as the cylinder rotates. The drying in will tend to occur in the shallower areas of the engraving and can usually be remedied by the addition of a strong solvent to redissolve the dried area or the judicious use of a suitable retarder to prevent the drying in.

Packing in

Extreme cases of ink drying in cells and not being resolubilized by the methods described above are referred to as packing in. It is a more serious fault and is likely to be caused by one of the following reasons.

- Similarly to drying in, the resin system lacks resolubility in the solvents. It may not be the total resin combination at fault, as it needs only one of the resins to have poor resolubility so that it lodges in the cells preventing ink transfer. Normally the cells require vigorous scrubbing to clear them. Retarding the ink or adding stronger solvents has no affect. Shellac is an example of a resin which behaves in this way.
- Equally troublesome, but fortunately not very common, is packing in due to excessive pigment flocculation. When the design has a solid area of print, pigment flocculation would show as unacceptable reticulation, but where there is only fine type, the lay characteristics cannot be assessed. Under these circumstances the flocculated pigment tends to pack into the cells, eventually completely blocking them. Pigments which have large irregular shaped particles are prone to this fault, e.g. phthalocyanine blues. Additions of retarder or any other additive has little effect. The cylinder has to be cleaned with a soft wire brush and the ink reformulated on alternative pigments.

Mottle (crawling)

Mottle is exhibited as a poor non-uniform lay of ink on a substrate, showing as a speckled, grainly pattern in the print. The ink may be too low in viscosity or unable to wet the substrate properly. Increase in ink viscosity can improve the appearance.

The fault is often excessive in inks containing pigments of widely different hues (e.g. browns). Sometimes additions of extender medium can improve the lay while the ink is on the press, however, the amount that can be added is limited by its effect on colour strength and gloss. Sometimes an alternative pigment combination will be required to overcome the fault.

Pinholing

This fault is characterized by the appearance of voids or holes in the ink film. Pinholing is the result of the inability of the ink to wet or to spread on the substrate surface. Adding a hydrocarbon solvent or a specifically formulated anti-pinhole compound usually overcomes the problem.

Due to the degree of agitation in a rotogravure ink duct, pinholing can occasionally be attributed to bubbles or foam trapped on the cylinder. Suitable foam suppressants should be used in such instances.

Skipping

This fault appears as elongated unprinted areas of work and is caused by either insufficient ink being supplied to the duct, air bubbles trapped under the doctor blade or water contamination in the ink with which it is immiscible.

Moiré pattern

In process printing a phenomenon can occur where a regular pattern appears across an area of tonal printing distorting the visual appearance of the reproduction. This is known as the moiré pattern and is due to the screen size and angle being identical on each of the engravings used to produce the print. Normally at least one of the cylinders has to be remade with a different screen configuration to overcome the problem.

Rewetting

When alcohol/ester-based inks with the same resin system, particularly nitrocellulose, are overprinted on each other on non-porous substrates there is a tendency for rewetting and hence screening to occur. Occasionally this can be overcome by thinning the base colour to a lower viscosity thereby reducing its solids. The screening results from the ink in the cells of the second colour rewetting and combining with the dried ink of the first colour. Thus at the critical moment of contact the ink in the cells increases in viscosity sufficiently to prevent flow out.

The problem is associated with the ease of rewettability and the rate of viscosity increase of the resin system. Often the problem can only be solved by using inks on different solvent/resin systems to avoid rewetting. Acrylic or polyamide resin-based inks can be used as alternative systems for the overprinting of nitrocellulose inks.

The problems discussed so far have dealt with faults which are evident at the time of printing. Unfortunately there are others which only manifest themselves at some stage after printing.

Blocking and set-off

When printing on non-absorbent, flexible substrates the formulator is faced with the problem of producing an ink which has good adhesion and flexibility, yet is sufficiently hard drying to prevent blocking in the reel.

Lightweight unsupported foil is a difficult substrate, especially if there is overprinting which leads to the build up of excessive pressure in specific areas in the rewind reel. Achieving adhesion on foil normally requires the correct balance of plasticizer and resin. Too little plasticizer and the ink will flake, too much and there is a danger of set-off or blocking. Particular care has to be taken to avoid set-off when the printed material is destined directly to wrap, confectionery or foodstuffs. Some converters use cotton wool soaked with acetone to check if set-off has occurred on the reverse side of the substrate.

Likewise there is a possibility of set-off or blocking with flexible films which have solvent sensitive coatings, as these can retain solvent absorbed from the ink.

Immediately after printing, coated films are usually heated to a sufficiently high temperature to ensure adequate keying of the ink, and to avoid blocking it is important that the chill roller effectively cools the print prior to re-reeling. Both blocking and set-off are serious faults, as they normally render the printed material unusable, leaving both the inkmaker and printer facing a difficult situation.

8.6 RECENT DEVELOPMENTS

Trends in the ink industry are often initiated by restrictions imposed by legislation or prevailing economic and ecological factors. For example, the USA governmental legislation to reduce solvent emission created the need for a water-based ink development programme which has stimulated interest not only in the USA but in virtually every country worldwide. In the seventies the effect of local ecological pressure was seen in the UK causing the withdrawal of lead and cadmium from printing inks, while in many other countries such pigments are still in common use.

New developments are also created as a result of novel approaches, either from the ink formulator himself or by the converter or end user of the final print. These ideas can lead to totally new technology or more often new applications of existing concepts. In the following sections various probable approaches for the future will be discussed, some of which may come to fruition while others may not prove to be technically feasible.

Economic needs

In discussing economic factors the aim is to reduce the cost of the end product and many different courses of action are available which inevitably affect the ink manufacturers' recommendations:

- ink raw material costs
- ink manufacturing costs
- higher printing speeds
- higher packaging speeds
- redesign to use lower cost substrates.

Raw material cost reduction is an on-going situation affecting all inkmakers and does not warrant further discussion in this section. However, reduction of manufacturing costs can be achieved not only by increasing throughput rates but in the case of inks by ensuring that the maximum strength is obtained in the dispersion process, since the pigment is usually the most expensive ingredient within the formulation.

Higher printing speeds often mean formulation changes to reduce tack levels, maximization of solvent release characteristics and lower effective web temperatures for adhesion (due to shorter time dwell in the drying oven). The significant increases in press speeds which can be achieved with modern-day presses can even require a complete change of the basic ink vehicle.

The constant demand for higher packaging speeds has seen the evaluation of first cool-seal wrappers and then cold-seal packages. The change from a heat-seal wrap to cold seal requires complete redesign of the ink system and future trends indicate more wrappers will move towards this technique. In addition, for smooth transit on packaging lines, slip levels of inks and varnishes are critical and the inkmaker will need to exercise even more control in future years in measuring the coefficient of friction values.

A reduction in cost of the end product can often by achieved by advancements made in ink formulation techniques. For example the use of an absorbent board instead of a more expensive coated board can result in savings even though a sealing coat provided by the ink manufacturer is necessary in the first unit of the press. Similarly savings could be made if the properties associated with a coated propylene film could be achieved with an uncoated film. Such challenges await the technology advance from the inkmaker and until these developments reach fruition the end user must wait patiently for his resultant cost savings.

Ecological considerations

In earlier sections of this chapter the trend away from hydrocarbon solvents and the elimination of certain glycol ethers has been discussed. Such moves have been required to produce a safer working environment for the printer. Further consideration of the solvent component of the ink system has now been made so as to produce a safer global environment. Legislation has been introduced in the USA and the EC to restrict the volume of solvent emissions. There are a number of options available to the gravure printer for reduction and abatement of his solvent emissions. These include:

- solvent incineration
- solvent recovery
- biological degradation
- alternative low or negligible solvent containing systems.

Each of these alternatives carries some disadvantages and the printer must decide which is most suitable for his business area.

Solvent recovery is widely used for gravure applications in Italy, where ink systems are generally monosolvent, e.g. ethyl acetate. This requirement for a single solvent system may restrict the use of some traditionally used resin systems. In addition the plant required to recover the solvent may be too expensive for the small printer.

Solvent incineration would require less basic equipment than a recovery plant and there is a potential for the heat energy generated during the process to be reused within the plant. However, it is possible that there may be further legislation restricting carbon dioxide emissions thus limiting the suitability of the process.

Biological degradation requires the use of selective bacteria to digest the solvent vapours. The process produces gaseous by-products and the bacteria must be 'fed' when solvent emissions are low, e.g. during wash ups and shut downs.

A more detailed discussion of these legislative measures and abatement methods is made in Chapter 17. Two types of new technology offer routes of obtaining solvent less gravure systems, these being water-based and UV -curing systems.

Water-based inks and varnishes

The technology of water-based inks has remained unaltered in its general principle of using an acidic resin which is solubilized in water by formation of an alkaline salt with a relatively volatile amine. On drying the amine is liberated to re-form the water-insoluble acidic resin. A comparison of water-based and solvent inks in relation to the gravure process can now be made.

In a normal solvent ink the solvents are balanced so as to be sufficiently non-volatile for minimal evaporation to occur from the time at which the doctor blade removes the excess ink from the cylinder to the point of impression. After impression, transfer of ink to the substrate and time to achieve uniform lay, the solvent must evaporate prior to rewind or stacking. By comparison, in a water-based ink the solvent is replaced not just by the water (which acts as the solvent) but also with the volatile amine required to neutralize the acidic resin. Water, while being relatively non-volatile and not causing problems due to evaporation from the cell, can create difficulties at the final drying stage dependent on the porosity of the substrate. Care must be taken to use paper substrates not prone to cockling as a water-based system presents the worst scenario for this type of problem.

In water-based inks the selection of amine and control of pH is as important as correct viscosity control. Premature release of the amine before the point of impression will result in a drop in pH and subsequent precipitation of the resin in the cell. Water-based systems are not as readily resolubilized as solvent systems and it is possible that precipitation will not be resolubilized within the next revolution of the cylinder in the ink duct. Therefore unless carefully controlled, fluctuation in pH can lead to packing in the etch and a screening phenomena on the print.

The surface tension characteristics of the water-based system must also be considered since they are generally totally different from those of conventional solvent systems. These differences can lead to poor substrate wetting resulting in ink reticulation and crawling, and non-uniform flow out of the ink from the cells. Surface tension can be reduced by careful selection of surface active agents or the inclusion of small quantities of alcohols. This latter addition also aids ink drying.

In order to provide water inks which will dry more quickly, resins have been developed which contain a dispersed resin phase within a solubilized polymer. This reduces the amount of amine to be released at the point of drying and increases the solids content of the ink and therefore reduces the quantity of water for evaporation. The disadvantage of such polymers, however, is that the risk of drying in the cell is increased and resolubility of the dispersed phase cannot be achieved.

A comparison of a water-based solution resin and a water-based dispersed phase resin is given in Fig. 8.12.

The drying speed of water-based inks can also be improved by formulating stronger inks which will print from shallower etch depth cylinders,

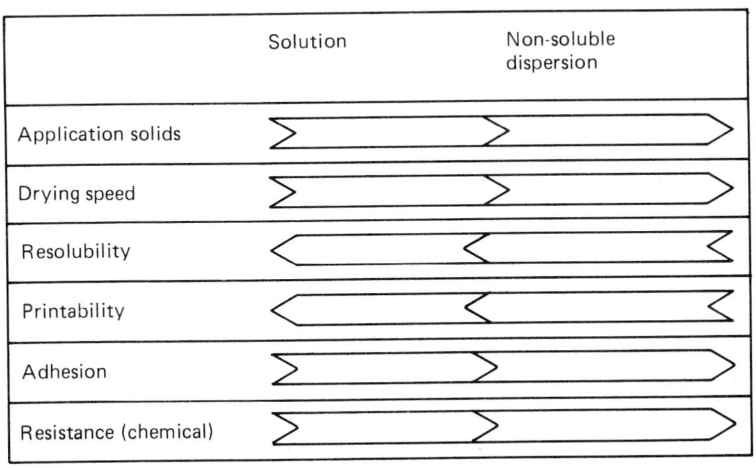

Fig. 8.12 Properties of aqueous resin systems.

Typical water-based ink for coated paper:

Phthalocyanine Blue (CI Pigment Blue 15:3)	15.0
Acrylic resin	32.5
Alkali	2.0
Water	30.0
Acrylic dispersed phase polymer	15.0
Antifoam	0.5
PE wax dispersion	2.0
Isopropanol	3.0
	100.0

Ink formulators are increasingly producing water-borne ink systems for a number of impervious substrates. The resins suitable for saponification and hence used in water-borne systems are more restricted than those available for solvent systems. Normally a combination of acrylic solution resins and dispersed phase resins are used to produce an ink with balanced properties.

Soft dispersed phase resins are used to impart adhesion properties to plastic substrates, although such polymers have low softening points and are therefore unsuitable for heat resistant applications. Gloss levels of this type of system are usually lower than comparative solvent systems.

Typical water-based ink for polyethylene/coextruded polypropylene:

Diarylide Yellow (CI Pigment Yellow 12)	15.0
Acrylic resin	10.0
Water	27.0
Alkali	2.0
Acrylic dispersed phase polymer	40.0
Anti-foam	1.0
PE wax dispersion	2.0
Isopropanol	3.0
	100.0

Harder styrenated acrylic dispersed phase resins can be used to promote adhesion and a degree of heat resistance on aluminium foil. The heat resistance can be further enhanced by the addition of zinc oxide.

Typical water-based ink for aluminium foil:

50% CI Yellow 83 acrylic chip	8.0
50% CI Orange 5 acrylic chip	2.0
Acrylic resin	5.0
Water	35.0
Alkali	2.0
Isopropanol	3.0
PE wax dispersion	2.0
Acrylic dispersed phase polymer	40.0
Anti-foam	1.0
Zinc oxide solution	2.0
	100.0

Ultra-violet curing of inks and varnishes

Ultra-violet-curing inks are now commonplace in letterpress and lithographic printing but have yet to make a significant impact in gravure technology. The difficulties facing the inkmaker will be examined in order to show the problems to be overcome before UV curing gravure systems are a practicable proposition.

Lithographic systems for UV curing inks contain the following ingredients:

- pigment
- oligomer
- diluent
- photo-initiator
- additives – waxes, inhibitors, etc.

The oligomer replaces the resin in a conventional lithographic ink and the diluent replaces the solvent. However, in UV-curing inks the diluent is also photoreactive and is cross-linked on irradiation with the oligomer and so the system is effectively a 100% solids ink. The photo-initiator is present to generate the free radicals necessary to start the polymerization process.

This technology is equally applicable to lithographic varnishes by omitting the pigment component and it has also been further extended to fairly low viscosity coating varnishes by inclusion of a low molecular weight diluent at the expense of some or all of the oligomer. Cure speeds are, however, slower since a greater degree of cross-linking is required to build a final polymer of sufficient molecular weight to achieve the necessary hardness and rub characteristics. The low viscosities required for gravure application can be achieved with some difficulty for lacquers, but for pigmented systems even greater problems arise in meeting the required rheological properties. Most low viscosity photoreactive diluents tend to have relatively higher toxicity and skin irritation figures. The Society of British Printing Ink Manufacturers (SBPIM) has recommended that a number of diluents and those with skin irritation values (Draize) above 3.5 are not used in ink formulations. Whilst these recommendations have not been universally adhered to, resin manufacturers have responded by producing a number of novel low viscosity oligomers.

In considering gravure application of UV curing systems an immediate advantage is that no evaporation will take place in the gravure cell since no volatile component is present. Hence, previous problems due to packing in the etch and resultant screening will be totally eliminated. However, even at low viscosities flow out remains a severe problem and while silicones can be successfully used in coating varnishes to reduce surface tension levels other problems of crawling and crazing are then often apparent if these additives are used in gravure.

New technology must therefore be developed before UV gravure-ink systems prove a viable proposition for the conversion industry and even then further challenges still have to be overcome. In lithographic printing for instance trapping of one colour over another in the wet state is achieved by tack grading of the inks (Chapter 6) whereas in the gravure process this is not feasible. The reasons for this are quite simply that one wet ink would pluck off the previous ink and that a web-fed gravure press requires the ink to pass face down over a turner bar prior to passing through the next unit. Interunit drying would therefore be necessary at least to tack dry each colour prior to a final curing process.

Ink film weight applied by the gravure process is higher than that achieved by lithography and in order to achieve full through cure at the commercial press speeds associated with gravure printing new polymers diluents and photo-initiators may have to be designed. In addition the problems of adhesion to the base substrate have also to be solved as well as satisfying the demands of the end user.

The quality of printed material produced using UV-curing flexo inks is very high and is very close to that of solvent-based gravure inks. This factor may result in the gravure printer reconsidering his process. The performance of the UV-flexo process and the problems outlined above suggest it is unlikely that UV-curing gravure varnishes for overprint applications will be an alternative to high gloss catalysed lacquers.

New concepts

Of all the areas of development, new concepts are one of the most difficult to predict since they are driven by a combination of the ink formulators' own ingenuity, the end users' packaging requirements, the printers' desire for improved performance products and not least by legislation, particularly environmental legislation. Solvent systems will still occupy a considerable amount of development activity with one goal; being to create novel effects to stimulate market appeal.

To reduce costs and the consumption of plastic packaging it is likely that the end user will demand the use of thinner gauge films. There should be no loss of bond strength in laminates containing such substrates and while pressure will primarily be placed on adhesive performance it is likely that inks with improved bond strength properties will be required. The use of multifilm laminates will also be questioned as the drive for total recyclability of packaging increases. Moves have already been made to produce labels and containers of like materials. It is likely that inks with either de-inking characteristics or compatible resin properties will be required.

The use of products containing potentially harmful components, e.g. chlorine-containing resins, hydrocarbon solvent, will increasingly be reviewed and the introduction of 'more friendly' alternatives will be made.

Water-based inks and coatings and monosolvent systems will continue to be introduced as part of the continued programme of solvent emission abatement. Further efforts will also be made to reduce retained solvent levels in food wrappers to reduce the risk of taint.

There is an increasing demand for package integrity. Non-reversible temperature induced colour change inks will be required for medical pouches and deep freeze packaging, whilst security marker inks may also be demanded

To sum up, whilst it is unlikely that major changes will occur to the gravure printing process, new ink formulations will be steadily introduced to meet the requirements of the changing market.

CHAPTER 9
Flexographic inks

The process of printing now known as flexography originated during the latter part of the 19th century. This process which was first called aniline printing used simple inks that were water or alcohol solutions of coal-tar dyes derived from aniline oil, the substrate being paper generally for bag making.

Commercial printing started in the early part of this century and the method was primarily used on the paper bag-making machine as a tail end printer, the print being hand synchronized with the bag-producing operation. Around the time of World War I press manufacturers began to make equipment to satisfy the needs of printers and paper converters. Since the equipment was somewhat basic at that time a number of users manufactured their own aniline printers to link up with bag machines.

The name flexography and its adjective flexographic were adopted much later in 1952 after the process had made considerable progress with improved print quality, press design, inks and a much wider range of work and substrates. The original name aniline together with aniline inks had chemical and toxicological association with coal-tar products which had been superseded. The process required a changed image not tied to its old bag-making origins. The original definition of the term flexographic became 'a method of rotary letterpress printing which employs rubber plates and rapid drying inks'. Since 1952 flexography has made further rapid strides and is used to print an increasing share of packaging and flexible packaging material to which it is ideally suited.

The range of materials for different end applications printed by the process is extensive. The principal materials are for packaging. In addition to the original paper bags the products include polyethylene bags and sacks, reels of paper, board, foil and films of all types. The reels can be converted either in-line as bags or boxes or subsequently used as slit or trimmed reels on various packaging machines to wrap or shape around food and other packaged items.

As well as packaging, the process is widely used to print stationery materials, tickets, coupons, forms, paper-backed books, children's comic papers and newspapers. Other applications are small labels on specially designed machines for adhesive, gummed and heat sealable types.

Disposable products are printed and examples of these are paper cups, kitchen towels, cloths and toilet tissue. Wallpaper and plastic wall coverings, gift wrap substrates, decorative materials, strappings and adhesive tapes are also printed together with a number of other products.

Process fundamentals

The flexographic printing process is described in detail in Chapter 2 of this manual. Basically similar to rotary letterpress printing it uses a raised printing surface made of a flexible material to transfer an ink image to the substrate. The flexible surface is able to transfer a good image even to rough substrates. The ink is contained in a duct and metered into a thin film often by the use of two rollers (Fig. 9.1). The first roller (A) called the duct or fountain roller can be metal or rubber and is mounted in the duct partly immersed in ink; this roller rotates normally at a slower speed than the second roller (B) called the forme or transfer roller. The second roller can be made of rubber or metal but not of the same material as the duct roller. On modern presses the transfer roller is an engraved anilox metal or ceramic roller and is likely to be fitted with a doctor blade for better ink film thickness control. Originally the pressure between these two rollers was the only means of film thickness metering.

The ink film this controlled is transferred by contact to the plate or stereo roller (C) which in turn transfers the ink image to the substrate web which is held against an impression cyclinder (D) normally made of rubber.

The improvements made by a number of different methods to control the film of ink, have over the years transformed the process. The use of engraved rollers, of trailing and reverse blades, the ability to vary the ratio of rotation of the duct and transfer roller, more refined engineering, improved register control for stack presses and the use of central

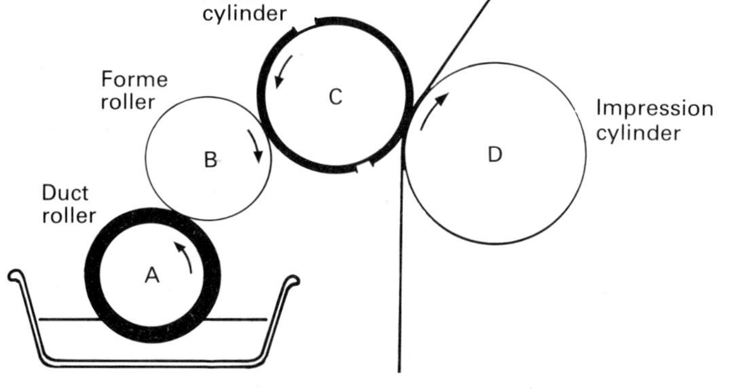

Fig. 9.1 A typical flexographic printing unit.

impression machines have all contributed. Added to these are developments in alternative stereo materials which have stimulated counter developments with rubber stereos, inks, drying and web handling, etc., all of which have allowed the flexographic printing method to continue to improve its quality and make the dramatic growth of recent years.

In general the improvements in design, particularly those which have affected ink application and drying, have allowed the inkmaker to offer a wider choice of inks which will print by the process and thus increase the possibilities of making suitable inks for difficult substrates with better end-use resistance properties. There are, however, still a number of flexographic machines of the older type giving good service that do not have many sophisticated features and the ink formulator should remember that some of the latest inks while perfectly satisfactory for a modern press may cause problems on these old machines. In general press design is as important to formulation as considerations of substrate and end use.

9.1 GENERAL CHARACTERISTICS OF THE INKS

There are a number of fundamental properties required of flexographic inks that are determined by the nature of the printing process and the conditions and forces met during printing; in addition there are the requirements of the printed article and the use to which it is put. The flexographic process did not originate and was not developed over the course of time in isolation. Considerations of ink performance, properties and limitations have played their part in the concept and improvement of the process alongside other products used by the printing industry.

Flexographic ink technology has also made considerable progress over the years and continues to advance. Ink consists of three main constituents – colourant, binder and solvent – and the variation and choice of these components, their properties and interactions will be covered in detail later, but for the moment, examination of the general characteristics of the ink and how these relate to the requirements of the process and its products will assist in an understanding of the basis of ink technology.

Fluidity

A necessary property of a flexographic ink on the press is that it be a free-flowing liquid and that it should remain in this state until deposited on the substrate. The ink as supplied may require reduction with solvent to adjust the viscosity for printing. On most presses the reservoir is sited below the level of the printing unit. The ink is pumped into the duct which is fitted with an overflow to keep the ink at a set level consistent with the position of the inking roller. The surplus ink is thus under constant recirculation by pumping and gravity fall, requiring a fluid state. The action of inking rollers and nips and the doctor blades also requires fluidity and on many presses the agitation can be considerable.

During recirculation of the ink through the unit, contact with the atmosphere in the duct and on the rollers causes some of the solvent to evaporate with a consequential increase of ink viscosity. This will cause an increase in film weight on the stereo as will be seen later, and it becomes necessary to replace the loss by the addition of a suitable solvent. This can be done manually by viscosity checks during the run adding solvent to correct or better still by the use of automatic viscosity control equipment.

Dye-based inks satisfy the fluidity requirement best of all since the solid content of the ink (colourant and binder) tend to be low when compared with other inks. Moreover, evaporation of the solvent with dye inks does not cause significant increases in viscosity (which is already close to that of solvent) and the losses are noticed more by an increase in colour strength of the print, due to a higher ratio of dye to solvent.

In contrast to organic solvent inks, water-based systems have the advantage of losing little solvent by evaporation and thus require less adjustment.

Viscosity

Ink viscosity control is important in flexography, particularly as a means of maintaining the print density during the run. The actual printing viscosity chosen for a particular job will depend upon a number of factors among which are press speed, substrate, type of metering, temperature, solvent mix and print thickness required. In practice satisfactory print can be achieved over a wider range of viscosity than is the case with the gravure process. The image as deposited on the substrate is faithfully reproduced from the stereo after contact resulting, within reasonable limits, in a satisfactory print.

Printers tend to use viscosity adjustment and the addition of solvent as a means of controlling colour strength of the print. As solvent is added to the ink there are two effects: first, a reduction of colour concentration due to the increased volume, and secondly, the lower viscosity of the diluted ink will give rise to lower sheer forces applied during metering and a decrease in film weight will result. This second effect is more apparent with simple roller nip metering.

Press speed will also influence the viscosity requirement of the ink. As speed increases sheer forces will also increase and the film weight of ink carried will correspondingly increase. A greater volume of solvent will need to be added to maintain the same print density. This is well known by printers and can cause problems since slow running during proofing at the start of a job will require the use of a thicker ink than is used on the run to give equivalent print density.

The control of colour strength and ink film weight by viscosity is a perfectly satisfactory method but care should be taken to avoid the extreme limits. Over-dilution may cause the print to suffer in appearance and lose abrasion resistance and the viscosity reading is of less significance as this moves closer to that of the solvent. At very high viscosities print definition and dirty working on fine type can occur and viscosity control is difficult due to wide variation as solvent is lost during the run. The use of a medium (non-coloured ink) to weaken excessively strong inks and the addition of a concentrated toner to a weak ink are ways of keeping the viscosity within the closer limits necessary for good quality printing, particularly with halftone.

Transfer

The more sophisticated ink metering becomes with reverse angle bladed anilox rollers, etc., the less do press speed and sheer forces affect film weight but with more simple metering the transfer properties of ink will alter the weight carried. Dependent upon the choice of ingredients used in the ink, tack properties and wet adhereance to rollers, stereo and substrate will vary extensively with each formulation and these will affect the amount of ink transferred to the substrate. Of the ink components the binder has the most influence on these properties and increases in concentration and molecular weight increase transfer.

Surface energy of the substrate can affect transfer and for good transfer this needs to be at a higher level than that of the ink.

The transfer properties of different formulations and resins can be compared in the laboratory by a side by side application of two inks using a hand anilox proofer, first ensuring that the strength and viscosity are equal. The transfer property of an ink has importance in formulation and the use of the hand anilox applicator for colour matching can be a better method of laboratory assessment than other hand applicators not related to flexographic ink metering.

Colour and strength

The colour and strength of a flexographic print will be determined by the ink film thickness applied to the substrate and the type and concentration of colourant used in the formulation. The wet ink film thickness will, in practice, vary over a wide range of approximately 2–15 μm, dependent upon press, metering, substrate and, not least, the working practice of the printer. On average the film thickness is lower than with gravure and a stronger ink is required to obtain a satisfactory coverage.

Since ink film thickness varies widely it is difficult to be precise when deciding colour concentration during formulation. The normal practice is to supply an ink at higher viscosity and of greater strength than is likely to be required on the printing press, thereby allowing solvent and reducing medium adjustments to be made. Experience and knowledge of the equipment and practice used by different printers can also be usefully considered when formulating inks of suitable strength and colour.

Stereo composition

Materials used to manufacture stereos have a considerable influence on the choice of solvent used in the ink. Those in use are various; natural and synthetic rubbers, photopolymer compounds and occasionally plastics have been utilized. In general, both water and ethanol will be suitable for all materials but modern flexographic requirements with many substrates and different and uses for the print lead to the use of many other solvents to widen the range of resins beyond those that are soluble in water and ethanol alone. A number of these solvents can swell or dissolve certain stereo composition materials. Table 9.1 gives an indication of the suitability or otherwise of different solvents for some of the more common materials in current use. The table should be only used as a general guide since both photopolymer and rubber compounds will vary in their solubility properties. Many inks are based on a blend of different solvents and a particular solvent may be used in a formulation at low concentration with no adverse effects, even though on its own it may be aggressive to the stereo. Stereo manufacturers are usually pleased to provide data on solvent resistance for their compounds.

The restrictions placed upon the selection of certain solvents, particularly with the use of the more soluble photopolymer materials, can create a limitation on the choice of binder for the ink. This may conflict with requirements for adhesion on difficult substrates or end-use resistance. These conflicting demands cannot always be simply resolved by ink formulation.

Table 9.1 Solvent resistance of stereo materials

Solvent	Natural rubber	Nitrile rubber	Butyl rubber	Photopolymer plate
Alcohols				
Ethyl alcohol	R	R	R	R
Isopropyl alcohol	R	R	R	R
n-propyl alcohol	R	R	R	R
Esters				
Ethyl acetate	N	N	R	N
Isopropyl acetate	N	N	R	N
n-propyl acetate	N	N	R	N
Ketones				
Acetone	C	N	R	C
Methyl ethyl ketone	N	N	R	N
Methyl cyclohexanone	N	N	R	N
Glycol ether				
Methoxy propanol	R	R	R	R
Ethoxy propanol	R	R	R	R
Aromatic hydrocarbons				
Toluene	N	C	N	N
Xylene	N	C	N	N
Aliphatic hydrocarbons				
SBP 3	N	R	N	N
Water	R	R	R	R

R = recommended; C = caution; N = not recommended.

Print characteristics

Ink will have a basic colour requirement and in addition may need to be glossy, matt, opaque or transparent dependent upon the design, substrate or job. The appearance of the print with good definition for halftone and type and good lay on solids are important considerations. The improvements in the process have helped ink printability although good print design, which takes account of the strengths and weaknesses of the process, are equally necessary. There can be conflicts on film weight requirements of solids and halftone and when these are on the same stereo, difficulties can result. Ink characteristics will also affect print quality and desirable features are good flow, wetting and good resolubility properties. Dye-based inks satisfy these needs and strong colours with high resolubility are easy to achieve.

Dye inks have limited fastness and most inks are now based on insoluble colourants and more care during manufacture and formulation is required to make pigmented inks with good printability. This is normally achieved but problems can be met when the formulation limits are stretched, examples being (a) excessive pigment loading reducing flow characteristics and (b) use of low solubility binders with high end-use resistance properties leading to dirty working.

After deposition on the substrate the ink is required to dry very rapidly this being most acute on a central impression multicolour press where the gap between print units is only a few inches and at high press speeds the time allowed for drying is a fraction of a second. Since flexography is unsuitable for wet-on-wet printing, drying, by penetration with absorbent substrates or, otherwise, by solvent evaporation has to be achieved instantly between colours. Thus modern drying equipment with high-velocity heated air has become normal practice.

Solvents used in the ink need to be volatile but a balance must be struck between satisfactory drying of the print and premature drying on the stereo which would otherwise affect print quality. Printing water-based inks on absorbent materials will most easily satisfy these requirements.

Overprinting colour on colour causes most drying problems as there is a tendency for solvent in the overprinting colour to resolubilize the first down colour and become trapped into the double layer which is less easily dried. With water-based inks overprint colour drying can also be a problem but the cause in this case is also due to the lower volatility of water.

The choice of solvent in the ink and the diluent used on the press have a significant bearing on drying and blends of solvent may be used to achieve the desired result. The binder used will also influence the evaporation of solvent as resins have different solvent release properties. Some drying will occur on the stereo and it is important that this redissolves on the next revolution or poor image definition and dirty working could result.

Adhesion and end use

Consideration of the substrate to be printed is of fundamental importance to formulation and this is particularly the case with non-permeable substrates where ink adhesion properties are critical. With paper printing adhesion is given by the penetration of the ink into the paper surface. In the case of non-absorbent substrates adhesion must be achieved by other means, therefore, and chemical and physical bonds and wetting between the surface of the substrate and the ink are important.

Of all the ink constituents the binder is the most important for adhesion, thus a suitable resin will be needed to confer adhesion to the particular substrate. Each type of substrate will present different problems and sometimes a combination of resins may be needed to give both adhesion and a balance of other ink properties. In general the more inert the substrate surface the more difficult it will be to obtain adhesion.

The end-use conditions of the print have a particular bearing on formulation when making inks for packaging. There may be requirements during the packing or processing for resistance to heat sealing, non-adherence to reverse printed lacquers or adhesives, or for particular surface slip characteristics. The print may have to resist packaged products during storage of the pack and these may contain various components such as water, oils, fats, soap detergent, etc. Particular conditions of storage such as deep-freeze may affect the print.

The choice of colourant will depend on the specification for bleed and fastness, and in general, dyestuffs and soluble pigments will be excluded from work with high requirements. The binder selected will need to have suitable resistance properties and many packaged products will soften certain resins resulting in a loss of adhesion or transfer of colour. Solvents will not normally affect resistance properties although resistant binders generally need stronger solvent mixtures for their solubility.

9.2 PHYSICAL PROPERTIES OF FLEXOGRAPHIC INKS AND THEIR MEASUREMENT

The measurement of the physical properties of flexographic inks and an understanding of how these properties are affected by the choice of ingredients and their interaction is a large part of ink technology. These measurements are ideally conducted alongside a study of ink performance on the press and in this way a relationship of properties and resultant performance can be built up by the ink formulator and subsequently used to improve inks and solve problems.

Measurement of print resistance properties must also be related to the conditions of end use, but care should be taken when setting specifications to avoid unnecessary margins of safety, particularly as this could lead to excessive restrictions on raw materials which may result in a lower quality of print.

Viscosity and dilution measurement

Inks are normally supplied at a viscosity which is higher than needed for printing, for two main reasons. First, to allow for adjustment on the press by the printer and secondly to prevent pigment settlement during storage which might occur with a thin ink. The importance of viscosity during printing has been discussed and the normal method for measurement is the efflux flow cup. This is a convenient method and is satisfactory for low viscosity liquids but can give false results with thixotropic inks. On the press it is helpful to allow the ink to circulate in the pump for a while before taking the final viscosity reading as this will tend to break any ink structure.

The temperature at which the viscosity is measured is important since the ink, in general, will decrease in viscosity with a rise in temperature.

For the ink formulator a solvent-dilution curve will yield more information than a simple viscosity measurement of the ink after manufacture. This can easily be carried out by taking readings at different dilutions and plotting the result as a graph. The graph shown gives typical results for some different ink types. Generally the steeper the curve the less solvent required to reach print viscosity and the stronger will be the resultant print (Fig. 9.2).

The tack of a wet ink can have a marked effect on printing properties and this will be more pronounced at high press speeds. While for oil inks it is relatively easy to take measurements with a tackmeter this is difficult with thin volatile flexographic inks. The effect of tack can be observed, however, using the hand anilox proofer and an ink with high tack will give more pull-though and strength.

Colour and strength testing

The colour and strength of a formulation must first match the agreed standard and this level must be maintained with subsequent batches. The ink film weight carried by the stereo is the governing factor affecting colour strength and some experience or estimation of the likely print weight will be needed when the colour is matched. A laboratory anilox hand applicator is normally used to simulate the printed result but less dilution with solvent will be required by hand proofing than with printing due to the relatively slower speed of hand application.

Once the colour and strength has been established maintenance of later batches to the same colour will be a more simple task. Side by side applications against the standard sample can be made and corrections carried out but viscosity differentials must be minimized or this will affect the result. Strength comparisons can be made by the 'white reduction' test – by diluting one part of the colour with 20 parts of a compatible white and comparing the result with the same test on the standard sample. In addition to colour strength, opacity must be considered and it should be noted that opaque inks tend to look stronger than those which are transparent.

Batch samples of flexographic inks can and do tend to deteriorate on standing particularly when they have been opened several times. It becomes difficult over a long period to maintain a master standard sample and rests tend to be carried out against the last batch sample. This can lead to colour drift and it will be necessary also to retain proofs to avoid this problem. Colour matching and batch passing can be carried out with the aid of a colour computer and, although it will be necessary to pay

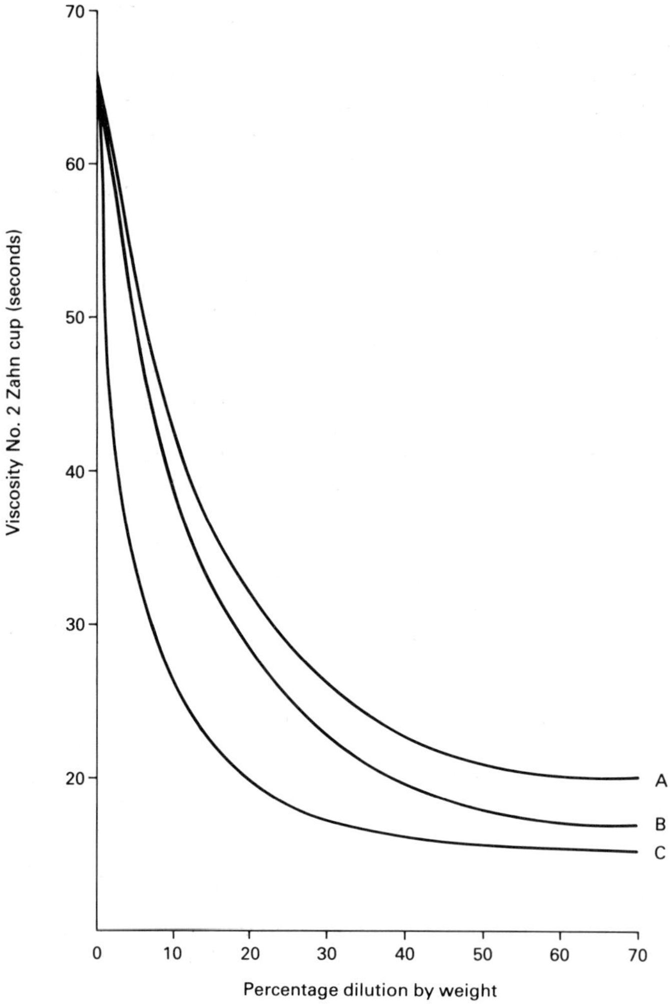

Fig. 9.2 Dilution–viscosity curve: (A) polyamide-based ink; (B) cellulose nitrate-based ink; (C) water-based ink.

attention to proofing control, the problem of deteriorating samples can thus be overcome.

Drying tests

With solvent-based inks there are a number of factors which will influence the drying, the most important of these being the evaporation rate of the solvents used and the solvent release characteristics of the binder. Some resins will release certain types of solvent more rapidly than others dependent upon their chemical make-up. A particular resin may release a slower evaporating solvent in preference to a faster solvent thereby reversing the expected drying rate of each. The hardness of the binder will also affect the point at which the ink feels dry and tack free, even though solvent may still be present in the ink film. High concentration of pigment and extender can improve drying, keeping the ink surface open without skinning to assist evaporation. Various additives and plasticizers can affect solvent release.

Drying properties can be compared in the laboratory by making comparative drawdowns and checking the relative time for the surface to become tack free but some solvent may still be contained in the ink. When testing it must always be remembered that on a press heat is normally applied, this reducing the time taken for a print to dry. Under laboratory conditions it is difficult to record results or make comparisons when drying is very rapid, as is the case with heat-assisted drying.

Testing of the final state of solvent retention after printing is normally carried out by the use of gas–liquid chromatography (GLC) and the results are expressed as the amount of solvent present in each square meter of print. The use of this equipment can be very informative in ink-drying research.

Prints which are poorly dried can give rise to serious problems in printing. First, the retained solvent can cause odour problems and secondly, the presence of solvent can leave the resin in a semi-solid state which is prone to block to the other side of the web. On certain coated films a ghost image may form on the unprinted side caused by solvent attack.

Water-based inks for paper and board dry mainly by penetration of water into the substrate and only partly by evaporation. The speed of drying is dependent upon the substrate absorbency and the type of resin/water dispersion used in the formulation. The inks dry too rapidly for this to be measured easily in the laboratory although a rough comparison can be made against an ink of known performance by testing a drawdown on the paper with one's finger as they dry.

With water-based inks for non-permeable substrates drying occurs by evaporation of the water, and any solubilizing amino compound contained in the formulation. Comparisons of drying can be made in the laboratory using similar methods given for solvent-based inks but again it should be remembered that on the press any heat applied will reduce the time taken.

Water-based inks – special tests

The properties of water-based inks are generally measured using similar methods to those used for solvent inks, as has already been discussed. Due to their special nature, additional tests for foaming and pH should be made on water-based systems.

Foaming can give rise to problems on the press, particularly when the design is such as to cause high agitation of the ink. A simple test of vigorously shaking or stirring a diluted sample of ink for a period of time will quickly determine any tendency to foam. Comparisons can be made with a satisfactory standard ink and the time taken for the foam to subside noted in each case.

Conventional water-based inks are alkaline, giving pH readings above 7. Using a standard pH meter inks can be tested and the normal satisfactory range will be in the region of 8–9. Problems of stability can occur with low readings, and poor ink drying may result if the pH is too high. The type of formulation used, however, will influence the range for a satisfactory pH and this can be determined by experiment.

Dispersion state

Pigmented inks exist as a dispersion and the state of dispersion is dependent upon the formulation and the process of manufacture. The normal method of measurement is the grinding gauge consisting of a metal wedge graduated from 0 to 25 μm. The ink is drawn down the wedge with a blade until pigment particles begin to form scratches. This is then taken as the particle size. Care should be taken that the test is conducted prior to the addition of waxes, etc. since these will interfere with the result.

Other indications of the dispersion state are gloss, colour development and transparency and these are monitored. Alterations in settings controlling the conditions on milling equipment will give rise to variations in quality of dispersion and careful monitoring will be needed to avoid batch variations. Chip-based inks are more consistent from batch to batch than with other methods of high dispersion. Chipping can, however, induce loss of flow due to the high dispersion state achieved.

While processing plays a major part in producing a dispersion there is a considerable influence and interdependence between the binder, pigment and solvent. All are equally important to the final result particularly with respect to the dispersion stability. Finding the optimum conditions and formulation balance for maximizing dispersion for a particular product can be time consuming involving many experiments. There is little doubt that bulk manufacture of single pigment bases, leading to fewer products milled, is to be preferred since in this way only a few products have to be studied.

Stability of the dispersion is important, both with neat ink as supplied and diluted ink as used on the press. Storage stability can be assessed by an accelerated test at 50°C for one week and dilution stability can be tested by comparing a diluted sample left overnight with a freshly diluted sample.

Adhesion and end-use tests

The adhesion of an ink to the substrate being printed must be of a satisfactory level for the life of the printed article. Normally the level of adhesion will not change after printing but occasionally on some substrates certain chemical type inks systems will need a short time for full adhesion to develop. On plastic films adhesion is often improved if some heat is applied to the film while drying; heat may also aid wetting or cause fusion of film coatings and ink resins. Once adhesion is achieved it is normally stable and should not deteriorate with time. On the rare occasions when this does happen it is usually associated with a change of state of the substrate or ink components.

The most common test for adhesion is the adhesive tape test and although it has the virtue of being simple to carry out there are a number of variables. Attempts have been made to standardize the tests and test conditions but it should be remembered that the test is dependent upon the adhesion of the tape to the ink surface as well as adhesion of the ink to the substrate. The result should therefore be viewed alongside other tests, i.e. scuff resistance and scratch resistance, before a final conclusion is reached.

With most presses being reel-to-reel, prints tend to be subjected to very high pressure after rewind. With non-absorbent substrates where the ink remains on the surface this pressure can cause blocking and it is necessary to ensure that film inks have good block resistance. There are a number of tests in the laboratory to simulate reel pressures by placing a test print under a hydraulic ram, screw clamp or weight for a period of time and observing if the print will block. Sometimes an elevated temperature is used and this will increase the severity of the test but the temperature chosen must not affect the substrate of its coating otherwise the results will be invalid. In practice, 50°C with most coated films would be a limit.

Much packaging is heat sealed and sealing is often applied in the printed area of the pack. With reverse printing there will be no direct ink-to-heated-jaw contact but, with surface printing, ink must not melt or stick to the heat seal jaw. Test heat sealers are available and the formulation must be such as to give the required properties. Surface-printed ink is also required to have good rub resistance as packaged goods will be handled and tend to rub together in transit. There are a number of standard rub testers available.

Flexographic printing is widely used for packaging applications and the properties required of the print to resist packaging, packaged products and storage conditions can be diverse. It is common for the ink to be printed on the outside of the substrate without the protection of an overlacquer. Many packaged products can have a powerful solvent or chemical action on the print and not all packaging materials are good barriers to migration or seepage of wrapped substances through to the ink layer.

Ink manufacturers will select their raw materials and formulations to give the necessary properties being sought, but it is important that the inkmaker be advised of the intended use of the ink and of any special conditions or specifications. Data will have been previously assembled by ink technicians of the resistance properties of raw materials and particular formulations but it is often necessary to carry out specific tests involving the substrate and the packed product or its active ingredients. On this basis recommendations or advice on ink selection will be made.

Tests can be devised for ink and print to simulate the various conditions of use and under ideal circumstances a specification is drawn up between the packer, converter and inkmaker to agree on test methods and standards which are acceptable to all. Within this specification the ink, print and package can be tested at all stages of manufacture and problems avoided. There are cases where the degree of resistance provided is left to the judgement of only one of the parties involved. It should be remembered that the final performance of the pack will be dependent upon all of the component parts and the conditions of printing and conversion.

The measurement of the physical properties of ink are clearly of prime consideration during batch testing of inks with quality control and assurance testing. All inks will be tested for colour, strength, adhesion and dispersion and dependent upon the ink product type specific fastness or end-use tests can be included. The decision regarding the extent of testing will be governed by the importance of the test, the associated risks and the time available during the manufacturing cycle. Every ink cannot be tested for all possible conditions and a sensible choice will need to be made of what constitutes a reasonable cover for good control. Systematic manufacture of a rationalized range of known specification products makes for a satisfactory resolution of this compromise.

Testing forms very much a part of the inkmaker's responsibility and is carried out to ensure that mistakes during making and blending are detected. If the final stage of manufacture is a mixing, blending, operation on the premises of a printer similar testing will be necessary for the same reason. Normally a simple test procedure can be recommended.

9.3 FORMULATING PRINCIPLES

As previously mentioned, flexographic inks consist of three main ingredients – colourant, binder and solvent – each having a different function but dependent on the others. Additives of various types are also used to impart other properties and as modifiers for the basic components to widen the range of properties available.

In considering the selection of raw materials to manufacture different inks the properties of each raw material must be known. Manufacturers will supply data of fastness properties, solubility, compatibility and general chemical constitution and this will assist the selection. In practice the properties of raw materials will be affected by the other constituents in the ink. For example, the fastness of a pigment may be improved by the presence of certain resins used in the formulation. It will be necessary to compile data for different ink systems which will show modified properties from that of the raw materials themselves.

The formulation of the ink will be influenced by a large number of factors associated with the use of the ink and the printed job. These would include: colour standard, finish required, type of press, ink metering, drying equipment, stereo composition, substrate, print design, end use, after processing of print and any specification requirement. In addition there will be considerations of ink manufacturing equipment, batch size and cost.

It is normal to classify inks into different named product types according to formulation, substrate usage and properties rather than all inks being bespoke for each job. To avoid the proliferation of many ink types held in stock most users will want the inks they use to have the widest possible use covering several substrates and applications. Distinct product types with known and defined performance properties and limitations will make the selection of the correct ink easier without misunderstandings. Products can also be manufactured in a planned manner and with larger batches of fewer inks, better control can be exercised both with raw material stocks and QC testing. The printing and packaging industry, however, is subject to rapid changes and these may require modified ink properties. Therefore, flexibility must also be preserved.

In order to consider the principles of flexographic ink formulation an examination of the basic raw material ingredients and their behaviour in the inks will be covered. Details of the properties, origin and chemistry of raw materials will be found in Chapter 4.

Colourant – dyes

The most widely used dyes in flexographic inks are the basic dyes which are salts of cationic dye bases (commonly hydrochlorides, sulphates, etc.). Without modification they are very soluble in water and have some solubility in alcohol, the alcohol solubility being improved considerably when the dyes are 'laked' with tannic acid or synthetic mordants. Colour strength is improved by laking and, depending upon the choice of mordant, solubility in water is greatly decreased. Laking of the dye will improve lightfastness, although this remains relatively poor, and bleed resistance to wax, oil, water and glycerine will also improve. Dependent upon the choice of mordant the optimum ratio to the dye will vary to achieve maximum colour strength development.

Price considerations are important but with wide strengths variations the lowest-cost mordant will not necessarily be the most economic. Phenolic and other high acid-value resins are also used to lake dyes and these can reduce the amount of mordant required together with improving transfer properties of the ink. Other modifications are made for inks used on kitchen paper rolls to improve resistance to domestic chemicals.

Dye inks have near Newtonian flow and give low viscosity bright colours with high strength and, on the press, printing is clean with good drying at high speed. The inks are used for paper bags, waxed bread wrappers and decorative wrapping paper. Semi-pigmented inks are made by mixing dye with white or coloured pigment. The applications of dye inks are for prints where lightfastness is not required and their tendency to migrate and bleed will cause no problems. Generally these inks are not used on plastic films. The basic dyestuffs most commonly used are:

Basic Yellow	CI 37
Basic Violet	CI 14
Basic Green	CI 1
Basic Red	CI 1
Basic Blue	CI 26
Basic Violet	CI 10
Basic Violet	CI 1

Lightfast dyes which are mainly metal complexes are available that have improved fastness to light, fats, water and other chemicals. Solubility and strength are much lower than with the basic dyestuffs and the manufacturers' literature must be consulted with respect to solubility and compatibility. Inks for foil and labels are made with these dyes and they have been used for plastic films but there is a tendency to migrate under some conditions. It should be remembered that their lower strength can require higher concentration of dye and being soluble the dye will become part of the binder so that the film-forming characteristics must be considered.

Examples of these dyes are:

Solvent Yellow	CI 19
Solvent Red	CI 102
Solvent Orange	CI 45
Solvent Blue	CI 37

Colourant – Pigments

Note that the Dyes are listed here as their CI references.

Pigments used in flexographic inks have similar specification requirements to those used for other processes. Irrespective of the properties required by the end use of the print, suitable pigments will be chosen for their wettability and dispersion characteristics in the various solvent and resin systems which are used. Generally what is being looked for is ease of dispersion with stability on storage, good flow at high pigment concentration and maximum colour strength at reasonable cost. Properties of gloss, opacity and transparency will be important for certain applications. The flow and dispersion characteristics will be dependent upon pigment type but pigments of similar chemical constitution vary considerably depending upon the processing conditions used by the pigment makers and the various additives and treatments of the pigment during manufacture.

Flexographic inks will vary widely in the solvents used (water to aliphatic hydrocarbon) and in the resin types for the different ink ranges. While it is desirable to use the same range of pigments for all inks the dispersion and rheological behaviour will differ with each solvent/binder system. In order to optimize the properties the pigments selected will tend to vary somewhat for each range. Other inks may contain additives and materials which are unstable in combination with certain pigments and these factors will also have to be considered.

The following is a list of the main pigments which are particularly suitable for flexographic inks.

Pigments	*Colour index*
4B toner	PR 57
2B toner	PR 48
Lake Red C	PR 53
Lithol Red	PR 49
Iron oxide	PR 101
Permanent Red R	PR 4
Permanent Red 2G	PO 5

Pyrazolone Orange	PO 13
Diaryl Yellows	PY 12, 13, 14
Monoazo Yellows	PY 3, 5, 98
Phthalocyanine Green	PG 7
Phthalocyanine Blue (β form)	PB 15
Ultramarine	PB 62
Permanent Violet	PV 23
Titanium dioxide	PW 6
Carbon black (furnace/channel)	PB 7
PMTA pink, green, blue, violet	PR 81, PG 1, PB 1, PV 3
Copper ferrocyanide dye complexes	PR 169, PG 45, PB 62, PV 27
Extenders	
Blanc fixe	PW 21
Calcium carbonate	PW 18
China clay	PW 19

Lead chromates are not listed on grounds of toxicity. Pigments other than those listed would be required for increased specifications, for example, soap wrappers, so that in general the choice of pigment will be influenced by the end use of the print. Many inks will be made using a combination of different pigments in order to achieve the desired shade. It is important to remember that the resistance of the resultant mixture tends to be governed by the least resistant component.

Extenders are used in flexographic inks particularly when tints or weak shades are required but, in general with the low film weights carried by the process, they are not extensively used.

Bronze and aluminium powders are also used to make metallic shades and sometimes these are blended with colourants to produce particular effects. When aluminium powder is combined with coloured pigments or dyes the resultant mixture is described as a polychromatic. It is more convenient for the printer that metallic inks be supplied as a ready made mixed dispersion but care will need to be taken with the choice of resin, solvent and additives to avoid instability problems which, at their worst, can involve evolution of gas and heat generation. Where there is any doubt, inks are supplied as a two-part mixture, powder and varnish or paste and varnish, to be mixed together just prior to printing.

Resins

The function of the resin is to act as a carrier for the pigment or dye and bind the colourant to the surface being printed giving adhesion. There are a large number of resins available and the main restriction is that any resin used should be soluble in solvents which do not adversely affect stereo materials. This can be a severe restriction with some photopolymer stereos.

With development, the number of resins available within each chemical type has increased in recent years. The ink technician needs to keep up to date with these developments and to be familiar with the properties of the variations available. In general the requirements for the resin are solubility in flexographic solvents, adhesion to substrates, good pigment-wetting, good solvent release, good film-forming properties, low odour, pale colour and compatibility with other resins. All resins vary in their fulfilment of these properties and a resin may have good adhesion properties but be a poor film former while another may have poor solvent release but improved pigment wetting. To overcome these difficulties and, because the perfect resin does not exist, blends of resins are commonly used in ink formulation and in this way a balanced range of properties can be incorporated into the ink. The increasing number of resins used in a formulation will, however, increase the possibilities of incompatibility and often increase the need for powerful solvents to bridge, the solubility requirement of all the resins used. Compatibility and solubility tests will be essential during formulation.

Shellac

Early inks were restricted to a very limited range of natural resins and shellac became the principal one used. Shellac has good pigment-wetting properties, solubility and adhesion to a large number of substrates. It is also compatible with dyes and will make good water-based inks when the resin is saponified with a base. Shellac has, however, slow solvent release properties and low heat resistance and is now very expensive. The resin, although still used, has largely been replaced by alcohol-soluble maleic and phenolics in pigmented paper and dye-based inks, although on other substrates the film-forming properties of these substitutes are not as good as shellac and other resins are used.

Nitrocellulose

The most common resin employed in solvent-based flexographic inks is nitrocellulose which is cheap and readily available and also has many other desirable features including good solvent release, low odour, good pigment wetting, heat resistance and very wide compatibility. This compatibility allows the possibility of making inks for nearly all substrates by the addition of other resins and plasticizer to the nitrocellulose. The properties of nitrocellulose inks can thus differ widely, dependent upon the particular modification. In these formulations nitrocellulose will normally confer pigment-wetting, drying, heat resistance and film-forming properties while the modifications improve gloss and adhesion. Inks made from pigmented nitrocellulose chips give further improvements in gloss which is necessary for film and foil printing

Nitrocellulose is available in different grades and in flexographic inks. It is normal to choose the alcohol-soluble grade, i.e. low nitrogen/low viscosity. This will make it possible to use less ester which is damaging to photopolymer plates.

The main limitation of the resin is firstly a tendency to react with bronze powders with the evolution of hydrogen and, secondly, its sensitivity to water contamination which can cause blushing of the print whilst drying.

Ethyl cellulose

Ethyl cellulose has similar properties to nitrocellulose in respect of heat resistance, solvent release, film formation and odour. Unfortunately however, the resin has a number of disadvantages and among these would be included poor flow, high cost and high viscosity. This has resulted in the use of ethyl cellulose being mainly confined to dye systems with which it has good compatibility and gives improved drying with low tack.

Cellulose acetate propionate

The third member of the cellulose family used in solvent inks is cellulose acetate propionate (CAP). It is similar to nitrocellulose in many of its properties but compatibility is more limited and more ester is required for solution. The primary use is in inks for PVdC coated films where it is used or in combination with acrylic resin to give low odour, heat resistant inks with good adhesion. There are two main versions available, one being more alcohol soluble than the other. CAP resin has a low tack level and this gives inks which print best at the upper speed range. Chips are available and with careful formulation free-flowing inks with good gloss can be made.

Polyamide resins

These are widely used in flexographic inks and they have good adhesion to many substrates. Pigment wetting is good and glossy inks can be made without chips by judicial selection of pigments and processing. The resin has a slight odour and low heat resistance.

Combinations with nitrocellulose are common. Many grades are available and these vary in solubility and compatibility and in general the more alcohol tolerant the resin is, the softer it becomes. Early polyamide inks required the use of aromatic solvents and the resins were described as cosolvent types. Polyamide inks can now be made with ethyl alcohol as the main solvent and these are suitable for photopolymer plates. They dry to give a slippery waxy surface. The resin is not suitable for laminating inks due to the inability of adhesives to adhere to the ink surface.

Acrylates and methacrylates

These resins in recent years have found increasing use in flexographic inks over a wider area of formulations. The range of resins in the acrylic family is large and the diversity of properties in both solvent and water inks is wide.

In solvent inks the resins are primarily used for their good adhesion properties and are also suitable as binders for metallic-pigmented inks. Acrylic resins in combination with cellulose esters or nitrate are used in inks for PVdC lacquer applications, certain lamination inks, PVC film inks and inks for adhesive tapes. The resin requires a careful balance of solvent as solubility can be difficult to achieve. Solvent release properties can vary with resin type and allowance must be made for this when formulating. Care must be exercised with the higher molecular weight acrylics as these tend to string causing dirty working.

Polyacrylic and polymethacrylic acid resins were first used in conventional water-based flexographic as a substitute for shellac (which had become too expensive). The resin is used either as a sodium or amino salt which is soluble in water or alternatively in an emulsion form. Most inks for paper sacks and corrugated cases are of this type.

In recent years there has been much development of water-based inks for film substrates and particular water-soluble acrylic resins are a basis of much of this work. With films penetration drying is not possible and the inks dry by evaporation of water and any volatile amino compound present. Given the lower volativity of water the inks tend to dry more slowly than solvent-based types. With careful formulation adequate results can be achieved, as will be explained later, although the problem is less easy to resolve when several colours are superimposed.

Styrene acrylics are used in their solvent soluble form in nitrocellulose overprint varnishes to provide good gloss. In water-based inks these resins may be used either as a solution or in a non-convertible emulsion form where their good gloss and heat resistant properties are utilized.

The use of acrylic resins in inks and varnishes has become extensive and furthermore somewhat complex, in both solvent and water systems. The technician may find it an advantage to catalogue properties, performance and modifications of the many different types of these important resins in order to optimize their use in formulations.

Polyvinyl butyral

Improved polyvinyl butyral resins have become available in recent years. These have good film-forming properties, low odour level good adhesion to some film substrates and solubility in alcohol. These properties make the resin suitable in formulations for a number of inks used for lamination, cold-seal packaging applications and metallic pigmented systems. The poor resistance however to heat and water prevents the resins being used in a wide range of applications.

Other resins

Ketone resins are used in flexographic inks to assist gloss or aid adhesion as they have good solubility and are inert. Low-odour versions are available although the film-forming properties of ketone resin is poor.

Other resins are polyvinyl acetate, polyvinyl chloride and the copolymers of these which find their main use in PVC film inks and other speciality applications. There are also a large number of additional resins used as minor components of inks to impart special properties.

Solvents

The two other main constituents of the ink, colourant and binder are both solids and therefore the prime function of the solvent is to convert the ink into the fluid form which makes it capable of being printed. The solvent is thus a carrier for the other ingredients but must be easily removed from the print by evaporation and by penetration into the substrate.

The choice of solvent is governed by a number of factors. The resin used in the ink will have specific solubility characteristics and, as is often the case, mixed resins with different solubility parameters will demand particular solvents or solvent mixtures. With flexography there will always be a restriction of choice of solvent determined by the nature of the stereo materials used. Some printers may specify the use of certain solvents which, of course, may restrict resin choice. With inks for food packaging certain solvents may be prescribed, particularly as retained solvents in print after drying may represent an unacceptable risk of odour or taint even in small quantities.

Solvent choice can also be influenced by the colourant. In the case of dye inks there is a requirement for the solvent to dissolve the dye. With pigmented ink however, flow characteristics are affected by the choice of solvent, and a particular pigment may have better flow in one solvent and worse in another. An example of this is the behaviour of diarylide yellows which while satisfactory in alcohol solvents give very poor flowing inks when based on ester solvents. Dispersion stability and colour strength development may also vary with the choice of solvents used in the ink.

Another property of inks which is affected by the solvent is adhesion. On coated films or those which are soluble, certain solvents may partly dissolve the surface of the coating or film and this will tend to assist adhesion. With non-soluble substrates and uncoated films, solvents can also affect adhesion by increasing or decreasing wetting of the surface. Adhesion is normally at its best when wetting is of a high order.

The balance of solvents used in an ink will be determined by a number of factors already mentioned above. During printing, however, more solvent will be added by the printer to adjust the ink to a suitable viscosity. The volume used may be equal to that contained in the undiluted ink. Thus the total composition must be considered when formulating and recommending diluents. Solvent will be lost by evaporation from the unit during printing and depending upon factors such as vapour pressure, temperature, resin constitution, etc., the composition of the evaporating mixture will vary. Should the evaporating mixture be at variance with that contained in the diluted ink an imbalance can occur with a possible result of instability of the ink causing precipitation of the resin and a number of side effects. This can be a particular problem on a unit with a small area of type or halftone where the same ink recirculates through the unit for hours without constant replenishment with fresh ink.

Solvent added during the run should therefore be of the same constitution as the evaporating mixture to avoid evaporation imbalance, particularly with inks for halftone printing (a classic case of low usage application). The best solution in extreme situations is to simplify formulation drastically and use only one solvent provided that a satisfactory ink for the substrate and end use can be made with these restrictions.

The principal solvents used in flexographic inks are water, ethyl alcohol (methylated spirit), isopropyl and *n*-propyl alcohols, ethyl, isopropyl and *n*-propyl acetates, methoxy propanol and ethoxy propanol. In certain cases aliphatic and aromatic hydrocarbon (SBP, toluene) or ketone solvents (acetone, MEK, MIBK) may be used for some types of ink but the suitability of the stereo composition must be considered when choosing these (Table 9.1).

Additives

These materials are incorporated into inks at different stages of manufacture to improve performance and to impart specific properties which are essential to the requirements of use and specification. The quantity of an additive used is normally small, and with the exception of plasticizer, rarely exceeds 1% of the formulation. There are many additives available and it is necessary to understand their function, limitations and any side effects they may produce, if the best use is to be obtained.

During manufacture of ink the use of surfactants will assist the dispersion of pigments, which will improve gloss, colour development and reduce the tendency of pigments to flocculate when inks are diluted. The phenomenon of flotation (which is a tendency of different pigments used in the same formulation to separate, causing streaks or marks on the print) can be overcome by careful selection of surfactant. Many pigments contain surfactants which are added by the pigment manufacturer but some surfactants do not give good results in all ink types.

Plasticizers are incorporated into many inks particularly those to print film and foil. Generally, plasticizers are used to give flexibility and adhesion to the ink film and are essential with cellulose-based inks, otherwise these would be too brittle. Certain resins do not require plasticizer since they are inherently flexible by their nature. These resins however, are not normally resistant to heat and tend to be thermoplastic. Heat resistant inks are made using cellulose materials and, as stated, it would be difficult to make such products without plasticizer. Excess plasticizer should be avoided because the ink film may thus become soft which could give rise to the risk of blocking.

With inks containing plasticizers it is possible that plasticizer, by being a liquid, may migrate through certain film substrates which have low barrier properties. Furthermore, studies have shown that plasticizer can transfer by set-off from the printed side of the web to the unprinted side during storage in the reel. Given that the pack consists simply of a single layer of outside-printed film it can readily be seen that plasticizer could find its way into the pack contents through the routes described. Should the pack content be a foodstuff the contamination possibilities are obvious, and although the amounts found during the studies were extremely small and are dependent upon many factors, this can have serious implications. Inks for food packaging should therefore only contain plasticizers which have been well researched in respect of their toxicological properties and found to be of low toxicity. The MAFF [1] Food Surveillance Papers No. 21 and 30 should be consulted. Finally, the responsibility of the packer for the choice of materials and design of the pack should not be overlooked.

Various stabilizers, which prevent reactions between ink components and inhibit breakdown of resins and reduce corrosion of metal containers, are also used in inks and intermediates. Anti-oxidants can also prevent oxidation of resins such as polyamide.

The most common additives for inks are waxes, and these will be present in most inks and varnishes, the principal exception being inks for lamination. Waxes which are normally added as powders or compounds, give improved rub, scuff and scratch resistance and also slip properties to the print. Polyethylene and polypropylene waxes are the most popular, as these do not inhibit drying to any great extent since they are not soluble in solvents. One disadvantage of wax addition is a reduction of gloss.

Adhesion promoters are important additives used in certain inks, particularly those for film substrates which have surfaces that present difficult adhesion problems such as treated polypropylene. There are a number of these materials available and some cause side effects with the use of inks such as:

- ink viscosity build due to reaction with resin or pigment;
- loss of heat-sealing properties of some coated films;
- colouration effects with certain resins or coatings;
- odour development.

It is essential, therefore, that careful testing of these materials, in the formulation contemplated, should be carried out prior to their use.

There are a number of other additives used in ink formulations which, like waxes, improve scratch and rub resistance and also reduce the coefficient of friction. Among these are fatty amides, silicone fluids and stearates. Most of these materials also migrate to the surface of the print and this can reduce the tendency of inks to block. It should be noted, however, that there can be side effects under particular circumstances. Examples of these are loss of seal properties with film, reduction of cohesion with cold-seal adhesives, and wetting and adhesion problems with overvarnishing and lamination.

Water-based inks may contain particular additive types. Anti-foams, fungicides, waxes, surfactants, wetting agents, transfer aids, coalescing agents, anti-blocking compounds and slip additives are all used to achieve different properties. Anti-foam is usually required as foaming can be troublesome both during manufacture of the ink and on printing presses particularly those with high agitation. Foaming may be difficult to overcome and there is a temptation to use excessive anti-foam agents which can cause craters and uneven print. Some anti-foams contain hydrocarbons which will attack stereo rubbers and therefore care should be taken in their choice.

9.4 INKS AND VARNISHES FOR SPECIFIC PURPOSES

Flexography originated as a process to print packaging materials in the form of paper bags. While the scope of the process has widened considerably with many substrates and products being printed, packaging has still remained the most significant area of printing for the process. Given the relative ease and low cost of flexographic plate origination, the possibilities of design changes at short notice and improving standards of printing and ease of handling of flexible materials, it is not surprising that flexography attracts an increasing share of packaging work. As previously mentioned there are, however, a large number of non-packaging applications, most of these being for paper products of many kinds. In recent years wallcoverings have been extensively decorated by flexography and newspapers are also printed by the process.

The formulation of inks and varnishes is normally governed by the needs of the substrates being printed and the end use of the print. Each class of substrate and its end use will be considered in respect of the formulation of inks and it will be seen that at present these are normally based on pigments and organic solvents. Initially, and because of their special nature, two types of inks will be considered: dye-based inks and water-based inks.

Dye-based inks

The dyes can be dissolved by stirring, and solubility and laking are improved if some heat is used. Denatured ethyl alcohol (methylated spirit) is the solvent and dependent upon the mordant which is used this can contain up to 6% water, thereby improving solubility although an excess of water will precipitate the laked dye. After the inks are made and allowed to stand at room temperature there will be a tendency for a small deposit to settle which should be discarded as this may cause problems on the press when printed.

Additions of resin solutions may be made, e.g. shellac or alcohol-soluble maleic, to improve transfer properties and wax or glycerine resistance (glycerine being used to improve the flexibility of paper for sweet wrappers). If further resistances are required phenolic resins and others can be used. Ethyl cellulose in small quantities is sometimes added to reduce tack and also to increase drying speed. Glycol ethers may be incorporated to retard the ink and because they are good dye solvents they improve clean working on the press by keeping the type open during printing.

A typical formulation for a paper ink is:

Basic Yellow 37 dye	8.0
Basic Blue 26 dye	4.0
Tannic acid (or other mordant)	20.0
Varnish (60% alcohol-soluble maleic)	5.0
Glycol ether	4.0
Ethanol	59.0
	100.0

Most dyes require a mordant level equal to their own weight but Basic Violet B and Basic Yellow 37 need twice this quantity to give maximum water and wax resistance. Colour matching with dye inks is similar to other inks but the formulator must remember that high dilution during printing is common and proofing with the anilox hand roller is difficult due to the low viscosities of these inks, therefore a knife-down or squeegee applicator gives better results dependent on the substrate and its absorbency.

Dyes may be blended with white to make a semi-pigmented ink but the use of a pigment will require more binder to stabilize the dispersion and prevent settlement. However, simple bag machines are unable to cope with the resulting increased viscosities and tack. Dyes can also be incorporated into coloured pigmented inks to increase colour strength but the dye may interact with some pigments and binders which can cause instability. Pigmented water-based inks can also be strengthened using dye solutions but the alkaline nature of the inks precludes the use of a mordant and the print is prone to bleed under damp conditions. Some of the dyes also have poor stability and compatibility with alkaline resin suspensions.

Basic dyes are not normally used in inks for film printing because of their bleed and migration properties although some early polyethylene inks were made with dye and titanium dioxide pigment. These inks were, however, superseded by pigmented polyamide inks which have better fastness.

Flexographic inks based on lightfast dyes are employed for aluminium-foil printing and these will give bright transparent effects. Inks for metallized substrates and coloured lacquers for overprinting silver inks are also made with lightfast dyes. These dyes do not require a laking agent and by consulting the manufacturer's data a selection of suitable dyes can be found that are compatible with several binders. Lightfastness can be good and resistance properties are better than with basic dyes although there is still a tendency for migration.

Disperse dye inks

Flexographic printing for the heat transfer process has decreased in recent years although it is still carried out for particular textile applications, mainly for synthetic fabrics. The inks are printed on a carrier web (usually paper) and are formulated with disperse dyestuffs which sublime at high temperature onto fabric held in contact with the web. Print quality has to be of a high standard and both solvent- and water-based inks can be used. The print on the carrier web must not rub and the resin used should not interfere with the subliming properties of the dyestuff. The selection of the dyestuff and binder is influenced by the type of fabric to be transfer-printed and the specification of the final print.

Water-based inks

Paper and board

The traditional water-based ink for paper and board consists of a pigmented resin system, the resin being first dispersed in water. While a number of natural resins exist which are soluble in water such as casein and gum arabic they will also remain soluble after printing, which is not acceptable. It is therefore necessary that the resin is capable of being dispersed but then become insoluble once printed. One method of achieving this is to add an alkali to a suitable resin and converting this to a soap or resin/water dispersion. Traditionally shellac was used but has been largely superseded by carboxylated acrylics and, at times, other resins.

Water-based inks have the advantage over solvent inks that due to their low evaporation rate the inks are stable on the press without drying on the stereo or in the duct giving constant viscosity. Water does not attack stereo materials including photolymers and the inks present no fire hazards and give economies by the use of water as a reducer and cleaning solvent.

A problem for the printer is the disposal of waste ink and wash-up solution which tend to be generated at a larger volume because water is freely available. Such wastes cannot be discharged into the drains because of pollution and have no recovery value, the alternatives being expensive water treatment plant or the cost of dumping.

The most extensive area for use of water-based inks is for multiwall paper sacks and corrugated case printing where non-flammability and fast drying on the highly absorbent substrates are an advantage. Good water resistance shortly after printing is obtained and also good level of rub resistance essential for the case- and box-making machines.

The choice of alkali for resin solution will be made depending on a number of factors – cost, equivalent weight, print waterfastness, press stability and drying speed. Properties for three alkalis are given in Table 9.2 to illustrate how these may vary.

Table 9.2 Alkali properties

Alkali	Cost	Equivalent weight	Water fastness	Press stability	Drying speed
Caustic soda	Low	40	Poor	Good	Fast
Triethanolamine	Medium	149	Fair	Good	Slow
Ammonia	Low	17	Good	Poor	Fast

Resins may be used in solution form. Acrylic resins are often supplied as a resin-in-water emulsion allowing a solution to be formed instantly by the addition of the selected alkali to the correct pH. Alternatively the emulsion can be pigmented without alkali, making an ink which will have faster drying and better resistance properties than the solution type. This will be at the expense of press stability and makes press wash-up difficult particularly when the ink is allowed to dry. Some inks are made as a mixture of the solution and emulsion form to try to arrive at a balance of properties. This may involve mixed acrylics and a limited amount of alkali.

Pigments have to be selected for stability in water-based inks and those which react in alkaline systems or show settlement must be rejected. To give improved rub resistance the use of a wax compound will be needed. Too much slip on inks for paper sacks may cause instability in the stack. Water-based inks are prone to foaming problems and anti-foam additives of the silicone type or organic defoamers are used although too much may cause poor lay, wetting problems and 'fish-eye' effects.

The manufacture of water-based inks is relatively easy and the three main methods are:

- milling of dry pigment into varnish;
- mixing of stabilized pigment water pastes with varnish;
- use of pigment press cakes with power mixing or milling.

A strength advantage is given by the use of pastes but the presence of wetting aids and glycols (which are present in the pastes) may be troublesome. A small quantity of organic solvent incorporated into the inks sometimes improves tack and lay but solvent content must be strictly limited in case of fire risk. Typical water-based flexographic ink formulations are as follows.

Red water-based ink:

Lithol red pigment (C.I. Pigment Red 49)	18.0
Acrylic/alkali water varnish	60.0
Polyethylene wax compound	4.0
Isopropyl alcohol	4.0
Water	13.9
Silicone anti-foam	0.1
	100.0

Blue water-based ink:

50% Phthalocyanine Blue water paste (CI Pigment Blue 15)	24.7
Acrylic emulsion	50.0
Water	20.0
Monoethylamine	2.0
Polyethylene wax compound	3.0
Organic defoamer	0.3
	100.0

Extender pigments may be incorporated into the ink to reduce cost and since many of these inks are used on natural kraft or coloured substrates, white pigment is added to improve opacity.

Newspapers

A number of newspapers are now printed by flexography using water-based inks. There are several reasons why the process is attractive, and among these are:

- lower installation costs – flexo presses can be cheaper than web-offset presses;
- less paper wastage – an ink applied via an anilox will result in less paper waste on start up;
- reduced cost of newsprint – water inks show less strike-through so that lower-grammage papers can be used;
- less maintenance and manpower;

- better rub resistance properties – gives a cleaner sheet to handle.

A major problem still associated with flexographic newsprint is plugging of the polymer stereo which then leads to deterioration of print quality. This can be caused by linting of some grades of paper and the effect can be exacerbated by the use of tacky inks which may pull fibres from the paper surface.

The ink requirements for printing newspapers are somewhat different from those for other applications. The main difference is the exceptional high speed of printing (up to 600 mpm).

High optical density is required with black ink used for printing type and the density should generally not be less than 1.15 Status T (USA standard – wide-band colour reflection densitometer). This must be achieved at very low printing viscosities and a thin film of ink, normally metered by a reverse-angle doctor blade and anilox roller. Typically the anilox is a ceramic type with 140 lines per centimetre at a depth of 35 microns. In order to obtain satisfactory optical density under these conditions, pigment:binder ratios for the ink will lie between 2:1 and 3:1.

Good resolubility of the ink is essential to prevent dried ink building up on the stereo and eventually filling-in the shadow areas of the halftone. Resolubility can be achieved by formulating inks based on acrylic resins or colloidal dispersions, solubilized with sodium hydroxide or liquid amines of low volatility.

Due to the very high press speeds, rapid-drying inks are required to avoid the problem of second impression set-off and marking on the turner bars. However, due to the adsorbency of the substrate this is only a problem when multicolour layers are involved. In fact the drying speed can be made too fast and, particularly if the dried ink is fairly brittle, this may cause the formation of a fine dust which can initiate plugging.

The way in which the inks dry is also critical. If too much tack is developed, during the drying phase, paper fibres might be picked off and these can swell in water and cause plugging. The correct choice of binder and amine is the best method of controlling drying speed. The use of alcohol should be avoided since this may soften certain photopolymer plates.

Due to the low viscosities and high press speeds, foam control of flexographic news ink is more critical than with conventional water-based inks. Powerful silicone-based defoamers are usually employed but care must be taken to choose one which does not swell the sensitive polymer plates.

A basic formulation for a News Ink Black:

Carbon black pigment C.I. 7	23.0
Solubilized acrylic resin varnish or emulsion	38.0
Anti-foam	0.5
Surfactant	0.5
Water	38.0
	100.0

Bulk storage of low viscosity, neat or diluted flexographic news inks can give rise to settling problems. These can be overcome by the correct choice of binder and pigment but a good degree of dispersion during manufacture is also required. The level of dispersion is often so high that microscopic methods of measurement are now replacing the traditional grinding gauge for quality control.

Impervious substrates

Water-based inks for impervious substrates is an area of technology which is currently being developed by ink companies. Although a number of successful commercial results have been achieved it is too early to provide guidance on formulation for specific applications. This section will therefore be confined to notes of the general principles which apply to these systems.

The requirements for these inks are quite different from those which are applicable to paper and board inks. The main problems are:

- drying
- adhesion/flexibility
- substrate wetting
- resistance properties
- printing conditions.

Because of the nature of impervious substrates ink drying is mostly by evaporation. Water has a much higher latent heat of evaporation than the commonly used organic solvents, meaning that it takes more heat for it to evaporate. It is desirable to keep the water content as low as possible and inks which contain high solids levels dry more rapidly than those which contain more water.

Generally, high solids are achieved by using non-convertible emulsion polymers. Emulsions when used alone, however, can give resolubility and drying-in problems during printing. It is therefore normal to include some resin, which has been solubilized with amine, to improve resolution. The inclusion of soluble resin will slow drying and a balanced blend becomes necessary to achieve a compromise of properties.

Addition of alcohol can speed up drying, although the level of alcohol will be limited to control flash point and furthermore to avoid compatibility problems with emulsions. A typical alcohol content is 5%.

The best way to speed up drying is by applying a lower film-weight of ink. Because water inks give rapid viscosity reductions on dilution (Fig. 9.2) this will allow high solids-strength at printing viscosities without compromising print strength.

Also important to drying is the choice of solubilizing agent (as previously mentioned) and of those available, ammonia gives more rapid drying due to its high volatility. The choice of pigment dispersion is also important to drying and those which contain excess glycol should be avoided.

The adhesion and flexibility of water-based inks to impervious substrates is controlled by the selection of polymer emulsion. These are usually based on acrylic resins and there are a wide range of types available. The softer, low-T_g (glass transition temperature) polymers normally give the best adhesion and flexibility properties when used on many film substrates, but these polymers have poor heat resistance.

On uncoated substrates water-based inks are far less tolerant to variations of treatment level and the presence of surface active materials, such as waxes, than is the case with organic solvent-based inks. Adhesion may suffer as the result of these variations. The minimum corona discharge treatment level for satisfactory adhesion to be obtained is normally 38–40 dynes/cm. Ideally, corona discharge treatment should be carried out in-line or shortly before printing.

Alternative resin systems which exhibit good adhesion to polyolefin films are water-based urethanes, polyamides, epoxy esters and saponified rosin. Unlike solvent-based inks it is not common to use plasticizers to increase flexibility.

Good substrate wetting is vital to ensure satisfactory printability. A major problem with aqueous systems is due to the high surface energy of water 73 dynes/cm. In contrast the surface tension of ethyl alcohol and isopropyl alcohol is 24 and 21.7, respectively.

In order to reduce the surface tension of water-based inks which will allow better wetting of substrates, surfactants and alcohols are used in formulations. The use of both anionic and non-ionic surfactants may cause problems of poor water resistance in inks and care in the choice of these must be exercised.

A further consequence of reducing surface tension is to increase the tendency to foam. The choice of defoamer is also critical to prevent poor wetting which may cause reticulation of the print. Good performance on impervious substrates using water-based inks usually involves a balancing act between surfactant and defoamer.

The end-use resistance properties of prints made using water inks is governed by the correct choice of resin/emulsion binder. With acrylic systems, emulsions tend to give better results. Emulsion-based inks are, however, more difficult to print due to their relatively poor resolubility and the ultimate resistance may have to be compromised by the addition of solubilized resins to achieve a balance.

In the past the addition of polyaziridines has been used as cross-linkers to confer high levels of resistance properties. Their use is, however, now in decline due to health and safety concerns.

It should be noted that, in general, water-based inks do not develop their maximum resistance properties as quickly as solvent-based inks. Resistance tests should therefore be only carried out after a period of time has elapsed.

Much has been written concerning how water-based inks should be handled on the press. Inkmakers will normally advise customers on the best use of their products. In general, the following rules should be observed:

- use the minimum reduction with water;
- apply a thin ink film with shallow anilox and a blade;
- print at a relatively high viscosity;
- do not adjust print strength with water, use medium;
- pH of the ink should be monitored (7.5–8.5);
- wash press immediately after use
- clean anilox and stereo after printing breaks;
- use a well treated substrate, free from slip additives if possible.

Solvent-based pigmented inks for specific substrates

Paper and board

In formulating pigmented solvent-based inks for paper substrates the main concerns are in producing inks having satisfactory colour and strength with good printing properties together with good drying and rapid solvent dilution characteristics. Adhesion to board and paper surfaces is not normally a problem with the possible exception of glassine and heavily coated papers. Most inks must be dry rub resistant unless an overvarnish is used.

Flexographic printed papers are used for labels, packaging and decorative wrappers, wallpaper and other applications already mentioned. Consequently the choice of pigment must satisfy the resistance requirements associated with the end use. General wrapping and bag inks do not demand lightfast pigments except the normal avoidance of those with poor lightfastness in pastel shades as this will reduce lightfastness even further. Pigment used for waxed papers and soap wrappers would need to be insoluble in the respective materials in each case.

The principal binder for these inks is nitrocellulose which has good pigment wetting and can be pigmented either by direct grind methods or, where increased gloss and strength are required, by using chip dispersions. Nitrocellulose resin gives good drying and since pigmented inks are not absorbed as readily into paper, as is the case with dye inks, faster drying is required. Maleic resin is often used with nitrocellulose in the inks and the main solvent is ethanol with esters or glycol ethers to aid solubility. Plasticizers and waxes may be incorporated to improve flexibility and rub resistance, and other binders or modifiers may be used for special applications or to give particular properties.

Typical formulation of ink for paper:

Phethalocyanine Blue (CI Pigment Blue 15:3)	14.0
Titanium dioxide (CI Pigment White 6)	6.0
Maleic resin varnish	16.0
Nitrocellulose varnish	38.0
Wax compound	4.0
Plasticizer	4.0
Ethanol	11.0
Isopropyl acetate	7.0
	100.0

A formulation using chips:

N/C 2B red toner chip (CI Pigment Red 48)	24.0
Isopropyl acetate	10.0
Ethanol	40.0
Maleic resin varnish	12.0
Polyethylene wax compound	4.0
Nitrocellulose varnish	10.0
	100.0

Film inks

Inks for films have different requirements from those for paper. There are two main differences. First, the substrates being non-permeable will cause the ink to lie on the surface. As adhesion is not assisted by penetration as is the case for paper, it will therefore be dependent on the physical and chemical attractions between the ink and substrate surface. In some cases, fusion of ink and film coating or solvent attack on the base film or coating occurs. Secondly, because the substrate is transparent it is possible to print on the face or reverse side of the film and it is normal to use a white underneath the colour in the former case and to back the colour with white in reverse printing. The white is used to opacify the film and increase the contrast of the print.

Reverse printing has the advantage that the colour is viewed through the film which increases gloss and brightness as well as giving protection to the print against abrasion and rub. Laminates will normally be printed this way if the lamination is carried out after printing. Films which are reverse printed need to be very transparent and cloudy substrates such as extruded polyethylene are unsuitable. In recent years there has been an almost total move away from reverse printing for non-laminated food packaging towards face printing to avoid any possibility of contamination contact of packaged foodstuffs by the ink or retained solvent. Reverse printing, although still carried out, normally requires an overlacquer or the use of an inner ply to prevent ink/food contact.

In discussing inks for films it will be assumed that these are face-printed unless otherwise stated or dealing with laminates. With face prints, adhesion of the white ink, if used, is paramount since this will be in direct contact with the substrate. It may be possible to use a common range of coloured inks on several substrates by taking advantage of the adhesion properties of different whites which are suitable for each substrate, the white being used as a key coat. The inkmaker will advise when such possibilities exist.

Cellulose films

Regenerated cellulose films were the first packaging films used but, although still extensively printed, they have given way to various plastic films in many areas of use for several reasons: price considerations, changing packaging demands and the more widespread use of deep freezing. Cellulose film is available in three main types: uncoated, nitrocellulose coated and polymer coated. Some of the more common films are listed below.

P and PF	Plain cellulose films, uncoated
PS and QMS	Nitrocellulose coated, semi-permeable, sealing
MF	Nitrocellulose coated, semi-permeable, moisture proof
DMS	Nitrocellulose coated one side with moisture proof, sealable

MS	Nitrocellulose coated both sides with moisture proof, sealable
MXDT	PVdC barrier-coated one side
MXXT/S	Solvent PVdC barrier-coated both sides, sealable
MXXT/A and MXXT/W	Aqueous PVdC barrier-coated both sides, sealable
WS	PVC/PVA coated both sides, sealable

Uncoated and single-side coated films

Uncoated P and PF film has reasonable fat and grease resistance but poor moisture resistance. It is not used to the same extent as coated films but is used for certain applications such as twist wraps for sweets and cellulose adhesive tapes. P film can present the inkmaker with a number of problems and among these are difficulties with adhesion at high film weights, blocking, solvent retention by the film and bleeding of certain pigments into glycol plasticizers contained in the film.

Inks for twist wrap may be based on nitro or ethyl cellulose (MF film is also used for twist wrap and the same inks can be used). For adhesive tapes polyamide inks are suitable, but care will be required with the choice of pigments to avoid those which bleed or react with the adhesives.

Single-side coated films DMS and MXDT are primarily used for polyethylene extrusion or adhesive laminates with printing being applied to the uncoated P-side of the web. Similar ink problems to those mentioned with P-film printing exist but adhesion to the substrate and compatibility with adhesives is of basic importance for a good laminate. Polyamide inks are unsuitable for adhesive lamination due to a tendency of polyamide to interfere with two-component adhesive reactants. Polyamide/nitrocellulose ink can however be used for extrusion laminates.

Nitrocellulose-coated films

There are a number of cellulose films coated with nitrocellulose most of which are heat-sealable. PS and Q films are permeable to moisture and are used for packaging bakery and meat products. Although the films are sealable, it is unusual for the seal to be in the printed area. Polyamide inks may be used which have the necessary deep-freeze resistance properties. Pigments which are prone to migration should be avoided as these may bleed.

MS film is designed to have heat sealing, moisture and gas barrier properties as well as being impervious to fats and grease. The film is used for packing products which require a moderate degree of protection from moisture. It has also been used for many years, as unprinted film for cigarette overwrapping. The main problem for the inkmaker is that wax contained in the coating causes wetting problems with the ink, which may crawl on the surface before drying. Pinholes which appear on the printed films vary according to the source of the film. Formulations to overcome this will contain both resins which wet the surface and anti-pinhole compounds to reduce surface tension. Ethyl cellulose helps as it sets the ink before crawling while non-alcohol solvents aid wetting. MS films vary in their tendency to give pinholing and over reduction of the ink with solvent will increase the problem.

Adhesion to MS film is less of a problem as it is possible to fuse the film coating and the ink together by heating the print while drying.

Inks which are heat resistant are, however, more difficult to fuse. Solvent retention can be a problem and alcohol solvents can be absorbed more readily than other solvents. Heat-resistant inks are often required for MS film if the print comes in the seal area and nitrocellulose is the normal binder used together with a modifying resin and plasticizer. It will be necessary to conduct tests to ensure that the formulation will not block and has satisfactory adhesion and flexibility particularly when inks are superimposed.

A typical formulation for an MS film ink is:

Calcium 2B pigment (CI Pigment Red 48.2)	14.0
Titanium dioxide (CI Pigment White 6)	4.0
Nitrocellulose varnish	40.0
Shellac varnish	9.0
Polyethylene wax compound	3.0
Anti-pinhole varnish	3.0
Plasticizer	3.0
Isopropyl acetate	7.0
Ethanol	17.0
	100.0

Alternatively the ink could be formulated from nitrocellulose pigmented chips to give more gloss.

PVdC copolymer-coated film

PVdC coated cellulose film has improved gas and moisture-barrier properties compared with MS film. The film is available in two main grades with the polymer coating being applied during manufacture from either a solvent solution (MXXT/S) or from an aqueous dispersion (MXXT/A and /W), the latter grade being less affected by strong solvents contained in the ink and having better barrier properties. MXXT films are widely used on food products which require its high barrier protection, e.g. biscuits and some snack foods, although coated polypropylene and also coextruded polypropylene films have replaced MXXT in many cases. The films are heat sealable and this will affect ink formulations since the seal is often in the printed area.

Polyamide resins have good adhesion to MXXT but poor heat resistance and although this may be improved by the addition of nitrocellulose it is difficult to achieve sufficient heat resistance for direct ink application in the seal area. A number of ink additives can damage PVdC coating either by affecting heat-sealing properties or moisture-vapour transmission rates. Tests will need to be conducted if additives of chemical active type are used.

Acrylic resins have traditionally been used to give adhesion to PVdC-coated films and when these resins are combined with cellulose esters or nitrate, heat-resistant inks can be formulated and since there are a large number of acrylic resins available the selection should be made on a balance of properties. These would include:

- high alcohol tolerance;
- low odour and a low solvent retention;
- good compatibility with cellulose-based resins;
- good pigment wetting;
- adhesion to PVdC surfaces.

The main problems with the use of these resins in flexographic inks is the relatively high ester solvent requirement for solubility and a tendency to string. Photopolymer plates are also difficult to use with the solvents needed. With surface printing it is possible to formulate a white ink on an acrylic/cellulose basis with good adhesion and use other heat resistance inks, for example nitrocellulose colours for overprinting.

A basic breakdown formulation for heat resistant white ink is:

Titanium dioxide (CI Pigment White 6)	33.0
Cellulose acetate propionate	9.0
Acrylic resin	8.0
Plasticizer	3.0
Ethyl acetate	11.0
Normal propyl acetate	10.0
Ethanol	25.0
Polyethylene wax powder	1.0
	100.0

Isopropyl acetate is avoided due to its tendency to be retained by PVdC coatings after drying.

Pinholing of the type associated with MS film is not a feature of MXXT films although occasionally wetting problems can be encountered which may be a function of the coating or its components. MXXT films (as is the case with other coated films) are subject to changes as manufacturers seek to improve their products for storage, sealing or packaging performance and these changes may influence ink behaviour in unpredictable ways. The ink manufacturer will need to keep abreast of such changes by constant liaison with filmmakers and users so that he can make necessary adjustments or recommendations as required, thus avoiding most problems before they occur.

PVdC overlacquering

Prints on MXXT films in common with prints on coated polypropylene and other PVdC coated substrates may be overlacquered with aqueous PVdC emulsions to improve sealability, barrier properties or give added protection and gloss to the print. The overlacquer can be applied in-line or as a separate operation and will be heated to coalesce the emulsified PVdC and to evaporate the water contained. With such applications both white and coloured inks need to have an affinity for PVdC. Adhesion to the film coating and receptivity of the print surface for the lacquer without poor wetting and crawling are necessary attributes. Additionally interlayer bond strength must be high to resist tape tests and possible breakdown of heat seals.

In these circumstances the overprint colours as well as the white need to be formulated with these considerations in mind and nitrocellulose overprint colours, as previously described for non-lacquered surface printing, are unsuitable. Colours similar to the white formulation given will be more suitable and tests should be made to check the performance of any formulation used. When considering the pigment choice for PVdC overlacquered colours, note should be taken of the highly acidic nature of these emulsions which may cause certain pigments to bleed.

Polyolefin films

There is a wide and expanding range of films in this group and together they now represent the largest use of packaging films. All of these materials are important with respect to the flexographic process. For the purpose of covering the ink requirement they are placed into three main categories: uncoated, co-extruded and coated. The following list gives the main films.

Uncoated polyolefins	
LD polyethylene	Low density polyethylene may contain additions of ethylene vinyl acetate (EVA) or linear low density (LLD) polymers
HD polyethylene	High density polyethylene
OPP polyethylene	Orientated polypropylene
Cast polypropylene	
Coextruded polypropylenes	
Coextruded polypropylene	Three-ply, polypropylene inner, modified polypropylene outer
Pearlized polypropylene	Three-ply, voided inner, modified polypropylene outer (may contain TiO_2)
Coated polypropylenes	
PVdC-coated polypropylene	Aqueous PVdC-coated both sides
Acrylic-coated polypropylene	Coated both sides
Differential-coated polypropylene	Acrylic one side, PVdC other side
Vinyl-coated polypropylene	PVC/PVA-coated both sides
Coated pearlized	Various acrylic or PVdC coatings

Polyethylene film together with cast and orientated polypropylene films require to be treated by corona discharge before printing. Without treatment, adhesion is difficult if not impossible with conventional inks. The treatment may be carried out on one or both sides of the substrate. Avoidance of ink blocking is more difficult on two-side-treated film, particularly if the film has a different level of treatment on each side, as there is a tendency of the print to be attracted towards the more highly treated surface.

Polyethylene and uncoated polypropylene films

Treated film will normally show a surface wetting value between 36 and 40 dynes/cm. Excessive treatment can result in a print with poor water resistance, due to improved wetting of the surface by water, which can penetrate under the print. With polypropylene films the effect of treatment is of shorter duration than with polyethylene.

There has been a steady move towards inks which are alcohol dilutable and hydrocarbon free and these have become the most common type in the UK. Most of these inks are still based on polyamide but of the alcohol-soluble type which is softer than cosolvent polyamide, therefore nitrocellulose is added to increase drying. Inks of this type can be formulated for polyethylene which have all of the desirable properties of the cosolvent systems but do not gell, are faster drying, suitable for polymer plates and require less solvent for dilution to reach press viscosity, thus giving stronger prints. Choice of resin, pigment and additives are all important in achieving a good result.

Polyamide inks can be modified with rosin-based resins to improve their adhesion and water resistance although compatibility must be considered. Waxes are used to improve scuff resistance and fatty amides may be incorporated to adjust slip characteristics. The film itself may contain several additives and the combination of these together with those in the ink will need careful control to achieve the desired result. Pigments should be chosen to give good flow in the binder and solvents used and for other properties such as water resistance (since these are affected by some pigments). Polyethylene is a low-barrier material and the use of pigments which may bleed or migrate through the film must be avoided or thoroughly tested. Some pigments may cause oxidation of polyamide and should be used only with a suitable anti-oxidant or avoided altogether.

Cast polypropylene film which is used for garment packaging because of its high clarity has somewhat improved barrier properties than polyethylene but because of the possibilities of lower treatment levels the alcohol-dilutable inks may need some modification to improve their adhesion. In other respects similar inks are used.

A typical formula for a cosolvent ink:

Organic pigment	12.0
Cosolvent polyamide resin	20.0
Maleic or phenolic resin	3.0
Normal propyl alcohol	22.0
Isopropyl alcohol	22.0
Aliphatic hydrocarbon	15.0
Polyethylene wax compound	5.0
Fatty acid amine	1.0
	100.0

A typical starting formulation for an alcohol-dilutable hydrocarbon-free ink:

Organic pigment	12.0
Alcohol-soluble polyamide resin	22.0
Nitrocellulose (dry weight)	4.0
Ethanol	29.0
n-propyl alcohol	18.0
n-propyl acetate	10.0
Polyethylene wax compound	4.0
Fatty acid amide	1.0
	100.0

Alternatively, chips could be used with both formulations although gloss is quite good by direct dispersion methods.

Polyethylene is also used in heavy gauges to make sacks to pack fertilizers, animal feeds and garden and agricultural products. Polyethylene sacks are stored in the open and the inks used must resist weathering, the chemicals which are packed, and have good lightfastness. A testing programme will have to be conducted over many months to ensure that both pigments and binders resist simulated weathering conditions before such inks are used. It should also be remembered that the substrate plays an equally important part in achieving its specified performance.

Coextruded polypropylene (COEX) film

Coextruded polypropylene films have in recent years become important for packaging of many products including crisps, snack foods, chocolate-coated products and confectionery. Moisture-barrier properties of these films are similar to coated films but gas barrier is inferior. COEX film consists of a central polypropylene core with the outside layers modified using polyethylene or other polymer groups to melting point, thus giving good sealing properties without film distortion.

Polyamide inks give good adhesion to the treated surface but, since the film is heat sealable, inks resistant to heat will be required for many jobs. Moreover, due to the nature of the products packed, grease resistance is also desirable. Because of the difficulties in satisfying these requirements, early systems consisted of a chemically reactive two-pot white or primer based on polyester–isocyanate chemistry, overprinted with conventional heat-resistant inks.

Some years ago these systems were replaced by nitrocellulose inks modified with urethane resins and adhesion promoters. These inks give satisfactory heat and grease resistance and have become the standard inks for printing COEX by flexo for snack foods.

Coextruded film may be used in cold seal applications and in this case non heat-resistant inks would be suitable; overprint varnishes may also be used and these applications will be discussed later.

Pearlized coextruded polypropylene is another coextruded substrate which has made fast progress in similar market areas to the plain coextruded films and also by the employment of special conversion methods used for biscuit packaging. These films have a voided polypropylene core which gives the pearl effect but also some loss of barrier properties and internal film strength. As much of the film is used on cold-seal packs, inks which have good adhesion but without heat resistance are normally satisfactory. The main concerns for the ink and overlacquer are that they should not adhere to the cold-seal adhesive. When heat resistance is required, similar inks to those described for plain coextruded films are suitable.

Pearlized films have found an increasing use for biscuit packaging because of their attractive appearance, but due to the low barrier properties of the film a PVdC lacquer is often applied over the print to help heat sealability; the lacquer may also be applied to the reverse unprinted side of the web. An ink will therefore be needed to adhere to the uncoated treated surface of the film and be receptive to the aqueous PVdC over lacquer. Conventional flexographic inks are not suitable and will only perform one of the functions required. One possibility is to use a suitable print primer on the substrate, overprinted with inks of the acrylic/cellulose ester type (already described for use with coated films) but in view of the many layers of coatings and inks an additional layer of primer is not desirable.

A method in use at present to solve these problems is to use a two-pot white of the epoxy type which will adhere to the base substrate and is also suitable for PVdC overlacquering. This gives good interlayer bond for heat sealing. More recently various one-pot white inks have been developed which will perform the same function, but these can have slightly lower bond to the base substrate when measuring heat seals made from the pack. Both the two-pot and one-pot whites can be overprinted with the PVdC lacquerable colours as just mentioned.

It should be noted that converter-applied PVdC overlacquering involves complex ink technology and certainly requires skills in printing and conversion. Several printing and coating operations are needed for a particular job.

Coated polypropylene film

PVdC-coated polypropylene was the first established and is still the most extensively used coated plastic film. Suitable for biscuit wrappers and a wide range of other packaging the substrate has excellent gas and moisture barrier properties and in common with other polypropylene based films has good low temperature stability.

Heat-resistant low-odour inks are used to surface print the coated film and when the print is overcoated with aqueous PVdC lacquers the ink must be receptive to this coating application. Inks previously discussed for MXXT films are also suitable for printing this film, although two main differences between the film types exist. First, coated polypropylene has lower heat resistance than MXXT and the inks therefore need to show good adhesion at the lower drying temperatures used during printing. Secondly, there is a tendency for the coated polypropylene film to absorb strong solvent such as hydrocarbons and esters and as well as being retained in the prints these solvents can reduce the bond between the base film and its coating.

Acrylic-coated polypropylene is used for similar packaging applications to PVdC-coated films. The film is a high-barrier sealable type and ink adhesion is not normally difficult so that more ink types can be used than with PVdC-coated films. Solvent retention by the acrylic coating and the possibilities of blocking do, however, demand care in the choice of solvents used in the ink, as esters and hydrocarbons will solubilize acrylic resins.

Differentially coated polypropylene films with an acrylic coating one side and PVdC the other have become more widely used in recent years, and both plain film and voided-core versions are available. These films have good barrier properties and make very good heat seals as both sides are compatible for sealing. The manufacturer recommends that printing is carried out on the acrylic side and similar ink recommendations are made as for acrylic-coated films, and of course the same precautions need to be taken with respect to solvents.

Other plastic films

Polyvinyl chloride (PVC) films

PVC films vary in their polymer composition and can be unplasticized or contain different quantities of plasticizer. Flexography is not the main printing process, although vinyl tapes, labels, wallcovering and some packaging are printed by the process.

The copolymer resins of polyvinyl chloride and polyvinyl acetate are difficult to use in flexographic inks because of the ketone and ester solubility requirement of this resin type. Compromise formulations of cellulose with acrylic resins or other binders have to be used and careful formulation with comprehensive testing should be carried out to achieve a balance of adhesion, plasticizer tolerance and non-blocking. Solvent retention can also be a problem particularly on the unplasticized grades.

Pigments have to be chosen for non-migration into the film and non-bleed in the film plasticizers. Adhesive tape inks should exclude colourants which cause breakdown of latex adhesives. With inks for wallcoverings, there are additional requirements of good lightfastness together with high temperature stability for heat embossing work.

Nylon films

Nylon films have high strength but poor moisture and modest gas barrier properties. They are used in laminates for thermoforming and in deep-draw vacuum packs and various boil-in-the-bag applications. Good adhesion is obtained with polyamide formulations (such as those already given for polyethylene) but inks for laminates are more difficult. Boil-in-the-bag work is also difficult as the substrate is permeable to water and two-pack catalysed ink systems may need to be used. Much film, particularly orientated nylon, is unprinted and retailers apply their own adhesive labels.

Polyester films

These films are tough, heat resistant, have very good barrier properties and are used in laminate complexes, pharmaceutical packaging, boil-in-the-bag and liquid pouches and for carbon-copy film.

In its uncoated form polyamide and vinyl inks have good adhesion and additional inks are suitable, if the film surface is corona-discharge treated by the converter prior to printing. Primed/treated versions of these films are available to which ink adhesion is less difficult and various modified nitrocellulose inks can be used for surface printing or with laminates. Polyester films are also coated with PVdC and this allows the use of similar inks to those described under MXXT film.

Metal and metallized substrates

Aluminium foil

This material has greater light, gas and liquid barrier protection properties than other flexible packaging materials. These properties, together with its reflective appearance, have led to foil being chosen where such requirements are of a high priority.

The surface of aluminium when exposed to air oxidizes rapidly, and once this occurs, ink adhesion becomes difficult. With virgin foil in the reel, the surface will not readily oxidize due to the exclusion of air and the presence of traces of lubricating oils left from the foil-rolling process during manufacture. When foil is laminated to paper or board however, it becomes necessary to wash-coat during lamination with a dilute lacquer containing various resins such as nitrocellulose or shellac. This wash coat acts as a primer for ink adhesion, in addition to preventing oxidation.

In the virgin or washed form, adhesion with polyamide, nitrocellulose or specialized formulations is not difficult. Inks are required to be flexible and resistant to the packed products and with confectionery and chocolate wrappers low odour level is essential. Resistance to blocking must be carefully tested with virgin foil, particularly as pressure in the reel can be severe.

Bright transparent effects can be given using dye inks (as already mentioned) but for non-bleed results and high-resistance specifications pigmented inks, which are also transparent, can be formulated using selected pigments and chips.

A typical formulation for a transparent ink for aluminium foil:

Nitrocellulose pigmented chip (60% pigment)	20.0
Maleic or shellac varnish	10.0
Alcohol-soluble polyamide	4.0
Nitrocellulose varnish	10.0
Micronized polyethylene wax	0.5
Plasticizer	3.0
n-propyl alcohol	20.0
Ethanol	22.5
n-propyl acetate	10.0
	100.0

Metallized substrates

A number of substrates are metallized by the vacuum metallizing process.

The metal can be deposited directly onto certain base substrates or, alternatively, by the use of a primer which is applied before processing. Among the materials which are metallized for packaging applications are paper and board, cellulose films, polyester and coated and uncoated polypropylene. Metallization gives the substrate a similar appearance to aluminium foil, reduced light penetration and slightly improved barrier properties.

When metallized substrates were first introduced, it proved very difficult to obtain satisfactory ink adhesion on to the surface of the metal. Furthermore, as the substrate aged, adhesion problems increased, due no doubt to oxidation of the aluminium. A number of improvements have been made to these substrates and in general the useable period has been extended considerably. Nevertheless, it is prudent to carry out tests, with ink and substrate, prior to printing.

Metallized paper presents fewer adhesion problems than metallized film. Polyamide inks will give satisfactory results, as will other formulations. Depending on the specification required, it may become necessary to test the ink on the batch of substrate which is being printed to ensure satisfactory results.

Printing of film is invariably carried out on the metallized side of the substrate as this gives brighter prints. Simple polyamide formulations will be satisfactory for some substrates but if the print has to be heat resistant, more complex specialized formulations will be required. Again tests will be necessary prior to printing.

Because of the problems which are described above, printing on metallized film is often carried out using a print primer. A number of primers are available which are suitable for different metallized films, and in recent years water-based types have proved particularly successfully and are widely used. The use of a primer will normally overcome many of the problems discussed.

Inks for lamination

Laminates are formed from two or more substrates which are held together and thus combine the properties from each substrate. Although printing may be carried out on the laminate complex this section will cover only those laminates where the ink is sandwiched between two layers. The possibilities of film combinations which can be laminated together is extensive although the majority of laminates employ polyethylene as the inner layer for heat sealability. There are several methods of lamination and three main types will be considered:

- adhesive lamination
- extrusion lamination
- extrusion coating.

Adhesive lamination may be carried out using solvent-based or solventless adhesives. These may be one-pot non-reactive or two-component catalysed types. Water-based adhesives are also used. The adhesive is normally, but not always, applied to the printed web. Any solvent or water from the adhesive is first evaporated and the two webs are brought together between nip rollers under pressure. Sometimes, dependent upon the adhesive, these rollers are heated. Generally, a few days are required before the bond between the laminated layers is developed.

The main concerns for the formulator is the nature of the printed substrate, the properties of the adhesive and the laminating conditions. Substrates which can be printed and adhesive laminated include DMS, MXDT, MXXT, polyester, nylon and the different kinds of uncoated and coated polypropylene. The nature of the unprinted web in the laminate construction is generally not important to the properties of the ink.

The ink requirements are good adhesion to the substrate and compatibility with the adhesive. It is difficult to be too specific regarding ink formulation because of the many substrates involved and adhesives used, but, in general, resins which give poor bonds are polyamide, ethyl cellulose and PVA while those which find use in formulations are acrylics, PVB, vinyl and various combinations and modifications of nitrocellulose and cellulose esters. The choice of plasticizer is important and often chemical adhesion promoters are used. Wax compounds and materials which give slip or abrasion resistance to surface prints can cause delamination if used. Moreover, it is frequently found that matt inks are more receptive to adhesives than gloss inks. Pigments are not normally a problem although PMTA types may bleed in adhesive solvents.

With the many substrates and adhesives used there is an obvious advantage if one ink type is suitable for several different laminates. It becomes necessary for the inkmaker to be very specific in the recommendations made regarding which substrates and which adhesives have been tested and found to be satisfactory. Lamination bond failure is only discovered after printing is completed and the laminate is formed. Checks which ensure satisfactory final results are not easily made during the production stages.

Extrusion laminates are formed from two webs which are combined by extruding a thin film of hot polyethylene on to one web and nipping the films together while the polyethylene is still molten. The webs may be of the same or different composition, dependent upon requirement. Extrusion coating is made in a similar manner but, in this case, the molten polyethylene is extruded onto the printed side of a single web which is then passed over a chill roller.

Ink formulations need to have a good affinity for polyethylene. Polyamide and shellac inks in combination with nitrocellulose give good results. Bond strength is generally not as high as with adhesive lamination and the inks need good solvent release properties because retained solvent affects the strength of the bond.

Miscellaneous products

Metallic inks

These inks are made from aluminium or bronze powders mixed with a suitable resin solution. The particle size of the metallic powders used in flexographic inks are larger than pigment particles and normally are leafing grades which give maximum reflectance and coverage. When first mixed into a varnish, brilliance will be at its best but rub resistance may take a day to develop to its maximum and there will be some loss of brilliance due to 'wetting out' of the metal with resin and solvent. Inks may be supplied ready mixed but the binder choice will be restricted to those that are neutral and give storage stability. With other less stable binders the inks will be supplied as a powder or solvent paste together with a varnish and the mixing is carried out prior to printing.

Aluminium powders are more stable with resins than bronze powders but gold shades can be obtained with aluminium by mixing with a varnish which contains some dye or finely dispersed pigment. These mixtures are, however, less bright than a bronze powder-based ink but have the advantage of being copper free, which may be necessary for certain food wrapper specifications. The type of resin binders for metallic inks are dependent on the substrate and end use but good leafing of the metal flake has to be considered. Over concentration of the powder should be avoided as ink film strength can be affected together with the adhesion of any overprinting colours. High powder concentration should also be avoided in lamination inks. Suitable resins are polyamide, ethyl cellulose, PVA, acrylics and cellulose esters. Nitrocellulose can be stable in some aluminium-based inks but bronze powders are reactive and care must be exercised. Stabilizers can be added to help stability in certain cases.

Gold ink:

Bronze powder	30.0
PVA varnish (25% solids)	55.0
Wax compound	5.0
Ethanol	10.0
	100.0

Silver ink:

Aluminium powder	16.0
Polyamide varnish (40% solids)	60.0
Wax compound	4.0
n-propyl alcohol	15.0
Aliphatic hydrocarbon	5.0
	100.0

Halftone inks

The use of halftone and colour process printing has increased dramatically in recent years and is made possible by better printing equipment and printing plates. Good quality inks will give satisfactory results with halftone, but most manufacturers offer specialized inks for two main reasons:

- to ensure that each colour gives the correct hue and clean shade which are required for process work and, moreover, as a standard for making plates.

- to ensure that the ink has good solubility with few solvents (preferably one), to avoid solvent balance difficulties which can arise with low ink usage over long periods as can be the case with halftone printing.

High resistance specification inks with low solubility resins and complex solvent mixtures are unsuitable and will not give clean printing, which is essential for good reproduction.

Polyethylene-coated board

Polyethylene-coated board which is printed before extrusion has been discussed under laminates, but when printing is carried out on the outside extruded layer the polyethylene is first treated by corona discharge. Both printed forms are used extensively for milk and other liquid packaging to make cartons of different types.

Flexibility of the ink is only required to facilitate folding and the main concerns for the inkmaker are abrasion, rub resistance and adhesion together with meeting the specification regarding resistance to the liquid packed. Suitable inks are made from nitrocellulose which has the necessary toughness and these are modified to improve adhesion and rub properties. Pigment choice is governed by the requirements for non-bleed in milk and other drinks.

Two-component reactive inks

The properties of prints produced from conventional evaporation-drying inks are related to the properties of the resins employed. With flexography the choice of resin which may be used is restricted by the limited range of permitted and acceptable solvents. In recent years these constraints have become more severe. Systems which incorporate a chemical reaction after the solvent has evaporated can give a 'cured' film which shows improvements in heat, solvent and product resistances with more gloss and adhesion than a conventional ink, thus offering the possibility of overcoming difficult print specification problems. The principal types are now usually based on epoxy–amine chemistry and isocyanate-cured systems are no longer in favour.

These inks also have disadvantages:

- the need to mix two components together before printing;
- limitations of pot life after mixing which can create waste ink;
- formation of an insoluble deposit on the stereo or part of the press if allowed to dry out;
- in certain cases odour may be a problem if the print is not properly cured.

Given these difficulties, two-component inks are used for certain applications where conventional inks will not satisfy the resistance or adhesion requirements. One way of reducing the problems is to use a two-component white overprinted with conventional colours and under these conditions the resistance properties of the colours are also improved.

Daylight fluorescent inks

Flexographic printing with daylight fluorescent inks has not increased in recent years, although some of the problems with poor strength have been resolved with the availability of better pigments and soluble fluorescent toners.

The inks can be used on paper substrates but when film prints are made it is better to print over a white as the inks are transparent and contrast is improved. Single-colourant formulations are best, as mixing two colours has the effect of quenching the fluorescence. Both organic solvents and water-based inks can be made using the binders contained in the toners or with simple resin systems. By using the coloured inks to overprint suitable white inks different substrates can be accommodated.

Overprint varnishes

The majority of flexographic printing on paper, foil and film is carried out without overvarnish, since this is an additional cost and requires the availability of a spare unit on the press. Certain work will specify or need a varnish to enhance the print or perform a certain function. Varnishes may also be used to increase resistance performance of the print which cannot be achieved any other way and examples of these are paper cups and plates and also some labels.

The most recent application need for varnishing is the increased use of cold-seal adhesives for packaging. These adhesives are based on latex and are applied to the reverse side of the web and function by contact. Since adhesion to the face side of the web is to be avoided, a release varnish is applied over the print. Although it is possible to formulate inks which perform a similar function, total ink coverage is required because any area left uncovered would allow the adhesive to stick to the substrate. Total coverage is the exception and over-varnishing has thus become normal.

Polyamide resins are used in these varnishes and the aim is to achieve good release, lowest odour and good drying. Tests will need to be made over a period to ensure that release properties are maintained. The use of additives for slip and abrasion resistance will need to be thoroughly tested to ensure that they do not interfere with the properties of the adhesive.

Specialized varnishes may be used to give particular properties. Among examples are varnishes applied over a print to aid heat sealing and barrier varnishes which will require high resin solids to obtain maximum film weight. Two-pot varnishes give good barrier properties.

Water-based varnishes are increasingly being used to replace organic solvent-based systems and in many cases similar results can be obtained. Some of the problems that are experienced using water-based pigmented systems are easier to overcome with varnishes and it is expected that the trend will continue.

9.5 INK-RELATED PRINTING PROBLEMS AND POSSIBLE SOLUTIONS

A number of problems can occur during printing and these may be a consequence of many factors including substrate, press design, ink and operating conditions. Most problems will be overcome by experienced printers making adjustments as necessary, although sometimes it may not be possible to do so with the equipment or materials available.

The following problems are discussed which may have an ink origin but it should not be assumed that these are always ink related as usually there are a number of possible causes, not all of which are discussed here. An ink change to one with different properties may overcome a particular problem even though the fundamental cause may lie elsewhere.

Colour and print appearance faults

Colour and strength drift may occur during a printing run. Viscosity or machine variations may be the cause, but a badly dispersed or formulated ink may be unstable after dilution giving rise to pigment settlement or flocculation. Loss of gloss may be noticed and the condition may be aggravated if the solvent balance becomes upset due to evaporation from the duct leading to resin precipitation. Good processing, care in selection of materials, binders with good mutual compatibility and solvent blends with controlled evaporation to retain balance should avoid such problems.

Mottle and poor lay can be the result of over-reduction of the ink but formulations which have poor flow tend to produce prints with lay faults. Over-pigmentation or poor pigment/vehicle wetting may be a cause and this should be corrected during formulation. If the problem is caused by the ink not wetting the substrate the addition of an alternative solvent or a better wetting binder may improve the lay properties.

Inks with poor flow or bad wetting may cause a 'ghost image'. This is the appearance of a fainter unwanted image in a printed area and is the result of local ink starvation on the transfer roll from the previous impression.

Printability faults

Dirty working is probably the most common printing fault with flexography. When the problem is due to ink, poor resolubility on the edges of the stereo is the main reason, causing dried ink to build up, thereby enlarging the stereo and giving it a ragged thick edge which shows on the print. The use of a slower solvent or retarder will minimize the problem, although binder solubility and low speed of solution are fundamental causes. Inks that need high resistance properties may necessitate the use of a binder which has poor solubility in flexographic solvents and this may give rise to dirty working.

Pinholing can occur on coated films, particularly MS film, due to poor wettability of the surface giving rise to ink crawl after impression and before drying. The addition of anti-pinhole compound may overcome the problem by improving wetting and sometimes an alternative mix of a diluent which includes a good wetting solvent will also help. Surface properties of substrates can vary and the printer may increase the ink film weight or pre-heat the web.

Pinholing or bad wetting may also be evident on overprinting colour on colour, due to the non-wetting of the dried ink surface by the next colour. Some ink additives and waxes can cause this problem.

Stereo swelling is normally caused by solvent attack and absorption on stereo materials. This problem is fundamental to the choice of both solvent and stereo composition. Changes consistent with the properties of these are required as soon as the problem is detected.

Foaming of water-based inks may be attributable to the press recirculation system causing high turbulence of the ink and thus aerating the liquid at a faster rate than it will break down. Some resins and formulations are prone to this problem, although the addition of a small quantity of an anti-foam compound will normally overcome the foam.

Drying problems

Solvent retention of prints is of great concern with food packaging. Given the low ink film weights deposited by flexography and the good drying equipment available, solvent retention is not normally a problem but occasionally on some substrates retention is encountered. A good solution is to increase the strength of the ink, as this will allow a thinner film of ink to be carried which is more easily dried. Choice of binder and solvents used in the ink will also determine the drying characteristics and with film inks the type of substrate will influence the solvent choice, it is best to avoid those which soften films or coatings.

Pick-off is a possible fault when overprinting colour, and is caused by the first-down ink not being sufficiently dry and remaining tacky when overprinted. Increasing the drying speed of the diluent mix on the first colour should help to overcome the problem.or, alternatively, applying a thinner film of a stronger ink.

Screening, which is the appearance of a screen pattern on the print, can be caused by ink which dries too rapidly giving evaporation during transfer from the anilox to the stereo. The engraved pattern thus shows on the print. Retardation of the ink is the obvious solution as the ink may be formulated with solvents too fast for the press. The use of dual purpose flexo/gravure inks can produce this problem because the choice of solvents is governed by the fast solvents required for gravure.

Problems in the reel

Blocking of print in the reel is mostly a problem when printing on non-absorbent substrates, and appears as sticking or transfer of the image to the unprinted side of the web. Solvent retention can be a cause since this will make the ink soft and liable to transfer under pressure. Ink formulations which contain too much plasticizer or low molecular weight binders can also be soft and sticky after drying, and these will cause blocking. Blocking tests are a standard procedure during ink formulation and these problems should be avoided at source.

Solvent retention may cause 'ghosting' in reels of printed coated films, and this phenomenon appears as a faint cloudy image on the reverse side of the web due to a partial solution of the coating. The presence of traces of slow evaporating powerful solvents in the ink or diluent is the most common cause.

Odour in the print may be caused by the substrate or by retained solvent, but certain resins used in ink have some odour which may be worse in the presence of retained solvents. All materials used in ink formulations require careful selection and screening.

9.6 RECENT TRENDS

Process developments

Notable flexographic process developments of recent times which are widely used are reverse-angle doctor blades, ceramic anilox rollers and improved rubber and photopolymer plates. Although in general these changes have assisted the process by a better control of the ink film, it has become necessary to produce stronger inks with more flow, in order to preserve sufficient colour intensity given the lower ink film weights used.

Because of the growth of halftone printing on polyethylene and other films, press improvements were necessary to achieve accurate impression and good register. Inks were also developed with the better resolubility properties which are essential to colour process printing.

Another press requirement for the future, given the likely increase in the use of water-based inks, will be improved drying systems if existing running speeds are to be maintained.

Substrates

There have been few fundamental changes in substrates in recent years although variations in film coatings and laminated complexes have posed new problems of adhesion for the inkmaker. There has also been an increase in the use of white-pigmented films which do not require the use of a backing white ink. Under these circumstances, and where crimp jaw heat sealing is in the colour printed area, it has proved difficult to avoid jaw marks on the print surface.

Water-based inks which are extensively used for paper printing will need to have improved drying properties as the use of coated papers becomes more commonplace.

Market changes

Changes in the market, and requirements of the market, for flexographic printed materials have traditionally been a large stimulus for change in inks and associated products. Market changes however, have in the past been somewhat difficult to predict.

More predictable is the continuing increased speed of packaging machines and the relentless search for lower cost printed packaging materials. The former necessitates inks with higher slip, and the latter requires inks which have lower cost or greater coverage.

The flexographic process continues to gain an increasing share of the market over some other processes. One example of this is the flexographic printing of newspapers. A further example is the wider adoption of the process to print packaging materials, given the trend towards shorter runs and more design changes which are more economic by flexography than by gravure.

Market changes driven by environmental considerations are likely to affect ink requirements considerably in the future and this is discussed later in this section.

Ink development

Raw material developments have continuously aided the progress of ink development, and, by better communication, the suppliers now have a greater understanding of the needs of the ink industry. Resin and pigment developments have continued to make new products available to the industry. It remains to be seen, however, if an increased priority placed on safety and environmental requirements results in any reduction of the pace of these developments.

Many new ink products result from development work to create improve performance products, or more rarely, from new and original concepts. Successful new developments tend to be more rapidly concluded when the work is conducted in cooperation with a converter or end user.

Two fields of work which are being intensely studied by various ink makers are those concerned with water-based inks and radiation-curing inks.

Considerable progress has already been achieved with new water-based inks for certain substrates used in particular applications, although these products have not, as yet, been generally adopted by printers and converters. There is much work which remains to be done if these products are to give similar results to those achieved after decades of development, with organic solvent-based systems.

The technology of radiation-curing inks for flexographic printing is being studied. This research is generally being carried out in conjunction with equipment and press manufactures because the problems to be solved involve printing press design and UV-lamp technology as well as ink chemistry.

UV flexographic inks are already in use for certain applications and early indications are that more rapid progress is likely with flexography than with the gravure process.

Environmental considerations

Society has become increasingly concerned about the impact on the environment by industrial processes. These concerns inevitably force governments to introduce legislation to control pollution of different kinds. The type and the extent of legislation tends to vary from country to country and at a higher level in the more developed part of the world.

The printing and ink industries are not normally considered as highly polluting industries. Nevertheless, their activities will be controlled by a number of anti-pollution measures which, when fully enacted, will have a severe impact on the way they operate. In the remainder of this section three areas where regulations will directly affect ink and print technology will be discussed. For full details of all current regulations see Chapter 17.

First, restrictions on the emission of solvent into the atmosphere are already in force or being progressed in several European countries. The EC Directive[2] is currently at the draft stage and national regulations such as the air pollution controls which are part of the UK Control of Pollution Act 1990[3] are currently being enforced. Solvent emissions from print installations will be required to be at much lower levels than are currently the practice using solvent-based inks.

Along with other processes, flexographic printers will need to consider the various options to reduce their emissions to such levels that will be allowed. The following options might be considered:

- incineration
- solvent recovery
- biological treatment
- water-based inks
- radiation-cured inks.

Of the above only incineration has fully proven technology. All options are likely to more expensive than current practice. The dilemma for the industry is which to choose, and will any option chosen be technically feasible within the time allowed for implementation.

Of the options given, the first three are largely the province of the printing industry, but with solvent recovery, suitable inks will be necessary dependent upon the type of equipment used. The last two options (as has previously been stated) are currently the subject of development by inkmakers, but as yet, these methods are not fully proven for all current applications.

Finally, potential new regulations on waste management are currently under debate. These are likely to have an impact on packaging and the materials used for packaging. Re-use and recycling targets for packaging waste are likely to be set. Furthermore, controls are likely in regard to limits on how much packaging waste can be disposed of into landfill. As yet, however, the final outcome is not clear.

The reuse and recycling of packaging may favour certain materials over others. In general it is more difficult to recycle packaging made from mixed materials, unless the mixture is compatible to enable manufacture of a product without separating the components. Recycling often involves extrusion and moulding and studies of ink components are being made to see if these cause problems during the respective process. Waste regulations could affect the choice of packaging materials and in turn the types of inks with which to print them.

There is no doubt that the impact of printing ink on the environment has become an increasing part of our thoughts and actions. It follows that this will lead to continuing and increased development within the industry, aimed at environmental improvements in the use of the ink itself.

BIBLIOGRAPHY

1. MAFF. Survey of Plasticiser Levels in Food Contact Materials and in Foods. Food Surveillance Paper No. 21 (1987) No. 30 (1990), HMSO.
2. EC. Draft of Council Directive on the limitations of emissions of organic solvents from certain processes and industrial installations.
3. UK-DOE. Environment Protection Act 1990, Part 1. (Processes prescribed for air pollution control by Local Authorities), HMSO.

CHAPTER 10
Screen inks

Screen printing is a stencilling process initially described as silk-screen printing or serigraphy. The process has grown considerably in importance during the last few years, particularly in the industrial area. Specific characteristics of the process have made it attractive outside the more established display and poster industries.

Screen printing on materials other than textiles began to attract attention in the 1920s, especially as an economic method of producing short-run posters. The process developed slowly, finding applications mainly in the display and point-of-sale market. The simplicity of the process combined with the thick ink film deposited governed its applications. These characteristics were particularly advantageous in the printing of fluorescent posters, water-slide transfers, display showcards, traffic signs and PVC window stickers.

During the 1950s screen printing was found to be a useful printing process for the decoration of polythene bottles and for the production of printed circuits. The growth of the plastics and electronic industries provided new substrates and techniques ideally suited to screen printing.

New applications continue to be found for this versatile printing process. Examples of this have been the incredible growth of the membrane touch-switch industry and the printing of rub-removable metallic inks for lottery tickets.

As a result, screen printing must be placed alongside the other main processes of lithographic, letterpress, flexographic and gravure in its importance.

10.1 IMPORTANT CHARACTERISTICS OF SCREEN INKS

The characteristics of screen inks become clear when one considers the basic characteristics of the process.

The stencil-like image is supported on a fabric mesh stretched across a rectangular frame. The ink is forced through the openings in the stencil

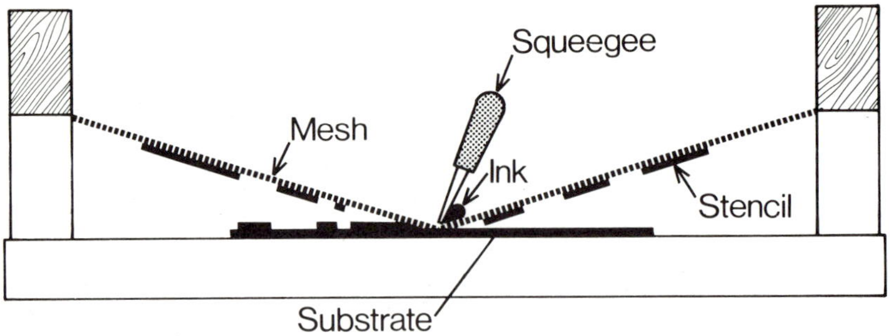

Fig. 10.1 The principle of screen printing.

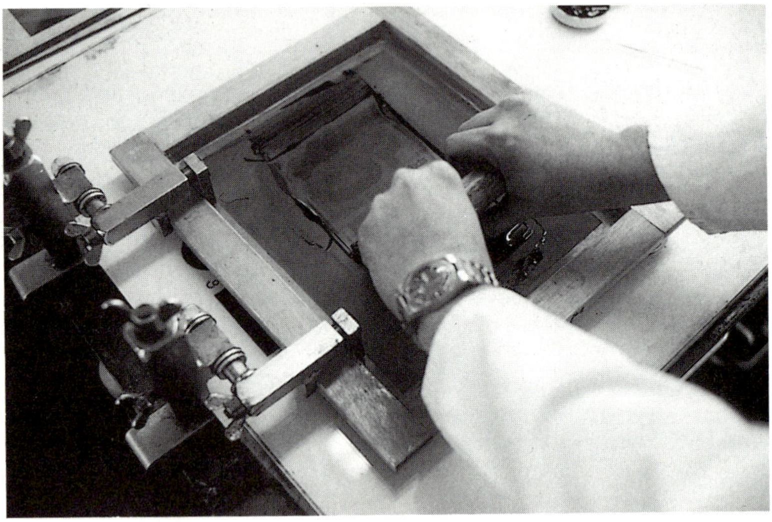

Fig. 10.2 Simple hand-operated screen printing. (Courtesy of Coates Lorilleux Screen Products)

on to the substrate by drawing it across the mesh with a rubber bladed squeegee (Fig. 10.1). Whether the printing machine is simple and hand operated (Fig. 10.2) or large and automatically controlled (Fig. 10.3), the principles remain the same.

First, during printing, the ink is held in a printing frame which is open to the atmosphere. The solvents must not be so volatile that there is excessive solvent loss during printing.

The ink solvents must not cause the squeegee rubber to swell or crack, nor remove the stencil film. An ink must have sufficiently low viscosity to pass easily through the mesh, but not so low as to give slurred printing.

It is undesirable that an ink has excessive flow; shortness is an advantage up to a point since it minimizes the tendency for the ink to string when the screen is pulled off the printed surface.

Once the ink has been applied to the substrate, it must dry, within a

Fig. 10.3 A fully automated screen printing press fitted with a jet drier. (Courtesy of Coates Lorilleux Screen Products)

reasonable period, depending on the application of the ink and the drying equipment available.

Screen and stencils

Although screen printing is often described as silk-screen printing, other mesh materials are employed, e.g. nylon, polyester and metal mesh. Monofilament polyester and nylon have many technical advantages and have largely replaced silk. Stainless steel mesh is used in the glass and printed circuit industries. Modern meshes are available in a wide range of mesh counts and qualities depending on the nature of the ink, printing application and end requirements.

There are two main classes of stencil, hand-cut and photographic. Photographic stencils can be further divided into direct emulsion, indirect film, indirect/direct and capillary film. Hand-cut stencils have limited application as they are less solvent resistant and less hard wearing, and, of course, restricted to hand-cut designs.

The squeegee is produced from either rubber or polyurethane. The latter material maintains its sharp edge profile on long runs and is, therefore, often used on automatic printing press. In practice, there is often some swelling of the rubber in contact with screen inks but this is not normally a serious problem.

Film thickness

Screen printing has been traditionally characterized by thick, opaque ink films. However, various developments in mesh materials, stencils and ink technology have made it possible to achieve a film thickness approaching that obtained with gravure. On the other hand, solvent-free inks, such as plastisols, are printed at a remarkably high film thickness.

Typical film thicknesses are as follows:

Printing process	*Thickness (microns)*
Offset lithography	2
Letterpress	3
Flexographic	3
Gravure	7
Ultra-thin film screen	8
Conventional screen	12
Thick film screen	30

The low film weight achieved with ultra-thin film screen inks is desirable to produce rapid drying and high mileage. Typical ethyl cellulose inks printed through a fine mesh (100 strands/cm) can give a total print area of 50 m^2/kg of ink and give drying times in the order of 10 seconds through a jet drier.

In other applications, high film thicknesses which cannot be achieved with any other printing process, are desirable, e.g. when printing fluorescent inks.

The pigment content of screen inks will vary considerably due to the wide range of film thicknesses that are possible. The coloured pigment content is still considerably lower than for letterpress or lithographic inks, although some screen inks contain a very high proportion of extenders.

Drying methods

Screen printing is almost unique in that it is possible to employ all available methods of drying. The very nature of the process allows the use of a wide range of solvents and a variety of catalytic-curing resin systems. However, the most important drying method is evaporation. Most screen inks contain volatile solvents, sometimes representing 70% of the formulation.

In the past, white spirit was the most generally used solvent, but this is no longer true. Printing speeds have increased to such an extent that this solvent is considered too slow evaporating. In addition, the applications of screen printing are so varied that a wider choice of resins is necessary than is possible with white spirit.

An interesting point about evaporation-drying screen inks is the wide range of driers that have been developed in an effort to remove the solvent. These include warm jet-air driers, wicket driers, simple drying racks, flame-driers, hot air blowers, oven driers and microwave driers.

Oxidation-drying inks are normally based on alkyd resins. Inks for use on decals and metal signs are normally of this type. Catalytic-curing systems are often employed for the more specialized applications of screen printing, e.g. polythene bottle printing and printed-circuit production.

Penetration drying is never used as the sole drying method; due to the heavy ink film, this method of drying is impractical. When printing an absorbent MG poster paper, penetration occurs but it is still essential to rely on solvent evaporation. Certain inks, especially those for printing PVC, incorporate solvents which penetrate the plastic sheet and promote chemical adhesion.

Ultra-violet curing screen inks are now firmly established for all aspects of screen printing. The advantages of rapid setting and space-saving curing units are utilized in the printing of plastic containers, circuit boards, flat plastic and general purpose display printing.

Viscosity/flow

The consistency of screen inks is often described as short and buttery. As discussed earlier, this is advantageous in giving clean, sharp prints. However, it is not always possible to meet this requirement because some inks rely on good flow to achieve a specific effect, e.g. gloss inks for showcards.

If a screen ink has excessive flow there will be a tendency towards bubbling on printing and slurred prints. There will also be a tendency for the paper to stick to the underside of the printing frame although the use of a vacuum printing base minimizes this effect. Although lack of flow is desirable, an ink with very poor flow will not flow out once printed to remove the pattern of the mesh from the printed ink film.

There is no one specific viscosity range which is suitable for all printing conditions. It is normal practice to supply screen inks at a higher viscosity than required; the printer adds solvent according to the nature of the design. Average printing viscosity of a screen ink would be 1.5–2.0 Pa s but may be much higher.

Drying and screen stability

The correct choice of solvent is the prime factor governing the balance between fast drying and screen stability. There has been an increasing tendency towards the use of faster evaporating solvents to meet the needs of fully automatic printing lines.

Solvents such as propylene glycol methyl ether or 160–180°C aromatic hydrocarbons typify the basic solvents of thin film inks.

As a screen printing frame is exposed to the atmosphere, there will always be a certain amount of solvent loss during the printing operation. Therefore, solvents such as xylene are generally unsatisfactory, due to poor screen stability, and attempts to enclose the printing frame have been unsuccessful in promoting better screen stability.

Adhesion

The screen process is frequently employed to print on widely differing surfaces; therefore poor adhesion can occur if the correct ink is not selected or the nature of the substrates proves particularly difficult. One of the problems is that the thick ink film is often less flexible than the surface on which it is printed. Poor adhesion will often be the result of this situation. A good example of this is that a good quality polythene bottle ink may be completely unsatisfactory when printed on polythene film. It is not recommended that films of less than 50–75 μm thickness be screen printed.

PVC is widely used as a display material and is the most common flat plastic to be screen printed. Migration of plasticizer from flexible PVC sheet will impair ink adhesion immediately after printing and on storage of the print. Therefore, printing of flexible PVC normally requires inks with a higher proportion of ketone solvent and, due to the variation in PVC sheets, it is advisable to test the ink adhesion before commencing a printing operation.

Due to the inert surface of polythene, even when pre-treated, good ink adhesion is often difficult. Also, certain additives in the polythene migrate to the surface and destroy the adhesion. Unfortunately this effect can take several weeks to manifest itself. Even correctly formulated polythene bottle inks will not adhere to poorly pre-treated bottles. This is probably the most common problem with the decoration of polythene bottles.

When printing screen inks based on curing systems, excessive stoving must be avoided to obviate poor intercoat adhesion. This applies, of course, only when a colour is being overprinted. A curing screen ink must not cure so rapidly that it is too hard to accept later colours.

End-use requirements

The end-use requirement has the greatest influence on the choice of pigments.

When printing paper, the selection of pigment depends on whether it is for outdoor or indoor use. A pigment would also need to be lightfast if it is to be used indoors but subjected to daylight in, for example, a shop window. For this type of application, pigments with Blue Wool rating of 5 would be acceptable. A distinction must be made between pigments with good lightfastness and weather resistance. For the printing of road signs, a weather-resistant pigment with lightfastness of Blue Wool 8 would be required, whether the road signs were made of plastic or enamelled metal.

The printing of polythene bottles does not require pigments of exceptionally good lightfastness; the most important factor is that the pigment is product resistant to the contents of the bottle, for example, detergent. This would also apply to the resin system.

Screen printing inks applied to printed circuits act as resists, therefore both resin and pigment must be functional. In the case of an acid resist, the pigment and resin must be selected to withstand the etchants. In addition, some resists are required to protect the circuit for many years.

10.2 REQUIREMENTS OF RAW MATERIALS

In common with most inks, screen inks are composed of resins, solvents, pigments and additives. However, many of the requirements are specific to screen inks.

Resins

The majority of screen inks that are for printing on paper dry by solvent evaporation. The first requirement is that resins should form a stable solution in acceptable screen solvents.

These resins should release the solvents readily to form a tack-free film and give the film adequate flexibility and prevent set-off in the stack. It is advantageous if the resin forms solutions of fairly high viscosity at relatively low solids contents.

Inks based on cellulose derivatives can be used without further resin modification but it is possible to add maleic or modified phenolic resins to improve the finish. These resins must be selected to provide good solvent release.

Screen inks for the decoration of flat plastics are almost exclusively based on acrylic or vinyl resins. Acrylic resins have excellent adhesion to a wide range of plastics and have good outdoor durability.

Catalytic-curing systems are frequently used for polythene bottle inks. These resins are selected mainly on account of their good chemical resistance.

Resins used in printed circuit inks perform a number of functions. The main requirements are that the resin should adhere to clean copper. It should act as a resist to various chemicals used in the involved processing technique of circuit production and, where necessary, be removed by a solvent or aqueous alkali.

Solvents

It is possible to use a wide range of solvents in screen inks and virtually all chemical types are employed. The limiting factor is the evaporation rate, as the printing frame is exposed to the atmosphere. Nevertheless, an adequate solvent range is available which, in turn, allows the use of most resin types.

The basic requirement of a screen ink solvent is that it provides adequate screen stability and will volatilize with the minimum expenditure of energy. This energy can be applied from a warm jet-air drier; therefore, the vapour pressure at the drying temperature is of significance. It is unlikely that a single solvent will provide the exact balance of screen stability, fast-drying print and cost. Two or three solvents blends are normal. Propylene glycol ethers are frequently used in mixtures with aromatic or aliphatic hydrocarbons and this has led to the generally accepted term 'cosolvent ink'.

Attention should be paid to the effect of solvents on both photographic stencils and polyurethane squeegees. Prolonged contact with specific solvents can destroy both stencil and squeegee.

Screen printing shops are often not well ventilated therefore health and safety considerations are of paramount importance.

Pigments

Three points are of importance when considering the pigmentation of screen inks, namely:

- the short soft desirable consistency:
- the hue of the ink which is mainly determined by its mass tone:
- end-use requirements.

The consistency of a screen ink can be controlled by the type and amount of extender; calcium carbonate or china clay are frequently used. In addition to controlling the consistency of the ink, the use of extender will lower the cost, which is particularly important for poster inks. The extender will also prevent penetration of the ink into absorbent stocks such as MG poster paper.

Screen inks with adequate colour strength can be formulated with as little as 5% of organic pigment. The mass tone of the ink largely indicates its hue but this is not strictly true with transparent and ultra-thin film inks.

It is very difficult to obtain a smooth solid print with completely transparent screen inks. Although these are available for special purposes, such as process inks and car indicator panels, they are not recommended for general-purpose printing. Transparent inks tend to highlight any imperfections in the mesh or squeegee and accentuate any variation in film thickness. Transparent pigments such as phthalocyanine blue and green should be used in conjunction with a small amount of titanium dioxide.

The end-use requirement is very important in selecting a suitable pigment. Due to the numerous applications of screen printing, the properties required of a coloured pigment can vary considerably. Much screen printing is for display work and therefore requires good lightfastness and weather resistance. Few organic reds are sufficiently lightfast; much use has been made of quinacridone red pigments. Phthalocyanine pigments are also suitable.

Poster inks for outdoor use do not require the same degree of lightfastness but fairly good lightfastness is still required. As a guide pigments with Blue Wool rating of 5–6 are acceptable.

When formulating screen inks containing solvents such as glycol ethers, ketones and esters, special regard must be paid to the solvent resistance of the pigment. PMTA pigments, for example, would be unsuitable for use with these solvents.

Where the printed ink film is subjected to certain processes or specific tests, this must be borne in mind when making pigment selection. Pigments with good resistance to bleaches will be required when printing polythene bottles to contain bleach. Naphthol red F2R and diarylide yellow are commonly used pigments in polythene bottle inks.

Additives

Screen inks require some form of control to avoid the tendency to bubble on printing. This can be achieved by the use of extender but where the use of extender is limited or impossible, then silicone fluids are frequently used.

Silicones of the polymethylsiloxane type are normally used in screen inks, usually in quantities less than 1%. However, inks containing silicone are not without problems. In certain circumstances, particularly curing systems, they lack intercoat adhesion or cannot be overprinted with an ink that does not contain silicone. If traces of silicone contaminate certain inks, then pinholing can occur on overprinting; this is most noticeable with metal-printing inks. Non-silicone defoamers are available but these have proved of limited value in screen inks.

While extenders are frequently used to control the consistency of screen inks, this can be achieved with structuring additives such as bentone. These additives are relatively expensive and can reduce gloss, and care should be taken to select the correct grade according to type of solvent being employed. The viscosity of screen inks is sufficiently high that pigment settling rarely occurs but anti-settle additives may be required where high specific gravity extenders such as blanc fixe are used in the formulation.

Wax pastes, such as micronized polyethylene wax, are added to formulations to promote good rub resistance. These are normally not required in the harder catalytic systems but are used in evaporation drying inks, particularly when the printed matter concerned is likely to be scuffed after printing.

Driers and anti-skinning agents are, of course, used with oxidation drying inks. As most oxidation drying screen inks are based on long-oil alkyds, the most commonly used driers are calcium, manganese and cobalt naphthenates. Due to the heavy deposit achieved with alkyd inks, it is important to ensure good through-drying.

10.3 INKS FOR PAPER AND BOARD

The early screen inks for printing on paper were oil based. They usually contained a high proportion of coarse extender, such as whiting, and the vehicle was based on highly bodied linseed stand oil in white spirit.

Drying was by evaporation of the solvent followed by oxidation of the oil. As might be expected, such inks suffered from a number of disadvantages; they were slow drying, had poor scuff resistance, low mileage and were very coarse. Rather surprisingly this type of formulation was used for many years for the printing of posters and showcards but has now ceased to be of commercial interest.

The first type of screen ink to dry entirely by evaporation was the early 'cellulose ink'. This description refers to nitrocellulose-based inks although the formulations were different from modern nitrocellulose inks in that they were similar to cellulose paint. In fact, they were developed by industrial paint manufacturing companies. The disadvantages of such inks were objectionable odour, fire hazards associated with nitrocellulose and high cost of the solvents.

Thin film screen inks

The most significant development during this early period occurred with the introduction of EHEC (ethyl hydroxy ethyl cellulose). This allowed the formulation of the first thin film screen inks. Inks based on EHEC were a significant advance in that they overcame the problems associated with 'cellulose inks', with the added advantage of being white spirit reducible. The significance of this latter point is that all types of stencils are suitable and the cleaning solvent and thinners are relatively inexpensive.

A typical formulation for a thin film screen ink for posters would be:

Toluidine Red (CI Pigment Red 3)	5.0
China clay (CI Pigment White 19)	36.0
Low viscosity EHEC	6.0
Penta-ester gum	10.0
White spirit	30.0
Aromatic hydrocarbon (160–180°C)	10.0
Propylene glycol ether	3.0
	100.0

The function of the ester gum resin is to increase the finish and assist pigment dispersion.

This type of ink would be suitable for printing on most paper stocks. Typical applications would be poster printing both for indoor and outdoor use, showcards and point-of-sale displays. In comparison to oil-based inks, EHEC inks offer better rub resistance, higher mileage, more finish and faster drying. Air drying time of the formulation given above would be approximately 20 minutes.

Ultra-thin film screen inks

The increasing use of automatic screen printing machinery brought a demand for even quicker drying inks. This demand led to the development of the ultra-thin film inks. The technique here is to use a cellulose derivative to give low solids and high viscosity vehicles. The introduction of higher speed printing machines permitted the use of more volatile solvents without problems of drying in the screen. The combination of lower solids vehicle and fast evaporating solvents formed the basis of fast drying, ultra-thin film inks.

Typical formulation for ultra-thin film screen ink:

Primrose Chrome (CI Pigment Yellow 34)	20.0
Ethyl cellulose	15.0
Dioctyl phthalate	5.0
Propylene glycol methyl ether	25.0
Dipropylene glycol methyl ether	5.0
Aromatic hydrocarbon (160–180°C)	30.0
	100.0

An aromatic hydrocarbon solvent would be a suitable reducer for this ink.

Apart from the obvious advantages of being fast drying, the ultra-thin film ink has superior rub resistance, increased mileage and more finish. A higher standard of pigment dispersion is essential with this type of ink; this point and the lack of extender are the reasons for the increased finish.

The trend towards the ultra-thin film screen ink has, to some extent, departed from the accepted image of screen printing. These types of inks no longer produce the characteristic strong, opaque prints. In addition, there is a greater tendency for the ink to penetrate absorbent stocks.

It is possible to formulate an ultra-thin film ink using nitrocellulose in place of ethyl cellulose. The high cost of ethyl cellulose has encouraged this trend, and it now forms the basis of many high quality screen inks which are capable of jet drying and adhering to a very wide range of substrates.

A typical formulation for a nitrocellulose-based ultra-thin film ink for printing paper, board, foils, hardboard and stove enamelled metal signs would be:

Phthalocyanine Blue (CI Pigment Blue 15.3)	5.0
Titanium dioxide (CI Pigment White 6)	3.0
Nitrocellulose	14.0
Dioctyl phthalate	7.0
High acid value maleic resin	3.0
Propylene glycol ethyl ether	38.0
Ethylene glycol isopropyl ether acetate	20.0
Aromatic hydrocarbon (160–180°C)	10.0
	100.0

However, a more expensive solvent reducer is required. The solvent balance given in the ink formulation would be a suitable reducer.

Oxidation-drying gloss inks

Oil-based gloss inks are used to achieve a thick glossy film not unlike the effect achieved with diestamping inks. This type of finish has only limited applications, for example special-effect showcards and labels.

The use of oil-based gloss inks has decreased in popularity for two reasons. First, the inherent slow drying of the very thick film and, secondly, the improved finish of evaporation drying inks. The higher standard of pigment dispersion achieved with ultra-thin film inks has resulted in improved finish. On coated paper, a semi-gloss finish can be obtained with a well-dispersed ethyl cellulose ink.

The slow drying of oil-based gloss inks, usually in the order of 6–8 hours, limits the production to the printing of one colour in a normal working day. A multicolour run would take several days to complete.

A typical formulation for an oil-based gloss ink:

Diarylide Yellow (CI Pigment Yellow 14)	4.0
Blanc fixe (CI Pigment White 21)	40.0
Linseed stand oil	10.8

Modified phenolic stand oil varnish	40.0
White spirit	4.0
Naphthenate driers	1.0
Methyl ethyl ketoxime	0.2
	100.0

Blanc fixe is chosen for its good gloss and levelling properties.

10.4 INKS FOR IMPERVIOUS SURFACES

Metal signs

Metal signs which are screen printed with non-ceramic inks are intended to have a life of 3–5 years. This is considerably more permanent than the relatively short-lived paper poster but not as durable as a vitreous enamelled sign. Typical applications of metal signs would be traffic signs, semi-permanent window displays, fire-exit notices (Fig. 10.4). In view of the long life of metal signs, the main requirement of a screen ink is durability. This refers to lightfastness, weathering and adhesion.

It is normal practice to use long-oil alkyds as the basis of suitable screen inks. These alkyds are ideally suited in that thick deposits are possible, which is essential for outdoor durability. Alkyds are also capable of being air dried or stoved and have a well proven history of reliable weathering properties.

Fig. 10.4 Screen printing on metal, used for a road sign.

The inks are normally applied over a stoved enamelled background; they may be stoved or air dried. Once the design has been printed it is then protected with a spraying or roller coating varnish. The completed sign is then subjected to a final stoving.

It is important that the screen inks are sufficiently dry to withstand attack from the solvents present in the spraying varnish. The base enamel, screen ink and protective varnish must be tested as a complete system. There should be good intercoat adhesion between the three systems. Formulation for a lightfast, durable, metal printing red ink is given below:

Quinacridone Red (CI Pigment Red 122)	6.0
Calcium carbonate (CI Pigment White 18)	24.0
Long-oil linseed alkyd	50.0
Mixed napthenate drier	4.0
White spirit	16.0
	100.0

Drier content should be calculated on alkyd resin solids. The amounts and type of drier to be used is dependent on the specific alkyd used; however, it is normal to use calcium, manganese and cobalt naphthenate with long-oil alkyds.

Inks similar to those described above are also used for the printing of nameplates. There are a large number of dials, clockfaces, instruction panels for household appliances which are screen printed. No general formulation can be given for these inks as there is such a variety of applications. Finish can vary between full gloss and dead matt. Resins used are those which give tough, chemically resistant finishes when stoved, such as melamine–alkyd, polyurethane–alkyd, thermosetting acrylic and epoxy–polyamide.

Metal containers

It is often convenient to add identification or other marking to drums. This can, of course, be done by normal stencilling methods but a neater result can be achieved with screen printing. If a large number of drums are to be printed, then screen printing is more convenient. The inks are not usually applied to bare metal but drums enamelled with a background colour. The screen print is usually air dried, particularly if the drums have a large capacity. The requirements of such inks are that they exhibit good adhesion and are not removed by the chemicals packed inside the container.

Since the printing process is liable to be slow and intermittent, it is desirable to use high-boiling solvents in the inks to avoid drying in the screen. It is also necessary to restrict the flow of the ink; this is particularly important with screens that are physically carried from drum to drum.

A suitable ink would be:

Carbon Black (CI Pigment Black 7)	6.0
China clay (CI Pigment White 19)	30.0
Medium long-oil alkyd	45.0
Naphthenate driers	3.0
Anti-oxidant	0.1
Structuring additive	3.0
Dipentene	12.9
	100.0

Inks for sheet plastic

The decoration of sheet plastics is now a very important part of the screen industry. Many traditional materials have, for economic and technical reasons, been replaced by sheet plastics. A good example of this is the increasing number of traffic signs produced in acrylic sheet, to replace metal signs.

The variety of plastic substrates that exists presents problems to the inkmaker in that each substrate may require special consideration. An ideal situation is where one screen ink is suitable for all substrates: in practice this is not possible but an ink formulation should be as multipurpose as possible. Possibly, the nearest approach to this is a screen ink based on acrylic/vinyl resins.

Polyvinyl chloride (PVC) sheet is probably the most frequently printed sheet plastic. This is available in rigid unplasticized sheet, self-adhesive sheet and heavily plasticized 'self-cling' PVC. Other important sheet materials are acrylic, polycarbonate, polyester, polystyrene, cellulose acetate, corrugated polypropylene and spun polyolefin.

Relatively thick sheets of acrylic or PVC are frequently used for display signs. These displays can be in the form of an illuminated light box which functions day and night. Inks for illuminated signs must be translucent: if the inks are too opaque, the sign will not be visible at night. Translucent screen inks contain approximately half the normal pigment content, this provides an ink which is neither too transparent nor too opaque.

An ink for a road sign must be very weather resistant and lightfast: pigments with Blue Wool ratings not less than 7–8 should be used. In order to achieve a high degree of weathering, the ink must have excellent adhesion to the substrate. Some printed acrylic sheets are subjected to thermoforming, and the ink must be sufficiently flexible to withstand this process. However, if an ink has a low softening point the prints will show mould marking if contact is made with the mould.

A lightfast red ink suitable for outdoor use on acrylic or PVC sheet would be:

Quinacridone pigment (CI Pigment Red 122)	10.0
Ethyl methacrylate copolymer	25.0
Vinyl resin	7.0
Propylene glycol monomethyl ether acetate	20.0
Aromatic hydrocarbon (186–214°C)	20.0

Cyclohexanone	10.0
Diacetone alcohol	7.0
Silicone anti-foam	1.0
	100.0

Although 1.5–3.0 mm PVC sheet is used for roadsigns, the vast majority of PVC sheet screen printed is considerably thinner, i.e. 0.13–0.24 mm. The main application of this thinner grade of PVC is in the form of self-adhesive labels.

There are two main methods of producing a self-adhesive printed label. The more conventional method is to screenprint directly on the surface of self-adhesive PVC with a gloss ink. The second method is to use transparent PVC and screenprint the legend in reverse (viewed through the plastic), apply a backing white ink over the whole print area and finally screen print a pressure-sensitive adhesive. The adhesive can be permanent or removable type but as it is tacky it must be protected with a release paper.

The main point of interest to the ink formulator is that the first method requires gloss inks and the second method, matt inks. Either system may be used to produce a double-sided window sticker, i.e. where the image is viewable from both sides. With such stickers 10 superimposed ink layers are not uncommon and this presents a problem to the ink formulator. A high degree of flexibility is required which cannot simply be achieved by plasticizer modification. In the case of gloss inks this is normally achieved by careful selection of methacrylate resin. A wide range of resins are available which can produce screen inks with either excellent flexibility or hard drying properties.

The most common problem with printing PVC is obtaining consistently good ink adhesion and it has been the practice to use strong ketonic solvent to promote adequate adhesion. This presents some problems: the inks have an unpleasant odour and cockling can occur on some thinner grades of plastic. The basic resins used are normally polyvinyl choride/ polyvinyl acetate copolymers and recent developments with these resins have led to inks with milder solvent balances with a high proportion of aromatic hydrocarbon solvent.

Pigment selection is important; ketone-resistant pigments must be selected. Even the slightest contamination with a non-resistant pigment can cause staining of subsequent overprinting inks. Pigments must also be resistant to bleed into plasticizers because the high proportion of liquid plasticizer in flexible PVC can cause certain pigments to migrate into the plastic.

Typical formulations for a matt and gloss ink for either plasticized or rigid PVC are given below.

Matt ink for PVC sheet:

Permanent Yellow H10G (CI Pigment Yellow 3)	5.0
Titanium dioxide (CI Pigment White 6)	7.0
China clay (CI Pigment White 19)	30.0

PVC/PVA copolymer	15.0
Aromatic hydrocarbon (160–180°C)	20.0
Pentoxone	16.0
Cyclohexanone	7.0
	100.0

Gloss ink for PVC sheet:

Rutile titanium dioxide (CI Pigment White 6)	27.0
PVC/PVA copolymer	19.0
Methyl/butyl methacrylate	7.0
Cyclohexanone	20.0
Aromatic hydrocarbon (160–180°C)	18.0
2-butoxyethyl acetate	8.5
Silicone anti-foam	0.5
	100.0

Good lightfastness and weathering properties are not essential for all vinyl inks; certain PVC articles, e.g. sachets, require only minimum lightfastness. On the other hand, permanent labels must conform to a higher standard. Basic requirements for the more permanent labels are detailed in BS 4781: Part 1: 1973, specification for self-adhesive plastic labels for permanent use.

Inks for glass

Inks for screen printing on glass fall into two main categories:

- conventional screen inks
- ceramic inks

The latter are normally produced by specialized pigment manufacturers and are outside the scope of the conventional ink manufacturer.

The application of conventional screen printing inks is mainly in the decoration of mirrors. Conventional inks have also been used on glass bottles, laboratory glassware, pinball machines and glass objects such as ashtrays and ornaments.

Printed mirrors are often used as advertising displays in shops. The screen ink can be applied to the front of the glass or printed in reverse on the back. In the case of reverse printing, the glass is screenprinted before the silvering operations. The ink must therefore withstand the mirroring process of degreasing with abrasive whiting, immersion in solutions of stannous chloride and hydrochloric acid and the final deposition of silver from a solution of silver nitrate.

For surface printing, an ink based on hard drying alkyd, e.g. vinyl toluene-modified alkyd, of medium oil length would be suitable. However, when reverse printed, this type of an ink would not always withstand the mirroring process.

A two-pack epoxy/polyamide, screen ink would be suitable for this application; a typical starting-point formulation would be as follows with a mixing ratio 100 parts of paste to 15 parts of catalyst.

Paste:	
Titanium dioxide (CI Pigment White 6)	45.0
Epoxy resin	35.0
2-butoxyethanol	20.0
	100.0

Catalyst:	
Reactive polyamide	70.0
2-butoxyethanol	10.0
Aromatic hydrocarbon (186–214°C)	20.0
	100.0

Mirror pictures, i.e. where the silvered area covers only part of the glass usually in the form of a simple picture of pattern, are produced with the aid of screen printing. An acid resisting ink is printed directly onto the silver. The mirror is subjected to an acid bath which removes the unprotected silver. This leaves a mirror pattern against a clear glass background. A suitable acid resisting ink can be formulated using isomerized rubber resins in aliphatic hydrocarbons.

Ceramic screen inks are used where a permanent mark is required on the glass container, e.g. reusable drink bottles, glassware, pottery and laboratory glassware.

Ceramic screen inks can take a variety of forms and may be applied by direct printing, decalcomania or hot-melt screen. Regardless of the type used, the principle is the same: the ink is fired into the glass and becomes an integral part.

A ceramic ink consists of pigment, frit or flux and a varnish. The function of the varnish is simply as a temporary carrier of the other components. After the ink is applied to a bottle it is fired at temperatures in the order of 800°C the pigment and frit, known as the enamel, fuse into the glass and the varnish completely volatilizes at these temperatures.

Formulation for a ceramic screen ink for direct printing is as follows:

Ceramic enamel pigment	52.45
Thixotropic alkyd	26.23
Hydroquinone	0.07
Naphthenate drier	0.13
White spirit	21.12
	100.00

10.5 INKS FOR PLASTIC CONTAINERS

Polythene containers

There are a number of alternative methods of decorating polythene bottles but, in the UK, the most popular process is screen printing (Fig.

10.5). The alternative processes are dry offset printing, paper labelling, hot melt transfer printing hot-foil stamping, in-mould labelling and shrink labelling.

Before discussing suitable ink formulations, it must be stressed that correct treatment level of the surface of the polythene is absolutely essential. If the pre-treatment is incorrect, this can result in poor ink adhesion, poor flexibility and lack of chemical resistance. A further problem is that certain additives included in the polythene can migrate to the surface and destroy ink adhesion. Inks must be tested on the specific polythene being used and the adhesion tested over an extended period.

One of the main requirements of a screen ink is that it is resistant to the products that are being packed inside the containers. The most common products are detergents, bleaches, lubricating oil, household cleaners and cosmetics (Fig. 10.6). With such a variety of products it is unlikely that one ink system will be suitable for all bottles.

Good flexibility is required as some bottles, particularly those containing detergents, are subjected to frequent flexing when in use in the house hold

The early screen ink formulations were oxidation drying; these were normally dried in an oven at 80°C for 8–10 minutes. In many ways, oxidation drying inks have not been improved upon: they are simple to use, relatively inexpensive and give thick glossy films with fairly good chemical resistance. Their main drawback is the necessity to stove each colour for several minutes. Although oxidation-drying inks cannot meet the demands of a high-speed bottle-producing plant, these are still used in certain circumstances.

The original polythene bottle inks were based on long-oil linseed alkyds

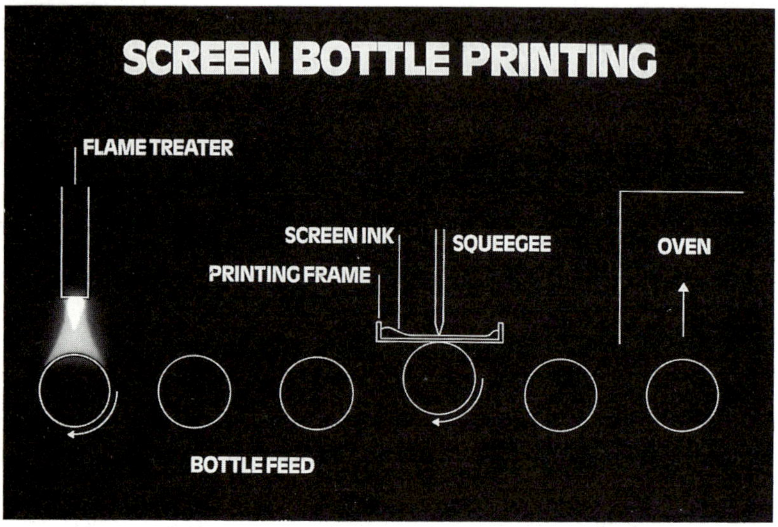

Fig. 10.5 Diagram of screen bottle printing. (Courtesy of Coates Lorilleux Screen Products)

Fig. 10.6 Examples of screen printing on polyurethane containers.

but in modern times better chemical resistance is required. An oxidation drying ink would be formulated by using epoxy ester resins as follows:

Phthalocyanine Green (CI Pigment Green 7)	6.0
Titanium dioxide (CI Pigment White 6)	10.0
Long oil epoxy ester	82.2
4% calcium naphthenate	1.0
6% cobalt naphthenate	0.6
Silicone anti-foam	0.2
	100.0

With the introduction of high-speed printing equipment it was essential to provide faster-drying inks, i.e. evaporation drying types. Traditionally polyamide resins have formed the basis of gravure polythene inks and these resins have also been found suitable for screen inks. Polyamides provide good adhesion to polythene but require some modification with a film former to improve solvent release and chemical resistance.

Polyamide inks can be developed which have fairly good resistance to oils, detergents and bleaches but are not resistant to cosmetics or aggressive products containing solvents. In addition, polyamide screen inks tend to lack gloss and are not very scuff resistant.

Where good chemical resistance is required, it is possible to use epoxy polyamide based inks. These are two-pack systems with good chemical resistance but they are not particularly fast drying. There is a need, therefore, to provide an ink that combines the speed of drying of evaporation inks, the gloss of oxidation inks and the chemical resistance of epoxy polyamide inks.

The inks which have most closely approached these requirements have been formulated using acid-catalysed amino resins. Unfortunately these inks suffer from the disadvantage of being two-pack systems with limited pot life.

Considerable development work has taken place to produce a formulation that is one pack and with an extended shelf life. Very sophisticated one-pack catalytic curing inks are available which rely on chemically blocked catalysts. The action of stoving the printed bottle at 80°C unblocks the catalyst and allows a high degree of polymerization. Excellent chemical resistance and fast production is achieved with the added advantage of ease of use. The one-pack catalytic ink is widely accepted as the most superior system for polythene bottle decoration.

PVC containers

Containers made from PVC have, to a limited extent, replaced glass bottles. Typical products that are packed in PVC bottles are soft drinks, disinfectants, shampoos, and sun-tan oil. The vast majority of PVC bottles are, in fact, paper labelled, not screen printed.

As PVC is being considered as a replacement for glass, most bottles are made from a transparent grade of PVC. This presents some problem to the ink formulator because it is essential to formulate inks with good opacity. If a dark-coloured liquid is introduced into a printed bottle, the legend or design must remain clear. Where possible, inks should contain titanium dioxide or contain a higher proportion of pigment necessary to achieve good opacity.

Similar requirements to the polythene bottle inks are necessary, i.e. chemical resistance, adhesion and fast drying. A problem in the decoration of PVC bottles is that of stress cracking. Printed bottles can split very easily or have an unacceptable low impact strength. The source of the problem is in poor blow moulding technique. PVC bottles moulded in a highly strained condition will stress crack in the presence of ink solvents.

Although the problem can be avoided by moulding technique or the use of impact modifiers the ink formulator needs to minimize the effect by careful selection of resins and solvents. Formulations given under PVC sheet inks would be suitable but these would have a tendency to scuff; PVC bottle inks should be formulated using harder resins. A typical mar-resistant PVC black bottle ink would be:

Carbon black (CI Pigment Black 7)	5.0
Hard acrylic copolymer	30.0
Dioctyl phthalate	6.0
PVC/PVA copolymer	8.0
Methyl propoxol acetate	20.0
Aromatic hydrocarbon (186–214°C)	20.0
Diacetone alcohol	5.0
Cyclohexanone	5.0
Polyethylene wax	0.5
Silicone anti-foam	0.5
	100.0

10.6 TEXTILE INKS

It was noted in the introduction to this chapter that screen printing was first used in textile printing, primarily for the application of dyes and dye-resists for subsequent chemical and physical treatment. Large textile producers normally blend these inks for themselves from dyes and bases.

However. there is considerable interest in the screen printing of pigmented inks, e.g. printed clothing, pennants, bags, etc. This work is normally carried out by the established screen printer rather than the textile manufacturer, although many specialist firms exist who produce printed sportwear such as sweatshirts and T-shirts (Fig. 10.7)

An ink should be capable of being applied to a variety of synthetic and natural fabrics and be washfast and dry-clean resistant. Once appllied the colour should not fade or impart a 'feel; or 'hand' to the material, in other words, when printed, the ink film should not alter the texture of the fabric.

There are two main types of ink used, pigmented emulsions and plastisol inks.

The emulsion inks are used for direct printing of material; they are suitable for all types of fabric. They may be allowed to dry by evaporation at room temperature but to achieve resistance to washing and dry cleaning, they must be cured at 160°C for 2–3 minutes.

Suitable inks are based on aqueous dispersion of an auto-crosslinking acrylate copolymer.

Fig. 10.7 Screen printing on textiles. (Courtesy of Coates Lorilleux Screen Products)

A typical formulation would be:

Water	10.0
Emulsifier	1.0
Thickener	4.0
White spirit	62.0
Catalyst soln	3.0
Binder	15.0
Pigment dispersion	5.0
	100.0

This type of ink is an oil-in-water emulsion where the oil is, in fact, white spirit. The high proportion of solvent, in a basically water-based system, gives an ink with easy printing properties and a low filmweight. As a consequence the print does not impart any feel to the fabric.

It is sometimes desirable to produce an ink that is entirely water-based. This can be achieved by replacing the white spirit in the above formulation with water. The resultant ink will give less bright colours and a higher hand.

The basic medium, described above as a binder, could be an acrylate copolymer which is capable of cross-linking in the presence of a catalyst and under heat. Catalyst, used are ammonium salts of inorganic acids, e.g. diammonium phosphate or ammonium thiocyanate.

The addition of proprietary chemicals such as definition improvers, lubricants, accelerators, thickeners, humectants and softeners is possible to achieve specific properties or variation in performance. However an improvement in one area is usually at the expense of another property. For example the addition of 1 or 2% accelerator will reduce the stoving schedule but have a dramatic effect on the pot life of the ink.

Pigmentation is achieved by adding pre-dispersed pigment pastes. These are normally organic pigment dispersions specifically formulated for textile inks. The lightfastness and washfastness are usually very good.

The second type of screen textile ink is based on plastisol, i.e. vinyl resin dispersed in plasticizer. This type of ink consists of virtually 100% non-volatiles; no solvent is present. As a consequence it applies an extremely thick opaque ink film and therefore is frequently printed on black or dark-coloured garments. Plastisols may be applied directly to the fabric or used as a transfer.

When used as a transfer the ink is screen printed onto a release paper, e.g. vegetable parchment. The ink is subsequently cured to a dry film and the print is stored. The plastisol, although partially cured, remains thermoplastic and can be transferred to a fabric simply by ironing or another heat-transfer process.

Plastisol inks are basically formulated from PVC homopolymer dispersed in phthalate plasticizer. The vinyl resin should be high molecular-weight emulsion polymer with fine particle size, i.e. paste-making polymer. The liquid plasticizers frequently used are dialkyl phthalate and di-iso-octyl phthalate.

The PVC resin requires stabilization against degradation by heat and light. The liquid barium/cadmium/zinc stabilizers combined with epoxy plasticizers are particularly useful.

A high proportion of extender is desirable to reduce the raw material cost and to improve the wet-on-wet printing properties. Printing on textiles is probably the only area of screen printing where wet-on-wet printing is common practice.

10.7 TRANSFER INKS

Transfers are sometimes described as decals, decalcomania or water-slide transfers. They are used where screen printing an object is impracticable, e.g it is far easier to apply a transfer to a large object like a compressed air cylinder than to print directly on the cylinder. The screen printing is carried out on a flat sheet of transfer paper which has the ability to release the ink film completely.

A transfer can be defined as a printed image being removed from a substrate (transfer paper) and adhered to another surface (usually a large object).

The simplest type of transfer involves the use of a transfer paper which is an absorbent paper coated with a water release agent and adhesive. The gum serves as a release agent and adhesive for the transfer (Fig. 10.8) There are variations in this process but the ink requirements remain the same.

The main point of interest to the ink formulator is that an ink film must be sufficiently strong and flexible completely to support itself at the time of transfer. This flexibility must be retained over many years; some transfers are stored for long periods before being used.

In order to achieve a tough, flexible transfer it is necessary to apply a very heavy deposit of ink. Coarse meshes are commonly used, particularly for printing the background colour. A typical transfer could consist of a background white, two or three colours and a clear overprint varnish.

Transfer inks are usually based on long-oil modified alkyds and are thinned with white spirit. The choice of alkyd and the manner in which it is modified is critical in achieving the desired flexibility. A resin should be selected with high solids content to achieve a thick ink film. Choice of pigment is dependent on the end use of the transfer.

Common criticism of transfer inks is the long drying times: evaporation-drying inks can be formulated and are used for less critical work. Highly plasticized nitrocellulose usually forms the basis of these inks but it is not possible to achieve the flexibility and opacity of alkyd inks.

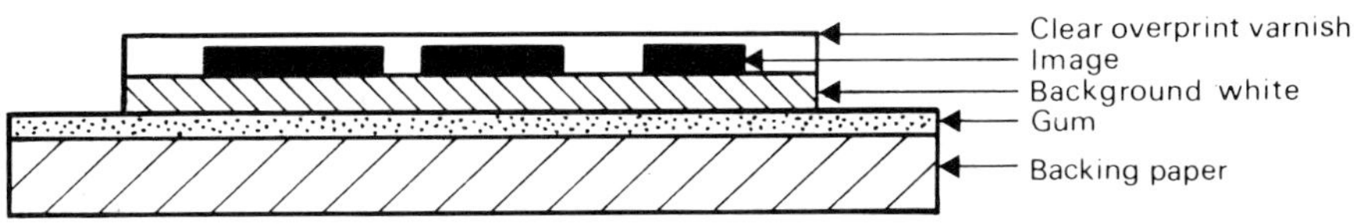

Fig. 10.8 A typical transfer print on simplex paper.

10.8 OVERPRINT VARNISHES

Overprint varnishes applied by screen printing often have to compete in terms of technical performance, cost effectiveness and ease of application with roller coating varnishes and film lamination.

The simplest application of a screen varnish would be a point-of-sale showcard requiring an overall gloss finish. Traditionally this type of varnish would be based on resin-modified nitrocellulose, or for higher gloss, a non-yellowing medium-long oil alkyd. To a large extent these decorative overprint varnishes have moved towards UV-curing formulations. These offer extremely high gloss finishes with the advantages of almost instantaneous drying.

It must not be assumed that the role of all overprint varnishes is to provide a gloss decorative finish. Many speciality screen varnishes exist which have other functions and indeed are not always full gloss. A few examples are given below with some indication of the chemical type.

- Alcohol-resistant varnish for labels on spirit bottles – chlorinated rubber.
- Sealing varnish for rub-removable lottery tickets – UV or modified nitrocellulose.
- Thermoplastic varnishes for blister packs – various.
- Barrier coatings for reducing gas permeability of plastics – PVdC.
- Alcohol and petrol resistant varnish for vehicle marking – vinyl/acrylic.
- Texturing varnishes for membrane touch switches – mainly UV but acrylic possible.
- Protective varnishes for PVC self-adhesive exterior stickers – vinyl/acrylic with UV absorber.
- Laminating varnish for PVC credit cards – thermoplastic vinyl/acrylic.

10.9 DAYLIGHT FLUORESCENT INKS

Fluorescence is a process of photo-luminescence by which light of short wavelengths, either in the UV or visible regions of the electromagnetic spectrum, is absorbed and reradiated at longer wavelengths. This re-emission occurs within the visible region of the spectrum and is manifested as colour. Fluorescent pigments are transparent organic resin particles containing dyes in solid solution.

Although fluorescent inks can be printed by most processes, the screen process gives incomparable results. The thick film of screen printing is the obvious method of printing fluorescent inks as fluorescence increases with film thickness.

Two main types of inks are available, evaporation drying or oxidation drying. The latter type gives thicker films and more brilliant results but is slower drying. In fact, oxidation-drying fluorescent inks, due to the pigment loading, are touch-dry relatively quickly but then take several hours to become hard.

Two formulations for a fluorescent ink suitable for posters are given below.

Evaporation dry:

Fluorescent pigment	46.0
EHEC	4.0
Penta-ester gum	16.7
White spirit	29.0
Dipentene	3.0
2-butoxyethanol	1.3
	100.0

Oxidation drying:

Fluorescent pigment	48.00
Long oil soya alkyd	46.00
6% cobalt octoate	0.25
Zirconium naphthenate	0.75
White spirit	5.00
	100.00

It is also possible to formulate screen fluorescent inks for printing PVC sheet. This type of printing is not very popular as the fluorescent pigment has limited outdoor life, while a PVC sticker may be regarded as semi-permanent.

A formulation for a screen ink for printing PVC sheet is as follows:

Fluorescent pigment (solvent resistant grade)	27.0
PVC/PVA copolymer	19.0
Methyl/butyl methacrylate	7.0
Cyclohexanone	10.5
Aromatic hydrocarbon (160–180°C)	32.0
Aromatic hydrocarbon (186–214°C)	4.0
Dioctyl phthalate	0.5
	100.0

When using fluorescent pigment in conjunction with ketone solvents, attention must be paid to the grade of pigment. Solvent-resistant grades are available which give complete stability with vinyl/ketone solutions.

It is essential to print all fluorescent inks over a white background because of the transparent nature of the pigment. It is possible to include some titanium dioxide in fluorescent inks but there is an inevitable loss of brilliance.

10.10 PROCESS INKS

The quality of screen process printing lines has improved dramatically, over recent years. This may be attributed to developments in inks, meshes, stencils, printing machinery, separation techniques and printing skills.

Printing of large posters would normally use rulings of 30-line screen but the average separation may be 65 to 85-line screen. It is not uncommon to use 120-line screen which produces superb quality that is sometimes difficult to distinguish from litho printing. Achromatic reproduction techniques are applicable to screen printing and can further enhance the quality of print.

Unfortunately, there is no British Standard that is truly applicable to screen printing. It is therefore normal practice to quote screen process inks as being matched to BS 4666 or DIN 16539 but strictly speaking this is incorrect.

As with other printing processes, suitable screen inks must be transparent and give good dot reproduction. Capability to jet-dry quickly is nearly always required to allow all colours to be superimposed rapidly.

Screen process inks should be formulated on low solids varnishes to avoid a rough appearance on build up. They should be structured to give dot reproduction and contain a proportion of transparent extender to avoid penetration into the stock and also lower cost. Pigmentation is lower than normal screen inks but must be maintained around 2–3% to produce good colour strength through fine meshes. A considerable amount of process work is produced on self-adhesive vinyl. In this situation pigment selection is more critical, only lightfast, solvent-resistant pigments may be used. As four-colour superimposition is inevitable in process printing a degree of flexibility is required, particularly on self-adhesive vinyl which is subject to embrittlement with heavy ink deposits.

A good starting-point for formulating an ultra-thin film process ink for poster printing would be:

Organic pigment	3.0
Alumina hydrate	8.0
Ethyl cellulose	15.0
Dioctyl phthalate	6.0
2-butoxyethanol	33.0
Structuring additive	2.0
Aromatic hydrocarbon	33.0
	100.0

Note that higher boiling point solvents are preferred to improve screen stability. For economic reasons and where appropriate it is possible to formulate either EHEC process inks or nitrocellulose-based inks.

10.11 METALLICS

The use of metallic inks in screen printing covers a very wide range of applications. Metallic effects are possible in the printing of T-shirts, point-of-sale showcards, self-adhesive vinyls, plastic containers, but possibly the largest application is in the printing of rub-removable lottery tickets and promotions.

Rub-removable tickets have gained worldwide popularity both as a lottery or as a promotional tool (Fig. 10.9). In situations where large prizes in cash or kind are at stake, security is essential. Screen printing offers the means of applying heavy ink deposits: a film thickness of 12–15 μm is required for complete obliteration.

More than one printing process is used in the production of lottery tickets. The graphics are litho or gravure printed and the metallic ink is screen printed. The ticket is a four- or five-part construction, base board, printed design including symbols, sealant, obliterating ink and an optional litho print to add to the personalization and security of the ticket.

The purpose of the sealant, which normally is litho printed UV-curing varnish, is to prevent screen ink penetration and assist removability. It also serves to combat fraudulent abuse of the ticket.

The rub-removable screen ink is the most important component of the system. It must give total obliteration and give controlled rub-removal characteristics and must remain removable for many months after printing.

The selection of the aluminium powder is critical; both leafing and non-leafing grades can be used but they must be of fine particle size to give adequate opacity. If they are too fine this can result in finger staining or board staining. Coloured rub-removable inks can be achieved by dispersing 2% of organic pigment into an aluminium-based ink. The rub-removable characteristics are achieved with the use of pliable resins. The inks are often described as latex inks but modern inks are based on synthetic 'plastic' resins. No precise formulations can be given but the importance of thorough print testing particular on ageing must be emphasized.

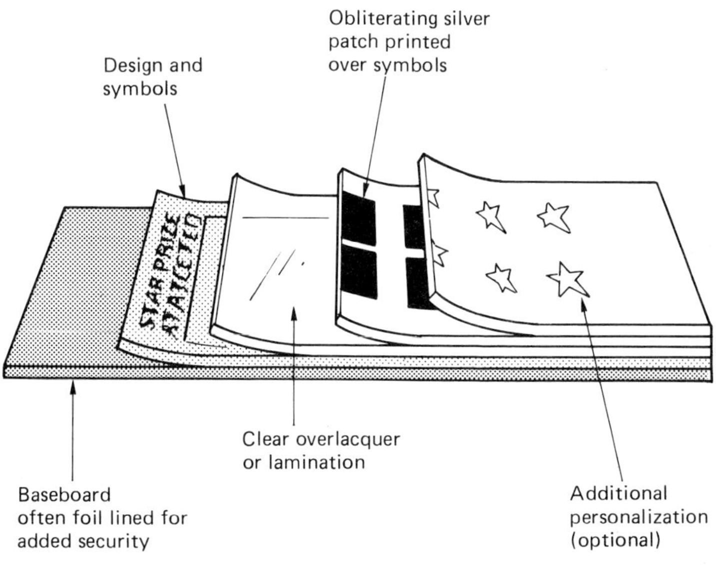

Fig. 10.9 The structure of a lottery ticket.

Metallic inks are used frequently in the printing of double-side self-adhesive labels. The aluminium screen ink provides complete obliteration between both sides of the printed legend. The metallic content should be limited to avoid poor intercoat adhesion with overprinting colours.

Very attractive glitter effects can also be achieved with screen printing. Glitter powders are actually particles of highly polished aluminium foil which have been coated with a transparent pigmented lacquer. Various grades of glitter are available differing in colour, particle size and solvent resistance.

The coating lacquer is usually nitrocellulose, vinyl or epoxy and the solvent balance of the ink must be considered before selecting the grade of glitter. An average particle size of 200 μm is common and therefore may only be printed through coarse meshes.

Glitter particles may be dispersed in almost any medium providing the solvents are taken into consideration. Plastisol glitters are popular for the printing of T-shirts.

10.12 SPECIALITY SCREEN INKS

Screen printing is a versatile technique which allows all sorts of jobs to be undertaken that are impossible by any other means. In recent years, a number of speciality ink systems have been developed to extend this versatility even further.

Encapsulated perfume inks

The technique of encapsulating is not new – it has been used for many years for 'no carbon required' multi-part stationery. The most recent application of this technique is in the encapsulation of fragrances which are then screen printed on to a paper substrate. They have the advantage of offering a stable fragrance in the form of a dry print, which is only released when pressure is applied. The fragrance remains locked in until the capsule is broken by scratching, unlike a conventional perfumed ink which will lose its fragrance rapidly.

The capsules are incorporated into a water-based medium which can be screen printed in the normal manner and jet or air dried. It is important that the capsules are not broken during printing and, as they are relatively large, a coarse mesh (77–90T) must be used. It is normal to print solid blocks in selected areas of preprinted material. The potential range of fragrances is enormous, although the most popular tend to be flowers and fruits.

Thermochromic inks

Thermochromic inks are based on liquid crystals that change colour as the temperature varies and have a number of novelty and practical applications. The inks may be used to indicate very specific temperatures or produce a vivid change of colour over a selected temperature range.

Specific temperature inks are used in the production of thermometers or fever strips, whereas inks with a colour play of several degrees are used for novelty products and promotional material. Thermochromic inks can be formulated on liquid crystals that react to any specified temperature or temperature range between – 20 and 50°C. They can be printed on a wide range of plastics, including PVC and polyester, and can find application on some papers and boards. For best visual effects, the inks must be viewed against a black background.

The use of the term 'liquid crystal' is somewhat curious since the molecules of a 'liquid' generally exist in a haphazard configuration, whereas a 'crystalline' structure implies an ordered arrangement of molecules. Such regular geometry normally means that crystals are solid and their molecules are unable to move. Liquid crystals exhibit the same ordered geometry but, because they are fluid, their molecules are able to twist and move in relation to each other. This can be brought about by gentle heat and as the disturbance takes place, a change in the wavelength of reflected light occurs – the crystal changes colour. On cooling, the crystal reverts to its original colour. Thermochromic inks are almost colourless but as temperature increases, the sequence of colour play is red/brown, yellow/green, blue and violet. Apply more heat and the ink becomes colourless again. The colour play is continuously reversible.

Phosphorescent inks

There is sometimes confusion between luminosity, fluorescence and phosphorescence. A phosphorescent material is defined as one that has the ability to emit light in total darkness, following excitation by various light sources.

Feature	*Activated by*	*After-glow*
Phosphorescence	Any light source	2 hours
Fluorescence	UV light	None
Luminosity	Radioactive chemicals	Long

It is now possible for inkmakers to incorporate phosphorescent pigments into screen inks for both promotional and industrial applications. Currently available is a phosphorescent green which in normal light has the appearance of very pale green, but appears yellow-green in the dark after excitation by daylight, UV or artificial light. The resultant light emission gradually fades but can be perceptible to adapted eyes for 1–2 hours. Re-exposure to light will reactive the print.

Phosphorescent pigments are activated zinc sulphide and as they contain large particles can only be printed through very coarse meshes. As the duration of the after-glow is dependent on the filmweight of ink deposited, it is normal to carry out two or three workings. Suitable substrates are acrylic, unplasticized PVC, polystyrene and polycarbonate films, paper and some enamels.

Phosphorescent inks have a number of applications. Warning signs can be produced in the case of light failure; light switches and keyholes can be identified in the dark; other applications include protective clothing accident prevention, dial illumination, toys and novelties.

10.13 INKS FOR THE ELECTRONICS INDUSTRY

Printed circuit products

Despite the fact that the term 'ink' is rarely used nowadays to describe products used in the manufacture of printed circuit boards, it is not surprising to find that a few ink manufacturers, with their specialist knowledge of coating technology, have taken up the challenge offered by the electronics industry. The term used for such coatings is 'resists' and the three types most commonly used are etch resists, plating resists and solder resists.

Strictly speaking, we could not use the description 'printed circuit', as this implies that the circuit is applied by a printing process. This is not the case and the pattern of the electrical circuit is achieved in two ways.

- Screen printing an etch resist on a copper clad base board in the image of the circuit required. The unprotected copper is removed by treating the board with an etchant solution. The etch resist is then removed either by spraying with a suitable solvent or a mild alkali. In the latter case, the acid resists are based on high acid value maleic resins.

  Alkali-soluble etch resist:

Phthalocyanine Blue	2.0
Extender	46.0
High acid value resin	30.0
Glycol ether	20.0
Levelling agent	2.0
	100.0

  The pattern of the circuit is inspected after printing and, to assist visual inspection, the resist should be matt and contrast with the background copper. As the requirements for higher printing definition increase, the flow characteristics of the resist become more important and excessive flow must be controlled. This can be achieved by using extender or structuring additives.

- Screen printing a plating resist to a copper clad board in the areas where the circuit is not required, leaving the circuit pattern exposed. After cleaning, the board is placed in a plating bath containing a metal salt which may be gold, nickel, silver, tin or lead (or a combination of the last two). By electrolysis of the solution, a thick layer of the relevant metal is deposited in the exposed, circuit pattern areas. After plating, the resist is removed by solvent or alkali, depending on the type used, and the copper is selectively removed using an etchant that does not attack the plated metal.

  Such boards are very expensive to produce and are therefore only used for the highest quality circuits, for example in military equipment. A plating resist must have a high standard of print definition, must resist the plating solution and must be easily removed in solvent or alkali. As with acid resist, finish is matt for inspection purposes, but the proportion of extender to control flow must be minimized to avoid excessive sludge formation in the removal solutions.

Once the circuit pattern has been established by one of the above methods, the electrical components can be soldered into place. Solder resists are applied to the board to protect those areas where solder is not required. They remain on the board as a protective coating and a high degree of adhesion is therefore required. They must be heat and solder resistant and must not be attacked by the solvents used to remove excess solder flux. They must also be electrical insulators. They are lightly pigmented, usually green, and transparent to allow inspection of the underlying circuit. The two most popular forms of screen-printed solder resists are conventional two-pack curing products based on epoxy/polyamide and UV-curing resists based on expoxy acrylates.

Two-pack epoxy solder resist:

Phthalocyanine Green	2.0
Extender	30.0
Epoxy resin	50.0
Glycol ether	16.0
Levelling agent	2.0
	100.0

Aromatic adduct amine hardener mixed 2:1 resist:hardener. Stove at 120–40°C for 30–45 minutes, convection oven.

UV solder resist:

Phthalocyanine Green	1.0
Extender	25.0
Thickening agent	3.0
Epoxy acrylate	30.0
Reactive monomer	37.0
Levelling agent	2.0
Photo-initiator	2.0
	100.0

The print definition of solder resists has become extremely critical as circuit tracks become narrower and narrower, particularly for professional boards. As the screen process is being pushed to the limit of its definition capabilities, a new technique using a photoimaging system has been developed. Here, screen printing applies a solid all-over coating of photodefinable resist which is then exposed through a photographic negative. The exposed area hardens to give the resist design and the unexposed areas are washed off. Definition is only limited to the accuracy of the negative, not the printing process.

The final printing operation in printed circuit board manufacture, before the components are added and the solvent is applied, is a board marking ink used to identify component serial numbers, etc. These are truly inks, as they are fully pigmented and the only element in PCB printing that is used for a decorative function. They must, however, resist solder at high temperature and the flux removal solvents. Conventional two-pack curing inks based on epoxy resins are used as well as UV-curing products.

Thermal-curing notation ink:

Titanium dioxide	20.0
Extender	20.0
Epoxy resin	40.0
Glycol ether	18.0
Levelling agent	2.0
	100.0

Amine hardener mixed 6:1 ink:hardener. Stove at 120°C for 30 minutes.

Much of the technology of resists and board-marking inks is now based on UV curing because of the following advantages.

- Dimensional stability of the boards is maintained. The temperature of the board does not rise above 50°C.
- Resists and inks are solvent-free and thus have excellent screen stability. The flammability hazard is removed.
- Productivity is increased. The boards can be processed very rapidly.

Membrane switches

Membrane switches are a relatively new technology and are finding widespread use in place of push button and other mechanical switches. As there are no moving parts, a membrane switch is extremely reliable and could easily outlast a domestic appliance. They are easy to clean, are attractive to look at and are relatively cheap to produce.

A membrane switch uses a number of different types of screen printing inks and is typically constructed with a plastic graphic overlay, top circuit, spacer, bottom circuit and a mounting panel. Finger-tip pressure on the top circuit brings it into contact with the bottom circuit completing the electrical connection (Fig. 10.10).

Each of the circuits is screen printed with a conductive ink based on silver, gold, nickel, copper or graphite, depending on the degree of electrical conductivity required.

Silver gives the best conductivity (lowest resistance) but at a premium price. Low-cost graphite is often used for less critical applications and it is quite common to blend silver and graphite to provide a cost/resistance compromise.

The graphic overlay can be either polyester or polycarbonate; the latter is popular because it is relatively easy to produce a textured, non-reflective surface without loss of clarity. Polyester is cheaper and has better chemical resistance for applications such as laboratory equipment, where it might come into contact with industrial cleaning compounds. The clarity of polyester suffers when producing a textured finish grade

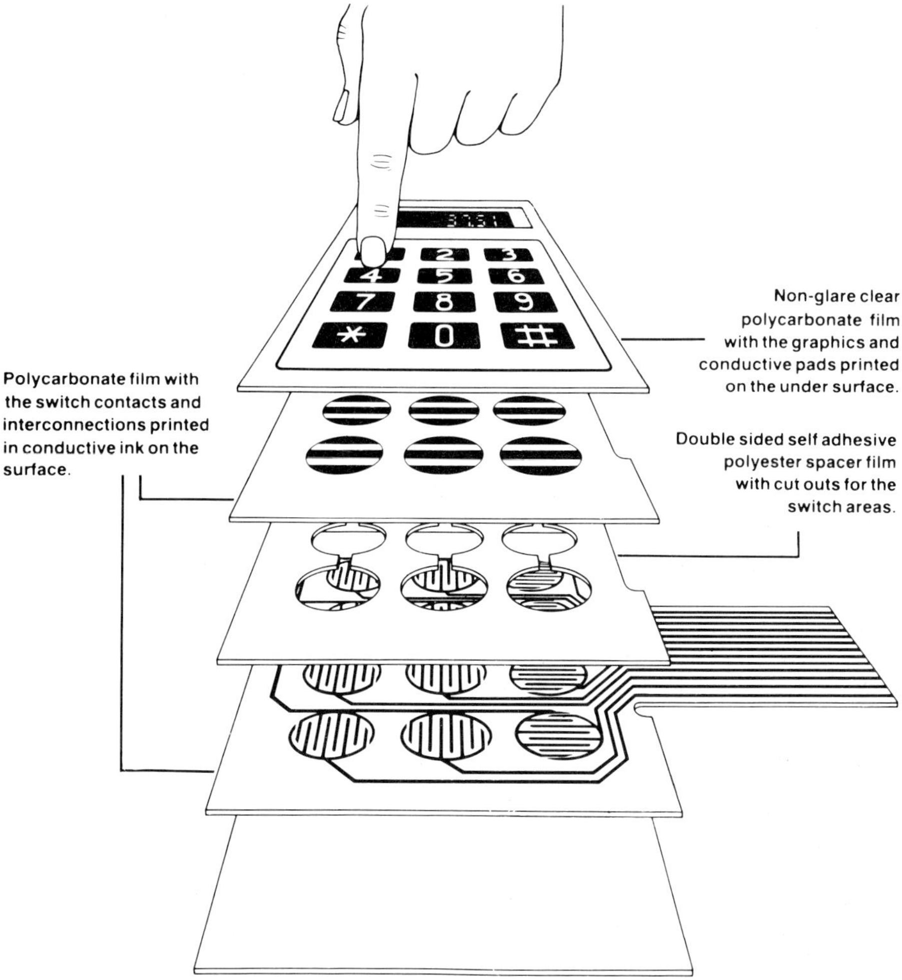

Fig. 10.10 The structure of a membrane switch. (Courtesy of Autotype)

and special matt coatings are used to give a non-reflective finish and maintain clarity.

The overlay may also incorporate transparent window areas through which a light emitting diode (LED) can be viewed. For maximum clarity a transparent ink is essential and a high gloss overlay is desirable. This has the disadvantage of creating a highly reflective surface on the overlay, making it difficult to read.

To counter this, a screen-printed anti-glare varnish is applied over the surface, but leaving the LED windows exposed. The switch then has a matt, non-reflective finish with patches of glossy transparent windows. These varnishes are available in a variety of finishes and in UV-curing grades. Nitrogen-blanketed UV equipment is available that can produce a variety of finishes from just one varnish by altering the conditions of cure. These are described as texturing varnishes and produce a grain finish.

The windows are coloured using highly transparent, dye-based inks which have good lightfastness in order to survive the working life of the switch.

10.14 INK-RELATED PRINTING PROBLEMS

Most problems that occur in screen printing are often not directly attributable to the ink formulation. Problems occur in misuse or incorrect application of the ink or the final print. Listed below are common problems and suggested solutions based on manipulation of ink formulations.

Adhesion

As screen printing is such a versatile process it is possible to screen print on a very wide range of surfaces and thus each substrate can present its own ink adhesion problems. When screen printing on some plastics it is possible to base formulations on resins that bear similarity to the substrate, e.g. acrylic resins for sheet acrylic, amino resins for melamine formaldehyde mouldings, vinyl resins for self-adhesive PVC sheet. Resin selection is the most important factor in achieving good adhesion. Indifferent adhesion can sometimes be improved by solvent selection but this is not always reliable. A technique frequently used when printing on PVC sheet is to include a higher proportion of cyclohexanone. This has the effect of etching and improving chemical adhesion. This technique is never successful if the basic ink formulation is unsuitable and has poor initial adhesion.

Inadequate flexibility can result in poor ink adhesion; in this situation modifications with plasticizer, or flexibilizing resins will improve matters. In thermosetting systems attention should be paid to the reactive component or catalyst, e.g. increasing polyamide content in an epoxy–polyamide screen ink will improve flexibility and adhesion.

Bubbling

This is a very common problem in screen printing. It is usually attributable to long flow or the inability of an ink film to break cleanly between mesh and substrate and may be overcome by the use of silicone defoamers, non-silicone defoamers, increasing extender content, structuring agents or increasing solvency.

Crazing

This has the appearance of hairline cracks in the print. Normally it only occurs on plastic and is believed to be attributed to shrinkage of the ink film. Crazing can arise due to the ink solvent being withdrawn rapidly from the film. It is most commonly seen when printing on solvent-sensitive plastics such as polystyrene or overprinting heavily pigmented inks. Crazing can be eliminated by reducing solvency, lowering pigmentation, correct resin selection or increasing plasticizer content.

Drying

Inks formulated for display application, i.e. printing of paper, board and self-adhesive plastics, are expected to jet dry rapidly in 20–30 seconds. It is obvious that in these conditions the evaporation rate of the solvent is critical. However, this alone does not govern the speed of drying. Most screen inks require a balance of solvent and diluent to achieve optimum drying. Attention should be paid to solubility parameters and solvent release of various resin systems. Poor solvent release can result in blocking and rewetting in the stack.

Intercoat adhesion

Poor adhesion between ink layers often appears in thermosetting inks, e.g. epoxys, amino and UV-curing inks. Overcuring of the first ink layers is the obvious cause but this sometimes can be overcome by reducing the reactivity of the system, e.g. lower catalyst content or in the case of epoxy systems using a resin with high epoxide equivalent. If possible it is best to avoid the use of silicone defoamers in thermosetting inks.

Feathering

This appears as an unsightly spray pattern around the print. Although very often caused by static electricity charges, it can be attributed to poor ink formulation. Feathering can be overcome by increasing extender content, improving solvency, retarding solvent evaporation, inclusion of anti-static agents.

Mesh marking

A pattern of the mesh appears in the dried ink film. Always due to poor flow or overpigmentation. May be overcome by increasing resin content.

Pinholing

This is sometimes described as cissing or reticulation but there is a subtle difference. Pinholing is nearly always due to high surface tension. It is often caused by silicone contamination but can be eliminated by increasing silicone content. Restricting the ink flow will also assist.

Pick-up

When overprinting screen inks it is possible to redissolve the first colour down with the overprinting colour. The fault shows itself as staining of the mesh, loss of finish and mesh marking. The problem is caused by using a high solvency blend and therefore may be overcome by including diluent or weaker solvents. Also selecting a less soluble grade of resin will assist.

Rub resistance

Unfortunately it is true to say that many screen inks give some degree of rubbing; this is not unexpected from the heavy ink deposit. It will be found that pigment selection and pigment/binder ratio is important. The use of polythene and fluorocarbon waxes will assist with this problem.

Stress cracking

This occurs when printing on plastics moulded under strain. Contact with ink solvents will release the strains and cause embrittlement and low impact strength. Apart from moulding considerations the problem may be minimized by weakening the solvent mixture or increasing the flexibility of the ink. In the case of printing on PVC it is necessary to reduce the ketone and ester solvent content and with polycarbonate and polystyrene reducing the aromatic hydrocarbon content.

10.15 RECENT TRENDS

Environmental/health and safety

Legislation continues to become more demanding in all areas of the world. It is therefore most important when formulating inks that local legislation is observed. This, both today and more so in the future, affects raw material selection and is as equally important as ink performance characteristics. Environmental considerations such as airborne emissions and the effect on both land and water pollutions are also necessary to take into account. This will challenge the ink formulator progressively to use safer and more environmentally acceptable solvent, resin and pigment systems and in the case of UV inks, monomers of lower skin irritancy. Lead chrome pigments, whilst still currently used in screen process inks, will no doubt be progressively phased out. Further details on health and safety and the environment are given in Chapter 17.

Water-based inks

Water-based evaporation drying inks will be of increasing interest for all screen printing applications in order to minimize solvent emissions and to satisfy health and safety requirements. Water-based systems formulated from either emulsion or solution polymers can now be formulated that will provide good printing characteristics and adhesion onto paper and board together with many 'plastics' including PVC, polystyrene, polycarbonate, top-coated polyester, etc. Solution polymers in general appear to provide superior printing characteristics. Current technology has certain limitations with regard to drying speed, particularly on multicoat printing applications which at present limit printing speeds.

Water-based UV-curing inks will also no doubt increase in use for both line and halftone printing applications. In four-colour process printing they show significant benefits against conventional UV-curing inks in both improved print quality and reduced film build. Similarly to water-based evaporation drying systems the new generation of UV-curing solutions of polymers/monomers offer superior stability characteristics during printing and also appear to be less sensitive to changes in relative humidity.

Ultra-violet-curing inks

UV-curing inks and varnishes continue to grow in popularity and can still be regarded as new developments within the screen process ink market. While formulation details are not provided, the chemistry is similar to that described in Chapter 11.

Inks for printing flat sheet plastics

The demand continues for inks that will dry faster yet still maintain good stability on the machine. This feature is most important when printing on the increasingly popular high speed cylinder screen machines. Modern formulations based on hard acrylic resins offer good adhesion, fast solvent release and good film flexibility. UV curing inks are also gaining in popularity in this field.

Lightfastness and weatherability is also most important and many signs and self-adhesive PVC decals are required to withstand 5 to 10 years exterior use.

Plastic containers

The movement towards UV-curing inks continues to grow. These inks offer exceptional gloss and the ability to print up to five colours on a bottle without large interstage drying units. Ink formulations will need to be fast curing to match the high speed production rates and maintain good chemical resistance.

Textile inks

Water-thinnable inks should be capable of curing and developing good washfastness without the need for long curing schedules.

Plastisol inks should improve in cure speed to allow the printer greater safety margin in production. There is an increasing requirement for plastisols with soft hand and therefore attention should be paid to plasticizer content, selection of plasticizer and grade of vinyl resin.

CHAPTER 11
Ultra-violet and electron-beam curing systems

Printing inks based on ultra-violet and electron-beam curing technology can be applied by most of the print processes. The characteristic which is common to these inks, and which divides them from other ink systems, is the ability to change almost instantaneously from the fluid phase to a highly cross-linked solid phase by means of a chemical reaction initiated by ultra-violet light or by high energy electron-beam radiation.

The fluid state of the ink must share the rheological characteristics of the type of ink normally used in the application concerned. The curing process must bring about a dry state within the printed ink film which is necessary for end print application. In this chapter we will use the term 'cure' to describe the formation of new chemical bonds under the influence of radiation.

As described elsewhere in this manual the dried state of 'conventional' inks is brought about by a combination of physical and chemical processes. Solvent-based gravure and flexographic inks for example, dry largely by evaporation of the solvent, together with a limited degree of penetration of the substrate. The solvents will also confer properties of adhesion on a particular substrate. Water-based inks for the same process dry partly by evaporation followed by a chemical reaction initiated by this evaporation.

Typically, all the above inks may contain high levels of solvent and correspondingly low solids, and are applied at relatively high film weights.

Oil-based paste inks, by contrast, tend to contain much less volatile solvent and dry initially by a combination of evaporation and penetration, followed by oxidation reactions which depend upon drying and semi-drying oils. These may well take a period of days to complete their reactions. The ability of radiation-cured systems to reach the same dry state almost instantly provided the initial incentive for their development. More recently the reduction of volatile organic compounds (VOCs), now required by law in some countries, has created further interest in development in areas traditionally confined to solvent-based inks.

11.1 RADIATION-CURING PROCESSES

Whether effected by chemical means, physical means or by a combination of both, ink drying requires at least an initial input of energy. There are essentially three ways by which the necessary energy can be applied to a wet-printed substrate. The printed matter may be passed over a heated surface and the thermal energy conducted through the substrate into the ink layer. Secondly, warm air may be blown across the surface of the print, a method particularly effective for drying solvent-based ink via forced convection currents. Finally, the energy may be transmitted across space from a suitable emitter and absorbed directly by the molecules that comprise the ink vehicle.

Radiation is the term used to describe the passage of energy from a transmitting source to an absorbing body without interaction with any intervening matter. Four categories of radiation have found prominence in the field of ink curing; these are listed in Table 11.1 but only two will be discussed in this chapter.

Table 11.1 Radiation in ink curing

Radiation type	Radiative particle
Microwave and radio frequency	Photon
Infra-red	Photon
Ultra-violet	Photon
Electron beam	Electron

UV radiation has been known to initiate curing for a very long time, although results reported before 1960 may depend upon other mechanisms accelerated by heat produced along with the UV radiation. The development of UV-curing inks had to wait until resins which both cured and were capable of participating in the print process were developed. Later in the chapter (Table 11.3) the advantages of using UV inks which have ensured a steady growth rate in ink usage are listed. Lamps and installations which have become more simple, relatively inexpensive and more versatile, have supported this growth. UV-cured inks and lacquers provide the largest volume of the specialized radiation-cured market. In lithographic and letterpress areas, the market is now mature in Europe and growth is slowing, but the introduction of environmental legislation is about to sponsor a rapid growth in solvent-free systems for flexographic and gravure applications. Finally, electron-beam (EB) curing will be examined. It is mainly limited to long run web work because of the cost of installation and the safety problems associated with the use of high energy electrons. EB has, nevertheless, established certain niche markets where complete cure and the deep penetration afforded by electron beams give a considerable technical advantage. Essentially, the inks resemble the UV inks without photo-initiators and thus may produce prints of significantly lower odour. It is now appropriate to consider each of these systems in greater depth. Firstly, however, it is useful to consider the electromagnetic spectrum and the origins of the radiation that will be discussed.

11.2 ELECTROMAGNETIC RADIATION AND ELECTRON BEAMS

Electromagnetic radiation

Electromagnetic radiation is a particular type of radiation for which a fundamental entity called a photon is the vehicle for energy transfer. Photons are massless particles of subatomic matter which transverse space at a velocity of ca. 3.13×10^8 m/s.

A characteristic of photons is that while certain aspects of their behaviour can be explained by considering them to be small particles of matter, other features can only be interpreted in terms of a travelling spatial wave possessing an electrical and a magnetic component. Hence the term electromagnetic radiation is used.

The peculiar behaviour of matter and energy at atomic and subatomic levels is the province of quantum mechanical theory. Although of mainly indirect interest to the ink technologist some appreciation of the concepts involved are useful in gaining an understanding of the principles which underlie radiation curing. Much of the terminology used in radiation curing technology has its origin in nuclear physics and spectroscopy.

The standard form for wavelength is to give values in units of nanometres (nm). However, spectroscopists, by tradition, also use the micron (μm) and the ångström Å as units.

$$1\ \text{Å} = 10^{-10}\text{m}$$
$$1\ \text{nm} = 10^{-9}\text{m}$$
$$1\ \mu\text{m} = 10^{-6}\text{m}$$

A further unit in popular use is the wavenumber. This is the reciprocal of wavelength expressed in centimetres.

For electromagnetic radiation a fundamental relationship exists between the wavelength λ, the frequency υ, and the velocity of propagation in vacuum C:

$$\upsilon = C/\lambda$$

Thus, regions in the electromagnetic spectrum can be designated in terms of frequency. The standard frequency unit is the hertz; 1 Hz = 1 cycle/second.

The minute quantity of energy $\triangle E$, carried by an individual photon is proportional to its frequency:

$$\triangle E = h\upsilon$$

where the proportionality constant h is Planck's constant. Hence energy units are acceptable in defining particular bands in the spectrum of electromagnetic radiation. It is customary to define the energy of radiation in terms of a mole of photons, called an Einstein. Thus units for energies may be expressed as kJ/mol or kJ/Einstein. Note that from Avogadro's number there are 6×10^{23} photon to the Einstein.

For convenience the spectrum of electromagnetic radiation can be divided into regions. Although there are no clear boundaries, each region defines a frequency range for which distinctive physical and chemical effects are observable when the radiation is absorbed by matter.

These bands of radiation and their inherent effects on matter are summarized in Table 11.2.

While all electromagnetic radiation involves the emission of photons, the most efficient type of emitter varies with each particular frequency band. Thus a hot ceramic surface is an effective source of IR radiation but radiates insignificant amounts of UV, while collisions between excited vaporized mercury atoms can emit substantial amounts of UV. This topic is further developed in later sections.

The physiological effects of a particular band of radiation are important as they dictate the degree of precaution that is necessary in order to operate a radiation curing system safely.

Electron beams

For EB radiation the entity conveying energy across space is a free electron.

Electrons are particles of atomic matter which like the photon exhibit a particle–wave duality. In contrast to photons, however, electrons have a significant, albeit small, mass (9×10^{-28} g at rest) and carry a negative electric charge.

Usually in matter, electrons are bound by the positive electric force from the nuclear protons in regions of space close to the nucleus.

In a metal there are certain electrons which are not bound to any specific nucleus, but are generally associated with all the atomic nuclei that comprise the metal structure. These so-called conducting-band electrons are responsible for the electrical conductivity of the metal.

If energy is supplied, by say heating the metal, some of the conducting band electrons may become sufficiently energetic that they temporarily escape through the metal surface to form a cloud of effectively 'free' electrons. Thus around a hot metal a dynamic equilibrium develops with electrons continuously leaving and drifting back.

An electric circuit can be made under conditions of high vacuum, such that the hot metal forms the cathode, and a positive electrode, the anode, placed a short distance from it, attracts the 'free' electrons away from the cathode.

Taking advantage of either the electric charge or the magnetic moment possessed by the electrons, suitably positioned electrostatic or magnetic fields can be used to manipulate this current of free electrons.

In an EB generator such fields are used to concentrate the electrons into a narrow band, accelerate them to the desired potential energy and if necessary deflect the beam to allow scanning across a substrate.

Electron-beam radiation is usually characterized qualitatively by the average kinetic energy possessed by the electrons on leaving the region of acceleration.

Table 11.2 Significant regions in the electromagnetic spectrum and their principal effects

Region	Radio	Microwave	IR	Visible and UV	X-ray
Frequency (Hz)	3×10^{10}		3×10^{12}	3×10^{14}	3×10^{16}
Wavelength	1 cm		100 μm	1 μm	10 nm
Energy (kJ/E)	7.1×10^{-4}		7.1×10^{-2}	7.1	7.1×10^{-2}
Effect on matter	Subtle effects on the electrons and nuclei of atoms	Changes in rotation and spatial translation of polar molecules	Changes in the extent of vibrations between atoms in molecules	Changes in the distribution of electrons bonding atoms in a molecule	Electronic effects as for UV but involving electrons more closely bound to the atom
Observable consequence	Requires specialized equipment	Increased evaporation of polar molecules	Increased evaporation; chemical reaction if intra-atomic vibrations lead to bond disruption	Disruption of chemical bonds with consequential chemical reactions	Ionization with subsequent molecular disruption and chemical reactions
Physiological effects	None at normal intensities	Deep tissue burns	Experience as heat; surface skin burns	400–700 nm radiation experiences as visible light; UV can cause damage to skin tissue	Tissue damage due to ionizing radiation

For convenience, the energies are often expressed in units of electron volts (eV), 1 eV being the kinetic energy increase acquired by electrons when accelerated across a potential of 1 V:

$$1 \text{ eV} = 1.602 \times 10^{-16} \text{ J}$$

Electron beam systems used in the printing industry provide electrons with energies of between 150 and 300 KeV.

In contrast with electromagnetic radiation where only photons with wavelengths shorter than ca. 100 nm have sufficient energy to ionize molecules, electron beams are essentially an ionizing radiation. When an electron with high kinetic energy enters matter, collision-induced disruption of the atoms and molecules occurs, resulting in the generation of ionic species. The subsequent formation of radical species which are responsible for the initiation of cure in inks and varnishes is discussed in section 11.5.

11.3 INTRODUCTION TO FORMULATION

Inks for radiation curing must function in the same way as their conventional counterparts on a printing press. The formulation, therefore, must be broadly similar to other inks, i.e.

- pigment
- vehicle
- solvent
- additives.

While the pigments remain similar to those used in conventional inks, the other parts of the formulation differ significantly chemically, although fulfilling similar roles. The solvents in these systems are low-viscosity monomers, capable of reacting in their own right. The vehicle is composed of oligomers, usually acrylates. For a UV ink the 'additives' contain a large amount of photo-initiators which respond to the photons of UV light to start the system reacting. These are omitted from an electron-beam ink where the electrons initiate the reaction themselves.

A UV ink formulation may be generalized as:

Pigment	15–20%
Prepolymers	20–35%
Monomers oligomers	10–25%
Photo-initiators	5–10%
Other additives	1–5%

Detailed examination of each component is given in later sections. At this point the formulation may be contrasted with that for a standard oleoresinous ink.

The low viscosity monomers, sometimes termed diluents, play the same rheological adjustment and pigment-wetting role as do the solvents in conventional inks. Usually, however, these monomers are capable of chemical reactions which result in their becoming fully incorporated into the ultimate polymer matrix. Selective use of non-curing solvent is possible in a UV vehicle formulation but this practice is limited.

In the main the very low viscosity reactive monomers are precluded by their volatility and, more importantly, their toxicology. Those which can be used in the printing industry have, if compared with conventional solvents, relatively high viscosities.

This factor has previously restricted the development of UV- and EB-curable inks for flexographic and gravure printing, although the use of heated ducts and cylinders to lower viscosities has eased the problem recently.

The oligomer provides the 'hard resin' portion of the formulation. Typically these are derived from conventional synthetic resins, urethanes, epoxides and polyesters are examples, in which residual hydroxyl groups have been reacted with acrylic acid to yield products possessing reactive ethylenic unsaturation.

The radiation-sensitive part of the formulation is provided by the photo-initiator package. Only small amounts of the total photo-initiator present are used up in the curing reaction. Residual initiator and photodegradation products are in the main not bound into the cured polymer. Indeed their presence may be detrimental to the overall properties of the film, a fact that has encouraged the investigation of EB curing.

Ultra-violet and EB-curable varnish formulations essentially consist of the vehicle as used for the pigmented ink systems. Appropriate adjustments are made in the selection of the prepolymers and monomers used in order to achieve the required viscosities for the different methods of application.

11.4 ADVANTAGES AND LIMITATIONS FOR ULTRA-VIOLET AND ELECTRON-BEAM INKS AND VARNISHES

The specialized nature of the raw materials used in the technology tends to make the formulations significantly more expensive than their conventional counterparts. Equipment to generate the radiation is sophisticated and particularly in the case of EB initially involves a considerable outlay of capital. Nevertheless UV printing has become well established (Fig. 11.1). Table 11.3 summarizes the principal advantages which can offset the additional expense incurred by a printer using a UV-curable ink.

Advantages which are specific to particular printing processes and those to be gained by the use of EB curing will be discussed in later sections. Disadvantages inherent in the use of UV technology are also noted.

11.5 CHEMISTRY OF ULTRA-VIOLET INITIATION AND CURE

Rupture of chemical bonds

A covalent chemical bond supplied with sufficient energy may dissociate to give either ionic or free radical intermediates.

Fig. 11.1 The UV curing unit of a Steinemann varnishing machine at Glossifilm Ltd.

$$A - B + \text{Energy} \quad \begin{matrix} A^{+}\ B^{-} \text{ ionic pathway} \\ A^{\bullet}\ B^{\bullet} \text{ radical pathway} \end{matrix}$$

The ionic pathway is favoured if A and B are atoms of significantly different electronegativity and if the environment is conducive to stabilizing charged ions. Generally speaking the raw materials in EB or UV inks and varnishes are comprised of molecules with limited polar functionality which make them comparatively poor ion-solvating species. Consequently reaction mechanisms involving free radical pathways are favoured.

Overall, the process of cure may be divided into the three stages of:

- initiation
- polymer chain propagation
- termination.

Initiation

Initiation is a term covering the early part of the curing chemistry and includes the generation of free radicals and their subsequent reaction with chemically active sites on the vehicle molecules. In most instances these sites will be alkenic double bonds, usually activated towards polymerization by the chemistry of their neighbouring groups.

If irradiated with shortwave UV, alkenes will undergo polymerization. This reaction is, however, usually too inefficient to be of use in practical

Table 11.3 Summary of advantages for printing with UV-curable formulations

- Dry prints, off the press, allow in-line post-print processing
- Ink system remains open almost indefinitely during press stoppages
- Elimination of spray powder leads to smooth prints and a clean press environment
- No solvent emissions
- Very high gloss prints achievable, in some cases able to replace lamination
- Elimination of stack airing reduces 'holding' space requirements
- Consistent, low odour level: no odorous species generated by postcure chemistry
- Compact lamp systems contrast with large IR or hot air drying ovens. Consequently considerable space savings are achieved, particularly for decoration of non-porous substrates such as plastics and metals

ink and varnish systems. Additives, commonly termed photo-initiators, have to be employed to provide an effective source of photochemically generated radicals to start the polymerization.

Although there are many chemical species that are potentially capable of generating radicals photochemically, UV-curing technology has almost exclusively employed the chemistry of the carbonyl function.

The carbonyl group may absorb a photon of electromagnetic radiation whose energy lies in the UV region with the consequent elevation of one of its electrons from a non-bonding orbital into the lowest anti-bonding π-orbital. A so-called $n \rightarrow \pi^*$ transition.

Although the carbon to oxygen bond is weakened by this increased anti-bonding character it remains intact. The spin of the electron does not change during the transition and the resulting species is said to exist as a singlet excited state, $S_0 \rightarrow S_1$. The lifetime of the singlet state is very short and the chances of the uncoupled electron reacting via a collision encounter with a nearby molecule is very low. A variety of decay modes from this excited state are possible. The electron may return to its ground state with re-emission of the absorbed radiation. Some or all of the excited state energy may be dissipated into the adjacent bonding system with a consequential increase in the vibrations between atoms. If sufficient vibrational energy becomes concentrated in a particular bond, homolytic dissociation into a free-radical pair may result.

Another possible fate for the singlet excited state is for the electron to change its spin, via a quantum mechanical process called 'interstate crossing'. This is symbolically written $S^*_1 \rightarrow T^*_1$.

The resulting 'triplet' excited state molecule may decay back to the ground state via either the re-emission of radiation or by increasing the interatom vibrational energy, with possible molecular fragmentation. However, the lifetime of the triplet state is long enough to allow reactions between what is essentially a diradical and surrounding molecules, during collison encounters. For excited state carbonyl, generated from a photo-initiator in an organic resin system, a hydrogen abstraction is the most likely subsequent reaction.

These theoretical ideas become clearer on examination of the proposed mechanisms for some of the initiators currently in use.

Benzil dimethyl ketal provides a good example of an initiator which undergoes a so-called 'Norrish I' molecular fragmentation on irradiation:

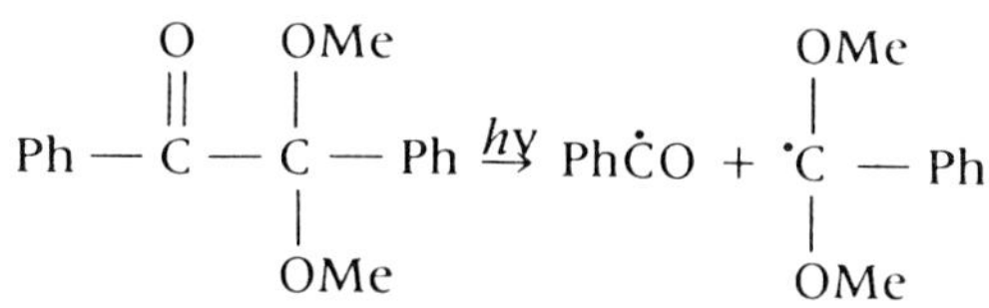

Benzophenone and its derivatives, on the other hand, owe their effectiveness as photo-initiators to the Norrish II hydrogen abstraction mechanism:

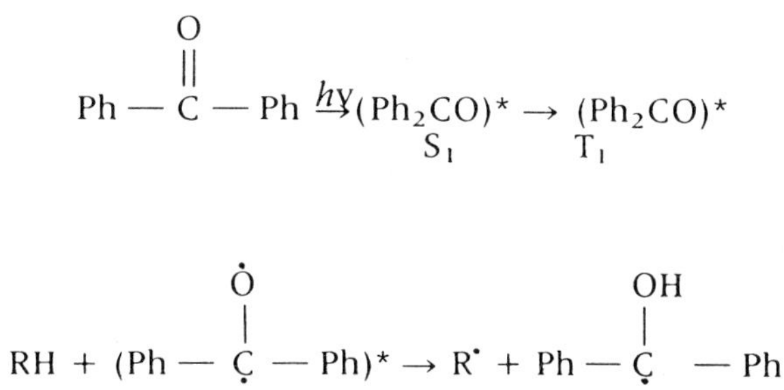

Both mechanisms may be operative simultaneously, as for example with diethoxyacetophenone (DEAP).

Fragmentation;

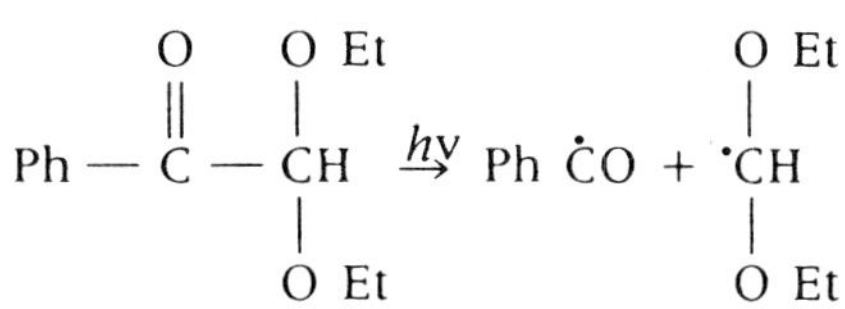

and

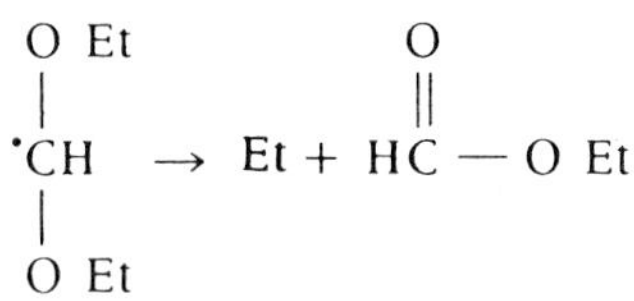

Intramolecular hydrogen abstraction:

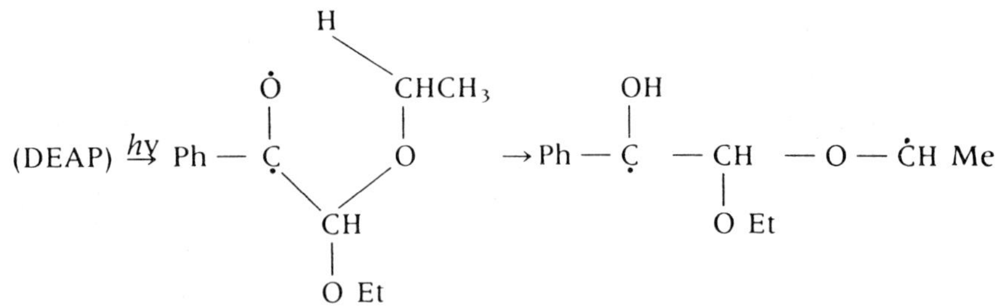

Table 11.4 lists some of the popular initiators that are available.

Photosynergists

For a Norrish II initiation mechanism to operate effectively a H-donor is necessary. Abstractable hydrogen is usually readily available from suitable sites on the vehicle molecules. Methylene groups adjacent to oxygen in epoxides and polyethers provide examples. It is thus usually unnecessary to add specific hydrogen donors to the formulation.

On their own, the curing performance of the ketone initiators is usually too slow to be of practical use for printing applications. Addition of amine containing molecules has been found essential if ink formulations are to have acceptable cure speeds.

Tertiary and secondary amines are found to be more effective than primary. Table 11.5 lists a number of the amines which have proved popular in the industry. These amine-containing molecules have sites with easily abstractable hydrogen, but this alone is insufficient to explain their enhancing effect on the photoformation of free radicals.

A mechanism involving an encounter complex between the amine and excited state ketone has been proposed. Benzophenone and triethanolamine, a commonly employed initiator package, provides an example:

$$Ph_2CO \xrightarrow{h\upsilon} [PhCO_2]^*$$
$$[Ph_2CO]^* + (HOCH_2CH_2)_3N \rightarrow [Ph_2CON\ (CH_2CH_2OH)_3]^* \equiv A$$

It is probable that charge transfer takes place. The resulting excited state ion pair complex is called an exciplex.

$$Ph_2CO^{\cdot -+}N^{\cdot}(CH_2CH_2OH)_3] \equiv B$$

In either case the excited state species readily decomposes into free radicals:

$$A \text{ or } B \rightarrow Ph_2C^{\cdot}OH + HOC^{\cdot}H\ CH_2N\ (CH_2CH_2OH)_2$$

An initiator combination which is extremely effective and has been extensively used in UV-curable formulation is that of benzophenone and Michler's Ketone. Containing the aromatic carbonyl group Michler's

Table 11.4 Some photo-initiators in common use

Compound	Structure	Comments
Benzophenone	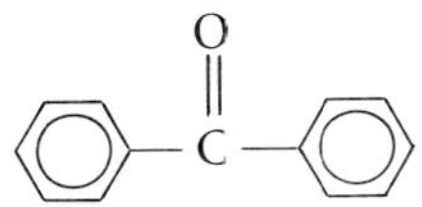	Readily soluble water-white solid. Functioning via a H-abstraction mechanism. A cheap moderately effective initiator. Also provides plasticizer and pigment-wetting effects.
Benzophenone derivatives (a) *p*-chloro benzophenone (b) *p*-phenyl benzophenone	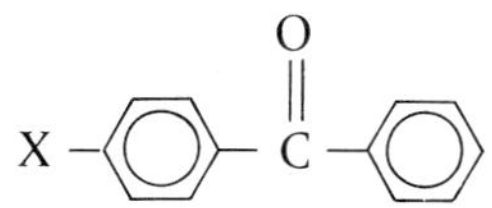 (a) X = Cl — (b) 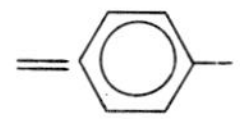	Generally more effective than benzophenone. Functioning via a H-abstraction mechanism. Derivatives are more expensive than benzophenone and are often coloured.
Acetophenone derivatives (a) 1-Benzoyl cyclohexan-1-ol	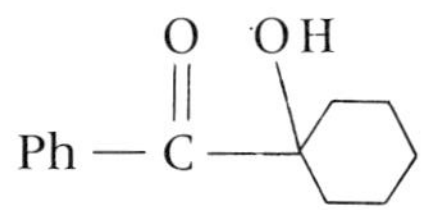	
(b) 2-Hydroxy-2.2-dimethyl acetophenone	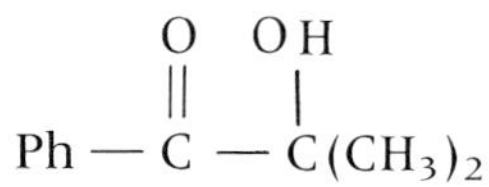	Water-white liquids or readily soluble solids. Initiating mechanism may be via H-abstraction or fragmentation or a combination of both. Highly effective initiators but are often expensive. Photofragmentation products may impart an odour to the cured ink or varnish.

(c) 2.2-Dimethoxy-2-phenyl acetophenone	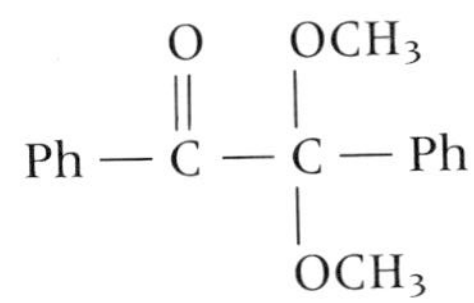	White, reasonably soluble solids. Functioning via a fragmentation mechanism. Initiation is not sufficiently fast for sole use in ink and varnishes, but this class can provide good through-cure if used in conjunction with other initiators.
	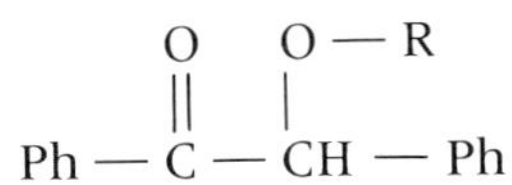	
Benzoin ether derivatives		
(a) Methyl benzoin ether	(a) R = CH_3,	
' 3 (b) *n*-butyl benzoin ether,	(b) R = $CH_3-CH_2-CH_2-CH_2-$	
Thioxanthones		Very effective initiators. Sparingly soluble but forming effective dispersions. Functioning via a H-abstraction mechanism.

(a) 2-chlorothioxanthone, (b) 2-isopropylthioxanthone, (c) 1-chloro-4-propoxythioxanthone,	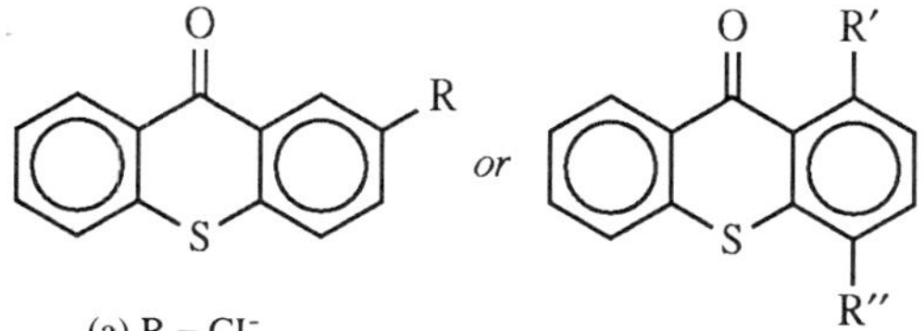 (a) R = Cl^- (b) R = $(CH_3)_2$–CH– (c) R^1 = Cl^- (d) R^{11} = O–C_3H_7	These initiators tend to be expensive. They are free from odour but their distinctive yellow colouration precludes their use in whites, varnishes and some pale tints.

Table 11.5 Photosynergists

Compound	Structure	Comments
• Aliphatic amines		Very effective liquid coinitiators; water solubility precludes use in lithographic inks; odour can be a problem; many cause yellowing of prints on ageing.
(a) N-methyl diethanolamine (b) Triethanolamine	$(OH–CH_2–CH_2)_2$ N–CH_3 $(OH–CH_2–CH_2)_3N$	
• Aromatic amines		Effective liquid coinitiators; insoluble in water; low odour; more expensive than the aliphatic species.
(a) Ethyl-4-dimethyl aminobenzoate	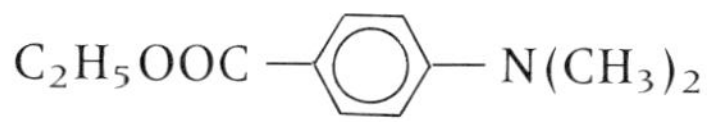	
(b) *n*-butoxy ethyl-4-dimethyl aminobenzoate	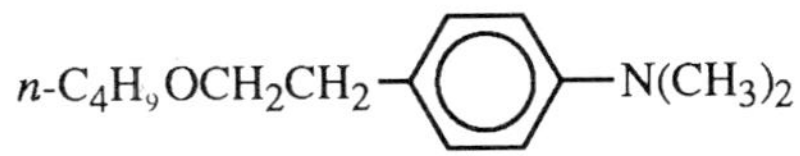	
(c) 4,4′-diethyl-amino benzophenone	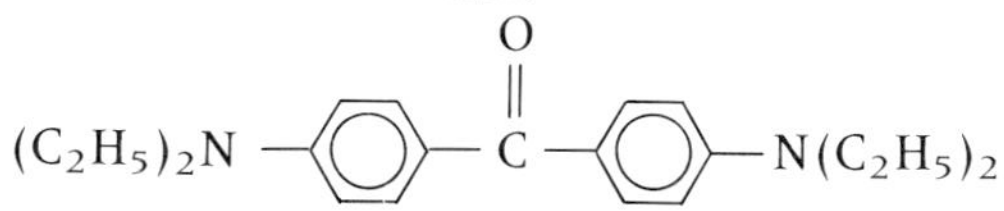	Very effective coinitiator for use with benzophenone; highly coloured solid; very effective with dark or highly pigmented inks.

Ketone is a photo-initiator in its own right. Its amino groups are also available to form complexes with excited state species from other Michler's Ketone molecules and also with benzophenone.

Unfortunately this ketone is toxic and is no longer used. The diethyl derivative is less hazardous and, despite its pronounced yellow colour, is widely used by the coatings industry.

Free radical generation by electron beams

Electrons from the beam accelerator bombarding an ink or varnish film have initially very high kinetic energies. The primary process is collision interaction with electrons from the vehicle molecules. This results in the ejection of secondary electrons with reduced kinetic energies and the formation of molecular ions.

$$e + MX \rightarrow MX^{+} + 2e$$

Further ionization reactions occur as the ejected electrons collide with other vehicle molecules. Successive collisions result in a cascade of electrons with progressively decreasing energy. Ultimately the electrons have insufficient energy to effect ionization and subsequent collisions with molecules lead only to the formation of excited states.

$$e + MX \rightarrow MX^{*} + e$$

Under favourable conditions some excited state molecules dissociate into free-radical pairs which may initiate a polymerization process.

$$MX^{*} \rightarrow M^{\cdot} + X^{\cdot}$$

The fate of the ions generated by EB radiation is not well understood. Certainly some molecular fragmentation and rearrangement occurs. This accounts for the greater degree of branching found in EB cured systems when compared with the same vehicle after cure by UV initiation using photo-initiators. It is probable that some electron capture occurs and this provides a further source of free radicals:

$$e + MX^{+} \rightarrow MX^{*} \rightarrow M^{\cdot} + X^{\cdot}$$

Propagation

Once generated, a number of reaction pathways are open to the free radical. Of primary importance is the process which results in chain polymerization of the prepolymers.

Where $\dot{X}$ is any free radical species, generated via an initiation reaction, polymerization may be summarized as follows:

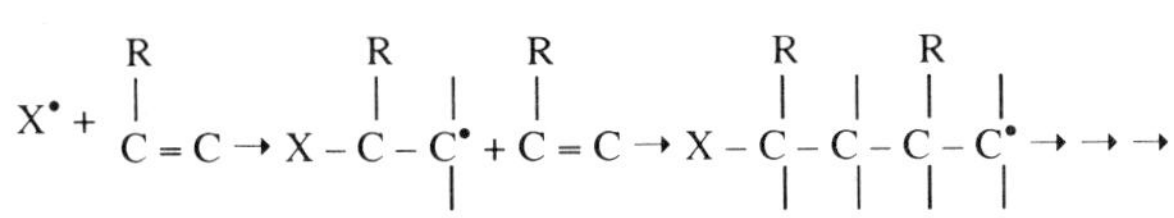

If the vehicle contains species with two or more ethylenic bonds per molecule the possibility exists for the growing chains to cross-link. As the chain lengths and the degree of cross-linking increases the viscosity of the ink or varnish film rises until a densely interlinked solid matrix is attained.

Termination

The rate at which an ink film is cured is dependent upon such factors as the number of initiating radicals generated, the number and reactivity of the unsaturated sites on the molecules of vehicle and the mobility of the growing polymer. Competing with the propagation reaction are any number of so-called termination processes which remove free radicals from the systems.

Ultimate termination occurs as the rate of polymerization is reduced to zero by a depletion of the unsaturated reaction centres and also by a decrease in molecular mobility as the viscosity of the vehicle increases. Termination can result from dimerization reactions, e.g.:

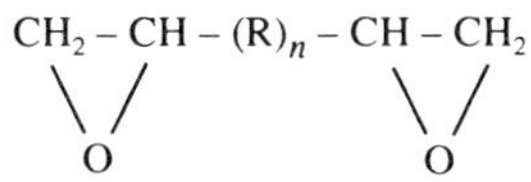

where P = growing polymer and In = initiating species.

Analysis of a cured coating containing benzophenone, for example, shows that a significant quantity of benzopinacol is formed:

$$2\ Ph_2\dot{C}OH \rightarrow Ph_2\underset{}{\overset{OH}{C}} - \overset{OH}{C}Ph_2$$

A serious termination process from the viewpoint of curing thin coating films arises from reactions with atmospheric oxygen. A free radical centre from either initiator of the growing polymer may react with oxygen to form a relatively stable peroxide:

$$X^{\cdot} + O_2 \rightarrow XOO^{\cdot}$$
$$XOO^{\cdot} + RH \rightarrow R^{\cdot} + XOOH$$
$$XOO^{\cdot} + R^{\cdot} \rightarrow XOOR$$

In addition to retarding the rate of cure, peroxides can be detrimental to the chemical and physical properties of the cured film.

Ultra-violet and electron-beam cure in thin films

With the exception of silk screen application printed ink and varnish films for UV or EB curing are essentially 'thin'. Being under 10 μm in thickness, surface effects become significant in relation to the bulk properties. The formation of peroxides in the upper surface layer has already been discussed.

Oxygen is an effective deactivator or 'quencher' of the triplet state and this process will further inhibit surface cure. With UV-curable formulations two factors are active which counteract to some extent the oxygen inhibition. First, the preparations usually contain an amine as a photo-synergist. Although the mechanism has not been clearly established there is evidence to suggest that the effect of oxygen inhibition is substantially reduced by the presence of amine.

A second and more significant feature is the high density of free-radical formation near to the surface on UV irradiation. The rate of formation of free radicals, and consequently the rate of cure, is proportional to both the intensity of the radiation and the concentration of initiator.

For an unpigmented, relatively thick film the Lambert–Beer equation provides an approximate description of the change in intensity with depth and concentration.

$$I = I_o e^{-kcl}$$

where

I_o = the intensity radiation at the surface;
l = the intensity at a depth l from the surface;
k = the absorption coefficient of the radiation absorbing species;
c = the concentration of the absorbing species.

This simplified picture shows that the level of radiation falls off rapidly away from the surface of the film. Adjustment must be made to I_o for partial reflection at the surface. For a thin film the situation is further complicated by multiple internal reflections from the substrate and the underside of the top surface. For a pigmented ink layer, scatter, as well as absorption by the pigment particles, has to be taken into account.

The photo-initiator itself is an absorbing species. Thus a high level of initiator while enhancing the formation of radicals near to the surface, reduces penetration of the radiation and can be detrimental to through cure. Indeed over-use of initiators typically results in soft, 'cheesy' films.

Electron-beam radiation is more penetrating and results in the generation of proportionally more initiating radicals deep in the ink layer.

Through-cure, even with thick highly pigmented inks is seldom a problem for EB systems. However, a comparatively low density of free radicals is created in the top surface where oxygen inhibition effects are dominant. Consequently inert gas blanketing is essential for effective EB surface cure.

11.6 PREPOLYMER RESINS FOR ELECTRON-BEAM AND ULTRA-VIOLET FORMULATIONS

Activated carbon–carbon double bonds can be introduced into a resin structure via a number of readily unsaturated compounds. However, the rate of reactivity for an alkene group in a free-radical polymerization reaction is very sensitive to the nature of its neighbouring groups.

Broadly, reactivity follows the order:

acrylate>methacrylate>allyl>styrenyl>vinyl

As far as the printing industry is concerned, it has been chiefly resins with acrylate functionality that have provided the necessary reactivity to bring about adequate cure at the throughput rates demanded by modern high-speed presses. Table 11.6 lists the three main resin types that have provided the prepolymer for EB and UV formulations.

Epoxy acrylates

Epoxy acrylates are prepared by the reaction of an epoxy group with acrylic acid. Generally the reaction produces medium to high viscosity fluids which have a fast cure rate and will act as the main vehicle of a lithographic ink. The most widespread type of resin used for inks is prepared by acrylating the diglycidyl ether of bisphenol A. Often, to ease handling during processing, this will be supplied diluted in one of the commonly used monomers to reduce viscosity.

Many epoxy resins may be acrylated, the important reactive group being the hydroxyl group(s). If we generalize an epoxy as:

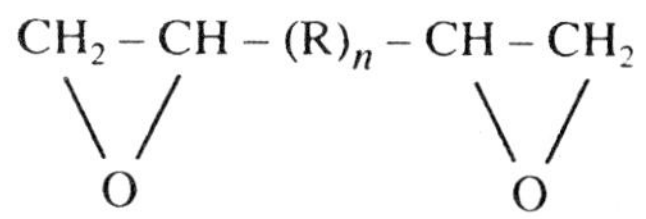

then each R contains one hydroxyl group, and in the general formula there will be n hydroxyl groups. For the reaction we must use n equivalents of acid, together with a small percentage of catalyst (e.g. triethylamine) and an appropriate inhibitor. These will be reacted together with heating until the acid value has reached the appropriate low level. Care must be exercised throughout to avoid gellation which can occur very rapidly.

A disadvantage of epoxy acrylate resins is their comparatively poor pigment-wetting ability. Consequently, UV inks based entirely on simple epoxy acrylate resin tend to have low gloss and poor lithographic performance.

Improved pigment wetting, and at the same time enhanced substrate adhesion, may be obtained from resins with residual hydroxyl functions.

Table 11.6 Some acrylated prepolymers

Resin type	Typical properties
Epoxy acrylate	Inexpensive, fast curing
Polyurethane acrylate	Toughness, chemical resistance
Polyester acrylate	Low viscosity, good pigment wetting

Such functionality is achieved by opening an epoxide ring directly with acrylic acid:

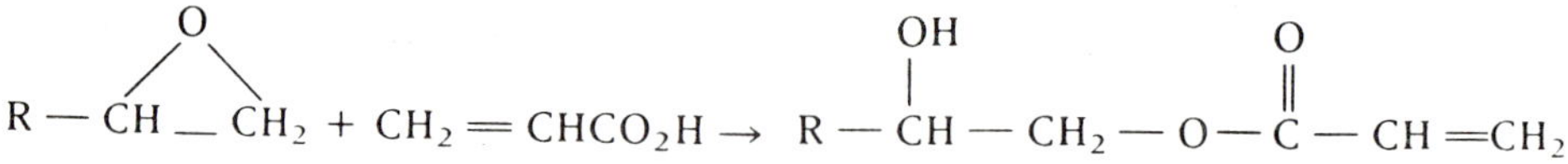

where R is the resin backbone.

Typical aromatic resin backbones are those derived from bisphenol A epichlorohydrin condensates:

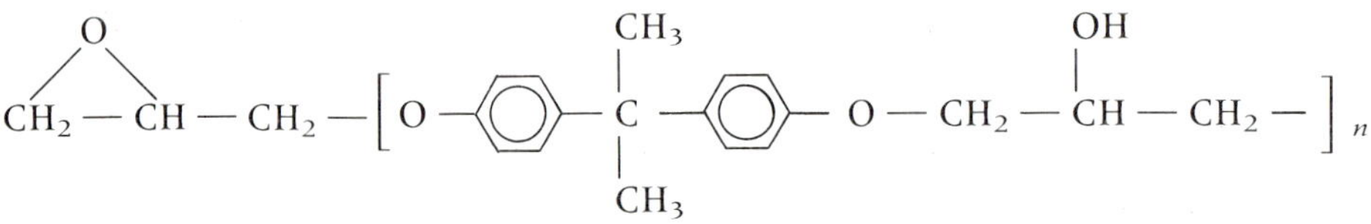

Notable features of such a structure are the rotational flexibility imparted by the bisphenol A group, the chemical inertness of ether linkages and the hydroxyl sites which enhance adhesion and wetting while providing centres for further reaction if required. For example, to improve lithographic performance further, these residual hydroxyl functions are sometimes esterified with long-chain fatty acids.

Unsaturated natural oils such as soya bean and linseed can be readily epoxidized. Subsequent reaction with acrylic acid yields radiation curable resins with good inkmaking characteristics. Unfortunately their reactivity is low and their use in formulation can lead to a retardation in cure speed.

Polyurethane acrylates

Inks and varnishes formulated on urethane resins are noteworthy for their toughness and chemical resistance. Urethane acrylates provide a class of resins for UV-curable products which more or less retain these desirable characteristics.

Products derived from resins based on aliphatic isocyanates have good colour retention on exposure to light or on stoving. These resins, however, are rather expensive. The somewhat cheaper aromatic urethane acrylates suffer the drawback that they tend to yellow significantly on exposure to light or heat.

A urethane acrylate prepolymer can be prepared by reacting isocyanate groups from a urethane resin with, for example, hydroxyl ethyl acrylate:

$$\mathrm{OCN-R-NCO + 2(HO{-}R') \rightarrow R'OCONHRNHCOR'}$$

where R is the urethane backbone.

For the simplest case where R is an aromatic di-isocyanate, such as toluene di-isocyanate (TDI), a lack of molecular flexibility leads to high-viscosity resins. Generally, therefore, prior to acrylation the isocyanate is chain extended by reaction with a polyol or alternatively with a polyester or a polyether which itself has residual hydroxylic functionality. In this way complex mixed prepolymers often, with extensive branching, can be constructed.

Polyester acrylates

The wide range of comparatively low-cost polyesters that are available as precursors make these an attractive group of radiation-curable materials.

Chemical and physical properties are very varied. General polyester prepolymers provide good pigment wetting and consequently make good inks with, if required, excellent lithographic performance. Their adhesion, particularly to non-porous substrates such as tinplate and plastic, is especially noteworthy.

The intrinsic ester linkages are more labile than are ethers thus when contrasted with epoxy acrylate the polyester acrylates tend to give products with inferior chemical resistance, particularly towards alkali. Furthermore, especially for the lower molecular weight polyester acrylates, reactivity can be poor and surface oxygen inhibition effects can be a serious problem.

On the positive side an advantage with many polyester acrylate prepolymers is that they have a low or relatively low viscosity. They may thus be used to replace part or all of a diluent monomer in a formulation.

Preparation is either by reaction of acrylic acid with residual hydroxyl groups from a polyester precursor or by reacting residual acid groups from a polyester resin with a hydroxyl acrylate such as 2-hydroxy ethyl acrylate. For example:

$$ROH + CH_2{=}CH{-}\overset{\overset{\displaystyle O}{\|}}{C}{-}OH \rightarrow CH_2{=}CH{-}\overset{\overset{\displaystyle O}{\|}}{C}{-}OR + H_2O$$

where R is the polyester condensation product from a polyol and a polybasic acid and contains one or more residual hydroxyl or acid functions. However, in practice the esterification process is by no means simple.

The volatility and instability towards spontaneous polymerization inherent with acrylic acid and its derivatives limit permissible reaction temperatures to under 100°C. To drive the esterification towards completion strong acids, for example methane sulphonic acid, are required. These have to be removed from the final product together with any excess acrylic acid.

To remove the water formed, azeotroping solvents such as benzene have to be introduced and subsequently removed. Also inhibitors such as nitrobenzene and phenothiazine are added to prevent free-radical polymerization.

In addition to all these potential contaminates, the reaction conditions are such that saponification of the ester linkages in the resin backbone are possible. Acrylation of the original polyols can then occur to yield low molecular weight products with possible toxicity hazards.

Acrylated polyols and polyethers

Resins which contain ether linkages that derive from polymerization of epoxides are customarily called epoxide resins by reference to their origins. Of considerable importance are the polyether acrylates derived by chain extension of simple polyols, prior to acrylation. Usually extension is limited to only a few ether units and these products are thus often termed oligomers.

These polyether acrylates and the products derived by direct acrylation of polyols are often comparatively low viscosity liquids. As such their use is considered in more detail in the section covering diluents. It is possible, particularly in a varnish formulation, for the vehicle system to be comprised almost entirely from such molecules.

Water-reducible resins

The wood coatings and furnishing industry has made considerable use of UV-curable water-based systems. Usually these have been formulated around emulsions of polyester acrylates. Such resins have so far attracted little interest in the printing field, being unsuitable for pigmentation and also lacking the gloss necessary for varnish formulations.

A few experimental water-thinnable resins have been developed. Despite the potential advantages of producing low viscosity inks cheaply these have so far met with very limited success. Possibly this is because their film-forming properties have not been adequate.

However, water-reducible systems have found acceptance in the field of silk screen printing where there is ample opportunity to remove the water from the system. Development is still progressing with water-reducible resins and, because water is capable of fast viscosity reduction, there is considerable potential for such resins in low viscosity systems.

Other resins

In the coatings industries early UV-curable formulations were based on blends of styrene and unsaturated polyesters, the unsaturation being introduced into the polyester from such precursors as itaconic acid and maleic anhydride. These systems generally had too slow a rate of cure to be of use for printing applications. Such resins can, however, be used in conjunction with the more reactive acrylated resins and diluents. Indeed it is possible to formulate a UV-curable ink containing portions of fully saturated polyester, triazine or epoxy resins. These become 'occluded' in the polymer matrix but do not necessarily copolymerize with the acrylated species.

Some of these, and other types of resins, have been included in formulations to improve pigment wetting and to improve adhesion to a specific substrate. Such uses tend to be the 'trade secrets' of individual companies and are many and varied.

11.7 DILUENTS FOR ELECTRON-BEAM AND ULTRA-VIOLET FORMULATIONS

Diluents are materials used to reduce the viscosity of formulation to levels suitable for the required method of printing. They may also be needed to solubilize a solid prepolymer and they certainly contribute to pigment wetting.

A diluent can be a non-reactive solvent which is either ultimately lost from the formulation by penetration and evaporation or else remains in the cured film as a plasticizer. In UV and EB technology, however, diluents are more frequently species which have acrylate functionality and are thus capable of copolymerizing with the main resin system to provide a 100% solids formulation. Potentially there are a large number of molecular species which could function as reactive diluents. While these may vary in the number of acrylate groups they possess and can be complex branched structures, they are usually distinct from resins in that they are discrete molecular units. Hence the term 'monomers' is often applied to them.

In practice, the number of monomers that are acceptable for use by the printing industry are comparatively few. Many good viscosity-reducing acrylates are eliminated simply because they present a toxic hazard, but odour and volatility can also give problems.

Monofunctional monomers

Monofunctional monomers lack the ability to cross-link and, although good viscosity reducers, their excessive use may lead to poor film properties. An important use can be to impart flexibility to the print film. In this respect they may be thought of as reactive plasticizers. Two good examples in this respect are isodecyl acrylate (IDA) and phenoxy ethyl acrylate (PEEA).

In the coatings industry generally, 2-hydroxy ethyl acrylate (2EHA) is very widely used but like many of the acrylated low molecular weight alcohols, toxicity, volatility and odour restrict its use in printing applications.

Isobornyl acrylate and *N*-vinyl pyrrolidone are two other examples of monofunctional monomers with excellent viscosity-reducing properties, and which have good compatibility with multifunctional resins. However, neither are accepted by the printer as both have very pungent odours and there is some doubt over their toxicological properties.

Difunctional acrylates

Most of the popular difunctional monomers are derived from simple diols. In this class may be included the low molecular weight species derived by acrylating the product of bisphenol A condensation with ethylene oxide:

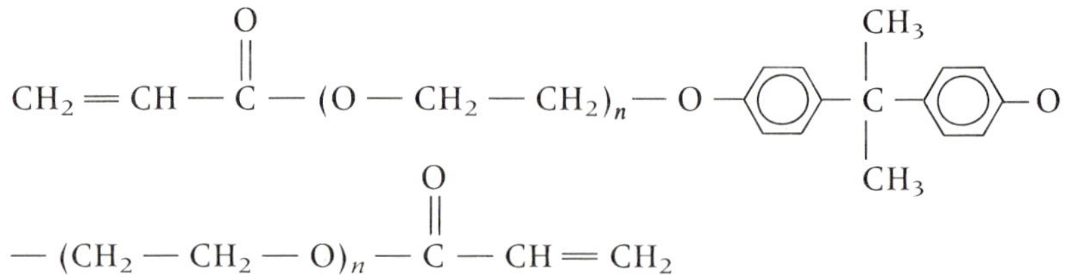

where $n = 1$ or 2.

The viscosity of these molecules is not particularly low but this is compensated for by their good pigment-wetting and inkmaking properties.

Table 11.7 lists a selection of the more important of the available difunctional acrylate monomers. Manufacture of this type of product is usually by reaction of a diol with acrylic acid.

Toxicity factors, in particular skin irritation and the potential sensitization, limit the use of many difunctional diluents in ink and varnish formulation. Many of these chemically aggressive materials are, however, employed in the coatings industries and also for silk screen ink production. In both these areas, adoption of safe handling procedures by the user permits the more liberal approach to formulation.

Of the species listed in Table 11.7 all except TPGDA and DDA are classified as severe irritants. TPGDA remains as the best difunctional viscosity reducer. PEGDA, for molecular weights of about 500, is classified as only a moderate irritant, but usefulness is restricted by the comparatively high viscosities prevalent at these molecular weights.

Table 11.7 Difunctional acrylate diluents

Abbreviation	Chemical Name	Viscosity at 25°C (cp)
BDDA	1,4-butane diol diacrylate	6
HDDA	1,6-hexane diol diacrylate	7
NPGDA	Neopentyl glycol diacrylate	7
DEGDA	Diethylene glycol diacrylate	8
TEGDA	Triethylene glycol diacrylate	25
PEGDA (*n*)	Polyethylene glycol diacrylate	10–30
TPGDA	Tripropylene glycol diacrylate	16
DDA	2,2-dionol diacrylate	ca.1000
ABPE2	Bisphenol A diacrylate	1000–1500

Trifunctional acrylates

The three most important polyols that have provided the industry with useful trifunctional monomers after acrylation are penta-erythritol, trimethylol propane and glycerol.

Pentaerythritol triacrylate (PETA)

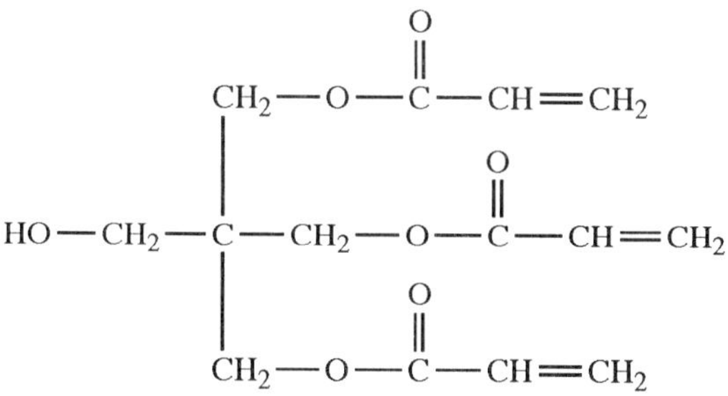

and trimethylol propane triacrylate (TMPTA)

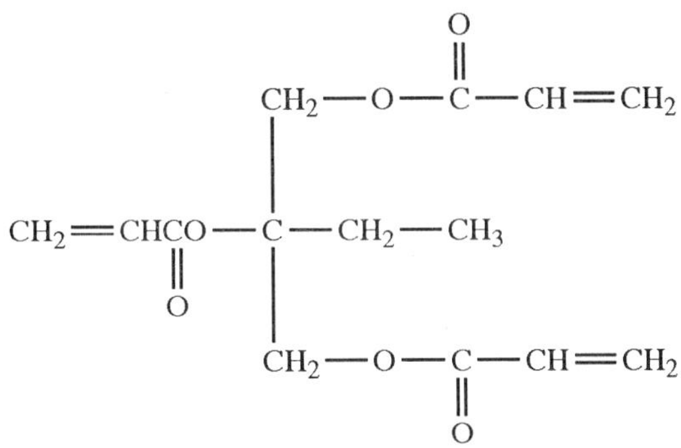

combine good viscosity reduction with fast cure and the ability for extensive cross-linking.

Unfortunately, PETA is a severe irritant and TMPTA, while classified as a moderate irritant, is a potential skin sensitizer and its use is declining.

To overcome these defects a new generation of trifunctional acrylates have been developed for which the hydroxyl functions have been ethoxylated or propoxylated prior to acrylation. This procedure has provided a number of proprietary diluents, the precise composition of which is often undisclosed by the supplier.

Propoxylation of glycerol with subsequent acrylation provides an example:

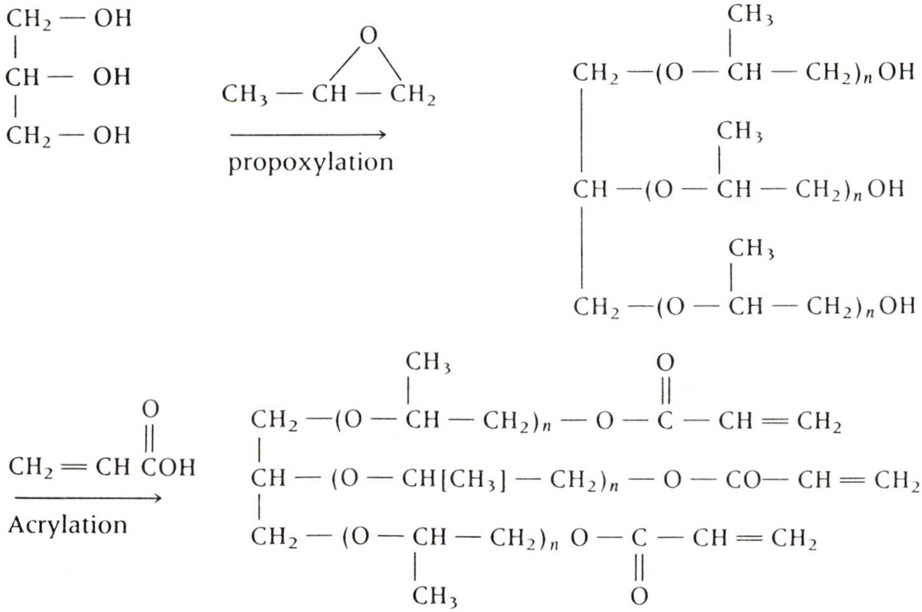

Typically $n = 1$ although the chemistry involved obviously leaves scope for variation in the molecular structure.

These products have viscosities lying in a range from about 0.1–1.0 Pa s. In combination with high molecular weight prepolymers they exhibit reasonable viscosity cutting power. The potential for high cross-link density and good cure speeds, characteristics of the non-chain extended forms, are retained.

Of special importance are the very low skin irritation indices and the negative results in sensitization studies that have been found for this class of materials.

High functionality monomers

Until recently very few acrylate monomers with functionality four or more have found favour in the ink industry. However, a number of companies have recently introduced tetra-acrylates variously known as PPTTA or ATTA, of which the full structure has not yet been disclosed.

These materials are claimed to have extremely low irritancy coupled with other beneficial properties which resemble those of the irritant TMPTA – good reactivity, low viscosity with good cutting power.

Hexa-functional acrylates are now readily available giving fast cure rates.

Non-reactive plasticizing diluents

Any of the typical plasticizing molecules can be used to improve flow-out and impart flexibility to the cured film. High boiling esters, highly branched alcohols and alkyl phosphates have been used. Benzophenone is interesting in that in addition to its photo-initiation properties it can reduce viscosity, aid pigment wetting and is a powerful solvent.

It is necessary for the diluent to be compatible with the ink, or varnish, vehicle system.

11.8 FORMULATION PRINCIPLES FOR ULTRA-VIOLET-CURABLE INKS

The formulation of any printing ink has to take account of:

- the method of application
- the substrate
- the end-use requirements.

In these respects UV-curable inks are no exception.

However, as the major portion of the vehicle system has to be comprised of acrylated species the range and number of combinations of raw materials available is more restricted than would be the case with conventional oleoresinous inks.

The required body and tack of a UV-curing ink or varnish is primarily achieved by balancing carefully selected prepolymers and monomers. Addition of diluent to a formulation usually results in a lowering of tack, yield value and plastic viscosity. This is in contrast to experience with conventional formulation where it is often possible to adjust the rheological parameters individually by choosing the right type of solvent. Further, the viscosity reducing power of a reactive monomer is invariably less than that of a typical petroleum distillate.

Most pigment types can be used in UV ink formulation, but some caution in selection is needed for three reasons. First, as a class, acrylated resins do not possess particularly good pigment-wetting characteristics. Consequently some pigments that might produce reasonable ink in a conventional vehicle are not suitable for UV or EB formulation. Secondly, some pigments are prone to initiating 'dark' polymerization reactions in the inks during storage. Such pigments must either be avoided or else a suitable stabilizer must be incorporated in the formulation. As these reactions may only manifest themselves after several months storage, evaluation of individual pigments and stabilizers can be a problem. Acceleration oven tests may be useful but are not always reliable. Finally, some pigments inhibit cure. This may be the result of chemical reaction with the initiating system, or it may arise because the pigment has significant UV opacity in critical special regions. Particular care is needed in formulating black inks.

All in all, formulating effective UV and EB ink probably calls for more skill and a greater understanding of the physics and chemistry involved than is necessary to achieve satisfactory results for corresponding inks based on conventional resins.

Formulation for ultra-violet-curing lithographic inks

By far the largest application area for UV-curable inks is sheet-fed lithographic printing of paper and board. Formulations of inks for this type of printing will thus be considered in some detail. Many of the principles introduced, however, apply generally to UV ink formulations regardless of the method of printing.

A typical UV-curable dark blue lithographic ink is:

Phthalocyanine Blue (CI Pigment Blue 15:3)	16.0
Carbon black (CI Pigment Black 7)	4.0
Epoxy acrylate resin	30.0
Fatty acid modified epoxy acrylate	25.0
Monomer viscosity modifier	8.0
Benzophenone initiator	8.0
Coinitiator	3.0
Aromatic amine photosynergist	4.0
Stabilizer, waxes, etc.	2.0
	100.0

The choice of resin blend reflects the need for achieving good lithographic properties while maintaining an effective rate of cure. If the ink is to be applied to a substrate lacking in absorbency, metal, plastic or polyethylene coated board for example, it may be necessary to use a portion of a polyester or a polyurethane acrylate to provide improved adhesion; albeit at increased expense and possibly a sacrifice of cure speed.

A first choice for the photo-initiator package would be benzophenone combined with an aromatic amine. However, the example chosen is a comparatively dark colour and a coinitiator such as 2-chlorothioxanone would probably be required in order to attain a satisfactory curing speed.

Alternatively the use of a cleavage initiator such as dimethoxyphenyl acetophenone might be considered. This latter choice would be important in the case of a white ink where the yellow tint from a thioxanone would be unacceptable. However, the ultimate end use must always be kept in mind. A side reaction of the above initiator leads to the formation of methylbenzoate:

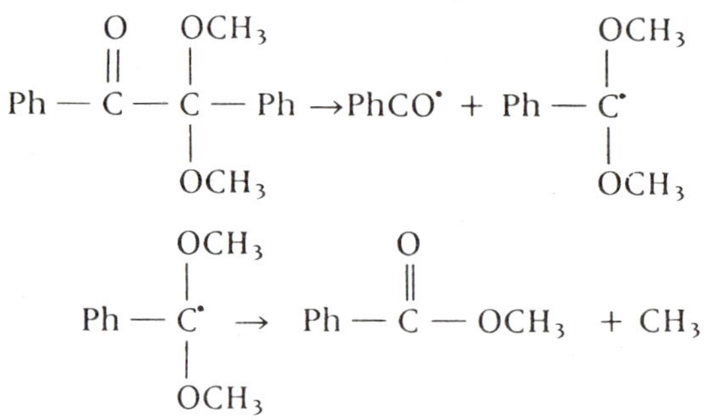

The odour and tainting risk introduced into a cured ink by this compound might be unacceptable if the end use involved food packaging.

Even when one of the more exotic initiators is selected, it is customary to leave some or all of the benzophenone in the formulation to retain the advantage of its pigment-wetting and plasticizing properties.

As the lithographic process is involved, use of water-miscible amines should be avoided. Ignoring the insignificant effect that these compounds may have on the emulsification of the inks, there would still be a risk of them leaching into the fount solution with a subsequent reduction in cure efficiency. Thus the more expensive, but water insoluble, *N,N*-dialkyl aniline derivatives are preferred as photosynergists for litho inks.

The body of a UV ink is controlled by the rheology of the prepolymer, the level of pigmentation, and the pigment type, adjustments to the tack and viscosity being made by the addition of monomer until a suitable printing rheology is attained. The selection and concentration of the monomer can best be made on the basis of experiments.

For process colour printing, tack grading of the inks for each different hue may be required to ensure satisfactory trapping. However, when process printing with UV inks it is popular, and relatively easy, to fit a lamp between print units, thus partially curing each colour before the next is applied and reducing the risk of backtrapping.

A typical formula for a letterpress ink:

Pigment	23.0
Diluent monomer	20.0
Resin prepolymers	40.0
Benzophenone	8.0
Triethanolamine	3.0
Waxes	3.0
Additives	3.0
	100.0

The relatively short-bodied inks required by the letterpress process are usually readily achieved from the available prepolymer and diluents. As the letterpress process does not require the sophisticated rheology demanded of litho inks the formulator may economize and select the relatively cheap epoxy acrylates as the total prepolymer component, provided that substrate and end-use requirements do not dictate otherwise.

The considerations for the selection of an initiator system for a letterpress ink are similar to those that apply to litho inks. However, as there is no involvement of water, cheap and highly effective alkanolamines, such as triethanolamine, can be used. While obtaining an adequate cure for blacks and very dark colours is a general problem with UV formulations, the existence of abnormally thick ink films in the regions around the image perimeter where squash may occur, aggravate this difficulty for letterpress formulations. In such situations the use of a combination of initiators, to take full advantage of all the available radiation, may be the only means of attaining a satisfactory cure.

Ultra-violet labelling inks

A rapidly growing market area for UV letterpress inks is in reel-to-reel printing of a narrow width web for the production of labels. Where paper substrates are employed a standard formulation based on, for example, epoxy acrylate prepolymer can be used. Increasingly, however, this type of label printing is turning to PVC substrates. To formulate inks which give adequate cure and adhesion on this type of substrate the use of more specialized resins, for example the polyester and polyurethane acrylates, are needed. The selection of monomer can contribute greatly to the success of this type of formulation. Unfortunately, most of the monomers which would provide a bite into the substrates and hence provide good adhesion may be restricted in formulations on grounds of irritancy

Inks for dry offset application on plastics and metal

Closely allied to the formulations for letterpress printing on impervious flexible substrates are inks for the decoration of rigid performed-plastic and metal containers. Inks are required that will offset from a relief zinc or polymer plate to a blanket prior to transfering the image to the substrate, such as those mentioned in Chapter 6.

A typical formula for such an ink would be:

Lithol rubine pigment (CI Pigment Red 57.1)	10.0
Talc (CI Pigment White 8)	2.0
Epoxy acrylate	14.0
Polyester acrylate	35.0
Polyurethane acrylate	20.0
Monomer, e.g. TPGDA	8.0
Initiator package	8.0
Waxes, surfactants, etc.	3.0
	100.0

Such an ink can be formulated using resins that are similar to those suitable for a letterpress ink. Rheologically, however, the inks require body and flow characteristics, more aligned for a lithographic application. The inks have to transfer well and have adequate flow.

Of prime consideration is adequate adhesion and abrasion resistance. Curing is often carried out 'on mandrel', where despite the limited space for lamps and the short dwell times, a sufficient cure has to be achieved to withstand almost immediate handling and stacking. Both the resin and initiating system have to be chosen with this in mind. Often a blend of a number of different initiators is used to optimize the rate of cure.

Selection of the monomer can be critical to the success of these formulations, particularly with respect to adhesion. Ideally the diluent should slightly attack the substrate, so providing points of anchorage. Unfortunately, materials which behave in this manner are largely precluded from use in the ink industry because of their associated toxicity. The mechanical resilience of these inks can be enhanced by the careful use of selected surfactants and waxes and this, at least in part, can counteract their relatively poor adhesion.

Ultra-violet curable web-offset inks

Despite the advantages of their being free from solvent emission UV web inks have made very little impact in the industry. Successful formulations have been run using inks based on those raw materials that are available for use in lithographic and letterpress inks. A few prepolymers are marketed which are specifically recommended by the manufacturers as suitable for use in web inks. The primary requirement is to make inks possessing good lithographic performance at high running speeds. Most currently available resin systems fail fully to meet this specification. Furthermore, because of the relatively poor pigment-wetting properties inherent with acrylated resins, the gloss levels fall somewhat short of the high standards set by conventional web-offset ink.

Web UV remains an area of challenge for the coated paper industry, although there is a limited use of UV in newspaper printing

Ultra-violet-curable silk screen

Silk screen ink films are typically between 20 and 60 μm thick. Although they can be much thinner, there is still a substantial depth for the radiation to penetrate when compared with the 3–5 μm thickness of a letterpress ink. It might be expected that this would present insurmountable problems for formulating an efficiently curing ink.

However, silk screen technology has two characteristics which have favoured the development of successful UV-curable products.

- The pigmentation levels are low and while high proportions of extender are used this is invariably transparent to UV radiation.
- Printing speeds are relatively slow, e.g. 130 m/min.

The wide spectrum of applications for silk screen printing, ranging from printed circuitry to glass-bottle decoration, has provided many areas for the successful introduction of UV inks.

A typical formula for a UV-curable silk screen ink suitable for decorating a plastic container would be:

Pigment	8.0
Calcium carbonate extender	20.0
Epoxy acrylate prepolymer	25.0
Monomer, e.g. TMPTA	27.0
Polyester acrylate	5.0
Initiator (1) 2-chlorothioxanone	4.0
Initiator (2) Benzophenone	5.0
Amine, e.g. TEA	3.0
Waxes, surfactant, etc.	3.0
	100.0

There is increasing use of water-dilutable resins in this field which has shown a rapid growth in recent years.

UV inks for flexographic and gravure applications

Until recently little commercial interest was seen for UV inks within these technologies. Available, safe monomers already are as viscous as a conventional solvent-based ink before pigmentation is considered. However, as emission controls become tighter, the options available to the former solvent-based printer narrow to include only water-based technologies and radiation-cured inks. Currently there is considerable interest in UV-cured flexographic systems and major printers are using UV commercially. It must be assumed that the advantages are great enough to warrant such a major change in technology. This being the case, it is certain that this will provide a new growth area for radiation curing during the 1990s.

11.9 ULTRA-VIOLET-CURABLE VARNISH AND COATINGS

Varnishes and coatings comprise a major portion of the market of UV-curing products. Varnishing of printed material is carried out to provide protection and promote gloss. The ultimate target in both respects is the quality set by film lamination. Ultra-violet roller coat varnishes almost achieve this criteria for gloss and lay, while the UV printing viscosity products although an order of magnitude lower in gloss are still superior to conventional oleoresinous or emulsion varnishes.

Table 11.8 lists the various popular methods of application together with an indication of the viscosity range most suitable for each technique.

A UV varnish may be required to coat any of the multitude of substrates in general use in the industry. Problems with soak-in can arise if low viscosity varnishes with very little body are applied to porous material.

Further considerations are imposed by the need to overprint inks. A spectrum of possibilities exist, some of which are summarized in Table 11.9.

Obtainable gloss levels usually decrease as the varnish viscosity increases. Indeed lithographic and letterpress printing varnishes can give disappointing results so far as appearance and gloss are concerned. Such products are mainly used to give a seal over the inks in circumstances where it is essential to leave certain image areas uncoated, for instance where clean flaps for subsequent glueing are necessary.

The majority of UV varnishing is carried out by a dry-offset method. Application via the damping system can give excellent results, but control of film weight is not easy and incorrect press conditions may result in

Table 11.8 Application methods and categories of UV-curable coatings and varnishes

Method of application	Approx. range of viscosities
Out of line: roller coating	Under 0.5 Pa s
In-line coating head	0.8–1.2 Pa s
Dry-offset via the damping system	2–3 Pa s
Dry-offset via a printing plate	4–15 Pa s
Lithography and letterpress	10 Pa s upwards

Table 11.9 Substrates for overprinting with UV varnishes

Substrate	Comment
Unprinted substrate	Soak-in on porous substrates can occur
Part printed: part unprinted	Disparity in gloss levels may be a problem
Wet conventional inks	Gloss levels often poor and variable; adhesion may be poor; special ink formulations advised
Dry conventional inks	Interfacial adhesion may be poor; reticulation can be a problem; low wax inks advisable
Wet UV inks	Results usually good; gloss levels may vary over different shades
Dry UV inks	Results usually good

severe 'orange-peel' effects. Application from the duct via the inking system is the more popular process. Film weight is readily controllable and, if desired, glue flaps can be left unvarnished by means of cutting out appropriate areas of blanket.

The above techniques suffer the disadvantage of utilizing one of the printing units. A recent trend has been the introduction of in-line end-of-the press coating heads. Such an operation, of course, only allows an allover varnish application. Despite the commercial advantages of an in-line single pass varnish procedure, a substantial amount of UV varnishing is still carried out off-line with trade varnishes via traditional single or double head coating machines.

Turning to end-use properties, the frequent need for adhesive receptive regions has already been mentioned. For an all-over varnish coating the question of its potential glueability thus arises. Experience suggests that this is very dependent on the type of adhesive that is used and also on the conditions of temperature, pressure and dwell time employed in its application. Such factors are usually beyond the control of the formulation. Often the best that can be said is that if formulated with minimal levels of surfactants and waxes then the varnish is potentially glueable. Similar considerations apply to the question of foil blockability; another end-use feature often demanded from a UV varnish.

Scuff and rub resistance for UV-curable varnishes can be excellent but to optimize these properties requires a correct balance of the prepolymers and the photo-initiators. Waxes will improve rub resistance, but gloss is sacrificed. Similarly silicones will provide desired levels of surface slip, but incorrect use may lead to reticulation and poor interfilm adhesion.

Because each varnish has to be matched to its proposed mode of application as well as its end-use requirements, it is far from easy to give a generalized formulation. For application at around 1 to 5 Pa s the following would provide a starting-point for an acceptable product:

Bisphenol A epoxy acrylate resin	13.0
Trifunctional polyether acrylate resin	60.0
Tripropylene glycol diacrylate	5.0
Amine acrylate	12.0
Benzophenone	6.0
2,2-dimethoxy-2-phenyl acetophenone	3.0
Silicone slip additive	1.0
	100.0

For a higher viscosity product a quantity of prepolymer resin would be needed. For a lithographic or letterpress varnish calcium carbonate would be required to provide body and other modification might be necessary. For example 2–3% of talc would probably be added to such a formulation to reduce misting; albeit with a sacrifice in the gloss levels.

The choice of initiator is usually restricted by the need for water whiteness. Thus use of the thioxanones with their yellow colouration is limited.

As with inks any end-use constraints on the allowable levels of odour and taint must be borne in mind when selecting the raw materials and in particular the photo-initiators.

The use of an acrylated amine is popular when formulating varnish. Free amines have a tendency to give rise to a surface bloom, which may not manifest itself until several hours after curing.

11.10 CATIONIC-CURING SYSTEMS

Most of the inks and lacquers described elsewhere in this chapter are cured by free-radical mechanisms, and are based on acrylated materials. This is now a well established technology and is dominant in the ink market place.

It is worth discussing here the other available technology, which, after several years due to patent problems, is now undergoing a resurgence in certain specialized areas.

Cationic cure systems differ totally in their components and chemistry from free-radical systems. They differ in their properties equally widely. They are not inhibited by oxygen. The reaction will continue even without the presence of UV light, once initiated, providing the possibility of a more complete cure which is superior to normal UV systems. The formulations exhibit much lower shrinkage than acrylated systems, which helps provide better adhesion on to many 'difficult' substrates.

Mechanisms

The reaction route of cationic curing is as follows:

$$\text{UV light} \xrightarrow{\text{Absorption}} \text{Initiation} \xrightarrow{\text{Proteolysis}} \text{Lewis / Bronsted acid}$$

$$\xrightarrow{\text{Initiation}} \text{Cationic species} \xrightarrow{\text{Propagation}} \text{Polymer}$$

Because the cationic species are long lived, i.e. not easily quenched, the reaction continues even if no more UV light is available, which does not happen to the same extent with a free-radical system, although some post cure does take place. The electron-deficient boron trifloride can initiate polymerization acting as a Lewis acid:

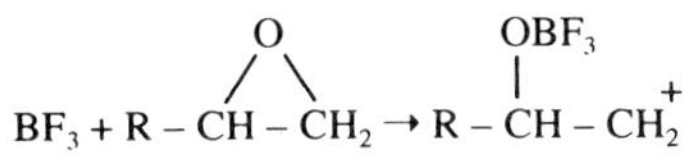

However, the evolution of nitrogen, which forms disruptive bubbles in the cured film, together with poor thermal stability has restricted the use of this system in printing technology.

Recently, considerable development effort has been directed towards photo-initiator having a general formula $Ar_nX^+Y^-$ (where Ar denotes an aryl group). Typically, X is an atom of sulphur, iodine or phosphorus and Y is an anion of the type BF_4, PF_6 or AsF_6. Examples of such molecules are:

$$Ph_3S^+ \; PF^-_6, \; (CH_3 - O - \bigcirc)_2 \; I^+As^-F_6$$

Formulations

A lacquer for cationic curing may be formulated quite simply. It may be composed of a backbone resin, for example an epoxide, together with vinyl ethers to enhance cure speed, or polyols to aid flexibility, or possibly phenoxy resins. To these, surfactants may be added, as usual. The photo-initiators, however, are extremely specialized salts.

Well known photo-initiators for this type of system include triarylsulfonium salts of hexafluorophosphoric acids from which acid is regenerated by photolysis, or by ion arene complexes.

The acids generated are, however, remarkably easily poisoned by bases, which stop curing completely. This has led to cases, for example, of lacquer curing on one side of a print but not the other. Also, since most free-radical systems contain amines as oxygen inhibitors, the two systems may not be used together without considerable care, and risk of failure.

While discussing problems it should be obvious that any basic pigment should be avoided at all costs. However, it is possible to formulate around this problem, and at least one manufacturer has produced a range of predispersed bases in an epoxy system.

11.11 PRESENT AND FUTURE USES

While initially the two systems, i.e. UV (traditional) and UV (cationic), were seen as rivals for the same market, the present view is that they are complementary technologies and can be intermixed. Until recently cationic systems have required a substantial degree of postcure or a large thermal bump to take cure past the tack-free stage. This problem is diminishing, however, and the relatively low viscosity of these systems is generating much interest for systems for flexographic and gravure application. The good adhesion qualities and the potential for very low odour mean that cationic lacquers are already used for specialized applications, a role which can only grow with the move away from solvents. Finally, it is possible to use a cationic lacquer as a fully cured, low-odour barrier coating over a free-radical-cured print, reducing the fear of acrylate migration.

11.12 ELECTRON-BEAM-CURABLE INKS AND VARNISHES

The essential difference between a UV-curable system and one designed for cure by EB radiation is that the latter does not require any photo-initiators. Thus the 8% or so of the UV formulation which comprises the initiator package is available for replacement by prepolymer or monomer.

The removal of this non-curing portion from the ink or varnish formula offers two significant advantages. First, the 'purer' vehicle would be expected to provide enhanced mechanical and chemical resistance properties after curing. Secondly, there is no risk of any odour, taint or toxicological contamination that might arise from either unreacted photo-initiator or its photodegradation products.

In all other respects an EB formula can be identical to one designed for UV curing. Some manufacturers of acrylated propolymers and monomers have made specific recommendations concerning which of their products are best suited to EB curing. However, current practical experience has not shown any outstanding distinctions between EB and UV performance for most of the popular resins in use and selection is probably best based on their potential printability and end-use performance. An exception to this, however, is the possibility that some resins are less prone to surface oxygen inhibition and could thus give a greater guarantee of satisfactory cure in the event of a partial failure in the inert gas blanketing.

Electron-beam radiation is sensitive to the density of the formulation but not to its optical properties. Thus the formulation is free from the constraints that exist which strongly absorb dark colour in UV ink technology. This leads to consideration of a further advantage for EB curing, namely that the technique gives greater assurance of in-depth cure of the print film. Again risk of uncured, migratory, odorous or tainting material in the cured film is minimized.

These factors are important, particularly since the major use of such inks at the moment is in aseptic packaging for fruit juices, wines, etc. Because these are printed on the web and re-reeled, the risk of set-off on to the inside of the package is very high if anything less than full cure is achieved. The possibility for taint in these circumstances has led to commercial pressure for acrylate materials with very low levels of residual solvent or acrylic acid. The availability of such materials has been of benefit in general usage for UV and EB curing.

The ability of electron beams to cure the ink film from the back of the substrate has proved advantageous in producing 'mirror finish' materials. Ink or lacquer is applied as the substrate passes around a polished drum, on the face presented to the drum.

The major limitation of the process is the necessity to print on a web since metal gripper bars might cause the generation of X-rays, and maintaining an inert atmosphere. This, however, need not impede the development of the process in flexo or gravure areas.

11.13 RADIATION-CURING EQUIPMENT

Ultra-violet-curing equipment

Sources

The most widely used, practical, source of UV radiation is the mercury vapour lamp. Commercially available lamps can be divided into three categories (1 Torr ≡ 1 mm mercury):

- low pressure (10^{-3} – 10^{-2} Torr);
- medium pressure (*ca.* 10^2 Torr);
- high pressure (10^3 Torr).

Typically the standard design is a linear tubular envelope of quartz glass with electrodes at either end. The envelope retains a charge of mercury and an inert gas, usually argon.

Energy is applied, which vapourizes and subsequently ionizes the mercury. Collision interaction between mercury atoms, mercury ions and free electrons possessing high kinetic energies, results in the population of excited states in some of the mercury atoms and ions. Decay transitions, back to the ground state, occur with the emission of radiation. The dominant transitions, and hence the spectral emission profile, depend upon the temperature and pressure prevailing in the tube.

Low-pressure lamps emit UV predominantly at two wavelengths, 185 and 254 nm. Penetration into ink or varnish films is relatively poor for radiation of these wavelengths. This fact, together with the low radiative intensity, only 10–15 W/cm, has limited the use of low-pressure lamps in the printing industry.

High-pressure mercury lamps emit a continuum throughout the UV and visible into the IR wavelengths. These lamps have relatively short lifetimes, are necessarily small in dimension and are little used in UV-curing technology.

Linear medium-pressure lamps with lifetimes in excess of 5000 hours can be constructed in lengths exceeding 1 m. They have efficient spectral output with two dominant UV bands at 365 and 366 nm, and are thus ideal as a press-mounted UV source.

Two types of medium-pressure mercury lamps are used by the industry. The most popular, the electrode lamp, utilizes a high voltage discharge to excite the mercury atoms. A warm-up period of 10–15 minutes is required with this type of lamp, over which time the mercury vapourizes and full spectral output is developed. If switched off, condensing mercury prevents restrike for several minutes.

The effective life of this lamp is limited by the slow erosion of the hot electrodes. Also, vapourized tungsten from the electrode is deposited on cool inner wall of the quartz envelope gradually reducing transparency. These problems can be overcome by eliminating the internal electrode completely.

In the so-called electrodeless lamp, microwave radiation is transmitted through the quartz envelope to excite the mercury atoms. Such lamps claim to offer a longer life, rapid attainment of their full spectral output, and almost instant restarts. They suffer the disadvantage of being short, 250 cm is a typical maximum length. Also the necessary microwave source is expensive.

Heat output and shuttering

The medium-pressure mercury source has a significant output in the IR. This can be an advantage to curing as the polymerization reactions proceed faster at elevated temperatures. However, the lamp envelope requires some means of cooling, which is usually achieved by mounting the tube in the air stream of an extraction system. Air or water cooling is also needed for the reflector and any other parts of the press exposed to incidental radiation.

In the event of temporary line stoppage electrode lamps are normally left on to avoid incurring a restrike delay. To handle such events without the risk of fire, from the over-irradiated substrate, interlocked shutter assemblies are generally fitted.

Ozone generation

Ozone is generated from atmospheric oxygen on exposure to UV radiation in the region of 185 nm. In addition to being an irritating gas, ozone may damage polymeric parts of the press and can reduce cure efficiency by absorbing the longer wavelength UV. The use of an effective extraction, venting into the outside atmosphere, is therefore essential.

Lamps are available for which the quartz has been doped to filter out short-wave UV and hence eliminate ozone production. However, under practical conditions such lamps are found to show a significant loss of cure efficiency.

Press assembly and reflectors

Lamps are generally fitted a few centimetres above the substrate at some convenient position on the press. Typically, two or three 80 W/cm full length electrode type lamps might be fitted at the delivery end of a four-colour sheet-fed press. However, on a multicolour press it is not unusual to find interdeck lamps sited between each printing unit. The high voltage transformer and controls for electrode lamps can be situated away from the press if desired.

Microwave excitation systems often have the lamp and power source housed in a single module. Multiple units are mounted across the press with overlap as required, to provide a continuous field of radiation.

Reflectors are obviously needed to utilize all the radiated output from a tubular lamp. Three basic configurations are in common use:

- elliptical
- parabolic
- planar (non-focusing).

Elliptical reflectors are the most popular design in use, particularly on web or sheet-fed presses where the substrate is flat (Fig. 11.2). The lamp may be positioned so that a partially focused beam is obtained and hence a compact band of UV radiation is applied across the substrate.

With the lamp positioned at the focus of a parabolic reflector, a partially parallel beam is achieved. This configuration is often recommended for the curing of printed performed containers and other non-flat substrates.

Non-focusing reflectors are less popular. There is however, some evidence that greater cure efficiency is achieved with the longer exposure at a lower flux density which occurs with planar reflectors as compared with an equal dose of high flux exposure from a focused system.

Cool ultra-violet lamp systems

The need arises for partial, or total, removal of the IR component from a UV lamp to protect sensitive substrates such as plastic films from overheating and deforming. Additionally some authorities suggest that lower substrate temperature leads to improved handling and stacking immediately off the press because the cured cold ink film carries less risk of set-off.

Filtration of the output by way of doping the quartz used to construct the lamp envelope has been mentioned. Dichroic reflectors, which absorb IR while reflecting the short UV, provide another possible means of

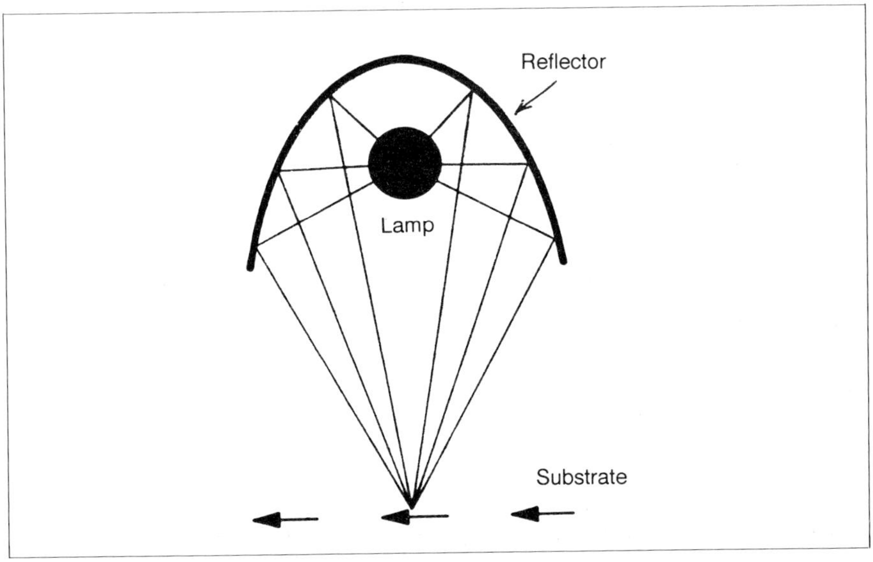

Fig. 11.2 An elliptical reflector.

partially removing the unwanted radiation. Alternatively a number of designs provide so-called cool UV, by circulating water in a quartz jacket either around the lamp or immediately above the substrate.

In all cases a significant loss of UV radiation also occurs, together with reduced thermal polymerization.

Electron-beam curing equipment

Introduction

In contrast to those for IR, UV and to a lesser extent microwave, systems for EB curing are bulky and expensive to install and run. Three factors contribute to these disadvantages. Firstly, the necessary high vacuum requires ancillary pumping and control equipment. Secondly, wherever the high-energy electrons impinge on a surface the risk of generating X-ray occurs. Unlike the electrons whose penetration in air is limited to a few centimetres, the X-ray radiation may present a health hazard several metres from the source. Thus extensive lead shielding in conjunction with a failsafe radiation detection system is necessary. Finally, for the reasons discussed earlier, inert gas blanketing is essential for efficient EB cure.

Installation is made at the delivery end of press with the beam window sited a few millimetres above the substrate.

Source design for electron beams

Two configurations of beam generator have been developed to the point of commercialization:

- the scanned beam type
- the linear cathode type.

In both cases the source of electrons is a hot metal cathode situated in a high vacuum.

For the scanned beam system the electrons originate from a point cathode and are accelerated towards a positively charged grid. Passing through the grid they are focused into a narrow beam by a magnetic lens. This accelerated pencil beam of high-energy electrons is deflected on its way to the emission window by a varying magnetic field which is applied via electromagnet coils positioned close to the beam pathway. In this manner the beam is scanned across the width of the window.

With the linear cathode design a filament cathode, which runs the effective width of the press, is suspended in a vacuum tube. Electrons leave the cathode and are accelerated towards a positive grid placed between the cathode and the emission window.

A curtain of unfocused electrons pass through the window to bombard the ink or varnish film which travels a few millimetres below. Although the linear cathode assembly is unable to deliver the very high energies at high flux densities that can be achieved with the scanner type, they have a compact design and are more easily shielded against X-ray emission. They are thus the more popular model for fitting to a printing press or coating machine.

A variation on the basic linear cathode design gives enhanced output by having an array of parallel cathode emitters within a single vacuum tube.

The dose

The operational characteristics of an EB curing system are defined by consideration of two features: the dose delivered and the beam penetration.

The dose is measured in terms of the quantity of energy absorbed per unit mass of irradiated material. Units in general use are the megarad (Mrad) and the Grey (Gy). The relationship between these units is:

$$1 \text{ Mrad} = 10 \text{ kGy} = 10 \text{ kJ/kg}$$

Dose is directly proportional to the number of electrons in the beam flux as measured by the current I flowing in the cathode circuit. It is inversely proportional to the throughput rate S of the substrate undergoing cure. Dose may thus be defined by the relationship:

$$\text{Dose} = kI/S$$

The proportionality constant k is determined experimentally and is dependent upon such factors as the machine geometry and the efficiency of electron transmission through the foil window.

Penetration

The depth of material that can be penetrated by an electron beam radiation is directly proportional to the energy processed by the accelerated electrons E and is inversely proportional to density, ρ, of the irradiated material, i.e.

$$P = K\,E/\rho$$

Penetration is essentially independent of the chemical nature of the material.

For a given accelerating voltage the penetration can be expressed in terms of the weight of material irradiated per unit area. Usually this is given in units of grams per square metre. The energy of an electron entering the material to be cured is the final accelerating voltage less potential losses which occur at the window and across the gas-filled gap between the window and the substrate.

For a given system it is possible to plot profiles displaying the fraction of surface dose delivered against the penetration for a given accelerating voltage and substrate density.

Shielding

Penetration distances for free electrons are relatively short. For example, a 200 keV electron will penetrate only about 40 cm of air. However, when arrested by collision with matter, X-ray radiation can be generated.

With the scanner type quite extensive overall shielding is required, which may include several feet of external concrete. For the linear cathode machines, which operate at voltages below 300 keV, stray electrons and the associated X-rays are less penetrating and can be contained by only a few millimetres of lead shielding. This is usually incorporated as a part of the unit housing.

Press installations include extensive electro-mechanical safety interlocks coupled to radiation monitors to ensure that an operator cannot enter an irradiated zone while the beam is on. With EB equipment there is no radioactive source and thus no risk from residual radiation when the beam is not operating (Fig. 11.3)

Inert gassing

The generation of ozone from atmospheric oxygen on interaction with an ionizing radiation and the effects of oxygen on cure efficiency have been discussed earlier. The gap between the window and the substrate has to be purged free of oxygen. This is usually carried out by maintaining a positive pressure of nitrogen gas in the region of the window. After a period of purging the curing region becomes virtually free of oxygen. Maintenance of a slow stream of nitrogen prevents ingress of more air from the slipstream above the moving substrate.

Maintenance

While a 'hard' vacuum has to be achieved and maintained during running it is customary to retain a rough vacuum throughout any stand-by period.

Over a period of time the cathode filament will evaporate and have to be replaced. However, the more usual limitation on a filament's life is imposed by mechanical shock arising from heating and cooling on start-up and shut-down.

The window which is constructed from an extremely thin foil of metal,

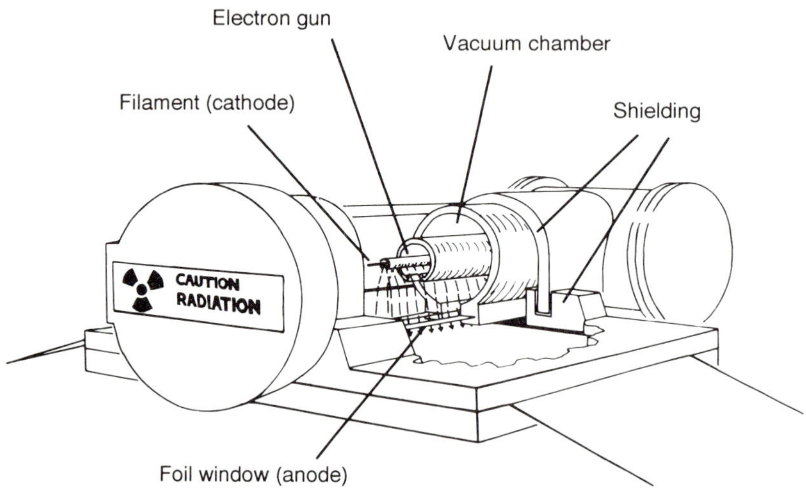

Fig. 11.3 Diagram of an electro-curtain, linear cathode type, EB system.

typically aluminium or titanium, is also subject to long-term thermal damage and thus requires periodic replacement.

A supply of water is usually required to remove excess heat, particularly from the window and the filament mountings. Additionally, a source of inerting gas, usually nitrogen, has to be supplied whenever the system is in operation.

BIBLIOGRAPHY

Holman, R.J. (Ed.) (1986) *UB and EB Formulations for Printing Inks and Coatings*, SITA, London.

Roffey, C.G. (1982) *Photopolymerization of Surface Coatings*, J. Wiley and Sons.

Phillips, R. (1983) *Sources and Applications of UV Radiation*, Academic Press.

A six volume set of books covering the entire field of radiation curing is available from SITA publications. They cover Chemistry and Technology of UV and EB Formulations for Coatings, Inks and Paints: Vol. 1 *Markets and Curing Equipment*; Vol. 2 *Prepolymers and Reactive Diluents for UV and EB Curable Formulations*; Vol. 3 *Photoinitiators for Free Radical and Cationic Polymerisation*; Vol. 4 *Formulation*; Vol. 5 *Photophysics and Photochemistry*; Vol. 6 *Specialised Applications for UV and EB*.

Printing Industry Advisory Committee (PIAC) (1993) Guidelines *Safety in the Use of Inks, Varnishes and Lacquers Cured by Ultra Violet Light*.

CHAPTER 12
Ink-jet inks

12.1 INTRODUCTION

Unlike traditional printing processes ink jet is relatively young. It is also fast emerging as a high technology and a practical solution, well suited for many diverse printing applications within the industrial and office sectors. Ink jet is a versatile, non-contact printing technique which has grown dramatically in popularity over the last ten years. In the industrial sector, it owes this popularity to an ability to print fixed and variable data at high speeds (up to 2100 characters per second) to virtually any surface, porous and non-porous, including those that are irregular and fragile.

Broadly speaking ink jet technology divides into two distinctive techniques for generating droplets namely, continuous ink-jet printing (CIJ) and drop on demand printing (DOD). By and large each technique reflects the market it serves. CIJ is the fastest form and is highly demanding of the properties of an ink for successful operation. Although commercial realization of continuous ink jet is recent, the scientific principles were established over a hundred years ago. Lord Rayleigh's work in modelling the instability of a capillary jet forms the basis of the technology. It involves exciting a jet of fluid by destabilizing it in a controlled fashion to produce droplets. It is used mainly for industrial applications, where it is ideal for printing information such as batch numbers and date marks onto products on a rapid production line, especially if the products are rough or do not have flat surfaces. Increasingly, CIJ is finding favour in printing on-line naming and addressing on high speed printing presses (up to 500 metres per minute). It derives its name from the nature of the process where droplets are constantly produced and recycled when not required for printing.

In contrast, DOD printing is characterized by the fact that droplets are only produced when required for printing and, in general, droplets are produced by an ejection mechanism rather than by destabilizing a jet. Several methods are currently in use to do this, including valve-jet, impulse, piezo, thermal and hot-melt printers. Again the type of technique employed tends to depend on the application, and these range from industrial large character outer carton coding through to high quality office printing.

The fundamental physical and chemical requirements of an ink are common to most ink-jet systems and the factors influencing performance such as solution stability and jet break-up characteristics are similar. DOD formulations tend to be simpler only by the virtue that the type of inks used is constrained by the inherent design of the technology. In most cases the applications are restricted to printing on absorbent materials such as paper and board. On the other hand, applications for continuous ink-jet inks are more diverse and a wide range of solvent-based systems have been developed that can be applied to more difficult substrates such as metals, plastic and glass as well as withstanding pre- and postprinting processes such as steam sterilization and solvent resistance.

Currently, ink jet represents only a few percent of the European ink market but it is predicted by the end of the century it will grow to about 11% which becomes comparable with today's flexographic ink market.

12.2 CONTINUOUS INK-JET INKS

Principle of operation

In continuous ink jet, a jet is deliberately made unstable and is broken into discreet droplets in a controlled manner by application of an externally induced vibration, typically a piezo crystal. The physical and chemical properties required of an ink for consistent droplet formation and subsequent charging are very demanding. To understand this it is worth considering the process an ink is subjected to in an ink-jet printer.

The principle of operation, typical of a commercial printer, is shown schematically in Fig. 12.1. At the printhead, a fine jet of ink is directed out of a nozzle which is the size of a human hair or less. The jet is travelling at about 60 mph and the droplet is broken within 1 mm. In most commercial printers the piezo crystal oscillates at 64 kHz producing 64 000 droplets per second, although in one commercial design up to 1 million droplets per second are produced. Break-up is controlled by varying the voltage applied to the piezo crystal so modifying the amplitude of vibration. The term modulation is often used to describe the range of voltage applied to the piezo crystal at which reproducible droplet formulation and printing can be obtained. The mechanical shear forces developed within the ink as a result of squeezing ink through the nozzle are very high, in the region of 10^6 s^{-1}. These forces are equivalent to those generated by a high revving motor car engine and experienced by the lubricating oil in the regions of the crankshaft journal bearings. The rheological properties of an ink during this process are complex. Ink-jet inks are generally non-Newtonian and although it is true to say the rheological process is not well understood, it is known that precise control of the behaviour of the fluid at the nozzle is important and small variations can greatly influence the stability of the jet and droplet formation.

The droplets, about 1.5 nanolitres in volume, equivalent to 2 million per teaspoon, must be controlled to keep size, shape and spacing

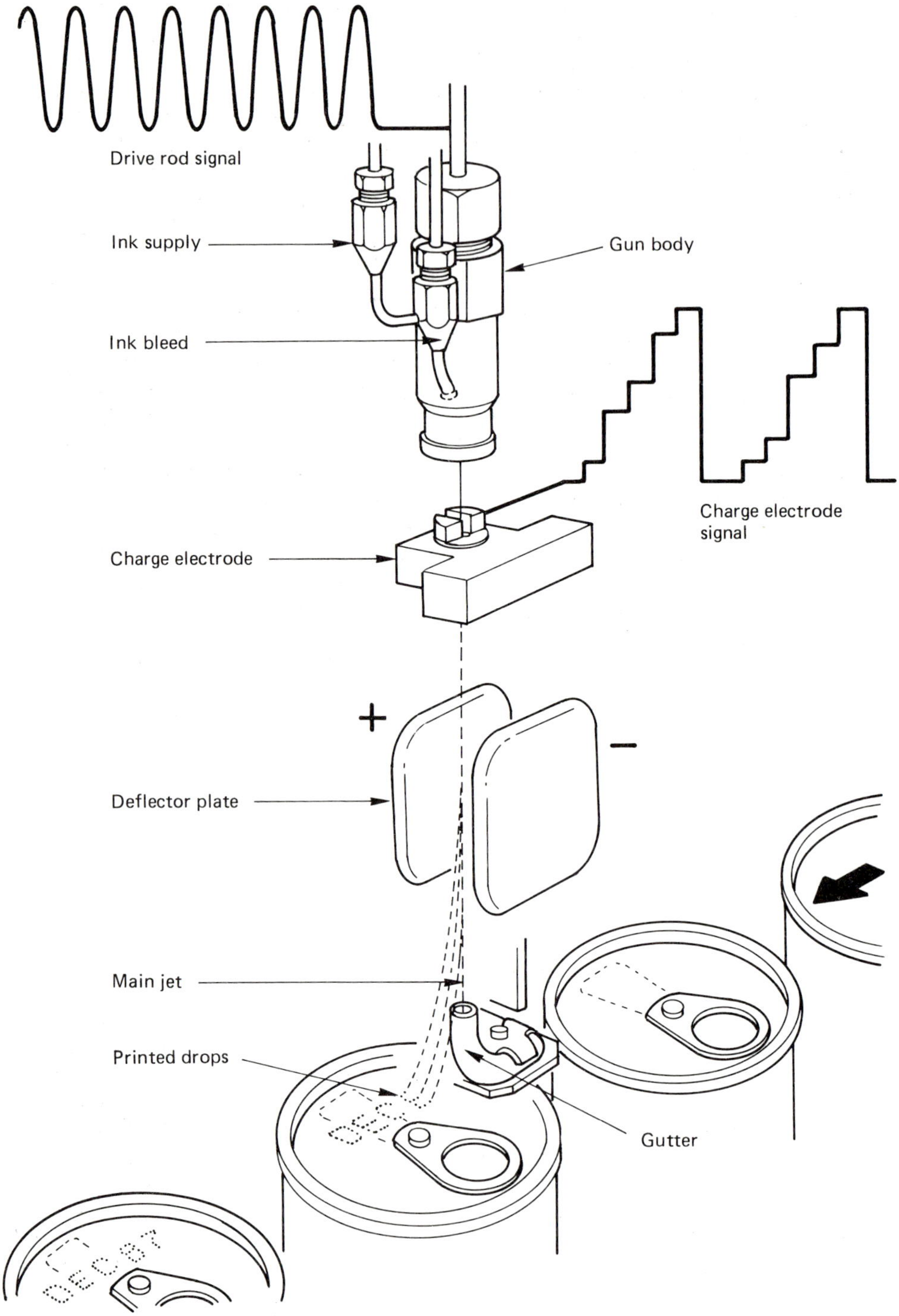

Fig. 12.1 The principle of ink-jet printing.

effectively constant. At the point of jet break-up a charge is applied to each individual droplet as it passes through the charge electrode. It must be capable of holding onto this charge during flight through the print-head. The size of the charge on each individual droplet will depend on the voltage applied. The charged droplets then pass between two deflector plates, which are maintained at high voltage. Ink which is not deflected and used for printing is caught in a tube under suction and returned to the ink container. This process goes on continuously even when the printer is not required to print. Ink is recycled many times and is exposed to oxidative, mechanical, thermal and hydrolytic processes.

The message of characters is built up dot by dot into a series of strokes to form a dot-matrix pattern. A whole host of different dot-matrix formats can be software designed and these formats are described as rasters. Rasters can range from simple characters comprising 7 drops high and 5 drops wide (7×5) through a multitude of variants up to 34 drop high, 26 drop wide (34×26) characters for high resolution multiline printing. For example, a letter 'E' using a 7×5 raster is generated according to the sequence shown below. The product travels on the production line in a perpendicular direction to the direction of the deflected drops. The first drop is placed on the product as shown in (a) of the sequence and the next is placed vertically above, shown as (b) in the sequence. This is repeated until the first stroke is completed, (c). Subsequent strokes are then built up such as (d) until the final character is developed, (e).

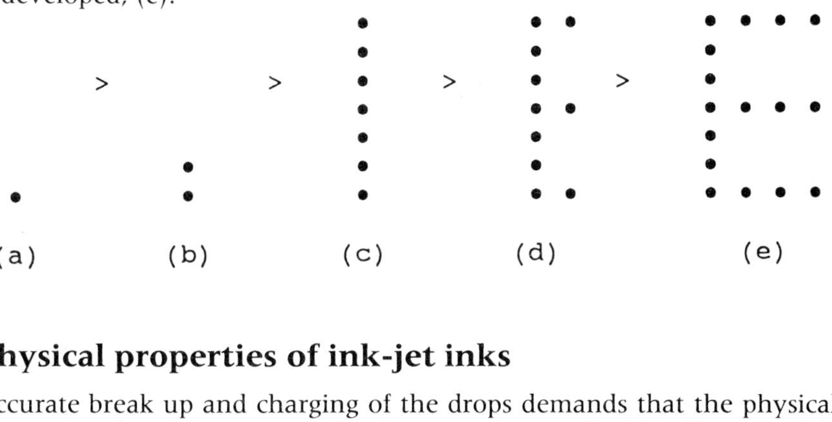

(a) (b) (c) (d) (e)

Physical properties of ink-jet inks

Accurate break up and charging of the drops demands that the physical properties of an ink are carefully chosen and controlled. To illustrate this Fig. 12.2 shows the formation of droplets from a continuous ink-jet printer. Good formation is characterized by the jet stream initially distorting to produce 'tadpole'-shaped patterns within the jet which grow to become more distinct until finally the jet breaks cleanly, rapidly forming spherical droplets of equal size, shape and spacing. Good droplet formation must be achieved over a wide operating temperature range, usually 5–45°C. Poor droplet formation, a result of inadequate control of the physical properties of an ink, causes the jet to break in either an erratic fashion or produce 'satellites'. Satellites are a result of the tail of the growth of the drop splitting to produce separate smaller droplets. Alternatively, erratic break-up is often a result of the jet distorting to form spheres connected to each other by very thin ink ligaments. Examples of these types of droplet formation are shown in Figs. 12.3 and 12.4 respectively. The reason for these types of behaviour can be subtle and complex but the effect of both results in unacceptable print quality.

For these reasons ink development cannot be carried out in isolation of printer design and success can only be achieved by working closely with the ink-jet physicist. Design of the system can greatly influence droplet formation. Nozzle size, jet velocity and pressure represent a few parameters that need to be precisely controlled. Aerodynamic and coulomb forces also affect the flight of the droplets and as a result uncharged droplets are employed to the shield, charged droplets making the distance between charged droplets larger and so reducing coulomb and repulsive forces, allowing more accurate drop placement.

The important bulk physical properties of an ink-jet ink are shown in Table 12.1. In particular, the viscosity and surface energy must be held within tight limits and the liquid must be electronically conductive to allow drops to be charged.

An ink-jet ink is designed to meet the specification shown in Table 12.1 to ensure satisfactory performance over the equipment operating range. However, this specification may vary from one commercial ink-jet printer to another, since the formulation is optimized or tuned to the printhead design. In the case of conductivity an ink is specified to have a higher conductivity than necessary, 1000 μS/cm at 25°C because the effect of temperature on conductivity is often pronounced and at either extreme of temperature the conductivity can be considerably lower. The relationship between conductivity and temperature is not easy to predict and depends on the particular formulation.

Any trace of particulate matter may cause jet instability, misdirect the jet or even clog the nozzle. Filtration of an ink jet is a vital property to control. Typically 100 millilitres of ink must take less than 30 seconds to filter through a one-micron filter and it is necessary on occasions to filter to 1 micron during the manufacture of an ink-jet ink.

It is difficult to separate the physical properties of surface tension and viscosity from each other, when describing the nature of jet break-up and the contribution of each towards the spreading and splattering characteristics, when the droplet impacts on the surface. Surface tension plays a critical role and maintains the size and shape of the droplet. An ink-jet ink can be successfully designed over a wide surface tension range from 22–45 N/m, although for solvent-based systems optimum performance is obtained between 25 and 30 N/m and controlled to ±3 N/m. Water-based systems tend to have a higher surface tension but suffer from a greater degree of splattering. Reducing the surface tension of a water-based ink often leads to an unacceptable level of foam so a design compromise is reached. Foam or dissolved air is an important consideration, especially for water- and alcohol-based systems at low temperature.

Fig. 12.2 An ink-jet ink exhibiting good drop formation.

Fig. 12.3 An ink-jet ink showing smaller satellite drops.

Fig. 12.4 An ink-jet ink producing droplets connected with thin ink ligaments.

Control of the viscosity of an ink-jet ink probably makes the greatest contribution to achieving good droplet formation. It is the rheology of the fluid in the nozzle that governs droplet formation and not the low shear viscosity value described in Table 12.1. This value is used as a starting point for formulation purposes. Currently the rheological process is not well understood but the choice of polymer type and molecular characteristics is very influential. Over the next few years, the unravelling of these parameters controlling the rheology of ink-jet inks represents an exciting scientific opportunity and an ability quantitatively to describe the performance of ink-jet inks will be a major step forward. The molecular weight of the polymer, chemical type and the polymer solvent interactions are all known to affect performance from a pragmatic approach. For example, Fig. 12.5 shows the effect of molecular weight on break-up performance for two inks meeting the specification shown in Table 1 and of the same solvent base, differing only in molecular weight. In this case the ink containing the polymer with a higher molecular weight has a greater modulation voltage range and is likely to display a more beneficial effect on performance over a wide temperature range. In a similar fashion two inks formulated to meet the specification shown in Table 12.1 and differing only in the ratio of solvent blend of methyl ethyl ketone to alcohol show marked differences in performance, as highlighted in Fig. 12.6.

Composition of a continuous ink-jet ink

The flexibility of continuous ink-jet printing places strong requirements on the properties of the ingredients used in an ink formulation. The choice of raw material and control of the specification is critical bearing in mind even small changes can cause jet instability with catastrophic effects. Currently there are few tailor-made materials for ink jet and this can often lead to a conflict in design to match printer performance with

Table 12.1 Typical properties of a solvent-dye based ink-jet ink

Viscosity	3.75±0.25 cP.
Surface tension	27±3 N/m
Conductivity	>1000 μS/cm
Particle size	<1 micron
Insoluble matter	<0.02%

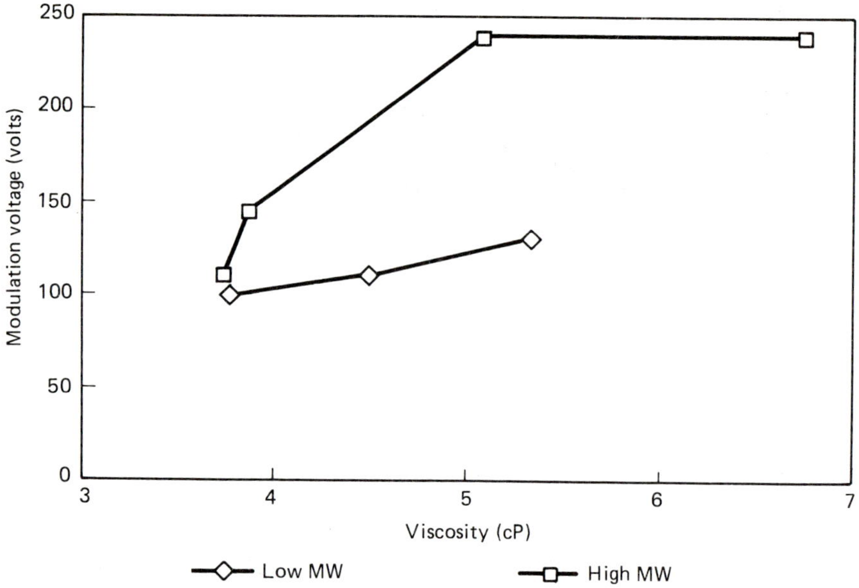

Fig. 12.5 Modulation voltage range for two cellulose derivative MEK-based inks.

the end-use application performance. This, in turn, presents the ink-jet chemist with both a challenge and opportunity and in practice a compromise is sought, resulting in a range of formulations to meet wide and diverse needs. With the current and predicted growth of ink jet it will be a surprise if the raw material supplier does not respond to this and swiftly design their products with ink jet in mind.

For simplicity, ink-jet formulations, like traditional inks, may be described as consisting of a solvent blend, colourant, binder and additives. It is worth considering these in turn to understand the role of each within an ink-jet formulation.

Solvents

Typically, the solvent blend may comprise as many as eight volatile solvents. The solvent blend acts as a medium to transfer ink from the printer to the surface and carefully controls the drying time. Modern high speed production lines require most inks to dry or at least be smudge free within one second. Varying the solvent blend will influence the drying time greatly. Clearly, if the ink contains only highly volatile solvent, it may evaporate too quickly at the surface which can lead to the formation of a skin and prevent or slow down the drying of the bulk of the droplet. In contrast, the inclusion of too much humectant or solvent of low volatility will simply slow the drying process to an unacceptable level. To achieve fast drying times the blend of solvent is designed to allow the ink to evaporate at a controlled rate.

Traditionally, common solvents such as ketones, especially methyl

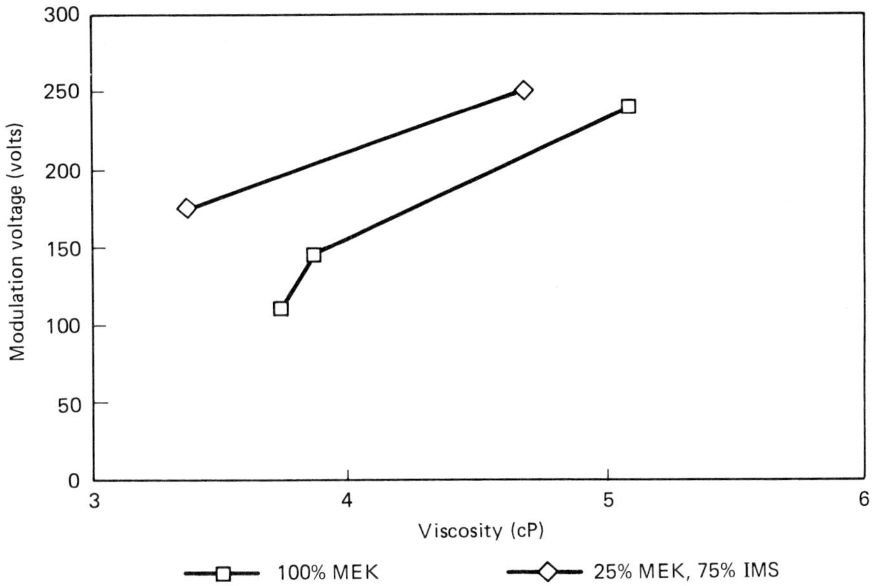

Fig. 12.6 Modulation voltage versus viscosity for two cellulose derivative solvent-based inks.

ethyl ketone (MEK), acetates, glycol ethers and alcohols have been used. Such solvents present little hazard to the operator, providing good practices are employed, because commercial printers of good design tend to be enclosed units with sealed ink containers and so the level of solvent within the vicinity of the printer is very low, typically, 10 ppm for MEK. Despite the safe use of these solvents, rightly the ever increasing environmentally conscious user is pushing harder for even safer solvents. Much work within the industry is focused on responding to these demands, reflecting a trend in improvements to equipment design and in the development of a range of new ink formulations. For example, recently there has been significant progress in the area of faster drying ethanol-based systems, solvent-free UV-curable inks and faster drying water-based inks. The advantage of moving to ethanol-based systems is the relatively low risk of operator exposure to a solvent vapour hazard. The remaining limitation, common to all current high performance ink-jet systems, is the flammability of the solvent. This can be a real hurdle for ink jet to overcome in some applications and progress in developing safe, non-flammable solvent-free UV-curable inks and water-based inks may be an interesting solution to these objections in the future.

Dyes and pigments

Predominantly, dyes are preferred to pigments as colourants in ink-jet formulations. Pigment systems can be successfully designed but the particles must be less than one micron in size and must be held in a homogenous suspension in the solvent blend at all times. In the past, control of particle size and stability of the suspension has often been difficult to achieve, especially for high density titanium dioxide-based systems. Pigments by their nature are abrasive and this fact is not popular with ink-jet designers who view this as equivalent to putting fine sand into a lubricating oil of a racing engine. If white or truly opaque inks are needed, there is no alternative pigment to titanium dioxide. Much research work is in progress to improve the chemistry of pigment dispersions to allow greater compatibility with ink jet. More recently, transparent organic pigments have been successfully used and such systems now provide the possibility of process colour cyan, magenta, yellow and black (C.M.Y.BK) ink-jet printing.

On the other hand, dyes are much easier to handle. They are selected to show high solubility in the solvent system and possess low levels of impurities with minimal insoluble matter, less than 0.02%. Ink-jet dyes must demonstrate high thermal stability and must not deteriorate on storage or during operation since the temperature of the ink within an ink-jet printer can exceed 60°C for long periods of operation. Thermal degradation products of dyes invariably lead to poor jet break-up. Similarly, solubility of the dye at temperatures as low as 5°C is also required for successful operation. Lightfastness is increasingly becoming important. In general, traditional printing dyes are unsuitable for ink jet and it is in this area where there has been, over the last few years, the introduction of high purity dedicated ink-jet dyes, reflecting advances in the reliability of ink-jet systems over a wide temperature range. Increasing legislation has also placed greater restriction on the use of certain heavy-metal complex dyes from a general user toxicity and packaging disposal point of view. As a result non-metal complex dyes are currently being developed.

Binders

The binder of an ink-jet ink may contain one or more polymers. The function of the polymer is manyfold. It controls the viscosity of the system and promotes droplet formation. This subject is explored in a succeeding section. The polymer also serves to bind the colourant and provide the adhesion to the surface. It is also responsible for the properties of the application of the ink such as steam sterilization, high gloss, good print definition and alkali solubility. In common with other ingredients in an ink the binder must exhibit good stability upon storage and during use.

The range of chemistry of polymers used within an ink-jet formulation is large. If a detailed analysis of the modes of failure of an ink-jet ink design was carried out, the choice of the polymer would figure high as a principal factor. This would not only be due to rheological properties on the influence of jet break-up but it would also be due to the dichotomy of design for performance in the printer and design for performance on the product to be printed. The amount of polymer to achieve high end use performance, such as good adhesion, is constrained by the low viscosity requirements of the fluid required for printer operation. When this is combined with the application of a thin film in the form of a dot it proves to be an extremely demanding set of properties required of the polymer. The ink mark on the product is the whole purpose of the technology and care is needed to ensure the fulfilment of an end-user property does not compromise performance in the printer. There is no better example to illustrate this point than the use of phenolic type resins in ink-jet formulations. Phenolic resins provide extremely durable coatings on a range of hard non-porous materials such as metals and glass, yet most phenolic resins are inherently thermally unstable and with constant recycling can undergo oxidative cross-linking and cause jet instability or, in severe cases, clogged nozzles. In the early days most ink-jet formulations were based on phenolics and were acceptable to a customer experiencing a new and innovative technology. Today despite their excellent end-use properties which are difficult to match, even with modern block copolymers, few survive to meet the quality and reliability standards of maturing and more demanding ink-jet users.

Additives

Additives impart specific functional properties and like traditional printing-ink formulations it is often the additive, typically at a level less than 0.1%, that will determine the difference between success and failure for a particular application. Many additives are used in an ink formulation and it is not unheard of to contain up to a dozen or more in a formulation. Again, similar to traditional printing ink formulations they are used to promote flow and adhesion, reduce tack, provide stability to oxidation, reduce foam, modify surface energy and plasticize the binder.

The most important and essential additive is the conductive salt which provides the electrical conductivity to the ink for charging the droplet. There is a limited range of organic and inorganic salts that may be successfully used. On occasions it is possible to exploit trace-metal salt impurities of the dye to provide the necessary conductivity. Much attention must be paid to the possible effects of corrosion to the printer and compatibility with other ingredients, in particular dyes and polymers. The success in choosing a conductivity salt for a new design is highly dependent on a thorough understanding of the chemistry of the formulation, experience and extensive testing. A simple CIJ formulation is shown in Table 12.2.

Designing and testing an ink

The design of a range of inks to withstand the rigours of ink jet and produce a wide variety of properties once printed is no ad hoc process. Experience, good scientific understanding and investment into research are all necessary. Reliability is a key requirement for all equipment on production lines and ink jet is no exception.

Once the design of the formulation has carefully been screened in the laboratory it has to go through extensive tests to ensure it keeps its properties in many difficult environments and after long periods of recycling through the machine. In particular, attention during the design of an ink is focused on the degradation process and the relationship between the changes in the chemical properties and printer performance. This must be well understood and controlled to ensure reliable perfor-

Table 12.2 A simple CIJ formulation

Component	% By weight
Methyl ethyl ketone (MEK)	40–60
IMS	20–30
Water	5
Ethylene glycol	5
Soluble black dyestuff	5–10
Cellulose derivative	5–15
Plasticizer, e.g. dibutyl phthalate	1–5
Defoamer	1
Anti-oxidant	1
Conductive salt, e.g. lithium nitrate	2

mance during service life. In general the need to establish reliability early in the design usually leads to an approach combining accelerated ageing trials of the ink and extended real-time printer trials.

Control of the lifetime of the ink receives much attention during the design stage such that it is stable and does not undergo deterioration due to ageing. Accelerated ageing trials, usually carried out at 60°C for periods up to one year, provide information on the degradation process and the stability of an ink.

More often than not a printer trial will involve monitoring the performance of ink for up to 1000 hours over a wide temperature range from 5°C to 45°C for continuous, recycled and start/stop modes of operation. The effect of tolerances of component design of the printer must also be taken into account to ascertain the robustness of the ink to system performance. Throughout the testing process, physical, chemical and end-user properties are measured and checked against the required specification. Chemical techniques including gas chromatography, Fourier-transform infra-red spectroscopy, and gel permeation chromatography can be used to elucidate any chemical changes occurring and allow for correction if necessary. It is inevitable that more than one design cycle of formulation, test and reformulation may be needed before a fully reliable ink is achieved.

To appreciate the degree of testing required it is worth considering some common modes of failure in the design of an ink-jet ink. These include some obvious failures of raw material instability and raw material incompatibility, but also subtle modes, including:

- polymerization, leading to cross-linking due to thermal instability;
- chain scission of the polymer due to oxidation;
- poor print quality;
- high humidity environments can lead to an accumulation of water content in an ink;
- polymerization due to catalysis of metal complex dyes;
- chemical reactions between dyes and conductivity salts;
- salting out of the dye by the conductivity salt;

- degradation of end-user performance during printer trials;
- agglomeration of pigment particles;
- incompatibility with components or glues in the printer;
- foaming;
- suppression of conductivity due to complexing of the conductivity salt with the binder or dye;
- solvent degassing of dissolved air;
- raw material batch variations of molecular weight properties of the binder, leading to a loss in modulation performance at extreme temperatures.

End-user properties describe the performance of an ink on the substrate. These are the characteristics of prime interest to the user.

Typical of these are:

- adhesion to various surfaces
- resistance to wet and dry abrasion
- colour and colour fastness
- opacity
- print quality
- resistance to process treatments such as steam sterilization, washing, heat and solvents.

During the design of an ink up to 30 different packaging films, common plastics, rubbers, metals and glass are evaluated, reflecting the wide variety of applications of ink jet and the versatility required from an ink. The variety of surfaces within packaging films include coated and uncoated celluloses, treated and untreated polyester, and different types of polyethylene and polypropylene. Common plastics include plasticized and unplasticized PVC, polycarbonate, polystyrene, polyamide and acrylics. Most general purpose ink-jet inks would be expected to provide satisfactory print performance on the majority of these materials.

Applications

There is a complex and diverse range of applications for continuous ink-jet printing. The growth in variety of these applications has promoted significant advances resulting in a vast array of commercially available inks. The formulations of these inks varies markedly to provide unique features and end-user properties. Traditionally, ink jet has found favour in the food, beverage, dairy and pharmaceutical markets. Over the years these industries have been subjected to increasing legislation, mainly from the EC, to put 'best before dates' on the packaging of their products. Ink jet has been the ideal solution because of the three key features and advantages it offers over competitive techniques. It is a non-contact printing process, simple to install, occupying minimal production line space, situated on line which means it does not affect the production process. It is capable of printing at high speed so does not slow the production line down. Finally, it can print variable, real-time and fixed data on each product.

A trip to the local supermarket will reveal that today virtually all packaged food products now have some kind of ink-jet code. The code can be a simple manufacturing batch number, expiry or best-before date. Alternatively, there may be additional lines of print detailing to an informed observer, the time the product was filled, the production line number and other production codes. Notably, inks based on methyl ethyl ketone (MEK) solvent have been a popular choice for the majority of these general coding applications. These inks are very fast drying, typically less than one second and show high performance on a variety of surfaces including waxed and varnished coatings, glass, metals and a host of common packaging films and plastics. A range of common applications using these inks is shown, Fig. 12.7.

In dairy applications, such as the one shown in Fig 12.8, where coding on direct filling lines is carried out the risk of taint from MEK, although minimal, means alcohol-based inks tend to be preferred. These inks are slower drying, approximately 4 seconds, and in the past have not achieved the high performance associated with MEK-based inks on a wide range of surfaces. More recently significant progress has been made in developing ethanol-based systems which not only match the versatility and performance of MEK-based inks but also achieve very fast drying times of approximately one second. The first generation of this type of ink has been commercially introduced and represents a major step forward in the development of more environmentally friendly inks. It is anticipated over the next few years these systems will see widespread use in many applications, not only in existing markets such as the food, dairy and pharmaceutical but are likely to open new opportunities for ink jet.

It is outside of these general coding and marking applications where the progress in the development of inks has come into its own and it is only possible to present a summary here of the range of applications. Many of the formulations have been developed to withstand pre- or postprinting process treatments. For example, inks have been formulated for the beer and beverage industry whereby the droplet of ink is required to penetrate damp surfaces and once dry become waterfast to refrigerator condensate. This is a particularly difficult and variable application where the surface can range from damp to soaking wet. Varying degrees of success have been achieved to date and there is scope for improvement in the future. In the canning industry the requirement is often for an ink to withstand steam sterilization, typically high pressure steam at 125°C for 25 minutes. In addition to these conditions the ink must also be resistant to high chlorine contents and anti-corrosion additives usually present in most steam retort processing chambers.

In more specialized applications the ink is also required to be thermochromic and change colour during the steam sterilization process, allowing an easy visual check that the product has been processed. In all cases the print of the processed ink must be of high quality and not suffer fading or spreading. Conversely, in some applications it is necessary for the ink mark to be easily removed by washing upon return to the factory and an example of this is alkali washing of coded returnable beer kegs or bottles. In this case the printed code must be waterfast during the service life of the product.

Fig. 12.7 A range of typical ink-jet applications.

Fig. 12.8 Ink-jet printing in the dairy industry.

Ink jet has found many applications in the electronics, cable and wire, automotive and aviation industries. The requirements across these markets show some similarity. Commonly used is a range of white and coloured pigmented inks providing high opacity, especially printing onto dark surfaces such as highly coloured cables and components. Likewise, a range of inks have been developed to withstand solvent washing of components, a common practice with many applications in these industries. These inks may either be air drying or UV-curable formulations and today it is possible to formulate solvent-free UV-curable inks.

An early concept for ink jet was to print on line variable, random or sequential data onto bingo or lottery tickets on high speed printing presses. This concept has grown over the years to encompass some interesting applications from printing lucky numbers in newspapers, postal and bar code printing, naming and addressing at speeds of up to 500 metres per minute and security printing. A typical application on a high speed printing press is shown in Fig. 12.9. With this growth it probably comes as no surprise of the requirement for high print definition inks and in some cases with specialized properties. The case of postal barcode printing has brought the need for easily machine readable inks such as optical character recognition (OCR) and fluorescent inks and, more recently, the Royal Mail has introduced phosphorescent ink-jet coding to aid sorting of letters.

The nature of the design of ink-jet printers and their ability to computer control the output data, combined with inks of special properties makes ink jet an attractive proposition for security applications. In the spirit of security it is not always possible to describe the function of these inks in the open literature apart from in general terms. Inks have been developed with UV readable, infra-red readable, fluorescent, biological, tamper proof solvent-resistant and photochromic properties.

For a long time Henry Ford's famous phrase 'you can have any colour you like so long as it's black' has often been adapted to describe the colour range of ink-jet inks. Today, this phrase is rapidly becoming outdated as

Fig. 12.9 Ink-jet printing on a high speed printing press.

inks with colour processing capability are introduced. Indeed there is a well established application where ink jet in conjunction with digital colour scanners is used to prepare pre-press proofing images. In another example ink jet has been used to print outdoor billboard posters and indeed the largest illuminated outdoor billboard in the world, measuring 8 m × 40 m, has been printed using ink jet.

Finally and almost in complete contrast a number of novel applications are evolving in the development of food grade and edible inks, enabling an ability to print indirectly and directly onto food products such as eggshells, citrus fruit skins, biscuits and confectionery. An example is shown in Fig. 12.10. The list of applications continues to grow daily and with it the requirements to develop new and even more specialized inks, sustaining the challenge and opportunity for the ink-jet chemist.

12.3 DROP ON DEMAND INK-JET PRINTING

There is a vast array of patents describing different designs of drop on demand ink-jet printers. This is mainly in the office printing sector and as

Fig. 12.10 An example of an ink-jet printing on eggs.

a result there are a number of commercial designs on the market. Most share a common principle of operation and similar requirements of an ink. The design of a DOD printer consists of series of nozzles in the printhead and droplets are ejected in a parallel flight to each other, producing a stroke of ink at a time. Characters are subsequently built up from a number of strokes. This is in contrast to CIJ printing where a stroke is built up from a series of droplets which are deflected one by one to form a single stroke.

Drop on demand printing methods

Designs differ mainly in the mechanism used in the printhead to produce droplets. The main techniques employed commercially include piezo impulse, thermal jet, valve jet and hot melt. The versatility of DOD printing is not as broad as CIJ printing. There are a number of reasons including speed, throw distance and ink performance. Speed is many times slower than CIJ and is dependent on the technique used. Throw distance, a term used to describe the distance the drop can be propelled without deterioration to print quality, is much smaller, approximately a millimetre or so compared with over one centimetre with CIJ printing. Although print quality can be superb with some systems ink performance limits the use to printing mainly on absorbent materials.

Piezo impulse ink jet is one of the simplest forms of DOD and consists of a printhead with open nozzles and channels. Droplets of ink within the capillary channels are ejected by the action of an oscillating piezo crystal. In the case of thermal ink jet or bubble jet as it is commonly known, the piezo crystal is replaced by a heater element located on the wall of the capillary channel. The heater is activated by an electrical signal, delivering energy to the liquid, creating a bubble which grows rapidly causing a droplet of ink to be displaced from the nozzle.

In valve jet DOD printing a valve is located at the nozzle which seals the ink in the capillary channels. The ink is held under pressure and when the valve is opened via an electrically operated solenoid a large droplet of ink is ejected.

Hot-melt systems, dubbed phase-change systems, have been applied to most systems and more recently introduced commercially in combination with a piezo crystal. Ink is heated within the printhead to a temperature above the melting point of the ink, typically between 100 and 150°C and when an electrical pulse is applied to the piezo crystal the wall of the capillary channel deforms to displace a droplet from an open nozzle.

Ink requirements

Piezo and bubble-jet inks

In most cases piezo and bubble-jet inks are used in office printing applications. The properties of an ink to operate in these printers revolve around meeting three key requirements. The first is to ensure the ink within the open capillary channels and open nozzles always remains fluid and is not allowed to dry. The second property which seemingly contradicts the first is for the ink to dry quickly on the paper once printed. Finally, the ink must also provide good print resolution on a wide variety of papers in use today. In the case of bubble jet the ink must also be capable of boiling rapidly to produce bubbles and the ink must be stable to withstand repeated and rapid heating and cooling cycles.

The approach taken to prevent ink drying in the nozzle has been to use a range of traditional humectants such as glycols and polymeric glycols in water-based inks. These are effective at mopping up moisture from the atmosphere surrounding the nozzle. The impact of using humectants to prevent clogging adversely affects the drying time of the ink on the surface. Reasonable drying times can be achieved by choosing a solvent blend that promotes wicking and adsorption of the ink into porous and coated paper. However care has to be taken to ensure the process of wicking does not lead to too much spreading, reflecting a deterioration in print resolution. In practice the balance is achieved with the aid of good equipment design. As well as automatic printhead cleaning the paper handling system, post printing, coupled with the low print speeds achievable, allows the ink to become smudge free upon stacking. Alternatively in some larger office systems heaters are used to speed up the drying process. Print quality on most papers is high but there is some variability and the price of the paper is no indicator of performance.

The limitation of ink performance is a direct result of not being able to use more powerful fast drying solvents such MEK. Much work is focused today on alternative approaches to ink design to provide the capability of printing on a wide range of surfaces with faster drying times. Currently the best formulations exhibit a drying time of about 10 seconds and in others a drying time of more than 30 seconds is more typical. High print quality is achieved by designing formulations with relatively high viscosities, up to 20 cP and a surface tension of about 45 N/m. The pH of an ink is usually between 8 and 9 and in some cases this is utilized to enhance waterfastness since most papers are acidic. Currently waterfastness of most systems is limited and new developments are expected to produce much improved waterfast inks.

High quality and stable dyes are required to ensure reliable performance. This is especially the case for bubble-jet systems. Much of the success of ink jet used as an office printer has been due to the effort expended in the synthesis of the dyes. Elaborate techniques including ultrafiltration, ultracentrifugation and reverse osmosis have been employed to purify the dyes and remove trace salt impurities. The reason is that in a bubble-jet printer the temperature at the heater element can be as high as 350° C and repeated heating and cooling can cause the dye to degrade to produce particulate matter. Over time this particulate matter can produce a hard residue which can block the capillary channels or hinder the efficiency of heat transfer to the ink. This effect is known as 'kogation' and it has been shown that dyes susceptible to this behaviour contain inorganic salt impurities. There are very few commercially available dyes that meet the stringent requirements of a bubble-jet ink.

Valve-jet inks

Valve-jet printers are used in industrial applications for large character printing onto outer cartons on production lines. Unlike other DOD ink-jet systems, the design of using a valve to seal the nozzle enables highly volatile solvents such as MEK to be used, since there is no longer a requirement to prevent drying at the nozzle. This allows inks to be formulated to print on a wide range of surfaces from paper through to packaging films. Drying times can be less than 10 seconds but print quality and speed is low due to the time taken to open and close valves. Formulations are relatively simple although the optimum viscosity is as low as 2 cP and surface tension must be greater than 23 N/m to prevent wetting of the nozzles.

Hot-melt inks

Piezo hot-melt or phase-change inks share similar properties with other DOD systems as far as the principle of operation is concerned. In addition they have a new set of features. The ink is solid at room temperature and is heated above its melting point and maintained at a temperature of approximately 150° C where it exhibits the fluid physical properties required for ink-jet printing. Again like other systems no evaporation must be allowed to occur in the capillary channel or nozzle otherwise this may lead to clogging the nozzle. Good performance is achieved when the formulation is designed to give a surface tension of 40 N/m and a working viscosity of 40 cP. Heat stability of the system is of paramount importance and the ingredients selected must be protected against oxidation, corrosion and colour fading.

Upon impacting on the surface the liquid cools and becomes solid to produce an image of very high print quality. The print gives an embossed or raised image and this is susceptible to damage through scratching. In the design of a hot-melt ink, care is taken to overcome this by blending materials to control the cooling rate as the ink impacts on the surface. Control of the rate of cooling allows the ink to be adsorbed or wick into the paper. A balance is needed to prevent too much spreading which can lead to deterioration of print quality and strike through the paper. Again like most DOD systems the performance of the ink is limited to absorbent materials, although a commercial hot-melt printer used for industrial applications has met with some success in printing onto non-porous materials.

The formulation of a hot-melt ink does not contain organic solvent and this provides some key advantages of being safe and non-flammable. Essentially, formulations comprise two types of components, namely, colourant and vehicle. The vehicle often consists of a blend of polymers to control the viscosity temperature profile and balance the performance in the printhead with the performance on the surface. Polymers used tend to be based on fatty acids, natural and synthetic waxes. The molecular weight distribution of the polymer is important and will determine the sharpness of the melting point. A blend consisting of a brittle, sharp melting point, low melt viscosity fatty acid combined with a tough, film forming thermoplastic is used to provide excellent print quality on many papers.

Conclusion

Up until now the majority of applications for DOD printing have been in office printing. Simple water-based inks in impulse systems are currently used in industrial applications for printing information on outer cartons on production lines. Specialized valve-jet systems allow the use of higher performance inks, based on MEK, for industrial applications. The systems used in industrial applications are in general slow and of low print quality.

Print quality of recent office ink-jet printers is superb but ink performance limits the use to printing on absorbent materials. Over recent years, much progress has been made in printhead technology, improving speed, reliability, ease of nozzle manufacture and print resolution which is fast approaching laser quality in office applications but the ink technology is currently lagging behind. This is similar to the position CIJ technology found itself during its early years of commercial development.

Progress in improving speed and print resolution is likely to continue and new techniques such as microjet technology are likely to emerge to ensure this is at a fast rate. The printhead in microjet technology consists of flexible capillary channels, manufactured from PZT wafers (Lead Zirconate Titanate), which distort compressing adjacent capillary channels and displacing droplets from the compressed channels. Ink is probably the key to the future growth and suitability to a wide range of applications for DOD ink-jet printing. Much attention is being paid to alternative approaches to ink design and to synthesizing tailor made materials. If progress can be made the future promises to be exciting and we can expect to see DOD printing in widespread use in office printers, facsimile printers, wide area office plotters as well as opening up opportunities in the industrial sector.

CHAPTER 13
Manufacture of inks and varnishes

The manufacture of printing inks is not a complicated process. Most printing inks consist of pigment dispersed into varnishes and manufacture is mainly concerned with the mechanical means of producing these dispersions efficiently, economically, reproducibly and, often, very quickly. Usually two techniques are employed, mixing and milling, which are briefly defined in respect of inks as follows.

Mixing is the mechanical agitation of pigment and vehicle/varnish together and is usually continued until no dry pigment is discernible and an agreed level of dispersion has been achieved. This is followed by milling, the process by which the pigment/vehicle mixture is refined and the pigment particles are reduced to a size suitable for the intended printing process. In some cases these two processes are combined in the same item of plant.

The emphasis of the last few years and for the foreseeable future is on speed of production and the ability to reproduce a consistent quality, meeting the delivery time and the print requirements of the customer. To achieve these objectives, the operation of an ink manufacturing plant demands an efficient and flexible organization since the number and size of orders to be processed can vary considerably. The equipment in common use is relatively simple.

13.1 GENERAL REQUIREMENTS

Printing ink production units are today becoming more specialized, organized on mass-production lines making a single product or a limited number of products of similar type. The manufacture of rotary newsinks, heat-set web-offset and sheet-fed process inks, flexo white, gravure process colours and the bulk manufacture of blending scheme bases or concentrates are examples of this change of emphasis. However, the inkmaker must cater for a wide variety of requirements, which explains why an ink factory, although with specific lines of bulk manufacture, has to retain the ability to be flexible. This has led to the blending of bulk lines for small quantities, a process which is costly and can be time consuming. The real cost of small batch production should always be borne in mind. In general, it is far easier to select equipment and to design and operate a plant for large-scale production of a few similar products than to operate a plant manufacturing many products in varying batch quantities.

Apart from the manufacturing capabilities of the equipment, there are other considerations to be borne in mind when designing an ink plant. Among these are the services which may be required, including electrics, health and safety, flame-proofing, water supply, drainage and cooling, steam supply and compressed air. Space requirements and access to the machinery, ease of cleaning, wear rate and maintenance (see later for possible control methods), cost of replacement parts, ease of feeding the equipment and transferring the finished products, are all items to be considered before the equipment is purchased. While the rate of production is of great importance, the nature and quality of the material produced and the likely profitability of the unit concerned have also to be considered.

13.2 THE MANUFACTURING PROCESSES

A study of the individual chapters on the inks for different printing processes will have shown that the majority of printing inks fall into two categories; basically these can be defined as:

- oleo-resinous systems for letterpress and lithographic printing;
- volatile solvent/resin systems for flexographic, gravure and most screen printing.

The former are formulated on non-flammable solids dispersed into vehicles containing mineral or vegetable oils or high boiling point solvents. Ultra-violet curing inks although not strictly oleo-resinous systems may be considered similar for the purposes of their manufacture.

The latter types of ink can be subdivided into those based on water-reducible vehicles, and those based on highly volatile and flammable solvents, which require special buildings and plant with flameproof electrics for their manufacture.

The manufacture of inks can be broken down into the following operations:

- varnish manufacture, i.e. the vehicle;
- the manufacture of additives;
- dispersion of pigments into vehicles using mixing and milling techniques.

Providing that the formulation has been carefully considered and appropriate raw materials have been selected, the process of dispersion is fairly straightforward. To understand the process it is necessary to review the materials and the routes by which dispersion can be achieved.

Primary dispersions for ink production demand that the pigment particles are thoroughly wetted by the liquid phase and that the particles in the finished ink for most processes should be in a range between 5 and 1 μm. If this level of dispersion is not achieved, printing problems will arise which are considered in other relevant chapters.

Pigments are generally precipitated from aqueous solutions. At the time of the chemical reaction causing the precipitate, these particles are molecular in size. However, the molecules of the pigment are strongly attracted to each other by physical forces and immediately form larger pigment particles, the size of which will depend on the nature of the material and the physical conditions existing at the time of reaction. These primary particles cluster together during processing to form agglomerates. The particles in an agglomerate are strongly bound together and high energy is required in the dispersion process to break them down. As the pigment is filtered from the liquid phase, the agglomerates tend to bind together more loosely but in larger groups known as aggregates. If the subsequent press cake is dried, the resulting solid will contain a mixture of primary particles, agglomerates and aggregates.

The pigment is finally crushed to reduce the size of the largest agglomerates, giving what appears to be a relatively fine powder. However, it is an essential part of the dispersion process to break down these large particles exposing a greater surface area of pigment to the wetting medium in order to produce a finished, evenly wetted product.

There are several routes by which ink dispersions can be produced. In some cases, pigments can be purchased or manufactured in a form which makes the inkmaking process, rather than the dispersion process, much easier.

While there are similarities in the way the individual classes of inks are made and, since the introduction of bead mills, much of the machinery involved is also similar in concept, there are distinct differences between them and they will be considered separately.

The manufacture of oleoresinous systems

Varnish manufacture

Varnishes have three primary functions. First, to wet the pigment particles and thereby to assist the process of dispersing the pigment in the vehicle system; the better the wetting property of the varnish, the easier it is to disperse the pigment thoroughly. The second function is to incorporate printability into the system. The system must be stable enough to perform on the press and must ensure faithful reproduction of the printing surface on the printed material. Thirdly, the varnish must bind the pigment to the printed surface so that the final print dries to an acceptable film which is suitable for the use for which the finished print is designed. Thus, varnishes impart such characteristics as gloss, rub resistance, product resistance, flexibility and adhesion to the finished print.

In the case of oleoresinous inks, these functions will be carried out by the use of varnishes based on drying oil–resin combinations with, or without, the addition of high boiling solvents. Vehicles based on hard resins dispersed in these solvents will also be used, and if the latter have a low aromatic content and are refined to give a narrow boiling point range material, the varnish may require the addition of 'bridging' solvent to give the required stability and to avoid the resin separating out.

The correct drying oil, resin and distillate combination in the finished ink may be achieved by combining the varnish raw materials at the manufacturing stage, but in practice is more often carried out by the inkmaker, by combining one or more varnishes in the formulation. Where practical, the varnishes may be preblended to create a composite vehicle system, thereby simplifying the weighing or metering stage operation.

Oleoresinous varnishes are produced in closed kettles, of, mainly, up to 5 tonnes capacity, and these are heated by oil heat exchangers or, on older plants, by gas-fired furnaces. Cooling of the larger sized vessels is also carried out by heat exchanging oil from a primary source. Relatively high temperatures are involved in the processing, usually in the range 120–260°C, and complex chemical reactions take place during this cooking process. These conditions require strict control of the procedures used, especially the rate of temperature increase and decrease, the maximum temperature achieved and the length of time for which the varnish is held at this temperature. After the initial cooking of the oil–resin mixture has been carried out, the structure of the varnish can be further modified by the addition of gelling agents or 'chelates' which are based on metallic molecules held inactive in an organic complex. The method of addition of such compounds and the control of the temperature at which they are added is critical if a stable varnish, and hence ink system, is to result.

Since many of these varnishes are formulated to dry by oxidation polymerization, it will be clearly seen that, at all stages of their manufacture, every effort should be made to exclude atmospheric oxygen. For this reason, the varnish is usually cooked under a nitrogen 'spurge' or 'blanket'. The use of a condenser helps to prevent the loss of distillate and 'bridging' solvents.

In modern varnish plants, all functions are controlled and monitored from a control panel (Fig. 13.1) and temperatures, times, heating and cooling rates, can all be recorded on flow charts.

A typical cooking cycle takes the following progression during the manufacture of a high gloss varnish used in a lithographic ink:

- distillate metered from tank farm to clean vessel under nitrogen flow;
- preheated alkyd charged by back weighing;
- charge heated to required temperature;
- hard resins added slowly maintaining temperature;
- resins cooked 1/4–2 hours between 160 and 240°C to achieve solubility/ modification;
- batch cooled to required temperature for addition of gelling agent – this is specific to the agent used;
- gelling agent addition made by spraying a solution in distillate;
- batch reheated to the reaction temperature to achieve optimum structure/modification – temperature and time specific to the gelling agent used;

Fig. 13.1 Varnish manufacture control panel. (Courtesy of BASF (Coatings + Inks) Ltd)

- batch tested during cooling stage with adjustment of distillate, as necessary;
- when completed, batch filtered (or strained) through fine filter into tarred 205-litre drums;
- vessel distillate washed for next batch;
- final sample of varnish product tested prior to use or storage.

Any combination of these operations may be used but once established as the procedure must be adhered to, to achieve continuity and consistency of product. Figure 13.2 shows a typical 'cook cycle' achieved during the production of a batch of oleoresinous varnish which has been 'structured' by chelate addition, while Fig. 13.3 shows the arrangement of a varnish kettle with the ancillary equipment required for the manufacture of this type of vehicle. The careful recording of previous experience is necessary to scale up the production of a batch of varnish, from a laboratory pilot plant size to that of full-scale works production, if the most efficient manufacturing process is to be devised. For consistent results, the raw materials used should have been checked carefully to ensure that they were up to the certificate of conformity agreed with their supplier. From previous records a working specification can be set up, to be used during manufacture, which may be modified after the initial batches, so that a final vehicle formulation is evolved where any adjustments are kept to a minimum. These adjustments will usually be small additions of the high-boiling petroleum distillate used to cut the viscosity of the varnish. Besides the final viscosity measurements, earlier checks for clarity and resin dispersion will have been carried out during the cooking cycle, while checks for tack and colour will usually be carried out prior to filtering and running off to storage.

Development work on varnishes requires a combined knowledge of resin and ink technologies, and expertise, gained over many years of varnish manufacture. This has led many users today to choose not to make their own vehicles, but to purchase proprietary varnishes from the specialist manufacturer. There are arguments for and against this policy but an essential part of ink manufacture is a knowledge of the vehicle and what it will do, when used in ink formulations.

Where smaller quantities of specialized varnishes are required, a 'cold cut' technique is sometimes employed. The term 'cold cut', in this context, describes the process of dissolving soluble resins in oils or high boiling-point solvents at temperatures between 160 and 180°C which are not far above the melting points of these resins. Modern techniques for cold cutting use high-shear, high-speed mixers, which can be used to achieve the temperatures required by friction to get the resins used into solution. External heating can be used to assist the speed of temperature increase and hence the solubilizing process.

The resulting oleoresinous varnishes can be considered as ultrafine particles of resin which are dissolved in solvents which have just sufficient strength to hold them in solution. Although clear and stable varnishes are produced in this way, their resultant properties do show differences from a varnish of similar formulation produced by the cooking method, because the higher temperatures employed (230–265°C) by the latter allow ester interchange reactions and then chemical complexing of the resin and oil. The latter method also permits the use of less soluble resins.

The 'cold cutting' method of varnish production is much more relevant

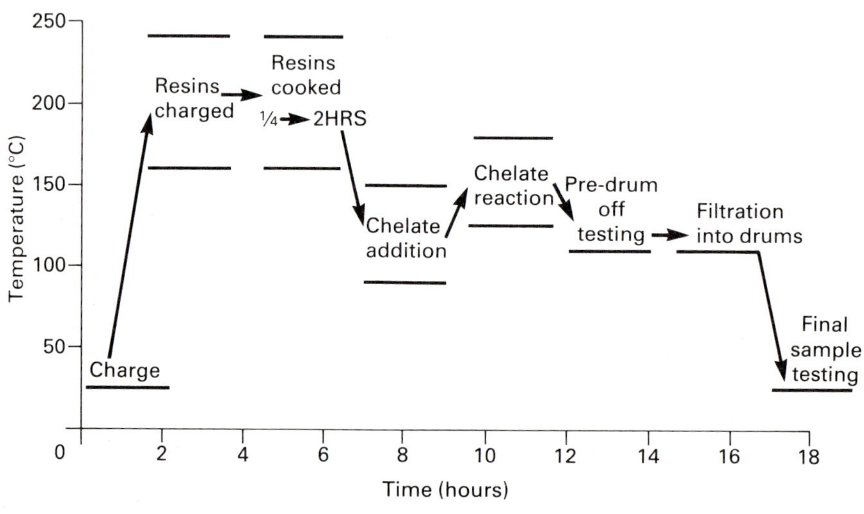

Fig. 13.2 Temperature/time related cooking cycle. (Courtesy of BASF (Coatings + Inks) Ltd)

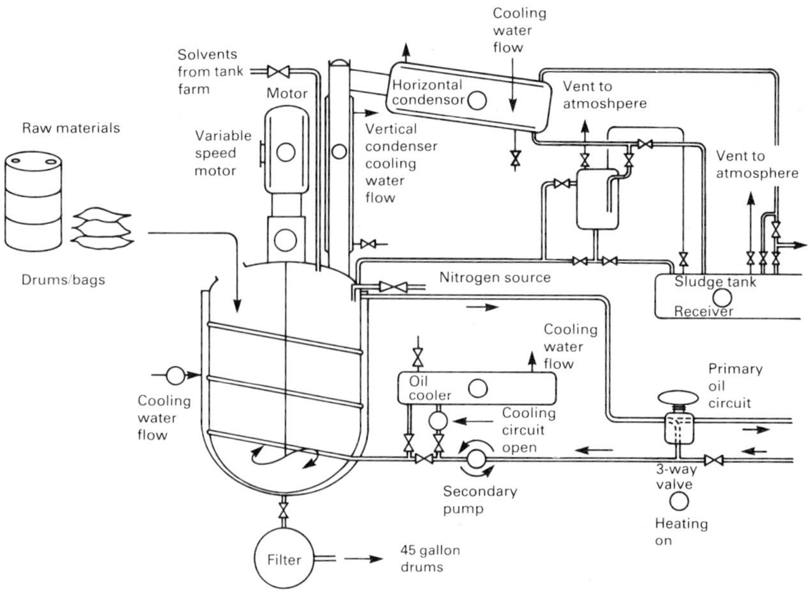

Fig. 13.3 Flow pattern in practical terms. (Courtesy of Coates Lorilleux/A Process Systems)

to liquid ink vehicle production and will be discussed further in that section.

Summary

Producing paste ink vehicles by cooking at elevated temperatures under carefully controlled conditions gives the following benefits to the formulator and manufacturer:

- good batch consistency;
- economy in large batch sizes;
- low oxidation (pale) colour when nitrogen 'blanket' used;
- good lithographic performance due to high temperature solubility process;
- good gloss;
- uses gelling agent to maximum effect giving consistent structure where required;
- controlled solubility leading to crystal-clear filtered product.

Although the oils and resins used are classified as non-hazardous, at these elevated temperatures great care must be taken to minimize the potential fire risk and operators should be appropriately trained in the safe handling of the hot varnishes to prevent any skin contact with the product. They must also adhere to all current health and safety requirements.

Bulk varnish storage and handling

How this is achieved depends on certain criteria, and although the list may seem long, all of the following aspects have a bearing on the type of storage and handling employed.

- Is the varnish made on the same premises as it is to be used?
- What is the temperature at completion of the manufacture and can the residual heat be used to aid transmission of it to the point of use?
- Are there any special requirements for special materials to be used for the tanks and pipework construction? Is the varnish oil based, or glycol based for water-reducible inks?
- Does it need further processing?
- What is the frequency of use or turnaround of the varnish? Long-term storage at elevated temperatures can lead to changes in the characteristic of the product.
- Is the varnish compatible with the previous contents of the tank?
- If the products are bought or made at a distance from the point of use, do they need special transport facilities and storage in a warm area to enable them to be pumped or used effectively?

In principle, liquid materials, whether at 10 or 1000 poise (1 Pa s or 100 Pa s), flow more efficiently if at an elevated temperature. The head of storage is also most significant, so that gravity feed can be used wherever possible, and surprisingly heavy materials flow with the aid of small amounts of pressure or warmth. To retain the heat, lagging of pipework and in some cases trace heating of valves to encourage this flow is well worth while and has a short payback period. The pressure can be achieved easily by compressed air or hydraulic pressure. The latter has a history of being more accurate across the wide range of viscosities which may be encountered.

Fail-safe devices need to be incorporated into the design of any system to avoid discharging into vessels with valves in the open position or filling beyond the capacity of the container.

Manufacture of wax pastes and other additives

The waxes used in the construction of printing inks are as various in their type as in their physical appearance, and have been described in another chapter.

The latest mode is to incorporate micronized versions but these do not always give the desired or expected results. There are several reasons for this; for example, the particle size may not be large enough to provide the abrasion resistance required, or the material may not be uniformly dispersed. The wax content of a finished ink is small, and must be thoroughly mixed to give the necessary protection to the ink film. As a result, waxes are, more often than not, added as a compound containing between 30 and 40% wax. The word 'compound' in this context needs to be further defined as the waxes are normally neither soluble nor compatible with the vehicle or solvents into which they are dispersed, which will be chosen so that the final paste is compatible with as wide a range of inks as possible. The only purpose of making a compound is to reduce the particle size of the wax to a suitable level for incorporation into the ink, and reduce the chances of localized additions of large agglomerates of wax formed by the attraction of smaller particles due to static forces.

Dispersion methods are crude and are normally by use of a star or rotor high speed mixer or, alternatively, the wax is heated as a mixture to a temperature around softening point and refined through a heated three-roll mill until an acceptable reduction in particle size is achieved.

Although waxes are the largest group of materials which need to be added to oleoresinous systems in small amounts, similar techniques can be used for other minor additions of solid raw materials which an ink technician may wish to include in his formulation.

With the use of highly accurate electronic balances or those systems linked to a computer base, the weighing of liquids as ink additives becomes a simple operation, however, dilution with some of the vehicle, or the solvent used in the formulation, aids transfer of the material and helps ensure a homogeneous mix. For example, see the addition of gelling agent, during oleoresinous varnish manufacture. If the reduction is done in bulk, a simple paddle mixer should be sufficient, but care needs to be taken to ensure that the reduced additive does not form layers of separated material and it should be homogeneous before use.

Manufacture of oleoresinous inks

Until the early 1960s the production of paste inks had followed the traditional path of mixing the items just sufficiently to wet out the dry pigment and then passing this rough mixture over a three-roll mill until sufficiently dispersed, i.e. most of the work was done at the milling stage. With the introduction of faster running printing presses there was a greater need to reduce more efficiently the particle size and to cost effectively develop the colour potential. As a result the required levels of dispersion meant that the ink had to be passed over the three-roll mill several times and this was time consuming and very expensive.

Combined with this, the increasing use of heat-set inks meant that a totally enclosed system was needed to overcome evaporation of the lower boiling point solvents being used in increasing quantities. Both these factors led to the introduction of bead mills, which were not only enclosed but also gave improved rates of production and permitted the easier handling of larger batch sizes. This method of manufacture is efficient for dispersing most pigment/oleoresinous varnish mixtures, although it would be premature to suggest that all the problems such as machine downtime, temperature control and maintenance of gloss in the final inks have been overcome. In fact, some companies use bead mills to disperse the premix and follow this by passing the ink through a three-roll mill to improve gloss and remove air.

Whatever the equipment being used, the following processes will be carried out when making a paste ink from dry colour and varnish:

- weigh out pigment and vehicle and pre-mix;
- disperse using a three-roll mill or a bead mill;
- incorporate additives and solvents;

- transfer to bulk or pot-off into tins, buckets or drums, once passed by quality control.

Using his experience of the pigments being used, the ink formulator should have broken down his formulation so that the initial stage consists of weighing out, or metering, sufficient vehicle to 'wet-out' the pigment. There is a vast range of equipment which can be used for this purpose, with a gradual move from mechanical scales to the load cell principle. The readings are either digital or make use of an optical print-out. Electronics in this type of equipment can be complex and if it is computer linked, boosters along the line may be necessary. Since these are costly, the location of the weighing and weighing-related activities is very significant. The addition of the dry colour needs to be carried out with care to avoid airborne contamination of the working place. To ensure thorough wetting of the individual particles, the pigment may be added in stages and thoroughly dispersed between each addition. It may be necessary to adjust the viscosity of the pre-mix, at this stage, to achieve the maximum working pressure and hence the most efficient condition for shearing and dispersion action in a particular mixing. This is especially important in the early stages of mixing, using a high-speed disperser, where under the desirable dilatant conditions, brought about by pigment agglomerates locking, the variable impeller speed permits extremely high forces to be applied to the mixture. When the resistance is overcome, the agglomerates are pulled apart far more effectively than by very high speed alone in a semi-paste or fluid state.

This type of machine could very well switch the emphasis of ink manufacture away from roller milling to high-intensity change pan mixing, at a much reduced capital and operational cost. Short predispersion cycles would be followed by a light milling.

Harsh pigments, which are difficult to disperse, will require extra work done on them to break down the agglomerates and this will be carried out on the roll mill. However, the wide use of easily dispersible (ED) pigments in modern oleoresinous formulations has speeded up production rates and enabled premixes to achieve very low readings on a grinding gauge. These pigments are treated, usually in aqueous solutions, with a wide range of surface-active material and resins, so that the pigment surface is compatible with the medium into which the pigment is to be dispersed. The offset ink formulator needs to ensure that the surface-active wetting agents do not react with the fount solution and leading to printing or plate problems.

Following the mixing stage, the viscosity of the ink formulation may be further adjusted depending on the ink type and the nature of the mill to be used for the next processing stage. For three-roll milling, the mixture should have sufficient structure to ensure shearing takes place, but should also have sufficient vehicle present to complete the wetting out of the pigment particles. For bead milling, there is an additional requirement to keep the temperature in the chamber low enough to avoid changes in shade and opacity with certain yellow and red pigments.

Once the ink is fully dispersed, final additions can be made and the ink technically approved. It may then be canned manually by passing through a three-roll mill with the roller gap set to give a high rate of throughput without aerating the ink. Alternatively, a modern ink-filling line may be used (see the section on ink packaging).

An alternative method of manufacture may be used for small to medium sized batches (say up to 100 kg) or when using flushed pigments or special concentrates:

- blend the predispersed base (made from dry colour and varnish, as above) or the flushed pigment or the concentrate, with the extra varnishes and additives required;
- refine the mixture over a three-roll mill;
- transfer to bulk or can.

The range of bases available to the ink formulator will depend on the type of inks being supplied, but ideally should contain a high pigment concentration and should be as versatile as possible. A simple mixing process using a paddle or 'butterfly' type mixer is normally used.

Flushed colour

Pigments had been flushed into various vehicles from the 1860s, but it was not until 1914 that pigment flushing, as it is known today, was invented by Robert Hochstedder.

The original flushing technique was developed to produce finished inks in a more efficient, less expensive manner than by three-roll milling, since the pigment is transferred directly from the water phase to the vehicle without the necessity of drying, crushing, blending and milling. Since the water paste avoids the drying process, agglomerates and aggregates are not formed. Much of the early flushing was done by inkmakers themselves.

The heart of the process was the use of vacuum to remove all traces of water that could not be poured off. The double arm, jacketed mixer used was equipped with a cover so that the system could be put under vacuum (Fig. 13.4).

Nowadays, the organic pigment manufacturer has greatly extended the process to produce highly concentrated bases which are in turn sold to the inkmaker for formulating into finished inks.

Whereas a few years ago the quality of flushed organic pigment was recognized to be superior to that of dry colour, nowadays, with the improvements in dry pigment technology, this difference has been narrowed significantly. Why then use flushed colour?

By far the most important property of flushed colour is that it contains fully dispersed pigment, free from any agglomerates which require grinding equipment. Moreover, in spite of the significant improvements in dry colour, especially in terms of dispersion efficiency, the basic need to grind out pigment agglomerates formed during the dry process is still the greatest limiting factor and the cause of high cost ink production.

The justification for flushed colour as an alternative method of manufacture becomes more apparent from the line diagram for the production of a theoretical 500 kg of ink (Table 13.1).

Clearly, with existing three-roll mill equipment, using flushed colour in place of dry, production can be somewhat more than doubled, resulting

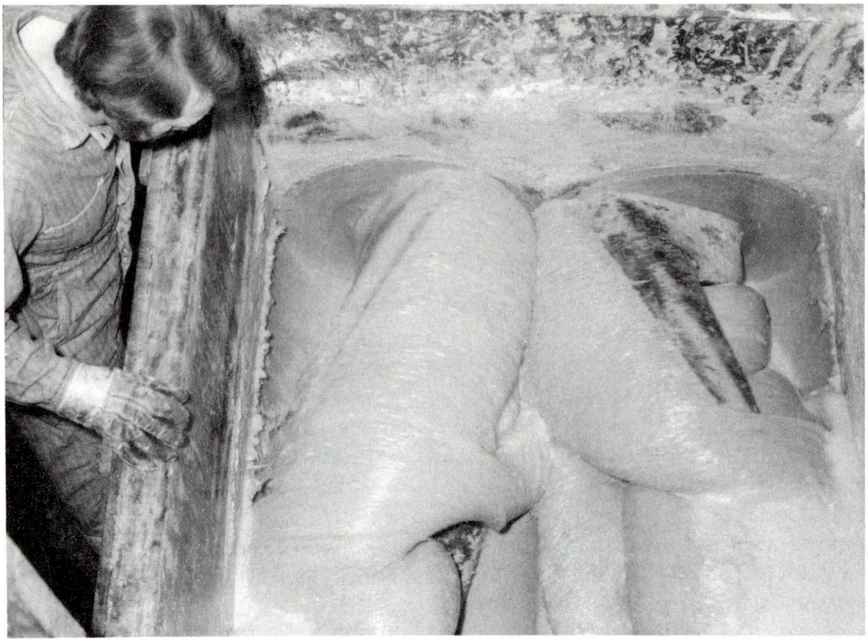

Fig. 13.4 Z-blade mix of flushing. (Courtesy of Sun Chemical Ltd)

in output significantly greater than that achieved by the introduction and increased capital cost of introducing shot mills. Inevitably, the economic viability of using flush depends very much on the price relationship between pigment powder price and flush, the savings in capital outlay and the increase in plant production, all of which are best assessed by the individual inkmaker.

Flushed pigments do offer significantly cleaner working possibilities to the ink manufacturer who does not wish to install sophisticated dust extraction systems. But if the inkmaker has a large capital commitment to shot and/or three-roll mills, it is as well to recognize the fact that flush is unlikely to figure in the immediate future as a major process; change to this philosophy has to be planned and may be introduced in part as a result of plant breakdowns or a sudden surge in demand.

One method for the effective handling of flushed pigment pastes is to have air lines with free air, to connect to the base of the containers, thus ensuring easy and economical emptying. Alternatively, under the action of high shear and the temperature induced, the flushing can be made sufficiently mobile to transfer to the blending plant. Whichever method is used, some care needs to be taken to avoid affecting the dispersion when added to a large volume of cold solvent and varnish.

Summary

Figure 13.5 illustrates a few of the possibilities from the multiplicity of methods and individual machines all with their own conceptual design style and benefits. Routes (1) and (2) could only be used for the lower

Table 13.1 Comparison chart of manufacture of 500 kg of finished ink (courtesy of Sun Chemical Ltd)

Powdered pigment 3-roll mill 1 grinding pass/ 1 processing pass	Powdered pigment shot mill/3-roll mill	Flushed colour	
1. Weighting – 15 mins	1. Weighing – 15 mins	1. Weighing – 30 mins	
2. Mixing – 30 mins	2. Mixing – 30 mins	2. Mixing – 30 mins	
3. Mill 1 pass – 4 hours	3. Shot mill – 2 hours	3. Control adjustment – 15 mins	
4. Control adjustment – 15 mins	4. Control adjustment – 15 mins	4. Processing to can:	
5. Mill pass processing – 2.5 hours to can	5. Mill pass processing and to can – 2.5 hours	*3-roll mill* 2.5 hours	*filter* 1 hour
6. Control during processing run – 15 mins			
7.75 hours 6 stages	5.5 hours 5 stages	3.75 hour 4 stages	2.25 hour

viscosity types of paste ink and (3) to (8) are by far those in most common use.

The individual pieces of equipment can be put together to form an online unit but, before purchasing, any machinery under consideration should be evaluated for throughput, quality and general handling conditions. This is particularly important with envisaged automatic controls, either remote or through a computer, as the program can only be written when the aspects just mentioned are known. A simple example of this would be to have a mixing capacity of 1000 kg per hour, when the product can only be refined at 500 kg per hour, which would mean the purchase of another refiner or using the mixer for only half the time. It is foolish to dismiss such simple examples because they do occur, and all too frequently.

Although a fairly rigid set of parameters is needed to program in-line production, the constituent parts need to be flexible to accommodate changes in the products' physical characteristics. For example, if, due to temperature, the viscosity of a heat-set colour is liable to change, it is important that the three-roll mill being used has some facility to adjust the rolls to enable a similar throughput even though, due to the viscosity drop, the cohesiveness of the product has reduced. With the exception of the major bulk lines, some ability to vary certain dwell times in the production line is useful, remembering at all times to reprogram the microprocessor, if one is being used.

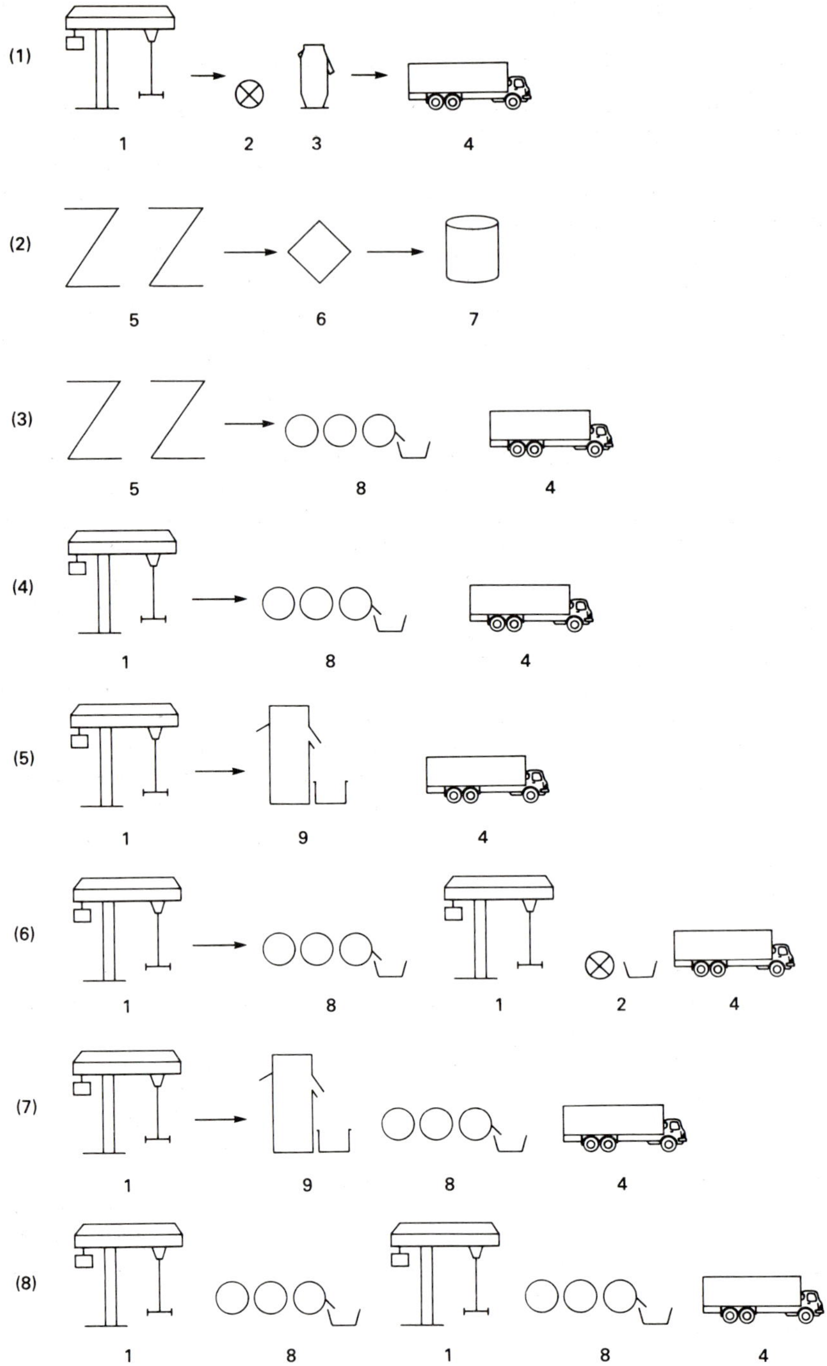

Fig. 13.5 Thumbnail sketches of production routes: 1 variable speed mixer; 2 pump; 3 filter; 4 distribution; 5 Z-blade mixer; 6 de-aeration; 7 container; 8 3-roll mill; 9 beadmill. (Courtesy of Coates-Lorilleux Ltd)

The manufacture of polymer/solvent systems

This section covers the manufacture of flexographic, gravure and most screen inks. As noted previously, there are two main types of formulation – water based and solvent based. The volatile nature of the liquid phase of both solvent- and water-based inks requires similar manufacturing techniques but they are often produced in separate units to avoid contamination. The highly flammable nature of a large number of the solvents used requires special precautions to be taken to prevent the risk of fire and explosion. Solvent extraction and containment along with fresh-air ventilation are both used to keep the level of solvents in the atmosphere below the threshold limit value (TLV). Emissions to the atmosphere need to be studied before plans are finalized on this subject because in principle they should be avoided.

The area where highly flammable and highly volatile products are manufactured is normally classified as a Zone 1 flameproof area and every effort must be made to eliminate the risk of fire; manufacturing equipment and containers must be earthed to eliminate the generation of static electricity; all motors, lights and switches must be to Class 1 flameproof standards and any hand tools used must be non-sparking. Further, the clothes worn by operatives should be so manufactured that overalls do not cause static and protective footwear does not cause sparks.

A number of the materials used in liquid ink formulation could cause a health risk if adequate precautions are not taken. The extraction systems should keep the overall levels of solvent in the atmosphere below the 8-hour weighted maximum allowed concentration (MAC), but care must be taken not to allow pockets of solvent fumes to build up.

Face filter masks must be worn if there is a risk of powder in the atmosphere and gloves are essential to reduce the risk of skin contact related health problems.

Varnish manufacture

It is possible when producing liquid inks on a ball mill to load pigment, resin, solvent and other ingredients to produce a finished ink from basic raw materials. However, it is more common to produce varnishes using the resins and solvents before the pigment dispersion process.

Liquid ink varnishes are formulated using a very wide range of resins such as nitrocellulose, ethyl cellulose, maleic resins, metallic resinates acrylics, polyamides, etc. The solvents used range from aliphatic and aromatic hydrocarbons, to esters, alcohols, ketones and water. Varnishes are normally produced on high-speed stirrers. In view of the volatility of the solvents used and the heat generated, it is essential that the stirrers are totally enclosed and, where the design allows, water cooled.

All the following types of mixer are employed.

Cavitation mixer

The cavitation mixer is a driven shaft fitted with a saw-tooth disc impeller which, combined with a variable speed drive, provides for flexibility of tank size and volume of charge. The impeller acts to keep the whole mass of the charge in motion and its size and velocity is determined in direct relation to the tank diameter.

This type of mixer is better suited to products in the higher viscosity range and, in this respect, process time can be considerably reduced if a proportion of the total liquid for the batch is omitted in the initial stages of the mix, to be added once the resin solids have been dissolved. This addition can be used to advantage as a coolant to the batch where heat-sensitive materials such as wax compounds are part of the formulation.

Rotor/stator mixer

This type of mixer has a fixed rotor/stator size and is normally operated at only one or two fixed velocities. The charge size and vessel size is therefore limited. The overall agitation is comparatively less vigorous than that produced by the cavitation mixer and viscosities must be kept correspondingly lower. The reduction of solids particles is achieved by their subjection to shear forces as they continuously pass through the narrow gap between the rotor and stator.

The application of heat by using a jacketed vessel will accelerate both processes although this is not always advisable when volatile flammable solvents are used unless strict temperature controls are used.

If multi-solvent systems are used, it is prudent to dissolve the resin in true solvents before adding the diluents. The normal procedure used is to agitate the liquid while adding the resin gradually to it. After the addition is complete, a further period of agitation dissolves the resin. For efficient dissolution the agitation must be of sufficient velocity to maintain resin particle suspension and separation. The liquid will thus act upon the whole surface of the particle, giving maximum opportunity for the solubilizing process to take place.

Storage and handling

In addition to the questions raised about storing oleoresinous varnishes, the following important points must be considered.

The low viscosity, possible flammability, and solubility characteristics of liquid ink varnishes requires pipework and storage containers to be constructed of materials adequately resistant to attack by them. A study of relevant legislation covering the storage of inflammable products is essential. There is more on this subject later in the section on dispense systems.

Manufacture of additives

When a formulation requires very small amounts of material, i.e. anti-foaming agents or ingredients that need pretreatment to highly controlled levels, i.e. wax pastes, a separate preparation is normally manufactured for incorporation into the formulation.

When the additive is a dilution of the active ingredient, it only requires a simple mixing of the ingredients but where predispersed waxes are required, these must be separately manufactured on normal liquid ink equipment.

Liquid ink manufacture

Referring back to Fig. 13.5 which gave optional manufacturing paths for oleoresinous inks, routes (1) – variable speed mixer followed by filtration – and (5) – variable speed mixing followed by bead milling and filtration – are also applicable to polymer/solvent systems. In addition, liquid inks can be made by charging a ball mill with either the whole mixing or a pigment vehicle base which can be thinned *in situ* once the initial dispersion has taken place but plant design and factory layout have both changed in concept during the early 1990s to ensure higher specialization and efficiency. Other significant factors are the on-going and ever demanding aspects of health, safety, welfare and the protection of the environment all of which are subject to changing and current new legislation.

As a result, new inkmaking facilities normally have the assistance of consultant or project management expertise to accommodate the demands sought by a tightening and better informed legal framework. Custom-built and designed factories are emerging yielding better working conditions and lower production costs.

We can see from the flow chart dealing with the manufacture of newsinks that it is or can be a continuous system and this is now also easily achievable in both liquid and paste ink plants. In each case the storage facility, which is dealt with separately, forms the buffer between manufacturing capacity and customer needs. This allows production to make in cost-effective amounts and to a programme.

The principle is the same for both types of ink. Obviously the equipment and facilities vary by necessity (e.g. viscosity and explosion-proof electrics, etc.) but the flow details are the same. The only predeterminate is whether the ink is made from prime materials or from a blending scheme.

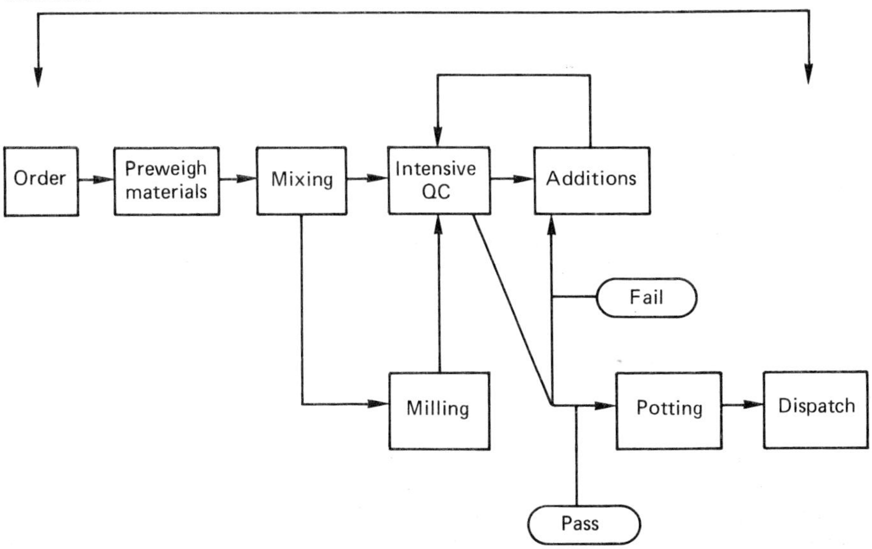

Fig. 13.6 Batch production flow chart: order to dispatch time. (Courtesy of Coates-Lorilleux Ltd)

From the flow charts (Figs. 13.6 and 13.7) can be seen more clearly the significance of whether there is a storage phase or not.

Ball mills

The traditional method of manufacturing low viscosity inks during the last 50 years was in ball mills. A ball mill consists essentially of a metal or ceramic cylinder rotating on a horizontal axis. This cylinder is partly filled with steel or porcelain balls or pebbles. During the rotation of the cylinder at the correct speed, the grinding balls rise and cascade over each other subjecting the ink components to crushing and shearing stresses.

The voids between the balls in a normal charge are estimated to be 40% of the charge volume, and this is the amount of ink material charge recommended for general work. In practice, good results are obtained when the ink charge is slightly above the ball charge level. If, as recommended this is 45%, it should make up the additional volume to a total ball and ink charge of 50%, thus a half-full mill will contain 23% of ink by volume. If the pigments are very easily dispersed, then figures of up to 65% total volume of ball and ink charge are suitable. In this case, the mill is acting mainly as a mixer.

The choice of pigment/vehicle ratio has a considerable influence on the performance of the mill and it may need experimental work carried out to determine the charge for maximum efficiency. It is claimed that the best result is obtained from a high viscosity mix, allowing for the limitation that the cascading effect of the balls must not be impeded.

Higher production speeds may be obtainable by milling the pigment in solvent or in a low viscosity first and then adding the higher viscosity medium later, but more wear takes place. A rough guide is that the volume of the liquid should be 40% of the apparent bulk of the pigment; however, the viscosity of the system will change during mixing. Thixotropy should not cause any problems while the mill is rotating but discharging the mixture may be a problem.

Bead mills

While the ball mill may be considered as the original totally enclosed system for manufacturing polymer/solvent type inks, the introduction of

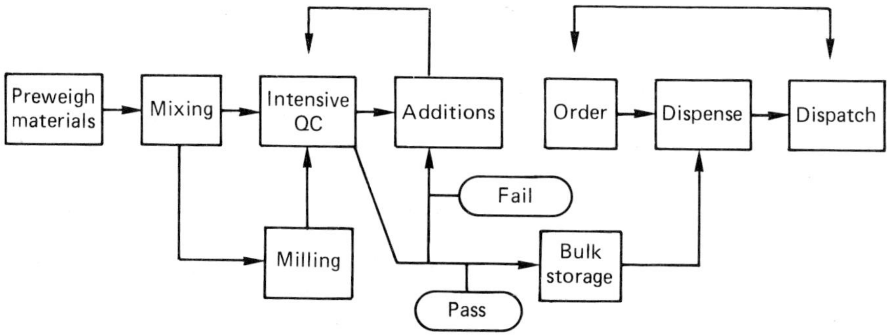

Fig. 13.7 Dispense production flow chart. (Courtesy of Coates-Lorilleux Ltd)

bead mills gave opportunities for introducing faster and continuous production methods. Bulk amounts of concentrates and inks can be manufactured from dry pigment and varnish using a combination of mixers and bead mills by the following processes:

1. weigh out pigment and vehicle and premix on a stirrer;
2. disperse on a bead mill or other continuous dispersion mill;
3. add additives and solvents;
4. filter and pump to the storage tanks or to containers for delivery, once passed by a quality control.

These processes are very similar to those used for making an oleoresinous ink from pigment and varnish. Decisions about the amount of vehicle required to wet-out the pigment at stage (1) and the correct pigment/vehicle ratio for optimum dispersion at stage (2) have to be taken to give the conditions for maximum shear. The increasing use of easier to disperse pigments make the grinding stage a much simpler operation in polymer/solvent types of vehicle and gives the level of dispersion and high colour strength required for quality inks with less energy.

Stage (3) consists of the addition of the remainder of any vehicle not previously used, also solvents to achieve the correct viscosity, together with other preparations not suitable for inclusion in the original grinding phase.

To prevent any chance of any of the bead charge or other adventitious material getting into the final product, the ink is sieved through a fine mesh filter. This will often be a multistage process, especially if the ink is held in stock for any length of time.

Chips

An alternative method of producing liquid inks with exceptional colour strength, transparency and high gloss properties, is by the use of the pigment 'chip'. Chips are produced by dispersing pigments into solid resins which may also include some plasticizer. Various wetting agents and anti-oxidants can be added to improve the manufacturing time and prevent instability on storage.

Basically, a chip is a solid ink which needs the addition of liquid to make it useable. Care needs to be taken to enable the chips to be dispersed in the solvents used in the formulation without flocculation. Various methods may be used.

- Soak the chip in diluent until soft and then use a high speed mixer to dissolve in the true solvent. Varnish and additives are then added.
- Disperse the chip in high resin content varnish and add solvent slowly, avoiding shock-cooling.
- Add the chip slowly to the varnish/solvent component of the formulation. When dispersed, add the other ingredients.
- It is also possible to use a ball mill although care has to be taken to ensure that the formulation of the initial charge does not cause the chip to form a conglomerate with the ball charge.

Figure 13.8 gives a sketch form of the manufacture and use of the pigment 'chip' and more detail regarding their production will be found later. However, the use of chip has diminished considerably over the last five years, mainly due to the cost of manufacture, while improvements in pigment manufacture and dispersion equipment have offered alternative production methods giving inks of nearly equivalent properties. For top quality results and when using high cost pigments the process has cost-effective advantages.

Other types of pigment dispersion, such as extrusion, can be handled in the same way as chips in the preparation of inks. These simply require mixing with the rest of the formulation, again avoiding flocculation, and refining in the normal way.

Pigment-chip manufacture

The hazards associated with the use of these materials are encountered mainly during their manufacture, especially where nitrocellulose, chlorinated rubber and oxidizing pigments are employed. After manufacture, although nitrocellulose and the oxidizing pigments remain flammable, they are generally less hazardous and more stable than a liquid ink being solids. This gives the product a greater shelf life, improved environmental tolerance and allows easier storage. The oxidizing action of specific pigments can be overcome by the use of inhibitors in the formulation.

The cycle of manufacture usually involves four separate stages, namely premixing, dispersion, dry/cooling, final size reduction.

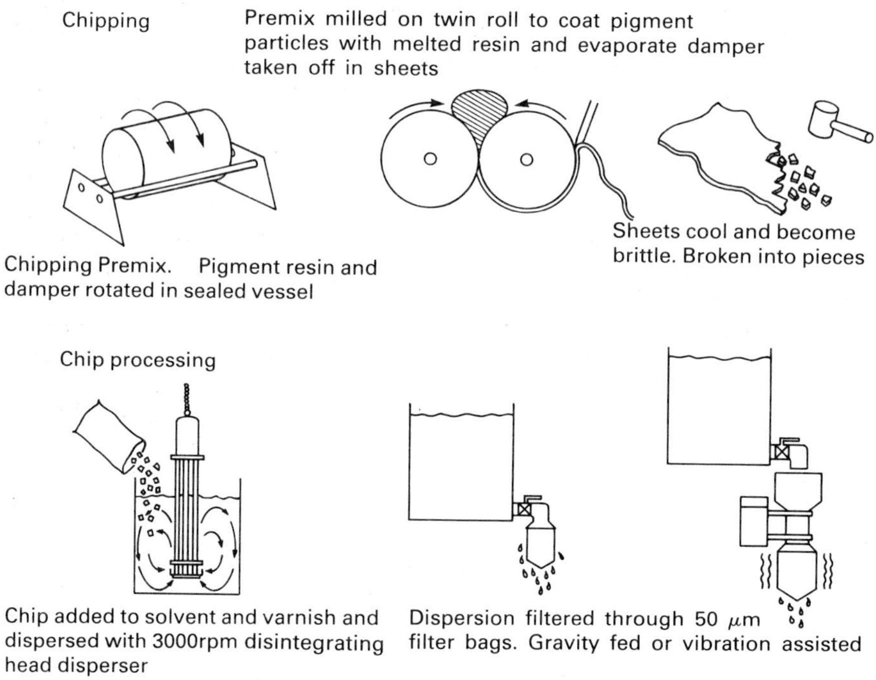

Fig. 13.8 Liquid ink manufacture from chips. (Courtesy of Coates-Lorilleux Ltd)

The initial premixing of pigment and resin can be carried out using a Z-blade mixer, solvent and plasticizers being added to give cohesion, producing a viscous paste. This method is commonly used for nitrocellulose systems as it contains any potentially dangerous nitrocellulose dust, at an early stage. Alternatively, a comparatively dry, although homogeneous mix, can be prepared either by tumbling or with a planetary type mixer, using a minimum of solvents/wetting agents. In both cases, the levels of solvent, plasticizer or other wetting agents must be strictly controlled as excessive use will prolong eventual milling operation and could lead to loss of dispersion. Heavy duty Z-arm mixers are available from several manufacturers, some featuring bottom unloading by an extrusion worm-screw fitted to the base of the mixing vessel. These machines should also incorporate a water jacket for cooling around the mixing pan.

After premixing, the material is transferred to a two-roll mill and placed in the nip. There it forms a plastic band around the rolls which is repeatedly cut off, turned and replaced into the nip until dispersion is complete. When using resins which have low softening points, for example polyamide, close control of roll surface temperature is desirable, as the mass can, if allowed, become very plastic, and therefore difficult to work, which in turn reduces shear efficiency. It is also advantageous to maintain a differential in roll surface temperatures as the mass (during its final stages) can be confined to one roll allowing easier removal. The nip setting varies with materials used, although it is usually between 1.0 and 1.6 mm. Some machines have a mechanism for changing this setting while the process is carried out. Shear is produced in the mass by the roll speed differential. The pressure applied during the milling operation is around 50 tonnes per square inch; power input is usually between 5.9 and 8.9 kW (8–12 hp) per kilogram of product. The mills are the types used by the plastics and rubber industries and are generally between 44.7 and 111.8 kW (60–150 hp). The rolls must have a speed differential, the ratio being around 1.1:1 and will be either cored or drilled peripherally to allow for cooling by water. The rolls have an adjustable nip and should incorporate a removable blade across one or both rolls to facilitate removal of the finished product.

When milling is complete a plastic sheet is produced (Fig. 13.9) which is allowed to cool during which any residual solvent evaporates off. The time taken for this varies with the resin system employed; for example, polyamide or proprionate require little drying time, while nitrocellulose systems can take several hours. Some nitrocellulose dispersions are made with a water-dampened resin which can present difficulties when drying. The use of a dehumidifying system can be contemplated if large volumes of this chip are required.

The sheeted dispersion becomes brittle when dry and is ready to be reduced to a convenient particle size. This is often carried out mechanically through a high-output cutting mill. Such machines are supplied with a selection of graded screens which determine the particle size; care should be taken to select a screen which produces a minimum of dust as this is detrimental to the quality of the finished ink.

Extreme caution is required when reducing dry nitrocellulose disper-

Fig. 13.9 Chip manufacture. (Courtesy of Coates-Lorilleux Ltd)

sions as any dust created during the process can be highly explosive, requiring purpose-built plant to be carried out safely, An alternative dispersion technique is to blend pigment and resin in a Banbury mixer. This machine produces shear under conditions of high pressure; the ram applies and maintains the pressure while the rotating arms provide the differential moving surfaces. Within the chamber the temperature rises rapidly and control is necessary to prevent the subsequent rise in viscosity and eventual loss of shear.

Manufacture of dye-based inks

Dyestuffs are soluble in the solvents and diluents used in the formulation and the method of manufacture is normally to dissolve the dye in the solvent part of the formulation with constant stirring.

The application of heat will increase the solubility of the dye and therefore lessen the time taken for the process to be completed. When the dye is dissolved, varnish and other additives are added to complete the formulation. Once again, care must be taken not to shock-cool the system, causing the dye to become insoluble.

13.3 MIXING EQUIPMENT

The dispersion of the dry materials, pigments, fillers, etc. is normally done in a mixing vessel. Whether they are dropped in from the top manually, or by means of a screw feed or, alternatively, drawn in from the bottom using vacuum in the mixer, is academic. The function of the mixer at this stage, is to achieve an accepted level of dispersion before it is further refined or filtered, also to facilitate the addition of liquid reducers and other additives to already dispersed material. Paddle mixers find a wide use for mixing all types of ink. They are also used to agitate or keep products in solution or suspension.

The types of mixers available are numerous but, in principle, they fall into fairly simple groups, mainly determined by the type, viscosity and volatility of the ink to be made. High-viscosity oleoresinous formulations are best dispersed using a low shear premixer, but as the ink viscosity falls, higher-shear machines give better results. Low-shear equipment employs slow speed mechanical movement, coupled with a high viscosity product to maximize the shearing effect.

Vacuum premixing has become general practice in the last few years, especially where oil inks are concerned. The main reasons for this are twofold. First, it is known that air dispersed in inks can be as high as 40% of the mix and this reduces the efficiency of the premixer considerably. Secondly, dispersed air can lead to problems in subsequent processing. Heat transfer within the mill is reduced considerably, leading to high processing temperatures which would affect the milling of some yellows and other heat-sensitive pigments.

Z-arm mixers have been used in the production of viscous pastes for many years and their use in manufacture of flushed pastes has already been mentioned. Modifications can be made to the basic mixer, such as the provision of heating/cooling jackets, while vacuum or pressure can be applied to the mix. In general, they are of heavy construction with relatively powerful motors and a low operating speed.

The mixer contains two gear-driven Z-shaped blades rotating, usually at different speeds, the blades almost scraping the bottom and sides of the mixer. The blades rotate in opposite directions, both towards the centre of the mixer, and the whole unit can be rotated through 90° to empty the contents for further processing. Usually, part or all of the varnish is added to the mixer and then the pigment and/or filler, added in stages, allowing each addition to be fully wetted out before the next one is added.

As the blades rotate, the mix is kneaded and thus strong internal shear forces result in a tearing action within the mix. Further shearing occurs between the blades and the sides and bottom of the mixer.

Butterfly mixers (change pan)

In some cases, combined with a fixed speed shearing head, these mixers are best suited for middle to high viscosity materials, such as paste inks. They produce a folding movement, where flow is restricted, but a kneading effect is achieved although the shear effect is limited. It is in fact a compromise between the planetary-type mixer and dissolver and in its field is a widely used machine.

High-speed impeller mixers

For fast, efficient dispersion of middle to high viscosity materials, this type of mixer is gaining in popularity today because of its speed, reproducibility, ease of cleaning and general flexible efficiency (Fig. 13.10).

High speed impeller-type mixers originally used a serrated blade (Cowles type) which creates high turbulence and intense shearing action. Generally known as dissolvers, they can be of fixed type, sited on large tanks or pedestal type which rise on a column allowing portable vessels to be used. The speed is generally variable allowing for a range of viscosities and loadings.

If the serrated blade is aligned on the central axis of the vessel, dispersion is achieved by the adhesion of the material to the disc surface combined with the configuration of the teeth. The teeth force the material radially outwards to the side of the vessel where they are deflected upwards causing a vortex effect. Additionally, the teeth act in breaking down clusters or lumps of solids. Peripheral speeds of such discs are between 20 and 25 m/s.

An infinitely variable, hydraulically driven disperser is an accepted processing tool in the surface coatings industry and has been supple-

Fig. 13.10 High-speed hydraulic mixer. (Courtesy of Buhler Ltd)

mented with other similar units such as the high torque mixer and the twin shaft disperser (Fig. 13.11).

For the manufacture of liquid inks, the mixers described below find particular use.

Rotor and stator high speed mixers

In machines like the Silverson, a stator is introduced between the rotor and vessel wall, in order to increase the shear. The disadvantage of this type, however, is that the radial velocity beyond the stator is reduced and it is therefore limited to low viscosity materials and chip dispersion (refer also to the previous section on varnish manufacture).

The 'star'-impeller type

This represents one of the latest developments where contrary to the Silverson type head, the rotor is the outer element and the stator the

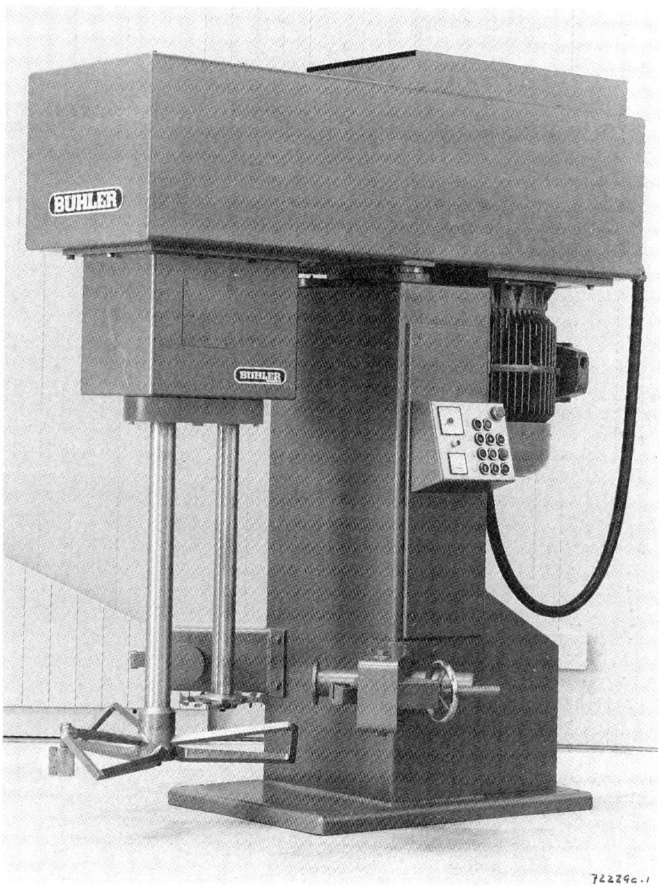

Fig. 13.11 Twin shaft mixer. (Courtesy of Buhler Ltd)

inner. This combines the advantages of the toothed blade and the combined rotor/stator system. A comparison of the disc impeller with the star head shows that, for a Newtonian fluid, the reduction of flow to the side of the vessel is linear, in the case of the disc. The star head, however, based on similar speeds, shows similar velocity parameters as the disc from rotor to vessel wall, but exhibits very high shear rates between rotor and stator, without loss of vortex.

The high-speed disperser

The field of use for these mixers is in the premixing of low to medium viscosity inks, such as the range of the various liquid ink formulations and web-offset inks for fast presses.

The apparatus used in high-speed dispersion usually comprises a cylindrical tank in which is operated a dispersion disc or head, driven from a central shaft, normally capable of variable speed operation from zero to approximately 1440 rpm (different speed ranges are usually necessary for high torque mixers).

There are numerous variations with respect to both dispersion disc design and position of the central shaft, but the commonly accepted formula for operation is as follows:

Diameter of turbine	D
Diameter of vessel	2–3D
Batch level	2–3D
Height of turbine from bottom	0.6D
Peripheral speed of turbine	1100–1500 m/min.

It should be emphasized that these operating conditions are merely those which are most commonly used. Obviously, it is necessary to vary, for example, dispersion disc diameter, batch depth, etc. in accordance with the requirements of the particular process under consideration.

The ratio of mixing tank diameter to disc diameter is an important factor, in that, as scale-up occurs to larger batch sizes, the actual distance between the dispersion disc and the inner wall of the mixing vessel increases and problems may sometimes occur with lack of circulation of material, particularly in those systems of unusual rheological properties, of thixotropy or dilatancy. It should always be remembered that the objective is to create a suitable condition of flow for the batch as a whole, when the intense shear of the high speed dispersion disc can be imparted to create acceptable dispersion of the solid phase.

If processing difficulties causing loss of material flow are encountered, then, should it not be possible to adjust the formulation, attempts may be made to operate the mixer at a higher peripheral speed. This may be achieved either by increased rotational speed or the fitting of a dispersion disc of different diameter. This can result in excessive temperature levels and the loss of volatile solvents, and indeed, colour change.

It should be mentioned at this stage that the raising or lowering of the dispersion head during the initial wetting out and mixing phase can be a most useful processing technique, and while this is often carried out manually in the high-speed disperser, there are other machines available with an automatic process control function, which will raise and lower the dispersion head in accordance with the degree of turbulence created within the batch and hence will greatly influence mixing and dispersion efficiency.

Another principle using the technique of asymmetric agitation is also widely used in the liquid ink field. This system relies on the intensive flow patterns achieved by placing the agitator shaft off-centre, which initiates high shear forces between adjacent streamlines having different speed patterns. Particle collisions are enhanced by such mixing and providing this can be maintained this is a highly energy efficient pre-mixing system.

In summary, while the classical dispersion technique of using a high-speed disperser does offer certain advantages, some outstanding problems remain. The most important of these is with materials having unusual flow characteristics, particularly in larger batch sizes, when adequate flow and circulation of material are not maintained at the periphery of the vessel. Hence, efficient dispersion is not obtained or, alternatively, time cycles can be extremely lengthy. In such cases, a slow scraper on the periphery of the container can be attached to rotate at a speed of 8–12 rpm wiping clear the inner surface and pushing the dead areas back into the vortex.

High-speed mixing

Over recent years, high-speed mixing techniques have shown little change as far as low viscosity mixtures are concerned. In the highly viscous paste field, however, significant technical advances have been made, and more regard is being given to the importance of good wetting out and better premixing. Few product manufacturers when purchasing or using a mixer, such as a Trifoil type, realize how much efficiency and money may be lost between the motor drive and the blade. The blade, after all, is the only working part in contact with the materials being mixed, yet its inefficiency is obscured by the mixer horse power; even so, between 30 and 50% power losses have been recorded.

To obtain the optimum results, it is essential to achieve the correct balance of torque, shear and vortex and then programme the conditions for repeat manufacture. The use of a hydraulic drive allows the development of these requirements but mixer manufacturers have been reluctant to develop this equipment, which is fully programmable and is currently proving its new technical advantages within the industry. To maintain torque and shear, the hydraulic system has been specially designed, balancing the pump, transmission lines and motors, so ensuring a minimum loss of power at the point of mixing.

To impart shear with particle reduction, the relation of the Trifoil blade to the bottom and side surfaces of the mixing containers is crucial, and the vortex should develop to ensure that the batch passes through the high energy zones.

As an example, the conventional high-speed mixer fitted with a 30.5 cm diameter rotor and revolving at 1440 rpm, should obtain a peripheral speed of 1379 m/min when used to disperse a low viscosity ink. A 70 cm diameter Trifoil rotor at 600 rpm will have a peripheral speed of 1320 m/min. It can be seen that they are almost equal, but if the Trifoil blade is being forced through a high viscosity oleoresinous ink, cooler conditions prevail because the work is done much nearer the container sides and bottom. This means that the work is far more effective in the earlier stages at a reduced speed, the force being very much higher with a shorter distance being moved, resulting in less heat. A higher speed is desirable as the viscosity drops and the resistance falls as the state of dilatancy changes to pseudo-plastic. With high-speed open rotor systems the movement near the container sides is slow with most of the work being done in the centre when heat exchange is made less efficient due to the lagging effect of the slower moving outer mass.

Having achieved a uniform and acceptable quality it now needs refining by filtration.

Filtration has its own parameters; its standards are normally measured in microns and its restrictions are viscosity and pressure. To filter a liquid requires some pressure, even if it is only gravity as in the case of many of the thinner liquids and varnishes. However, in the paste ink and liquid ink toner areas, cartridge filtration has made very considerable advances. Caution is needed here to avoid filter bag and cartridge blocking and the choice of material and pressure resistance is paramount because if either causes deterioration the batch is spoilt, but just as important is the danger if undue pressure builds up in the line. A pressure gauge is essential in such lines.

13.4 MILLING EQUIPMENT

Three-roll mills

These have been the mainstay of the paste ink industry for many years, but the machines today bear little resemblance to the early mills, being much more sophisticated.

Modern hydraulic three-roll mills, such as the 1100–1300 mm range, while being considerable improvements over their predecessors, are costly and, relative to output, very expensive to run.

In order to optimize output, the major mill manufacturers have developed sophisticated systems of gearing ratios, specific roll cambers and automatic ink feed, all of which have improved the specific results but have done nothing to assist the flexibility of these machines, which was once one of their greatest assets.

The mill is in fact a three- (or in some rare cases, a five-) roll disperser made to very exacting engineering standards. Roll widths are between 150 and 1300 mm with almost any gear ratio and roll camber to accommodate the vast range of products which are refined in these machines. A word of caution concerning the operation of these machines. They have two in-running nips which in the UK have to be electronically guarded with a fail-safe device to avoid the entrapment of fingers or any other appendages. They also have an extremely sharp 'doctor' take-off blade which must be handled with respect. Should the machine run dry, the roll surfaces will be damaged and this will necessitate regrinding. To avoid this, an automatic feeding device such as the SAT control system which both regulates the feeding of the mill and protects their rolls from running dry can be used. This does not predetermine the method of filling the hopper as the system can accommodate tub tilting, pressurized pan emptying or pumping, allowing the system to work automatically without the need for manual control. As can be seen in Fig. 13.12, a float or ultrasonic sensor scans the product level in the hopper of the roller mill. As soon as the product arrives at level I, II or III, preselected at the transmitter, this level will be kept constant between the limits 2 and 3 by switching the feeding mechanism on or off. If material supply fails and the hopper empties, both the roller mill and the feeding device will be automatically stopped at point 1. At the same time, this state will be announced by an optical signal. For cleaning purposes or when grinding small batches, the automatic device may be put out of operation and the float removed. A monitoring light indicates the state of supervision of the mill.

As has been said, this type of roller mill has been with us for some years but the degree to which it has been refined is most impressive, as also has the feeding of the hopper; see Fig. 13.13 illustrating press-out feeding.

The SDV-1300-E is now available in a programmable format offering clear advantages in factories where optimum output with uniform quality is required even with wide and frequently changing production programmes. The advantages are that the preselected values are electronically controlled and thus kept constant avoiding continuous attendance by the operator or influence from unauthorized persons with the knowledge that the machine's safety is assured. By recording and evaluating the production data the results can be used to optimize the efficient operation of the mill and enable the product to be made consistently from batch to batch. This is a fail-safe piece of equipment where roll temperatures and hydraulic pressures are automatically adjusted and there is electronic

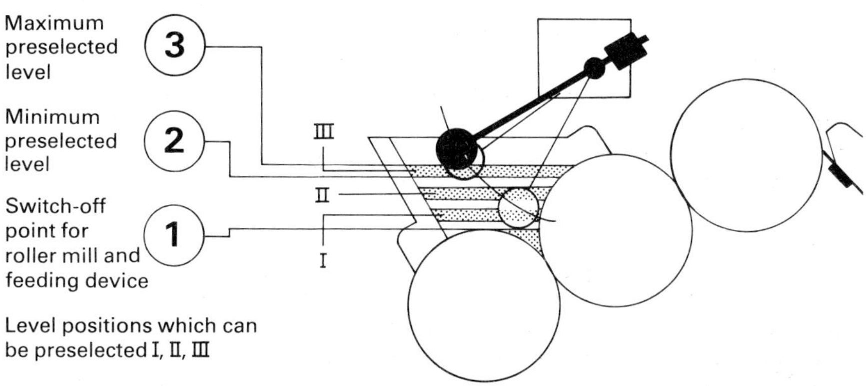

Fig. 13.12 Hopper with float. (Courtesy of Buhler Ltd)

Fig. 13.13 Press-out feeding. (Courtesy of Buhler Ltd)

locking of selector switches, to avoid error, even in the limited operator related aspects of the process (Fig. 13.14).

These machines and their performance figures are not comparable with the machines built 10 years ago or earlier. A modern roller mill is capable of meeting the following requirements with regards to a rational manufacture of quality products:

- high throughput;
- optimum level of grinding with good wetting;
- optimum and at all times uniform dispersion;
- simple and safe operation;
- continuous operating process without the supervision of an operator;
- high space utilization factor;
- rapid and simple colour change;
- robust and maintenance-free design.

Fig. 13.14 Saphir triple-roll mill. (Courtesy of Buhler Ltd)

Three-roll mill in operation

It is often incorrectly assumed that during the dispersion process on the three-roll mill, a reduction of solid particles occurs. It is more correct to say that the passage through the mill breaks down the pigment agglomerates formed during production, storage and transport of the dry colour into primary particles which may be smaller than the required terminal fineness of the dispersion as well as being smaller than the roll gap (Fig. 13.15).

Figure 13.16 shows how the levels of dispersion are improved by the increasing amount of energy used. All of these micrographs were taken using a Zeiss Universal Light microscope with bright field illumination. The overall magnification is 200 × and the practical limit of resolution is estimated to be around 4 μm.

The pigment particles are coated with a layer of binder replacing the skin of moisture and air previously present – this is equivalent to a microflushing process, otherwise designated as wetting. The breakdown of these agglomerates, achieved by the high shear forces present in the nip areas between the rollers, causes an increase in the pigment surface area. This, in turn, leads to the absorption of more binder, during the wetting process, producing an increase in viscosity for a given level of vehicle. The improvement in the degree of dispersion is normally accompanied by an improvement in the level of colouring power.

An important factor is to obtain, by the whole working process, a uniform and at all times reproducible depth of colour. This means that all physical parameters concerned in the dispersion process, namely roll pressure, speed ratios and dispersion temperature, should be held automatically constant and should be as reproducible as possible.

Fig. 13.15 Roll pressures. (Courtesy of Buhler Ltd)

Today it is possible by means of the centrifugal casting process to cast a 10–15 mm thick wear-resistant surface layer on the rolls; the hardness and elasticity of which is determined by the composition of the metal and can be adapted to meet requirements. With statically cast rolls, on the other hand, the surface hardness is obtained by quenching, thus preventing graphite formation on the surface so that the hardness falls rapidly with a decrease in the diameter.

The centrifugal casting process causes diffusion between the inner bearing cast-iron layer and the surface. Shrinkage cavities and hardness stresses are virtually excluded. All rolls must, nevertheless, undergo ultrasonic testing. A well-designed structure and maximum care during machining, influence the good and uniform temperability of the rolls and guarantee vibration-free running even at high rpm.

Automatic roll cooling

The importance of the correct roll temperature during the processing of pigmented systems is often overlooked.

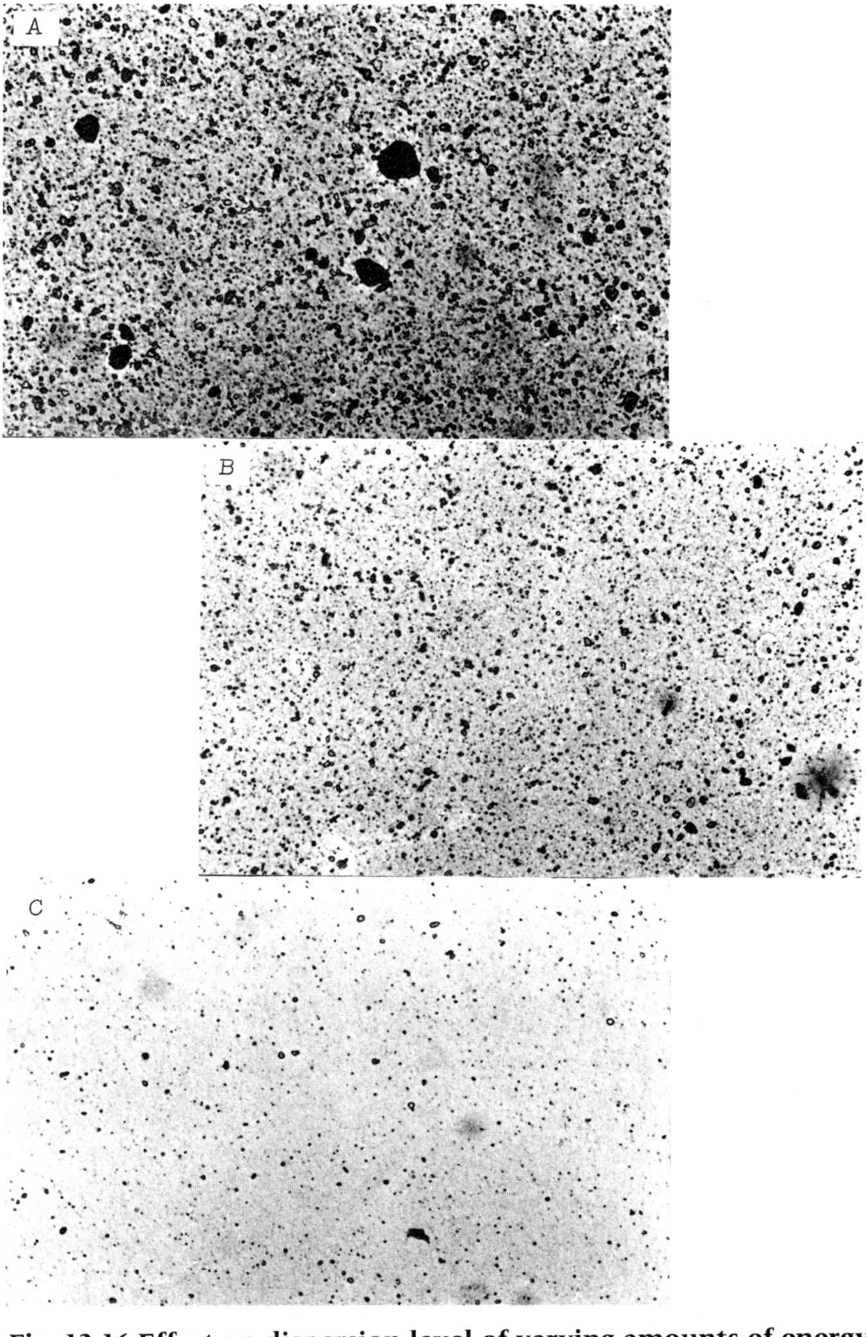

Fig. 13.16 Effect on dispersion level of varying amounts of energy used: A, Rubine PBC 15 mins MM; B, Rubine PBC 30 mins MM + 1 LP; C, Rubine PBC 30 mins MM + 1 LP + 1 HP. (Courtesy of Ciba-Geigy Pigments)

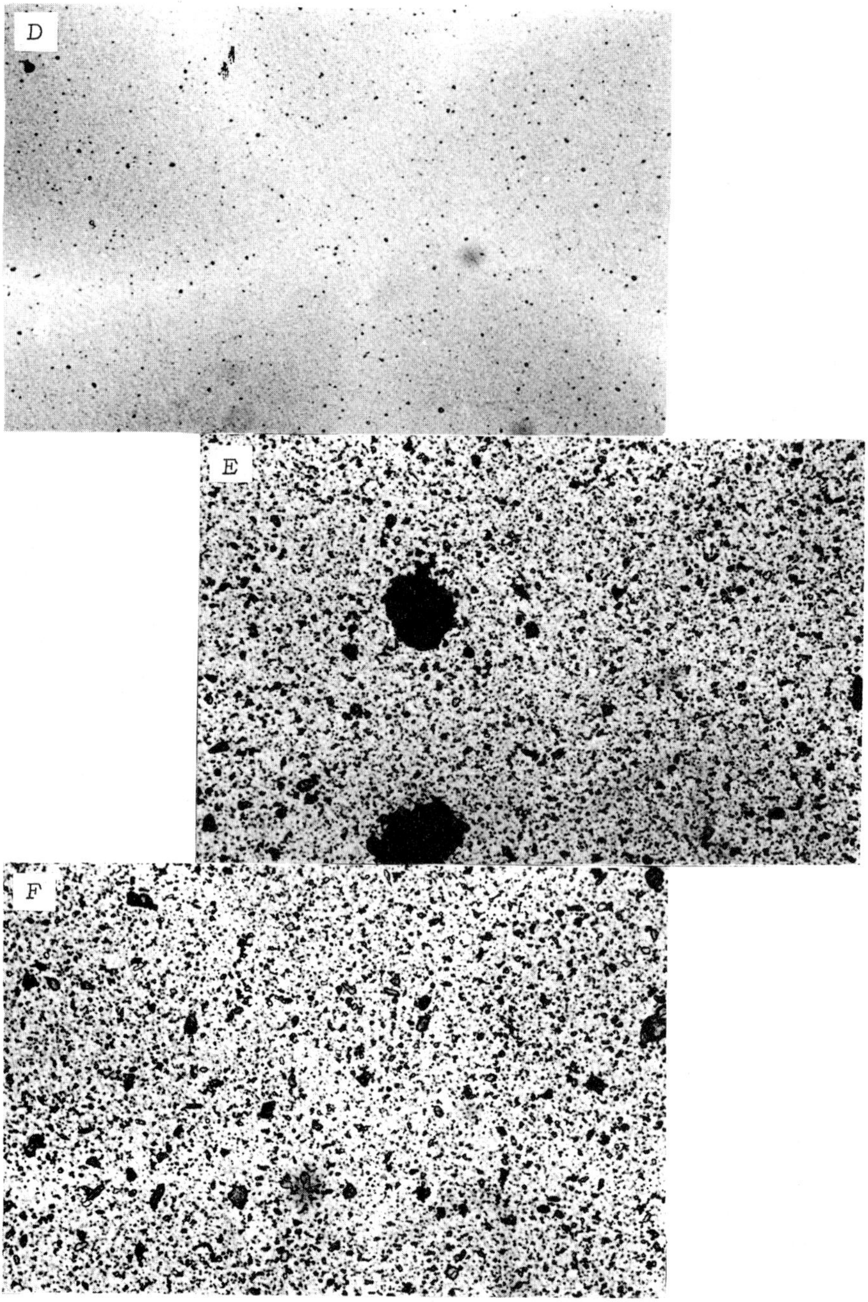

Fig. 13.16 ***cont.*** **D, Rubine PBC 30 mins MM + 1 LP + 2 HP; E, Yellow BL Premix 15 mins; F, Yellow BL 2 mins Eiger. (Courtesy of Ciba-Geigy Pigments)**

Fig. 13.16 *cont.* **G, Yellow BL 5 mins Eiger; H, Yellow BL 10 mins Eiger; I, Yellow FG4 flocculation. MM = laboratory mastermix, LP = light pass over three-roll mill, HP = heavy pass over three-roll mill. (Courtesy of Ciba-Geigy Pigments)**

Since the heat output is not identical on all rolls, each roll should be fitted with its own thermostat system. The temperature gradient for each roll is produced as a result of friction heat. If the actual temperature, measured in the water bath in the rolls, exceeds the preselected required figure, automatic regulation of the cooling water feed is effected. This ensures the uniformity and reproducibility of all the required working temperatures. It also provides an estimated 30% saving in water consumption.

The same automatic system includes a magnetic valve which permits the inflow of water only while the main motor is running – this again saves water and obviates the formation of condensed water on the rolls, when the machine is not running. Should the cooling water supply fail, for any reason, it stops the machine.

Hydrodynamic roll pressure

Dynamic roll pressure involves a system consisting of a pump, maintaining an incompressible or compressible liquid or gas under a constant pressure under all circumstances, irrespective of external factors such as temperature, time, etc.

The mechanical and static hydraulic systems are still widely used; in such systems the transfer of force is taken over by an incompressible liquid or paste. Any desired transfer of force is possible by selection of the actuating and working cylinder system.

The dynamic system guarantees a constant pressure while the working distance resulting from the movement of the hydraulic cylinder (the quantity of oil in the cylinder and thus the roll gap) can be varied independently of the pressure.

The position is exactly the reverse with the static system – any variation in the roll gap changes the contact pressure.

This means that in relation to the three-roll mill, only the dynamic pressure system is capable of taking into account variations due to temperature differences or changes in the rheology of the product, without corresponding changes in the pressure. This is, however, critical for the reproducibility and uniformity of the operating process and the performance of a roller mill will be decisively affected thereby. Although the dynamic system involves a slightly greater technical and financial outlay, this will be rapidly recouped by the resulting decisive advantages, outlined as follows:

- reproducibility of all operating processes;
- increased output;
- constant pressure;
- uniformity over the whole operating process and thus a uniformly high quality of the end product;
- extremely simple operation, the pressure being preselected and subsequently controlled merely by the movement of a lever;
- ancillary assemblies, such as hopper cheeks, stripping blades and tub-tilting devices, operate at constant oil pressure and can be very simply regulated and automated.

Floating-roll system

In the floating-roll system, which also has considerable advantages from the point of view of operation, the third roll (delivery roll or cutter roll) is mounted in a fixed position. The high rpm of this roll (in excess of 400 rpm) is thus achieved without vibration. The angle of the blade to this roll also remains constant without subsequent adjustment. Pressure is applied to the first roll (feed roll) with the middle roll floating between the outer rolls. Correct adjustment of the milling gaps between the roll is thus automatic. Any spraying is avoided, since the first milling gap only releases as great a quantity as the second milling gap is capable of processing.

Bead mills

Bead mills or stirred ball mills have been known for almost 60 years although their use in the ink industry has been limited to the last 30 years, the breakthrough in liquid inks being the introduction of glass media rather than 'Ottawa sand' which, until the early 1960s, had been the only medium available. Oil inks, due to their viscosity, required further technical developments in mill design and the use of bead mills for their manufacture came considerably later.

History

The first stirred ball mills were basically, fixed vessels with an internal rotor, unlike the standard ball mill which rotated about its own axis. The grinding in the tumbling ball mill is that of impact and compression where one can only affect the efficiency by mill diameter and speed – its limit being that where centrifuging takes place. Grinding in the case of bead mills is by compression and shear where accelerations of more than 50 times that of gravity are normal.

The original open top 'attritor' type mills used slow speeds and large beads. The restriction of this type of mill was that there was a limit to energy input and hence grinding efficiency. These mills were followed by the open top Dupont sand mills using 'Ottawa sand' as the medium, which became a standard production unit in the paint industry until the early 1960s, when they were replaced firstly by mills using glass beads and then by a multiplicity of closed mills which are today's standard. These operate at high speeds using very small grinding media – as small as 200 μm as against a minimum size of 13 000 μm (1/2 in) in the tumbling ball mill.

There are many ways of grinding/dispersing solids in liquids. A particularly efficient means is by using the combination of the dissolver (impeller mixer) and bead mill, the first part of the process being batchwise, the bead mill continuous.

A dispersion produced from a mixture of liquids and solids needs a certain given energy input to produce the required quality. This total energy input can be reached by sharing the work between the dissolver and bead mill. However, the more efficient the dissolver, less is required by the bead milling process and vice versa. It is therefore important to achieve a high degree of predispersion both to maximize the bead mill and increase product quality. It also avoids the production of agglomerates within the premixing process itself.

Recent developments have now made it possible to avoid pre-dispersion by using a new direct dispersion process where the dry and liquid phase are fed directly to a bead mill and this process will be described later.

The following are types of bead mill, in chronological order, which have contributed towards its development.

Open-sieve mill (Dupont and Sussmeyer)

This was the original sand mill and was basically a vertical narrow cylinder with an inner shaft to which were attached agitating discs shaped as a ring supported by bars to the centre. Subsequent shapes of disc have been: flat cam type, a plate with inner holes, etc. Separation of the medium from the product was by a wedge wire screen at the top of the grinding pot, the medium being Ottawa sand. Peripheral speeds were in the range of 9–11 m/s. The major development in this type of mill was the introduction in 1962 of the use of glass media by Draisewerke GmbH in what became known as the Perl Mill.

The various types of media now in use will be discussed later.

The open sieve type of machine was mainly used in the pigment, dyestuff and paint industries and was manufactured in sizes of 1 litre to 1000 litres. It still remains in use today, mainly in the production of titanium dioxide and calcium carbonate based coatings. However, its drawbacks are that the viscosities of feedstock are limited due to bead lift and overflow in the sieve area. Thixotropy can also be a limiting factor as the lack of movement around the discharge area can result in a build-up of ground material. This could be moved by vibrators but they were not particularly successful.

Closed-sieve mill (Dupont and Draiswerke)

The design is basically as the open sieve type but with an enclosed housing around the sieve area and the addition of a scraper to clean the sieve discharge. This machine enabled products such as water-based pigment pastes, which tend to foam, to be handled as once the air had been purged, the mill operated in a totally closed condition. Viscosity limitations, however, appeared to hamper further development of this mill.

Gap separation mill (Draiswerke GmbH)

Developed by Herr Durr and Herr Engels of Draiswerke GmbH, this major development enabled higher viscosities to be handled without the discharge problems experienced with the sieve designs.

The gap separation consists of a rotating disc fixed to the agitator shaft and an annular ring fixed to the grinding pot. The device is adjusted so that there is only a few tenths of a millimetre between the ring and rotor which retains the medium but allows the product to be discharged through what basically is a self-cleaning gap.

With this development, and its ability to handle pastes, a change of medium was required due to the low specific weight of sand/glass relative to the paste viscosity. Steel beads (ball-bearing quality) were introduced, mainly 2 mm size being preferred. These easily destroyed the available agitator discs due to the turbulence they produced and a 'finger' design of disc was introduced. These discs were stellite-coated to resist wear and created a 'rise and fall' effect rather than turbulence. Such mills were first used for the production of rotary newsinks but then moved quickly into the cold- and heat-set inks field.

Limitations of this mill were mainly those of temperature. High energy inputs were necessary and the heat transfer through the only cooling surface, the pot wall, was limited. Blacks are not affected by the high temperature; in fact, dispersion is improved at temperatures of 80–90°C, but pigments such as yellow and some reds cause problems through changes in appearance.

John mill (Netzsch GmbH)
Developed by Willi John of Netzsch in the mid-1960s, for the manufacture of high quality paste inks, it came to prominence in the early 1970s because of its superior cooling system. Using an annular shaft gap consisting of an outer sleeve over the shaft, the gap between creating the separation zone, the John mill was able to cool the central shaft, increasing the cooling surface area by more than 30%. A further difference was the use of agitator pins in place of discs and counter pins in the pot wall creating increased shear. The relative increase in energy input required was countered by the ability to extract the heat more efficiently. This development enabled the heat-sensitive pigments to be milled.

Tex mill (Draiswerke GmbH)
This was the first totally closed grinding chamber, and the first step towards the development of what we now know as the high duty agitation mill (Fig. 13.17). By virtue of the closed chamber, the medium is forced to keep a 'mean distance' apart irrespective of other influential factors. The separation of the medium from the product is by means of two sieve-type cartridges which fit from the pot wall into the centre of the chamber and are therefore in the actual medium agitation zone, discharge being through the wedge wire cartridge into the outlet pipe. It is also necessary to use a double-acting mechanical seal on the shaft to avoid wear by the medium and also counter the inner chamber pressure which is developed. The benefits of the closed system are considerable:

- higher throughput
- ability to handle higher viscosities
- elimination of evaporation losses
- better environmental control
- full use of the grinding chamber volume
- foaming avoided
- possibility of vacuum operation.

Dyno mill (Willy A Bachofen Ag)
This machine changed the parameters of existing milling technology by incorporating a number of already known features into a totally new package (Fig. 13.18). Milling to this time had been based on vertical mills

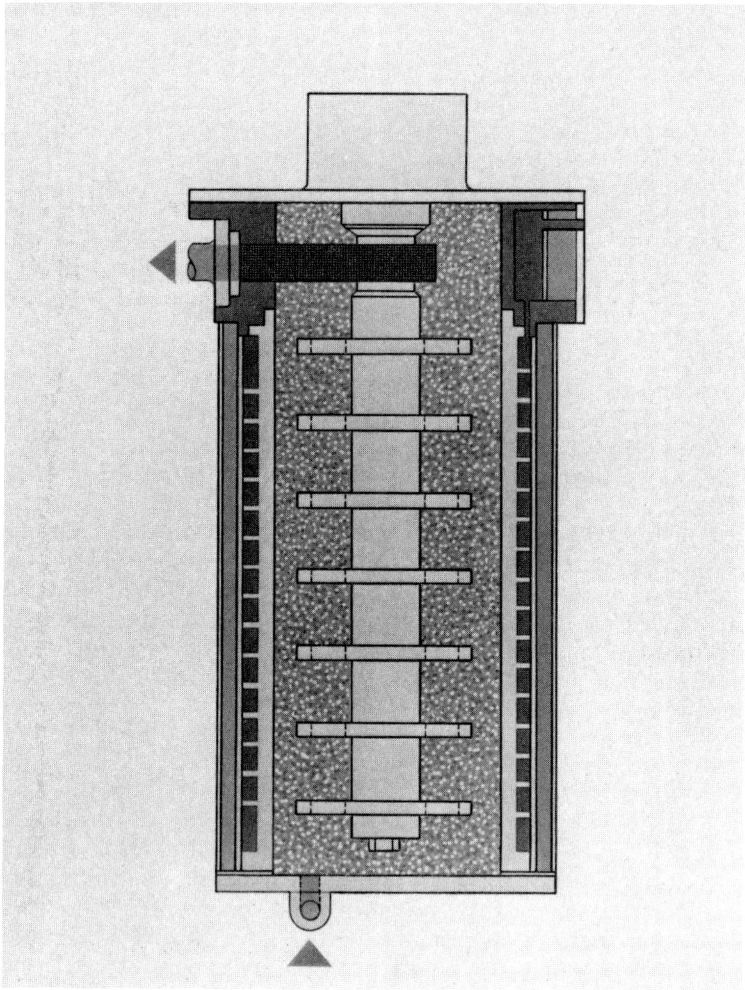

Fig. 13.17 Tex mill. (Courtesy of CIMACO Draiswerke GmbH)

with narrow grinding chambers and relatively slow peripheral speeds. The Dyno mill is a horizontal mill with a large diameter to length ratio and peripheral speeds ranging from 13 to 18 m/s. Separation is either by wall-mounted sieves or gap separation depending on viscosities and product need.

STS mill (Draiswerke GmbH)

The STS is a supercooled mill similar to the John system previously described but with considerably increased cooling area (Figs. 13.19–13.21). In this machine not only the shaft is cooled but also the discs are cooled enabling even more useful energy input for grinding. Although mainly used for paste inks, the STS can also be used for liquid ink

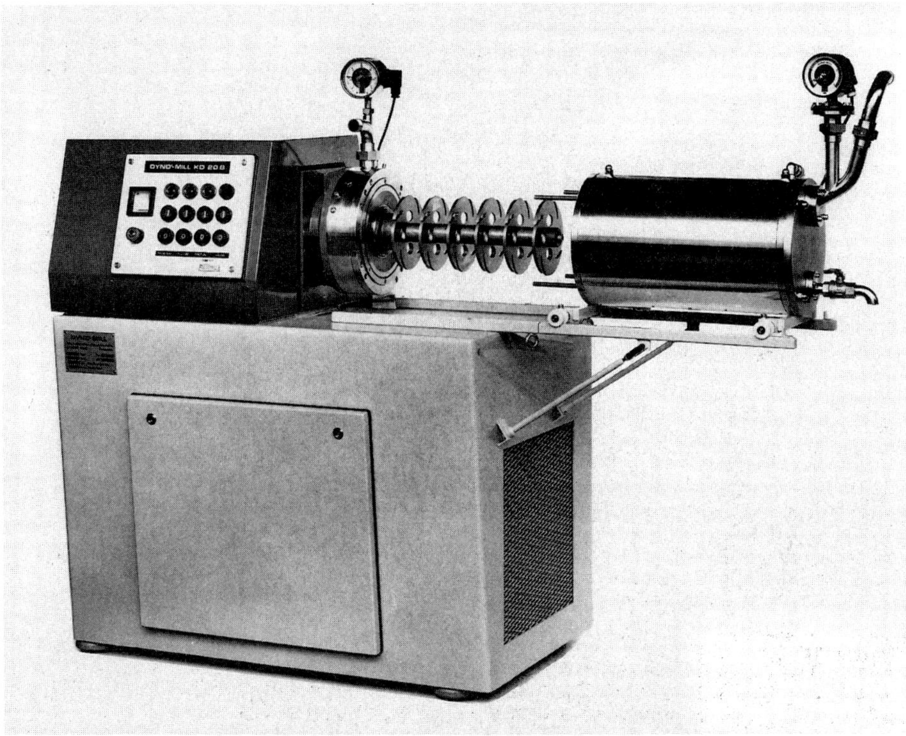

Fig. 13.18 Dyno mill. (Courtesy of Glen Creston Ltd)

production, particularly solvent based, but high input energies lead to high medium wear.

Electronically controlled Cobra mill (Buhler Brothers Ltd)
An addition to the bead mill scene is the Cobra mill, using a known system of variable bead charge which can be controlled during milling. This enables an easily ground product such as a white to be milled at a low bead loading and a subsequently more difficult pigment at the full loading without the need for medium addition.

Boa 251 mill (Buhler Brothers Ltd)
This is a mill with a horizontal ring chamber which can handle a wide range of viscosities, with simplified cleaning facilities (Fig. 13.22). For product re-production the settings can be preset to obtain batch to batch comparity and feature a variable-volume chamber, allowing effective bead change variation for optimum throughput.

Co-ball mill (Fryma Ag)
A novel approach to dispersion is the Co-ball mill which uses the ink plus grinding medium without any moving agitator. Though still in early development, it is finding acceptance in liquid dispersion. It still, how-

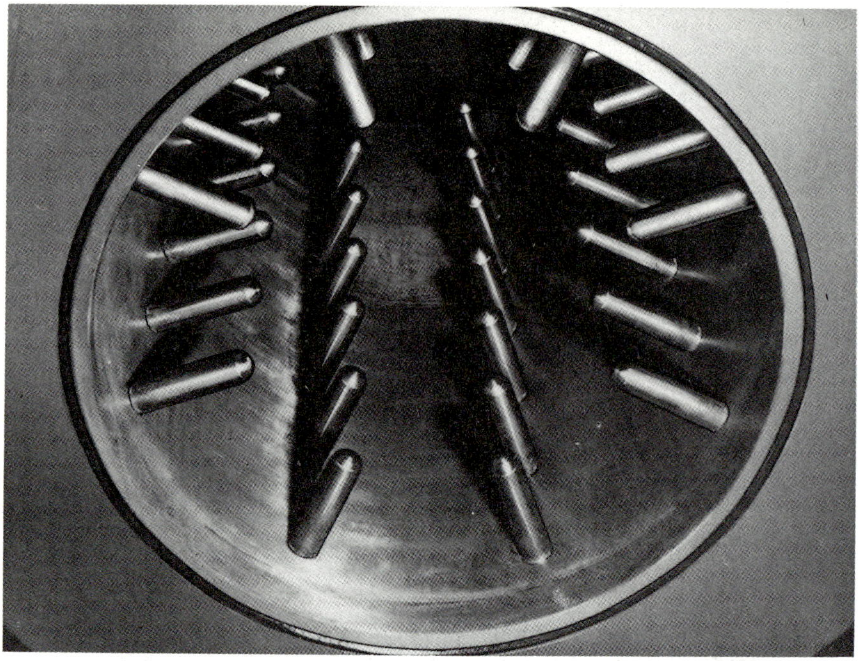

Fig. 13.19 STS chamber. (Courtesy of CIMACO/Draiswerke GmbH)

ever, has a number of problems to solve regarding flow, heat transfer, viscosity changes, etc.

There are many other types of mill on the market incorporating features first developed in the previously discussed mills but none have particularly contributed to the further enhancement of the dispersion technique, although they play a large part in ink production.

Table 13.2 Materials used as grinding medium in stirred ball mills (courtesy of CIMACO/Draiswerke GmbH).

Material		Density (g/cm^3)	Diameter (mm)
	Ottawa sand	2.8	0.4–0.8
	Hard glass, unleaded	2.5	0.3–3.5
Siliquarzite*			
	Hard glass, leaded	2.9	0.3–3.5
	Aluminium oxide	3.4	0.5–3.5
Draisal*			
	Zirconium silicate	3.8	0.3–1.5
	Zirconium oxide	5.4	0.5–3.5
	Steel shot	7.85	1.0–12.0

* Registered trade mark for DRAISWERKE.

Fig. 13.20 STS shaft. (Courtesy of CIMACO/Draiswerke GmbH)

Grinding media

The original medium, Ottawa sand, used in the Dupont mill is still in use today but to a very small extent. It has been replaced by glass and oxides for liquid inks and steel for paste inks. Table 13.2 is a list of the various media available and their densities.

Horizontal versus vertical mode

In the past few years, there has been considerable discussion as to the relative merits of using mills of horizontal or vertical design (Fig. 13.23). Before any direct comparison can be made, it must be based on like parameters and it is not possible to compare make X horizontal with make Y vertical where both have differing features, speeds, etc.

It is known that, to achieve the optimum conditions for grinding/dispersion, it is necessary to achieve a maximum of energy input while obtaining a state of bead flotation. This varies from product to product and also with temperature, surface area, behaviour under shear, etc. Some inks are particularly difficult to disperse and throughputs are of

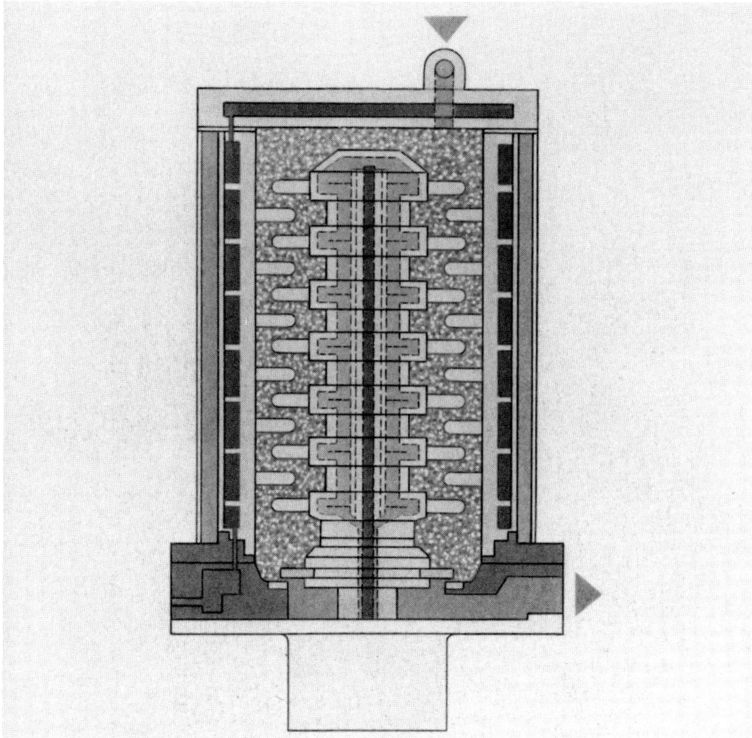

Fig. 13.21 STS sketchform. (Courtesy of CIMACO/Draiswerke GmbH)

necessity very low. In the vertical configuration it is not possible to lift the beads with such low throughputs and so the horizontal mode is superior. The converse is true with higher throughputs. However, there is no hard and fast rule which can be applied. When one refers to oil inks or higher viscosity liquid inks, then the horizontal mode is not recommended due to compaction of the medium and product at the outlet. This leads to flow reduction resulting in high temperatures and excessive wear. The decision, therefore, as to whether a machine should be horizontal or vertical is not just a matter of design or fashion but it should be related to achieving correct bead flotation in the mill, while processing a particular grade of ink.

A mill exists which is able to swivel through 90° during production in order to attain the most satisfactory mode relative to the product being processed. It has yet to realize its potential.

Direct dispersion

All the machines discussed so far are based on the standard two-stage process of pre-dispersion and subsequent bead milling. This process has drawbacks in that it is labour intensive, and requires high capital outlay. Research has therefore been conducted by a number of companies to

Fig. 13.22 BOA 251 mill. (Courtesy of Buhler Ltd)

realize a totally enclosed process where the solid and liquid phases can be fed to a single machine without the need for predispersion or premixing, and the required fineness of product achieved in a single pass.

The Direct Perl mill from Draiswerke GmbH was the first to reach this requirement and is now in use in the ink industry, mainly for the production of rotary and offset newsinks. This in due course may well spread to liquid inks with the introduction of coloured granular pigments.

A standard bead mill is used but with the addition of an integral screw feeder which ensures, by means of a positive sweeping device, that the unmixed materials are transferred from the feed hopper to the bead mill.

The feed screw itself does not undertake any dispersion, although a certain amount of pre-mixing and wetting occurs. A major problem was the avoidance of agglomerate formation. This was solved by the use of an annular nozzle system which sprays the liquid phase to the sides of the hopper while the solids are fed through the centre.

It is claimed that, as no pre-mixing is made, agglomerates which tend to be produced in standard dissolvers are not evident and therefore even higher production rates can be obtained than by conventional methods of production, namely dissolver/bead mill.

Fig. 13.23 Horizontal versus vertical. (Courtesy of Coates-Lorilleux Ltd)

Feeding of the two streams is made through a loss in weight feeding system controlled by computer giving accuracies of ±0.5%.

Bead mill/roller mill – paste inks

Bead milling is the random disintegration of primary particles by the use of a retained grinding charge the important word being 'random'. It cannot be guaranteed that all particles will be subjected to shear within the mill and therefore the process is liable to produce a 'tail' of stray large particles. It is also not a self-sieving device so that these particles will be discharged with the majority of product which is already ground to the required standard. Any air which is also in the feed will also be finely dispersed within the product.

Bead milling also produces strong boundary line attraction between the ground particles which in itself restricts flow of the ground product. In the early days of bead milling, much work was necessary to formulate inks which would retain flow after passing through the bead mill. Even so, it is not always possible to achieve the required rheology in spite of formulation changes.

The three-roll machine is still widely used after beadmilling to 'finish' or polish the ink. Basically, it breaks down the molecule attraction boundaries by rolling shear and not acting simply as a deaerator. In fact, ink passed over a three-roll mill may have more entrained air than after bead milling. Developments to date have still not been able to eliminate the use of the three-roller.

Conclusion
Modern high-speed bead mills are very versatile devices. The number of operating parameters and the variations possible allow for optimized grinding processes over a wide range.

The requirement for finer grinding and higher quality inks will demand even more from the bead mill which will inevitably lead to further developments of the process. The full potential of the bead mill, however, has still not been fully developed.

Ball milling

This has diminished considerably in the last few years, but for reference and history, as they are still common in other parts of the world, we recommend the section on page 327 of the third edition of the *Printing Ink Manual* published in 1979, which is précised here.

Advantages of ball mills:

- after installation cheap to run as they run through the night on low tariff electricity;
- they act as mixer as well as mill;
- low maintenance costs.

Disadvantages of ball mills:

- cumbersome and bulky;
- noisy;
- only 50–60% of the volume can be used for the batch or the cascading is influenced detrimentally;
- cannot be speeded up as demand increases as there is a set production-time cycle;
- emptying can be tedious particularly if the product has any thixotropic tendencies;
- can only handle low viscosities;

13.5 HANDLING, STORAGE AND MANUFACTURE OF UV INKS

It should be understood that the manufacture of UV inks and lacquers is an entirely different proposition from making the equivalent product using conventional materials. In fact this type of manufacture is probably more closely related to the pharmaceutical industry where cleanliness and handling care are of paramount importance. These comments apply not only to the way the components are put together to make the final product, but also to the way the ingredients are stored at controlled temperatures. Skin contact is to be avoided as certain of these materials are complex chemicals which, although very carefully selected, have been known to cause irritation to some people. The materials used are also considerably more expensive than conventional materials and as a result justify the extra care required.

Since these materials are generally poor wetters of the dry phase of the ink, the mixing procedures are critical for satisfactory dispersion. They do not differ in principle from conventional operations but enforced cooling is a prerequisite to avoid the temperature of the mix exceeding 55–60°C. Certain of the ingredients can be crystalline so these have to be added at a crucial temperature during manufacture or the product seeds.

Milling is best achieved on modern three-roll mills with comprehensive temperature control on the individual rolls which means that they can be warmed or cooled individually or *en bloc,* as required. This is an essential requirement as certain UV products are not enhanced by the shock-cooling which can occur during the start up period of milling. Handling of the finished product is less sensitive but it must be stored in opaque containers and not exposed to extremes of temperature change or sunlight.

13.6 MANUFACTURE OF NEWSPAPER INKS

Newspaper inks lend themselves to automated manufacture primarily because large volumes of standard products are required and secondly because a relatively few basic materials are involved in their formulation.

Nevertheless, because of the critical nature of newspaper manufacture, these inks must be of high and consistent quality to perform on long runs on high-speed presses. These inks are mainly cold-set web-offset. Reference to Chapters 5 and 6 will show that these are mainly produced from mineral oils modified by small additions of performance-enhancing varnishes with carbon as the pigment. Additives include wetting agents and anti-misting additives such as clays and extenders, while oleoresinous varnishes are added to cold-set litho inks to improve rub and lithographic properties.

Furnace black is normally used in newsink manufacture in the form of oil and water-wet beads.

The advantage of the beaded grades over the fluffy material is that they are cleaner to handle and feed more easily in bulk plants. The beaded form is also more compact and because the specific gravity is almost twice that of the fluffy grade (hence less storage space), is easier to handle in bulk. Beaded carbon is supplied in bulk tote bins, big bags and tanker vehicles for filling static silos, whereas fluffy is only supplied in 12.5 kg sacks. The disadvantage of beaded carbon is that the bead is more difficult to disperse and small seeds of cokey material form a residue which is unacceptable even in low concentrations (below 0.1%) in the finished ink.

The handling of carbon in any quantity can give rise to significant dust levels in the workplace, and manual handling of bags should be carried out under effective extraction; even under these conditions it is almost impossible to prevent some contamination, particularly with fluffy grades. Most modern bulk plants (Fig. 13.24) incorporate some sort of mechnical bag slitter, these machines not only open the bags efficiently but also compact the empty bags and automatically wrap the scrap in polythene film. In many plants without this facility the most significant carbon contamination arises from the handling and disposal of the empty paper sacks. Once the sacks have been opened the carbon is processed in a totally enclosed system.

The main problem in bulk handling fluffy carbon is the difficulty of getting the product to flow from bulk containers, and it is primarily for this reason that it is supplied in 12.5 kg sacks. Even when the sacks have been emptied into a suitable hopper it is difficult to transport the powder into the mixer or mill. Without some form of mechanical assistance fluffy carbon will block pipes and valves and will feed irregularly in the system. Several mechanical techniques have, however, proved quite successful. The best of these is fluidization, which is achieved by blowing low pressure air into the carbon. The air separates the light carbon particles and the resulting fluidized carbon will flow through the system satisfactorily. The air is diffused into the system through sintered pads, usually placed at the base of all hoppers and cones in the plant where blockages are likely to occur.

The use of live-bottom hoppers is also a satisfactory technique in keeping fluffy carbon mobile. This process involves vibrating the base of hoppers or cones at high frequency, preventing a build-up of carbon at these points, by agitating and thus separating the carbon particles.

Other mechanical means of agitating the carbon include mechanically driven rotating paddles of various designs located at the base of the hoppers.

Screw feeding has been used for fluffy carbon but has two major disadvantages. First, because there is a gap between the screw and the feeder body, the carbon tends to be compacted against the vessel wall. This compacted carbon will flake off in the screw feeder and the resulting material is more difficult to disperse. Secondly, the screw itself tends to aerate the fluffy carbon which, if the feed is needed to lift the carbon through the system, limits the angle of elevation of the screw feed by the tendency of the carbon to flow back through the screw.

The most satisfactory method of transporting fluffy carbon through the system is by pneumatic transport. The carbon is blown through smooth-bore PVC pipes using pulsed air supplied from a Roots blower. The carbon is carried to elevated day bins by the air which then escapes from the system through reverse air jet self cleaning filters mounted on top of the day bins and storage silos. A transfer rate of over 5 tonnes/hour can be achieved easily.

In modern plants the flow of carbon is controlled by microprocessors. The day bins and storage silos are filled automatically and the level of carbon in these vessels is monitored by level probes.

The above methods of handling fluffy carbon are equally applicable for handling beaded product. However, because of its higher density and because of the bead form, this product flows perfectly well through the system without fluidization. The screw feed method is quite effective for this grade of material.

Oil handling

For economy, oils are usually purchased in bulk from the refiners and delivered in tankers of up to 20 tonnes capacity. The oils are supplied in a

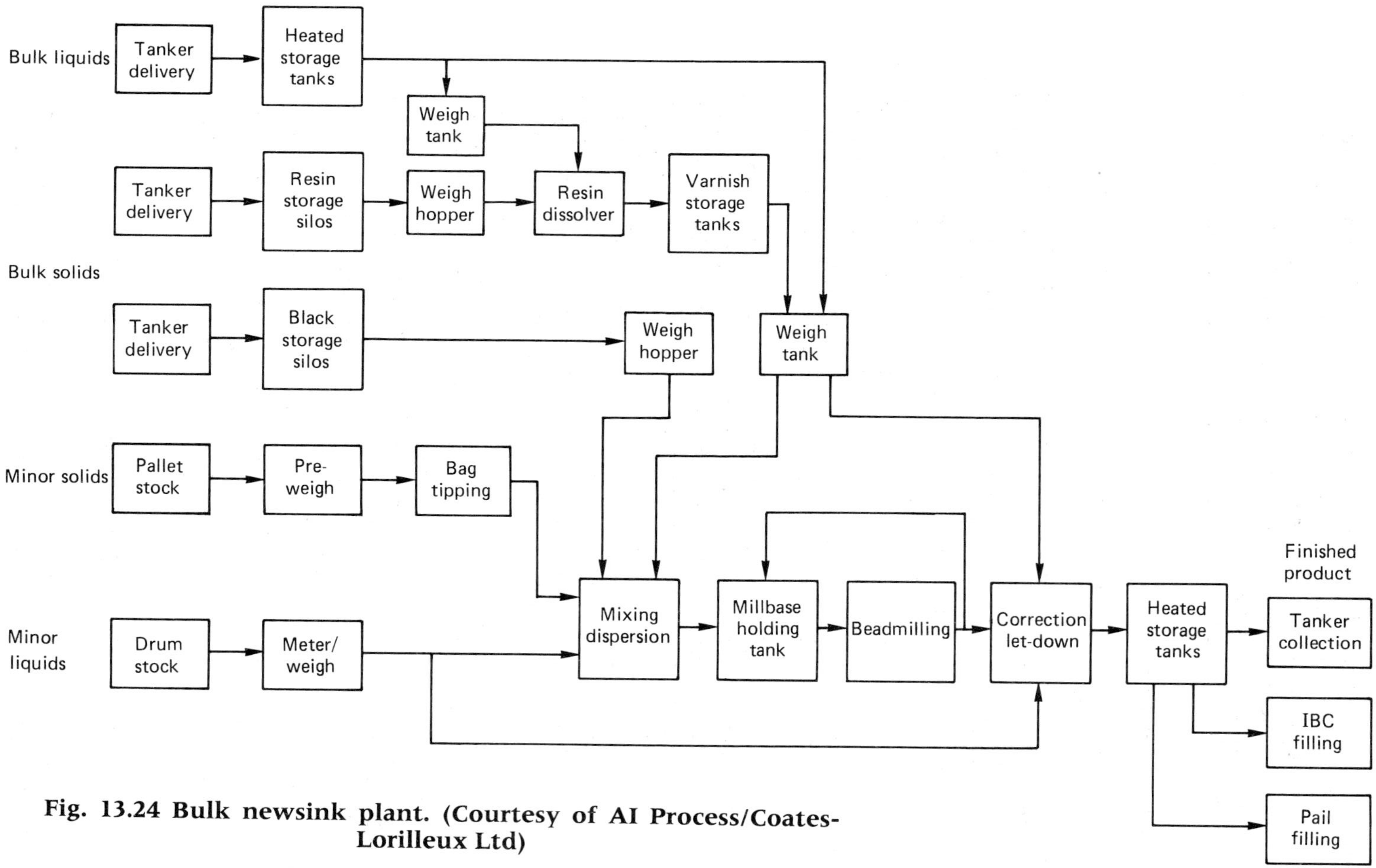

Fig. 13.24 Bulk newsink plant. (Courtesy of AI Process/Coates-Lorilleux Ltd)

range of viscosities but those in the range up to 10 (1 Pa s) poise are the most widely used. The heavy newsink oils are pumped into large holding tanks. These tanks should be sited as near to the production area as possible to facilitate ease of pumping. The holding tanks for the heavy oils need to be trace-heated by steam coils or immersion heaters in order to keep them within a practical viscosity range for pumping and metering.

Production method

Bulk production can be undertaken by either batch process or by continuous process.

In the batch system the process oils and varnishes are metered into a large mixing vessel. While these are being metered the carbon is weighed in a weigh hopper usually situated above the mixers. In practice half the oil is metered into the mixer together with the varnishes, wetting agents and extenders, then the preweighed carbon is dosed into the mixer at a controlled rate, together with the remainder of the oil. Controlling the rate of carbon addition at a speed which is compatible with the ability of the mixer to incorporate the pigment into the oil is important as it gives more efficient predispersion and, particularly with fluffy carbon, takes advantage of the fact that the carbon particles have already been separated by the fluidizing air.

The mixers used are usually high shear mixers or of the cavitator type, fitted with additional slow speed stirrers to increase mixing efficiency. When the predispersion is complete the premix is processed by high dispersion bead mills, using 2 mm steel beads as the grinding medium. The resulting ink is filtered both mechanically and magnetically before being pumped into bulk ink storage tanks. The magnetic filters remove any ferrous particles which may be generated in the bead mill, while the mechanical filters remove any cokey undispersed carbon particles. This attention to filtering is necessary because even low concentrations of undispersed matter can cause press problems due to the high volume of ink which is used and stored in press installations over long periods in enclosed systems. Web-offset inks are generally stored in heated and lagged tanks and are kept stirred to break down the natural thixotrophy of the system.

Continuous production is more easy to achieve with beaded carbon due to the relative ease by which the carbon feed can be controlled. The granular material can be fed by a screw-feed mechanism directly into the mill. Oil is fed by means of a loss-in-weight system giving a controlled and consistent flow of product together with the dry pigment. The loss-in-weight system is a microchip system constantly monitoring the weight of the hopper, in which the oil is held, together with the control of the carbon flow.

The manufacture of cold-set colours is very similar to that of heat-set (Fig. 13.6) and it can be from either source, but is mainly 13.7)

13.7 HANDLING AND STORAGE OF INKS

The adoption of bulk handling or tanking makes it possible to ensure better use of available space and appropriate controls on ink stocks. If the tanks are linked with an in-house blending scheme then the whole process is subject to a degree of simplification because it is possible to operate from a reduced number of materials, in greater bulk and with less frequent deliveries. The inkmaker has the ability to control or to make up special colours when they are needed.

There are a number of things which have to be considered when deciding whether tanking is an appropriate method to adopt. In storage, there can be a gradual decline in quality as pigments flocculate, settle out, or the binder systems can separate. It is therefore recommended that the tank size be related to approximately 2 months' consumption. This ensures that stocks are turned over in a tank at least six times a year (in the case of newspapers, etc. this is measured in days). Furthermore, agitators need to be installed which are time controlled to give periodic stirring. With regard to the movement of liquid materials, experience has shown that the rectangular tanks do promote better agitation in bulk, without undue turbulence.

The tanks, having been equipped with automated agitation, have to be provided with a means of charging and discharging, level control, venting and sometimes temperature controlling. Changes in legislation have led to the simple venting valve on the top of each tank being scrapped and fixed piping conducted to external atmosphere so that the factory is free of solvent vapours (Fig. 13.25).

A well-designed tanking system will make maximum utilization of the available space. It will also provide a very much safer and cleaner working environment. Finished inks are blended by mixing the single colour bases held in the tanks. In the case of lower viscosity bases dispensing is by means of gravity. As the viscosity increases it becomes increasingly necessary to provide some other force. An air driven, low geared or diaphragm pump is the ideal method of providing that extra pressure behind the fluid to ensure smooth and adequate dispensing rates.

With a driven delivery system, it is possible to incorporate the dispensing valves from individual tanks into automated dispensing stations. The valves are mounted on a frame which is powered backwards and forwards to bring each individual nozzle, as required, into the appropriate dispensing position over the centre of the receiving tank. Normally, these heads would be arranged over dormant mounted scales (Fig. 13.26).

The sophistication of the system can vary from punch card control to computerized controls.

The agitators which would normally be used for mixing the blend are powered by electric motors. Air-operated agitators are used, but the electric agitators are very much more practical, not only for periodic timings, switching on and off automatically, but also for overcoming the varying forces met by the impellers as the level of the material in the tank drops. A constant speed is maintained with an electric motor, whereas the 'revs' of air-driven versions can vary appreciably.

Holding the materials in bulk necessitates efficient stock control. Using steel tanks, an effective form of level indicator is required. This is one area in which equipment has gone through quite dramatic changes, recently. The development has been a continual process. The preferred method now is the ball-float linked to an external indicator needle and scale. This

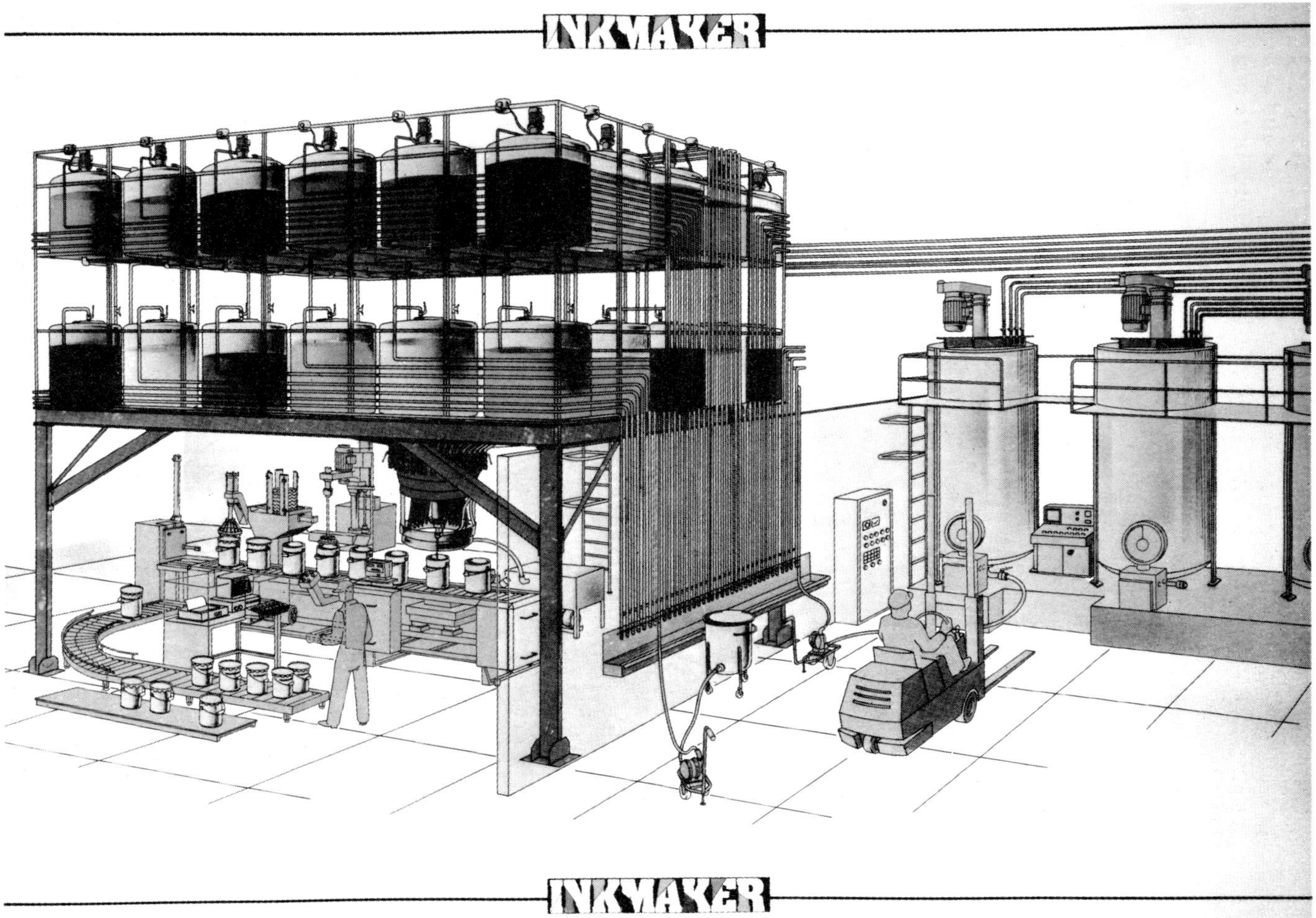

Fig. 13.25 Typical liquid ink storage. (Courtesy of Inkmaker/Coates-Lorilleux Ltd)

Fig. 13.26 Flexible dispersing end for filling varying container heights. (Courtesy of Fullbrook Systems Ltd)

mechanism has been subjected to very rigorous testing on the part of certain of the EC countries, and has been given safety certificates for its ability to resist ingress of solvent and hazardous or corrosive materials. It is now shown to be a very safe application. By using this system of content indicator, it is possible to link it with end-contact switches. These come into play when filling the tanks, prevent accidents caused by overfilling and, at the opposite end of the scale, they will switch off the agitators when the level of the contents is too low. The latest thinking on this matter is the Weka gauge which is electronically operated by pressure. The selection of the pipework, the fittings, the glands, is a specialist operation (Fig. 13.27).

Charging the tanks can be done in one of two ways, either with an external barrel pump which will be linked to the coupling pipes at the base of the tank or with slow-moving geared ink transport pumps. With appropriate valving these units double as dispensing and filling pumps, thus simplifying the system along with the introduction of pouring valves with a two-stage flow rate. These flow rates can be controlled to such accuracy as to permit the delivery of as little as 75 g per charge. These valves are PTFE seated right at the exit point of the nozzle. Conical seating obviates the problems that would otherwise occur with ink drying into the valve thus causing clogging, jamming and leakages. The unit has been developed precisely to overcome those problems. The valves are spring loaded so that it is a fail-safe system, whether it is a hand or powered

Fig. 13.27 Tanking system filling end. (Courtesy of Coates-Lorilleux Ltd)

operation. The valves, hand or electronically operated, can be coupled to the transport pumps giving maximum safety and avoiding blow-backs.

The application of automated weighing does minimize waste in that the accuracy is guaranteed if one is using an electronic data processor or a punch card-controlled system. The accuracy of the dispensing is such that shading is normally unnecessary.

What are the economies available from bulk tanking? The obvious one is the limitation of stocks to a number of singly pigmented toners for blending, or finished inks sold in bulk. This is a very real asset to the ink manufacturer who can plan his production so much more effectively. The fact that there is no need to fill shelves with special shades surplus to requirement, thus tying up vast quantities of money, has to be another effective way of making economies.

Much of the above is also true for paste inks, but here one is dealing with materials of a much higher viscosity, which may very well flow quite adequately just after manufacture with the internal heat processing has produced, but when stored at ambient temperature in the UK between say −5°C and +30°C the rate of flow is slow. The methods to move such products at 250 poises/20°C are gravity, or gear pumps which are accurate and instant, or air-driven displacement pumps ideally suited for 200-litre straight-sided drums, and finally and more sophisticatedly, hydraulic pressure, for example Schwerdtel (Fig. 13.28). The air or hydraulically driven following plate method is now commonplace. In the higher viscosity ranges if distances are to be covered using pipework the bore and jointing are of great significance and must be calculated very accurately, with the same care being given to the driving (moving) force to cause the product to be mobile through small pipes and valves. Thorough heating of these pipes is essential to ensure mobility, particularly if any of this pipework is outside.

These storage systems can be linked to a dispense system, which has been briefly referred to, so it is important to know what a dispense system is. Essentially this is a system designed to dispense blends of inks from a range of single-pigmented bases, varnishes and solvents.

The main elements of a dispense system are:

- a range of storage containers, either tanks or barrels;
- a central dispense point (the dispense head);
- a system of pipes (and sometimes pumps) to deliver ink to the dispense head;
- a control system, either manual or automatic;
- a balance to weigh the components.

Tanked systems

For volumes greater than 100 tonnes per annum the larger tank system is normally more advantageous. Here the system components are stored in

Fig. 13.28 Canning line. (Courtesy of CIMACO/Ludwig Schwerdtel)

tanks which are sized according to need, from 500 to 5000 l capacity for large usage items.

These systems can be either pumped to dispense or (which is preferable) gravity fed. As for barrel systems they can be manual or automatic although automatic is by far the most common.

Advantages of tank systems

- Greater storage capacity means less handling, higher efficiency;
- Storage in tanks allows for agitation of the product, maintains homogeneity;
- Higher number of components increases system flexibility;
- Opportunity to use 'gravity' dispense, greater accuracy, enhanced reliability;
- Can incorporate automatic pails, handling devices, higher efficiency, reduced labour costs.

Why install a dispense system?

In house

The manufacture of printing ink is a service industry and is judged by both the quality of the product and the efficiency of the service. Increasingly, the production of a high quality product is asked for at very short notice, on occasion even same day delivery. This of course places great demands on both production and technical personnel to meet these targets with traditional batch manufacturing techniques.

Batch manufacture

It is necessary with batch manufacture from base components to concentrate the quality control effort at the post-milling or mixing operation stage when the finished product is ready to be potted into the delivery container. Because the formulation has usually been blended by a manual method it is normal to make adjustments/additions at this stage. Experience has shown that in some cases the incidence of adjustment can be as high as 60% of all batches produced. All this effort is often taking place against a deadline for delivery and if things go seriously wrong it could mean that the order is delivered late.

After the batch has been produced there is also some effort required to clean containers and machinery in order to begin the next process, all of which is labour intensive and time consuming.

Effective dispensing is only achieved if a high standard of repeatability is maintained, the gravity system offers the greatest opportunity to control these factors through its inherent simplicity.

Characteristics of solvent- and water-based systems

The major differences for the two applications are in the materials of construction and the degree of electrical protection required.

For a typical water-based system:

- tanks or barrels in plastic (HDPE)
- valves in stainless steel or PVC
- pipework in PVC or ABS
- electrical equipment protection normally IP55 for the UK
- pumps in polypropylene.

For a typical solvent-based system:

- tanks or barrels in stainless or carbon steel
- valves in stainless steel or PTFE
- pipework in stainless steel
- electrical equipment normally 'explosion proof'
- pumps in cast iron.

Dispense valve operation is by compressed air which is compatible to both applications.

Gravity versus pumped systems

This is the other option. Due to operational experience with dispense systems it has been possible to evaluate the merits of both methods of delivery.

Pumped system

Where pumps are utilized to dispense ink it is essential that frequency variator/pressure variator control devices are fitted to adjust both pump speed and pressure. This is necessary to maintain the system accuracy.

The addition of this equipment increases the overall complexity of the system which then requires a greater maintenance effort.

Where diaphragm pumps are used it is also normal to fit pulse-dampers to reduce the variance in flow inherent with this type of equipment.

Gravity system

Here pumps are used only to load the tank initially and require no special control devices. Ink is then dispensed by gravity at atmospheric pressure which remains constant.

Flow is determined by calculating the correct pipe diameter and the manometric head of the storage vessel.

Other benefits

- Dispensing directly into the delivery container eliminates container cleaning and potting operations (increases efficiency, reduces labour).
- Reduced stock-holding through greater flexibility.
- Rapid manufacture, e.g. typically a 20 kg batch of four components would take approximately 4–5 minutes to dispense (200–300 kgs/hr).

Systems installed for printers

Systems have now been installed for printers by inkmakers. For the printer there are many advantages to having a dispense system on-site.

- No delay in delivery – even with in-house systems at inkmaker's own plants, delays can be caused by vehicle failure, weather, distance, etc.

- Elimination of over-orders – it is common for printers to order in excess of what is actually required.
- Re-use of press-returns – many systems are now operated along with a colour computer and contain tanks specifically for the collection of press-returns. This feature can be of tremendous benefit to the printer both economically and environmentally.

Ink packaging

During the last 20 years, considerable strides have been made in packaging and ink supply to the printer.

For lithographic inks, slip-lid tins of various sizes, particularly those capable of holding 1 and 3 kg of ink, are still widely used. However, with the advent of larger web presses, 10 or 15 kg capacity polythene buckets are now widely used. For still easier handling, 200 kg straight-sided drums can be used, where air pumps with a follower plate, such as the Graco pump, are installed on press lines. In cases where the ink consumption justifies it, such as on the larger heat-set web-offset presses, supply by use of 800 kg or 1 tonne drop tanks of each colour is becoming more widespread.

The main change in oil-ink packaging was the appearance of the vacuum tin developed by Gustav Gruss of Dusseldorf. It was of three-part construction. Michael Huber of Munich carried out trials using the original can and although these were accepted by printers, the leakage rate of 0.5% made the system unreliable. It was evident, however, that the can had many advantages over the standard slip-lid can such as saving, storage space, easy emptying, no necessity for the inner paper seal, etc. Investment for a single deep-drawn can, however, was too high for a single ink outlet and so Huber were joined by Kast & Ehinger and the can we now know was produced by Gruss in 1970.

Paste-ink filling lines were already being made by Schwerdtel GmbH of Munich (Fig. 13.28), but no machine was available to seal the vacuum can automatically: this Gruss built themselves. It became evident, as the use of the vacuum can grew, that a specialist machine company was needed to take the development of this vacuum closing unit further and in 1973/4 Schwerdtel made the first such unit, the S10-ZV.

Since that time, the vacuum can has become a standard in Europe, but still has to be accepted in the USA.

There is a growing trend to supply the larger flexographic and gravure printers with a limited range of concentrates, supplied in 205-litre drums, which are pumped into individual tanks for use in blends as required. For smaller quantities, polyethylene containers, of various sizes, can be used for water-based inks but the generation of static electricity and the latest packaging regulations makes them totally unsuitable for solvent-based inks and sundry items. Steel cans and drums with an internal coating of lacquer are most frequently used for the latter type of ink.

Bulk cold-set web-offset blacks and colours are usually delivered in custom-built tanker vehicles. Tankers for offset inks need to have the facility to be heated and insulated to facilitate efficient pumping at the customer's premises, owing to the higher viscosity involved. Attention should be paid to the type of pumps and the design of the tank bottom or pumping problems may occur, particularly in winter or on longer journeys.

13.8 MODERN PRODUCTION TRENDS

Computerization

The computer has now become an established tool in ink manufacture. The routine procedures of preparing manufacturing instructions, progressing of orders, preparation of labels, stock control, etc. are carried out at high speed and high efficiency. Furthermore, the recent advances in electronic, computer and microchip technology are changing the face of ink manufacture. It is now theoretically possible to design a completely automated inkmaking factory in any sector, from gravure through to screen process (whether this concept would, in the short term, be cost effective remains to be studied, but in the long term it looks very attractive). This has been made possible not just because of the increasingly greater scope of computers and their on-line electronic attachments, but because these previously cumbersome installations have been streamlined, to the point where the most complex electronics are pocket-sized. Such units can be used to actuate valves, monitor the weighing (whether mechanical or via load-cells) and flow-rates of liquid products. (Fig. 13.29 for paste products and Fig. 13.30 for liquid.)

Costs of production and related subjects

In manufacturing operations it is essential to know the costs of labour, be it direct or indirect, power and utilities, rent, rates, maintenance costs, depreciation costs of plant and machinery, etc. because the product being made has to bear an apportionment of all these costs in relation to its real cost of manufacture so that they can be recovered with a realistic selling price being constructed. One of the major problems facing the ink industry is the pricing of small batches of ink which rarely bear any relation to their real costs and can be more expensive in a production sense than batches many times their size.

Maintenance strategy in the printing ink industry

A maintenance strategy, where this exists, usually embraces one of a number of separate philosophies. There are those who consider maintenance to be the repair or replacement of items which have failed, a necessary evil and an undesirable overhead, and consequently such maintenance is frequently given a low priority within the management strata. In some organizations, sufficient spare capacity is available and a failure may not adversely affect production. In this case the 'on failure' approach may appear to be adequate. This is not often the case, however, as most companies have to operate as near to maximum plant utilization as possible and a prolonged failure normally has serious consequences for production and profitability.

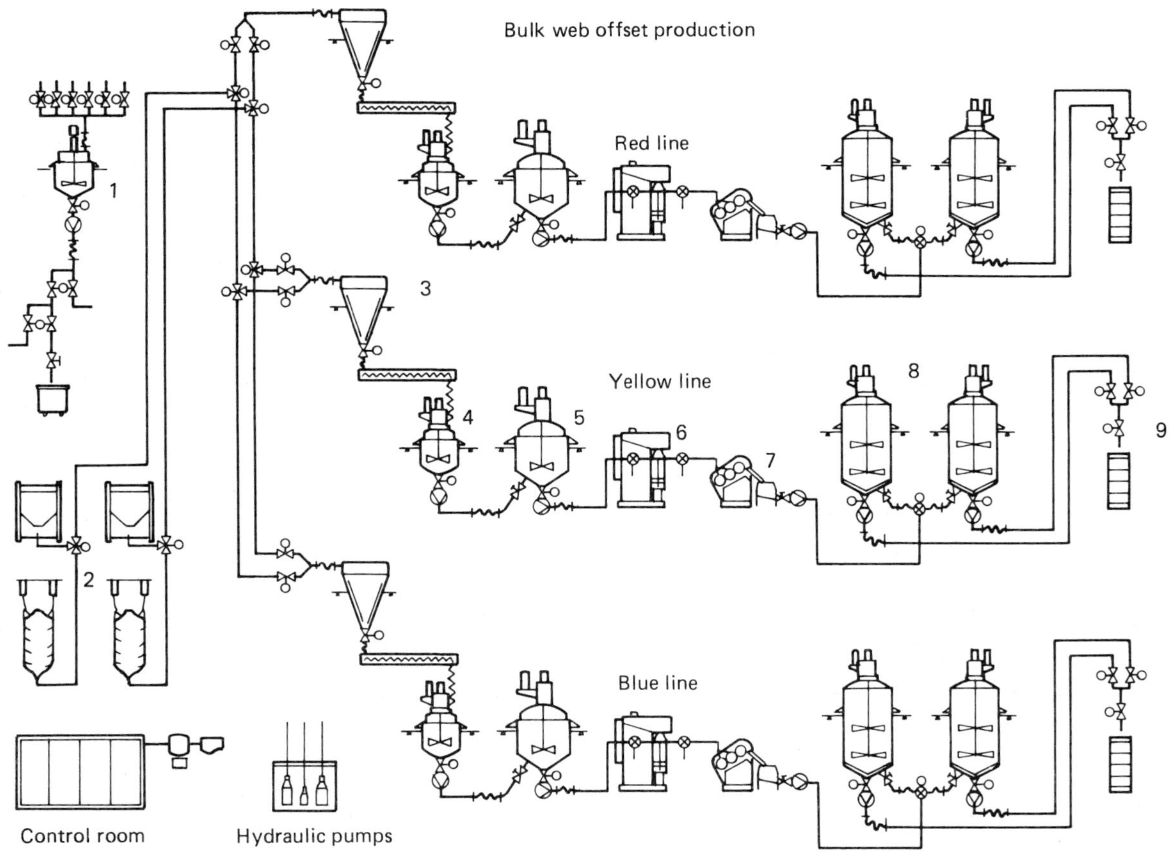

Fig. 13.29 Bulk web-offset production: 1 pre-blending and weighing of liquid components; 2 bulk pigment infeed from IBCs or big bag; 3 pre-weigh and metering of pigments; 4 premixing tank; 5 holding tank; 6 COBRA 501 bead mill, milling and dispersion of pigment; 7 SDVE 1300 triple-roll mill to finish product; 8 let-down tanks; 9 final product discharge to filling line. (Courtesy of Gebruder Schmidt)

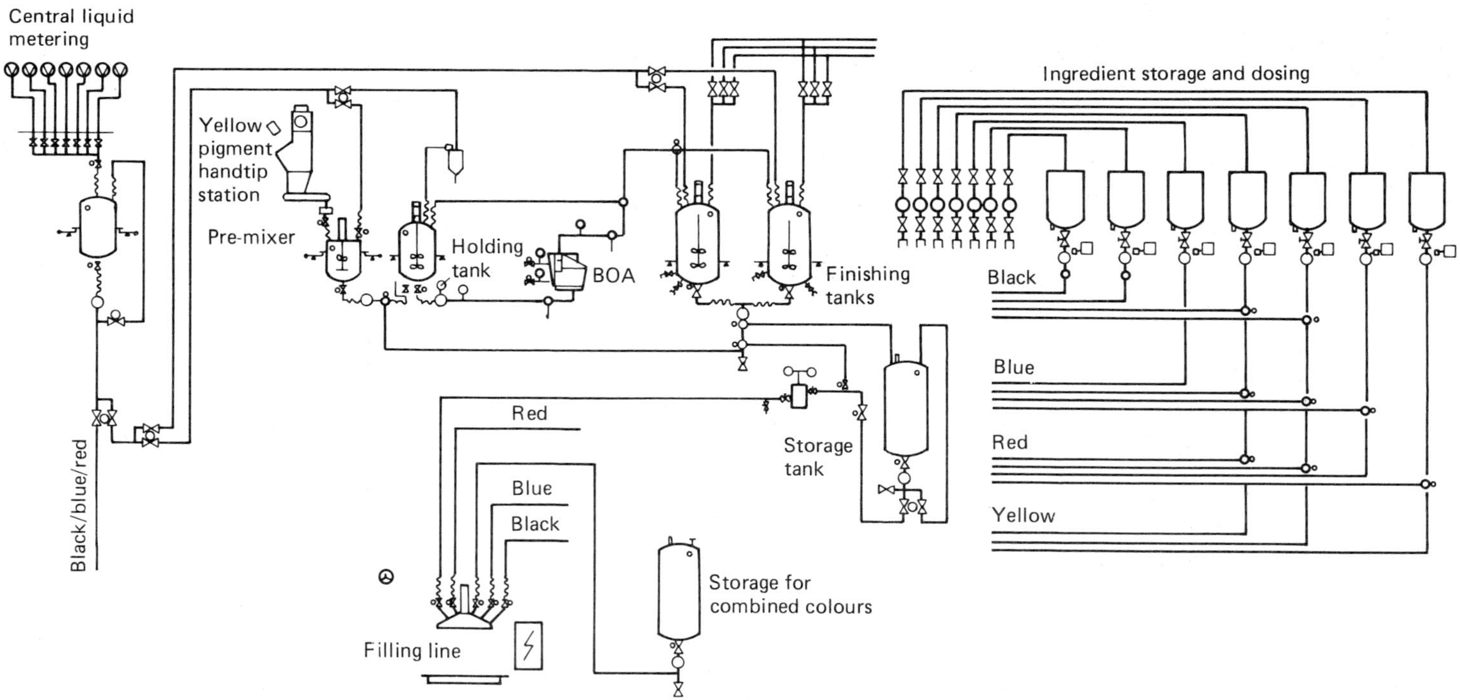

Fig. 13.30 Publication gravure ink production. (Courtesy of Buhler Ltd)

13.9 THE FUTURE

A scheme for a modern ink plant designed for the automatic production of publication gravure and heat-set inks gives an example of current thinking about production and indicates the trend towards automation and computerized control of ink manufacture. This is self-generating as the computer gives more information, the information creates new opportunities to improve the efficiency.

Plant control system

A modern plant control system ensures smooth operation of the production plant. The control system is designed so that each production line is allocated its own field on the panel.

The formulae are stored on disc. Each line is equipped with a reader and a printer for recording the actual weights, batch sizes, production sequence, bead-mill data, and alarms. Digital displays continuously show information on the formula number, the gravimetric proportions of ingredients and the major data concerning the bead mills.

Further plant features

Great importance has been attached to safety and environmental compatibility:

- the entire electrical system is of explosion-proof design;
- clean pigment intake system with the application of fabric dust collection filters;
- entirely enclosed system, hence no loss of solvent vapours;
- gas displacement device with flame arresters on all production vessels; exchange of air with atmosphere through activated carbon filters.

Other developments

It is difficult to see any further major developments in the basic bead mill other than continued refinement. The next steps could well be towards the direction of the Co-ball type mills where only the medium and product are in movement but many problems have yet to be solved to produce a universal machine, suitable for wide viscosity ranges. Ultrasonics have been tried and tested, as yet without success, and technology such as the use of laser beams is in the very early stages.

In the next few years the more efficient handling and automatic control of inkmaking will be normal. It will no longer be possible or viable to produce small batches without a blend scheme as the reduction of costs and delay become paramount in the constant push to state-of-the-art thinking.

CHAPTER 14
Rheology of printing inks

Rheology is the study of the flow and deformation of matter. The aim of the science is to define and evaluate such words as consistency, tack, length of flow, stiffness and body, in terms of the physical properties of the materials. These words are used in everyday language and with special significance in many sections of industry. At first sight such terms may appear to be self-explanatory and simple to define; in fact the converse is true and the greater the need for precision the greater is the complexity of separating the factors involved.

In the printing industry there are two main areas of interest for the rheologist.

- Lithographic and letterpress inks where the ink is required to be ready for the press without addition or modification.
- Gravure and flexographic inks where the inkmaker supplies ink at a greater viscosity than that required for use on the press and the printer accepts the responsibility for adjustments in viscosity to satisfy his printing conditions.

In lithographic and letterpress inks the need for precise evaluation on numerical scales arises because of two factors which are diametrically opposed. For good printing a lithographic ink should be as stiff as possible but when a printing machine is run at high speed the stiffness of the ink must be sufficiently low to avoid the paper surface being torn away from the body of the paper (i.e. it will not pluck) and to avoid the generation of too much heat on the machine.

With flexographic and gravure inks there are two reasons for the inkmaker to supply ink at a viscosity much higher than required by the press conditions. First, it overcomes the tendency of the pigment to settle to the bottom of the container used for transport. Secondly, it allows the printer to thin the inks to a viscosity suitable for the press and running speed and enables him to adjust the balance of solvents to achieve the desired solvent release and drying rate.

Most laboratory workers concerned with printing ink do not require a specialist knowledge of rheology. They will be better served if they acquire a more general appreciation of this subject and refer problems to the specialist for explanations when the need arises. The depth of knowledge required may be stated as follows.

- In branch operations where the laboratory worker is called upon to mix inks to pattern, an understanding of 'tack' and an elementary appreciation of flow should suffice.
- The chemist responsible for ink development should understand tack, Newtonian and non-Newtonian aspects of flow and gain experience of how these change in relation to the choice of pigment, pigment concentration, the ink media and their composition, particularly with regard to molecular size and pigment media interaction. A consideration of these parameters will lead to a greater appreciation of the visco-elastic nature of most ink systems under the high shear stress and shear rates encountered on a printing press.
- The specialist rheologist will have an appreciation of the flow of liquids and pastes which will extend far beyond the information recorded here. Such knowledge may be gained by a study of the extensive literature. Membership of the British Society for Rheology would be useful for their literature surveys and conferences on a wide range rheological topics.

14.1 FLOW IN IDEAL SYSTEMS

In commencing this study it is convenient to begin with an examination of flow in an ideal or simple system. Fluids which show ideal viscous behaviour are known as Newtonian, named after Isaac Newton (1687) who was responsible for the original theoretical work on the flow of liquids.

Consider two parallel planes, abcd and efgh of area A, in a fluid as in Fig. 14.1. Holding efgh stationary and subjecting abcd to a force F results in a movement of velocity v, the intervening fluid may be looked upon as having many planes sliding over one another. Consequently there is a velocity gradient between the planes abcd and efgh:

$$\text{velocity gradient} = \frac{dv}{dx} = D \text{ (rate of shear) } (s^{-1})$$

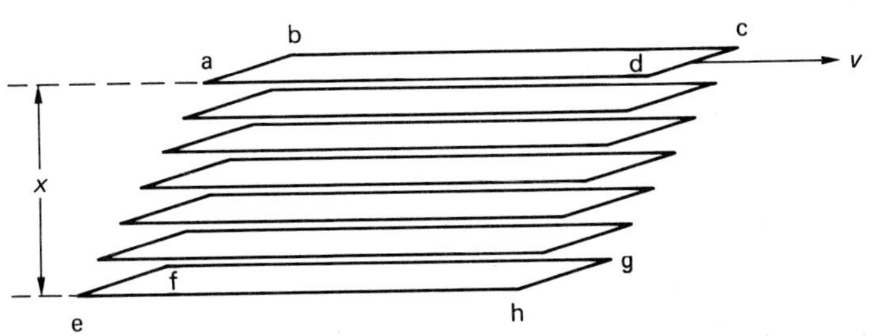

Fig. 14.1 Newtonian flow.

The shearing stress is given by the force per unit area:

$$\tau = \frac{F}{A}$$

For a Newtonian liquid τ is proportional to D, i.e.

$$\tau = \eta D$$

where η is a constant for the liquid under examination and is known as the coefficient of viscosity. The unit of viscosity is derived as follows:

$$\begin{aligned} \text{shearing stress} &= \text{force per unit area} \\ &= \text{newtons/m}^2 \\ &= \text{pascal (Pa)} \\ \text{shear rate } D &= \text{velocity gradient} \\ &= \frac{dv}{dx} = \frac{\text{ms}^{-1}}{\text{m}} \\ &= \text{s}^{-1} \end{aligned}$$

$$\begin{aligned} \text{coefficient of viscosity } \eta &= \frac{\text{shear stress}}{\text{shear rate}} \\ &= \frac{\tau}{D} \text{ (Pa s)} \end{aligned}$$

1 Pa s = 10 poise
1 m Pa s = 1 centipoise

European literature is now predominantly using the pascal . second (Pa s). However the poise is still in wide use in the ink industry.

Many fluids, such as the solvents used in inks, e.g. light mineral oils, ethanol and water are Newtonian liquids, so also are light-bodied stand oils, low viscosity alkyds and hydrocarbon-resin solutions. Changes from Newtonian behaviour occur with pigmentation of the ink medium and by chemical modification of the medium (Vincent, 1984; Love, 1983).

Flow curves or rheograms for two Newtonian liquids are shown in Fig. 14.2, the rheograms are straight lines passing through the origin, the steeper the line the higher the viscosity, so in the diagram liquid A is more viscous than liquid B. With a Newtonian liquid, measurement of shear stress and shear rate at a single point on the curve is sufficient to define the viscosity at all shear rates and shear stresses.

14.2 DEVIATIONS FROM NEWTONIAN BEHAVIOUR

The introduction of pigments into Newtonian liquids produces considerable deviations from Newtonian behaviour. This is due to particle association, by chemical bonds and physical interaction during flow. The following sections describe rheograms illustrating this non-Newtonian or anomalous flow.

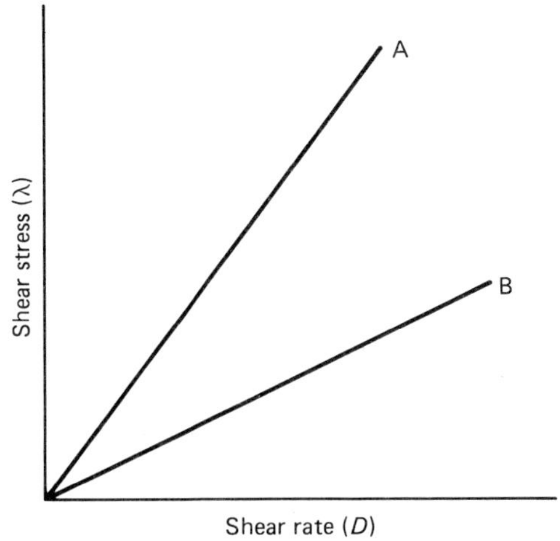

Fig. 14.2 Rheogram for Newtonian liquids.

Ideal plastic substances

The rheogram (Fig. 14.3) illustrates the ideal plastic substance; here flow commences only when the shear stress reaches a specific value, in this case τ_o.

Once flow starts the substance shows a simple relation between τ and D resulting in a straight line plot. The plastic viscosity U is given by

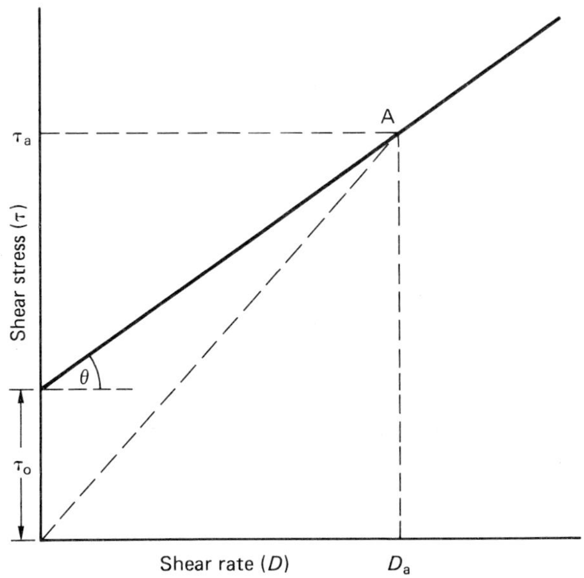

Fig. 14.3 Ideal plastic substance.

$$U = \tan\theta$$
$$= \frac{\tau - \tau_0}{D}$$

This value is a constant, however, if measurements are obtained at only one point on the curve, e.g. point A, the resulting calculation for viscosity

$$\eta = \frac{\tau_A}{D_A}$$

will give a different value. This is called the apparent viscosity and will be different for different shear stresses and shear rates. At high shear rates and shear stresses the apparent and plastic viscosity values will approach one another. This type of rheogram is typical of highly pigmented low-viscosity media, an example of which is toothpaste. When the toothpaste tube is squeezed the shear stress at the edge of the hole exceeds the minimum stress needed for flow to occur and thus the cylinder of toothpaste is extruded. This minimum value τ_0 for flow to occur is known as the yield value.

Pseudo-plastic substances

When the rate of shear increases more in proportion to the shear stress the material is described as a pseudo-plastic or shear-thinning material. At the higher shear rates the curve becomes approximately linear and it

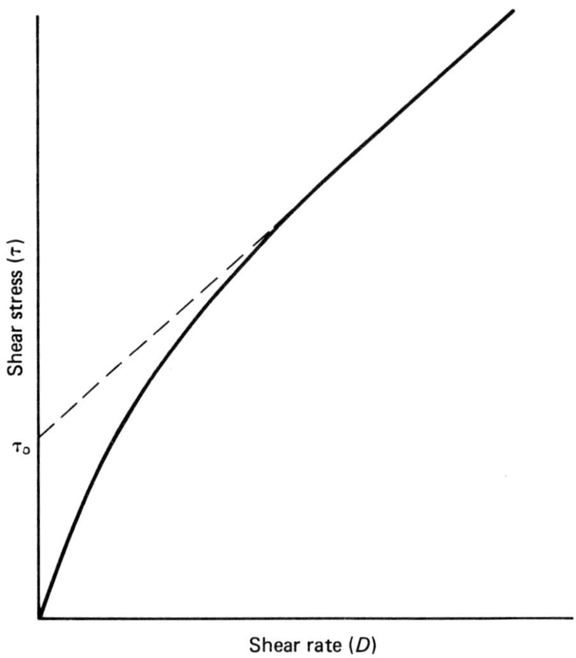

Fig. 14.4 Pseudo-plastic substance.

has become a practice in the ink industry to extrapolate this portion of the curve to the stress axis, as shown in Fig. 14.4, and record this value of shear stress as a yield value. This is not sound practice as all pseudoplastic materials will give apparent yield values using this method.

Figure 14.5 shows the flow curve of a pseudoplastic material with a yield value. Flow does not commence until the yield value stress has been exceeded. Once this has happened the resulting shear stress, shear rate curve is similar to that in Fig. 14.4.

Dilatancy

When the rate of shear decreases more in proportion to the shear stress the material is known as a dilatant or shear thickening substance and gives a rheogram of the type shown in Fig. 14.6. Dilatancy occurs when particles in the fluid tend to pack and form rigid structures. Consequently, dilatancy is associated with inks having a high proportion of pigment or extender.

Substances which are dilatant would be useless as lithographic or letterpress inks because at the high shearing stresses encountered with fast running presses, the ink would fail to distribute properly. Dilatancy may occasionally be found in stiff pastes of very high pigment concentration designed to function as the base of inks to which varnishes are subsequently added in preparing an ink for printing. Dilatancy is also sometimes specifically formulated into intaglio inks for copper and steel engraving to assist in clean wiping of the plate.

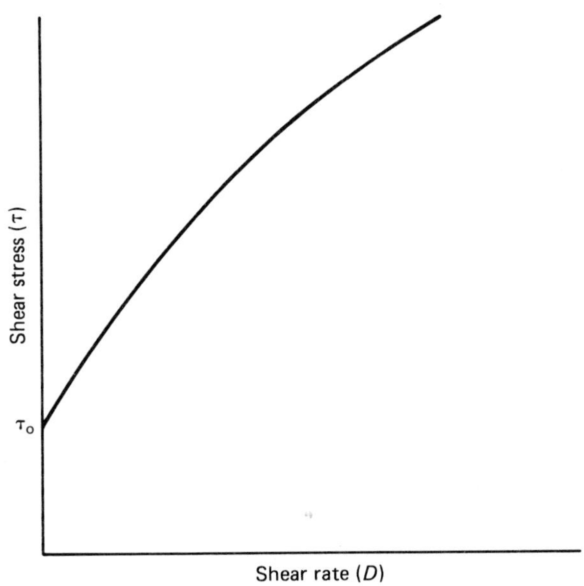

Fig. 14.5 Pseudo-plastic substance with yield value.

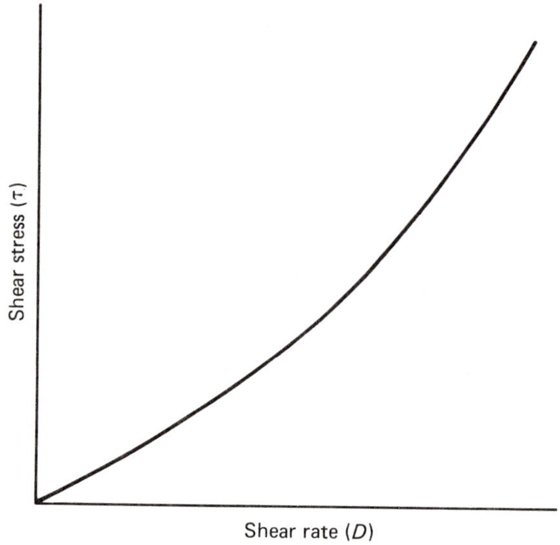

Fig. 14.6 Dilatant substance.

Thixotropy

Most printing inks thicken on standing and the gel structure can be broken down by stirring, a phenomenon generally known as thixotropy.

When a thixotropic substance is stirred at a constant shear rate the apparent viscosity falls with time and eventually attains a constant value. At this point the breakdown and buildup of gel structure are balanced for a particular shear rate. Graphs of viscosity against time at three different shear rates are illustrated in Fig. 14.7.

Thixotropy is a time-dependent phenomenon. It is assumed in flow curves, such as those for plastic, pseudo-plastic and dilatant substances in Figs. 14.3–14.6, that the breakdown of structure and its re-establishment when the shear stress is removed are instantaneous. This is true within the limits of measurement for some materials, and curves such as these could be followed upwards and downwards an indefinite number of times without any variation. For many other materials, however, including printing inks, the process of breakdown and especially set-up takes time and this leads to further complications.

In the flow curve for a thixotropic material, readings taken at advancing shear stress differ from those where the shear stress is being reduced (Fig. 14.8). As the stress is increased, the structure in the ink is reduced and the curve AB results. If after B the stress is progressively reduced, the curve BO results because structure recovery takes much longer than the time taken to measure and record the results.

The area in the hysteresis loop may be taken as an indication of the degree of thixotropy existing in the material under test. Care should be taken to establish that this thixotropic loop is in fact real because fluid systems (even if Newtonian) which are subjected to stress will become

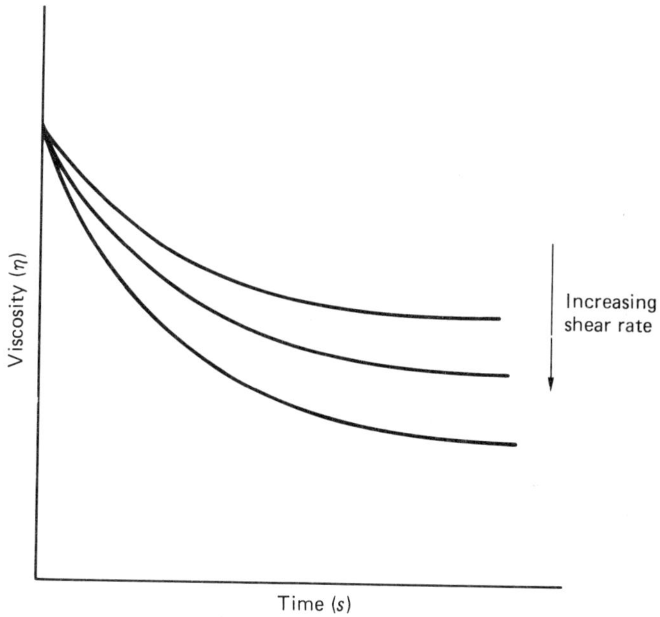

Fig. 14.7 Thixotropic substance at different shear rates.

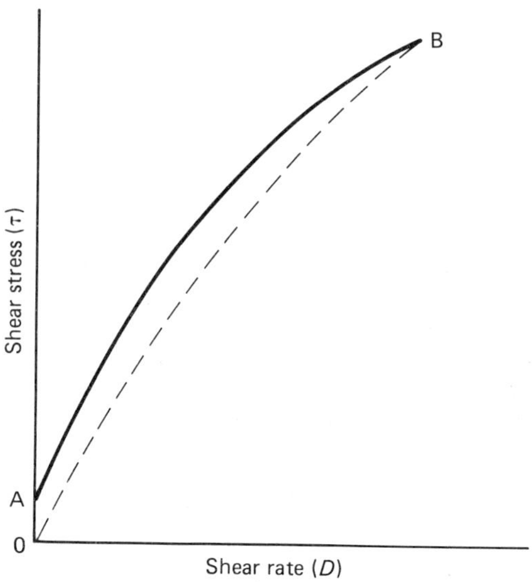

Fig. 14.8 Thixotropic material.

heated and yield similar hysteresis loops if the thermostatic control system is inadequate.

Influence of temperature

Flow behaviour is strongly dependent on temperature which means that during measurement great care must be exercised to ensure that the temperature is kept constant. The viscosity of liquids is always reduced by increasing the temperature – for example, a rise of a few degrees Centigrade transforms sugar syrup from an immobile liquid almost to the fluidity of water.

Alternative methods of expression of ink flow

The use of the rheogram is the classical approach to representation of the flow of a material and most who are familiar with this from a graphical expression of rheological behaviour are happy with it. However, for students who seldom refer to a rheogram, this method of expression may be confusing particularly as the important interest of the inkmaker is to understand the apparent viscosity on application of differing shearing stresses (i.e. the structure when in the process of being broken down). On the rheogram viscosity is equal to tan α which is inconvenient in making a quick appreciation (α is the angle between the tangent to the curve and the shear rate axis). An approach similar to that of Goodeve and Whitfield (1940) in expressing the apparent viscosity against reciprocal shear rate may perhaps be more readily understood.

Figure 14.9 shows the apparent viscosity of printing ink plotted against reciprocal rate of shear. This means that the rate of shear at the origin is infinite and the residual viscosity of the ink when the structure is completely broken down may be obtained by extrapolating to infinite rate of shear at A. In the graph, the point A is found in printing ink to be slightly in excess of the viscosity of the medium. In addition this value at A may be found to bear some relationship to values ascribed to the tack of an ink.

It is also found that this value at A is in general conformity with the Einstein equation for a system of particles suspended in a liquid:

$$\eta_\infty = \eta_o \, (1 + \phi c)$$

where η_∞ is the residual viscosity, η_o is the viscosity of the medium, c is a constant and ϕ is the volume of particles in the system per unit volume.

Visco-elasticity

A purely viscous liquid will move on application of a minimal stress, the deformation remaining on removal of the stress. An elastic solid, on the other hand, will recover its original form immediately on removal of the stress. No substance is either totally elastic or viscous and so most materials contain both characteristics, the extent of each depending on the molecular structure, the presence of particles, any particle to medium or interparticulate interactions.

These materials are called visco-elastic and most inks come within this classification. For example, an ink which has a yield value will behave as an elastic substance at shear stresses below the yield value, i.e. the stress

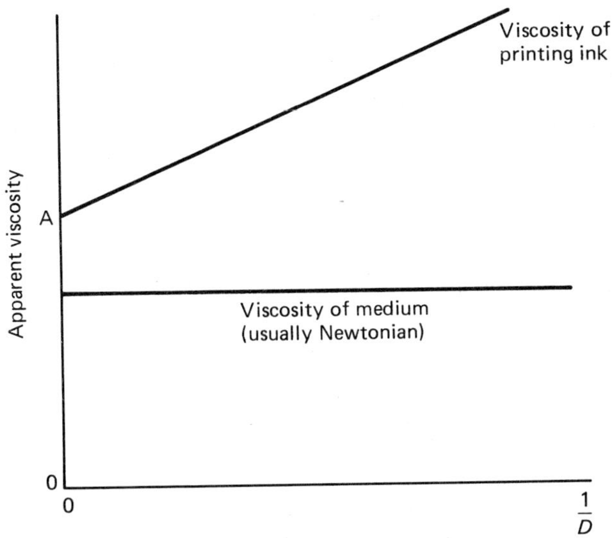

Fig. 14.9 Apparent viscosity.

will cause a deformation and on release of the shear stress the original form will be regained. On exceeding the yield value the ink behaves as a liquid.

These considerations are important in the conditions which occur on printing presses where the response of the visco-elastic material will depend on the rate of application and release of the shear stress on the distribution rollers and in the separation of paper and blanket.

The visco-elastic nature of inks becomes more significant with higher pigmentation. The choice of medium with respect to its molecular nature can have considerable influence, rosin-based derivatives will behave differently to long polymer chains based, for example, on linseed oil.

14.3 APPARATUS FOR THE MEASUREMENT OF THE VISCOSITY OF NEWTONIAN LIQUIDS

The range of liquids which can be examined with this equipment includes mineral oils, solvents, drying oils and varnishes (other than those with a gel structure). The rheology of Newtonian liquids does not require the establishment of a rheogram because the curve is simply a straight line passing through the origin. The slope of the line is therefore the only matter of interest, this being the viscosity.

Some instruments, e.g. Ostwald U-tube type, are essentially single-point instruments. That is to say they have the capability of making a measurement at one set of conditions of shear stress or shear rate. These viscometers cannot be used to obtain pairs of values of t and D necessary to plot a rheogram. However, when a liquid has been confirmed as being of Newtonian behaviour a single point apparatus can be used and several are available.

Reference should be made to the determination of the viscosity of liquids in CGS units BS 188:1977(1988) issued by the British Standards Institution which provides information on capillary viscometers and falling sphere viscometers. Details are given therein of the design, use and calculation of results of these two pieces of apparatus which in the printing industry may be used for solvents, oils, etc. and varnishes other than gel varnishes.

Capillary viscometers

These are used for liquids of low viscosity. Since the stress involved in this apparatus is that due to the difference in liquid level between the two arms of the 'U', the density of the liquids enters into calculation and the result is expressed as a kinematic viscosity, the unit of which is the stoke. The relationship between the stoke and the poise is:

$$\eta = \delta \, \varepsilon$$

where δ is the viscosity (stokes), η is the viscosity (poise) and ε is the density of the liquid.

The working formula for capillary viscometers is as follows:

$$\delta = Ct - \frac{B}{t}$$

where t is the time of efflux; and C and B are constants.

The second term can be neglected if it is greater than 100 s, BS 188:1977(1988).

Falling sphere viscometers

This method is applicable to Newtonian liquids although some concept of non-Newtonian behaviour may be established by dropping spheres of varying size through the liquid contained in the tube of the viscometer. It is essential however that the Faxen corrections are applied in any attempt to establish non-Newtonian behaviour.

Again this method gives the result as a kinematic viscosity centistokes.

Dependent on convenience, either formula (14.1) or (14.2) may be used:

$$\text{viscosity} = Mg \frac{\delta - \varepsilon}{0.03\pi v d \varepsilon \delta} \times F \text{ (centistokes)} \qquad (14.1)$$

or

$$\text{viscosity} = \frac{d^2 g\, (\delta - \varepsilon)}{0.18 v \varepsilon} \times F \text{ (centistokes)} \qquad \{14.2)$$

where M is the mass of the sphere (g), d = the diameter of the sphere (cm), δ = the density of the sphere (g/cm^3), ε = the density of the liquid

(g/cm^3), v = the velocity of fall (cm/s) and g = the local acceleration due to gravity (cm/s^2). F is known as the Faxen term and is a correction for the effect of the wall of the fall-tube on the motion of the sphere:

$$F = -2.104d/D + 2.09d^3/D^3 - 0.9d^5/D^5$$

where D is the diameter of the fall-tube in centimetres and d is the diameter of the sphere.

This Faxen correction should be used unless D is so large that d/D can be neglected for the largest diameter of sphere employed. British Standard 188:1977(1988) gives the range of viscosity measurements employing the capillary viscometer as 0.01–1500 stokes for clear liquids, and 0.02–4500 stokes for opaque liquids. When using falling sphere viscometers for clear liquids it is 20 stokes and upwards.

Use of falling sphere viscometer for the measurement of the viscosity of opaque liquids

For opaque mineral oils the tube can be filled with water or glycerol (dependent on the specific gravity requirements) to the lower mark and between the marks, with oil. The time of fall of the sphere is taken between its disappearance at the surface and its emergence into the water layer. The equation normally applied may be used in these circumstances but with the obvious error associated with the sphere not having achieved uniform velocity. The method is convenient and the error is small.

Flow cups

Flow cups should be used only for the measurement of the viscosity of Newtonian fluids and can be calibrated by means of an oil of known viscosity. In practice where the fluid is quite thin the deviation from Newtonian behaviour is ignored for practical purposes and so flow cups may be used for the control of varnishes and gravure and flexographic inks. However, it is essential to be alert to those circumstances where the structure in the fluid is such that the flow cup will produce any result dependent on the method of operation. BS3900 Part A6 1986 (ISO 2431–1984) gives a complete description of flow cups and their use and lays emphasis on the care and checking of the cup if erroneous figures are to be avoided.

The results are expressed as the flow time from commencement to the instant at which the stream first breaks into droplets. They are not recommended for use where the time of efflux is very short. Normally the results are expressed as the flow time, although conversion to stokes is possible by reference to conversion factors.

Flexographic and gravure printers frequently use Zahn flow cups for routine control. The cups have a handle and can be conveniently dipped into tanks. The inkmaker usually supplies ink at a viscosity which the printer can reduce by solvent addition to achieve the optimum printing conditions for the job in hand. Some presses have automatic control which adds solvent to maintain a predetermined viscosity. The formulator should anticipate such additions to ensure that the flow properties remain satisfactory.

Further studies have been commissioned by NAPIM on the measurements of the flow properties of flexographic inks using the Ford, Shell, Zahn and ISO flow cups (Lavelle, 1988). Laboratory experiments showed the limitations of flow cups and the careful control needed to obtain correct data. The ISO cup was found to be the most precise.

Flow cups have been reviewed by Stockhauser (1984).

14.4 PRACTICAL MEASUREMENTS FOR NON-NEWTONIAN SYSTEMS

Most printing inks are non-Newtonian because the element of structure is involved. In the case of lithographic and letterpress inks, the pigmentation and extender particles may occupy 25% of the volume. The particles of pigments and extender are therefore sufficiently close together to exert an attractive influence by the van der Waals forces so that in time some orientation may be produced, giving an appearance of solidity in the ink. Such systems are normally said to be thixotropic because they will break down with agitation, but when the agitation is removed the structure will set-up again. The time of setting-up continues for many days. In rheological measurements this means that depending on the shearing stress or the degree to which the structure is broken down, the apparent viscosity will vary. In other words, the greater the shearing stress the greater becomes the fluidity of the system, i.e. the viscosity falls.

The rheogram for the typical ink system is frequently that of an ideal plastic substance (Fig. 14.3) or pseudo-plastic (Fig. 14.4). It will be seen that the ideal plastic substance does not move until the shearing stress reaches a critical value. This value is described as the yield value, which needs to be measured.

Since non-Newtonian substances, in which many printing inks are classed, vary in apparent viscosity with increase in rate of shear stress, it is obvious that a picture of rheological behaviour must include measurements taken at varying rates of shear to enable a graph to be produced relating shearing stress to the shear rate.

Rotational viscometers

The most generally used of these is the coaxial cylinder type, shown diagrammatically in Fig. 14.10. The material under test is sheared in an annular gap between a driven outer cup and a cylinder bob, suspended from a wire or spring. The viscous drag tends to rotate the bob, until it is balanced by the torsion in the wire or spring. This type of viscometer was first employed by Couette and numerous modifications of it have been devised, including some in which the bob is rotated and the viscous drag measured at the cup. Rotational viscometers have the particular advantage that a sample can be subjected to a known rate of shear for any chosen period and a graph of the shearing stress against the shear rate may be obtained (i.e. a rheogram).

The rate of shear will obviously be proportional to the speed of rotation of the cup or the bob. Since it is desirable to measure printing inks at high rates of shear comparable to those on a printing machine, this leads to the choice of instruments with narrow gaps to avoid the use of very high speed of rotation of the cup or bob.

It should be noted that the rate of shear in the Couette viscometer varies across the gap. However, if the gap is sufficiently narrow the mean value may be taken. With large gaps care must be taken in comparing results obtained with instruments of different geometries.

A further reason to favour a small gap is that it helps temperature control. At high rates of shear considerable heat is developed which must be conducted away if the temperature is to be held constant. In practice this is difficult to achieve. Printing inks are generally poor conductors of heat and the best results are obtained by shearing thin films between metal surfaces with thermostatic control. The practical disadvantage of using narrow gaps may be the difficulty, particularly with stiff materials, of getting the ink into and out of the cup.

The temperature effect is not the only limitation in the use of coaxial cylinder viscometers at high rates of shear. Above certain values some materials, including printing ink, tend to 'climb' out of the annular gap producing anomalous results. Unfortunately, these limitations occur at or below rates of shear (of some thousands of reciprocal seconds) known to occur in the manufacture of the ink and also during the distribution on the printing press. This phenomenon is connected with the visco-elastic properties of the materials and is known as the Weissenberg effect. Visco-elastic materials have some of the characteristics of both elastic

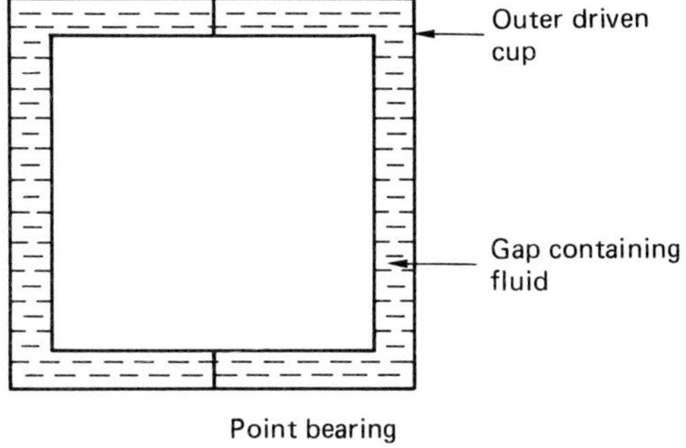

Fig. 14.10 Schematic diagram of a rotational viscometer.

solids and viscous liquids and such properties can be investigated by more complex instruments called rheogoniometers which measure forces at an angle to the plane of shear.

A modified version of the rotational viscometer provides convenience and speed of operation for the evaluation of gravure and flexographic inks. It is particularly useful when high pigmentation gives rise to structure which makes the use of flow cups inappropriate. The outer cup is a standard can which thereby avoids transference of the ink to a specific cup for the instrument. The shearing force is applied to the inner cylinder which, by means of holes drilled through it, produces a vigorous stirring action in the liquid. The force produced on the outer cylinder (i.e. the can) is opposed by a spring and the apparent viscosity may be read in poises on the dial.

Cone and plate viscometers

A variation of the rotary principle is found in the cone and plate viscometer, the Ferranti-Shirley instrument being one of the first to be used in the ink industry. The principle is shown in Fig. 14.11. The torque required to rotate the cone is measured and by using different sized cones and a variety of speeds a range of measurements is possible. The angle AOB between the cone and plate is normally so acute, approximately 1°, that the rate of shear is almost constant across the gap. Cone and plate viscometers are relatively easy to fill and clean and therefore very convenient to the inkmaker's requirement.

The introduction of inexpensive powerful computers has given rheologists the opportunity to control the variables of shear stress and shear rate to a much greater extent than ever before. Complex shear stress–shear rate programs can be applied to the test substance under strictly controlled and repeatable conditions.

The resulting rheograms can be analysed to see how closely the data fit to Newtonian, pseudo-plastic, Bingham, Casson (see section on falling-bar viscometers) dilatant or thixotropic behaviour and the best-fitting curves found. An estimate of the thixotropic nature can be found by

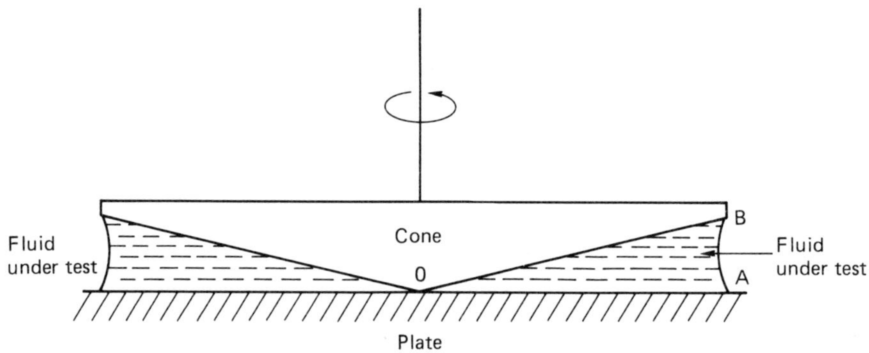

Fig. 14.11 Schematic diagram of a cone and plate viscometer.

measuring the area in the hysteresis loop on the rheogram. All these curves can be displayed on monitors or *x–y* chart recorders and compared with established standards. For routine quality control methods batch details can be stored on discs together with all the relevant rheological data.

Several companies are now producing a wide range of instruments for investigational work on ink flow. These include both concentric cylinder and cone and plate systems which are most convenient for measurement on offset inks.

Haake, Contraves and Carri-med produce a range of computer controlled viscometers suitable for making viscous and visco-elastic measurements. The Haake and Contraves systems control the speed of rotation of the system, as already described, whereas the Carri-med viscometer controls the shear stress applied to the sample.

Although expensive these viscometers are being used for quality control work in the ink industry where their simplicity of use, combined with the advantage of computer control should produce benefits in accuracy and time saving.

Simple versions of the cone-and-plate system are available for routine work for paints and similar substances with apparent viscosities of between 0.1 and 1 Pa s (1 and 10 poise). Their employment in the ink industry for flexographic and gravure ink is limited by the fact that the solvent evaporation rate is too rapid. However, for inks with solvents evaporating at the rate of white spirit, this instrument is most useful.

The principles of this apparatus are described by Van Wazer *et al.* (1963).

Controlled stress rheometers

Until recently rotational viscometers of the cup-and-bob and cone and plate types were shear rate-controlled instruments. Such instruments cannot measure yield values since by definition they must rotate at some speed, no matter how small, thus flow is always occurring.

The controlled stress rheometer, such as the Carri-Med CSL, can apply very small stresses, similar to those exerted by gravity up to those encountered on the rollers of high speed presses.

The central core of the CSL rheometer is the air bearing and torque motor assembly. The drive shaft, which forms part of the torque motor, is supported on an air bearing which can provide virtually frictionless support since it is 'isolated' from mechanical friction by a thin film of compressed air. The low level of friction enables the instrument to operate at very low shear stresses. Mounted on the drive shaft close to the bearing is an optical encoder which can detect movements as low as 2×10^{-5} rad and angular velocities up to 50 rad/s.

The drive shaft can be connected to any of three measuring systems, cone and plate, parallel plate, or concentric cylinder. The mechanical components are directly controlled by a dedicated single-board computer which enables close and accurate feedback between the desired and actual measurements. The standard temperature control system employs this feedback to enable a unique and rapid method of heating and cooling, controlled directly by the computer. The Peltier system provides a temperature range of 5 to 60°C, which may be extended to cover the range – 10 to 99.9°C if required.

If there is a need to test under controlled shear rate conditions this can be done using the appropriate software.

The data obtained from the controlled shear rate or controlled shear stress experiments can be fitted by the flow curve analysis programme to various mathematical models including the well known Bingham, Casson and power-law equations. The results can be plotted out on a printer or plotter in almost any desired combination.

Conditions, such as the settling of pigment suspensions, which are stress controlled can be examined by the creep-mode software. In this technique a single small stress is applied to the sample and the resulting strain (deformation) is monitored. Visco-elastic materials will give a characteristic creep curve which will either reach an equilibrium value, if the measurement is below the elastic limit after which no further flow occurs and any strain is completely recoverable, or it will reach an equilibrium shear rate, beyond the elastic limit, from which a viscosity value at very low shear stress can be obtained. The creep curve can be analysed to give retardation times which characterize the material tested, revealing its rheological properties.

Subjecting the sample to a sinusoidal stress or strain wave form will give an insight into the visco-elastic characteristics of a substance. If the material is purely elastic the stress and strain waves will be in phase. If, however, the material is purely viscous, then the stress and strain waves will be 90° out of phase. Most materials show an intermediate phase angle which changes with frequency. The analysis software can interpret the results in terms of the usual visco-elastic parameters.

Controlled-stress rheometers are also available from Bohlin Rheologi.

Falling-bar viscometers

For routine control in a printing ink laboratory the falling-rod viscometer is a widely used and convenient method of determining the rheology of lithographic and letterpress inks. In common with cone and plate and rotational viscometers it is possible to evaluate the plastic viscosity, and the yield value. The numerical values obtained depend upon the precise method of use, but once that is established repeatable results can be achieved. This makes the instrument suitable for control of batches and also for the characterization of inks which have been found particularly satisfactory on the printing machine for some specific process.

The apparatus consists of a vertical rod which passes through an annulus. A sample of ink is placed between them, after which the rod is forced through the aperture by applying weights to it (Fig. 14.12). The time of fall is measured at each load. The readings can then be converted into shearing force (load) and shearing rate (time) by applying appropriate viscometer constants.

A series of pairs of values may be plotted to obtain a rheogram. The apparatus is particularly useful in the viscosity range of letterpress and lithographic inks and the concentrated pastes from which they are often made (approximately 5 to 100 Pa s). Materials of this kind are

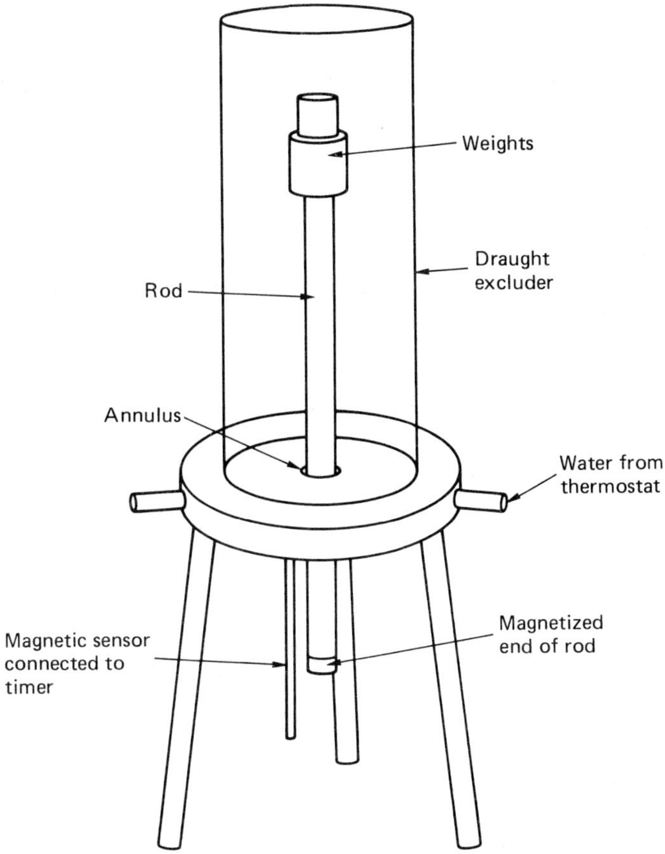

Fig. 14.12 Schematic diagram of a falling rod viscometer.

usually pseudo-plastic and many are very thixotropic.

The yield value of an ink is defined as the force required to produce flow and it is of interest in predicting, for instance, whether a given ink will feed readily from the duct where shearing forces may be small. However it is difficult to make reliable measurements at low rates of shear to see if a yield value is present. The difficulty is commonly overcome by the use of Casson's equation and charts based upon it are supplied by the equipment makers. Casson's equation is as follows:

$$\sqrt{\tau} = K_0 + K_1 \sqrt{D}$$

where τ is the shearing stress (N m^{-2}; Pa), D is the shear rate (s^{-1}), K_0 is the square root of the yield value and K_1 is the square root of the plastic viscosity.

The application of the Casson equation produces a straight line which can be extrapolated to zero rate of shear to determine the square root of the yield value. The slope of the line represents the square root of the plastic viscosity. A programmable calculator can be used to find the best least-squares fit for the straight-line extrapolation.

Manufacturers charts have been prepared so that the loads and their respective times of fall can be plotted and the viscosity and yield value directly read off (Fig. 14.13). It should be noted that these charts are prepared on the basis of the original dimensions of the instruments which change with wear. To check that wear is not excessive the instrument should be tested regularly with a standard viscosity Newtonian fluid supplied by the viscometer manufacturer and if the error exceeds a specified amount the bar and the annulus should be changed.

Frictional heat is generated when the rod passes through the annulus and cooling is provided by circulating water at constant temperature through a jacket surrounding the aperture.

The recommendation about the performance of measurements vary both in terms of choice of loads and whether they are applied in ascending or descending order. Generally, determinations using high loads produce relatively high figures for yield values and low figures for plastic viscosity (Turner 1973). This effect was found to be reinforced if loads were applied in descending order. An article prepared by the Experts Committee in Rheology of the European Association (1975) recommends a maximum load corresponding to a minimum falling time of 5 s (rate of shear 500 s^{-1}), or if an electrical timer is in use, this can be reduced to about 1 s (rate of shear 2500 s^{-1}). It is advisable not to use loads above 2000 g, because excessive heating will occur. The committee also recommends that the highest load is first applied, then the lowest, then the second highest and so on, in this way the influence of thixotropy is reduced.

Special attention is applied to the cleaning of instruments, because small amounts of dried ink on the rod or in the annulus will upset the results. It is also important to ensure that all solvent is removed before a new measurement is made.

Flow at low shear stresses

The shear stresses involved in producing ink flow in the duct of a press are due entirely to gravitation and are thus quite small. In these circumstances the rate of build up of structure is of considerable importance. If too rapid then the ink will not feed properly into the nip. A simple test to assess the build up of structure under low shear stress is to run the ink down a vertical or near-vertical glass plate. A sample of about 1 ml is agitated and collected on a palette knife and the sample drawn across the top of a glass plate. Timing of the rate of flow down the plate and the ultimate length of flow can give useful information on the low shear stress characteristics of the ink. Reproducibility will depend on the care taken to standardize the method of sample preparation and the adequacy of the temperature control.

Ranges of rate of shear

The rates of shear with which the inkmaker is concerned in lithography and letterpress range from zero to very great in the order perhaps of 10^6 s^{-1}.

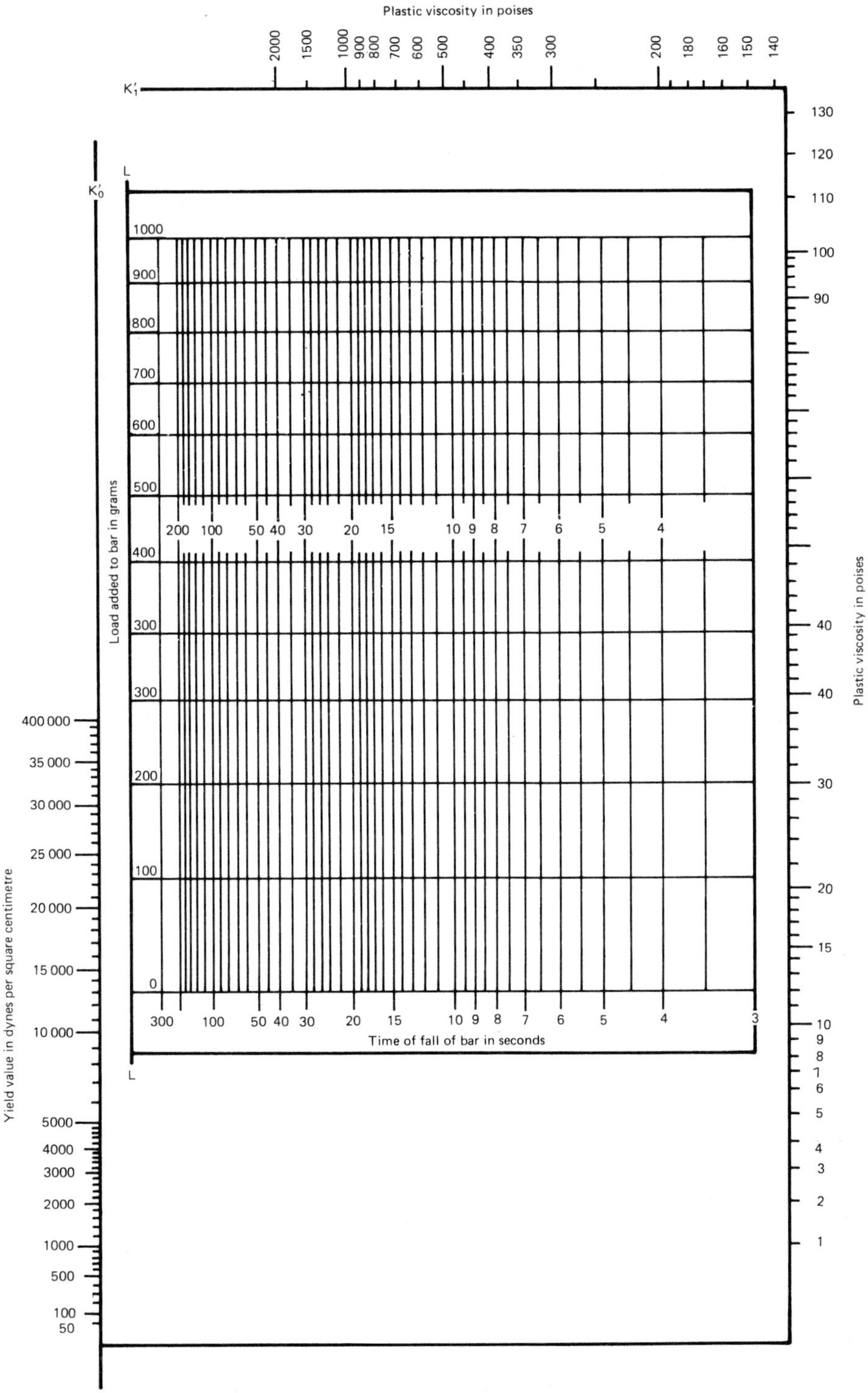

Fig. 14.13 Chart to determine the apparent viscosity from a plot from the weights and times on a falling rod viscometer.

This arises from the several situations encountered.

- Stiffness of ink in the can – inks kept in store can develop a degree of rigidity which gives inconvenience to the printer in removal from the can.
- The behaviour of ink in the duct of the printing machine under very light shearing stress. It is essential that the ink responds to the movement of the ink duct roller. If the shearing stress is less than the yield value the ink will fail to feed unless it is constantly agitated.
- The rollers of a fast-running press subject the ink to exceedingly high rates of shear.
- The capacity of ink to be pumped from large containers . The supply of inks by container is common in web-offset news printing and heat-set web-offset. These containers are delivered and located in the printing plant and ink is pumped from them direct to the printing machine. The rheology of the ink requires that such containers are designed so that the pumping operation is performed easily and efficiently. Recently, container supply has commenced in the field of offset four-colour sheet-fed printing which requires stiffer inks than those previously supplied on a commercial basis for pumping systems.

14.5 TACK

Essentially 'tack' may be defined as the 'stickiness' of a substance. Alternatively it may be looked upon as the force required to separate two planes of unit area at a set speed when they are held together by a liquid.

'Tack' would seem to be the factor governing the process of applying ink to paper. With fast running presses, particularly when printing coated paper, the tack of an ink must be below the critical point at which the ink will pick the paper at the required printing speed. Otherwise, tack should be as great as possible (allowing a reasonable safety margin) to produce the sharpest printing by the letterpress process and the cleanest working in offset lithography. With increase in printing speed it has become a matter of great importance both to the inkmaker and the printer that inks should be provided for the machine at a given tack (plus or minus a small tolerance), this implies the need for an acceptable standard scale of tack to satisfy these requirements.

Much has been written about this, notably by Voet (1952), Strasburger (1958), Myers, Miller and Zettlemoyer (1959), Mill (1964) Whitfield (1965) and in the European Study of Tack Measurement (1973). It is clear, however, that the theoretical side is much less developed than that of viscometry. It is probable that tack involves not only the viscous behaviour of fluids but also the elastic rupture properties of solids. Nevertheless, it is a property of great practical importance, and instruments have been developed to measure it, necessarily empirically.

The measurement of tack is usually carried out by evaluating the force exerted by a rotating roller on another via an ink film joining them but such a system is not amenable to a rigorous mathematical treatment. Tackmeters available to the printing industry therefore rely on empirical methods for evaluation and standardization.

The relationship between tack and viscosity

Whitfield (1965) found in a series of lithographic and letterpress inks that tack was related directly to the residual viscosity.

Mill (1952) noted that tack can be closely associated in some cases with viscosity and Turner (1976) also found a very rough general relationship between tack and plastic viscosity. Thus when extremes are considered high and low plastic viscosities are normally associated with high and low tack values respectively.

14.6 TACK MEASUREMENT

Originally, the evaluation of tack in printing ink was made by tapping out an ink with the finger on a glass plate. Particularly with the introduction of fast-running machinery, the need arose for a numerical scale and attempts were made to estimate tack by applying a force to separate two flat plates with an ink film between them. The result of such experiments did not lead to the design of any instrument of commercial acceptability.

Those tackmeters now commercially available measure tack under dynamic conditions at known ink film thickness, temperatures and speed. The basic principle of such tackmeters is that two rollers, one of which is driven, rotate with a film of ink of known thickness on them. The second roller may be supported in a manner which permits the measurement of the force required to hold it in a stationary position relative to the driven roller (Fig. 14.14).

An alternative system identifies the movement of the axis of the second roller to a position of equilibrium against a strain gauge. In all cases a third oscillating roller is employed but its function is solely to help in the distribution of the ink over the roller surface and to eliminate ribs which otherwise would form (Fig. 14.15).

At least five instruments designed to measure the tack of printing ink are available. While there are similarities in design there are also important differences particularly in the nature of the rollers. In the case of the LTF Inkometer and the Churchill, these instruments involve a metal roller operating a roller of some resilient substance employed in the printing industry for the manufacture of distributing rollers (Patra, 1963; Cartwright and Lott, 1962; Van Hessen 1963).

There has been discussion concerning the desirability of having metal on metal rollers or metal on resilient rollers. Metal on metal has the advantage of not having complications arising from the absorption of ink solvents into the roller. On the other hand, some observers believe that with well-seasoned rollers this difficulty is avoidable. The resilient roller has the advantage that uniform contact is more readily achieved throughout the length of the roller systems, although some workers in this field would not agree that this is so. A resilient roller will, however, make a contribution to the value obtained in a tack measurement.

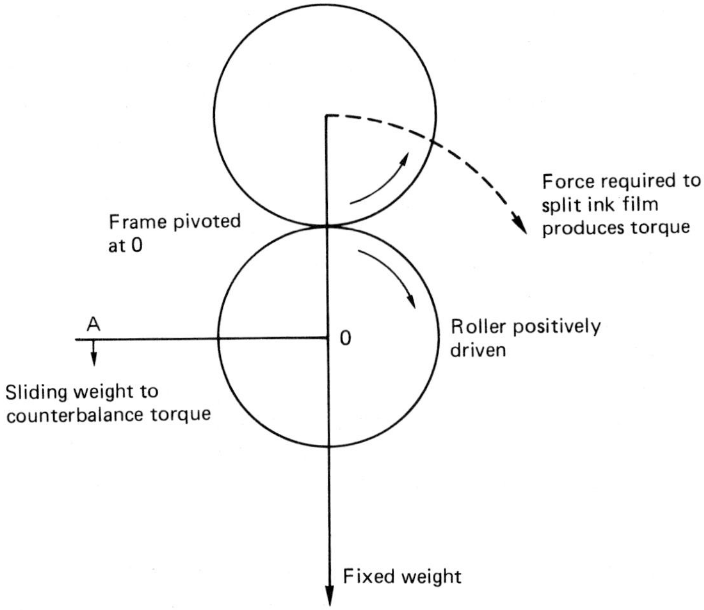

Fig. 14.14 Tackmeter principle (torque measurement).

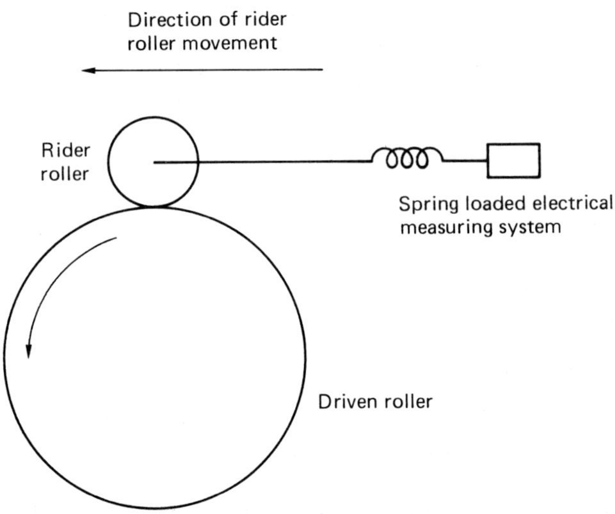

Fig. 14.15 Typical tackmeter (drag movement).

It has been stated that the two-roller system with an ink film between them is not susceptible to a rigorous mathematical analysis. Consequently, tackmeters need to be calibrated with a standard substance. The Society of British Printing Ink Manufacturers and PIRA were members of a European co-operative committee which studied tack measurement and a report was issued (1973), discussing the several devices available at that time. A recommendation was made that tack measurements should be made at a room temperature of 23°C and circulating water at 25°C.

The instruments under study by the European Committee involved the following:

- LTF inkometer (five instruments)
- PIRA (PATRA) tackmeter (five instruments)
- Churchill tackmeter (five instruments)
- Tack-o-Scope (four instruments)
- DIAF tackmeter (five instruments).
  See also Souty (1968).

In all, six samples of printing inks were examined. From the figures obtained it was apparent that in order to compare instruments of the same design it is first necessary to calibrate all instruments. This calibration should be carried out in accordance with the manufacturers' instructions. However, such calibrations may not take into account all the factors which affect the tack reading. Consequently any final correlation must be made with suitable inks and the use of available tack standards is to be recommended for this purpose. This procedure has the additional advantage of revealing any drift arising from any gradual deterioration in an instrument.

Grapho Metronic Mess.-und Regeltechnik GmbH & Co. have developed the LithoLab instrument which can simulate the procedure of water absorption by a printing ink at various roller speeds. In combination with the Grapho Metronic LithoMeter, the ink tack meter FZM and the combined ink film thickness and tack meter FSZM, the following investigations may be carried out under defined conditions with respect to the ink quantity, temperature, roller speed, etc.:

- the amount of both dampening solution and ink in the ink/water emulsion on the hydrophic area of the printing plate;
- the amount of dampening solution on the hydrophilic areas of the printing plate;
- the filling-in limit;
- the emulsification spread;
- the wet tack;
- the percentage of dampening solution in the printing ink.

Although in the LithoLab instrument (Fig. 14.16) there is no ink take-off, investigations are claimed to be useful, particularly where the state of emulsification of the ink corresponds to the situation on the press where there is little ink consumption. This is a situation where ink–water problems are prevalent (Rosenberg, 1986 a,b).

The fount solution is sprayed into the nip of rollers 9 and 10, a system which is said to be better than other laboratory testers in that overdampening of the ink is not possible.

The LithoLab offers the opportunity to investigate how the individual components of an ink affects its capacity to take up dampening solution. At the same time the tack behaviour can be recorded. Instabilities in the emulsification behaviour of inks and changes in the ink splitting may be

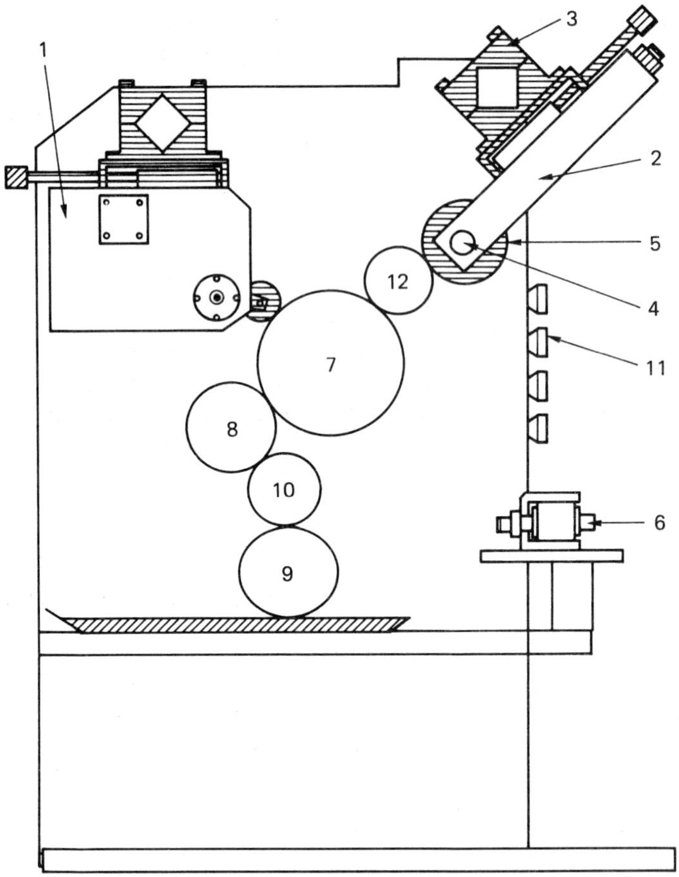

Fig. 14.16 Schematic diagram of LithoLab with measuring systems FSZM and LithoMeter: 1 FSZM measuring head; 2 LithoMeter measuring head; 3 measuring head bracket; 4 incremental encoder; 5 removable auxiliary roller with hydrophilic and hydrophobic areas; 6 sprayer device; 7 driving roller (option: w/temperature-control); 8 oscillation roller; 9 rubber-coated roller; 10 rilsan-coated roller; 11 controls to adjust contact among rollers; 12 rubber-coated roller.

detected. The LithoLab enables economical tests to be made on inks with regard to emulsification spread and their ability to absorb dampening solution, however the LithoLab does not replace the run on the litho press. The influence of dampening solution additives on the uptake of water by an ink may be investigated.

All the above information can be automatically stored and displayed on-line using the LithoComp computerized data acquisition system developed by Grapho Metronic.

The Lithomat, available from Dr Ing. H. Dürner has an Ink film thickness and tackmeter attached to the Prüfbau printability instrument. Thwing Albert, in the USA, produce a tackmeter giving automatic readout of tack readings.

Instruments of different design cannot be correlated without recourse to a series of ink samples covering a range of tack values.

Tack standards

Tack standards are manufactured from well matured printing ink pastes by adjustment to some specific figure on a research laboratory tack measuring device. Unfortunately such pastes are not absolute standards, but they can be regarded as convenient reference samples.

Further work on the use of poly-isobutene fluids as surface strength test liquids, the rheological properties of black letterpress and lithographic inks and tack graded inks has been carried out by Aspler *et. al.* (1982) using a Rheometrics mechanical spectrometer.

Tack testing

The equipment is widely used for routine quality control by comparison with a master sample. In combination with a printability tester such as the IGT or Prüfbau instruments it helps to avoid picking and provides some indication of the safe machine speed for any paper and ink combination. In high speed wet-on-wet printing, tack grading of letterpress inks in particular has proved beneficial to the attainment of good trap. It should be remembered, however, that the figure obtained will most closely relate to the condition of fresh ink in the duct, which may be different from the printing head.

Press stability can be predicted by running an ink for a set period and noting the change of ink tack with time. A graph having one of the general forms illustrated in Fig. 14.17 will be obtained. Inks having relatively steep slopes and high peaks when tested in this way will often have poor stability and may cause piling on the blanket due to incipient picking of the paper surface.

The equipment is also useful for assessing the tendency of an ink to mist, and the speed at which its onset is likely. The general method is to place a piece of graph paper adjacent to the rollers for a standard time, and then to count the number of ink dots over a unit area, using a magnifying glass. It is important to compare the ink under test with some reference ink whose practical misting tendency on a printing machine is already known, because small changes in atmospheric conditions may modify the results.

Testing of tack without recourse to an instrument

The experienced operator is able to produce results for comparative tack measurement by tapping the ink with a finger on a glass surface and appreciating by 'feel' the comparative stickiness. It has been suggested in

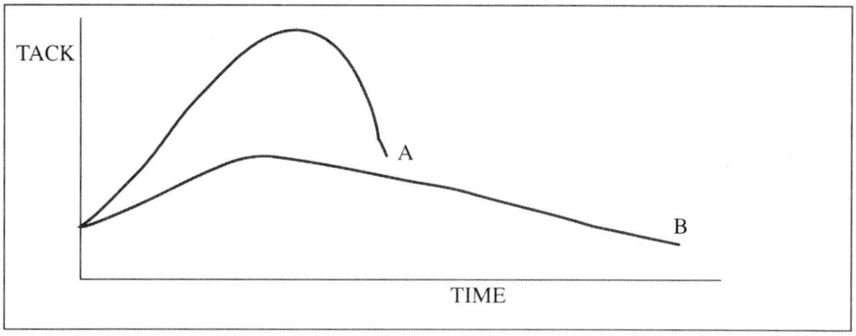

Fig. 14.17 Press stability prediction from inkometer readings. Curve B typifies an ink with better stability than the one giving curve A.

the past that such methods of measurement are unscientific and therefore must be inaccurate. In practice, the fully experienced operator will be found to have the accuracy of many approved tack measuring devices.

Such evaluations are only relative to known samples and unless a series of tack standards is available for which values have been obtained on a suitable tackmeter, it is not possible to ascribe numerical values to unknown samples.

14.7 INK DISTRIBUTION AND RELATED MATTERS

The rheological characteristics of an ink appear to determine its performance on the press, and many attempts have been made to relate the rheological parameters of viscosity, yield value, thixotropy and tack, both to press performance and to printing quality. So far, most of the relationships have defied a rigorous quantitative treatment, but rheological measurements can be useful for a semi-quantitative assessment of probable machine performance. Additionally, for quality control, they provide a helpful way of ascribing numbers to an ink whose working properties are known to be commercially satisfactory on the press, so that later batches can be supplied to a similar quality.

The qualitative properties needed in inks can be evaluated by some consideration of the inking system. In letterpress and litho for example, the ink is applied by a roller system from a duct. In the duct, the ink requires a consistency high enough to prevent leakage by flow between the duct blade and roller, but not so high that it will not feed, i.e. 'hang back' on the rollers. The structure must be broken down, and the ink must spread and transfer from roller to roller; it must therefore have an adhesion to the rollers greater than its own cohesion. However, the ultimate fluidity at the point of transfer must not be too great, otherwise the ink will flow laterally at impression and so not print the image accurately. However, if the ink is insufficiently fluid, the forces developed between the forme and the paper may damage the print.

The ideal rheological description for an ink therefore differs from stage to stage throughout the printing process and, clearly, practical formulations are a compromise.

'Hanging back'

The shear stress conditions in the duct are comparatively low, so good feeding will be consequent upon a low yield value and low viscosity. Inks having high yield values and high viscosities also feed because they have good cohesion and so act as one mass in the duct. It is an ink with a high yield value and a low viscosity which feeds badly, because its lack of cohesion means that action of the duct roller liquefies only a portion of it. Once this has been fed, the remainder of the ink in the duct becomes isolated from the duct roller and must be agitated to maintain printing (Fig. 14.18).

This phenomenon is likely to occur when the formulation of low to medium viscosity ink of high pigmentation is needed, such as might be desirable for adequate results on strong solids for instance.

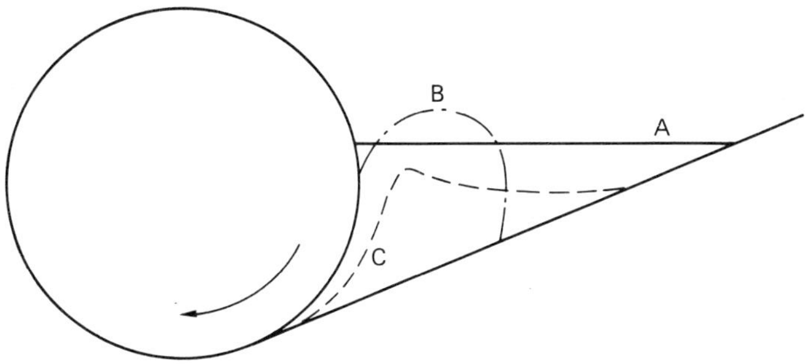

Fig. 14.18 Ink profiles in a duct: (A) low yield value, low viscosity – feeds; (B) high yield value, high viscosity – feeds; (C) high yield value, low viscosity – hangs back.

Distribution

It is evident that the ink must be fluid to distribute. However, deviation from Newtonian flow should not be considerable. Highly pigmented systems may exhibit visco-elastic or dilatant properties at the high shear rates encountered. This will reduce the ability of the ink to feed through the roller train and transfer to the printing surface.

Consideration should then be given to the reasons for poor distribution in terms of the effect of pigment concentration, pigment–medium interaction and medium type. The structure of inks can be controlled by the use of gelled varnishes which reduce in structure on agitation, i.e. they are thixotropic. This means that the structure will be reduced as the ink passes through the ink train. Thus an ink can have high structure in the duct, at low tack levels, and low structure and tack at the printing surface.

Ink stability

Printing inks are a paradox in the sense that they are required to dry as quickly as possible after printing, and yet remain fluid on the machine. This becomes a matter of considerable compromise in letterpress and litho inks, which must retain stability when spread into thin films for prolonged periods over the large surface areas of a roller system running at 30–40°C. The viscosity may increase by oxidation polymerization, but solvent losses by evaporation are usually particularly critical. On print, it is a great advantage to arrange for small losses of diluent into the substrate to produce a correspondingly large increase in viscosity; the ink left on the surface of the print is thereby rapidly stiffened or 'set' and this is one basis for quick-setting inks. However, premature loss by evaporation from the rollers must be suppressed by the choice of diluents having a limited volatility, otherwise an increase in tack may produce a vicious circle of heat generation and drying on the rollers and a tendency for the picking of the paper. Piling on the blanket sometimes arises because the ink acquires sufficient tack in this way to pull off a little more paper coating with each successive impression.

Occasionally ink starts to pile as a different manifestation of the basic problem of solvent loss. In this instance dilatancy may be induced. Some quick-setting varnishes consist of hard resins barely dispersed in diluent mixtures, so that very rapid phase separation occurs when the ink is printed on to paper. Should the separation happen through evaporative loss on the rollers, the resin seems to become part of the disperse phase instead of the continuous phase and the tack falls. It is the case in litho, water droplets also augment the disperse phase, the ink may become dilatant, lose its proper capacity to distribute and transfer and so pile up on the rollers and perhaps the plate.

Stability can be assessed by testing on a tackmeter and observing the increase in tack with time of running. Curves of the general shape shown in Fig. 14.17 may be obtained.

Transfer characteristics

The rheological characteristic ideal for printing solid areas are rather different from those ideal for printing halftones. Inks with high plastic viscosities usually do not lay well, particularly if the yield value is also high. This may be connected with the difficulty of 'bottoming' the low surface contours of the stock, particularly in letterpress; lower viscosities and yield values often achieve better lay on the ink by allowing it to flow into surface irregularities.

On the other hand, sharp dots and small dot gain is generally favoured by high yield values and high plastic viscosities, because lateral flow of the ink is thereby minimized.

It must be noted, however, that factors other than ink greatly affect dot gain and image sharpness. These include impression, press speed, temperature and the substrate. Consequently, it is not always possible, by adjusting the rheological properties alone, to overcome all difficulties which may need to be investigated in a logical manner.

J.M. Fetsko's (1989) survey of the published research since 1965 on the role of emulsification on ink transfer through the press and on final print quality provides a number of theories on the phenomena encountered. A failure to transfer properly on the ink distribution rollers produces a build up of ink known as piling and this may be associated with inks which retain water and/or exhibit shortness of flow.

Transfer of publication gravure inks is a low shear process and investigations by Schubert *et. al.* (1984) have shown that, in this area of 0.01–300 s^{-1}, pseudo-plastic flow can greatly improve transfer and give better reproduction. Demonstrations of these results are given in photomicrographs and viscosity studies on the Haake PV 100/CV 100 viscometer.

Jeske (1990) has described the simulation of the nip region on a high speed rotogravure press by modification of a hammer and anvil type impulse press and showed that ink transfer increased with increased pressure, contact time, and reduced ink viscosity. Capillary forces had no significant effect at high speeds.

Mottle

When ink is the cause of mottle (sometimes it might be due to an inconsistency of ink acceptance by the paper) it is thought to be connected with the ink being drawn out into strings as the film splits (Fig. 14.19). The strings then collapse onto the print, producing a characteristic mottled appearance due to slight differences in film thicknesses all over the surface. It often appears to be aggravated by some combination of a heavy impression, a paper of high surface absorption (which may prevent the strings of ink flowing out when they relax back to the print surface) and inks of low viscosity and low yield value. Mottle may be somewhat reduced by adding extender whose high oil absorption will reduce the flow of the ink, or a paste containing starch or wax which has a similar effect

Sirost (1990) has described the theoretical aspects of ink film splitting in offset inks. The laboratory experiments have been related to the actual printing process. In a further article Sirost (1991) examines the scission of the ink film during offset printing and describes a theoretical model of the process. The conditions of transfer and order of importance of the various

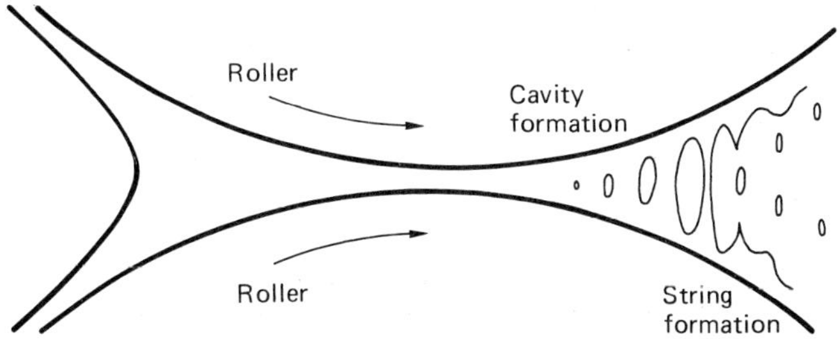

Fig. 14.19 Representation of an ink film splitting.

parameters are established. The influence of roller retardation, ink viscosity, printing temperature, excess water in the ink and printing speed are considered in relation to the laboratory tests and the practical limitations of the results are also discussed.

Ink-film splitting on non-porous substrates has been described by De Grace and Mangin (1988). The splitting is asymmetric, becoming more so as the film thickness and printing speed are increased. Asymmetric splitting can be explained on the basis of entrained air at the ink/substrate interface. A review of ink-film splitting mechanisms is examined.

Squash

When printing on a non-absorbent surface in particular, the impression must be kept as low as possible to reduce the lateral flow of ink which shows up as squash beyond the true image boundary. The extent will be determined by the printing pressure and viscosity; to avoid excessive flow, it is desirable to formulate for the highest possible viscosity. In lithography, physical squash is not the only factor concerned, because the ink and water balance keep the image sharp. Although the rheological properties may appear less critical in lithography, inks generally need a high viscosity in order to minimize the scumming and water marking associated with excessive take-up of water.

Misting

The tendency for inks to mist or fly follows from the formation of strings as the ink film splits on emerging from the nip. High speed photography has shown the formation of cavities in the ink as a prelude to splitting, the walls of which eventually draw out into strings; sometimes multiple rupture seems to occur and the free droplets produced in this way can form a stable mist which disperses into the machine room (Fig. 14.19). It is an objectionable factor at a time when attention is being increasingly focused on environmental issues.

The phenomenon appears to start at some critical speed for a given ink thickness and it is a particular nuisance in high speed rotary newspaper printing. Empirically, it can be suppressed by an addition of an extender having a high oil absorption which reduces the capacity of the ink to flow into long strings. Bad misting sometimes occurs even at moderate speeds on sheet-fed machines and the usual cause is the use of resins containing long linear molecules similar to those found in rubber. These materials are extremely soluble in the usual ink solvents and there are indications that very soluble resins and/or the use of strong solvents may increase the tendency to mist. Increasing the concentration of extender also increases the elastic nature of the ink. This property ensures that the ink snaps on elongation of the ink filament on leaving the nip of the ink rollers, thus preventing formation of ink droplets.

A tendency for misting can be assessed by running an ink through a series of increasing film weights and speeds on a tack machine and assessing the amount of mist drops which fall on to a piece of graph paper sited adjacent to the rollers.

Wet-on-wet trapping

In multicoloured printing, the degree of acceptance of an ink film on one previously printed but still wet, will depend on the relative tacks of the inks. Theoretically it is desirable for the first down colour to have a tack as great as possible so that the subsequent colours of lower tacks are pulled from their printing heads. If the reverse happens, that is, successive units partly scavenge ink already on the print, 'back-trapping' is said to occur, with a consequent loss of colour balance and printing control.

Tack grading is more desirable in high speed wet-on-wet letterpress printing than in litho, where the resilience of the blanket and lower ink film thickness seem to make trapping less critical. The nature of the paper also plays a very important role in the degree of acceptance of quick-setting inks. The first colour down, having lost solvent to the paper, presents to the oncoming colour an ink film with a tack greatly increased over its original value.

It must be remembered too that the tack of a set of printing ink measured on a tack machine are numbers derived under uniform and standard conditions. In the case of practical printing, the film weights of inks may not be uniform among the printing units; the losses of solvent on passage through the distribution system from duct to printing surface may not be consistent; and in the case of litho, the take-up of water may be different in each ink. Each of these factors would have made some considerable difference if they could have been allowed for during the original measurements on the tack machine, and together with the effect of paper absorbency, they suggest why the theoretical requirements may sometimes be of reduced significance. Nevertheless, the tack machine remains a convenient and most useful empirical method of sorting inks into their correct order for printing in succession, provided other factors of importance are taken into account.

On contact with the paper there will be a rapid loss of low viscosity diluent into a porous substrate. This will mean a considerable increase in the pigment concentration in the ink with a subsequent increase in the elastic component of a visco-elastic ink. This change should ensure that the first ink down is not pulled from the paper.

Ink uptake can be related to the following parameters; the thickness of the previous layer, the speed with which this first layer sinks in, the degree of tack and viscosity, structure and chemical composition of this paper, printing speed, excessive damping temperature, roughness of the plate, hardness and porosity of the blanket.

14.8 RHEOLOGICAL MEASUREMENTS AND MACHINE DESIGN

The mechanism of feeding ink on to the roller trains places demands on the inkmaker in providing inks with appropriate rheological properties to satisfy this aspect of machine design (Whitfield, 1970). It is suggested in that article that the design of the ink duct should not call for inks of defined consistency because the necessity of satisfying the requirements of the ink duct prevents the inkmaker from formulating an ink which would give the best results in the printing process.

Many letterpress and lithographic presses continue to operate with the ink duct roller and duct blade system. The degree of agitation with ducts of this type is frequently very small which means that structure in the ink must be reduced to a low level. More recently such ducts have been fitted with duct agitators which permit the use of structured inks.

In the field of web-offset news printing, ducts of new design have been introduced. Many machines are now fitted with a device operating on two rollers with serrated faces not in contact which meter the quantity of ink to the press. Such ink-feeding devices call for a reasonable degree of structure in the ink to avoid the inconvenience caused when the ink dribbles through the rollers onto the floor.

In some web-offset machines of very sophisticated design the ink is fed on to the machine through a series of jets producing an ink-jet feed on to the duct roller. With these systems the quantity of ink is metered for each point on the rail and control is by remote press button. This system calls for detailed study on the part of the inkmaker if the best results are to be obtained.

Measurement of flow properties

Descriptions of a device for measuring ink properties on the press are given by Ritz (1978), and Ritz and Rech (1978). The instrument consists of two rollers, the top of one of which oscillates from side to side to imitate the movement of a distribution roller. The flow behaviour of inks in the presence of water, the viscosity of the ink when oil is added and the influence of water in the press on the distribution capacity of the rollers were tested. The drop in shear between rollers is highest in the transverse direction.

The relationship between tack and viscosity has been drawn up by Bauer, Decker and Bosse (1977). The errors in rheological values are discussed and the determination of these values on the press with the help of a newly designed viscometer.

Rohn (1987) states that the steady shear viscosity method of testing printing inks is unsatisfactory as a method of predicting printing performance. Using a Rheometrics fluid spectrometer to determine visco-elastic properties enables an ink to be formulated for optimum printability. Tack tests can be performed on a Rheometrics mechanical spectrometer.

The non-linear visco-elastic properties of printing ink from the standpoint of strain, strain rate and strain history dependencies have been studied by Amari (1987–1988). Creep compliance and creep recovery were also measured.

Watanabe (1990) has investigated the visco-elastic properties of inks during the initial drying stages using various viscometers, a double band viscometer, an oscillating plate rheometer and a reflecting ultrasonic viscometer. The dry ink film was examined using creep and creep recovery methods.

Tasker *et al.* (1983) has devised three new laboratory tests for determining the water pick-up of lithographic inks. These involve emulsification of the ink in a Litho Break tester; the Delta plastic viscosity test compares the viscosity of emulsified and non-emulsified ink, the a–k fit test analyses the emulsion for water and the emulsified water particle test measures this parameter.

Surland (1988) has investigated the relationship of ink emulsification curves to rate of change of ink flow properties and ink tack increase, also the ink transference, misting and sling. Surland used a modified Sunbeam Mixmaster to emulsify fountain solution into the ink. An emulsification curve was found by measuring fount uptake at one-minute stages up to a total of ten minutes. The emulsification curve could then be used to predict the press performance of the ink. The method has met with some success but has probably been limited by the fact that the physical properties of the inks are changed when fountain solutions are emulsified into them and these changes have considerable bearing on ink performance.

The process of emulsification of fount into ink is carried out under low shear rate conditions both in the Surland test and under press conditions. This is not the case during the distribution and printing of the fount in ink emulsion. Chou and Fadner (1986) have used a controlled shear stress rheometer with a cone and plate system to investigate shear stability of fountain solutions and lithographic inks. Meaningful correlations have been obtained between the measurements made on the ink emulsions and press performance which were previously unobtainable to the industry. They also observed that emulsion inks are unstable at shear rates much lower than 2500 s^{-1}. Tasker *et al.* (1983) used a falling bar viscometer to examine inks at this shear rate. They noted that a large change in the viscosity at this shear rate due to emulsification could cause printing problems. It would appear that not only is the uptake of water important but also the manner of its separation and reincorporation during the printing process.

Chou and Bain (1988) compared keyless and conventional litho newsinks on a Carri-Med controlled-stress rheometer. Fitting the data to the Herschel-Bulkey equation enabled a shear thinning index, a consistency index and a dynamic yield value to be obtained. Together with a thixotropy index these factors indicate that from a rheological viewpoint keyless inks can be made using standardized measureable parameters without the need for on-press testing.

A controlled shear stress rheometer with cone and plate geometry and a controlled shear rate rheometer were used to measure the stability of emulsion inks. Both methods yielded consistent results, but quantitatively not the same for metastable and unstable emulsions. The shape of flow curves and viscosity versus temperature curves enables classifying the inks as stable, metastable and unstable. Flow curves of stable inks are similar to fresh inks without fountain solution, both typical of pseudoplastic systems, viscosity decreasing with increasing shear rate. Metastable emulsions exhibit irregular flow curves at higher shear rates, the viscosity changes randomly, water separates and the emulsion reforms. Unstable emulsions feature shear-induced gross water separation. The stability of emulsion inks depends on the physico-chemical nature of inks, fountain solutions, emulsion composition and temperature.

The effect on ink viscosity in relation to water absorption has been examined by Braun (1985). Other parameters using simulated offset conditions on a Lithomat tester from Prüfbau include the effect of pigments and the calcium solubility of lithol rubine pigment in acidified medium.

The rheological characteristics of inks have been described by Bassemir (1981) and the relationship between flow properties, emulsification rate and printability. Particular attention has been given to dot resolution; however, much work remains to be done in order to understand the phenomenon fully.

The need for instruments which will measure viscosities at both high and low shear in liquid inks is discussed by Anderson and Jensen (1982). At high shear rates the visco-elastic nature of the binder solution plays a major role, while at low shear rates the flow properties are controlled by the dispersion characteristics of the pigments. For measurements of the flow at high shear rates which are most interesting from the printing point of view the widely used flow cup will not give any definite impression of the flow properties. The viscometer must be able to achieve high shear rates and measure visco-elastic effects.

Babula (1983, 1984) has described the measurement of the rheological properties of liquid inks in relation to printing properties, e.g. ink split, dirty printing, pigment dispersion and rewettability.

The relationship between the flow properties of screen inks and their performance has been well described by Frecska (1982, 1984) and Messerschmitt (1980, 1982a,b,c, 1984). The effect of fillers on the flow properties of ink and their thixotropic nature is reviewed. The range of viscometers is discussed and their place in viscosity control is considered. A suitable viscometer for assessing performance of screen inks at a practical price is not yet available.

Ink control on the press and in tanks

The development of fully automatic ink pumping systems has been discussed by Kirby (1984, 1985) and Quilliam (1983). The pump equipment and its operation to pump ink from the bulk container meter the quantity and maintain levels in the ducts are described.

The viscosity of the ink will rise rapidly in the winter months and a knowledge of the ink flow properties at low temperatures will be required for efficient delivery. Storage tanks should be narrow and deep with a tapered bottom. Pipe diameter is important and there should be no tight bends which would reduce flow.

The importance of viscosity control in flexography and how its major effect on colour control has been described by R. Lichtenberger and G. Johnson (1991). Manual control is time-consuming and sometimes unreliable, however, simple mechanical controllers can detect and correct viscosity variations. Continuous monitoring by electronic viscosity controllers such as a rotating sensor, falling piston, vibrating reed or ball or an automated efflux cup can control solvent additions when readings deviate from a reference point.

The importance of automated viscosity control to the flexo and gravure printer with regard to reduction of variable costs and waste has been reported Anon (1988).

BIBLIOGRAPHY

Amari, T. (1987–88) Peculiar rheological properties of printing inks. *Gr. Arts Jpn.*, **29**, 14–20. (In English.)

Andersen, J.H. and Jensen, H. (1982) Rheological Properties of Liquid Inks. T2–82T Horsholm Denmark: Scandinavian Paint and Printing Ink Research Institute. (In Danish.)

Anon. (1963) PATRA tackmeter. *British Ink Maker*, **5**(3), 55.

Anon. (1975) Falling Rod Viscometers.*British Ink Maker*, **17**(4), 167-170, 171.

Anon. (1988) *Viscosity Control Increases Profits*. International Flexographic Printing Conference, London, 8–9 March 1988. PIRA and EFTA, Leatherhead, UK.

Aspler, J.S. *et al.* (1982) Rheological properties of new inks and surface strength test liquids. In: *Advances in Printing Science and Technology*, Vol. 16. Proceedings of the 16th International Conference of Printing Research Institutes, Key Biscayne, Florida, June 1981, pp. 235–51.

Babula, S. (1983) Viscosity is not enough–a study of the behaviour of ink in the distribution system of a flexographic press. In *Report of the Proceedings of the Colour Conference*, 6–7 Dec. 1983, Atlanta, Georgia, pp.61–4. New York Flexographic Technical Association.

Babula, S. (1984) Viscosity is only one aspect of rheology that determines ink distribution. *Flexo*, **9**(3), 21–3.

Bassemir, R.W. (1981) The physical chemistry of lithographic inks. *Am. Ink Maker*, **59**(2), 33, 36, 41–2, 44, 46, 98.

Bauer, M., Decker, P. and Bosse, R. (1977) *Relationship between Ink Viscosity and Tack; Theoretical Analysis and Laboratory Experiments Carried Out on an Offset Press.* Forschungsbericht 5.00G, Deutsche Forschungsgesellschaft fur Druck- und Reproducktions-Technik e v (Fogra). Munich. (In German (1977))

Braun, F. (1985) Studies of offset inks–the effect of pigments on the formation of emulsions. *Am. Ink Maker*, **63**(2), 26–48.

Cartwright, P.F.S., and Lott, A.D. (1962). The Metal Box/Churchill tackmeter. *British Ink Maker*, **4**(4), 224, 225, 227–237.

Chou, S.M. and Bain, L.J. (1988) Rheological characteristics. Keyless versus conventional litho news inks. *TAGA Proc.*, 354–386.

Chou, S.M. and Cher, M. (1989) Rheological studies of emulsion ink stability. *TAGA Proc.*, 257–280.

Chou, S.M. and Fadner, T.A. (1986) Shear stability of fountain solution emulsified in lithographic inks. *TAGA Proc.*, 37–62.

De Grace, J.H. and Mangin, P.J. (1988) A mechanistic approach to ink transfer. Part II–*The Splitting Behavior of Inks in Printing Nips*. 19th International Conference of Printing Research Institutes, June 1987, Eisenstadt, Austria, pp.146–161. IARIGAI 1988.

Douglas, A.F., (1989) Lewis, G.A. and Spaull, A.J.B. (1971) The investigation of the dynamic visco-elastic functions of printing inks. *Rheol. Acta*, **10**, 382–6.

Fetsko, J.M. Effects of emulsification on printability. *Inknews*, June–August, 21–25.

Frecska, T. (1982) The operating characteristics of screen printing inks. *Screen Print*, **72**(6), 62–5.

Frecska, T. (1984) Viscosity measurements. A challenge for screen printers. *Screen Print*, 56–9, 124.

Goodeve, C.F. and Whitfield, G.W. (1940) *Trans. Faraday Soc.*, **44**, 652.

Jeske, D.R. (1990) *Impulse Press Simulation of a Gravure Printing Nip*. International Printing and Graphics Arts Conference, Vancouver, Canada, pp.97–104. Canadian Pulp and Paper Association, Montreal, Canada.

Kirby, S. (1985) From tankers to tank. *Ink and Print*, **3**(1), 16–17.

Kirby S. (1984) Saving multicolour costs with automatic ink pumping systems. *Lithoweek*, **6**(35), 42.

Lavelle, J.S. (1988) NAPIM studies on flow cups. *Flexo*, **13**(6), 18, 20, 22, 24, 26, 28–29.

Lichtenberger, R. and Johnson, G. (1991) Viscosity stands as a keystone to quality print. *Flexo*, **16**(11), 27–28.

Love, D.J. (1983) The modification of lithographic inks with Manchem aluminium compounds. *Polym. Paint Col. J.*, **173**(4109), 837–8.

Messerschmitt, E. (1980) The release of printing ink from the screen. *Siebruck*, **26**(11), 702, 704, 706, 708–10. (In German.)

Messerschmitt, E. (1982a) Rheological considerations for screen printing inks. *Screen Print*, **72**(10), 62–5.

Messerschmitt E. (1982b) Rheological considerations for screen printing inks. Part II. *Screen Print*, **72**(11), 136–9.

Messerschmitt E. (1982c) Rheological considerations for screen printing inks. Part III. *Screen Print*, **72**(12), 80–3.

Messerschmitt, E. (1984) The continuing story on viscosity. *Screen Print*, 50–7, 60.

Mill, C.C. (1952) *Chemistry and Industry*. S.C.I. p. 159.

Mill, C.C. (1964) *British Ink Maker*, **6**(2),

Monk, C.J.H. (1966) Routine measures of the viscosity of paint samples. *JOCCA* **49**(7).

Myers, R.R., Miller, G., and Zettlemoyer, A.C. (1959) *F. Coll. Sci.*, **14**, 287.

Newton, I (1687) *Philosophiae Naturalis Principia Mathematica*, Book 2.

Pira International and Society of British Printing Ink Manufacturers (1973) European study of tack measurement – interim report. *British Ink Maker*, **15**(2), 57–58.

Quilliam, B.L. (1983) Development of a system of computerized automatic ink-flow control. *Taga Proc.*, 89–103

Ritz, A. (1978) Determination of the rheological magnitudes of inks in the inking mechanism of an offset litho press. *FOGRA Mitt.*, **27**(96), 10–11. (In German.)

Ritz, A. and Rech, H. (1978) *Determination of the rheological magnitudes of inks in the inking unit of an offset machine*. Forschungsbericht S. 204, Deutsche Forschungsgesellschaft fur Druck- und Reproducktions-technik e v (Fogra) Munich.

Rohn, C.L. (1987) Predicting the application behaviour of printing inks from dynamic rheological measurements. *Taga Proc.*, 536–559.

Rosenberg, A. (1986a) The part played by surface water and emulsified water in the lithographic process. In *Advances in Printing Science and Technology*, Vol.18, (ed W.H. Banks), Pentech Press, London.

Rosenberg, A. (1986b) *What Parts Surface Water and Emulsified Water Play in the Lithographic Process*. GATF Conference, Chicago, USA, August 1986.

Schubert, F.G. *et al.* (1984) Low shear viscosity of publication gravure inks and its implications. *Taga Proc.*, 219–29.

Sirost, J–C. (1990) Ink flow. *Caractère*, **288**, 58–60.

Sirost, J–C. (1991) Ink rheology during printing. *Nouv. Gr.*, **41**, (10), 16, 18–22.

South, G.R. (1968). A comparison of commercial tackmeters. *British Ink Maker*, **10**(3), 150, 152, 153, 155.

Strasburger, F. (1958) *F. Coll. Sci.* **13**, 218.

Stockhauser, N. (1984) *Cup viscometers: a testing method with limits*. Symposium Drucktechnik Forschungsergebnisse Fur die Praxis, Fogra, Munich 4–5 April 1984. (PM10263G.)

Surland, A. (1983) Factors determining the efficiency of lithographic inks. *Taga Proc.*, 191–236.

Tasker W. *et al.* (1983) Water pick-up test for lithographic inks. *Taga Proc.*, 176–90.

Turner, T.A. (1973) Use of bar Viscometer. *British Ink Maker*, **15**(2), pp. 59, 61, 63, 65, 67.

Turner, T.A.L. (1976). Rheology of ink in relation to printing properties *British Ink Maker*, **18**(2), 55–57.

Van Hessen, J.L. (1963) The Tack-O-Scope. *Am. Ink Maker*, **41**.

Van Wazer, J.R., Lyonsz, J.W., Kim, K.Y., and Colwell, R.E. (1963) *Viscosity and Flow Measurement*. Interscience, Chapter 6,

Vincent, J.E. (1984) Chemistry of aluminium–organics in the preparation of ink vehicles and gel varnishes. *Am. Ink Maker* **62**(10), 25–6, 28–30, 32, 34, 34B, 104.

Voet, R.E. (1952) *Ink and Paper in the Printing Process*. Interscience Publishers.

Watanabe, K. (1990) Physical properties of printing inks–dynamics of printing inks during drying processes. *J. Jpn. Soc. Col. Mater.*, **63**, (12), 717–727. (In Japanese.)

Whitfield, G.W. (1965) Rheology of printing inks – Letterpress and litho. *British Ink Maker*, **7**(2), pp. 100, 102–106.

Whitfield, G.W. (1970). *Lithographic Technology*. PIRA/IARIGAI

CHAPTER 15
Testing control and quality assurance

The inkmaker has at his disposal an extensive inventory of chemical materials from which to manufacture printing ink. It is unfortunate that the term, 'raw materials', is used for these components since few are naturally occurring substances intended for use after minimal processing. Modern pigments, resins, additives, solvents and oils are the sophisticated products of industrial chemical manufacturers.

A great many of these ingredients are specially made for the ink industry to meet stringent specifications. The final performance of printing ink relies on the selective choice of starting materials, on the formulation, and, of no less importance, the care with which the ink is manufactured.

Consistent product quality is the rightful expectation of an end user. To meet this objective, careful scientific testing is essential at all stages of manufacture.

Since the selection of starting materials is an expensive part of the ink development programme, it follows that having made a suitable choice of components, reproducibility is essential. This can only be ensured by the rigorous testing of each batch of the approved materials as they are made. Such testing must be carried out, and be guaranteed, within the manufacturer's domain. It should also be systematically checked by the user to ensure that compliance to the agreed standards is being attained.

The inkmaker must always agree with the supplier a full specification and the expected quality parameters for all starting materials. A similar expectancy arises between the inkmaker and the printing ink end user, the printer. To this end product specification and testing is a central activity of the printing inkmaker.

Testing of inks and starting materials falls into two main areas:

- qualitative methods which rely on visual and comparative techniques such as flow, consistency, shade;
- quantitative methods which give numerical results such as tack, viscosity, yield value.

With the development of newer methods of ink testing many subjective tests can be replaced by more scientifically based procedures.

This chapter provides a guide for both the testing of raw materials, and for the testing of finished inks or intermediate base components of ink. It is not possible within the confined space of the chapter to offer a complete account of all the possible tests since some are specific to definite products. Tests which are considered essential are covered in some detail. Where standard tests are included only the salient points are given and the reader is advised to refer to the published text for the complete details.

15.1 STANDARD TESTS

It is essential that standardization of test procedures should be established, so that results obtained in one laboratory can be directly compared with those from another laboratory. This is true both with raw materials (between the inkmaker and raw materials supplier) and also with the printed product (where an ink may have to comply with an end-use specification).

Some tests have been widely agreed and standardized by such authorities as:

- The British Standards Institute (BSI);
- European Confederation of Paint Printing Ink and Artists Colours Manufacturers Associations (CEPE);
- The Technical Association of the Pulp and Paper Institute (TAPPI);
- The American Society of Testing Materials (ASTM);
- The Institute of Petroleum (IP);
- Printing Industries Research Association (PIRA);
- The National Association Of Printing Ink Manufacturers, Inc USA (NAPIM).

15.2 SAMPLING TECHNIQUE

For raw material testing to be meaningful, the sample taken from a delivery of raw material must be representative of the entire delivery. A delivery may include more than one batch, and so batch numbers should be carefully checked when sampling. Suppliers should be asked to make deliveries which consist of as few different batches as practicable. This will reduce the amount of testing required. If suppliers can also provide predelivery samples, these can be tested well in advance of the delivery, and only an occasional delivery spot check is then necessary.

The inkmaker may choose to test all materials at the time of delivery – tanker quantities are an example – or to carry out regular spot checks to ensure that agreed standards are being achieved.

It is important that materials which are not tested on delivery or are only spot checked are provided by approved suppliers. Approved suppliers are those who have a proven ability to supply consistently materials of agreed quality. They should have a recognized quality standard, or be prepared to give certificates of conformity with each delivery. It may be desirable to visit the supplier and carry out an audit of their quality system to ensure that the required standards are being attained.

The following paragraphs are intended as a guide to sampling different kinds of materials.

Resins

Resins which are delivered in lump form may well contain some smaller pieces and fine powder. A composite sample should be taken which is representative of the proportions of each state present. It must also be remembered that there is usually a specific reason for resins being delivered in lump form. Usually this is to prevent oxidation. These resins should only be crushed immediately prior to the actual testing procedure.

Pigments

The vibration in transit of a cask of pigment may well have caused some segregation of particles of different size. Hence the cask should either be tumbled before opening, or several samples taken at varying levels in the cask with a special auger-like tool.

Liquids

Liquids should be thoroughly stirred or the barrels tumbled before opening, to ensure that no separation into distinct phases has occurred. If the material is being delivered by tanker then each compartment within the tanker should be sampled.

15.3 PIGMENT TESTING

Quality control of pigments is particularly difficult as a single pigment can be used in a wide variety of vehicle systems to give inks for many different end uses. Hence, the testing procedure must be selective and it can logically be divided into three sections. First, limited testing can be carried out on the dry pigment itself to ensure the batch meets the specification for shade, particle size, oil absorption, moisture and volatile content. The second and far more important consideration is to compare the inkmaking characteristics of the batch against the master standard in the inkmaker's own vehicle systems. Finally, provided the batch meets the specification, prints from the sample and standard inks are checked for their resistance properties.

In reality most inkmakers would be unlikely to carry out very intensive testing of incoming pigments, as most pigments originate from manufacturers who have recognized quality standards. Many of the resistance properties are a factor of the chemical nature of the pigment and are unlikely to vary considerably from batch to batch, or are related to the vehicle systems being used, and are hence tested for as part of the finished ink quality-control checks. Furthermore, testing for such parameters as particle size will involve tests and equipment which will generally be beyond the scope of the average quality control department.

Tests on the dry pigment

Particle size

Since the pigment particle size will affect such fundamental properties as strength, hue, rheology and stability of the finished ink, it might seem desirable to check particle size or at least the average size distribution, as part of the quality control procedure. However, in the size range of most pigments (less than 1μ m) the majority of the standard techniques for sizing, e.g. gravitational sedimentation and optical microscopy, are slow and inaccurate. Only two methods are suitable for particle size measurement in this range – electron microscopy and centrifugal sedimentation – both of which are generally beyond the scope of the average quality control department, and would not be considered as normal control techniques for inkmakers. However, even the inkmaker without analytical facilities can use a local university to assist in particle size determination if necessary.

Electron microscopy is an expensive method and requires an experienced operator, but is capable of producing very accurate readings.

The disc centrifuge (Fig. 15.1) is perhaps the preferred technique and results compare favourably with those obtained using electron micros-

Fig. 15.1 A Joyce Loebl disc centrifuge.

copy. However, as neither method is simple and quick enough to be used as a standard quality control procedure the inkmaker places greater emphasis on the dispersion of the pigment particles in the finished ink (discussed later). For further reading on particle size determination by centrifugal sedimentation the reader should consult *Paint Technology* (1965) **29** (5), 30 and J.Beresford (1967) *J. Oil Col. Chem. Assoc.* **50**, 594–614.

Oil absorption

The oil absorption of a pigment will depend on:

- the chemical nature of the pigment;
- particle size and size distribution;
- particle shape;
- surface characteristics of the sample, such as those obtained by treatment to improve wettability or even simply due to absorbed moisture.

If the oil absorption of the pigment varies from batch to batch, then formulation changes may have to be made to the inks, and clearly this is undesirable. Since the factors affecting oil absorption cannot easily be measured, the test is reduced to a purely empirical comparison between the batch and the master standard. The actual test method is given in BS 3483: part B7: 1974, the result being quoted as the mass (or volume) of oil required to wet 100 g of pigment. Alternatively, the sample and standard can be dispersed in a measured amount of oil and the relative consistencies of the two samples can be compared.

Moisture and volatile content

Absorbed moisture may affect the ease of dispersion of some pigments, and this can be determined according to the method in BS 3483: part B6: 1982. A temperature of 105°C is chosen for this test, as at higher temperatures volatile matter may be lost which is an essential part of the pigment. This is especially true of carbon black and pigments on an aluminium hydrate base.

15.4 INKMAKING CHARACTERISTICS

All pigments should be checked wherever possible in the vehicle systems in which they will be used in production, but in certain cases where the pigment is used in several different types of inks, this may be too time consuming and a selective range of formulations must then be chosen. This should include oil and liquid ink formulations where appropriate. The method of dispersion in the laboratory should:

- simulate as near as possible the production process for mixing and grinding;
- be rapid, comparative, and reproducible;
- utilize small quantities of pigment so as to retain a single master standard for testing as many batches as possible.

The first test must be to ensure that the batch of pigment disperses at least as readily as the standard and to an equivalent degree, so that costly production time is not lost. The degree of dispersion is measured using a fineness of grind gauge (Fig. 15.2). This is a stainless steal block containing a shallow groove which is graduated in depth from zero to, usually, 25 or 50 microns. A sample of ink is placed at the deep end, drawn down with a steel blade towards the shallower part, and the degree of dispersion is taken as the point at which the continuous ink film breaks down across the groove. However, care must be taken in interpreting the result if discontinuity in the ink film occurs higher up the gauge in the form of scratches due to the odd oversize particle. It is then important to determine whether these scratches on the gauge are caused by undispersed pigment agglomerates, or by, for example, undispersed paste or dried ink or varnish skin. Also, care should be taken that these odd scratches on the gauge are not merely caused by airborne dust particles that have settled on the gauge or contaminated the ink sample after it was taken. This will be more of a problem on the wider channel 50-micron gauges than the narrow channel 25-micron gauges which are more commonly used, as most quality control specifications will require a dispersion of at least 5 microns and below. The wide channel 50-micron gauge is more commonly used for examining the degree of dispersion at the mixing stage before the ink is ground, but can be very useful for checking for particulate contamination from skin, etc. which may be present but may not actually show up on a narrow 25-micron gauge. If unusual scratches are observed on a gauge, higher up than the overall degree of dispersion, then the examination of the ink under a microscope is recommended to determine the cause.

Once the correct dispersion level has been attained, the trial ink can be compared with the standard for strength, hue, gloss and flow (or consistency). First, the flow is checked as it is impossible to obtain meaningful comparative prints unless the batch and standard are similar in consistency. Liquid inks having near-Newtonian rheological properties can be compared by viscosity measurements using any of the standard techniques (e.g. Ford, Sheen and Zahn cups). Oil inks, however, are not Newtonian and require more difficult instrumental measurement such as a Laray falling bar or cone-and-plate viscometer. It is important that the viscosity of the trial ink matches that of the standard or comparative flow tests will not be meaningful. To match the flow alone by comparing their consistency with the aid of a palette knife, or by comparison of their relative flow properties on an inclined or vertical glass/perspex plate is not adequate as the flow properties of the pigment may vary from batch to batch and should also be checked. Similar flow properties may be obtained, but at different viscosities, which is obviously not meaningful as production batches will be primarily adjusted to a specific viscosity or tack (section 15.13), rather than being adjusted to a specific flow without regard to the latter parameters.

If the flow of the trial ink is different from the standard at a similar viscosity, then this may indicate that there is some variation between batches of pigment, or that the degree of dispersion of the two samples is not equal.

Fig. 15.2 A fineness of grind gauge.

The strength of the trial pigment is then assessed on a drawdown at full strength and at reduced strength as a 'bleach'. The drawdown is carried out by placing small quantities of the standard and test inks about one inch apart on the top of a securely held piece of paper, and then firmly scraping them towards the operator with, for example, a 3-inch wide blade. Initially the blade should be held nearly vertical to reproduce a film thickness comparable to a printed film weight and then finally the blade should be brought almost parallel to the pad to give a thick film weight – the masstone of the ink. The drawdown should be carried out on a semi-absorbent paper on which a broad black band around 19 mm wide has previously been printed, so that the opacity (or transparency) of the batch and standard pigments can also be compared.

The bleach technique involves dilution by mixing (not grinding) a coloured ink with a much larger quantity of opaque white, so as to eliminate any minor differences of gloss and hue. In the case of yellow inks, the eye cannot differentiate small strength differences, and so sufficient blue is added to the bleach to bring the hue into the blue-green region.

Hue can be measured by instrumental techniques (Chapter 3), but normally a visual comparison of a drawdown and print is sufficiently accurate. The undertone and toptone of the inks can be assessed using the thin film of the drawdown and the thicker masstone respectively. Gloss can also be checked on the drawdown, but more accurately by examination of a print of the batch versus standard on an appropriate stock.

Resistance tests

Provided the pigment matches the standard on the tests conducted so far, the appropriate resistance checks are carried out. Many of the characteristic properties of a pigment are a function of its chemical nature, and hence are unlikely to vary from batch to batch, unless an impurity is present. Pigments are, however, coated to provide easier dispersion in different vehicle systems and this can give rise to batch variation, especially when the pigment is obtained from an alternative supplier.

All the resistance tests are carried out on prints of the batch and standard inks, rather than dry pigments themselves.

It is important that a vehicle is chosen which is appropriate to the intended application of the final ink.

Lightfastness

As the lightfastness of a pigment is a function of its chemical nature, it is only necessary to carry out random spot checks, but attention should be paid here to pigments used in inks where lightfastness is an important criterion, e.g. wallpaper inks.

Pigments will show a higher lightfastness at full strength and therefore it is important that a realistic strength print be submitted for test. For example, a PMTA pigment should rarely be tested at full strength, since its main application will be as a toner pigment. As pigments for use on wallpapers could often be used in tints, it is advisable to test these at both full and tint strengths. Prints should be prepared on a good quality paper which does not yellow on exposure to light.

The exposure of prints to daylight is too lengthy a procedure to be suitable and the use of an artificial source is therefore adopted. The only light source to have an emission spectrum close to that of daylight is the xenon arc and provided a filter is used to reduce the amount of UV radiation, a result comparable to daylight exposure is obtained in a much shorter time. Other light sources such as the carbon arc and mercury vapour lamp are not recommended as the results bear little comparison to those obtained under daylight conditions.

Another factor which greatly affects the lightfastness figure is the effective humidity, which is dependent on the relative humidity and temperature of the surrounding air and the surface temperature of the pattern. The average effective humidity value for daylight exposure in Britain is 45% and therefore the test apparatus should be capable of approximating to this figure.

In order to classify numerically the lightfastness results obtained, the prints are compared with a set of eight pieces of wool, each dyed with different blue dye which has been carefully graded to fade after a set exposure to light. This is known as the Blue Wool scale, and the scale ranges from one (poor lightfastness) to eight (excellent) according to the number of the wool dyes which have faded under the same exposure as the test method. The procedure for mounting and assessing samples is given in BS 4321:1969, but it is worthy of note that the term 'fading' includes not only a reduction in strength but also changes in hue, brightness, darkening, or any combination of these factors.

Heat resistance

Heat resistance of pigments is a critical factor in the areas of tin-printing inks (and other metal coatings) and wallpaper inks, which will be subjected to hot embossing. The pigments are simply tested by placing prints in a suitably heated oven for a set period of time, the actual conditions depending on the end usage of the print. The prints are then compared with an unheated sample for changes in strength, hue and brightness. Again, prints should be at a realistic strength, as heatfastness, like lightfastness, will vary from full to tint strength.

Soap and detergent resistance

Pigments to be used on soap and detergent packaging should be checked to ensure no impurities are present which are liable to cause bleed problems. Although test methods for both soap and detergent resistance are given (BS 4321 test methods 6 and 7 respectively) the manufacturers of these products have their own test methods which are generally more severe than the British Standard tests.

A typical soap resistance test, as provided by a leading manufacturer, is as follows. A 25% soap gel is made up and allowed to set. The gel is moistened and a print of the pigment to be tested is first soaked in water to prevent curing (1 minute for paper and 5 minutes for board). The print is then placed face down on the gel and left for 15 minutes, after which time it is examined. After this period there should preferably be no bleed, although a slight bleed could be considered acceptable.

Testing for detergent resistance is becoming more complex with the advent of enzyme and 'solvent' powders. Provided a pigment passes the above soap test, it is unlikely to fail with the standard soap powders. However, enzyme-detergent resistance should be assessed with an enzymic powder. (To date no standard test has been published for resistance to 'solvent detergents'.)

A typical detergent resistance test is as follows. A 5% solution is made up using the appropriate enzymic detergent in water. Two filter papers are soaked in the solution, one is placed on a glass slab and the print to be tested is placed face down on the filter paper. The second filter paper is then laid on top and a further glass slab provides a sandwich effect. The print is then examined after three hours and although ideally there should be no bleed, a slight bleed could be considered acceptable.

The disadvantage in interpretation of these test methods is that the degree of staining is only expressed in vague terms such as 'slight bleed', whereas both the British Standard tests (numbers 6 and 7) quote a numerical means of assessment of staining by use of the grey scale (BS 2663C).

Alkali resistance

This test can be conducted either by streaking dilute sodium hydroxide solutions onto a print of the pigment (e.g. 2, 3 and 5%) and examining the dried streaks for discolouration, or as in the British Standard test (BS 4321 Test method 5) by sandwiching the print between filter papers previously soaked in a dilute alkaline solution. In this case, the filter papers are dried after a set contact period, and then the degree of staining is assessed with the aid of a grey scale.

Acid resistance

Acid resistance can be carried out in a manner similar to BS 4321 test method 5 above, but replacing the alkaline solution with either dilute hydrochloric acid or dilute acetic acid. The latter reagent is normally used where the print could come into contact with vinegar or pickles.

Wax resistance

Wax resistance is essential for inks on waxed paper wrappers, ice cream cartons, certain vending machine cups, and similar applications where the substrate is made more water resistant by immersion in molten wax (usually paraffin wax). The standard test (BS 4321 Test method 10) involves immersion of a print of the pigment in molten wax for five minutes, after which the wax is allowed to run over the margin surrounding the print. The margin and the wax in the basin are then examined for discolouration.

Plasticizer bleed

This method is extremely important for pigments that will be used in inks to be printed on plasticized PVC such as for vinyl wallpaper and shrink wrap applications. Prints of an ink made from the pigment and a standard ink on plasticized PVC are sandwiched between plain sheets of the unprinted material and placed between glass plates in an oven at 45°C for 3 days. A small weight is often placed on top of the sandwich to increase contact pressure. On examination of the unprinted substrate adjacent to the print there should be no evidence of colour bleed at the end of the test period.

Deep-freeze resistance

This test is becoming increasingly important as more and more products are being stored under deep-freeze conditions. On carton board, the vehicle system (or overprint varnish) will normally provide adequate deep-freeze resistance and testing of the pigment is not necessary. However, for polyethylene printing the deep-freeze resistance of the finished pack is a function not only of the vehicle (usually polyamide), but also of the pigment as well. The deep-freeze resistance of the pigment in these cases is not so much a function of its chemical nature, but of the pigment coating, and therefore quality control of each batch of pigment is essential for this application. The tests are carried out on prints made from the pigment under test.

The severest condition met by the print is not when the pack is actually under the deep-freeze conditions, but during the stage of thawing on removal from the freezer, and hence the test normally conducted is one of water immersion.

A typical test is as follows. The print is immersed in water for a period of at least 1 hour and a maximum of 5 hours. The print is then flexed between the knuckles ten times and should show no more than 5%

removal. A second portion of the print is then blotted dry and should show no more than 5% removal when checked by the Scotch tape test.

15.5 CHIPS, PREDISPERSIONS AND FLUSHED PASTES

In chips and predispersions the physical process of dispersing the pigment in the resin system has already taken place, and in theory an ink can be made simply by solubilization of the resin component in the solvent system. However, in production the dissolution usually takes place on a high-shear mixer and therefore the laboratory technique should simulate this method.

A useful comparative method of testing such predispersions is to solubilize the pigment 'chip' in the solvent by shaking in a sealed container on a Red Devil paint mixer for a set period. Finally, if the dispersion is less than adequate both batch and standard can be further processed for a short period on a high-shear laboratory mixer such as a Silverson blender. A comparison between batch and standard inks can then be made according to the test methods given in the section on pigment dispersion (section 15.3). The method of first shaking in a sealed container and then mixing under shear enables the tester to ensure that dissolution takes place without excessive solvent loss, as well as providing a technique for assessment of relative ease of dispersion of the standard and batch.

If the predispersion is in the form of a preground paste this can be tested as if it were a finished ink, such parameters as degree of dispersion, strength, hue, and rheology being examined (section 15.13)

Flushed pastes are slightly different in that the pigment is not ground into the varnish base, but 'flushed' into it as part of the manufacturing process. Flushing is the process of transferring the pigment from the aqueous phase into the oil phase without the normal conversion into dry powder. The pigment particles within the flushed paste are therefore fully dispersed to give maximum strength, gloss, transparency, etc. Flushed pastes are normally so highly concentrated and viscous that they cannot be tested as if they were a finished ink. The most convenient way of testing flushed pastes is to 'let-down' the paste with a precise quantity of a standard varnish to give an 'ink' for which a specification is set. As the varnish is standard any deviation from the specification has to be due directly to variations in the flushed paste.

15.6 DYE TESTING

Solvent dyes

Solvent dyes have a limited usefulness as an alternative to the more commonly selected pigmented chips. They are tested by ensuring that the dye is soluble to the required concentration in the appropriate solvents.

On standing for a further period (usually up to 48 hours) no precipitation should occur. The strength, hue and brightness of the resulting solution may be determined on a colorimeter or by making up a suitable ink and comparing with a master standard in the normal manner. Lightfastness and bleed tests can then be carried out on a print as described earlier in this chapter.

Basic dyes

These dyes have poor fastness properties, high tinctorial strength, and find their greatest use as toners in flexographic inks, after laking with, for example, tannic acid. They should be checked for solubility and resistance properties against a master standard in an appropriate formulation.

Acid, reactive and direct dyes

These are little used in the printing ink industry, but do find specific end uses, such as in security inks and inks for printing tissues and paper kitchen towels. Again these are tested by comparison with a master standard in a typical formulation.

When these dyes are to be used for printing tissues, etc. particular attention must be paid to the properties of the dried print (e.g. soap bleed resistance, ease of bleachability) since the dyes play an important part in determining whether or not the inks will meet the customer's specification.

15.7 RESINS

There are very few classes of ink which do not contain a resin of one form or another, thus before discussing the importance of testing and control procedures an appreciation of the many forms a resin may take is essential. Encompassed within the name 'resin' is a wide range of differing polymeric chemical entities ranging from low molecular-weight oligomers to high molecular-weight complex polymers, and from low viscosity mobile liquids to plastic solids. Beyond these boundaries, resins may also be chemically reactive or comparatively inert, thermally stable or sensitive to light and heat.

The properties of resins selected by the inkmaker are governed by their chemical and physical nature, which in turn are controlled by manufacturing procedure and the purity of the raw materials. Both of these factors may be influenced at any one time for any number of reasons, the economic climate and changes in health and safety legislation to name only two. To ensure that a consistent end product is manufactured, suitable control procedures must be applied to the materials used in manufacture. The simplest and most useful form of resin testing would be to incorporate it into an ink and then monitor the desired properties of that ink. This would be very costly in terms of manpower and time if resin used by an inkmaker had to undergo the same testing. It is therefore desirable to apply empirical tests which are rapid and will convey sufficient information to determine whether a batch of resin is acceptable or not.

The following methods are typical of those employed by an inkmaker to determine whether incoming materials are acceptable against a proven standard.

Acid value

The stability, printability, flow, gloss and pigment wetting are just some of the properties which are affected by the acid value of a resin. Changes in the acid value may not always be detrimental, a higher acid-value resin in an oleoresinous lithographic system may improve the water take-up and thus press performance of the ink. However, a reproducible product is essential and therefore this property must be controlled within set parameters.

The method for determining the acid value of a resin can be found in the section on analysis.

Hydroxyl value

Residual hydroxyl groups in a resin may influence different factors according to the end use of that resin. Hydroxyl groups may affect resin and varnish viscosity via hydrogen bonding; solubility and solvent release characteristics may be influenced for the same reason. The water resistance of a system may be impaired if the value is too high. Similarly, reactive systems may not cure properly if the value is too low. Pigment wetting and thus gloss, transparency and flow may be adversely affected and 'can' stability may suffer if the hydroxyl value is outside of specification.

The exact method for determining hydroxyl value for resins is described in the section on analysis.

Solubility

Resins are often formed into a varnish prior to incorporation in an ink and this may itself be used as a quality test on the resin. This is not always so and in the case of plastisols and some aqueous systems the resin is not in the solution in the finished ink at all. In plastisols the resin forms a plastic solution in plasticizer only when the final print is heated. The complex interactions which occur within ink systems are not as simple as this and many different solubility factors come into play, solvent on solvent, solvent on resin, resin on solvent, plasticizer on resin to mention a few of the many possibilities. If the resin is required to be compatible and stable within a defined system it is relatively simple to test this characteristic. Solubility is a good means of testing a resin's performance within a system and changes in the solubility of a resin are an excellent indication that there is some significant batch-to-batch variation. Differences in the solubility of resins manifest themselves mainly in a change in the solution viscosity. Significant variations in the viscosity of a resin solution indicate that either the chemical or physical nature of the resin is different. As the molecular weight of a resin increases so does its solution viscosity. Similarly, subtle changes in the solvent being used may have significant effects upon the solution characteristics being measured. It is therefore essential to either keep a stock of standard test solution, or compare each new batch of resin with a retained sample of standard material under identical conditions.

Melting range

The melting point of a resin is also characteristic of its molecular weight, thus batch-to-batch changes in melting point indicate a variation in molecular weight distribution which will directly influence solution viscosity, film hardness and gloss. The melting range of a resin is also a very useful test to apply when resins are being considered for certain applications. For example a resin of too high a melting range may have no application in heat-sealing inks. Similarly, if the melt range is too low, marking due to resoftening in reverse printing may prove a problem.

There are two standard methods for determining melting ranges, both of which can be found in the section on analysis.

Colour

Gross colour variations although immediately obvious may not always have a significant effect upon the finished product. Visual colour measurements of a resin are far more realistic when carried out in solution, and instrumental techniques such as a colorimeter or standard colour guides are normally employed.

One of the most important aspects of colour testing is the indication of either age or overheating that may have caused the darkening of a resin. If the colour of a resin is acceptable but outside normal standards, it would be advisable to test thoroughly the other resin properties discussed above.

15.8 VARNISHES AND OILS

Varnishes in this category may be simple solutions of a single polymer or they may be complex mixtures such as the 'cooked' varnishes encountered in paste-ink technology. Oils may be pure or partly polymerized vegetable oils or animal- and mineral-derived products.

Often the most successful test method for these materials is to fabricate them into a finished ink. However, this is time consuming and therefore would not generally be a realistic quality control procedure. Good quality control practice should make use of a sufficient number of reproducible, rapid tests to ensure that the product is the correct material of a suitable grade. Invariably economic pressures force the inkmaker to test only when necessary and thus to rely upon his supplier to submit raw materials of an acceptable quality. Many of the larger varnish and oil suppliers now have BS 5750/ISO 9000 registration, but those which have not will normally be prepared to supply certificates of conformity either prior to a delivery or with it. This goes some way towards giving the inkmaker confidence only occasionally to spot-check incoming material. This is never a wholly acceptable situation; the ideal procedure would involve comparing an incoming raw material with a retained sample of standard material before approval to use that particular batch.

General tests

General tests available to the inkmaker for testing incoming varnishes and oils are as follows:

- colour
- viscosity
- infra-red spectroscopy
- refractive index
- acid value
- iodine value
- diene value (i.e. degree of unsaturation)
- saponification value
- unsaponifiable matter
- hydroxyl value
- water content.

Colour

The colour of an oil or varnish may influence the final colour of an ink or print, but generally the colour is used as an indication of consistency of manufacture. For comparative purposes one can use standard colour tubes or colour comparators such as the Tintometer or Gardner tubes. The colour of an oil or varnish, if it has darkened, may indicate that the sample has deteriorated and, on oxidation, oils and varnishes may no longer have the same drying characteristics.

Viscosity

According to Newton, viscosity is the ratio of shear stress applied to a fluid versus the shear rate, and in an ideal system this ratio should be constant at constant temperature. It is rare that the inkmaker will encounter Newtonian liquids except when testing some oils, as the majority of varnish systems encountered will have some form of thixotropy depending on the resin systems being used. In some oil–ink formulations the resins within the varnish will have been 'gelled' with a chelating agent, and in this case the thixotropic nature of the varnish will be marked. For a Newtonian liquid the relationship between shear rate and shear stress is shown in Fig. 15.3. Deviations from this encountered in thixotropic systems would exhibit a curve as shown in Fig. 15.4. Figures 15.5 and 15.6 are typical viscosity/time plots for a thixotropic system. A suitable measurement technique must be chosen which will reflect the characteristic of the oil or varnish.

There are many types of viscosity measurement equipment available and the choice of which instrument to use will depend largely on the rheological nature of the material to be tested. Newtonian oils and varnishes can be tested using efflux or orifice viscometers such as the Ford, Shell, Zahn and Sheen cups, which are the most basic types of viscometers. For varnishes which exhibit thixotropic properties rotational viscometers such as the Brookfield or Stormer may be used, but other methods such as a cone and plate, or Laray falling rod viscometer are ideally suited, particularly where the yield value of the system is critical (Fig. 15.7).

When testing viscosity specifications it should be noted that viscosity is temperature dependent and all measurements should be made at a specific temperature. Also it is very difficult to obtain correlations between instruments which measure the viscosity in different ways, such as a falling bar versus a cone-and-plate rheometer. Furthermore, it is extremely unlikely that the viscosity measured on a similar instrument will be identical to that specified by the supplier unless the material is Newtonian. If this is a problem the inkmaker must establish their own specifications on the instrument being used, or work with their supplier to establish a correlation between their own instrument and that of the supplier.

The remainder of the tests listed, although not necessarily lengthy, are likely to be beyond the average quality control laboratory and more applicable to an analytical laboratory. However, this will vary between inkmakers and the facilities available to them, and for inkmakers who either wish to carry out some of these tests within their quality control laboratories, or have separate analytical laboratories, the test methods can be found in the section on analysis.

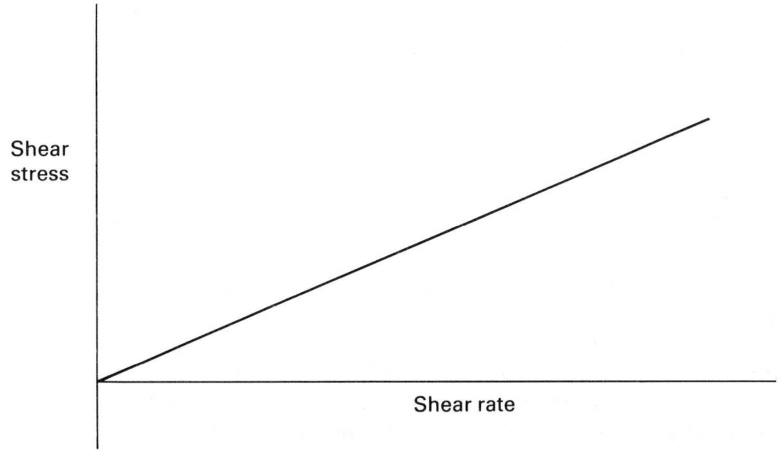

Fig. 15.3 Shear stress versus shear rate for a Newtonian liquid.

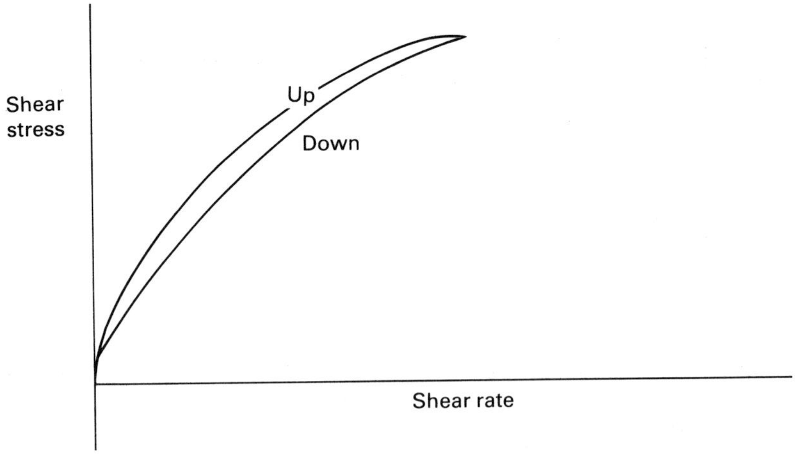

Fig. 15.4 Change of shear stress with varying shear rate for a thixotropic system.

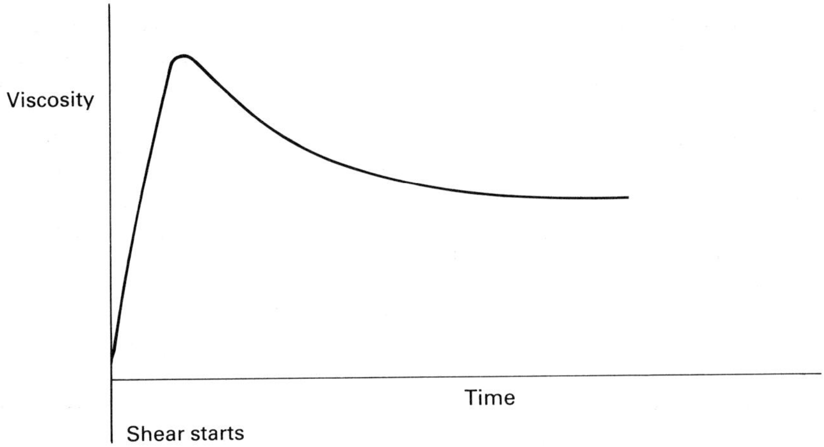

Fig. 15.5 Change of viscosity with time at constant shear rate for a thixotropic system.

15.9 SOLVENTS

Solvents are all too often considered to be discrete chemical compounds, yet even highly refined single compound solvents contain low levels of impurities. Solvents commonly used in the printing ink industry are invariably a technical grade and therefore contain residual impurities resulting from their method of manufacture. Most solvents have a characteristic odour which may be used as a first indication of the nature of the solvent, and can also be indicative of contamination. Odour from ester or ketone contamination in mineral-oil solvents is easy to detect and

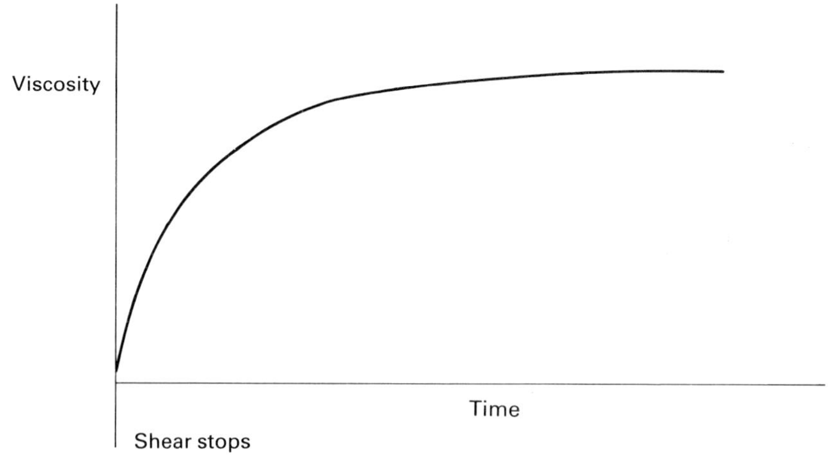

Fig. 15.6 Re-establishment of original viscosity with time of a destructed thixotropic system.

can easily occur if a road tanker is inadequately cleaned between deliveries. Quality control of solvents involves a blend of modern instrumental techniques such as IR spectroscopy and gas–liquid chromatography (GLC) with traditional testing procedures.

Infra-red spectroscopy may be used to give a chemical 'fingerprint' of the solvent under test. No two different compounds will give exactly the same spectrum. The IR spectrum of the test sample should be exactly superimposable over that of the standard. Variations from this would indicate the presence of impurities. It should be noted that solvents of large extinction coefficient (absorbance in the IR region) may well conceal significant levels of low extinction-coefficient impurities.

Gas–liquid chromatography, a vapour phase separation technique, may be successfully used to identify the test material, its boiling range and its purity. Spurious chromatographic peaks obtained from a single compound solvent may automatically be labelled as impurities. It is not often possible to do this for solvents derived from petroleum as the chemical composition will vary with the raw material supply.

Petroleum solvents may be usefully 'fingerprinted' using GLC and significant changes in composition or aromaticity may be detected.

High performance liquid chromatography (HPLC) and column chromatography are also finding application in solvent control. A particular application of these techniques is the determination of aromatic content.

Boiling range

The boiling range of a solvent may be cited as the literature value provided that the level and nature of impurities is within the acceptable level. The boiling range may be readily determined by distilling a known volume of the solvent, recording the initial boiling point, dry point, and

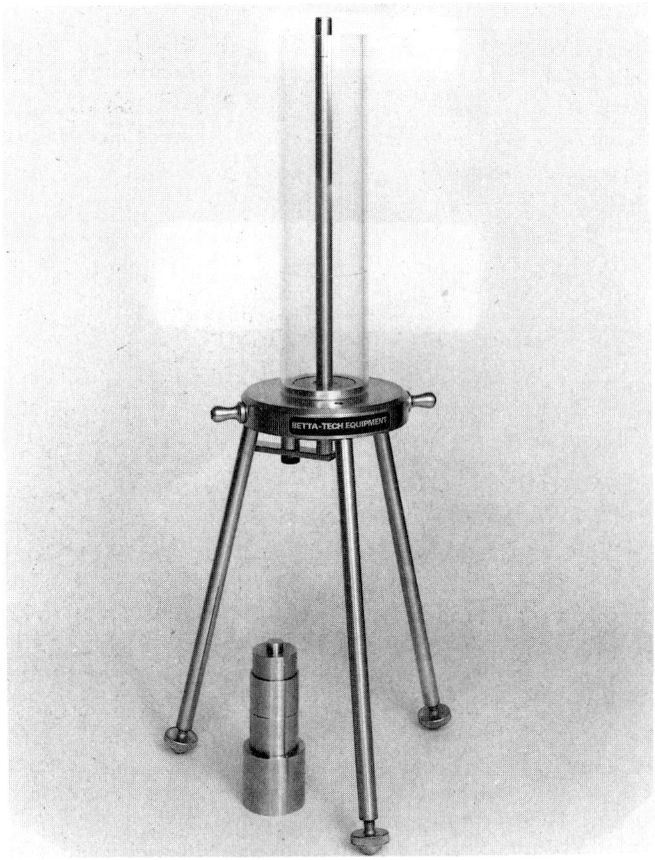

Fig. 15.7 Falling rod viscometer with temperature-controlled waterbath. (Courtesy of Betta-Tech Equipment Ltd)

the temperatures at which specific volumes of the solvent have distilled over. This method is covered by BS.638: 1962.

Simulated distillation by GLC may also be used especially when utilizing modern very high performance capillary columns with computerized data handling.

Relative density

The relative density (or specific gravity (SG)) of a liquid is the ratio of the density of the liquid and the density of air-free distilled water at the same temperature. The density of a solvent is commonly determined using a hydrometer or by weighing using a calibrated density bottle. Useful information about the purity of a solvent can be obtained from density information. Aliphatic and aromatic solvents of the same boiling range may be distinguished as the aliphatic hydrocarbons normally have a lower density.

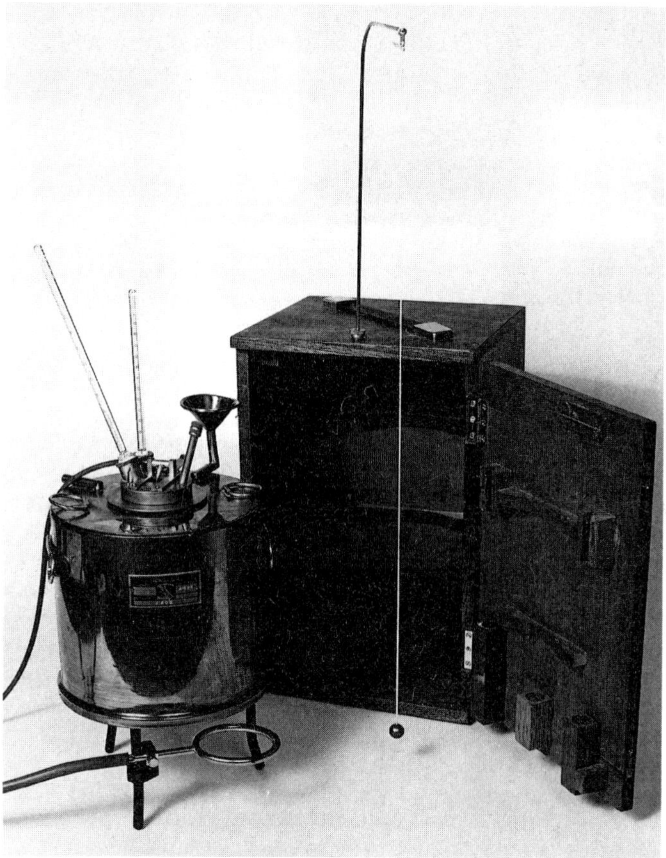

Fig. 15.8 Abel apparatus for measurement of flashpoint.

Flash point

The flash point is defined as the lowest temperature at which the vapour above a liquid will ignite when brought into contact with a flame or spark. The flash points of the more volatile solvents encountered in flexographic and gravure ink technology are usually very low. Ethyl acetate has a flash point (closed cup) of –4°C. The hazards involved in handling, transporting and storage of flammable solvents have led to definitive legislation covering all aspects of exposure, handling, transporting, packaging and labelling. There are three approved methods for flash point determination using the Abel apparatus (Fig. 15.8), Pensky-Martens apparatus and Cleveland open cup. The open cup method is preferred for solvents of flash point greater than 80°C. The details of these methods may be found in BS 2000: Part 4: 1983, IP 34, 35 and 36.

The flash point of a mixture of flammable solvents can usually be taken as that of the lowest. If any doubt exists the flash point of a mixture should be determined according to one of the approved methods.

There has been a review of current legislation to allow for mixtures which have a flash point when measured using one of the approved methods, but which do not support combustion. Any solvent with a flash point below 21°C is deemed highly flammable whether it supports combustion or not. Any solvent with a flash point between 21 and 55°C and which supports combustion at 55°C is deemed to be flammable.

Typically an aqueous emulsion containing a low percentage of isopropanol may have a flash point between 21 and 55°C but would not support combustion. Such a mixture does not require a 'flammable' label.

Aromatic content

Commercial grades of petroleum hydrocarbons contain varying proportions of aliphatic, naphthenic and aromatic compounds. The levels of these vary according to the source of the raw materials and the methods of refinement. The presence of aromatic compounds leads to greater solvency, a tendency to swell the compounds used in producing litho blankets, increased odour and toxicity. The effect of small differences in aromatic content on the solubility of resin systems should not be underestimated, particularly when being used to produce varnish systems using resins with limited solubility characteristics. The aromatic content is normally measured by aniline point or kauri-butanol value, and details of how to carry out these tests can be found in the section on analysis.

Water content

The presence of water in solvents invariably leads to problems such as poor adhesion, lack of gloss and flow in inks. Details of how to determine water content can be found in the section on analysis.

Refractive index

The refractive index of a solvent can be used as an indication of its purity, the method for determining refractive index being found in the section on analysis.

Sulphur content

Petroleum products may be contaminated with sulphur compounds from two routes. Sulphur compounds occur naturally in crude oils and may not be removed by distillation or purification, and may also be introduced when aromatics are reduced by treatment with concentrated sulphuric acid. Sulphur compounds lead to tarnishing and odour problems.

The usual test for sulphur consists of immersing a polished copper strip in a known volume of the solvent at elevated temperature.

Compounds of this nature may also be detected by GLC using a flame photometric detector which is very sensitive to sulphur compounds.

Odour

Odour, although a subjective parameter, is becoming of increasing concern to the inkmaker, particularly in the field of food packaging.

Very simple odour tests such as allowing solvent to evaporate slowly from a filter paper and then examining for residual odour may be employed.

Odour tests are normally carried out using a panel of testers and comparisons are made against proven standard materials using a simple numerical scoring system.

Particular attention should be paid to bulk tanker deliveries, and each tanker compartment should be tested for odour, small amounts of residual ketones or esters in a compartment leading to major odour problems.

15.10 RADIATION-CURING PRODUCTS

Products which are polymerizable by the action of UV light, or by the bombardment of electrons in the case of EB curing, are by their very nature reactive and unstable. Any quality control procedure or testing method must be concerned with the unique properties of these compounds. The compounds used in radiation curing inks fall into four categories.

- Diluents – these consist of reactive monomers generally based on the reaction product of a polyol, or alkoxylated polyol with an ethylenically unsaturated carboxylic acid.
- Resins – this category consists of reactive polymers of the acrylate type, varying in molecular weight from viscous liquids to solid resins.
- Photo-initiators – compounds which form free radicals when exposed to UV radiation.
- Synergists – a class of compounds, generally amines, which promote the action of photo-initiators.

Diluents and resins

The role of the diluent and resin is to produce a curable film via a printing process, therefore all the usual tests for colour, viscosity, curing speed and odour may be applied.

Some reactive monomers are a blend of isomers whose reactivity may vary. It may therefore be of use to examine this distribution using gas chromatography or HPLC.

Residual acrylic acid in either the monomer or polymer can affect the curing rate, odour, shelf-life and skin irritancy of the product and the acid value should therefore be determined.

The unstable nature of these reactive products require that the shelf-life should be examined. Stabilizers such as methoxyphenol are incorporated into diluent monomers to prevent premature reaction leading to viscosity increases, gelation and darkening, and it may be advantageous to monitor the levels of these. Inhibitors can be readily determined using HPLC. If instrumental techniques are not available the conventional method for examining shelf-life by accelerated testing may be applied. Accelerated testing for stability may be done by placing samples of a standard material and the test material in an oven at temperatures ranging from 30° to 60°C and then comparing viscosities and colour after suitable time periods.

Photo-initiators and synergists

These classes are, with the exception of amine acrylates, invariably pure compounds and as such may be tested using analytical methods. Infra-red spectroscopy can be used to generate a spectrum of the sample which may then be compared with that of the standard. The majority of these compounds may also be examined using GLC or HPLC. All of these techniques can be used to identify the compound and its purity. The only other applicable test is to fabricate the initiator into a standard vehicle and assess its performance under fixed curing conditions.

15.11 MISCELLANEOUS MATERIALS

Driers

Driers are oil-soluble metallic salts or soaps in which the metal is rendered soluble by reaction with a long-chain acid (e.g. octoates) or complex mixture of acids (e.g. naphthenates). Variation from batch to batch can therefore be associated with the organic portion or with the type and quantity of metal ion.

Comparison of the batch against a known standard using IR spectroscopy will characterize the organic portion of the drier. The type and quantity of metal can be accurately determined by either atomic absorption (AA) or HPLC. All these techniques will be discussed in the section covering analysis.

Plasticizers and pure chemicals

These materials can be simply tested by IR spectroscopy. The spectrum of the test material should be exactly superimposable on that of the standard.

Waxes

Waxes fall into two main categories for raw material testing purposes. The simple chemical types such as fatty acid amines can be treated as pure chemicals and analysis by IR spectroscopy as indicated in the previous section. The more complex polymeric waxes (e.g. paraffin and polyethylene waxes) cannot be identified by this analytical method and the following more traditional methods have to be employed:

- melting point determination similar to that used for resins;
- solubility characteristics at room and elevated temperatures;
- empirical end-product testing.

Others

Many miscellaneous items are often included in small amounts in printing inks for specific uses, e.g. pigment-wetting aids, flow additives, anti-foams, and these can only usually be tested by assessment against a standard in a typical formulation.

15.12 INK QUALITY CONTROL

Once there is absolute confidence that all raw materials being used for manufacture are within specification, then the primary objective of quality control is to ensure that all finished products manufactured consistently conform to that specification. Confidence in raw material specification is gained, as discussed previously, by either using quality-assured suppliers, insisting on certificates of conformity, by inspection, or by using a combination of all three.

Quality control of the finished product alone is inadequate and quality control should be applied throughout the processes used for manufacture. This is to ensure that there is no deviation from the formula during manufacture and that the final product will match the master standard in all respects.

Testing of the product can commence as early as the mixing stage, it often being easier and more cost effective to correct major variations in strength or rheology while the batch is still on a mixer. If a batch is not tested until extensive processing has taken place additions of dry pigments or pastes can necessitate expensive reprocessing to reach the same stage of manufacture, which is obviously undesirable.

In both oil and liquid ink manufacture there are fundamentally two types of grinding apparatus, i.e. continuous and batch mills. Continuous mills present a problem to the quality control laboratory, since while the sample is being tested the mill, if left running, could be producing ink inferior to the standard, or alternatively if stopped will result in lost production time. Hence a series of quick tests must initially be carried out, followed by a more rigorous test procedure when the milling operation is complete. Batch mills present the same problem but to a lesser extent, it often being possible to stop batch mills whilst testing is being carried out. Triple-roll mills present special problems for quality control as they have a sieving effect as well as grinding the ink. This means that although the product may be homogeneous when entering the mill, it could well differ during a run, larger particles being held back until later in the grinding run. The other problem with triple-roll mills is that they are very sensitive to minor alterations in mill settings, which could be inadvertently due to a change in mill temperature as it warms up, or due to an operator changing settings when a shift changes. As most of the parameters of the final product, ranging from strength to viscosity, depend very much on the degree of dispersion this can cause problems with lack of homogeneity throughout the batch, and regular tests for fineness should be carried out during a run to ensure that the degree of dispersion has not changed. Whichever form of mill is being used it is recommended that the ink is mixed before a final sample is taken for testing to ensure that it is representative of the whole batch rather than being just the last portion off the mill.

Care should be taken when finally approving a batch of ink to ensure that there will be no change in the ink between it being passed by the quality control laboratory and the ink getting into a package. Inks are often passed from post mixers when a further process such as loose milling into a package will take place, possibly affecting the physical characteristics of the finished ink. If a batch is to be approved but is still to undergo another process, even if that process is only running into a package through an 'open' mill, the sample being tested should have had the same processing, i.e. run through an 'open' laboratory-size mill prior to testing. It should not be taken for granted that further processes such as potting off will have no effect on the final product, every step being taken to ensure that what is passed by the quality control laboratory is in fact what reaches the customer.

In any ink quality control area, it is important to have good lighting which should be as near 'white' as possible. The more complex forms of lighting used for colour matching purposes are not necessary as all quality control procedures should be on a comparative basis of batch versus master standard, and any distortions of hue due to an off-white light source will be equal for both inks.

For convenience, the quality control of inks will be divided into four sections:

- short-term ink testing
- long-term ink testing
- press performance tests
- dry print performance tests.

Under each heading, tests common to both oil and liquid inks will first be considered and then specific tests on each of the two basic types of inks will be given.

15.13 SHORT-TERM INK TESTING

There are tests which can be carried out rapidly while the ink is still on the mill and includes checks for dispersion, flow, strength, hue, opacity (or transparency) and gloss.

Dispersion

A satisfactory level of dispersion (i.e. equal to the standard) is the fundamental requirement for any printing ink and until this has been achieved, any further testing is unnecessary. Dispersion can be adequately checked on a fineness of grind gauge as described earlier in this chapter.

Rheology

Before comparative prints can be obtained, to compare the strength and hue, the test ink must be of similar rheology to the standard. Provided the ink is not potted directly from the mill, it is not necessary at this stage to adjust the ink accurately for any small flow variations as this can be carried out after the ink has been taken off the mill.

Tests on the rheology of oil and liquid inks are quite different and hence will be considered separately.

Oil inks

Rheological parameters are extremely important in controlling the quality of oil inks. While many of these parameters can be assigned limits, others involve an assessment based purely on a comparison with an agreed standard. This subject has been discussed fully in Chapter 13 and only points pertinent to quality control will be further considered.

Feel on the slab

Observation of the body of an ink, the length of flow and 'stringiness' resulting from the rapid lifting of a flexible palette knife from the body of an ink on a slab can provide valuable information in a short space of time concerning certain vital properties of the finished product. Comparison against the master standard can be effected in a similar manner and substandard products can be readily identified even by the novice ink technician.

Ink viscosity

Ink viscosity can be measured in a variety of ways, giving values for both the plastic viscosity of the sample and its yield value. These two values will in many instances give information concerning the anticipated performance of a product. A lower ink viscosity will be required for web-offset ink, for example, than for a similar product for sheet-fed lithographic applications.

Certain applications will also demand a high yield value particularly in metal-decorating applications where varnishing on wet inks is desired. A low yield value, conversely, may be paramount in overprint varnishes where good flow and subsequently high gloss may be required.

Ink tack

There are several different instruments which will measure the tack of an ink. None of these gives an easily defined tack or stickiness value but are a related property which is dependent upon the design and materials of the instrument itself. The instruments measure the forces needed to overcome the resistance of an ink film to splitting between two moving rollers. This simulates the ink-film splitting which occurs during the distribution of ink on a printing press roller train.

The most usual instrument is the Tack-O-Scope tackmeter (Fig. 15.9), which can be found in the laboratories of the larger printer as well as the inkmaker. This instrument has the optional advantage of using a chart recorder. This gives a permanent record of the results obtained. A further use is checking tack stability over a set period of time.

Other instruments include the Thwing Albert Inkometer and the Churchill tack meter. Precise tack values are very important in obtaining the best results in multicolour wet-on-wet applications where the precise control of ink trapping is essential.

Ink flow

Ink flow has become a more significant test in recent years with the introduction of much higher press speeds, where transfer is important but the ink must not fly or drip from the duct. Flow characteristics can be measured comparatively by way of reference to a standard, but care must be taken if using this method as freshly manufactured ink may change considerably on standing, even overnight. For higher-viscosity products, such as sheet-fed inks, flow tests can be comparative by knifing both samples vigorously on a bench and then visually assessing the flow from a palette knife.

More precise tests can be carried out by applying 0.5 cc of ink, which has been worked on a bench for 30 seconds, onto an inclined or vertical plate using a narrow blade knife. By using this method numerical specifications can be set for the length of flow down the glass in a 15-minute period, and this is often used, in combination with a vertical glass or perspex plate, as a standard test within the newsink industry.

Liquid inks

Many liquid inks when first sampled from a mill will be above room temperature and may also be quite heavily aerated. All efforts should, therefore, be made to equalize as far as possible the batch sample to the

Fig. 15.9 The Tack-O-Scope tackmeter. (Courtesy of Testprint BV)

standard. The viscosity can then be checked with a Ford or Zahn cup, while the relative flow properties can be assessed by stirring the inks with a palette knife and then comparing the flow off the knife as it is withdrawn from each of the inks.

Strength, hue and opacity

The 'drawdown' technique as described earlier in this chapter is sufficient as a quick test for comparison of the relative strength, hue and opacity of the batch against the standard.

Gloss

Relative gloss is more difficult to assess on a drawdown and therefore prints of the test ink against the standard should be prepared. For liquid inks, a drawdown with a wire-wound applicator is suitable (Fig. 15.10). This technique gives an excellent comparison between two inks provided they have similar flow. The applicators are available for manual drawdowns. Automatic applicator models with a built-in applicator bar can be obtained so that a consistent pressure can be applied on consecutive prints.

Gloss may be assessed subjectively by visual means. However gloss meters with the capacity for variation of the angles of incident illumination and reflected light are in common use. These provide a numerical measurement of gloss under standardized and reproduceable conditions. The instrumental method uses a light source which can be directed so as to fall upon a surface to be tested at a known angle to the normal of that surface. The reflected light is then measured by a photocell set at the corresponding angle to receive the reflected light. The procedure is repeated with standard gloss and matt surfaces; these are usually tiles. The comparison of the readings is then expressed as a percentage reflectance. The results obtained do not necessarily agree with those obtained visually, since the eye will judge the relative gloss levels at a variety of angles and will also take into account other factors such as print imperfections and colour variation. Visual assessment is, therefore, also important since the printer is concerned with the total visual appeal of his finished print. Discrepancies in the gloss level of the batch sample could be caused by a misweigh of one or more of the ink components or, on the continuous type of mills, it could be due to an incorrect mill setting or condition. Hence, by checking the gloss level at this early stage, any mechanical adjustments necessary can easily be carried out.

Additional tests on inks cured by radiation

Inks 'curing' under radiation (UV or electron beam) vary in their rate of polymerization. The end-point, when the ink is 'dry' on the substrate, is somewhat more speculative.

For quality control purposes, it is necessary to ascertain whether an ink batch is curing at the same rate as the standard. This is normally carried

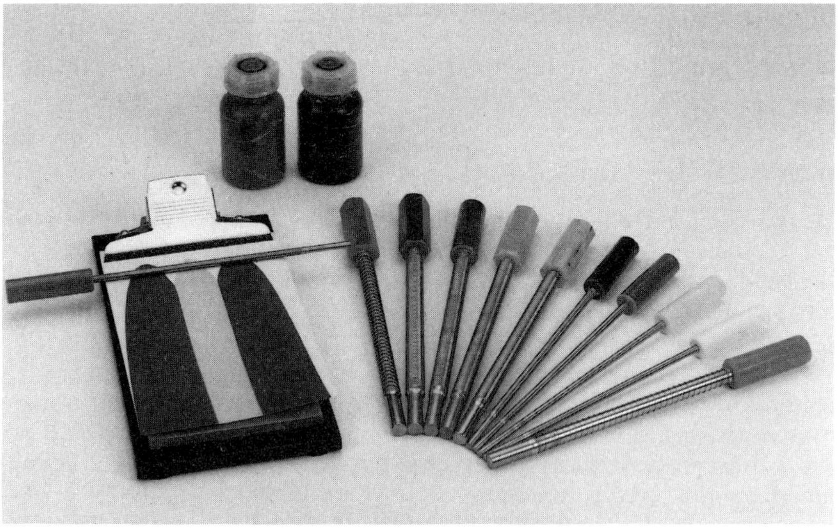

a

b

Fig. 15.10 A wire-wound applicator: (a) K bars; (b) applicator. (Courtesy of RK Print-Coat Instruments Ltd)

out qualitatively using a finger rub test at a given exposure time under the source of radiation (e.g. 2 lamps at 500 fpm at a film weight of 2 mμ). There are certain situations, however, when it is difficult to distinguish between cure and adhesion. An ink may be dry to the touch, but easily removed under the finger rub test. This can best be judged by experience.

Very slow rates of polymerization can still result in 'chalking' as in the case of rapid setting conventional inks on certain absorbent stock. Cure rate can also be measured by employing a rub test on the substrate after a given radiation exposure. Indeed, the toughness of a cured film, which may reflect the degree of cure, can be checked, for example, using the PIRA rotary rub tester. A cured ink can thus be described in terms of its resistance to rub under given conditions.

15.14 LONG-TERM INK TESTING

The tests already outlined under section 15.13 should first be carried out on a representative sample of the final ink. Alterations should be made as necessary, ranging from additional pigment to increase strength to diluents to lower viscosity.

Generally if an ink is too high in tack/viscosity then a diluent addition of a solvent or oil should be added to lower the solids concentration of the ink. If the tack/viscosity is low then it should be raised by the addition of more solids, be it pigment in the form of an extender, or resin in the form of powder or varnish. Generally, batches will be formulated so that the former scenario is experienced rather than the latter, as it is far easier to lower the tack/viscosity of a product than to raise it.

If the ink has a flow which is too short, this again suggests that the solids content is too high and a further addition of varnish/oil is normally used. If the flow is too long then the ink can be bodied by increasing the solids content or, if it is undesirable to raise the viscosity further, a rheological shortening agent can be used.

A more precise strength test can be carried out by assessing the batch against the standard using the bleach technique as described in section 15.3. Additionally, both oil and liquid inks can be tested for stability on ageing. This test is usually accelerated by placing the inks in a well-sealed container in an oven preset at between 35 and 50° C. After 24 hours the sample is allowed to cool to room temperature and checked for any separation and/or loss in flow characteristics compared with the bulk. A comparative drawdown of the heated and unheated sample is then prepared and examined for changes in strength and hue. Caution should be exercised with this test, however, as some pigment–vehicle interactions can cause a shade change to occur at these temperatures over a period of time.

Specific tests on oil inks

Drying time

The drying time of an oleoresinous ink is important to know for reasons of both press stability and print drying. Optimum drying rates are a compromise between these two extremes and will obviously depend upon the nature of the particular application. Before discussing methods available to measure ink drying times, it is important to note that this term refers to the time taken to dry a print of a particular ink film thickness on a certain substrate (e.g. paper/board/glass).

Factors affecting drying time of an oleoresinous system include:

- temperature
- humidity
- availability of atmospheric oxygen – a single laboratory print left on a workbench will have a greater access to the atmosphere than a production situation of printed stacks of varying pile heights.

In carrying out a manual assessment care should be taken to ensure all fingers are clean, as a build-up of deposit on the fingers can severely affect the test. In cases where powdering of the ink may be mistaken for drying failure, the print should be warmed in an oven for a few minutes. This will aid drying, but have little or no effect on powdering. For quality control purposes, one rarely needs to know the exact time of drying, but merely whether or not the print is dry within a certain time. However, since atmospheric conditions can vary quite widely, it is always imperative to carry out the test against a standard ink.

Automatic drying time testers

Exact measurements of drying times on prints is obviously both time consuming and laborious, and one to which automation has been applied. Unfortunately, no equipment exactly reproduces a real situation. Automatic test equipment can only be used to give relative drying times.

PIRA ink drying time tester

This equipment was described in a paper by Coupe entitled 'Ink and paper testing' in the Proceedings of the PATRA Letterpress Conference, 1955. Modifications to this equipment are currently available in other similar forms.

The equipment utilizes a glass upon which is printed the ink in question (or the standard) and which is subsequently rotated. A track is scored in the wet film using a metal stylus and as the ink dries, the smooth track pattern changes to a tearing pattern. At this point where the track changes its pattern, the ink is designated dry, and the time required to reach this point is classed as the drying time.

The main drawbacks to methods based on glass plate prints are the reproducibility (which can be poor) and the remoteness of the system from behaviour on absorbent substrates. It is, however, a rapid method suitable for quality control purposes and gives information concerning the ink drying time on non-absorbent surfaces such as rollers.

IGT drying time recorder

The IGT drying time recorder measures the drying time of an ink or the suitability of a particular paper for a given ink in two ways. The first is the set-off at preselected time intervals over a period of up to 24 hours and the second is related inasmuch as one examines the ink removed during the dwell time of a pressing-bar, over the same period. Various ink film weights can be examined on any paper. Whenever the moving bar is in contact with the wet ink film either disturbance of the test strip is observed or set-off on a backing strip is noticed.

BK drying time recorder

The BK drying time recorder is very similar in the method of measurement to the PIRA drying time tester, utilizing glass strips upon which the ink is printed. The glass strips are placed on a flat bed down which six hemispherical-ended needles travel, in a time between 6 and 24 hours. The recorder defines the following stages in the drying process.

- The first stage is a pear-shaped impression corresponding to the time taken for evaporation of the solvent.
- The second stage is the cutting of a continuous track, corresponding to a sol–gel transition.
- The third stage is an interrupted track corresponding to the surface-dry time.
- A fourth stage is where the needle no longer penetrates the film, corresponding to the final drying time.

The advantages and disadvantages of this system are similar to those given for the PIRA drying time tester.

Setting time

Ink setting time is governed by the choice of paper as well as the ink film weight used. A control should thus be run under similar conditions. An approximate qualitative test for setting time can be carried out by careful finger assessment checking surface tackiness. It is also noteworthy that setting is often accompanied by a change in surface gloss and the wet glossy appearance after initial printing is reduced as the ink sets.

A more quantitative test utilizes a 'back-up' technique on a laboratory proofing press or printing unit. Back-up in this context refers to the process whereby an extra sheet of paper is applied to the surface of the printed substrate and run through the press again. Wet ink will be transferred to the second sheet of paper. The time interval between printing and backing up is kept constant for a series of prints (e.g. 1, 3, and 5 minutes for quick-setting inks) and the amount of ink transferred to the backing paper is inversely related to the setting time. The time required for no back-transfer of ink is thus a function of the setting rate and gives an indication of the 'back-up' time when printing the reverse of the sheet.

Ink fly

The problem of ink 'misting' or 'flying' in the vicinity of a press is a serious nuisance which can largely be overcome by formulation changes. This phenomenon is often a function of the nature of the vehicle, but will obviously also depend upon a multitude of other parameters, e.g. pigment, extender type and level.

It is comparatively easy to monitor ink fly in the laboratory on a given batch of ink, given a two-roller system. It is preferable to have one roller oscillating to reduce the tendency of the ink to form ridges. A useful piece of equipment in the laboratory for this purpose is the tackmeter or inkometer.

Comparisons between different inks can be undertaken quantitatively by either:

- weighing the weight increase per unit area on the collection device;
- visually assessing the droplet distribution per unit area.

Any test for ink fly ideally should try to reproduce the most severe conditions likely to occur in the field on a given press. The parameters for consideration should thus be the following.

- Ink film thickness on the roller surface – generally ink fly will increase as the ink film thickness increases.
- The temperature of the roller surfaces – certain inks will dramatically change their rheological properties as the temperature changes and will generally decrease in viscosity as the temperature rises. As modern roller temperatures may be appreciably above the surrounding air temperature, this factor should be remembered when assessing ink fly.
- Roller speed – it is important to remember that ink misting primarily occurs in the ink-roller chain. It is worth noting that speed of separation is higher as the roller diameter decreases, hence a greater tendency for ink fly will arise on small diameter high speed rollers such as those in an inking-roller chain.
- Ink consistency or rheology – a low bodied ink will show a predictably greater tendency to mist than a high bodied ink.
- Siting of the target – in measuring misting levels of a variety of inks, it is important to site the target at a point which reflects this level. On a tackmeter for instance, a convenient point is at the same height as the nip between the two rollers, on the rear wall of the measuring chamber. It is perhaps surprising that the position of maximum fly should be on the opposite side to the parting nip of the rollers, where the filaments are formed and broken: the explanation appears to be that particles are carried around in a cloud which follows the roller surface until it collides with the cloud following the other roller which will be at this very position. It is also important to note that a percentage of heavy droplets, which are often termed 'sling', will be formed on the base of the measuring chamber. Although this may or may not be a nuisance in the same way as ink mist is (depending upon the design of the press), it may result in the necessity for constant press cleaning, and does not lead to a clean operation. This should also be monitored carefully.
- The complete enclosure of the test roller system is important to minimize operator exposure to ink mists, by restricting the spread of particles, and to exclude draughts which could affect reproducibility.

Specific tests on liquid inks

The drying rate of liquid inks is a critical factor since an ink which is too fast drying will cause printing problems, whereas a slow drying ink could give rise to excessive solvent retention. Also, an incorrect solvent balance, especially in a system where two or more resins are present, could lead to premature precipitation of one of the resins leading to poor printability and a drop in the overall gloss level. Tests on the drying rate can be carried out by drawing down side by side the test and standard inks on a grinding gauge and comparing the relative evaporation rates at different film thicknesses.

Pigment settling

This cannot easily be tested except by examination of a batch sample over a long period of time which is, of course, completely impracticable. However, the batch sample should be allowed to remain undisturbed for 24 or 48 hours before being examined for any signs of either a supernatant layer or pigment sediment on the bottom of the can.

Water-based inks

These should additionally be checked for foaming and pH variation. The batch and standard inks are either stirred or shaken for a set period of time and compared for the degree of foaming. Both inks are then allowed to stand for a few hours to check on the ease of dispersability of any foam formed on the above test. Excessive use of anti-foams should be avoided as they can lead to printability problems and stereo swelling.

The pH of a conventional water-based ink (i.e. one in which an acidic resin is solubilized with ammonia or an amine) is an important factor since if it is too high the ink would be too slow drying due to the excess of amine present. Alternatively, if the pH is too low, then large viscosity fluctuations can occur and the resin can even be precipitated out of solution. A check on the pH of the finished ink can simply be carried out by using a standard pH meter.

15.15 PRESS PERFORMANCE TESTS

Oil inks

The ultimate test for examining the performance of an ink is to run it on a press, and there is no laboratory test which can precisely predict the performance on press. However, there are some tests which can be carried out in the laboratory to predict, to a certain extent, the behaviour of an ink on press, although many would not normally be within the scope of a normal quality control specification. Quality control of most printing inks will rely on the principle that if the standard ink performs satisfactorily on press, and the batch being tested is made with the same materials, by the same method, and to the same specification, then it can be reasonably assumed that this test batch will also perform satisfactorily. Some inkmakers may have the facility to test products on presses identical or similar to those to be used by its customers, and in this case may choose either to spot check, or test every batch for press performance. However, this will involve considerable test time and expense and will tend to be limited to products which are extremely critical and where larger contributions can justify the resources needed.

Those inkmakers which do not have press facilities available will use one or more laboratory instruments and tests in an attempt to predict the final press performance of a batch of ink. These tests which range from those which will measure such parameters as transfer of the ink, to the water-uptake properties, the complexity of the instruments being used will be dependent on the resources available to the inkmaker.

The IGT printability tester (Fig. 15.11) exhibits certain features of the printing operation and is well used for both ink and paper testing. It consists of an ink distribution system and a print unit (Fig. 15.12). This printing unit is comprised of a 150° sector of impression cylinder suitably covered, on which a 1-inch (25-mm) wide strip of paper is placed. Ink is applied from the distributor to the rim of the black-centred wheels, which act as the rotary printing plates. Power is furnished by a pendulum or spring and in later models by electrical means. One has thus the choice of either having a constant speed or varying speed impression. In the case of the latter, the paper strip can be graduated along its length in terms of

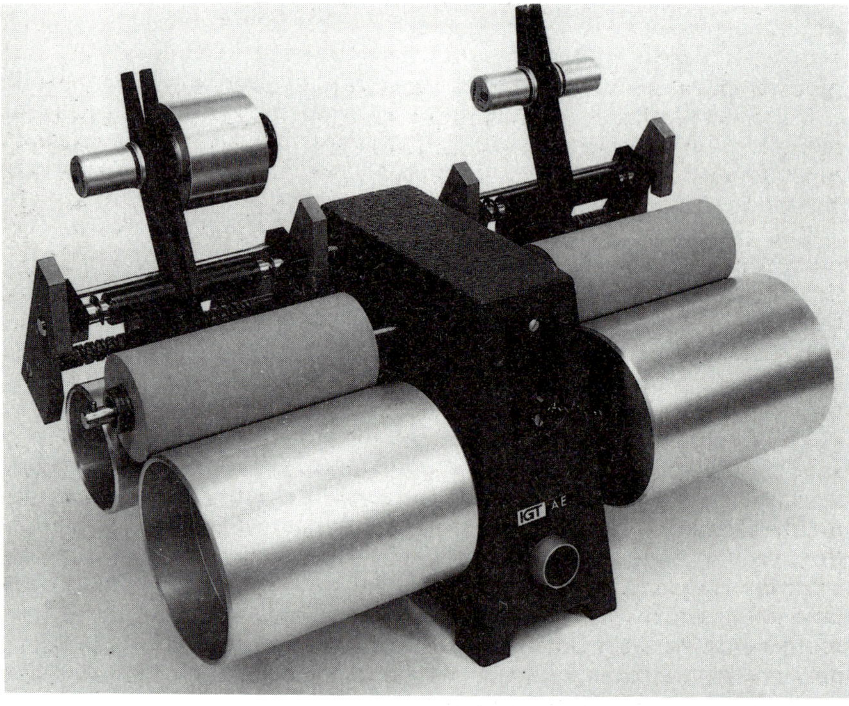

Fig. 15.11 Distribution unit of the IGT printability tester. (Courtesy of Testprint BV)

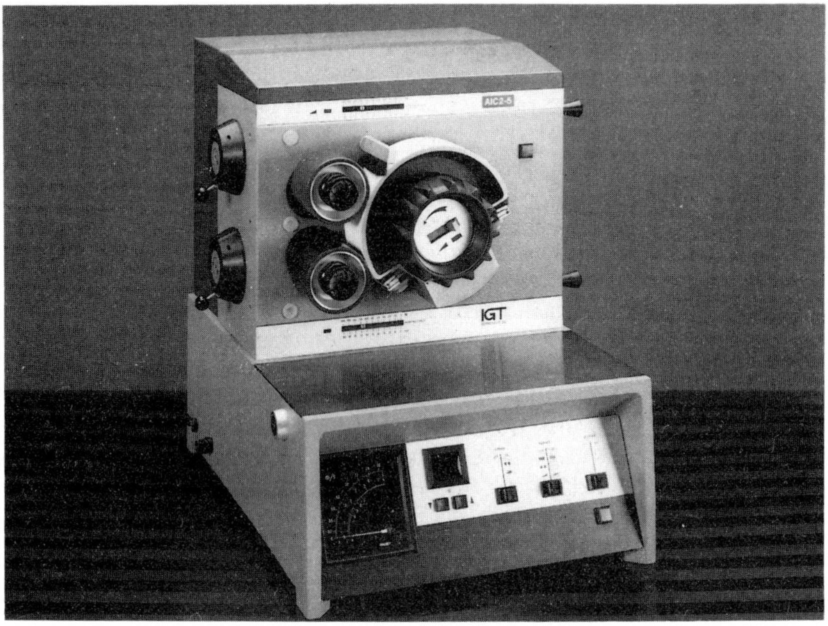

Fig. 15.12 The print unit of the IGT printability tester. (Courtesy of Testprint BV)

speed at given pressures, which themselves can be varied.

Under controlled conditions of humidity and temperature, this instrument can be used to examine a whole variety of printing properties together with comparisons of ink and paper. The manufacturer's information is recommended for a full list of these applications.

One property missing from the IGT list involves the lithographic properties associated with water–ink interfacial phenomena. Although the IGT is not designed specifically for examining water–ink relationships, there are pieces of equipment available to do this. The Pope & Gray Litho-Break Tester is perhaps the most simple and commonly used equipment, whereas the Prüfbau Lithomat can also be extended to measure changes in tack, film thickness and water take-up.

The Lithomat consists of a steel roller that revolves in the fountain solution. This roller rotates a rubber roller which presses against another steel roller that lies above it. Both steel rollers oscillate to ensure an even ink distribution. The effect of the fount solution on tack and thickness of the ink on the rollers can be assessed by combination with a graphometronic instrument which measures the tack as the force exerted on a roller driven by the inked roller. The change in tack and film thickness is recorded with time. The final water content take-up by the ink is measured by removing a sample from the centre rubber roller and analysing by the Karl–Fisher method.

The Tack-O-Scope tackmeter can also be used to assess the lithographic characteristic of an ink, in many ways being similar to the Lithomat. The

Tack-O-Scope has the facility to apply fountain solution to the main-metal roller using a shallow tray and fount roller, the effect of adding or removing the fount being monitored using a chart or graphometronic recorder.

For the small inkmaker who does not have such sophisticated equipment available there is always the manual Surland type of water take-up test, which is carried out by mixing fount solution into a quantity of ink in a beaker, and measuring the rate at which the fount is taken up. This may be a relatively primitive method of investigating the lithographic properties of an ink but can give useful clues where something is radically wrong with the lithographic characteristics.

As one moves away from conventional oleoresinous systems to, for example, ester chemistry, as in UV curing inks, it is important to ascertain the effect the ink may have on the rollers, blankets and plates of the press. While this is of paramount importance in introducing a new product, it is not necessary for routine quality control testing. The assumptions made earlier should hold.

Liquid inks

The majority of solvent-based inks are printed by either the flexo or gravure processes, although screen printing is also important in certain specific areas. Whatever the process, the performance of the ink on the press is obviously a critical factor and is one which should be carefully examined in the quality control procedure.

Gravure printing can be carried out in the laboratory by using a scaled down version of the production model, the rollers being either mechanically driven or even simply turned by hand. Alternatively, small gravure proofing presses are available (Fig. 15.13) and these, although ideal for quality control work, suffer from the disadvantage that only a single print can be obtained before cleaning up the rollers. All inks should be thinned where possible with the recommended reducers, although the use of slower evaporating solvents may be necessary to obtain clean prints at the relatively low printing speeds obtained in the laboratory.

Before any printing is undertaken, the proposed end use of the ink should be carefully considered and the following questions raised.

- Is the ink expected to be used for printing solids, tone areas, or a combination of both?
- Is the ink part of a process set?
- Is the ink to be used for surface or reverse printing?
- For coloured inks: is the ink to be used in conjunction with a white? Conversely, for whites: is there a typical coloured ink also used when printing this white?
- What substrate(s) should the ink be printed on?

The answers to these queries will provide the quality control tester with a guide in assessing the press performance of the test ink. It is important in this context that any inks to be used in conjunction with the test ink (e.g. for a process ink, the remaining colours of the process set) should also be

Fig. 15.13 A gravure proofer. (Courtesy of RK Print-Coat Instruments Ltd)

printed so that the effect of superimposition of the inks can be carefully examined.

Other problems associated with the gravure process, such as examination for cylinder wear and scumming, are more difficult to reproduce as a standard procedure and a great reliance must be placed here on having a good ink formulation. Several tests have been designed for measuring cylinder wear, but perhaps the most common uses a small trough containing the test ink in which is immersed a glass slide coated with a thin layer of chromium. A weighted arm with a brass weight on the end is then passed back and forth in the ink over the chrome-plated slide for a set time period. The test is then repeated using a control ink on a second chrome slide. A comparison between the degree of chrome abrasion on the trial and control ink slides is an indication of the wear potential of the ink.

Scumming, i.e. the formation of an ink film in non-image areas, can result either from an incorrect press setting or from an ink fault. Assuming correct mechanical adjustments on the press scumming can still be evident, in certain extreme cases from the beginning of a print run, and this is usually due to oversize particles in the ink causing the doctor blade to lift fractionally away from the cylinder surface. This fault should have been eliminated, of course, in the initial quality control, but in other cases where scumming is obtained only after a longer press run the cause is normally more difficult to ascertain and hence to rectify as a standard procedure.

In flexography, the film weight applied is a much less readily defined factor than in the gravure process. This is because the film weight carried can be varied simply by altering, for example, the pressure between the rollers, the speed of running, the ink viscosity, as well as varying the screen size of the anilox and the Shore hardness of the rubber rollers. Hence, flexographic proofing presses have a limited value, and a greater reliance is usually placed on the more versatile anilox hand roller (Fig. 15.14)

This consists of an engraved roller (termed the anilox) and a rubber roller. A print is obtained by placing the ink in the nip of the rollers and rolling the whole unit across the substrate. The film weight applied by this method should roughly parallel a typical commercial flexographic print and can be varied by:

- Altering the ink viscosity;
- Applying more or less pressure on the roll out;
- Using a different screen size anilox roller;
- Using a different Shore hardness rubber roller.

Some raw materials, especially some solvents and anti-foams, can cause swelling of the stereos leading to poor printability and in certain cases it may be necessary to specify the type of stereo material (i.e. natural or a synthetic rubber or a photopolymer plate) to be used with a particular ink. Tests can easily be carried out by immersing a sample of the stereo material in the ink for a short time and measuring the thickness of the sample before and after the test using a micrometer gauge and also the hardness before and after immersion.

Screen inks

Before the press behaviour of a screen ink can be examined, the formulation of the ink should be checked since inks for this process vary considerably in composition depending, for example, on the substrate to be printed and the drying facilities available. Generally, screen inks can be dried by evaporation, absorption, oxidation, polymerization or more often by a combination of these methods. Inks drying mainly by oxidation

Fig. 15.14 An anilox roller. (Courtesy of Weller Patents Development)

(e.g. by the incorporation of a drier) resemble a letterpress ink in many ways and would normally be manufactured in a similar manner (e.g. triple-roll mill). Alternatively, an ink drying mainly by evaporation (e.g. using a cellulose ester in solvent) is more akin to a high viscosity liquid ink and would normally be manufactured on conventional liquid ink milling equipment. Hence, in assessing the press performance of screen inks, many of the difficulties encountered may be a direct result of the formulation type and the reader should then refer to the appropriate section of this chapter.

In the actual printing of screen inks, the best results are normally obtained when the inks are soft and buttery to allow easy passage through the screen and short and thixotropic to give smooth lay and sharp definition. Testing can be carried out in the laboratory simply by drawing a squeegee over a hand screen, the film weight applied being varied by the use of different mesh screens (Fig. 15.15).

15.16 DRY PRINT PERFORMANCE TESTS

Paper and board

Although inks for paper and board printing may have different properties, many tests carried out on the final print have similar performance ratings.

Properties such as lightfastness, scuff resistance, soap and alkali resistance, etc. on the printed substrate are generally similarly tested in both paper and board printing and for convenience are treated together.

Fig. 15.15 A screen proofer.

Lightfastness

It is not normally necessary to check the lightfastness of each individual ink except for cases where contamination may cause a specific problem. Details of the test have already been described earlier in this chapter and are also covered in BS 4321:1969.

Chemical resistance tests

Several chemical resistance tests on finished prints have already been covered in the testing of pigments (section 15.3). These are:

- soap and detergent resistance
- alkali resistance
- acid resistance
- wax resistance.

Specific test methods

The following specific resistance tests are all fully described in BS 4321:1969 and are all self-explanatory.

Determination of resistance of printing inks and prints to cheese: Test Method 8, p. 27. Determination of resistance of printing inks and prints to edible oils and fats: Test Method 9, p. 29. Determination of resistance of printing inks and prints to spices: Test Method 11, p. 34. Determination of resistance of printing inks and prints to water: Test Method 3, p. 17. Determination of resistance of printing inks and prints to detergent: Test Method 7, p. 25. Determination of resistance of printing inks and prints to solvents: Test Method 4, p. 19.

Abrasion resistance

The resistance to abrasion is an important aspect of many printed articles, particularly on paper and board. Demands for improved abrasion resistance continue while the speed of packaging rises, giving greater possibility of severe scuffing from other printing surfaces or from contact with the production machinery itself. In certain cases (e.g. oleoresinous inks where excessive spray powder may give an abrasive surface) the package in itself can have a very rough surface and should be examined prior to any test. Failure on laboratory testing may result from incorrect or over use of anti-set-off powders. It is also worth noting that problems may arise in certain packaging areas where spillage of the contents can lead to an abrasive material being trapped between two areas of print. Any laboratory test should be designed to cover the severest conditions likely to be encountered.

Several instruments are available for checking rub resistance of prints and three such instruments are described in the British Standard *Methods for Measuring the Rub Resistance of Print*, BS 3110:1959.

While the simplest method described in BS 3110 is the Procter & Gamble tester, it is not used much outside the soap and detergent pack area for which it is specified. This rub tester is composed of a frame acting as a clamp to a print sample which is held against the base of the test instrument. The frame also acts as a guide to a block on which is attached a second piece of print in surface contact with the first sample. Hand movement of the block for a number of times at a steady speed (around 1 ft/s or 30 cm/s) gives direct comparisons between prints.

The Sutherland instrument and the PIRA instrument (Fig. 15.16) are discussed in BS 3110 which is recommended for further details.

It is important that the age of prints being tested is known. With conventional oleoresinous prints a week is the normal time allowed to elapse before rub testing. Even with radiation-cured prints 24–48 hours normally elapse before testing as inks dry to the touch can change with respect to their rub resistance properties within these times.

Crease resistance

In board printing the finished print is often required to be formed into a carton (e.g. a cigarette packet) and the ink must therefore have sufficient flexibility to prevent cracking when the board is folded. This test is particularly important where higher film weights of ink (and/or overlacquer) are carried as in gravure printing. Ideally, the test should be conducted using an apparatus such as the PIRA carton board crease tester (Fig. 15.17) which reproduces quite accurately production conditions. A cruder test can be carried out by either folding the board back and forth through an angle of 180° until the first signs of ink split is apparent or by

Fig. 15.16 A PIRA rub tester. (Courtesy of PIRA)

folding the board once through an angle of 180° and lightly scuffing the surface of the fold with the finger nail.

Film printing

Immediate adhesion of the ink to the substrate should normally be obtained in film printing to prevent set-off occurring in the reel. All film prints should be tested for adhesion, flexibility, scuff, slip and set-off as outlined below.

Scotch tape adhesion

A length of standard Scotch tape is placed over the printed areas, smoothed down with the ball of the finger and then pulled back slowly for half of its length and snatched back rapidly for the remainder. The print is examined for ink removal and the tape for any ink transfer. Care should be exercised with coated films so as not to remove the base coating and so obtain a completely false result.

Flexibility

The printed film is held firmly between the thumb and forefingers of each hand with about half an inch (13 mm) in between the hands and flexed 10 times. The print is then examined for any ink removal.

Fig. 15.17 A PIRA carton board crease tester. (Courtesy of PIRA)

Scuff

The print is repeatedly scuffed with the back of the fingernail, then examined for ink abrasion.

Slip

The friction of film to film and film to metal are important factors in processing packaging materials. If the coefficient of friction is too high, this can result in:

- movement of the packaging material over fixed elements becoming difficult;
- problems in controlling web tension;
- print which is stacked becomes 'adhered' and this can give rise to blocking;
- generation of a high static electrical change;
- excessive creasing during packaging.

Alternatively, if the coefficient of friction is too low, then:

- guiding rollers on the packaging line will not rotate and this results in irregular flow;
- web tension is difficult to adjust;
- Finished packages tend to slip during transportation.

Two types of friction can be measured, i.e. static and dynamic friction. In static friction there is no relative velocity between the boundaries of the materials and the static coefficient of friction can be described as the energy required to break down the localized 'adhesion' between the substrates in order to allow them to move freely. The static coefficient of friction can be measured using the tilting plane where a print is fastened face upwards to the plane and a metal sled is then placed on the film. The plane is slowly raised until the sled slowly and evenly moves down the plane and the tangent of the angle of inclination is equal to the static coefficient of friction. Alternatively, the static coefficient of friction between two printed surfaces can be measured by attaching a print to the metal sled and repeating the test as above.

Dynamic friction measures the friction between two materials which have a relative velocity to each other and is caused by molecules on the surface of the two materials colliding. These collisions are imperfectly elastic and hence energy is consumed. The dynamic friction can be measured on a flat bed by drawing a sled across a print at an even velocity (Fig. 15.18) or alternatively by the pendulum method where the print is held in a fixed position and the pendulum is rotated.

Blocking

The print immediately after drying is placed in contact with the back of the same film and subjected to a pressure of between 750 and 3000 psi for 16 hours in a hydraulic ram to simulate conditions produced in a reel of printed film (Fig. 15.19). The pieces of film are then slowly pulled apart and the test ink examined for degree of cling relative to the standard and any signs of ink transfer to the back of the plain film. Additionally, the

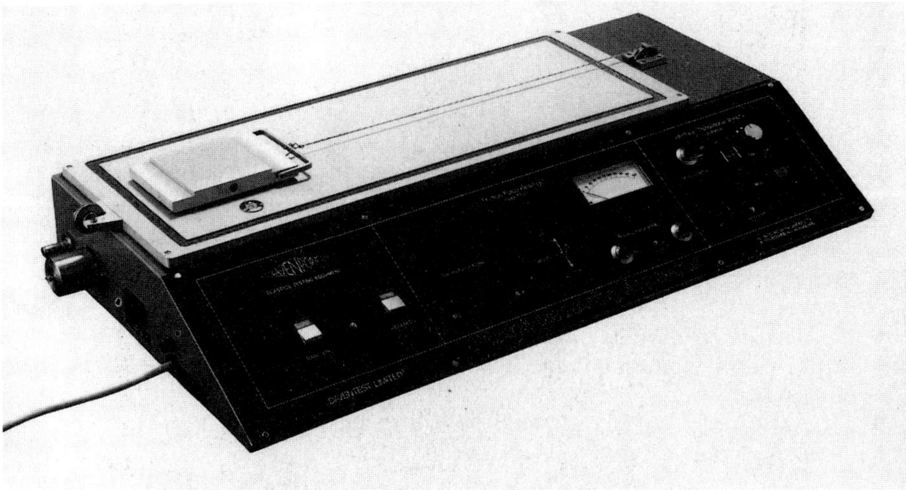

Fig. 15.18 Daventest dynamic friction tester. (Courtesy of Daventest Ltd)

blocking tests can be repeated at elevated temperatures if the facility exists for heating the platens of the hydraulic ram. This test can be used to simulate the effect of entering the re-reel stage while still warm or alternatively as a guide when film is used for packaging hot goods. In carrying out these tests, care must be taken in selecting a realistic temperature, time and pressures.

Additional tests for film

The following tests are dependent on the type of ink and its end application and need only be checked where stated in the ink specification.

Heat resistance

Heat-resistance inks are necessary where the film is printed in the heat seal area so as to prevent ink pick-off onto the heat seal jaws. An area of print in contact with virgin foil is placed between the heat seal jaws (preset at the heat seal temperature of the film under test) for a short time dwell (usually 1 second). After cooling, the print and foil are pulled apart and examined for ink pick-off and ink transfer on to the foil. Ideally, the print and foil should show no cling and there should be no ink transfer from the print to the foil. A further and more stringent test involves folding the print inside the foil so that the two ink surfaces are in contact. The heat resistance test is then repeated and when the ink surface are separated there should be no cling (Fig. 15.20).

Adhesive lamination

If necessary the adhesive is mixed according to the manufacturer's instructions and reduced to a suitable printing viscosity with solvent. The

Fig. 15.19 Hydraulic ram for blocking tests. (Courtesy of Specac Ltd)

print under test is set aside for 12 hours to allow for any waxes to migrate to the surface and then coated with the adhesive. After evaporation of the solvent, the tacky print is smoothed onto the laminating film (or foil) taking care that all air bubbles are removed. In order to allow for easy access into the laminate after the adhesive has cured, a strip of paper is wedged between the substrates down one edge of the adhesive layer. The bond strength of the laminate is tested after 24 hours and ideally also after 1 week, on a spring balance. Laminates for boil-in-the-bag work also require an additional bond strength test after being immersed in boiling water for 30 minutes.

PVdC coating

In certain types of packaging (particularly biscuits) the print is coated with a PVdC emulsion to allow for better seal integrity on A–B and A–A seals. To simulate this process, the PVdC emulsion is simply applied by a wire wound applicator over the print and the adhesion checked with Scotch tape. An additional problem can often occur after heat sealing, when interlayer adhesion between the PVdC and the ink film or the ink film and the base film breaks down. To ensure good interlayer adhesion is maintained the coated print is heat sealed (PVdC to PVdC) at 135°C/0.5 s/0.5 psi and the strength of seal tested on a spring balance as for a laminate

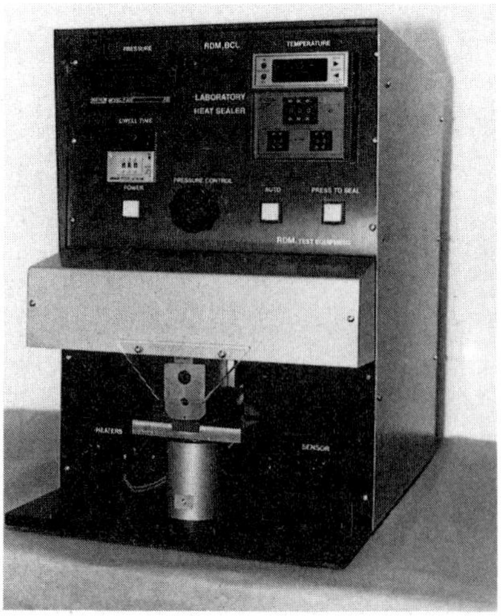

Fig. 15.20 A BCL heat sealer. (Courtesy of British Cellophane Ltd).

described in the preceding section. If poor interlayer adhesion occurs a poor seal strength will result.

Cold-seal packaging

Cold-seal packaging is being increasingly adopted to prevent heat contamination of sensitive products and for faster packaging line speeds. In this process a cold-seal adhesive is applied in line to the reverse side of the printed film so that on re-reeling the adhesive is in direct contact with the inks or release lacquer. It is therefore imperative that:

- no set-off of inks on to the adhesive occurs or blocking will result;
- pigments, dyes, waxes do not migrate into the cold seal and cause discolouration and/or loss of seal;
- no chemical interaction takes place between components in the inks or lacquer and the adhesive which might poison the adhesive.

Testing of inks and lacquers must be such as to show if any of the above situations can occur in practice. The normal test involves both short- and longer-term blocking tests with ink/lacquer in contact with the cold-seal adhesive and with a maximum pressure of 750 psi in the hydraulic ram. In addition to no cling being obtained on separation of the inks/lacquer from the cold-seal adhesive, the cold-seal adhesive must be sealed to itself to ensure no drop in seal strength is given compared with a print of adhesive which has not been in contact with the ink/lacquer.

Deep-freeze

This test was fully described earlier under pigment quality control (see page 813).

Odour and taint

Since many liquid inks are used on films for food packaging, it is necessary that they must have low odour and do not taint the packaged product. Samples of the test print, standard print and plain film (as a reference) are placed in separate tightly sealed containers, and heated in a 50°C oven for 1 hour. On removal the samples are tested for residual odour while still hot, then rapidly resealed and tested again after cooling to room temperature. A more elaborate odour test and taint test procedure is given in BS 3755: 1974.

Non-curl

A few films are especially prone to curl when printed with certain types of ink, causing great problems on the packaging lines. In cases where curl could be a problem, a sample of film, printed all over with the test ink, should be allowed to lie undisturbed on a bench. After a period of time (e.g. 1–2 hours), there should be no signs of the film curling at the edges.

Grease resistance

Printed film used for packaging certain foods, e.g. potato crisps, must be resistant to the oils used in the manufacture of these goods. A test can be carried out by rubbing the printed film a standard number of times with a piece of cotton wool saturated in the appropriate oil, followed by a flexibility test on the greased print. The print and cotton wool are then examined for print breakdown and stain respectively.

Alternatively, the grease rub test can be carried out using the SATRA rub tester by impregnating the felt pad with a measured amount of the oil and then rubbing the test print for a set number of revolutions with an agreed pressure. This test, although apparently more scientific, is subject to wide variations in results due to variability in the felt pads and an average result of three readings should be taken and compared with those of tests on a master standard print carried out at the same time.

Foil

The testing of foil prints closely resembles the tests already outlined for plastic films. Care should be exercised to ensure the correct type of foil is used whether it be, for example, the simple virgin foil (such as used on many chocolate wraps) or the paper/board-backed foil which has a surface wash (normally either nitrocellulose, shellac or vinyl) to prevent tarnishing and to improve print receptivity.

The procedure for film printing should be adopted, where appropriate, with the exception of the flexibility test. For virgin foil, flexibility is tested by folding the printed foil through a 180° bend and then with the thumb and forefinger at either side of the crease the foil is rolled back and forth through the fold ten times. The ink should then show no signs of breakdown at the fold. Paper- and board-backed foil is examined for crease resistance using test methods already outlined for paper and board (see page 845).

Metal

While some of the tests already described for paper and board prints will apply in the case of metal decorating inks, others are extensions of these (e.g. abrasion resistance), to take into consideration the tougher film involved, or the subsequent processing requirements of the metal container (e.g. high temperature process resistance), that is not possible with other substrates.

Abrasion resistance

The tougher film formed on metal surfaces, necessitated by the abrasion resistance requirements of the ultimate package, demands a different approach to testing the scratch susceptibility or hardness of that film. Several instruments are available for these tests.

Scratch testers

The apparatus consists of a weight-loaded needle arm which is placed in contact with the panel under test. Operation of the lever gently lowers the needle on to the test surface and slowly withdraws the panel so that the needle makes a scratch on the surface. As the load on the needle increases, a point is reached where penetration of the ink film is achieved and the needle strikes the metal substrate. This contact completes an electrical circuit as signified by the illumination of a lamp. Results are thus expressed as the maximum weight at which the film passes the test.

Hardness tester

Two types of hardness tester are used in the main: these are the Sward type and the pendulum type (DN 53–157). The Sward type utilizes a rocker action and the hardness is described as the percentage of oscillations made by the wheel type rockers on the test plate as opposed to the reference glass plate.

The pendulum type hardness tester is primarily used on the continent. The apparatus consists of a triangular shaped pendulum, free to swing on two ball points which rest on the sample. An electronic counting device records the number of swings made by the pendulum between the 6- and 3-degree marks on the scale. Comparison is then made between the number of swings on the glass reference plate and the number on the test surface.

Heat resistance

This test depends primarily upon the properties and concentration of the pigment, but may be affected by the medium.

While knowledge of the pigment will provide the basis for a selection, to be decisive, a test should always be carried out on the final print. Routine checking on ink batches need not be necessary for reasons outlined earlier. In the case of new products, however, this test must always be carried out where heat resistance is required.

The test itself, using a print of known film thickness on the metal to be used, employs a stoving operation in an oven at a suitable temperature for a realistic time. An average schedule is 150°C, for 5 minutes, but this can vary depending upon the type of container under test, and higher temperatures may be encountered for example on DWI can lines. There are specific applications (such as solder seaming) where much higher temperatures (e.g. 230°C) are encountered, albeit for much shorter times.

The stoved panel is then compared for hue, strength and brightness with the control panel, making due allowance for the wet state of the latter.

Flexibility or elongation properties

Simple flexibility tests can be carried out by bending the printed metal substrate and examining for failure, but the method can show poor reproducibility and often has little relationship to the stresses that the printed substrate is subjected to in commercial situations.

Reproduction of the actual stress situation on a tin-printing assembly line is not easy in the laboratory, but some equipment is available that can give information on the flexibility of a printed or coated film. One such piece of equipment is the conical bent test apparatus designed to measure the elongation potential of coated metal panels. The panels are clamped in position and formed round the conical mandrel by rotation of the roller frame. The 200 mm long mandrel tapers from 37 to 3 mm diameter along its length and after the test the panel is examined to determine the minimum diameter at which failure by stretching occurred.

Adhesion

Adhesion to the metal and good interfilm adhesion on overprinted areas are both important requirements for printing inks in metal-decorating applications. Failure of abrasion resistance tests may indicate that insufficient adhesion to the base metal or the underlying film is the cause. The principal test for poor adhesion is probably a simple scuffing test with a hard, blunt surface. Soft films or brittle films can be disturbed relatively easily. Poor interfilm adhesion can also be noticed in the same test as indicated by removal of surface layers only. The use of Scotch tape can also prove helpful, particularly for radiation-cured inks where adhesion has not been assisted by thermal treatment. It is particularly useful to predict the behaviour of cured ink films on interdeck drying as poor adhesion or interfilm adhesion can give rise to stripping of the ink from the sheet on to subsequent blankets.

Pasteurization and processing

Film cracking, softening or adhesion failure can also occur during the processing or pasteurizing of the printed metal. Where the ink is to be used in this application, the necessary laboratory tests should be performed.

Processing tests are normally carried out under steam pressure (15 psi) for various times, depending upon the particular application. The print is then checked for failure.

Pasteurization varies somewhat from laboratory to laboratory but an average test uses partial immersion of the print in a water batch at 150°F (66°C) for 30 minutes. Print failure tests are performed as for processing.

It is important when carrying out these tests to ensure that the print has been treated as would the finished article. Where for instance bending or stretching occurs this should be carried out under laboratory conditions. Also where a base coating or lacquer is present, it is important that the correct material is used in conjunction with the ink under test.

Plastics

The tests required for prints on polymeric surfaces (plastic containers, etc.) tend to lie between the paper/board and metal areas.

As many of these containers hold food, the main tests are contact tests with the material in question. These are similar to those described for paper and board, and are carried out by immersing the printed substrate in the material for a certain period and testing for film weakness or pigment bleed.

Ink adhesion and scuff resistance

Ink adhesion is tested using Scotch tape in a similar manner to that described for film inks (page 846). While complete resistance to Scotch tape may not be required, the inks must possess good scuff resistance. A fingernail test as performed for film inks generally suffices.

15.17 STATISTICAL PROCESS CONTROL

As mentioned in the previous section, statistical process control (SPC) is an important tool in the improvement of manufacturing industrial procedures. SPC is a technique for monitoring the quality of processes rather than products, which makes it different from the majority of other quality improvement strategies.

Failure to accept SPC as a quality tool is often due to the perception that a detailed knowledge of mathematics is required. This is not so, for although the derivation of the theory is complex, it is only necessary to have a basic understanding of mathematics to apply the technique in practice.

Unlike traditional quality control measurements, SPC monitors the performance and the stability of the manufacturing process, and is not dependent upon final product inspection. This means that it can be used as a tool for improving the quality of the manufacturing process and thereby reduce the time spent on final product inspection. If SPC is being used to monitor and adjust processes within manufacture, the products are more likely to meet specification with less final adjustment required. Furthermore, there will be a smaller incidence of product rejection. This gives a saving in time and materials which fully justifies the time required to set up and use such a system.

Before considering the details of applying statistical process control techniques, it is useful to have an understanding of the elements which go to make up a process. A process usually contains five distinct variable elements:

- people
- methods
- machinery
- materials
- environment.

These are all variable to some extent. When they are combined into a process there will be a natural variation in the process itself and hence in the resulting product. It is lack of understanding of these natural variations within the process that makes them difficult to control and minimize. Therefore, by understanding and controlling the causes of the variations observed, management can then concentrate on monitoring these process variables to ensure they are natural and that nothing is really wrong with the processes.

In practice SPC acts as a technique by providing a visual display of a process on charts which are marked with the limits of the process variations if they are occurring naturally. If results are obtained which fall outside the natural limits action can be taken to investigate the cause and possibly correct it. In traditional quality control techniques the effect, rather than the cause, is monitored and corrected so that the problem is observed again and again as the cause is never understood and removed.

To set up an SPC chart the process to be investigated needs to be monitored over a period of time to give data required for a statistical analysis. As already mentioned the resulting data will give a spread of results which will not all be the same. The data being collected may be such things as the time to run a product through a machine, the degree of dispersion obtained, the viscosity obtained, the results obtained with a calibration standard for an instrument, etc. – the possibilities being almost endless, as there are so many processes involved within the manufacture of printing inks. If the process is a true process involving the above five elements then the results obtained will not be the same. If the results appear to be the same it is more likely that the method of data collection is not sensitive enough to detect variations than there being no variation. However, many processes are not critical and small variations are not important for final product quality, the application of SPC having to be applied selectively to areas which are critical and will provide the greatest benefit.

The data obtained from monitoring a process can be used to produce a histogram, which gives a visual display of how the results were distributed across the range of results obtained. To produce a histogram the range of results is split up into small divisions or 'cells', and the number of results which lie within each cell is determined. For example, the following are results obtained from monitoring a process:

5.5	4.6	4.9	4.3	5.2	5.3	4.4	6.3	5.0	4.9
4.4	4.9	3.9	4.8	4.9	3.6	5.4	4.5	5.2	3.8
4.2	4.9	5.7	4.8	5.1	4.4	4.5	3.9	4.3	5.4
4.9	3.8	4.7	4.2	4.3	5.3	5.3	5.0	5.1	4.8
5.2	4.8	4.1	5.5	5.0	4.1	4.3	5.0	4.8	4.8

If the range covered by these results, e.g. 3.1 to 7.0, is split into equal cells of 0.5 units then a tally chart or 'frequency distribution' can be produced.

Range of values	Frequency of occurrence	Percentage
3.1–3.5	0	0
3.6–4.0	5	10
4.1–4.5	13	26
4.6–5.0	18	36
5.1–5.5	12	24
5.6–6.0	1	2
6.1–6.5	1	2
6.6–7.0	0	0
	50	100

The data are now easily plotted onto a bar chart or histogram which gives a visual display of the distribution of results obtained, with each cell representing one bar on the chart (Fig. 15.21)

The results obtained give a 'lumpy' appearance when plotted on a histogram, but this is only due to the small number of results obtained and the small number of ranges used.

As more results and ranges are used to plot the histogram the curve of the chart becomes smoother and eventually a smooth bell-shaped graph is obtained (Fig. 15.22).

The bell-shaped curve or distribution is known as a normal distribution and is very useful because there are various statistical facts that can always be stated for this type of distribution.

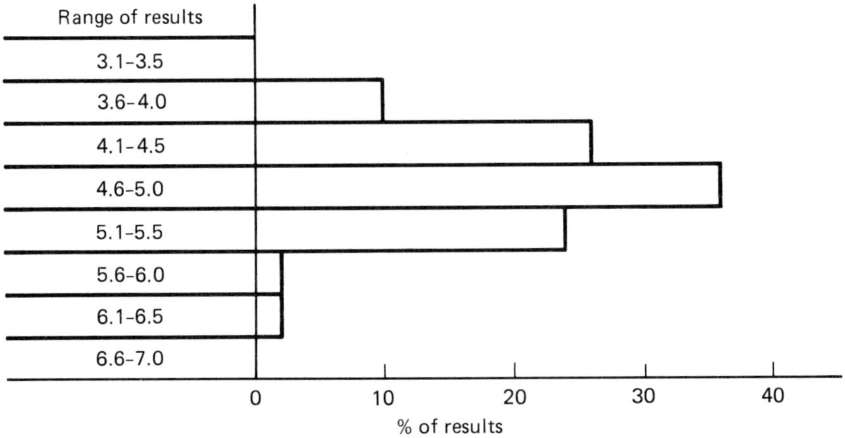

Fig. 15.21 Histogram.

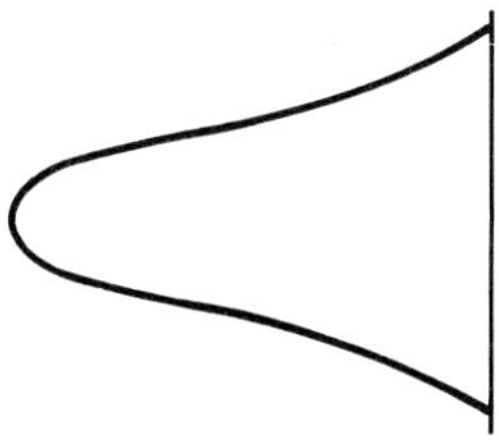

Fig. 15.22 Bell-shaped graph.

Mean and standard deviation

The mean of a sample, sometimes referred to as the average, is obtained by adding up all the results and then dividing by the number of results in the sample. For a normal distribution this is an important value, which should be found at the centre point of the distribution with the remaining results being distributed symmetrically about this centre point. The standard deviation of a set of results shows the dispersion of values around the arithmetic mean, the naturally occurring scatter either side of the average. The standard deviation of a sample can be calculated easily using a calculator with a button for standard deviation, but can also be calculated by hand using the formula:

$$\sigma = \sqrt{\frac{\Sigma fx^2}{\Sigma f} - \left(\frac{\Sigma fx}{\Sigma f}\right)^2}$$

where σ is the standard deviation, Σ is the summation symbol, x is the value and f is the frequency with which x occurs.

For example, if the following results are obtained:

8	7	9	5	8	12	16	9	11	12
7	7	10	6	8	12	13	14	14	11

then

x	f	fx	fx^2
5	1	5	25
6	1	6	36
7	3	21	147
8	3	24	192
9	2	18	162
10	1	10	100
11	2	22	242
12	3	36	432
13	1	13	169
14	2	28	392
16	1	16	256
	$\Sigma f = 20$	$\Sigma fx = 199$	$\Sigma fx^2 = 2153$

$$\text{arithmetic mean} = \bar{x} = \frac{\Sigma fx}{\Sigma f} = \frac{199}{20} = 9.95$$

$$\text{Standard Deviation} = \sigma = \sqrt{\frac{\Sigma fx^2}{\Sigma f} - \left(\frac{\Sigma fx}{\Sigma f}\right)^2}$$

$$= \sqrt{\frac{2153}{20} - \left(\frac{199}{20}\right)^2}$$

$$= \sqrt{[107.65 - (9.95)^2}$$

$$= \sqrt{8.6475}$$

$$\sigma = 2.94$$

This formula for standard deviation may initially seem complicated but can be useful for analysing data and is quicker than keying in each number into a calculator.

The standard deviation and mean of a normal distribution are important measurements for industrial processes, the mean being the centre point of the curve, while the standard deviation describes the actual shape of the curve. Statistically both figures in combination predict process performance accurately, and this fact is useful for process control purposes.

It is a statistical fact that for a normal distribution approximately 68% of all results will lie within one standard deviation of the mean (centre point) of the distribution. Similarly 95% of all results will lie within two standard deviations either side of the mean, and 99.8% within three standard deviations either side of the mean (Fig. 15.23).

These statistical facts regarding normal distributions are useful for process control purposes because after determining the mean and standard deviation of a set of process results the future behaviour of that process is completely predictable, provided that there is no change in the process. Indeed, it is the purpose of SPC to detect these changes in the process and determine whether they are within acceptable 'natural' limits, or whether they are outside natural limits and there is something causing the process to become 'out of control'. The natural limits for a normal distribution are therefore 3 standard deviations either side of the mean, and results obtained outside these limits may indicate that there is a problem with the process, and an investigation can be carried out to determine the cause.

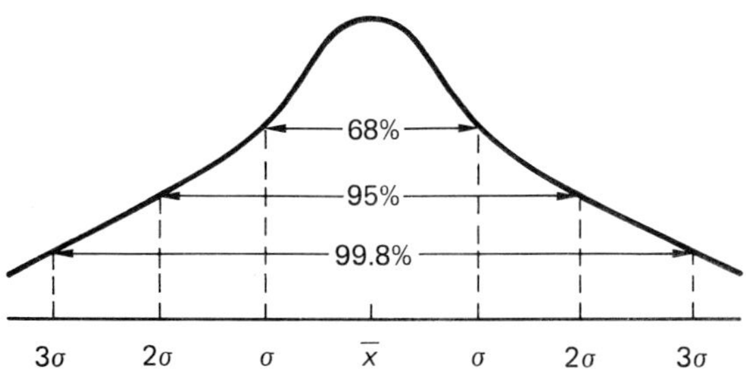

Fig. 15.23 Normal distribution curve.

Quality control charts

When using the mean and standard deviation of a process as a method of control it is normal to use a 'quality control chart'. A control chart is similar to a distribution chart which has been turned on its side, so that the value range is on the vertical instead of the horizontal axis. The horizontal axis is plotted from the arithmetic mean as calculated from the distribution, this line being plotted mid-way up the vertical axis. The standard deviation of the distribution is also calculated and horizontal lines are plotted at 2 and 3 standard deviations from the arithmetic mean, so that the final chart is made up of one vertical line and five horizontal lines as shown in Fig. 15.24.

The inner lines which are 2 standard deviations either side of the mean are known as the warning lines and the lines 3 standard deviations from the mean are known as the action lines. From the previous discussion regarding normal distributions it can be seen that 95% of the normal scatter expected from a process will lie within the inner 'warning' lines. Furthermore, it can be seen that 99.8% of results obtained will lie within the outer warning lines, and these two lines are important in the control of the process.

In practical terms we can use the chart to monitor process performance by plotting results from regular measurements onto the chart, and if all is well with the process there will be an even scatter of results either side of the mean point.

If the process is 'under control' we can predict that 95% (i.e. 19 in every 20) of results will lie within 2 standard deviations of the mean, and that 99.8% (i.e. 499 in every 500) of results will lie within 3 standard deviations of the mean. If more results lie outside the warning lines than expected then it is possible that the process is 'out of control' and the process should be investigated. If a result lies outside the outer action lines, because only one point in every 500 would be expected to fall outside these lines, the process should be investigated as a standard routine. In practice if a process is going to go out of control it will tend to drift slowly rather than suddenly jump outside the control limits. As the process drifts the points will tend to move towards one action line or the

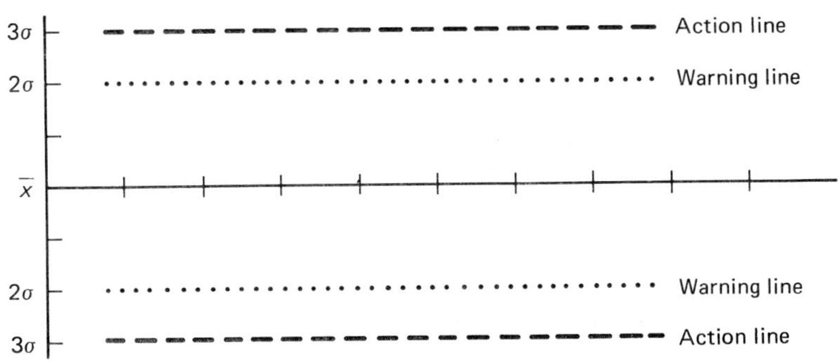

Fig. 15.24 Quality control chart.

other, but well before it reaches this point it will be clear to the observer that there is a trend towards one of the action lines and the process can be investigated before the process moves outside the control limits set.

Charts showing processes which are 'under control' and 'out of control' are given in Fig. 15.25.

The role of statistical process control is to ensure that process performance is such that no defective or non-conforming items will be produced. However, if the specification tolerance is set within the control limits for a process non-conforming items will result, and in this case the process is said to be 'incapable'. Tolerance limits should be set to excede the control limits for a process, so that the results obtained within the normal distribution will always be within the tolerance required. In this case the process is said to be 'capable', and the process will produce zero defects.

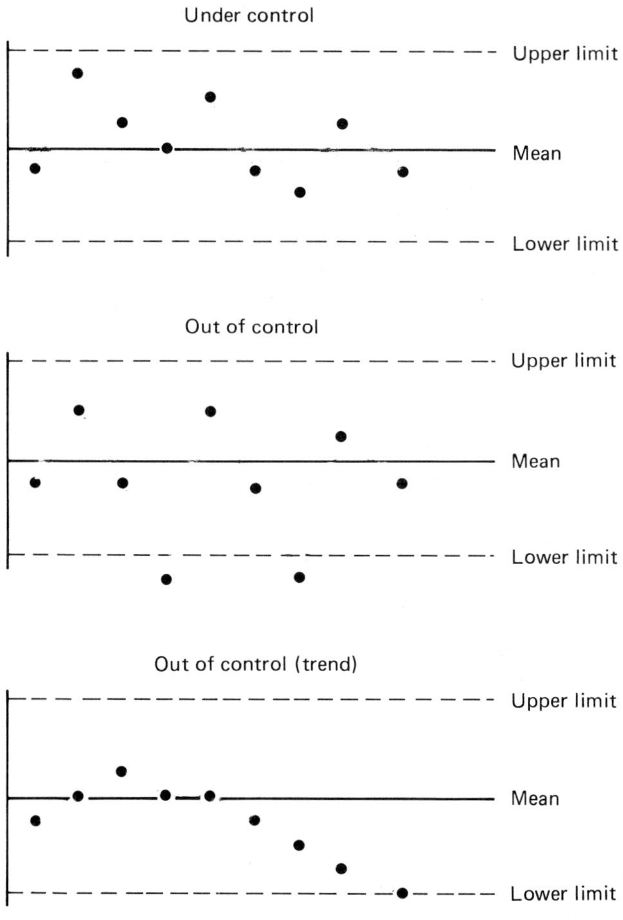

Fig. 15.25 Charts showing processes which are 'under control' and 'out of control'.

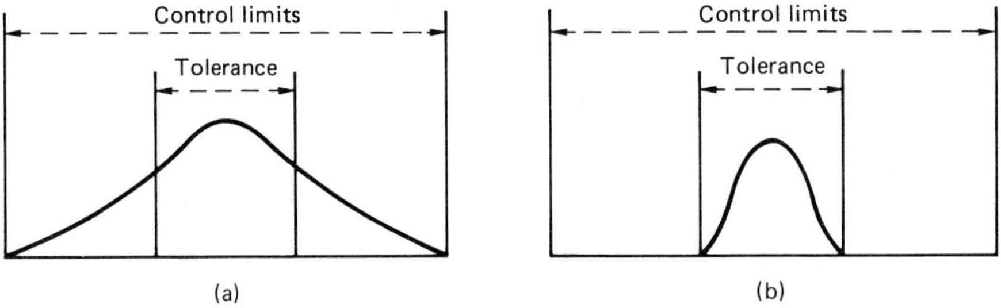

Fig. 15.26 (a) Incapable process; (b) capable process.

If the width of the distribution for a process is inconsistent with the tolerance limits being set then it is usually management's responsibility to improve the process by making changes in methods, materials, equipment, etc. to produce a narrower distribution which is well within the tolerance limits set.

In practice SPC is a method of identifying those variations in process performance which give rise to defective products. It will distinguish between problems which require operator response (out of control processes), and those which require management response (incapable processes), and enables both operator and manager jointly to assess and tackle the problem before them.

As with any aspect of total quality management (of which SPC is a part) it requires the right attitude and commitment from the highest level down if it is to be successful. As well as management commitment there must also be clear operating procedures so that all personnel involved know exactly what is required of them.

Although the practical operation of an SPC programme does not require a detailed mathematical knowledge, it is of benefit for some training in basic statistical techniques to be given.

The enhanced process control which SPC provides leads to reductions in rejects, waste and re-work, and hence much lower processing costs. Other benefits of SPC are associated with the implementation of any total quality programme which will naturally result in much greater employee involvement and closer management/shopfloor relations. The implementation of SPC will therefore yield both tangible and intangible benefits, the magnitude of which will depend on the commitment of both management and staff to achieving total process quality.

15.18 QUALITY ASSURANCE

One of the key features of the most successful companies is invariably their reputation for producing quality goods and services. As markets become more competitive and choice of suppliers wider there is little room in the marketplace for poor quality goods and services. Customers are insisting more and more on a supplier being able to demonstrate that their goods and services meet a required quality standard. Often this requirement may not only be related to final product quality, but companies may also be asked to demonstrate that the quality of their internal activities is also of a high enough standard for them to be considered as a suitable supplier.

For many companies reputation is no longer enough to be an automatic choice of supplier, and the primary requirement to be an acceptable supplier will be to have a recognized quality standard. The most common quality standard requirement is for a company to have BS 5750, which has various equivalents around the World which are listed below:

United Kingdom	BS 5750–1
Europe	EN29001
International	ISO 9001
Germany	DIN-ISO 9001
NATO	AQAP-1
USA	MIL-Q 9858-A
Australia	AS 1821
Canada	CSA-Z 299.1

BS 5750 is a national quality standard which is an indication that the service provided or product designed and constructed will satisfy customer needs. A company which has BS 5750 registration has had to establish, document and maintain an effective quality system, a system which is regularly audited to ensure that the standard is maintained. By specifying that only BS 5750 suppliers are used a company can reduce its inspection costs, and divert resources to more productive part of the business. Using only quality assured suppliers can be used in their own quality system to demonstrate to their customers their commitment to supplying satisfactory goods and services.

In practice some customers will insist on all suppliers having BS 5750 or equivalent registration, whereas others will be satisfied if a supplier can demonstrate that they have equivalent quality systems even though they are not registered. This is normally ascertained by sending out questionnaires with searching questions regarding the suppliers' quality systems, and for those companies without a recognized quality standard a visit to audit the suppliers' quality systems may be necessary. Those companies which fail to return the questionnaires or demonstrate that their quality systems are below the standard required are not considered as suppliers.

BS 5750 has three main constituent parts numbered 1, 2 and 3. As can be seen below Part 1 (ISO 9001) covers the other two parts also, being a specification for the design/development, production, installation and servicing of goods or services, i.e. when the requirements of goods or services are specified by the customer in terms of how they must perform and which are then provided by the supplier. Part 2 (ISO 9002) covers the manufacture, installation and servicing of goods or services, whereas Part 3 (ISO 9003) covers only final inspection and test procedures.

BS 5750

<table>
<tr><td>Design and development</td><td></td><td></td><td rowspan="3">Part 1</td></tr>
<tr><td>Production</td><td></td><td rowspan="2">Part 2</td></tr>
<tr><td>Installation and servicing</td><td>Part 3</td></tr>
</table>

Although it may be desirable for all suppliers to have Part 1 registration, in practice most of those which are quality assured will only have Part 2 registration. Part 1 registration is considered by many companies to be unnecessary as the primary concern when purchasing products is with the quality of manufacture and final inspection, rather than the actual design. Because it is only necessary to acquire Part 2 registration to become an approved supplier that is all that many companies choose to achieve.

BS 5750 is all about being able to demonstrate to a customer an ability to design and supply products to an acceptable standard, and the prevention of non-conformity at all stages from design to servicing, depending on which parts a company is registered for. This is achieved by having a quality system which is fully documented in the form of a manual, and is completely auditable by a third party who can verify that the procedures detailed in the manual are being adhered to. For a company to retain its registration it will have to be prepared to have its quality systems regularly audited, normally without prior notice, by an approved certification body, often the British Standards Institution (BSI) itself being used for this purpose. If a company is inspected and found not to be operating a procedure or system as detailed in its own quality manual this will be registered as a 'discrepancy' by the auditing body and the company given a period of time to correct the discrepancy or have its certification terminated.

Everything that is required to be included in the quality manual is detailed in BS 5750: Part 1:1987 and QAS 2552/312 which is a quality assurance schedule to BS 5750 Part 1 relating to the product development, manufacture and supply of printing inks and associated products.

Preparing a quality manual

BS 5750, and equivalent standards, require that all the necessary operational procedures relating to quality be documented, usually in the form of a manual, and more importantly be used. The writing of a manual is important as it gives a written and rational picture of all the quality systems in operation. Assessing the quality of systems is no different from assessing the quality of a product in that there must be a standard reference to make comparisons with. Without the systems being fully documented it is impossible to assess whether the systems are being operated correctly, or to make changes to the existing systems. In reality what will happen is that everyone within a department will operate their own systems and procedures as there is no precise description of their departmental functions and their execution.

The size and scope of the manual will depend on the size of company and the level of quality assurance required, whether the documentation of the quality systems is to BS 5750 standard or equivalent, or whether the company feels that a nationally recognized quality standard is not necessary. The quality manual should be broken down into clear sections and indexed, covering all aspects of the companies' operations which are related to the level of quality assurance recognition being sought. Everything which goes into the manual must be auditable, with the company being able to demonstrate that the systems are not only acceptable from a quality point of view but are also being operated as written. For a small company which does not wish to achieve a recognized standard the manual may be relatively short with contents such as:

- the company quality policy;
- the organizational structure of the company with regard to quality;
- all operational procedures relating to the control of the functions of the quality department.

Such a short manual would be unacceptable to many larger manufacturers who would have quality systems consistent with BS 5750 or equivalent. The contents of a manual for such a company would be far more comprehensive with typical contents:

- the structure of the manual, which would include:
  company name
  manual title
  manual issue number and date of issue
  person authorized to issue copies of the manual
  amendment record sheets, clearly indicating dates and nature of amendments
  manual contents page
  list of manual holders.
- a quality assurance policy statement;
- the organization of quality assurance;
- job descriptions of quality assurance personnel;
- quality assurance procedures;
- quality assurance audit procedures;
- documentation.

The requirements of BS 5750/ISO 9000 with regard to the quality manual are comprehensive and precise. Full details are given in the quality assurance schedules Part 1, 2 or 3. Part 0 gives guidance to companies seeking registration and also guidance in the writing of quality manuals. The full details are contained in the various parts of the Standard which should be studied in detail.

CHAPTER 16
Analysis of printing inks

16.1 INTRODUCTION

Modern analytical instruments are able to provide such a huge wealth of information that these techniques now play a vital part in every aspect of the ink business, from development through manufacture to after sales support.

Techniques such as Fourier transform infra-red spectroscopy (FT-IR), capillary gas chromatography (CGC) and gel permeation chromatography (GPC) are often used to quality-control incoming materials. With some of these instruments computers can now take part in the decision-making process between acceptance and rejection of materials.

In the ink development stage analytical techniques are used to study different ink properties such as solvent retention and tainting potential through migration studies.

The effect of the shearing forces on pigments and resins during ink manufacture are now studied with the aid of electron microscopy and molecular weight measurement techniques.

Powerful bench-top GC-mass spectrometers and GC–FT-IR instruments are used to study problems of taint and odour and environmental issues such as pollution and atmospheric monitoring.

The need for precise, accurate information has never been greater which can only lead to more and more reliance on the use of sophisticated analytical instrumentation in the ink industry.

Classical analysis

Despite the rapid development of instrumental methods of analysis more 'classical' approaches still have their place, and often offer an easier solution to a specific problem.

Solvent precipitation techniques are often used in the first stages of ink analysis to affect a semi-quantitative separation of an ink into its major component classes.

A typical scheme such as that shown in Fig. 16.1 may be used for variety of different ink technologies. This type of approach can be readily adapted by changing the sequence or choice of solvent according to the ink type. The resulting fractions can be weighed and examined using infra-red spectroscopy or other appropriate identification techniques.

The other classical approaches used are in the form of 'wet chemical' tests.

If an ink is considered, then a manual solvent fractionation scheme is often the simplest means of partial semi-quantitative separation of the ink components. A weighed sample of ink, usually 1 g, is sequentially washed with at least three 10 ml aliquots of solvents of increasing solvent power starting with a very weak solvent such as petroleum ether. At each extraction stage centrifugation is used to separate the solvent from the insoluble matter. The extracting solvent can then be removed from each fraction by heating and the residue weighed. Each of the fractions thus obtained are identified using an instrumental identification technique such as IR spectroscopy. The simple extraction scheme shown diagrammatically (Fig. 16.1) will give sufficient information on any ink system for an experienced analyst to select the optimum method of analysis.

The instrumental techniques available have enabled the analyst to advance and in many cases to combine the separation and identification procedures.

Gas–liquid chromatography is used to separate and identify quantitatively volatile materials, i.e. solvents, plasticizers, anti-oxidants.

Techniques such as HPLC and GPC allow us to separate soluble materials according to their chemical nature or molecular size respectively.

These techniques and others will be discussed in depth.

16.2 CHEMICAL TESTS

In spite of the invasion of instrumental techniques specific wet chemical tests still play an important role in the analytical laboratory. These are the tests which lend themselves to on-the-spot analysis by giving definitive reactions with distinct chemical entities. For example, the simplest way of determining if an amine has been used to produce an aqueous acrylic solution is by the distinctive odour obtained when the sample is heated gently after the addition of sodium hydroxide.

The following colorimetric tests are very useful when only small samples are available or if there is insufficient time to carry out a purification process prior to identification.

Diphenylamine test for nitrocellulose (NC)

This is a rapid sensitive test which may be used to identify whether NC contamination is present in an acrylic ink or if a print contains NC. Diphenylamine of 1–4% may be made by dissolving 1–4 g of pure diphenylamine in 70 g of concentrated sulphuric acid; this solution is then diluted by the very careful addition of 30 g of de-ionized water. When spotted onto the test sample a strong blue colour develops if NC is present.

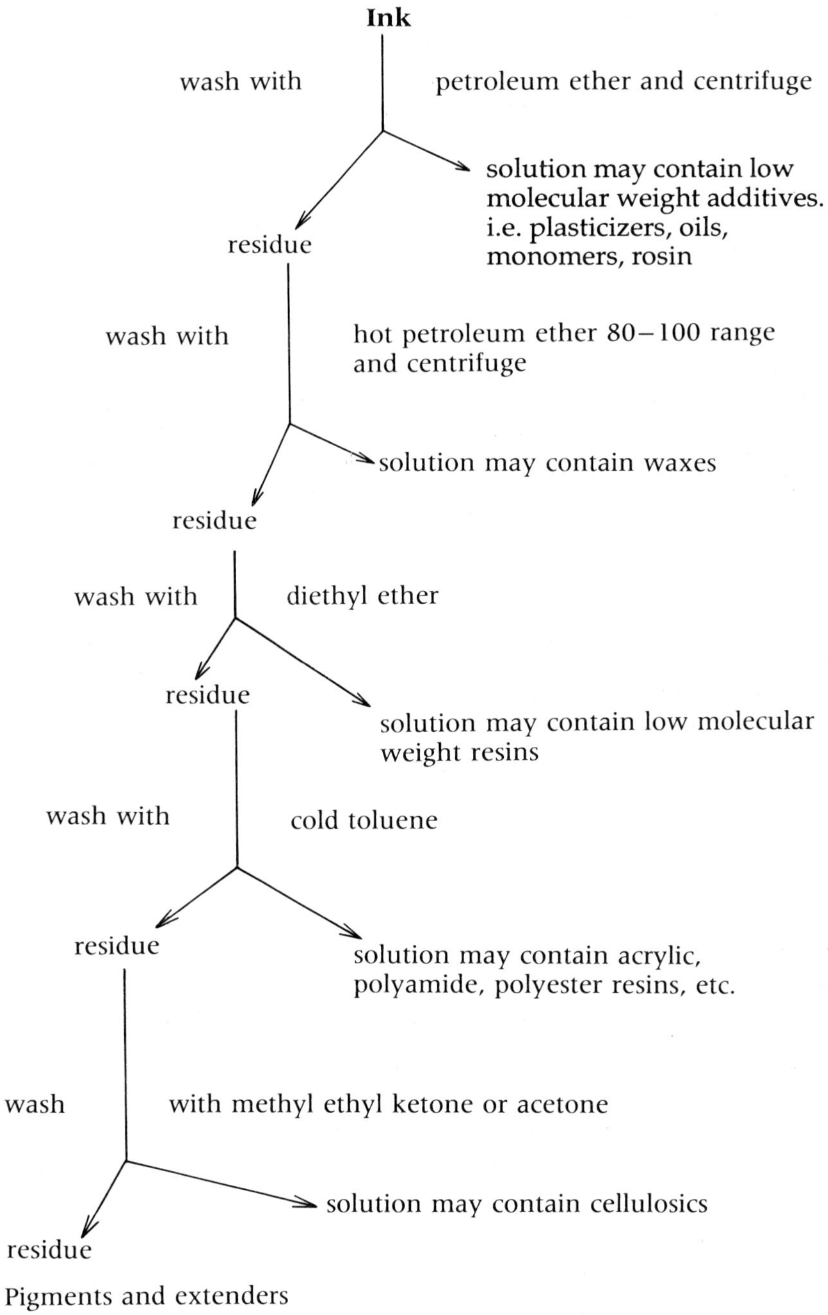

Fig. 16.1 Ink extraction scheme – can be adapted to suit the type of ink under investigation.

This reagent also gives a colour change when ketones are present; a weaker red colouration results. This may be very useful in distinguishing styrenated ketone resin from styrenated acrylic and styrenated allyl alcohol. It is also used as a confirmatory test for the presence of a ketone resin. Every precaution has to be taken to ensure no ketone solvent is present.

Liebermann–Storch test for rosin

A small amount of the sample is boiled with acetic anhydride. A few drops of this are spotted on to a porcelain spot test plate and concentrated sulphuric acid added. A violet colour indicates rosin. Other resins such as maleics and fumarics also react under these conditions.

Beilstein test for halogens

A halogen-containing material when ignited in a bunsen flame in the presence of copper gives rise to a strong green-turquoise coloured flame.

This is very useful for instances where the presence of vinyl chloride is suspected. The presence of chlorinated photo-initiators or amine hydrochlorides may also be tested in this manner.

Amine–formaldehyde resins

An amine–formaldehyde resin when warmed in a test tube with 70% sulphuric acid and one crystal of chromotropic acid will give rise to an intense violet colouration.

Urea–formaldehyde resins

A small portion of the resin should be boiled gently with a mixture of 20 parts acetone, 30 parts water, 10 parts of 30% HCl and 1 part of furfuryl alcohol. A red colouration indicates urea.

Epoxy resins

Either the resin or a print is dissolved in concentrated sulphuric acid with minimal charring. By means of a glass rod some of this solution is streaked on to a filter paper. A purple colouration indicates the presence of epoxy.

This is a very useful test for identifying whether a print is conventional litho or UV, assuming of course that the UV ink contains a bisphenol-A – epoxy resin.

Polyurethane resins

A simple test may be applied to polyurethane resins to determine whether they are TDI or MDI based. A filter paper previously impregnated with 4-nitrobenzene diazonium tetrafluoroborate is subjected to the fumes given off from a polyurethane resin when contacted with a red-hot glass rod. MDI gives a yellow colouration, TDI gives a red colour.

This test is specific for urethanes based on aromatic isocyanates and will also give a colour reaction with the less common 1,5-naphthalene di-isocyanate. A violet colour is obtained. No distinct, colour reaction occurs with aliphatic polyurethanes.

Phenolic resins

The most sensitive test for phenolic resins is the Gibbs Indophenol test. There are two ways of performing this test and although the former is quick and simple the latter has been shown to be considerably more sensitive.

A little resin dissolved in alcohol is treated with a drop of Gibbs reagent (1% solution of 2,6-dibromoquinone-4-chloroimine in alcohol). The solution is made alkaline with dilute sodium hydroxide. Phenolic resins generate a strong blue colour and a modification of this test can also be used to detect the presence of phenolic stabilizers in UV diluent monomers.

The alternative is to impregnate a filter paper with some Gibbs reagent and subject it to the fumes given off when a sample of resin is pyrolysed in a test tube over a bunsen flame. The filter paper may then be developed by the alkaline fumes from a bottle of 0.880 ammonia. Once again a strong blue colour is obtained.

Tests for PVdC

The presence of polyvinylidene chloride (PVdC) on the surface of flexible packaging film may be readily detected using IR spectroscopy but some simple colorimetric tests are often valuable.

Application of the Beilstein test will determine whether any chlorine is present. A hot piece of copper is briefly scraped across the surface of the film and then placed in a bunsen flame. A bright blue-green colouration confirms the presence of chlorine. A distinguishing test for PVdC may then be applied. A spot of morpholine dropped on to the surface of the film will darken through brown to black if PVdC is present. This reagent does not react to the other forms of chlorine which may be encountered in packaging films.

Traditional 'wet' chemical tests are frequently used by organic chemists to characterize the materials they make and use. The following tests are often applied to resins, varnishes, vegetable and mineral oils used in ink-making.

Acid value

The residual acid value of a resin will have a significant effect on its stability, printability, flow, gloss and surface wetting characteristics. It can also be a factor monitored during the production of a specific polymer. The acid value may be determined by the direct titration of a known weight of the sample dissolved in alcohol or a toluene/alcohol mixture. This is titrated with a decimolar alkali solution to a colorimetric end point using a suitable indicator such as phenolphthalein. Acid value is normally expressed in terms of the number of milligrams of KOH required to neutralize 1 g of the sample. This test is often applied to varnishes and oils also.

Hydroxyl value

The residual hydroxyl groups in a resin, varnish or oil may affect the ink due to intermolecular hydrogen bonding. This will influence a variety of physical properties, viscosity, solubility, solvent release and wetting, to name a few.

This value is also important if the hydroxyl group is to be used in a reaction mechanism.

Hydroxyl value is determined by refluxing a weighed sample with 10% acetic anhydride in pyridine. After a sufficient time the hydroxyl groups will have been acetylated and the remaining anhydride can be hydrolysed with water and titrated with standard alkali.

The hydroxyl value is expressed as the number of milligrams of KOH required to neutralize the acetic acid capable of combining with 1 g of the sample. Other less volatile anhydrides have been successfully used (stearic anhydride); however, acetic anhydride is still the most popular.

Iodine value

The oxidation drying potential of an oil, varnish or resin is related to the level of ethylenic unsaturation present. This may be determined by reaction with iodine which adds across the double bonds.

The reaction is not always directly proportional to the number of double bonds, for example the conjugated double bonds in tung oil do not all react with iodine and give an experimental value of 160 which is below the theoretical value. The values obtained are reproducible from one batch of material to the next and may therefore be used in a comparative manner.

A typical procedure for the determination of iodine value is the Wijs method which may be found in many organic chemistry textbooks.

Diene values

This is another method for determining the level of unsaturation of an oil and involves reacting a known quantity of the oil with an excess of maleic anhydride. The excess anhydride is hydrolysed to maleic acid and back-titrated with alkali.

Saponification value

The saponification value of an ester, whether it be a polyester resin or vegetable oil, varies in accordance with the molecular weights of the acids and alcohols involved. It is useful in assessing the purity of a vegetable oil and may be useful in the analysis of a polyester resin.

The test is done by refluxing the sample with an excess of alcoholic KOH and then titrating the excess alkali with standard acid. The result is expressed as milligrams of KOH per gram of oil.

Unsaponifiable matter

This test is normally applied to vegetable oils and is once again a measure of the purity of the oil. The sample is saponified with caustic alkali and the alkaline solution extracted with diethyl ether. The ethereal solution is evaporated and the residue dried at 80°C. The residue is the unsaponifiable matter.

Water content

The presence of water in varnishes, oils or solvents may lead to incompatibility, poor pigment dispersion, loss of reactivity or 'bumping' and 'spitting' during the cooking process.

The water content of varnishes or oils may be determined quantitatively by azeotropic distillation with xylene in a Dean and Stark apparatus. Alternatively, a Karl Fischer titration or gas chromatography with a Katharometer detector may be used (page 880). The Karl Fischer titration uses the following reaction:

$$I_2 + SO_2 + CH_3OH + H_2O \rightarrow 2HI + CH_3H\ SO_4$$

The end point is determined electrometrically. However aldehydes and ketones may give an erroneous result if determined in this way.

The last two tests in this section are specific to petroleum solvents which contain varying quantities of aliphatic, naphthenic and aromatic components. The aromatic content of these solvents is related to their solvating power. The aromatic content is normally measured by the aniline point or kauri-butanol value.

Aniline point

The aniline point of a solvent is the lowest temperature at which a clear solution is obtained with an equal volume of aniline. High aromatic solvents are miscible with aniline at lower temperatures than more aliphatic solvents. Equal volumes of the solvent and pure aniline are mixed and is then allowed to cool. The temperature at which it becomes cloudy is the aniline point. This method requires standardization with solvents of known aromatic content.

Kauri-butanol value

This test consists of adding the solvent from a burette to a standard solution of kauri gum in *n*-butanol until the mixture becomes turbid.

16.3 PHYSICAL TECHNIQUES

Physical separation techniques play an important role in the analysis of surface coatings. Manual solvent fractionation has already been discussed and instrumental techniques will be covered in the following pages. However, there are several other simple techniques of note. These are thin layer chromatography, paper chromatography and column chromatography.

There are two 'classical' analytical tests which may be classed purely as physical tests: melting point and refractive index.

Melting point

Organic chemists have used the melting point of substances as an indication of purity for many years. A pure compound exhibits a sharply defined, reproducible melting point.

This test is also applicable to resins although a melting range is normally specified and this relates to the molecular weight of a particular batch of resin.

There are two common methods for determining melting ranges. The standard method is the ball and ring method in which a steel ball is placed upon a cast resin film supported in a ring cradle which in turn is immersed and heated in a liquid bath.

As the temperature increases, the resin softens and the ball eventually passes through the film. The ball and ring softening point is the temperature at which the ball is clear of the film according to BS 2782 1970.

The other common method is the capillary melting point; this is widely used for pure compounds and may also be used for resins. The powdered sample is placed in a glass capillary tube and heated at a constant rate in a metal heating block, monitored by thermometer and viewed through a lens.

Both digital and automated melting point apparatus are now available. Reproducibility of this method for resins is not good and results tend to be lower than for the ball and ring method.

Refractive index (RI)

When a ray of light is incident at an interface between two media, i.e. glass/liquid, air/liquid, observation reveals that some of the light is reflected from the surface while the rest of the light travels in a new direction. On account of the change in direction the light is said to be refracted.

For two given media, the relationship between the incident light and the refracted light, known as the refractive index, is a constant. The value of this constant is also dependent upon the colour of the light and is commonly specified as that obtained for yellow light from sodium vapour.

A typical device for measuring RI is the Abbé refractometer. The liquid sample is squeezed into a thin film between two prisms arranged in such a manner that the RI may be read directly from a scale.

Thin layer chromatography (TLC)

The mechanism employed in TLC is such that a mixture of compounds is separated as it is moved through a thin layer of sorbent material by a migrating solvent. The most common sorbents are silica gel and aluminium oxide, the thin layer of sorbent may be mounted on either glass plates, aluminium or plastic sheet. Using these materials the separations are achieved by a process of adsorption, which is very useful for the separation of organic compounds of medium or low polarity. The sample being separated proportions itself between the developing solvent and the active sites of the sorbent. The polarity of the components determines what this proportion will be. Polar solvents cause the greatest movement of the sample. The following solvents commonly used in TLC are listed in order of increasing polarity: hexane, benzene, toluene, trichlorethylene, diethyl ether, chloroform, ethyl acetate, butanol, propanol, acetone, ethanol, methanol, water.

Paper chromatography is a form of partition TLC which finds application in separating water soluble inorganic compounds and polar organic compounds. Other forms of partition TLC utilize cellulose or silica containing absorbed water as the stationary phase. Reverse-phase partition TLC utilizes a non-polar stationary phase and a polar mobile phase. The non-polar stationary phase may be a coated silica. Identification of components in TLC is achieved by comparing the R_F value of the sample with a standard. R_F values are computed by dividing the distance the component travels by the distance the solvent front travels.

Some typical applications of TLC are:

- separation of the acids from 2B, 4B, Lake Red C and other pigments;
- separation of plasticizers;
- separation of dyestuffs.

Visualization of the different components may be achieved by a variety of methods including: exposure to iodine vapour, charring with concentrated sulphuric acid, colour reactions with specific reagents. More commonly TLC plates are manufactured with a fluorescent agent incorporated in them. Under a UV lamp the components appear dark against a light background.

Plate scanners are also available which read the density of the chromatographic 'spots'. This enables accurate quantitative TLC to be done.

Column chromatography is an extension of TLC often used as a large-scale preparative technique to isolate components of a mixture for further investigation. Once a separation has been achieved using TLC then a similar stationary phase packed into a hollow glass column and elute using the same solvent via gravity feed may be used to separate a larger quantity of the same mixture.

16.4 INSTRUMENTAL TECHNIQUES

Whatever the reason for having an ink or related substance analysed, be it a substandard product from the factory, a customers print requiring matching or, as happens on occasions, a competitor's product, the analytical results are always required within a very short space of time. Modern electronics have put sophisticated instrumentation within the reach of every ink manufacturer. A skilled analyst using such equipment to its limit can gain valuable knowledge concerning the sample under investigation very rapidly.

Most of the common analytical instruments discussed in the following pages are available at a price to suit the inkmaker.

Infra-red spectroscopy

This technique is undoubtedly the most important in a modern analytical laboratory. Infra-red spectroscopy is an identification technique used to characterize the chemical nature of the sample under investigation. Nearly all molecules absorb radiation in the IR region of the electromagnetic spectrum. The instrument (Fig. 16.2) is used to obtain spectra characteristic of the chemical nature of the sample under investigation. Any type of sample may be studied using appropriate accessories, solids, liquids, gases, pastes, powders, prints, etc. No two different pure chemical compounds will give exactly the same spectrum and even physical differences in the same compound, solid or liquid for example, will give rise to differences in the IR spectrum. Variations in the spectra of polymers of the same type may not be so obvious as polymers are complex mixtures and not pure compounds.

At temperatures above absolute zero molecules are in a constant state of motion, the vibrational energy of the molecule causing a change in its dipole moment which gives rise to an alternating dipolar electric field. It is this which interacts with the electrical components of electromagnetic radiation giving rise to an absorption of energy. The vibrational transitions are classified as stretching modes and deformation modes, generally occurring in the wavelength range 2.5–15 μm.

Spectroscopists commonly use the term wavenumber which is the reciprocal of wavelength and is expressed in units of cm^{-1}. The usual range of an IR spectrum is therefore between 4000 and 667 cm^{-1}. A complex molecule has a large number of vibrational modes which involve the whole molecule; however, most of these molecular vibrations are associated with the vibrations of individual bonds or functional groups. These vibrations are either stretching, bending, rocking, twisting or wagging. The fundamental stretching and deformation vibrations for the methylene group are:

asymmetric stretch	2926 cm^{-1}
symmetric stretch	2853 cm^{-1}
scissoring	1468 cm^{-1}
wagging	1305 cm^{-1}
twisting	1305 cm^{-1}
rocking	720 cm^{-1}

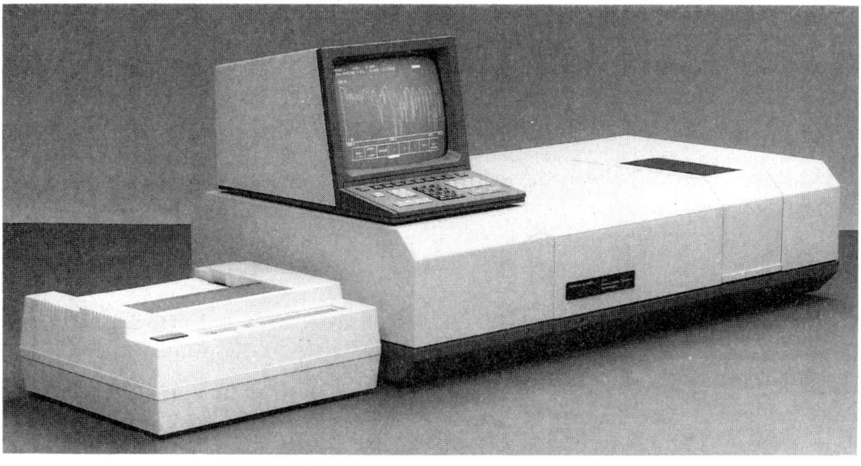

Fig. 16.2 A Perkin-Elmer infra-red spectometer. (Courtesy of Perkin-Elmer Ltd)

Infra-red quantitative analysis is made possible as the absorption of IR radiation takes place in accordance with the Beer–Lambert law $A = Ecl$, where A is absorbance, E is a constant for the sample under investigation, c is the concentration of the sample and l is the pathlength of the IR radiation.

The most common use of this relationship is in 'ratio' quantitative analysis. In this approach the strength of a peak in compound A compared with the strength of a peak in compound B is used to calculate the ratio of the two compounds in a mixture by comparison with a calibration table generated from known mixtures of A and B.

Quantitative and qualitative IR analysis are only successful when the sample is properly presented to the instrument. To present the correct quantity of sample to the instrument it is necessary to support the sample in the IR beam. Typical supports are constructed from the following: NaCl, KBr, CsI, KRS-5 (thallium bromide and iodide), AgCl or polyethylene. Liquids may be supported in cells of fixed or variable path length, the windows of which are usually NaCl or KBr.

Solids when they are film formers may be cast from solution as a thin layer on a salt plate. The most common technique for sampling solids is to prepare a paste (mull) of the solid in a viscous liquid.

Films and prints may be examined by placing them directly in the IR beam. However, a more common technique widely used in the surface-coatings industry is that of surface reflectance.

There are two useful variations of reflection IR spectroscopy. The first involves specular reflectance of the IR beam from a metal or metal oxide surface. An organic layer on this surface will selectively absorb IR radiation as the beam travels into and returns through the layer. The strength of the absorption is then dependent upon the material and the layer thickness.

The second variation is multiple internal reflection (MIR) spectroscopy. The surface of the sample to be investigated is clamped against a crystal of TiBr–TiI (KRS-5) or germanium. Infra-red radiation enters this reflection element and is totally internally reflected at the interface between the crystal and the sample. Some penetration of the sample takes place thus leading to selective absorption. Horizontal ATR accessories are now available in which the crystal is clamped against a trough allowing routine sampling of liquids and easier sampling of solids and prints. Fibre optics have made remote sampling with infra-red a possibility. A fibre optic ATR probe may be used to sample liquids in drums directly.

The following equation derived by Harrick gives the depth of penetration of the IR beam:

$$d_p = \frac{\lambda_\theta}{2\pi n_l\,[\sin^2\theta - (n_2 n_1)^2]^{1/2}}$$

where θ is the angle of incidence between the IR beam and the surface normal, n_1 and n_2 are the refractive indices of the crystal and the sample respectively and λ_θ is the wavelength of the radiation.

Utilizing reflection techniques information regarding the surface or coating on a sample as small as 1 cm^2 may be obtained. Infra-red microscopy extends this still further and very small samples such as print defects of 'spots' of only 5–10 μm may now be examined using infra-red spectroscopy.

In the analysis of printing inks interpretation of IR spectra is rarely done by identifying every peak. Analysis is normally done by comparing the unknown spectrum with known references. Most of the compounds used in the surface coatings industry are well documented. Typical examples are shown in Figs. 16.3–16.5.

The use of PCs to enhance the data obtained from IR spectrophotometers has improved every aspect of IR analysis. Modern IR software can manipulate the raw data (spectrum) in a variety of ways: baseline smoothing, expansion of weak spectra, spectral overlay, abscissa expansion, baseline flattening, conversion from transmittance to absorbance and vice versa, spectral subtraction, etc.

One of the standard routines, that of spectral substraction, may be used to illustrate the fact that this type of data handling brings its own unique problems. Firstly it should be noted that the absorption of IR radiation by a compound in a mixture is subject to matrix effects. These effects invariably lead to band broadening or even a slight shift in the frequency at which the absorption takes place. The result of subtracting the spectrum of a pure compound, e.g. plasticizer, from the spectrum of a mixture may result in the appearance of 'peaks' which do not truly exist.

Once the spectral data are in a satisfactory form the computer may be further used to interpret spectra, search software libraries to find the closest match and carry out automated quantitative analysis.

Fourier-transform infra-red spectrophotometers (FT-IR) have now

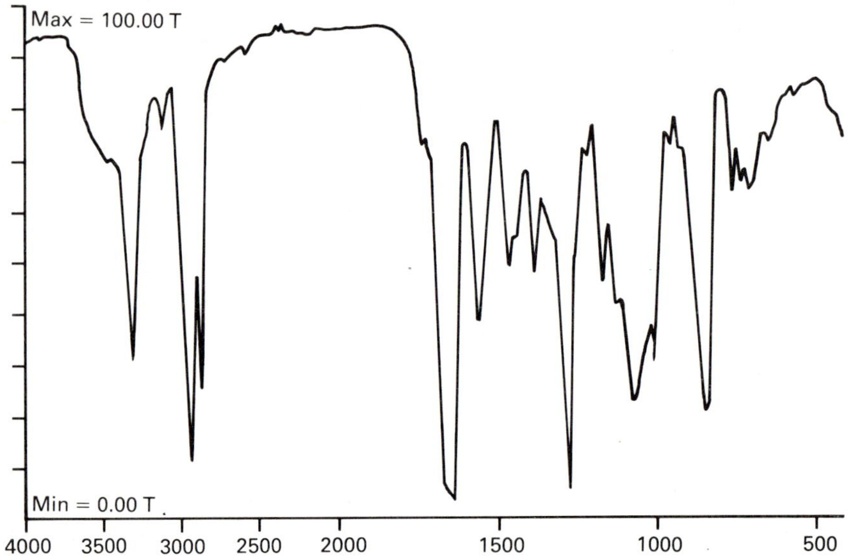

Fig. 16.3 Mixture of nitrocellulose and polyamide as a cast film on KBr.

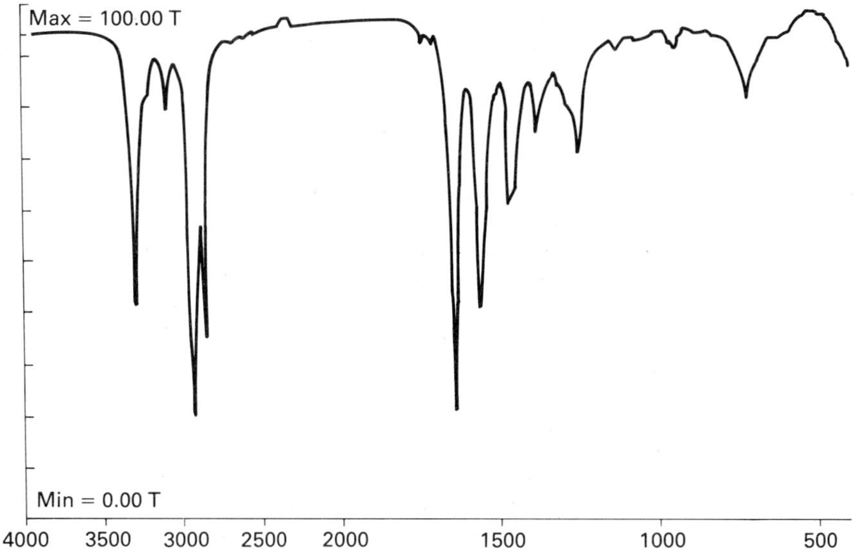

Fig. 16.4 Polyamide resin as a cast film on KBr.

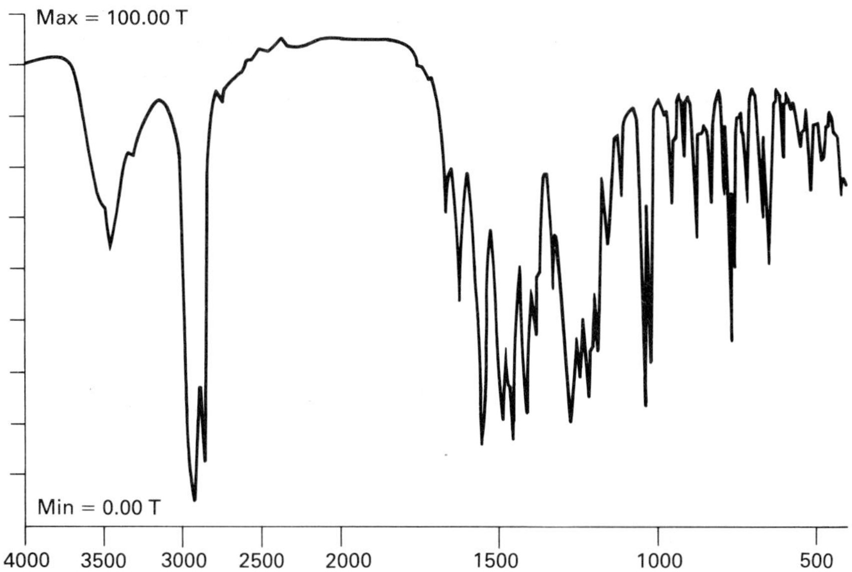

Fig. 16.5 CI Pigment Red 48:1 as a nujol mull.

replaced dispersive instruments as the normal choice. These instruments combine high scanning speeds, high signal to noise ratio, high energy throughput and accurate wavenumber measurement.

This brings significant advantages when investigating highly absorbing or scattering samples or when using accessories such as reflectance

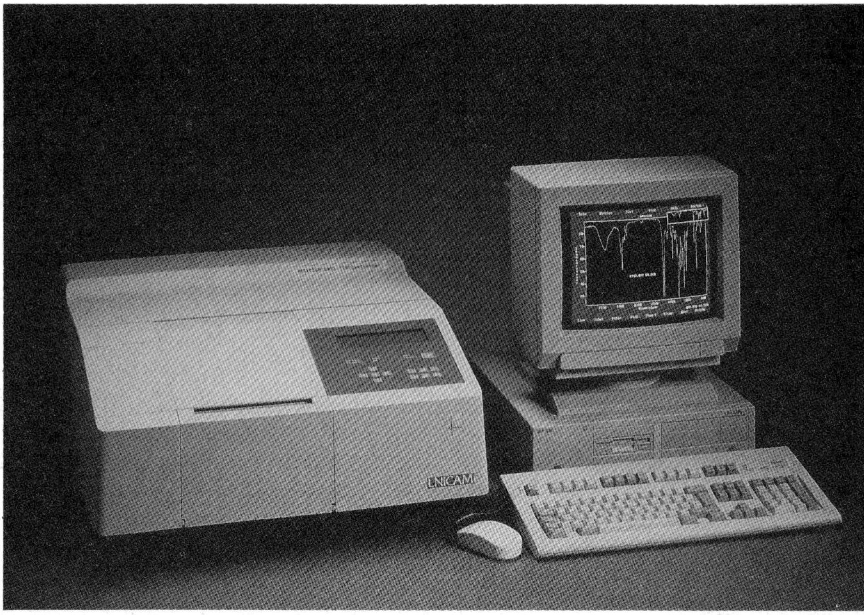

Fig. 16.6 Fourier-transform infra-red spectrophotometer. (Courtesy of Mattson Ltd)

devices and FT-IR microscopes. The advances in FT-IR spectroscopy have made hyphenated techniques such as FT-IR microscopy and 'on-the-fly' capillary GC–FT-IR measurements into routine laboratory techniques. The advantages of being able to obtain an infra-red spectrum from a spot on the surface of a print or from a well separated trace impurity as it elutes from a gas chromatograph are enormous (Fig. 16.6).

High performance liquid chromatography

This is now a well established instrumental laboratory technique (Fig. 16.7). It is once again a separation technique that when properly utilized may be used both to identify and quantify substances. A separation is achieved because the samples undergo a competitive distribution between a mobile liquid phase and a stationary phase. The stationary phase is contained within either a glass or steel column, the liquid phase is pumped through the column under pressure at precisely controlled flow rates. There are three different modes of separation employed in HPLC:

- adsorption
- liquid–liquid partition
- gel permeation or size exclusion.

The latter of these will be dealt with in detail in the next section.

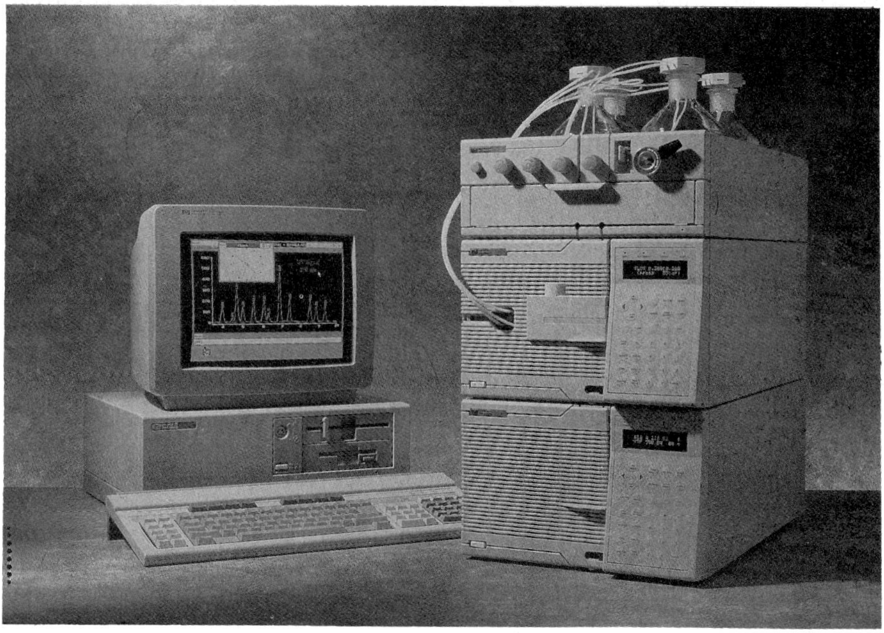

Fig. 16.7 High performance liquid chromatograph. (Courtesy of Perkin-Elmer Ltd)

Adsorption chromatography

This is also known as liquid–solid chromatography. Typical adsorbents are silica and alumina. Silica has a slightly acidic surface and preferentially retains basic compounds, alumina has a basic surface and as would be expected retains acidic compounds. For the purpose of determining the order of elution in liquid chromatography it is necessary to classify compounds according to their polarity. The following list depicts the polarity of compounds with respect to polar adsorbents starting with the least polar: fluorocarbons < saturated hydrocarbons < olefins < aromatics < halogenated compounds < ethers < nitro compounds < esters < ketones < aldehydes < alcohols < amines < carboxylic acids.

As HPLC is dependent upon a competitive distribution of sample between stationary and mobile phase, the polarity of the solvent used has a marked effect upon the separation achieved. In order to achieve optimum separation careful selection of solvents or even a programmed solvent gradient method may be required.

Liquid–liquid partition

The separation in liquid–liquid chromatography (LLC) is achieved by the distribution of the sample between two immiscible liquid phases. The stationary liquid phase is chemically bonded to the finely divided support particles. There are two distinct types of LLC: normal – where the stationary phase is polar and the solvents used are non-polar – and reverse phase – where the stationary phase is non-polar and the solvents polar. Solute elution in reverse phase is usually the opposite of that in normal LLC, i.e. polar compounds are eluted first.

In order to achieve a separation in reverse phase HPLC the components of the mixture submitted for analysis are encouraged to have a greater affinity for the stationary phase than the solvent. This is done by progressively increasing the polarity of the mobile liquid phase until the required separation is achieved. To ensure that all the compounds introduced on to the column are subsequently removed, gradient programming to a less polar solvent may be required.

One of the significant advantages of HPLC is that the detectors are mostly non-destructive and therefore specific detectors may be connected in series. Of the detectors available the following are the most common:

- refractive index detector
- UV detector – fixed, variable wavelength, or diode array
- fluorescence

Other available types are:

- Electrochemical detector
- conductivity detector
- infra-red detector
- mass specific detector.

Some of the most useful detectors in HPLC produce complete UV–Vis spectra of peaks as they elute. The two types of detector that can do this are diode-array detectors and rapid-scanning detectors. Both of these obtain spectra very quickly and can take many full spectra across the HPLC peak. This generates a very large amount of data compared with conventional fixed wavelength detectors. These data may be used to aid peak identification by comparison with spectral libraries and also in peak purity measurements.

The RI detector is widely used as the detector in GPC as gradient elution is not employed and it responds to any compound present in the solvent.

The possible applications of HPLC to the analysis of surface coatings are continually growing. Some typical applications which have already been investigated are:

- phthalate plasticizers
- surfactants
- polyols
- amines
- anti-oxidants
- polynuclear aromatics
- dyes
- fatty acids
- triglycerides
- solvents
- acids
- inorganic anions and cations (using ion exchange columns and buffered aqueous mobile phases)
- stabilizers

- UV monomers
- UV initiators.

Any separation which may be achieved by TLC may be achieved far better using similar HPLC conditions. Thus TLC is often used as a guide for the selection of a starting point in HPLC. Typically a system for examining polynuclear aromatic compounds would require a reverse phase liquid–liquid partition separation using acetonitrile and water as the mobile phase. Fluorescence detection in this case would give the greatest sensitivity.

Gel permeation chromatography

Gel permeation chromatography (GPC) is essentially a process for fractionating polymers according to their molecular size in solution resulting from permeation of the molecules through a porous matrix. Polymer solution is injected into a continuously flowing stream of solvent and passes through a bed of highly porous particles packed together in a column. If the pore size is so large that all the molecules can enter the pores or so small that none can, then no efficient separation can occur. The pore size range of the column packing must match the size of the polymer molecules under investigation. If this condition has been met, as the polymer solution flows through the column the smaller molecules will penetrate more pores than the larger and thus will travel a greater distance from injection to elution. As a result of this the large molecules elute first. GPC may be used as an analytical or a preparative technique.

One distinct advantage this has over HPLC is that everything put on the column assuming it is dissolved in the solvent and not of such large particle size that it physically blocks the column, will be eluted.

There are two detectors used in GPC which are considered to be virtually universal in their response, differential refractive index detectors and low angle laser light scattering detectors. The data from gel permeation chromatography may be presented in the form of a distribution curve on a chart recorder or numerically. The numerical data, M_w (weight average), M_n (number average), M_v (viscosity average), usually encountered are very useful for comparing bulk properties of resin samples. These data, however, often fail to account for the low and high molecular weight tails of the polymer distribution and thus a direct comparison of two polymer distribution curves may prove more useful. Overlaying curves in GPC is a widely used technique (Fig. 16.8) but molecular weight distributions are often expressed in terms of the numerical data, which can be obtained from the molecular weight distribution curve. These values are all different for a given polymer and were originally obtained using classical methods. M_n may be derived from osmotic pressure measurements, vapour pressure measurements or other colligative properties such as elevation of boiling point or depression of freezing point.

M_w is obtained from light scattering experiments and is a measure of the size and shape of the molecules present.

M_n is measured by examining the viscosities of dilute solutions of polymer. The actual viscosity obtained is dependent upon temperature,

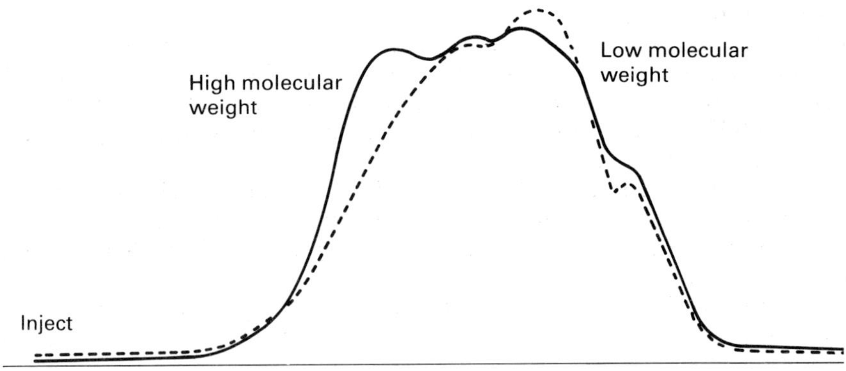

Fig. 16.8 Gel permeation chromatograms: (---) 'bad' resin; (—) 'good' resin.

the nature of the solvent and polymer, polymer concentration and the size of the polymer molecules.

The same numbers may be obtained with reasonably good agreement by using a calibrated GPC column set. Molecular weight standards are readily available for this. The elution time at any point on the GPC curve may be related to a molecular weight and the height of the curve when using a RI detector may be taken as proportional to the number of molecules of that molecular weight. By taking data from several points on the curve M_w and M_n may be obtained from the equations

$$M_w = \Sigma h_i \times M_i / \Sigma h_i$$
$$M_n = \Sigma h_i / \Sigma (h_i / M_i)$$

where h_i is the height of the GPC curve from baseline at elution time i and M_i is the molecular weight of species eluting at time i.

As previously mentioned, GPC can be used as an aid to the classification of a polymer. The molecular weight distribution affects many of the characteristic physical properties of a polymer. These may include:

- tensile strength
- brittleness
- melt viscosity
- softening temperature
- cure time
- coefficient of friction
- hardness
- gloss.

One very underrated application of GPC is as a separation technique. Even analytical grade columns may be used to fractionate a mixture of compounds although several injection/collection cycles may prove necessary in order to obtain sufficient samples to examine using other techniques such as IR spectroscopy.

Gas chromatography

Gas chromatography (GC) since its inception in the early 1950s has become a very powerful analytical technique. The instrument may be used to separate, identify and quantify almost any compound volatile below 450°C. The technique may be used either in a microanalytical mode or on a preparative scale.

The main principle of GC is that of a competitive distribution of a compound between a moving gas phase and a stationary phase.

There are two modes of GC, packed column and capillary column chromatography.

For many years packed column chromatography was the method exclusively used in ink analysis and this may be subdivided into two categories:

- partition – where the stationary phase is a non-volatile liquid;
- adsorption – the stationary phase is an active solid.

Partition or gas–liquid chromatography (GLC)

In conventional GLC the stationary liquid phase is coated onto a fine, particulate, inert support, which is packed into a tubular column, normally made of glass. The moving gas phase or 'carrier' gas is then passed through the packed column. The mixture to be separated is then introduced into the carrier gas and swept through the column.

If no partition of sample between the carrier gas and the liquid phase takes place then the column behaves like a distillation column and compounds are eluted from the column in boiling point order. To separate two different compounds of the same boiling point then a stationary phase must be selected which will have a greater affinity for one of the compounds. In order to separate a polar from a non-polar compound, a polar liquid phase would be chosen. The degree of partition for the polar compound would be greater than the non-polar and thus the polar compound would be eluted from the column last.

The degree of separation in GLC is influenced by the following factors.

- Column temperature – as temperature increases the partition equilibrium is shifted to the gas phase and the compounds elute more rapidly.
- Stationary phase – the greater the affinity for the stationary liquid phase the longer a compound will take to elute.
- Carrier gas flow rate – increasing this decreases the elution time from a compound: however, separation efficiency of the column also depends upon this factor.
- Particle size of support – this controls the area of stationary phase in contact with gas phase.
- Column dimensions – length and internal diameter may be adjusted to increase or decrease analysis time.
- The percentage loading of stationary phase on the support – this greatly affects the amount of sample that may be analysed and the retention time.

Careful control of these parameters eventually leads to the optimum separation of a mixture of compounds.

Adsorption or gas–solid chromatography (GSC)

The process of GSC is very similar to that of GLC; only the mechanism by which compounds are preferentially retained by the column is different. These are two types of stationary phases used in GSC: active adsorbents and molecular sieves. Active adsorbents, such as carbon and alumina, retain compounds according to the chemical/physical nature of the active sites. Molecular sieves retain small molecules longer and therefore separate on the basis of molecular size. Gas–solid chromatography is particularly applicable to the study of small molecules such as gases, typically H_2, He, Ne, Ar, Kr, O_2, CO, CO_2, N_2, NH_3, CH_4 to name a few.

Capillary gas chromatography

Capillary columns now dominate gas chromatography as the main choice of separating phase. Typical capillary columns are made of fused silica which when coated with a polyimide resin become very flexible. Dimensions of these columns are typically 0.1–0.3 mm ID by 10–100 m in length. The incredible length of these compared with a typical packed column of 2 m is possible because capillary columns have a hollow core and therefore generate very low back pressures.

The liquid phase is bonded to the inside wall of the capillary column in thickness of 0.1–5 μm. Capillary columns compared with packed columns (Figs. 16.9 and 16.10) give very efficient separations but have a very low sample capacity, limited normally to nanogram quantities per component. Column efficiency is measured in theoretical plates, a packed column may have 5000 to 7000 theoretical plates in total. However, a capillary column will have 1000–4000 plates per metre.

The very low sample capacity of capillary columns means that special injection techniques must be used to prevent overloading and any resulting loss of resolution.

Perhaps the most useful of the various types of capillary injector for the examination of inks and related products is the split/splitless system. This technique allows the reproducible introduction of very small samples into the capillary column.

Other factors influencing the performance of a capillary column are the carrier gas flow rate and temperature.

Of the carrier gases available, H_2 offers the fastest analyses with maximum resolution, the greatest number of theoretical plates being available when H_2 is used at a velocity of 40 cm/s. The flow rates through a capillary are of the order of 1–2 ml/min and thus difficult to control; flow rate is normally controlled therefore by pressure.

Capillary columns have very thin walls and being of low thermal capacity are very sensitive to temperature fluctuations.

A modern gas chromatograph has three main component parts (Fig. 16.11): an injection block, a column oven and a detector. The injection block is where the sample is introduced into the column without leakage. The block is heated in order to vapourize the sample. A septum injection technique is normally employed using a precision sampling syringe.

Column ovens are necessary for maintaining reproducible temperature conditions. Modern chromatographs allow almost infinite variation of

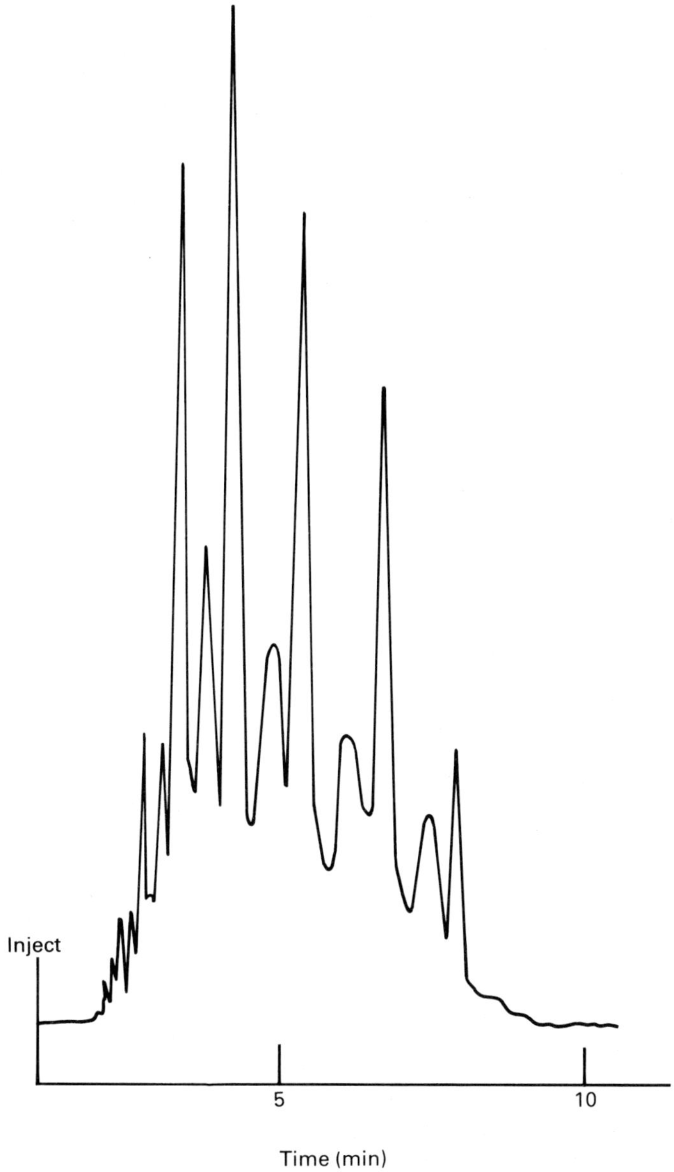

Fig. 16.9 Gas chromatogram of a distillate sample on a conventional packed column.

oven temperature with time and this is known as temperature programming.

Various detectors are available, flame ionization detector (FID), katharometer or thermal conductivity detector (TCD), being the most common. Other more specific detectors are available, electron capture, nitrogen/

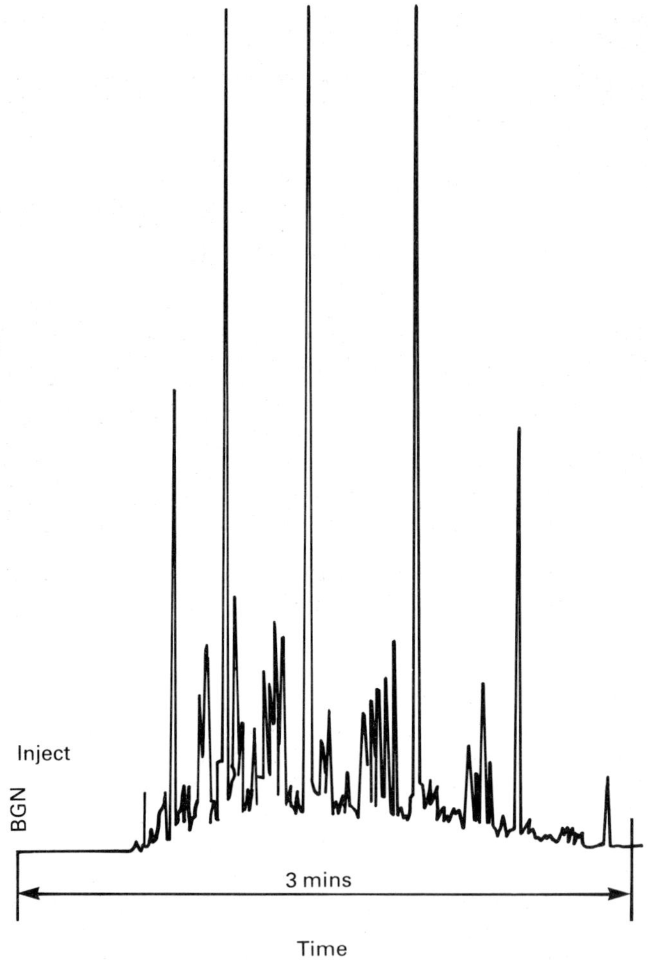

Fig. 16.10 Gas chromatogram of a distillate on a 25 m capillary column.

phosphorus and flame photometric to mention a few.

None of the detectors listed are able to identify the compounds eluting from the chromatographic column. Identification is made by comparison of retention times with those of known materials.

There are now several very specialized bench-top detectors which can be linked to a GC which use spectral information to provide a high level of information about eluting compounds. There is an atomic emission detector which is element specific, a bench top mass spectrometer detector and an FT-IR detector.

Gas chromatography/mass spectroscopy, or GC/MS as it is known, combines the separating power of a gas chromatograph with the identification power of a mass spectrometer to produce one of the most

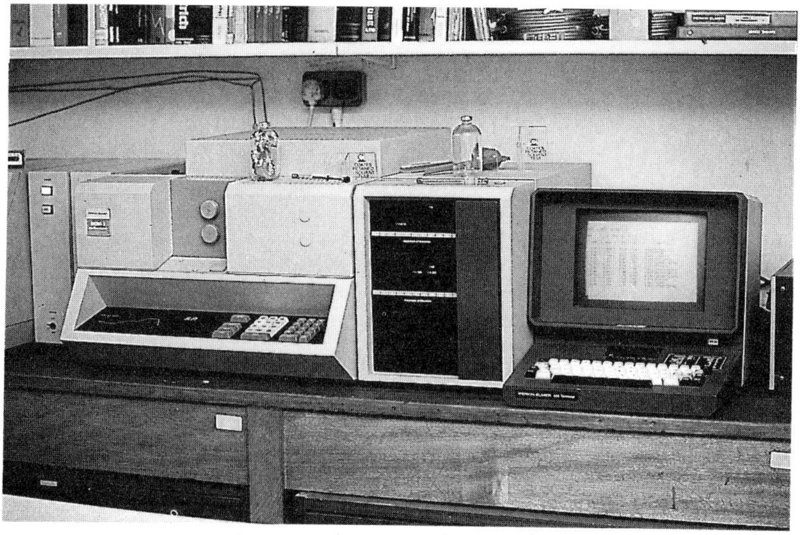

Fig. 16.11 A Perkin-Elmer chromatograph with data station. (Courtesy of Perkin-Elmer Ltd)

powerful analytical tools available (Fig. 16.12).

The components separated by the GC are eluted in turn into the MS. The component is fragmented into ions by an electron beam. The ions are then accelerated through a varying electromagnetic field formed by a radio frequency generator and a magnet in the form of a hollow centred quadrupole. The varying electromagnetic field serves to elute the ions in order of charge to mass ratio. Thus a mass spectrum is obtained for the component eluted. Every compound has its own unique mass spectrum, and therefore these instruments are equipped with computers and extensive data banks for the purposes of identifying the compound.

FT-IR detectors produce a vapour phase spectrum of the eluted components. These spectra are different from the normal condensed phase spectra of liquids and solids; however, many of the normal interpretation techniques may be employed.

The GC eluant is passed through a heated light pipe which is normally internally coated with a gold film. The FT-IR beam is 'bounced' inside this pipe to obtain maximum absorbance from the small amount of materials used in modern capillary GC. The FT-IR detector is not as sensitive as the mass spectrometer but is often better at differentiating between isomers (Fig. 16.13).

Whatever type of detector is employed, the output is proportional to the amount of component, and thus the technique may be used quantitatively. Two types of quantitative analysis are normally encountered, external standard and internal standard. The external standard technique involves calibrating the instrument by injecting a known quantity of reference material and relating this to a known quantity of the sample.

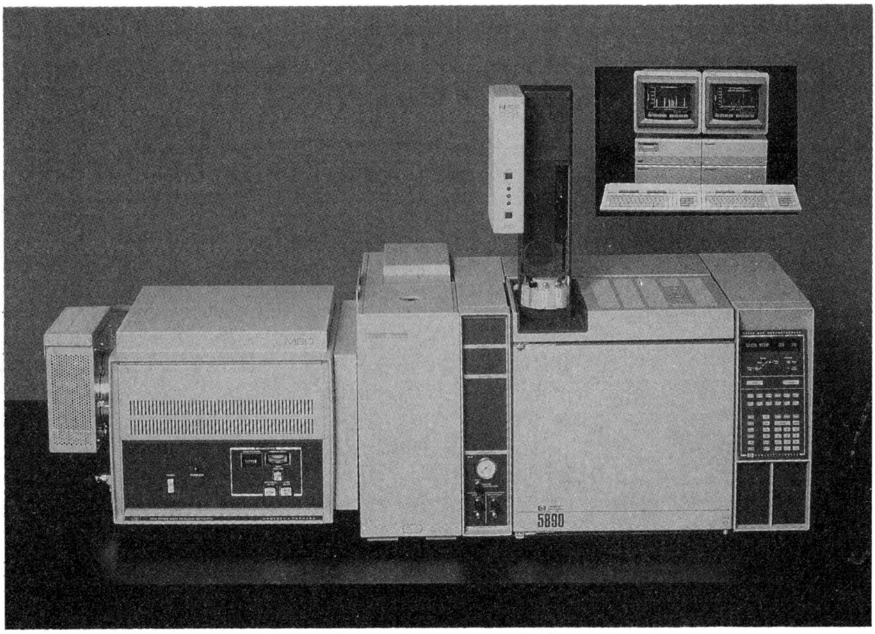

Fig. 16.12 Bench-top gas chromatograph with mass spectroscopy. (Courtesy of Hewlett Packard)

This method is reliant upon the operator's ability reproducibly to inject accurately known quantities.

The most common quantitative analysis method involves the use of an internal standard. With this technique a known quantity of a standard compound is added to a known weight of both the sample and calibration mixture. This amount then figures in the final calculation and eliminates injection errors and changes in detector sensitivity.

Applications

Gas chromatography may be applied to the analysis of the following compounds met in printing inks:

Solvents	Alcohols, esters, etc. (Fig. 16.14)
Distillates	Petroleum-based or synthetic
Plasticizers	Phthalates, adipates, citrates
Amines	Diethylamine, alkanolamines
Fatty acids	From oils or alkyd resins
Photo-initiators	Benzophenone, acetophenone
Anti-oxidants	BHT
Polyols	By prior derivatization
Resins	By prior pyrolysis or chemical cleavage followed by derivatization
Solvents retained in prints	By headspace techniques (see BS 6455: 1984) or discontinuous gas extraction

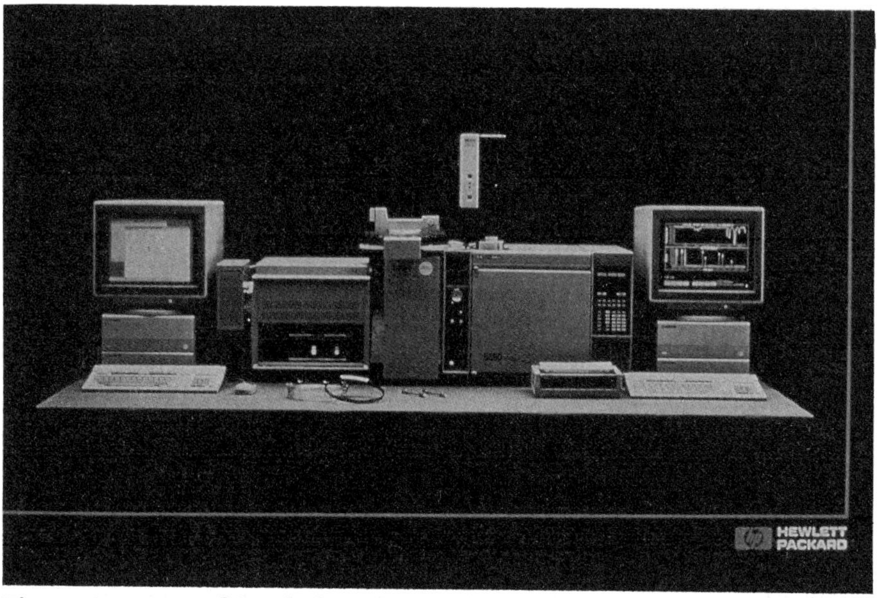

Fig. 16.13 Combined bench-top GC-MS–FT-IR. (Courtesy of Hewlett Packard)

Sampling techniques

Samples are normally introduced into the GC by the use of precision sampling syringes. Syringes are available for injecting liquids, solids or gases. Liquid samples lend themselves to precise repeatable injections more so than solids or gases. It is preferable therefore to have samples in the form of non-viscous liquids. This may be done by diluting the sample with a known solvent or extracting the desired components from the rest.

Resins may be studied directly by GC using a pyrolysis technique; this has been shown to be of particular use for acrylic resins.

Headspace gas chromatography is a technique which has found increasing popularity in the packaging industry as it enables the quantitative analysis of retained solvent levels in prints to be carried out.

The technique involves enclosing a known area of print in a fixed volume sample jar fitted with a septum cap. The jar is heated such that any solvent retained in the print is driven into the 'headspace'. A fixed volume of this is then sampled using a gas tight syringe and the quantities of each solvent determined chromatographically. The quantity of solvent retained in a print is normally expressed in terms of milligrams per square metre;' typical values for an acceptable print are 10–30 mg/m^2, depending upon the final use of the print.

The headspace technique is not the best quantitative method available but is the most appropriate for giving rapid, reproducible results and is a technique which has been standardized throughout the industry in BS 6455.

More recently automated thermal desorption has been applied to the

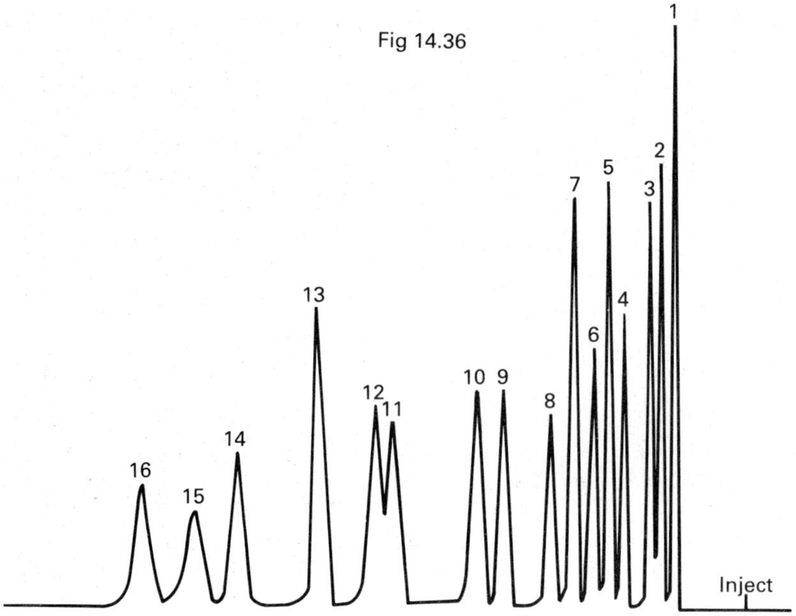

Fig. 16.14 Gas chromatogram of liquid ink solvents on a packed column: (1) acetone; (2) ethanol; (3) isopropanol; (4) ethyl acetate; (5) MEK; (6) isopropyl acetate; (7) *n*-propanol; (8) *sec*-butanol; (9) *n*-propyl acetate; (10) isobutanol; (11) propylene glycol ether; (12) *n*-butanol; (13) toluene; (14) *n*-butyl acetate; (15) ethylene glycol ether; (16) MIBK.

analysis of residual solvents in packaging film. Automatic thermal desorbers have been available for a number of years (Fig. 16.15) and work by heating the sample in a stream of inert gas. All the volatiles are removed from the sample, refocused on a cold trap which is subsequently heated very rapidly to effect a rapid transfer of the volatiles to the gas chromatograph.

The main advantages of this technique are that it does not rely on an equilibrium condition and, being automated, eliminates a lot of human error.

Gas chromatography calculations

The two most common calculation methods encountered are external standard and internal standard. With these methods the area under a given peak is related to the concentration of the component by comparison with a chromatogram obtained from a standard mixture.

The calculation used in external standard methods is the area of component A in sample/area of component A in standard times amount of component A in standard = amount of A in unknown.

Obviously with this method the amount of sample injected must be accurately known.

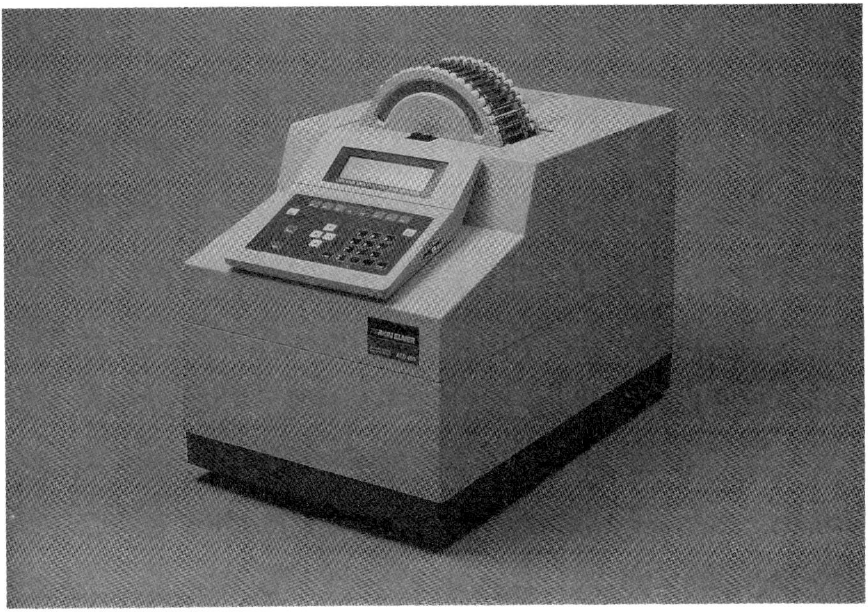

Fig. 16.15 Automated thermal desorber for analysis of residual solvents, etc. (Courtesy of Perkin-Elmer Ltd)

The calculation used in the internal standard method is quite different:

$$\frac{A_S}{A_C} \times \frac{AL_C}{AL_S} \times \frac{QL_S}{QL_C} \times Q_A = \text{amount of component A in sample}$$

where

A_s = area of component A in sample;
AL_S = area of internal std in sample;
Ql_s = quantity of internal std in sample;
A_c = area of component A in calibration mix;
AL_C = area of internal std in calibration mix;
QL_C = quantity of internal std in calibration mix;
Q_A = quantity of component A in calibration mix.

This calculation is far more precise as the quantities of the various components can be accurately weighed and by relating the calculations via the internal standard component the amount of either the sample or the calibration mixture injected no longer matters as long as the quantities of components are within the linear working range of the detector and do not exceed the capacity of the column used.

Atomic absorption spectroscopy

Since its introduction in the 1950s, atomic absorption spectroscopy (AAS) has proved itself to be among the most powerful instrumental techniques for the determination of metals (Fig. 16.16). Between 60 and 70 metals may be determined invariably without prior separation. The technique is not restricted to aqueous solutions as many organic and mixed phases are suitable.

Atomic absorption spectroscopy is the measurement of the absorption of photons by an atomic vapour. As an atom absorbs a photon one of the outer electrons is transferred to an excited energy level. The wavelength at which this occurs is characteristic of the element and the degree of absorption is directly proportional to the concentration of the solution.

There are two popular techniques employed to create the atomic vapour, flame and electrothermal. The most widely used technique is flame atomization, whereby a solution of the sample is sprayed via a nebulizer into an air/acetylene or nitrous oxide/acetylene flame. The light source for AA work is specific for the element being analysed and is normally a hollow cathode lamp. The lamp filament is coated with a layer of the element to be investigated, thus the light given out is at the appropriate wavelengths for absorption by the sample. The light given out is of several wavelengths and only one is used for the measurement; this is selected after passing through the aspirated sample by a monochromator. A photomultiplier detector amplifies the amount of unabsorbed light. The technique is applicable to solutions containing ppm levels of the element. Typical sensitivities are shown in Table 16.1.

Disadvantages of the technique are that the sample must be in the form of a solution. Only 5–15% of the aspirated solution reaches the flame and is then further diluted by the flame volume. The minimum sample size is 1 ml.

The majority of new equipment is available with flameless atomization.

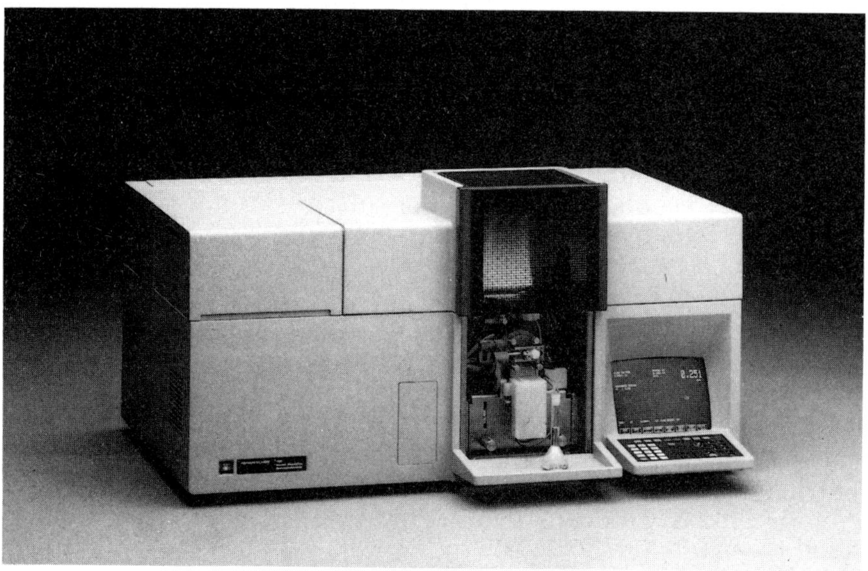

Fig. 16.16 An atomic absorption spectrophotometer. (Courtesy of Perkin-Elmer Ltd)

Table 16.1 Detection limit

Element	Flame atomization (μg/cm^3)	Electrothermal(μg/cm^3)
Aluminium	0.09	0.002
Barium	0.05	0.0045
Cadmium	0.003	0.00002
Chromium	0.006	0.0016
Copper	0.005	0.001
Iron	0.008	0.002
Manganese	0.003	0.0003
Molybdenum	0.15	0.004
Nickel	0.012	0.002
Lead	0.02	0.001
Silicon	0.3	0.02
Titanium	0.2	0.012
Vanadium	0.15	0.009

This technique employs the use of an enclosed furnace and the sample is atomized by a heated filament which may be carbon or tungsten. The furnace is flushed with an inert gas, i.e. argon, to prevent oxidation of the heating source. As the sample chamber is enclosed during the measurement stage this technique is far more sensitive. Solid samples may also be used, although solutions are often preferable. Typical sample size would be 50 μl.

Atomic absorption spectoscopy may be used for the following determinations typical of the type of analysis found in the surface coatings industry:

- determination of soluble metals from pigments;
- determination of metals from driers;
- metals in resinates;
- metals from leafing agents;
- metals from chelate type adhesion promoters.

The sample solutions may be prepared by either dry ashing followed by acid digestion of the metal oxides or salts, or by a wet ashing technique, depending on the volatility of the raw metal or its compound.

Dry ashing is carried out by heating a weighed sample until completely oxidized and the ash is then acid digested and diluted to a known volume.

Wet ashing involves completely charring a sample with concentrated sulphuric acid and then oxidizing with nitric acid. The resultant solution is again diluted to known volume.

One of the many uses of AAS is determining whether an ink meets the specifications listed in the *Toys (Safety) Regulations* 1989 BS 5665 Part 3 (EC EN71/3). This legislation requires that inks printed on toys do not exceed the following amounts of element available for migration (ppm of the dry film):

Soluble lead	90
Soluble arsenic	25
Soluble cadmium	75
Soluble barium	500
Soluble antimony	60
Soluble chromium	60
Soluble mercury	60
Soluble selenium	500

Ultraviolet–visible spectroscopy

Ultraviolet–visible spectroscopy is concerned fundamentally with the electronic transitions between energy levels within a molecule. The transitions are generally between a bonding or lone-pair orbital and an unfilled non-bonding or anti-bonding orbital. The wavelength at which this transition takes place is then proportional to the separation of the energy levels of the orbitals involved. Excitation of electrons from p, d and π-orbitals and π-conjugated systems gives rise to very useful spectra. But UV–visible spectroscopy is not confined solely to inorganic salts and conjugated organic systems, for even simple ketones will give rise to a weak spectrum: however, the former are the most widely investigated using this technique.

One of the most useful aspects of this technique is that the absorption is directly proportional to the concentration of the sample in dilute solutions. This relationship tends to break down in concentrated solutions. The absorption of energy in the UV–visible range obeys Beer–Lambert's law such that the absorption is directly proportional to the concentration of sample in solution. The relationship may be expressed by the equation:

$$A = Ecl$$

where A is the absorbance; E is the extinction coefficient (constant for the compound being investigated); c is the concentration, and l is the pathlength.

The UV–visible spectrum is usually taken on a solution of typically 0.001% concentration. A portion of this is transferred to a silica cell of pathlength 1 cm. An identical cell containing the pure solvent is also prepared and the two cells placed in the appropriate part of the spectrophotomer (Fig. 16.7). Two equal beams are then passed through these cells and the intensities compared over the wavelength range of the instrument. The spectrum of absorbance versus wavelength is plotted automatically. The excitation of electrons is accompanied by changes in both the vibrational and rotational quantum numbers giving rise to a broad peak. Further interactions with the solvent molecules blur this out and ultimately a smooth curve is obtained.

The most common solvent is 95% ethanol as the commercial absolute ethanol contains benzene. Fine structure due to vibrational and rotational absorbance may be revealed by using non-polar solvents where possible such as cyclohexane. Dipole–dipole interactions with the solvent molecules lower the energy of the excited state and therefore ethanol solutions give longer wavelength maxima than hexane solutions. This is known as a red-shift.

Fig. 16.17 A UV spectrophotometer. (Courtesy of Perkin-Elmer Ltd)

One important factor to note is that in UV–visible spectrophotometers the beam is split into discrete wavelengths by a diffraction grating before passing through the sample. This is because UV light is at a frequency and energy likely to cause decomposition of a molecule, therefore the minimum necessary is passed through the sample at any one time.

Despite having many applications in the ink industry UV–visible spectrophotometry is not a widely used technique due to the problems of sample preparation. The sample under investigation needs to be free from interfering compounds, thus the obvious application of UV-curing inks is probably one of the least exploited.

Typical applications of UV–visible spectrophotometry are:

- identifying and quantifying dye mixtures;
- identifying and quantifying optical brightness;
- determination of phenolic anti-oxidants;
- determination of aromatics in solvents.

Nuclear magnetic resonance spectroscopy (NMR)

This analytical technique depends on the fact that nuclei of certain atoms act like spinning bar magnets when placed in a magnetic field. The most commonly studied nucleus is that of the hydrogen atom, i.e. the proton, and in any organic molecule the various hydrogen atoms react differently to a magnetic field, depending on their position in the chemical structure.

In the study of complex molecules and macromolecules carbon-13 NMR is of even more importance. Correct interpretation of ^{13}C spectra can rapidly lead to a qualitative and quantitative analysis of relatively complex polymers. Although the abundance of ^{13}C is much lower than that of the proton the chemical shift due to different environments around a carbon atom are much greater.

The resolution in ^{13}C NMR is therefore easier to achieve than proton NMR.

A modern Fourier-transform NMR will produce high quality spectra in a relatively short time. Very high resolution proton NMR may still require several hours of scanning.

Figure 16.18 shows the spectrum obtained with ethanol using tetramethyl silane (TMS) as the internal standard and clearly identifies the different H atoms in the molecule.

Figure 16.19 shows the ^{13}C spectrum of a phthalate plasticizer; note the different scale on the *x*-axis.

For more detailed information on NMR spectroscopy the reader is recommended to read one of the specialist books on this subject.

16.5 SURFACE ANALYSIS TECHNIQUES

Earlier in this chapter we discussed testing procedures which may be used to identify a 'good' or 'bad' ink or an acceptable or unacceptable raw material. An ink, in order to perform properly on a press, must have particular rheological properties, flow, thixotropy, tack, etc. In order to perform well the final printed product must have good adhesion, slip, gloss, colour strength, flexibility, product resistance, and an assortment of other properties. Of all of these adhesion is possibly the least understood. Flame treatment of polyethylene bottles prior to screen printing for

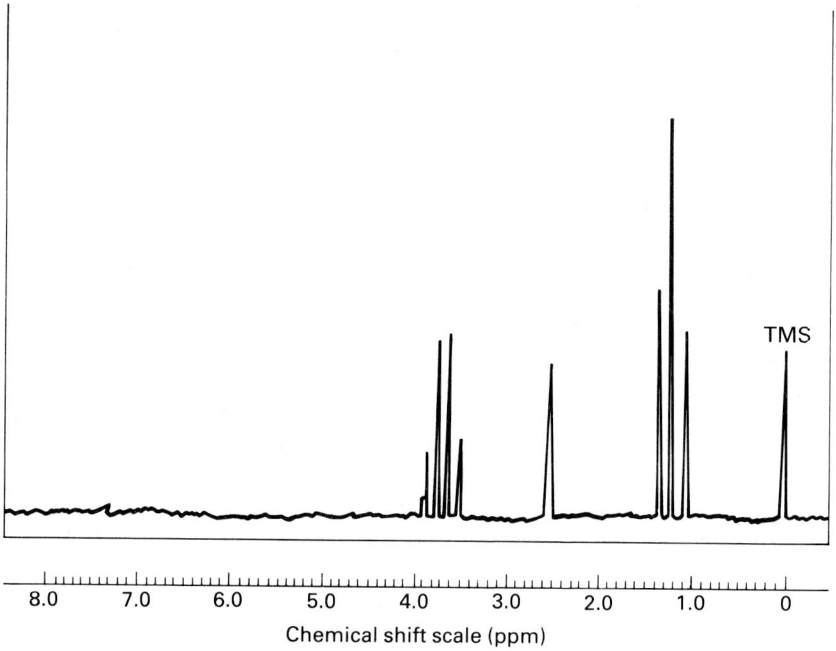

Fig. 16.18 NMR spectrum of ethanol.

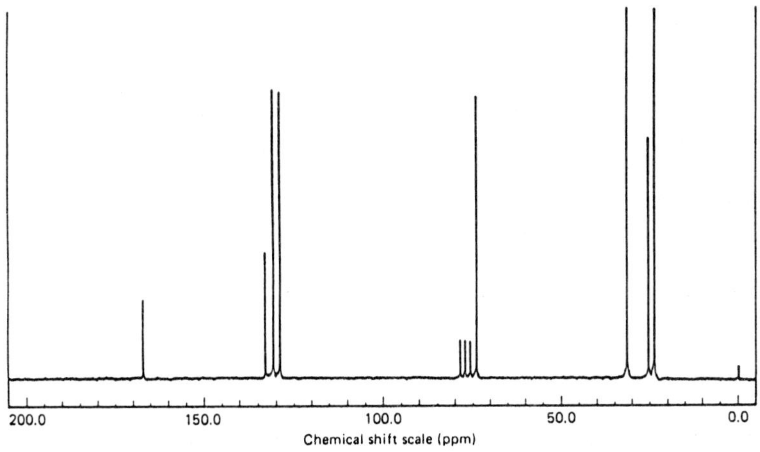

Fig. 16.19 ^{13}C NMR spectrum of DCHP.

example modifies the surface to give improved adhesion. Exactly what happens to the surface is only just beginning to be understood. In order to understand the phenomenon of adhesion or failure to adhere we need to study the chemical and physical nature of the surfaces at the interfaces between ink to substrate and ink to ink layers. There are several important new techniques which will be discussed. The instruments are prohibitively expensive to purchase but access to such techniques is available through research institutes and universities.

X-ray photoelectron spectroscopy (XPS) or electron spectroscopy for chemical analysis (ESCA)

In this technique the sample is bombarded with X-rays of typical energies 1253.6 and 1486.6 eV. Photoelectrons are emitted from the surface as a result of this bombardment and are analysed in terms of kinetic energy. Photoelectrons emitted from the atomic core, i.e. the carbon is electrons, are highly characteristic and allow the identification of all elements except hydrogen. The peak intensities at the emitted energy levels are proportional to the numbers of atoms present. But XPS is a surface technique despite the fact that X-rays may penetrate several microns into the sample the emitted electrons still have to 'escape' from the sample matrix. The maximum escape depth for this technique is typically 3 nm.

Most XPS equipment permits the variation of the angle at which electrons leave the surface and thus changes the escape depth of the electrons. The actual depth sampled is given by the equation

$$d = 3\lambda \sin \theta \ (\theta = \text{'take-off angle'})$$

A direct comparison of peak intensities at high and low angles enables a form of 'depth profiling' to be carried out.

Ion scattering spectroscopy (ISS)

This instrument bombards the sample with low energy (–2 keV) noble gas ions. A proportion of these are scattered by the surface atoms and their energies are analysed. This technique is not applicable to organic compounds because the ions are not scattered by hydrogen atoms.

Transmission electron microscopy (TEM)

This technique makes use of the fact that electrons are similar to light, having both a wave function and a particle nature. The electrons having passed through a very thin sample (10–100 nm width) are focused using electromagnetic lenses. Any investigation of an organic compound requires that the sample is coated with a thin layer of metal or carbon to prevent charging. Damage to the sample by the electron beam is common.

Scanning electron microscopy (SEM)

This technique is perhaps the most useful tool for gaining information regarding the topography of a surface. A finely tuned electron beam is scanned over the sample surface. The beam undergoes a variety of interactions leading to emission of electrons of varying energies, X-rays and light. The image in an SEM is produced by the detection of low energy secondary electrons. The scanning electron microscope has a depth of field some 10–100 times greater than an optical microscope would have at the same magnification, the resolution of an SEM is also of the order of 10 better. The only major problem in SEM is the charging of insulating materials, and this may be overcome by applying a 30–50 nm conductive coating by vapour deposition.

The use of energy dispersive analysis of X-rays (EDX) is a technique which may be incorporated in an SEM. This utilizes a high energy electron beam which induces the emission of X-rays. The frequency of the X-rays is very specific to the emitting atom and proportional to the number of similar atoms. Therefore, EDX may be used to identify the chemical nature of the surface under investigation. The X-rays generated are not confined purely to the surface layer but may originate from a depth of several microns.

Secondary ion mass spectrometry (SIMS)

The sample is bombarded with a beam of ions, e.g. Ar^+, O_2^+, of 2–20 keV. This bombardment gives rise to a few charged particles being emitted from the surface. These are then examined using a mass spectrometer from which secondary ion mass spectra are obtained.

This technique may be used in either one of two modes of analysis. Low energy or 'static SIMS' is used for gaining the maximum information from the surface being investigated. High energy 'dynamic SIMS' utilizes a high current density ion beam which may be used to give a depth profile of the sample.

16.6 ENVIRONMENTAL MONITORING

Historically the printing ink industry has used whatever materials are necessary to achieve the performance required. Solvents, catalysts and other additives were chosen purely for their characteristics. Ignorance of potentially unhealthy side effects led to a great deal of misuse. With the increasing awareness of the possible damage to both the environment and the health of workers, legislation has followed to ensure that standards of cleanliness are met. Factory effluents are monitored as is the working atmosphere. Atmospheric monitoring is that which greatly concerns this industry due to the very volatile nature of the solvents used.

A much greater emphasis on environmentally friendly products, recycling and low emissions are becoming the order of the day. New legislation limits the level of volatile organic compounds which may be emitted by ventilation systems.

It is important to monitor stack effluents regularly to ensure compliance with local government standards. Various on-line detectors are available for this, automated GC, FID detectors, IR detectors, and photoionization detectors may all be used continuously or for spot sampling.

The measurement of fundamental importance to the health and welfare of the employee is that of the average level of airborne contamination to which any one person may be exposed. The recommended limits for occupational exposure levels are available from the Health and Safety Executive. However, every manufacturer should attempt to eliminate the presence of contaminants from the working atmosphere.

There are several ways in which the atmosphere may be tested. 'Gas sampling' involves taking an atmospheric sample using a gas-tight syringe or sampling cylinder and then determining the contents of the device by gas chromatography. This technique as with sampling tubes such as the Draeger tubes gives a result only for the atmospheric conditions at the time of sampling. In order to obtain a time-weighted average, passive or active sampling devices must be worn by a worker for a significant time period. These devices contain absorbents which may consist of activated charcoal or porous polymers. Passive samplers absorb solvents from the atmosphere by diffusion. Active samplers require a pump to draw air over the absorbent. The absorbed contaminants may be desorbed by a suitable solvent or heat for subsequent analysis using gas chromatography.

Automated thermal desorbers as described for monitoring solvent in prints are also used for this type of work. It was in fact with this application in mind that the first automated thermal desorber was developed (Fig. 16.15).

Environmental monitoring is an ever-increasing, important aspect of an analyst's duties.

FURTHER READING

Feigl, F. and Anger, V. (1972) *Spot Tests in Inorganic Analysis*, 6th edn, Elsevier.
Grob, R.L. (1988) *Modern Practice of Gas Chromatography*, 2nd edn, J. Wiley and Sons.

Hachenberg, H. (1973) *Industrial Gas Chromatographic Trace Analysis*, Heyden and Son Ltd.

Hachenberg, H. and Schmidt, A.P. (1977) *Gas Chromatographic Headspace Analysis*, Heyden and Son Ltd.

Haslam, J., Willis, H.A. and Squirrell, D.C.M. (1972) *Identification and Analysis of Plastics*, Butterworths, London.

Mark, H.F., Othmer, D.F., Overberger, C.G. and Seaborg, G.T. (1984) *Kirk Othmer Encyclopedia of Chemical Technology*, John Wiley and Sons.

Vogel, A.I. (1968) *Elementary Practical Organic Chemistry*, Part 3, *Quantitative Organic Analysis*, Longman.

Willis, H.A, van der Maas, J.H. and Miller, R.G.J. (1987) *Laboratory Methods in Vibrational Spectroscopy*, 3rd edn, J. Wiley and Sons.

Williams, D.H. and Fleming, I. (1973) *Spectroscopic Methods in Organic Chemistry*, 2nd edn, McGraw-Hill.

CHAPTER 17
Health, safety and the environment

Although many substances may present hazards at work, they can still be used without risk to health and safety provided those hazards are understood and the appropriate precautions are taken.
(Health and Safety Executive, Substances for Use at Work, *1986)*

17.1 GENERAL INTRODUCTION TO UK LEGISLATION

This chapter is aimed at a review of UK legislation and other constraints affecting health, safety and environmental aspects associated with the manufacture of printing inks. It is not intended to set out in fine detail all the provisions of the regulations discussed but simply to highlight the existence of relevant legislation and note the provisions that impinge on the inkmaker's art.

The first section presents an overview of three principal areas of legislation:

- Health and Safety at Work, etc. Act 1974 and subsequent legislation;
- Environmental Protection Act 1990;
- Product liability legislation.

Brief mention is also made of a number of isolated statutory instruments made under other legislation.

The second section covers the handling of chemicals and mixtures (preparations) under the following headings:

- ECOIN/EINECS and notification of new substances
- Occupational exposure limits
- Control of substances hazardous to health
- Control of lead at work
- Fire hazards
- Data communications to the user
- The label
- The data sheet
- The control of industrial major accident hazards
- Transport of dangerous goods

The third section deals with mechanical and operational aspects of health and safety in the manufacture of printing inks.

The fourth section covers particular applications for printing ink which require more than general attention to health and safety matters.

The last section looks at some international constraints.

17.2 HEALTH AND SAFETY AT WORK, ETC. ACT 1974 (HSW ACT)

The quotation at the beginning of this chapter embodies an unexpressed principle of current UK health and safety legislation. With few exceptions, UK legislation does not seek to impose an absolute ban on the use of chemical substances; what it does seek to do is to ensure that those who decide to use chemical substances at work do so in the knowledge that they are required to make adequate provision for the health and safety of those expected to come into contact with chosen substances.

Prior to 1974, health and safety legislation was dispersed across a wide range of statutes. As a result of the deliberations of the Robens Committee of Inquiry [1], the Health and Safety at Work, etc. Act [2] set out to consolidate earlier legislation and established general duties on employers and employees. It also gave ministers the power to issue specific regulations to deal with specific problems. Every employer needed to consider, within the context of his own organization, what action was necessary to fulfil the duties imposed by the Act [3, 4].

Section 1 of the Act made specific provision to replace a range of earlier statutes dating back to 1875. Existing duties, spelt out by earlier statutes, remained in force until the latter were replaced by specific statutory instruments under the HSW umbrella. The Act had immediate effect on the method of enforcement. The gradual replacement of those earlier provisions is nearly complete.

A two-component governmental agency was charged with overseeing actions to promote health and safety at work and their inspectors were given the powers to issue improvement notices or prohibition notices. Improvement notices set a specified time within which remedial action must be taken; prohibition notices require immediate suspension of an operation until deficiencies are rectified.

The Act aimed to ensure that employers looked after the health, safety and welfare not only of their employees but also of those outside their employment who might be affected by their work activities. Employers were, and are, required to control the handling and storage of dangerous chemicals and the emission of noxious or offensive substances into the atmosphere (not just those which are dangerous to health).

The primary responsibility for accident prevention and combating occupational ill health, lies with those who create the risk. However, the Act aimed to involve everybody at the work place in achieving high standards of health and safety.

The employer's general duties were to be carried out so far as was reasonably practicable. Specific mention was made of the provision of safe plant and systems of work, safe use, handling and transport of raw materials and finished products. Necessary information, training and a safe working environment were to be provided and a written statement of general policy covering health and safety at work was to be issued where more than four people were employed. However, earlier provisions under the Health and Safety at Work Act have been largely replaced by the UK legislation implementing the EEC Framework Directive and the various 'Daughter' Directives which came into force on 1 January 1993.

When announcing proposals for what has now become the Management of Health and Safety at Work Regulations 1992 [5], the Chairman of HSE described the proposals as 'probably the most important addition to general health and safety law since the Health and Safety at Work, etc. Act 1974'. Nevertheless, it was generally acknowledged at the time that the proposals contained no principles not already implicit in the 1974 Act, but reflected the philosophy of the EC directive on which the proposals were based. They were characterized as more prescriptive and detailed (that is more explicit) than most regulations made under HSW. Much of the underlying philosophy had already begun to appear (especially from within the Control of Substances Hazardous to Health Regulations (page 910), i.e. assess, eliminate hazard, prevent/minimize risk, protect employees).

In particular, they brought into HSW the strongest move ever to place responsibility on all employers for the proper management of health and safety matters in their workplaces. They gave to the employer explicit responsibility to manage (plan, organize, control, review preventative and protective measures) health and safety. Again, in all but the smallest companies (four employees or less), the arrangement must be formally recorded. Finally, and probably most importantly, where the employer is not himself competent to undertake the task, he must appoint one or more competent persons to assist him to discharge his responsibilities under HSW. When more than one competent person is appointed, the employer must make arrangements for their adequate cooperation.

The Environmental Protection Act 1990

The Environmental Protection Act 1990 (EPA) [6] provides for improved control of pollution, re-enacts and modifies provisions in the earlier Control of Pollution Act 1974 (COPA) [7] (relating to waste disposal) and substantially changes the organization for waste management.

From the point of view of the surface coatings industry, the biggest changes which were introduced by EPA relate to air pollution and the management of waste.

Processes are either designated for central control (Her Majesty's Inspectorate of Pollution) (HMIP) or for local authority control (typically at District Council level). The manufacture of printing inks is subject to local authority control in respect of air pollution only. Those industries which are controlled by HMIP are subject to Integrated Pollution Control (IPC) and they are required to take account of the effects of discharge practices on all environmental media.

For the less polluting industries, which category includes the surface coatings industry, air pollution control is governed by reference to sector-specific Guidance Notes from the Secretary of State [8].

It should be noted that occupational exposure control continues to be the concern of the Control of Substances Hazardous to Health Regulations; both it and EPA contain provisions to avoid the transfer of problems from one jurisdiction to another.

EPA continues to define 'Controlled Waste' [9] and 'Special Waste' [10] as they were defined under the earlier Control of Pollution Act for the time being, though it seems likely that 'Special Waste' will eventually be redefined to bring it into line with the EC's definition of 'Hazardous Waste' [11].

Controlled Waste continues to include material, emanating from factories or laboratories, which is scrap material or effluent or any other unwanted surplus substance arising from the application of any process, or any substance or article which requires to be disposed of as being broken, worn out, contaminated or otherwise spoiled.

Special Waste is a sub-category of Controlled Waste and contains listed substances so as to render the waste dangerous to life or to have a flash point below 21°C. The latter criterion is more significant to ink manufacturers, as many ink solvents – particularly those used in gravure and flexo processes – have flash points below this temperature. The EC amendment which will probably modify this definition includes irritant substances and raises the flash point threshold to 55°C.

While COPA was concerned with waste disposal, EPA is concerned with waste management. Under the new Section 33, it is illegal to treat, keep or dispose of controlled waste in a manner likely to cause pollution of the environment or harm to human beings (33,1,c) or to treat, keep or dispose of controlled waste or knowingly cause or knowingly permit it to be treated, kept or disposed of, except in accordance with the conditions of a waste management licence (33,1,b). Any holder of controlled waste is under a duty of care [12] to:

- prevent any contravention by others of Section 33;
- prevent the escape of waste from his or any other person's control;
- transfer waste only to those authorized to deal with it
- provide a written description of transferred waste.

Product liability legislation

Product liability is that branch of the law requiring retailers, distributors, manufacturers, designers and installers to provide goods which are fit for purpose and are not inherently dangerous. Unfitness and danger are distinct but overlapping concepts. In general, the fitness for purpose is left to agreement between contracting parties in the UK. Only in the case of dispute is it necessary for the law to intervene in determining the meaning of a contract. However, safety is an area where the law sets itself a preventative rather than an arbitration role. Increasingly, safety has become the subject of criminal (as opposed to civil) sanction as under the Health and Safety at Work Act 1974.

In the UK, under the Sale of Goods Act [1], seller and buyer enter into a contractual agreement which guarantees the reasonable suitability and safety of the goods received by the buyer. Should the goods prove defective in either sense, the contract has not been fulfilled and the buyer can claim against the seller. During the course of any such claim, the law does not require that the buyer proves 'negligence' on the part of the seller. The seller remains liable to the buyer whether the defect is known or unknown. Also, the contract is between these two parties only; the buyer has no contractual arrangement with the manufacturer (if he is not the seller). The primary responsibility rests with the seller not the manufacturer. In legal terms, this is the concept of strict liability.

However, in the absence of contractual rights (for example, where the injured party was not the buyer) the injured party must prove negligence. In the USA and some European countries, this unreasonable situation where a third party has to prove negligence has been removed and there is considerable pressure on the European Community (EC) to adopt this broader application of strict liability. In the UK, there has been what might be termed isolated recognition in various Acts (e.g. in both the Toys Regulations [14] and the Pencils and Graphic Instruments Regulations [15] made under the Consumer Safety Act 1978) that this wider application of strict liability is appropriate and desirable.

As a result of this pressure, a European directive, and subsequently the Consumer Protection Act 1987 [16] have introduced the concept of strict liability, though it must be emphasized that 'defectiveness' in this context is with respect to safety only and not its fitness for purpose in a technical sense. Thus, the Act is not concerned if an electric fire, for example, fails to heat a space adequately, but is concerned if, in performing its function, the electric fire electrocutes the operator. Also, the seller is not under an obligation to provide an absolutely safe product, as some products (one extreme example might be dynamite) are inherently dangerous. What he must do is provide adequate information regarding the safe handling and use of such products. The legislation does not affect contractual obligations to provide products which meet the performance criteria agreed between buyer and seller.

In connection with the preparation of Safety Data Sheets, it is important to note that the criterion on which a producer's liability should be assessed is that of 'Safety which a person is entitled to expect . . .'.

The assessment of 'expectation' will be with reference to the information supplied in respect of foreseeable use; that information being in the form of labelling, warning notices, instructions and guidance, for example.

If a product is properly labelled and is supplied with adequate instructions for safe use and information about hazardous properties, this can protect the producer. The consumer is made aware of the risks being run and the degree of safety that the user can expect.

The Act provides for compensation, without the need to prove negligence, for death or personal injury resulting from defective products. Manufacturers need to check that their insurance arrangements cover these liabilities.

The Act (Section 36 by reference to Schedule 3) also amended the Health and Safety at Work, etc. Act to ensure that manufacturers exercise reasonable foresight in product design and construction.

Other UK acts

In addition there are a number of isolated statutory instruments issued under other acts which potentially impinge on ink manufacture. These will be dealt with in greater detail in sections 17.3 and 17.4 They are:

- Consumer Protection Act 1978: Toys Regulations, Pencil and Graphic Instruments Regulations;
- Factories Act;
- Food and Drug Act.

17.3 HANDLING OF DANGEROUS SUBSTANCES IN THE MANUFACTURE OF PRINTING INKS

The major emphasis of UK legislation at present relates to chemical exposure and the risk of fire, though this is now changing, following the adoption of the framework directive on minimum health and safety requirements and its subsequent daughter directives.

This section deals with the chemical and fire hazards associated with the handling of inks and their raw materials; the requirements of the new framework directive and its daughter directives are dealt with in the next section.

ECOIN, EINECS and notification of new substances

Several years ago, the EC set itself the task of establishing toxicity parameters for all known chemical substances in order to control the information made available to users through, for example, labels and health and safety data sheets. Relatively early on in this exercise, it was realized that this was a virtually impossible task.

The EC sought to overcome this problem by the simple device of establishing a definitive list of 'known' substances for which it is presumed that their manufacturers are sufficiently knowledgeable to provide relevant information on the basis of experience in use.

Thus, all chemicals demonstrated to have been on the market within the period 1 January 1971 to 18 September 1981 fall into this category of 'known chemicals' and were published initially as the European Core Inventory (ECOIN [17]) drafted by EC itself. Subsequently, industry was requested to submit suggested additions to this list and the updated list, known as the European Inventory of Existing Chemical Substances (EINECS), was published in the Official Journal in June 1990. It must be remembered that EINECS is simply a listing of chemicals and makes no attempt to provide information relating to the toxicity of the listed materials.

Any chemical substance which appears in EINECS is defined to be a 'Known Substance' [18] those which do not are, by definition, 'New Substances' for the purpose of notification legislation and as such are subject to pre-marketing notification requirements. Substances were not taken to include polymers which were deemed to be dealt with by virtue of the inclusion of their component monomers.

However, under the new Chemicals (Hazard Information and Packaging) Regulations [19], a definition of polymer is about to be introduced and this will have the effect of requiring registration of some oligomers previously regarded as polymers. This provision is not expected to apply retrospectively.

Any 'new' substance must undergo extensive physical and toxicological testing if a total quantity of 1 tonne or more is expected to be put on the market in a 12-month period. This is covered by the Notification of New Substances Regulations [20] which came into effect at the end of November 1982 and the requirements are amplified by a guidance booklet [21].

With effect from 17 August 1987, HSE adopted the Annex to EC Directive 84/449/EEC [22] as the source of its recommended test methods for the purposes of the Notification of New Substances Regulations.

The dossier of information included in the notification to the competent authority will include a proposed classification, and labelling recommendations.

At the time when registration for EINECS closed, all significant ink raw materials had already appeared on ECOIN or were the subject of notification by their manufacturers.

Much speculation has been voiced about the curtailment of commercialization of new chemical compounds but it remains to be seen what effect the 'Notification of New Substances Regulations' will have on the availability of new materials. Suppliers will need to assure themselves that likely financial returns on investment will justify the considerable expense involved in the toxicological testing required by these regulations.

Certainly, the number of new substances which have been listed in ELINCS (European List of New Chemical Substances) [23] since EINECS was closed gives some support to the contention that the rate of introduction to the market is slower than before (340 substances by 30 June 1990 and a further 140 substances by a year later).

Occupational exposure limits

On an annual basis, the Health and Safety Executive publishes guidance [24] on exposure limits for some 560 airborne substances including many solvent vapours and dusts which are of direct interest to the inkmaker. In the intervening period, quarterly amendments are noted in their periodical *Toxic Substances Bulletin*. The guidance is issued especially to assist manufacturers of substances to discharge their responsibilities under Section 6 of HSW. Overall the intention is that exposure to such substances should be eliminated or minimized.

Within the present regime, it is a manufacturer's responsibility to ensure that his processes are so controlled as to ensure that the concentrations of solvent vapours and dusts in the factory environment do not exceed any of the published limits. Measurement of solvent vapours can be either for single point, single substance using for example a Draeger tube, or for monitoring mixed atmospheres over an extended period of time using for example an ORSA gas badge.

The Draeger tube is similar to a breathalyser in operation as it samples a known volume of air and gives an indication of the atmospheric concentration of a single substance by means of a colour change. It is quick and simple to use and gives a reasonably accurate assessment although it is unable to take account of the possible cumulative effect caused by mixtures of vapours of a similar chemical type.

The gas badge, while much slower in operation, is capable of measuring the atmospheric concentration of mixed solvents by diffusion into an absorption tube under controlled conditions followed by measuring of the absolute quantity of each component by gas/liquid chromatography.

The published exposure limits take the form of a two-part listing, the first entitled 'Maximum Exposure Limits' (MEL) covering some 30 substances or groups of substances which the authorities believe require stricter control than the further 470 or so substances in a second listing entitled 'Occupational Exposure Standards' (OES). A further 12 substances were listed as subject to approved changes in the OES list and about 40 substances were under review. Collectively, MELs and OESs constitute Occupational Exposure Limits.

Both MELs and OESs are further subdivided into long-term and short-term exposure limits. The long-term exposure limit, or 8-hour time-weighted average (TWA) is that level of exposure to which it is believed that nearly all workers may be repeatedly exposed for 8 hours per day without adverse effect. The short-term exposure limit (10-minute TWA) is set for those substances for which there is evidence of a risk of acute effects occurring as a result of brief exposure.

Maximum exposure limits

The MELs derive from Regulations, Approved Codes of Practice, European Community Directives, or are by adoption of the Health and Safety Commission (HSC). They are deemed to be reasonably practicable on the basis of scientific and medical evidence. Failure to comply with control limits, or to reduce exposure still further where this is reasonably practicable, may result in enforcement action.

The lower glycol ethers and their esters are subject to maximum exposure limits, the limit for 2-butoxy ethanol (8-hour TWA) is set at 25 ppm/120 mgm^{-3} and has the skin (Sk) notation (indicating a material where absorption through the skin represents a significant health risk).

Occupational exposure standards

The OESs are simply recommendations of the Health and Safety Executive (HSE) on advice from HSC's Advisory Committee on Toxic Substances (ACTS). These limits are considered to represent good practice and realistic criteria for the control of exposure to the relevant substances. Inspectors will use the recommended limits as part of their criteria for assessing compliance with the HSW Act. Table 17.1 illustrates typical examples from the 'recommended limits' list of ink raw materials.

However, it is important to note that the Guidance Note from which they are taken is reviewed and reissued annually. **The most recent Guidance Note should always be consulted for the current figures**.

The HSE is at pains to point out that exposure to all substances, including those listed, should be kept as low as is reasonably practicable at all times. In addition, for listed substances, the exposure limits specified should not normally be exceeded.

The exposure limits should not be used as an index of relative hazard or toxicity. Also it should be noted that they apply to concentrations of single substances. Exposure to additional substances, either simultaneously or sequentially, could give rise to greater hazards to health. There is no universally applicable method for the derivation of exposure limits for mixtures. The suitability of the wide range of formulae, devised in recent years, for a particular mixture should be assessed by toxicologists or other competent persons.

While occupational exposure limits apply to solvent vapours and dusts, dusts in general are perceived to represent an additional hazard by virtue of their physical form, irrespective of their chemical nature.

Dusts in general should therefore be controlled [25] to the minimum that is reasonably practicable, although not all dusts have been assigned occupational exposure limits. The lack of such limits should not, however, be taken to imply an absence of hazard. In the absence of exposure limits for a particular dust, exposure should be controlled to the minimum that is reasonably practicable (a point which is emphasized in the COSHH regulations). In addition, where there is no indication of the need for a lower value, personal exposure should not exceed 10 mg/m^3 8-hour TWA total dust and 5 mg/m^3 8-hour TWA respirable dust (Table 17.2).

An official method for the measurement of total and respirable dust has

Table 17.1 Typical examples of recommended limits of ink raw materials

	Long-term exposure		Short-term exposure	
	ppm	mg/m^3	ppm	mg/m^3
Ethylene glycol (vapour)	–	60	–	125
Isophorone	–	–	5	25
n-Butanol	–	–	50	150 Sk[a]
Toluene	50	188	150	560 Sk
White spirit	100	575	125	720
n-Butyl acetate	150	710	200	950
Butan-2-one (MEK)	200	590	300	885
Methanol	200	260	250	310 Sk
n-Propanol	200	500	250	625 Sk
Ethanol	1000	1900	–	–

The Sk notation is explained above.

Table 17.2 Specific dust limits which are set for some powders

	Long-term exposure		Short term exposure	
	ppm	mg/m^3	ppm	mg/m^3
Calcium carbonate (total)	–	10	–	–
Calcium carbonate (respirable)	–	5	–	–
Carbon black	–	3.5	–	7
Mica (total)	–	10	–	–
Mica (respirable)	–	1	–	–
Silica, amorphous (total)	–	6	–	–
Silica, amorphous (respirable)	–	3	–	–
Talc (total)	–	10	–	–
Talc (respirable)	–	1	–	–

been published by the HSE in the Methods for Determination of Hazardous Substances series [26].

Similar guidance is published by many other countries. The ACGIH in the USA publish an annual listing of threshold limit values and biological exposure indices [27] together with an updated publication [28] which presents reviews of the documentary basis for the values assigned. Corresponding data, assigned in Germany, are referred to as maximum allowable concentrations (MAK).

In the particular case of asbestos dust [29, 30] specific regulations have been issued. Under these regulations any work undertaken which is likely to involve asbestos dust must only be done by licensed contractors.

Control of substances hazardous to health

The regulations were modelled on the earlier Lead at Work Regulations and came into force on 1 October 1989 [31]. They are designed to control the risks associated with exposure to substances which are hazardous in an occupational context, i.e. those which:

- are listed in the Authorised Supply List;
- are not so listed but present a comparable hazard;
- are assigned a maximum exposure limit (MEL);
- are micro-organisms which create a hazard;
- are a dust of any kind when present at a substantial concentration in air.

The regulations do not apply to materials which are dangerous solely by reference to their flammability properties, nor to lead or asbestos, all of which are subject to separate regulations.

The regulations are revised annually to give effect to revised MELs.

The employer must carry out an assessment before employees are exposed to hazardous substances. The assessment will be reviewed regularly and a review will also be carried out if the validity of the earlier assessment is in doubt or where there has been a significant change to the work to which the assessment related.

Exposure must be prevented (replacing the hazardous substance, altering the process so that the hazardous substance is not produced, or totally enclosing the process, for example) or, where this is not reasonably practicable, by removal from the atmosphere before it contaminates the workroom. A general review of exhaust ventilation has been published by HSE [32, 33]. Only as a last resort will personal protective equipment (PPE) [34, 35] be considered; HSE have issued a practical guide for users and, from time to time, publish an updated list of legislative requirements, approved standards and type-approved equipment. Very recently the UK implementation of the PPE directive [36] has been issued and it came into force on 1 January 1993.

Exposure to materials which have been assigned an MEL will be regarded as being 'adequately' controlled if the level of exposure has been reduced 'so far as reasonably practicable' and in any case below the MEL.

In the case of materials assigned an OES, 'adequate' control will be acknowledged if the OES is not exceeded or, when it is exceeded, the employer takes action to remedy the situation as soon as is reasonably practicable. ('Adequate' is defined to be adequate having regard only to the nature of the substance and the nature of the degree of exposure – no reference to the economics of control provision.)

Control equipment will be maintained in an efficient state and, in the case of exhaust ventilation plant, it will be subject to thorough examination at least once every 14 months. Other equipment will be examined at suitable intervals. These examinations will be recorded and the records kept for at least 5 years. Exposure will be monitored where it is deemed necessary for protecting the health of employees. Records of such monitoring will be kept (for 40 years in the case of personal exposure figures or for at least 5 years in other cases).

Employers are required to take reasonable steps to ensure that any protective equipment provided is properly used. Employees are required to make full and proper use of the equipment provided.

Health surveillance is deemed appropriate where there is reasonable likelihood that a disease or effect may occur and there are valid techniques for detecting indications of the disease or effect (personal records to be kept for 40 years).

Employees and their safety representatives will be made aware of the nature and degree of risk to which they are exposed, the control measures that have been adopted and the circumstances under which PPE will need to be used, and they will also be kept informed of the results of monitoring procedures and collective results of surveillance procedures.

Extensive guidance [37–39] on all aspects of COSHH procedures has been issued by HSC, including advice to occupiers of sites where other employers' workers may work (off-site workers) [40].

Site occupiers have a duty, under COSHH, to persons other than their own employees, who may be affected by site activities. Guidance is provided under the following headings:

- substances already on site;

- substances introduced by off-site workers;
- hazards associated with waste disposal;
- special problems for maintenance workers;
- attention to one-off tasks;
- sources of information;
- working in confined spaces, permit to work systems.

A growing awareness of the general need for the prevention of adverse effects of micro-organisms associated with static warm water (so-called legionnaire's disease) has led HSC to issue specific guidance [41] and an approved code of practice [42] for the control of legionellosis.

More recently, a draft ACoP [43] has been issued on the control of sensitizers.

Control of lead at work

It is generally recognized that lead is a significant cumulative poison and the HSC has issued specific regulations to control exposure to lead at work. The regulations are important for the ink industry to the extent that the screen ink sector continues to use lead pigments in limited quantities. SBPIM members outside the screen ink sector have voluntarily withdrawn all lead-based raw materials.

Requirements needed to control exposure to lead (including lead compounds) are spelt out in regulations [44] under the Health and Safety at Work Act, and a Code of Practice [45] details acceptable methods for meeting those requirements. At this stage it is worth noting that failure to comply with the provisions given in a Code of Practice is not in itself an offence, though such a failure may be used in evidence in criminal proceedings. It is open to the defendant to satisfy the court that he complied with the relevant regulation in some way other than that spelt out in a Code of Practice. In contrast, failure to comply with regulations can result, directly, in criminal proceedings.

Central to the control is the assessment by employers of the nature and extent of exposure. Under no circumstances should the lead-in-air standard (currently 0.15 mg/m^3 of air as an 8-hour TWA) be exceeded. To ensure this a further lower 'trigger' level has been set, above which exposure is defined to be 'significant' and regular air monitoring and medical surveillance are required. This second, trigger level is set at half the lead-in-air standard (i.e. 0.075 mg/m^3). At levels below that defined to be significant, only some of the regulations apply.

The underlying philosophy is one of control by prevention of lead releases; only as a last resort should protective clothing be relied upon.

The Code of Practice identifies some of the types of lead work where significant exposure will occur unless adequate controls are provided; these include work which gives rise to lead dust as exemplified by the manufacture of lead paints. Works in which compounds of leads are used in processes which give rise to dust have a statutory duty [46] to use the best practicable means for preventing lead dust emissions to the atmosphere. Air sampling techniques have been published in the Guidance Note series [47].

Fire hazards

Legislation

The Fire Precautions Act [48, 49, 50] aims to ensure that adequate fire precautions to protect people from fire risks are established in certain classes of buildings, now including 'places of work'.

All factories in which more than 20 people are employed, must possess a fire certificate. The certificate imposes requirements relating to the maintenance of fire exits and means of escape, fire warning systems and signs, fire-fighting equipment and the training of employees on a case-by-case basis.

Because of split responsibility for the enforcement of separate controls, especially in the area of flammable solvents (fire certificates and the Petroleum Consolidation Act 1928) there exists potential for conflict between enforcing authorities. Any overlap will be a matter for discussion between the licensing authority and the HSE.

Separate regulations define 'special premises' [51] where fire risk is deemed to require special attention but this is most unlikely to apply to printing ink factories.

Under the Fire Safety and Safety of Places of Sport Act 1987 [52], the fire authority is given power to serve improvement notices and prohibition notices (the latter in cases where there is serious risk to life) and to recover costs incurred in the issuing of fire certificates.

The 'Highly Flammable Liquids Regulations' (HFLR) [53, 54] are of more direct relevance to the printing ink industry. A highly flammable liquid is defined to be a liquid giving off a flammable vapour at a temperature of less than 32°C and which supports combustion (according to the prescribed test). (This definition should not be confused with the EC labelling criterion of 'below 21°C'.)

The regulations apply to all factories where any highly flammable liquid is present in connection with a business purpose, with the exception of those subject to a licence under other related acts.

The regulations control the storage, marking of storage facilities, precautions against spillage, source of ignition, prevention of escapes of vapours and their dispersal, explosive pressure relief, means of escape, solid residues, smoking, taking of samples, fire fighting and duties of employees.

The storage provisions [55, 56] are intended to ensure that highly flammable liquids, when not in use, are stored in a safe manner. The main purpose is to separate storage from process areas. Small quantities (total amount not to exceed 50 litres) may be kept in a work room in a suitably placed cupboard or bin of fire-resisting material.

Marking requirements specify the wording on store-rooms and containers (excluding small containers below 500 ml capacity) but the container-marking provision may now be discharged by labelling according to the Classification, Packaging and Labelling Regulations (CPL; page 920). However, the HSE still recommend that some indication be included in the labelling which shows the applicability of the HFLR or Petroleum Consolidation Act (whose flash point criteria are different from those of CPL).

Ideally, spills will be prevented by transporting liquids in totally enclosed systems incorporating pipelines. If this is not possible, conveyance vessels will be so designed as to avoid the risk of spills. Open buckets and tins are not acceptable.

Sources of ignition must be eliminated. Smoking must not be permitted and 'no-smoking' areas clearly defined. Escape of flammable liquids into the workplace must be prevented and, where this is not possible, safe dispersal arrangements must be made.

A general guide to safe practice in the storage and use of highly flammable liquids for the manufacture of paint and other highly flammable surface finishes has been published [57]. Two further booklets in the same series [58, 59] give basic guidance on the types of electrical apparatus available for use in flammable atmospheres.

In passing it should be noted the HFLR covers those nitrocellulose solutions which are dissolved in solvent systems having a flash point below 32°C which were previously covered by the Cellulose Solutions Regulations.

Fire and fire prevention

Fire theory

The tragedy of fire will fortunately have been experienced only by a minority of people, but with modern communications the horrific scenes at a football stadium fire were televised almost immediately illustrating how a match or lighted cigarette carelessly discarded into waste material caused a flaming carnage within a matter of minutes.

The theory of fire is very simple and can best be illustrated by the fire triangle. Fire is the result of the interaction of oxygen, heat and fuel. If any one of these factors is removed the other two are unable to support combustion and the fire will go out. Taking each of the factors in turn, oxygen is present in the atmosphere at approximately 21% v/v and this concentration is sufficient to aid combustion. Heat may be in the form of a naked flame, mechanical metallic spark, electric spark, spontaneous combustion, static electricity, or even generated by sunlight. The fuel can be paper, wood shavings, waste material, solvents or any other combustible material.

Solvents used in flexo and gravure inks give cause for concern because of their highly flammable nature. A flammable gas or vapour in air can only be ignited and sustain continuous combustion when its concentration is neither too low nor too high. The lowest concentration to support combustion is known as the lower flammable limit and the highest concentration to support combustion is termed the upper flammable limit; these are the limits of flammability (Fig. 17.1). All intermediate concentrations are said to be within the flammability range. Flammability limits are usually quoted as volume percentages but may be given by mass concentrations. The lower flammability limit can best be illustrated when a car will not start because of insufficient petrol getting through, and the upper flammability limit when the carburettor is flooded. (Flammability limits are sometimes referred to as explosion limits.)

Flammable limits usually relate to limits determined in air under a total

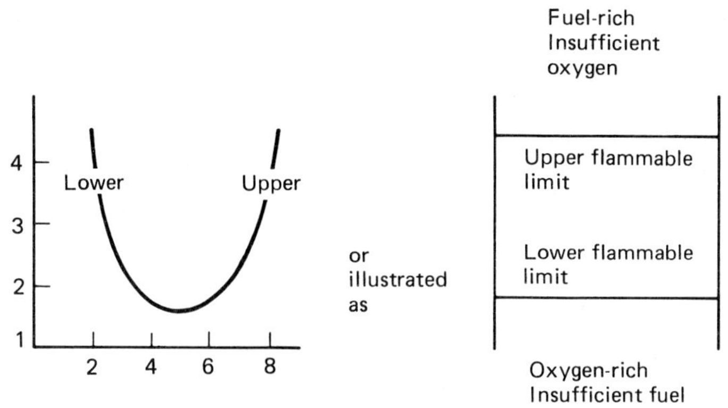

Fig. 17.1 Flammability limits.

pressure of 1 bar at 20°C. The theory of fire and flammable limits should be thoroughly understood.

As can be observed, from Table 17.3, most solvents used in the printing ink industry have a flammable range between 1 and about 12 vol. %, IMS having an upper value of 19%.

Sources of heat

Static electricity

The principles of static electricity were taught to us at school and we remember such experiments as those of rubbing ebony and glass rods with various materials and the effect the charges had on a gold leaf

Table 17.3 Flammable limits

	Lower vol.%	Upper vol.%
IMS (Industrial methylated spirit)	4.3	19.3
Propyl alcohol	2.1	13.5
Isopropyl alcohol	2.0	12.0
Butyl alcohol	1.4	11.2
Isobutyl alcohol	1.6	10.9
Sec. butyl alcohol	1.7	9.8
Ethyl acetate	2.2	11.0
Propyl acetate	1.8	8.0
Toluene	1.2	7.1
Xylene	1.0	7.0
Acetone	2.6	13.0
MEK	1.8	10.0
White spirit	0.9	8.0
SBPs	0.9	8.0
HBPFs	0.6	6.0

electroscope. Today these simple experiments are illustrated in our everday life. Shocks from static electricity often occur when walking over a synthetic fibre carpet and the body is earthed by the hand coming in contact with a filing cabinet or other fixture in an office or room. When a synthetic fibre jumper is taken off the body in the dark it is even possible to see the static electrical charges jump across from the garment. Clearly static electricity can be generated by very simple means and is not only restricted to man-made fibres. Static electricity can be generated by the flow of liquids and powders and therefore can be a potential danger in promoting a fire risk in an atmosphere of highly flammable vapours.

To prevent build-up of static electricity it is essential to earth all tanks, vehicles, drums, etc. before starting any operation. Such earthing systems must be kept clean and in good condition and regularly inspected. The faster solvent is pumped through pipes the more readily static is generated, so flow rates should be kept to a minimum; 1 m/s is the recommended flow rate. Static can build up in a system and special static footwear may be purchased to give good conductivity and allow a charge to dissipate. Ordinary leather shoes act as a good conductor and are much safer than many synthetic materials used in shoe manufacture.

Spontaneous combustion

If a vegetable drying oil such as linseed oil is spilled and cleaned up with rags and wipers which are then left in a bundle in a restricted space, oxygen and heat could be generated possibly leading to spontaneous combustion causing a fire. For this reason all soiled and dirty wipers should be stored outside a building in a closed-top metallic container and should not be left in a building overnight. It has been known that even dirty wipers in a closed top metallic container can cause an exothermic reaction, the heat generated may be detected by touch or vision. Should this occur the container must not be opened, as the flow of fresh air could cause immediate combustion. The closed container should be doused with water until the temperature is reduced to a safe level and then carefully investigated to ensure complete safety.

Fire extinguishers

While brief instructions for use are given on fire extinguishers, it is highly recommended that all personnel working in areas that present a higher than usual fire risk receive instruction on fire prevention and the use of extinguishers. It is common practice for the Fire Brigade and insurance companies to state the types and number of fire extinguishers to be placed in any location. The size, type and number of extinguishers will be assessed by the fire hazard and the volume of the room, laboratory or workshop. As far as possible the type of extinguisher placed in any location would be the type necessary to deal with the fire hazard; for example, it would be hazardous to use foam against fires involving electrical risk, therefore foam extinguishers should not be located near electrical equipment. Extinguishers should be placed in an accessible position and not be blocked in, should be regularly serviced and be in good working order.

Fire alarm procedure notices should be exhibited in all departments giving clear and precise instruction what to do in case of a fire. It is essential that all personnel be aware and know what exactly to do in case of a fire. Should a fire break out, the fire alarm procedure and drill must be strictly followed, but the following points on fire fighting are important if the fire can be tackled by local personnel.

The largest hand fire extinguisher has an active life of less than 3 minutes; if a fire cannot be brought under control in this period of time it is unlikely that the fire will be extinguished with just one extinguisher. Before tackling a fire it is essential that there is a safe escape route and if possible never tackle a fire alone. When attempting to extinguish a fire, aim at the base of the fire, and if there is excessive smoke keep down near the floor where the air is cleaner and breathing and visibility are easier.

Spillage of printing and auxiliary products should be cleaned up at once using sand or some other absorbent material. Most good fire prevention is simple and based on common sense, and is usually called good housekeeping. Fires do not occur – they are caused. One cardinal rule to remember with regard to fire is that life is infinitely more valuable than property.

Types of fire extinguishers

1. *Water – painted red.* Usually of 2-gallon (9-litre) size; discharges water under carbon dioxide pressure; although not suitable for chemical fires they are very effective against ordinary combustible materials such as wood, paper, textiles. Water must not be used where live electrical equipment is involved.
2. *Foam – painted cream.* Usually 2-gallon (9-litre) size; foam extinquishers are suitable for use on small fires in flammable liquids, especially where these are in a container or are overheated. A 2-gallon (9-litre) extinguisher can extinguish 10 square feet (1 m^2) of burning liquid. Foam must not be used where live electrical equipment is involved.
3. *Carbon dioxide (CO_2) – painted black.* Carbon dioxide extinguishers are suitable for use on small fires of flammable liquids and on electrical equipment, especially live equipment, where it is necessary to avoid damage or contamination by foam by dry powder. When using a CO_2 extinguisher never hold it by the horn as this will give severe freezing to the hand.
4. *Bromochlorodifluoromethane (BCF) – painted green.* Ideal for electrical, laboratory and computer protection. BCF is slightly toxic. Personnel should not be subject to a concentration greater than 4.5% and then only at this level for about 1 minute. The agent should not be used in a confined space and rooms should be thoroughly ventilated after use.
5. *Dry powder – painted blue.* These are useful all round extinguishers for fires of flammable liquids, especially where the liquid may be spilled over a large area. They are also effective on electrical fires. They have the disadvantage of leaving a very large deposit of dusty white powder. Ensure the gauge indicating the pressure of the extinguisher is in the charged position.

Figure 17.2 illustrates colour-coded fire extinguishers.

Automatic detection systems
Sprinkler installations detect fires and operate automatically to extinguish them, at the same time operating the fire alarm. Where a sprinkler system is not installed an automatic fire detection and alarm system should be considered as a means of an early warning.

Fire blanket
Non-asbestos fire blankets should be available to smother flames where clothing and other materials catch fire.

Sand buckets
Sand buckets are very frequently abused and used for waste, rubbish or cigarette ash trays. Dry, soft sand is a very effective means of extinguishing a small fire and can be used to prevent flammable solvent from flowing into public drains.

Zone classification and flame-proofing

The regulations which govern the use of highly flammable liquids set down a very strict code of practice for bulk use of such solvents to give the maximum protection against fire and explosion. Areas in which highly flammable solvents are used must be clearly marked 'No Smoking', and in order that highly flammable liquids are identified immediately by any persons using or coming into contact with them the sign 'Highly Flammable Liquid' should be clearly and boldly displayed. Highly flammable solvents, in this context, are those with a flash point below 32°C.

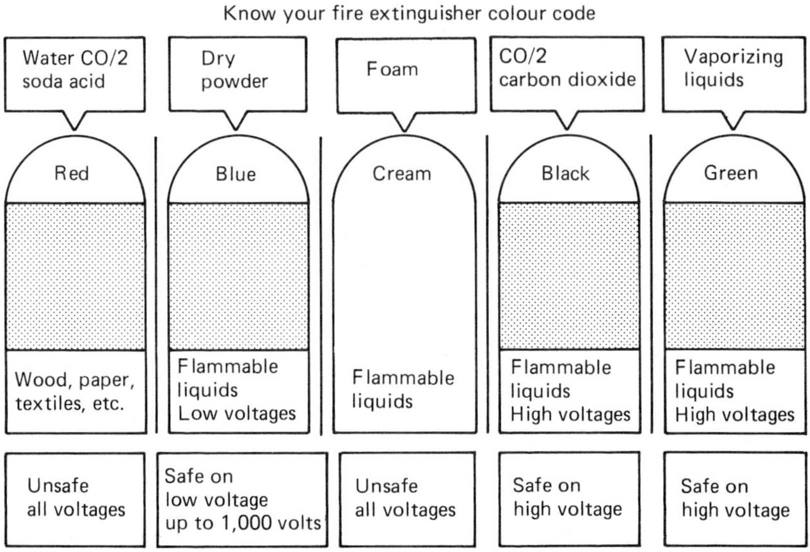

Fig. 17.2 Colour-coded fire extinguishers.

In dealing with the risk of fire or explosion in the presence of flammable liquids or vapour three sets of conditions are recognized.

Div. 0
An area or enclosed space in which any flammable or explosive substance, whether gas, vapour or volatile liquid, is continuously present in concentration within the lower and upper limits of flammability. These conditions do not occur in printing ink manufacture.

Div. 1
An area within which any flammable or explosive substance, whether gas, vapour or volatile liquid, is processed, handled or stored and where during normal operations an explosive or ignitable concentration is likely to occur in sufficient quantity to produce a hazard. A risk of the nature described under Div. 1 can be combated by the use of flameproof or intrinsically safe electrical equipment.

Flexo and gravure ink plants will come into this classification and all electrical equipment will be flameproof. However, in a modern ink plant with good extraction ventilation, volatile liquid concentrations should not accumulate in the workplace for significant periods.

Div. 2
An area within which any flammable or explosive substance, whether gas, vapour or volatile liquid, although processed or stored, is so well under the conditions of control that the production (or release) of an explosive or ignitable concentration in sufficient quantity to constitute a hazard is only likely under abnormal conditions. An area falling into the category of Div. 2 is sometimes known as a 'Remotely Dangerous Area'.

Div. 3
Electrical equipment and switches need not necessarily be flame proofed but situated in an area placed well outside the risk area or in an adequately ventilated compartment.

Data communications to the user

One of the intentions of Section 6 of the Health and Safety at Work, etc. Act 1974 was to ensure that adequate information about products, with respect to their safe use, was made available. This requirement was amended to ensure that users are provided with necessary health and safety information by Section 36 of the Consumer Protection Act 1987 [16] (as opposed to being simply 'made available'). In addition, suppliers are now required to take account of foreseeable operator error when compiling their information documents.

Despite this direct attention to product safety data sheets, it has long been recognized that their provision is not always the most effective way of alerting those most at risk to the hazards associated with exposure to substances at work.

Labels and data sheets are complementary; clearly, the separate messages must be consistent, but the one does not replace the need for the other.

Under the new Chemical (Hazards Information and Packaging) Regulations (CHIP) [19], the distinction between two target audiences is made clear.

The label communicates directly with those who come into contact with the container, though the information cannot be regarded as exhaustive; the user is being given first-line alerting advice each and every time the container is handled.

In contrast, the recipient of the data sheet (under CHIP) is the employer and/or the person in control of premises. The information is provided to him to enable him to take the steps necessary to protect health and safety at the workplace and in the environment.

The Chemical (Hazard Information and Packaging) Regulations

The overall objective of these regulations is to ensure the provision of information to users on the safe handling of dangerous goods and to supply those goods in appropriate containers. The key step is classification; once this is right the other steps follow relatively straightforwardly.

The purpose of classification is to identify the properties of substances and preparations that may constitute a hazard during handling and use. (It is important to note that, in general, CHIP is concerned with hazard; the topic of risk is the subject of the COSHH regulations – see page 910.)

The process of classification, which is more significant in CHIP than it was under its predecessor, Classification, Packaging and Labelling Regulations (CPL), must be completed before a substance or a preparation is supplied. It consists of placing the material into one or more of the categories of danger and assigning the appropriate risk phrases to describe the danger.

The categories of danger are more extensive than they were under CPL and now include carcinogenic, mutagenic and teratogenic as classes in their own right, distinct from toxic and harmful.

Classification is undertaken from three points of view:

- physico-chemical properties (assessed by testing);
- health effects (new test to be avoided);
- environmental effects (limited to substances only for the moment).

Classification – physico-chemical properties

Of the physico-chemical classes corresponding to the listed properties in Schedule 1 of the regulations, only Highly Flammable/R11 and Flammable/R10 are of real interest to inkmakers. The criteria and test methods remain as they were under CPL. That is to say, flash points may be measured by one of the listed non-equilibrium methods of Schedule 1, part III, unless the measured flash point is within ±2°C of one of the classification thresholds. In these cases, the flash point must be remeasured by one of the listed equilibrium methods.

Materials with a flash point below 21°C (with a boiling point above 35°C) will be classified as Highly Flammable; those that flash between 21 and 55°C will be classified as Flammable so long as the material supports combustion. For some single substances, their flammability class has already been assigned in the revised Authorised and Approved List[60].

Classification – health effects

Some 1400 substances have been officially classified with respect to their health effects in the revised Authorised List, prepared for CHIP[19]. These substances include all those previously appearing in the Solvents or Paints annexes (under CPL).

Other substances which are listed in EINECS[18] but not in the Authorised List will be classified by their supplier (so-called self-classification) with reference to the criteria listed in Schedule 1 to the regulations using available information.

Substances not listed on EINECS are by definition new substances and are subject to the Notification of New Substances Regulations. The suggested classification for these materials will be included by the supplier in his dossier of toxicological, physical and chemical data which he is required to submit to the competent authority before the substance is marketed.

Thus, the majority of inkmakers' raw materials will have been classified before they are supplied to the inkmaker.

The classification of preparations may be done by animal testing, but this is discouraged and is in any event subject to the Animals (Scientific Procedures) Act 1986. The preferred classification route is by the conventional (calculation) method following, essentially, the model originally established under CPL for the classification of irritant preparations, but now extended to cover toxic and harmful preparations as well.

Every substance will be associated with a set of assigned or default thresholds. These thresholds come from one of three sources; many but not all of the 1400 substances in the Authorised List have thresholds assigned to them; for the remainder, together with materials in EINECS, but not in the Authorised List, there are default thresholds applied by reference to the classification of the substance. Finally, the thresholds for new substances (subject to the Notification of New Substances Regulations[20]) will either be determined by the competent authority and will eventually appear in the Authorised List (via ELINCS initially), or will rely on default thresholds.

The data used for classification will be recorded and retained for at least three years after the dangerous substance or preparation is last supplied.

The label

The label for single substances will contain either the classification data, symbol(s) and risk phrases, together with the associated safety phrases given in the Approved Supply List or the data derived from the classification procedure (including any environmental classification data). A new requirement is the telephone number of the supplier to supplement the name and address requirement set out previously in CPL (Fig. 17.3). This telephone number does not imply a 24-hour emergency service; it, together with the name and address, is there to provide a contact point with no implication of an immediate response. As before, the substance label will carry the chemical name of the substance and, in addition, the EC Number, if known. If the label data derive from the Authorised Supply List, the label will finally carry the phrase 'EC Label' (to indicate to the reader that the information is the official, peer-reviewed information; other data may be/will be of variable quality).

The label for preparations will indicate the trade name of the preparation, the name, address and telephone number of the supplier and the chemical name of all significant dangerous components. These are, in the case of Very Toxic, Toxic or Harmful preparations, the name of each component above its **harmful** threshold.

For corrosive preparations, the significant components are those present in concentrations exceeding their lowest irritant limit.

In addition, the name of any component causing a preparation to be classified with one or more of the risk phrases R39 to R48 (except R41 and R44) will appear on the label, but under CHIP it is no longer necessary to name simple R36/37/38 Irritant substances.

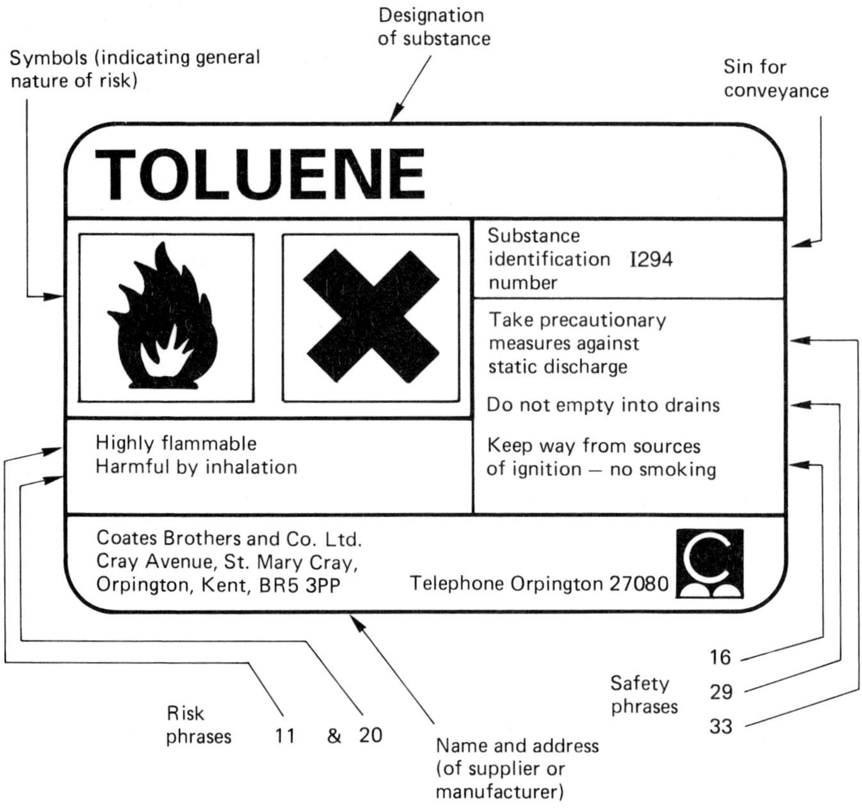

Fig. 17.3 Supply and conveyance label. (after CPL)

For the time being, the label for preparations will not carry reference to environmental effects (pending the development of classification criteria for these effects for preparations).

The final choice of risk and safety phrases to appear on the label follows the same principles established under the previous CPL Regulations; that is the choice is primarily governed by the need to give all the necessary information, but taking account of the clarity and impact of the final label. To assist in the goal of clarity, it continues to be the case that a maximum of four risk phrases is generally recommended; the same applies in respect of safety phrases.

The data sheet

The target audience of the data sheet is the employer in the user organization and it is intended to enable him to protect health and safety at his workplace and to protect the environment. A data sheet will be provided when the dangerous substance or preparation is first supplied to the recipient and revised data sheets will be provided to all those who have been supplied in the preceding 12 months, when significant new information becomes available, giving an indication of which text has been revised. The provision of a data sheet is required no matter how the product is delivered (package, tanker, pipeline, for example). Each supplier in the chain is responsible for the information that he passes on to his recipient, even in circumstances where he simply passes on data prepared by somebody else.

The information will be provided under 16 obligatory headings. Guidance on the information content under each heading is provided in the associated Approved Code of Practice. Ultimately, the adequacy of the information provided will be assessed by reference to the criteria set out in Regulation 6(1); that is to say, does the information enable the recipient to take the necessary steps to protect workplace health and safety and the environment?

Further guidance has been provided by CEPE in the form of a limited number of model sheets which are recommended by SBPIM. They depend on a belief that there is only a relatively small number of raw material types that contribute to the danger category of printing inks and, despite a wide range of individual raw material components, there is only a restricted number of basic sets of health and safety advice that will need to be provided.

On the basis of current estimates, it is not expected that the CHIP regulations will be enforced until the first quarter of 1994. Until then existing CPL[61,62] regulations will continue to apply; that is to say, the classification of harmful preparations by three different routes according to the source of the components (Solvents directive, Paints directive, Substances directive) with a further calculation type for Corrosive/irritant preparations. For data sheets, the HSE Guidance Note will continue to apply[63].

This transitional regime will be difficult for international inkmakers with a whole range of implementations across the 12 member states.

The control of industrial major accident hazards

The Control of Industrial Major Accident Hazards Regulations (CIMAH)[64–66] implemented the so-called 'Seveso' directive and set out to impose controls on sites which use or store significant quantities of toxic, explosive, reactive or flammable substances and represent an extension of the controls earlier presented in the Notification of Installations Handling Hazardous Substances Regulations (NIHHS).

The main aim of the new regulations is to ensure that adequate emergency plans are developed (by both local authorities and industrial organizations) for handling major accidents involving a number of specified dangerous substances. For each substance, there is a threshold quantity which must be present before the regulations apply.

The ink industry is unlikely to find itself within the scope of the 1984 regulation as amended. Nevertheless, it is prudent for all sites to have a contigency plan to deal with emergency situations.

The Notification and Marking of Sites Regulations 1990[67] require notification to the fire authority and the enforcing authority before the total quantity of dangerous substance(s) is 25 tonnes or more. These notification requirements are not required where the site is subject to either the Notification of Installations Handling Hazardous Substances Regulations, the Control of Industrial Major Accident Hazard Regulations or the Petroleum (Consolidation) Act 1928.

Sites carrying a total quantity of 25 tonnes or more of dangerous substance(s) will carry triangular safety signs strategically placed to warn fire personnel before the trigger quantity is reached. Mixed-hazard sites will carry the exclamation mark symbol on a triangular background.

The Planning (Hazardous Substances) Act 1990[68] came into force on 1 October 1992 to provide a back-up to pre-existing regulations such as CIMAH to protect those living in the vicinity of large quantities of stored dangerous substances. None of the categories in scope (either by chemical type or by total quantity) is of concern to inkmakers.

Transport of dangerous goods

Historically, the Tanker Regulations were the first of a group of three sets of regulations dealing with the labelling and conveyance of dangerous goods. These were followed in 1984 by the CPL Regulations (though these superseded the earlier 1978 regulations, which only covered the supply of single substances). The third arm of the group, dealing with conveyance of dangerous goods in packages, was published in November 1986.

Both the Packaged Goods regulations and the Tankers regulations have been updated and a separate requirement for driver training issued.

The Tanker Regulations[69] cover the transport of all dangerous substances in tankers and impose duties on the operator to ensure that the vehicle is fit for the purpose and is regularly examined and tested.

The operator must obtain information from the consignor to enable him to be aware of the risks involved and the driver must have sufficient information to make him aware of the identity of the substance and its

CEPE GROUP MATERIAL SAFETY DATA SHEET. "BASIC PREPARATION"

1. IDENTIFICATION OF THE PREPARATION AND COMPANY.

Product Name and/or Code :

Intended use :

Name, full address and tel. of company:

Emergency phone no : of the company:

and/or

of official advisory body :

2. COMPOSITION/INFORMATION ON INGREDIENTS

Substances presenting a health hazard within the meaning of the Dangerous Substances Directive 67/548/EEC.

Names	Conc. range	Symbol	R phrases
....	....	....	....
....	....	....	....
....	....	....	....
....	....	....	....
....	....	....	....
....	....	....	....

3. HAZARDS IDENTIFICATION OF THE PREPARATION.

4. FIRST AID MEASURES.

General:
In all cases of doubt, or when symptoms persist, seek medical attention. Never give anything by mouth to an unconscious person.

Inhalation:
Remove to fresh air, keep patient warm and at rest, if breathing is irregular or stopped, administer artificial respiration. Give nothing by mouth. If unconscious place in recovery position and seek medical advice.

Eye contact:
Irrigate copiously with clean, fresh water for at least 10 minutes, holding the eyelids apart and seek medical advice.

Skin contact:
Remove contaminated clothing. Wash skin thoroughly with soap and water or use recognized skin cleanser. Do NOT use solvents or thinners.

Ingestion:
If accidentally swallowed obtain immediate medical attention. Keep at rest. Do NOT induce vomiting.

Fig. 17.4 Front page of CEPE model safety data sheet.

risks. A copy of such information will be kept in the cab of the vehicle. The driver must receive adequate instruction and training. The parallel Carriage of Dangerous Goods in Packages Regulations [70] have extended the range of materials in scope; the new criteria include:

- any packaged quantity of flammable solids;
- 5 litres or more of packaging group I materials;
- 200 litres or more of packaging group II or III materials.

Special requirements apply to the carriage of peroxides.

The requirement to supply information in writing continues as before and the vehicles will be equipped with adequate fire fighting equipment. Vehicles carrying more than 500 kg of in-scope dangerous packaged goods will be marked with an orange reflectorized plate. A draft code of practice has been circulated, covering PGR, and includes extensive advice on the provision of information and 'adequate fire-fighting equipment' requirements.

Requirements for driver training (Packaged Goods and Tankers[71]) have now been issued as a separate Statutory Instrument.

Operators of relevant vehicles will ensure that drivers of such vehicles will receive adequate training so as to understand the nature of the dangers involved and the actions-required in the event of an emergency. The operator will keep records of such training. The drivers will be required to have a valid vocational training certificate (for tanker drivers, this came into force on 1 July 1992; for packaged goods drivers it will come into force on 1 January 1995).

Provisional vocational training certificates can be obtained to cover a transitional period to 1 January 1995 for those with 5 years' relevant experience prior to 1 January 1992.

The certificate must be available for inspection by a police constable or a vehicle inspector at all times when the driver is engaged in the carriage of dangerous goods.

The transport-label will carry an indication of the contact point in the UK from where expert advice may be obtained, the designation of the substance, the substance identification number (SIN) and the appropriate hazard warning sign. For conveyance in receptacles greater than 25 litres, supplementary information (nature of the danger and emergency action) will be provided either on the label or as a separate sheet accompanying the package (such as a Tremcard for example).

However, to avoid double labelling (Supply and Conveyance), CHIP permits the construction of composite labels under specified circumstances.

Most international rules relating to the transport of dangerous goods are based on UN recommendations[72]. Each substance or article is assigned a UN number (the basis for UK SINS). The primary listing, in UN number order, records the official description together with recognized alternatives, the assigned UN (danger) class and where appropriate the division number, subsidiary risks and any special provisions. The listing also assigns substances or articles to a packing group and gives recommendations for the packing method.

There are nine UN classes, the most important of which to the inkmaker is Class 3 (flammable liquids). Such liquids have a closed-cup flash point of not more than 60.5°C or an open-cup flash point of not more than 65.6°C. Having classified goods, subsequent chapters detail the recommendations for the transport of each UN class.

Substance identification numbers of importance to inkmakers together with their associated packing group, are given in Table 17.4.

Based on the UN recommendations, the International Maritime Organisation (IMO) has issued and updates the IMDG Code[73] on a regular basis to control the transport of dangerous goods by sea and provide readily accessible advice to ships' masters as to requisite action in the event of an emergency. Regulations for transport of dangerous goods by air similarly follow the UN recommendations, as published in the IATA regulations[74]. Not surprisingly there are restrictions on what can be carried on passenger or cargo aircraft. ADR[75] is an agreement between (19 contracting states whose purpose is to ensure that dangerous goods arriving at a frontier in or on a road vehicle have been suitably packed and are being carried safely. Under these circumstances the contracting states agree to allow the transport by road of dangerous goods in their territories.

ADR is currently subject to major review and will probably form the base of control of UK inland road journeys for dangerous goods in the foreseeable future.

Hazchem signs

Hazchem code signs[76] give information to the police and fire service about the action to be taken within the first 15 minutes in dealing with an incident. The sign gives an immediate indication of the hazard to

Table 17.4 Substance identification numbers and packing groups

SIN	Description	Packing group
1210[a]	All 'flammable' – only inks with a flash point below 60.5°C[b]	II or III (depending on flash point and viscosity)
1993	All 'flammable' – only thinners and diluents which are mixtures (single-substance solvents use their own SIN)	II or III (depending on flash point)
2059	Nitrocellulose solutions, less than 55% nitrocellulose of less than 12.6% nitrogen[b,c]	not assigned
2557	Nitrocellulose with more than 18% plasticizing substance and less than 12.6% nitrogen	II
1950	Aerosol-packaged materials irrespective of their contents	Not assigned

[a] It is fairly certain that 1210 includes inks containing NC.
[b] There are minor differences between authorities as to the exact flash point boundaries. The above are according to UN (transport) criteria.
[c] The original UK criterion was 12.3% nitrogen.

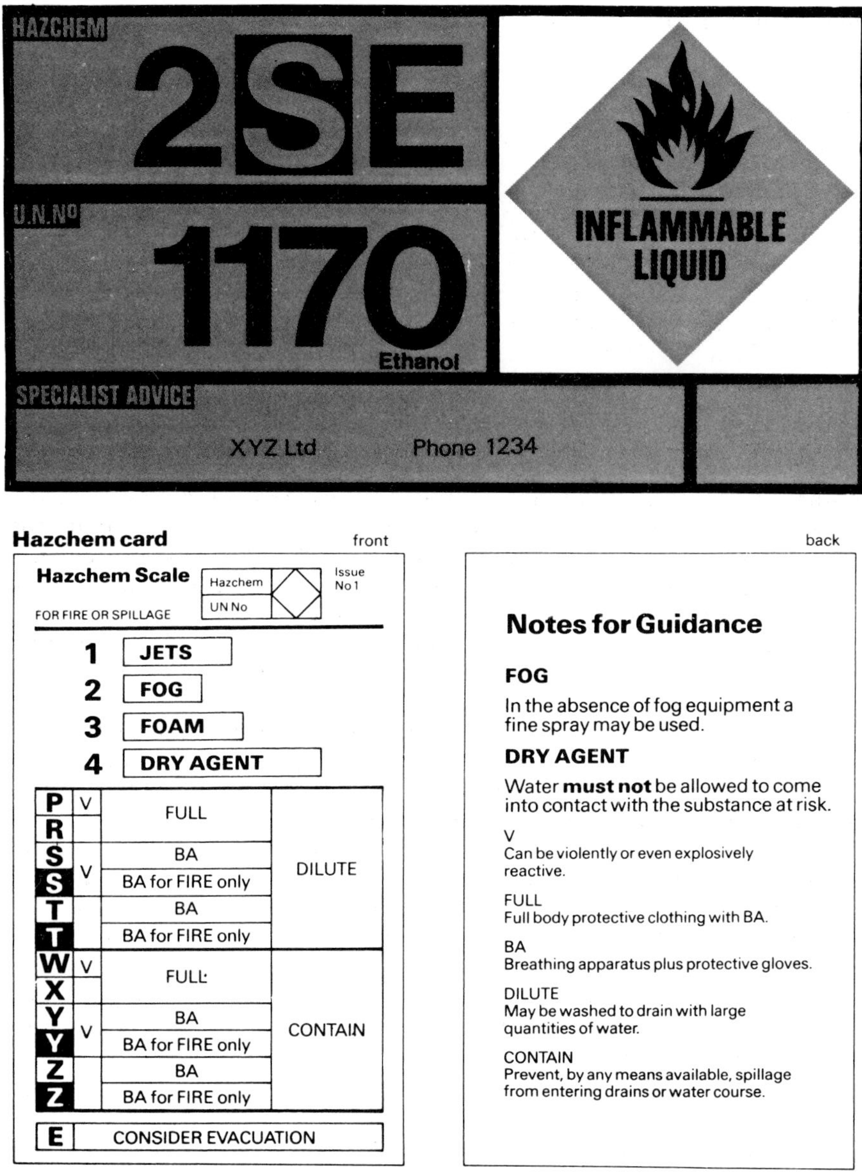

Fig. 17.5 HAZCHEM sign and card.

the public, immediate information on the action required by the emergency services and is simple to use. It does not need an elaborate reference text for interpretation but can be understood from a simple card of the type illustrated. The information provided covers fire-fighting media, personal protection, risk of violent reaction, spillage and evacuation. The Hazchem sign can be attached either to vehicles such as tankers or to the entrances of factories and other premises (Fig. 17.5)

Tremcards® (registered trade mark)

It was generally recognized that there is a need to provide different information to emergency services in the event of an accident during conveyance. Under the Tankers Regulations the driver of the vehicle is required to be given information in writing to ensure that he knows the identity of the substance(s) being conveyed, the associated dangers which might arise in an accident and the necessary emergency action. The CEFIC system of transport emergency cards is intended to meet the need for a uniform European system enabling the driver of a road vehicle carrying dangerous substances to have clear information in writing on immediate action that should be taken in the event of an emergency.

The European Agreement concerning the International Carriage of Dangerous Goods by Road (ADR)[75] at marginal 10385 of Annex B requires that, as a precaution against any accident or emergency, the driver shall be given instructions in writing in the language of the originating state and those of the transit and destination states. The instructions will be kept in the drivers cab and will indicate the nature of the danger inherent in the substance(s) carried, the safety measures needed to avert the danger, the treatment for persons who come into contact with the goods, necessary fire-fighting equipment and countermeasures in the event of spillage.

Two volumes of recommended Tremcards have been published, covering single substances[77]. Another covering groups of substances and mixtures (Group Text)[78]. These documents will find additional application with the Dangerous Goods in Packages Regulations.

After considerable discussions, SBPIM have developed three Transport Emergency Cards (to replace the CEFIC cards TEC(R)595 and 596), basing their text on that provided in four Group Text Cards (30G30, 30G35 and a combination of 30G34 and 30G36). It was not possible to get international agreement on these sheets and they were consequently issued independently by SBPIM outside the CEFIC TREMCARD system.

The SBPIM sheet covering nitrocellulose-containing inks is illustrated as Fig. 17.6. A typical CEFIC TREMCARD is illustrated as Fig. 17.7.

17.4 MECHANICAL AND OPERATIONAL ASPECTS

First aid at work

In the UK, the provision of adequate first aid facilities is one of the many aspects of the Health and Safety at Work Act. The First Aid Regulations[79,80] place a general duty on employers to make or ensure adequate first aid provisions for their employees if they are injured or become ill at work.

In determining the total number of first-aiders need in an establishment, account should be taken of all the relevant factors. These include

Class 3 ADR 2301
Item 5(b) or (c)
UN No. 2059
Packing Group II

TRANSPORT EMERGENCY INFORMATION

Cargo	FLAMMABLE PRINTING INK (containing more than 20% nitrocellulose) having a Flash Point below 55°C giving off toxic fumes in a fire. Liquid or paste with perceptible odour immiscible or only partly miscible with water [Lighter than water/Heavier than water]
Nature of Hazard	Highly flammable or flammable volatile. The vapour is invisible, heavier than air and spreads along ground. Can form explosive mixture with air particularly in empty uncleaned receptacles Heating will cause pressure rise with severe risk of bursting and subsequent explosion Decomposes in a fire giving off toxic fumes. The effect of inhalation of the fumes in a fire may be delayed The vapour may have narcotic effect. The liquid or vapour may cause irritation to eyes, skin and air passages.
Protective Devices	Suitable respiratory protective device Goggles giving complete protection to eyes Plastic or synthetic rubber gloves Eyewash bottle with clean water

EMERGENCY ACTION notifying police and fire brigade immediately

	Stop engine No naked lights. No smoking Mark roads and warn other road users Keep public away from danger area. Keep upwind Use low-sparking hand tools and explosion proof electrical equipment when available
Spillage	Shut off leaks if without risk Prevent liquid entering sewers, basements and workpits; Vapour may create explosive atmosphere Absorb or contain leaking liquid with sand or earth or other suitable material. Consult an expert If substance has entered a water course or sewer or contaminated soil or vegetation, advise police.
Fire	Keep containers cool by spraying with water if exposed to fire Extinguish preferably with dry chemical, foam or waterspray Do not use water jet Incase of fire warn everybody – explosion, toxic and corrosive hazard – evacuate if necessary
Fire Aid	If substance has got into eyes, immediately wash out with plenty of water for several minutes Remove soaked clothing immediately and wash affected skin with soap and water Seek medical treatment when anyone has symptoms apparently due to inhalation or contact with skin or eyes Depending on substance, persons who have inhaled the fumes produced in a fire may not show symptoms. Due to possible delayed effect of poisoning and for safety reasons they should be kept under observation In case of burns immediately cool affected skin as long as possible in cold water Additional information from:- March 1990

Fig. 17.6 SBPIM transport emergency card (for nitrocellulose-containing inks).

TRANSPORT EMERGENCY CARD (Road)

CEFIC TEC (R)-31
May 1971 Rev. 1
Class 3 ADR
Marg. 2301, 1°a
UN No. 1294

Cargo — **TOLUENE**

Colourless liquid with perceptible odour
Immiscible with water
Lighter than water

Nature of Hazard

Highly inflammable (flashpoint below 21°C)
Volatile
The vapour is invisible, heavier than air and spreads along ground
Can form explosive mixture with air particularly in empty uncleaned receptacles
Heating will cause pressure rise with risk of bursting and subsequent explosion
The substance poisons by absorption through skin and inhalation
The vapour has effects resembling alcoholic intoxication

Protective Devices

Suitable respiratory protective device
Goggles giving complete protection to eyes
Plastic or synthetic rubber gloves
Eyewash bottle with clean water

EMERGENCY ACTION — Notify police and fire brigade immediately

- Stop the engine
- No naked lights. No smoking
- Mark roads and warn other road users
- Keep public away from danger area
- Use explosionproof electrical equipment
- Keep upwind

Spillage

- Shut off leaks if without risk
- Contain leaking liquid with sand or earth
- Prevent liquid entering sewers, basements and workpits, vapour may create explosive atmosphere
- Warn inhabitants—explosion hazard
- If substance has entered a water course or sewer or contaminated soil or vegetation, advise police

Fire

- Keep containers cool by spraying with water if exposed to fire
- Extinguish with dry chemical, foam, halones or waterspray
- Do not use water jet

First aid

- If the substance has got into the eyes, immediately wash out with plenty of water for several minutes
- Remove contaminated clothing immediately and wash affected skin with soap and water
- Seek medical treatment when anyone has symptoms apparently due to inhalation

---

Additional information provided by manufacturer or sender

TELEPHONE

Prepared by CEFIC (CONSEIL EUROPEEN DES FEDERATIONS DE L'INDUSTRIE CHIMIQUE, EUROPEAN COUNCIL OF CHEMICAL MANUFACTURERS' FEDERATIONS) Zürich, from the best knowledge available; no responsibility is accepted that the information is sufficient or correct in all cases

Obtainable from **THE WHITEFRIARS PRESS LTD, MEDWAY WHARF ROAD, TONBRIDGE, KENT TN9 1QR.**

Acknowledgment is made to V.N.C.I. and E.V.O. of the Netherlands for their help in the preparation of this card

Applies only during road transport English

Fig. 17.7 CEFIC TREMCARD (for toluene). (Copyright CEFIC produced in England by Whitefriars (Tremcards) Sales)

not only the number of employees but also the nature of the work. As a guide, it is recommended that in establishments with relatively low hazards such as offices, etc. it would not be necessary to provide a first-aider unless 150 or more employees were at work – a ratio of 1 per 150.

In establishments with a greater degree of hazard such as factories, dockyards or warehouses there should be at least one first-aider present between 50 and 150 employees. When there are fewer than 50 employees at work in a factory then there may be no need for a first-aider, but an employer, if he is not providing a first-aider, has to provide an 'appointed person' at all times when employees are at work.

Employers whose establishment present special or unusual hazards will need to provide more than one first-aider. Where establishments have a larger number of employees and are divided into a number of more or less self-contained working areas, the employer will need to provide both centralized facilities and supplementary equipment and personnel in other locations.

A first-aider is a person who has been trained and holds a current first-aid certificate issued by a recognized authority. The certificate lasts for a 3-year period and then is renewed by examination. An appointed person is a person provided by the employer to take charge of the situation (e.g. to call an ambulance) in the absence of a first-aider. Soap and water and disposable drying materials, or suitable equivalents, should also be available. Where tap water is not available, sterile water or sterile saline, in disposable containers each holding at least 330 ml, should be kept easily accessible, and near to the first-aid box, for eye irritation. First-aid boxes should be clearly marked and made of materials to protect the contents.

A first-aid box should contain a guidance card, individually wrapped sterile adhesive dressing, sterile eye pad, traingular bandages, sterile covering for serious wounds, safety pins, medium, large and extra large sterile unmedicated dressing.

Personal hygiene

Where dermatitis is a hazard, proven barrier creams and hand cleansers, preferably antiseptic, are necessary and a weekly hand inspection by a factory nurse or a supervisor is advisable. Barrier creams are formulated for various operations, some water repellent, others oil repellent, etc. The correct types must be chosen in the light of the work to be performed.

Reporting of injuries, diseases and dangerous occurrences

The 1985 [81] regulations cover three specific topics and their reporting procedures. The requirements and implications are extensively reviewed in an associated HSE Guide [82].

Death or injury at work

This topic covers fatalities and is intended to apply, in practice, to any death which occurs either at the time of the accident or subsequently.

Specific injuries and conditions are illustrated, such as fracture of the skull, fracture of any bone in the arm (but does not include collar bone), fracture of the leg, amputations and injury from electric shock for example.

All accidents no matter how trivial or small must be recorded in the local accident book. A set procedure should be established to deal with all accidents and dangerous occurrences during and out of working hours.

Dangerous occurrences

Dangerous occurrences, having a high potential to cause death or serious injury, must be reported to the HSE. This reporting gives the HSE the opportunity to accumulate information on the circumstances under which such accidents occur, adding to the pool of knowledge which can be called upon in accident prevention.

Locations of occurrences which require reporting include warehouses, factories, fairgrounds, conveyance by road, etc. and cover collapse of lifting machinery, building structures, scaffolding, release of harmful substances into the atmosphere, explosions and fires.

A free leaflet covering reporting of injuries and dangerous occurrences is available from the HSE [83].

Reportable diseases

The diseases are grouped into poisonings, skin diseases, lung diseases, infections and a group of miscellaneous diseases which are associated with exposure to particular chemicals or physical agents. A free leaflet on reporting diseases is also available from the HSE [84].

Training in health and safety

Accident prevention can most effectively be achieved by a well-trained and able workforce. The HASAWA 1974 clearly states that instruction, training and good supervision are essential. Therefore all personnel must be thoroughly trained in all aspects of their work. The training should be performed by a competent person and a record kept of all training on the individual's personal file. The training syllabus for health and safety could be formulated from the objectives listed in the safety audits. Individual training programmes can be formulated using the whole of this chapter and should be integrated with general training programmes.

Lifting

Twenty-five per cent of all industrial injuries occur during the handling of materials. Many of these injuries are to the back causing much pain and the possible recurrence of the injury. The introduction of mechanical aids for lifting progresses at a steady rate but manual handling is employed at a very high level in industry and in the home. It is very difficult to legislate for which loads should be lifted by manual means; the size and shape of the object are never the same and the average worker does not have the capabilities of a trained weightlifter.

The Manual Handling of Loads Regulations (1992) [85, 86] came into force on 1 January 1993 and require employers to avoid, so far as is reasonably practicable, the need for employees to undertake manual handling. Where this is not reasonably practicable, employers will make an assessment of tasks and loads in the context of the working environment and the capabilities of individual employees.

Following the assessment, the risk of injury must be reduced to the lowest level reasonably practicable. The enforcement of general numerical weight limits has been avoided, but the guidance to the regulations offers numerical guidelines to serve as an initial filter in the identification of processes needing fuller examination.

The underlying philosophy continues that started in COSHH; that is avoid, assess, minimize and then provide mechanical aids.

Noise

The human ear does not judge sound powers in absolute terms but judges how many times greater one power is than another. A logarithmic scale is used to measure sound level. The decibel (dB) is the unit used for sound intensities. Noise can be divided into two different groups, environmental and occupational noise.

Environmental noise would be that of a nuisance noise audible outside a factory, and can be illustrated by the running of ball mills overnight causing disturbance to a neighbouring residential area. If the level of noise from a mill could not be reduced to a satisfactory level it could result in an order from the Public Health Department of the local authority forbidding the running of the ball mill overnight.

Occupational or work area noise has frequently been associated with industry and although recognized as a hazard it is only in recent years that work has been performed on deafness inflicted by industrial noise. In 1972 a Code of Practice [87] for reducing exposure levels of noise in industry was published.

The Noise at Work Regulations came into effect on 1 January 1990 [88]. They follow the now traditional COSHH pattern; that is to say, employers are required to assess the exposure, maintain records of the assessment, reduce the risk to the lowest level practicable and provide protective equipment where this proves necessary.

Three action levels are set with respect to noise:

1. the first action level (daily personal exposure 85 dB (A));
2. the peak action level (peak sound pressure of 200 pascals);
3. the second action level (daily personal exposure 90 dB (A)).

If there is a likelihood of employees being exposed to levels in excess of levels 1 or 2, ear protection equipment will be provided at the employee's request.

If there is a likelihood of levels 2 or 3 being exceeded, employers must reduce, so far as is reasonably practicable, the employees' exposure to such noise, personal ear protection will be provided (and must be worn), and the area will be designated an ear protection zone.

Accompanying general guidance has been published by HSE [89, 90] and PIAC [91] have issued guidance on noise reduction at web-fed printing presses. Table 17.5 gives a guide to some comparative sound levels.

Lighting

The intensity or brightness of light is expressed in lux and some idea what this term means can be gained by noting that an overcast summer day is estimated to be between 30 000 and 42 000 lx and a midwinter day 10 000 lx. BS 8206 Part 1 deals in general terms with the code of practice for artificial light. The following gives some general guidance for the light requirements in the workplace:

General offices, laboratories, kitchens, medium bench work	500 lx
Drawing office	750 lx
Gauge and tool rooms, retouching paintwork	1000 lx
Inspection of graphic reproduction	1500 lx

In the laboratory artificial daylight is most important for colour matching. This subject is covered in BS 950, *Artificial Daylight for the Assessment of Colour*.

The significance of lighting and how it affects the safety, health and welfare of people at work, is dealt with by one of the booklets in HSE's HS(G) series [92].

Guidance is given on the provision of appropriate lighting for a wide range of task-types by reference to required average luminance and minimum measured luminance and the need to avoid glare. Although not restricted to advice on the office environment, the booklet will be particularly useful when carrying and out assessments under the Display Screen Equipment Regulations.

Working with VDUs

Display Screen Equipment Regulations [93] were published on 16 November 1992 and came into force 1 January 1993. They implement EC Directive 90/270/EEC.

Table 17.5 A guide to comparative sound levels

Situation	Sound level (dB)
Quiet room	30–35
Noisy office with typewriter	60–70
Saloon car at 40 mph	60–70
Paste ink manufacturing mill room	70–80
Liquid inks manufacturing plant	80–85
A shouting person	95–100
Riveting hammer on steel	110–125
Four airliner jet engines	160–170

Employers will analyse the health and safety risk to which their employees are exposed as habitual users of workstations (display screens and their immediate environment) and assessments will be reviewed as necessary. Identified risks will be reduced to the lowest extent that is reasonably practicable.

Workstations installed after 1 January 1993 will meet the requirements of the schedule. Existing workstations will be brought into compliance by 21 November 1996.

Work activities of such habitual users will be so arranged as to provide periodical interruption of the screen work. On their request, employees will be provided with appropriate eye and eyesight tests by a competent person, to be repeated at appropriate intervals. The cost of special corrective glasses, where normal glasses cannot be used, will be met by the employer.

Adequate health and safety information and training in connection with the use of workstations will be provided.

The Guidance Note [94] expands on the concept of a habitual user and provides useful illustrative examples.

Those responsible for the assessment should be familiar with the main requirements and have the ability to assess risks from a checklist of activities. Assessment can be made by health and safety personnel or line management after the appropriate training. It may be necessary to seek outside expertise.

Machinery protection

It is estimated that over half the annual 16 000 UK accidents involving machinery could have been avoided by the adoption of well-established safeguarding standards. Legislation stemming from the Health and Safety at Work, etc. Act and the Factories Act state that plant and machinery must be safe and the employer must provide a safe system of work. To achieve this, dangerous parts of every machine shall be securely protected with properly constructed guards. The guards must be inspected, maintained and remain firmly in position while the machine is in motion and in use. Manufacturers of machinery have a responsibility to ensure that the products they supply comply with legislation regarding safety. If there is doubt on the safety protection of a machine there are BS and HSE publications covering most machinery. If doubt exists on the best method or if it is necessary to protect a piece of machinery the local factory inspector will offer advice.

The HSC proposed to replace the existing complex provisions, relating to machinery guarding, by a single set of new regulations as part of a package to implement the 'Provision and use of work equipment' directive [95, 96a]. Amongst other things, they will replace Section 14 (1) of the Factories Act 1961.

Types of guard

Fixed guards

These should prevent contact of any part of the body with the dangerous part.

Interlocking guards

These should ensure that an opening is automatically closed before the machine can be operated, and will remain closed until the operation is completed.

Trip guards

These should ensure that, if the position of the operation is such that the operator is liable to be injured, the guard will automatically stop the machine.

Automatic guards

These are usually of the photoelectric type for power presses.

Mills, mixes and stirrers

The types of mills, mixers and stirrers used for the manufacture of printing inks fall into the following categories: mixers, ball and bead mills and multi-roll mills.

Mixers

A large number of different types of vertical shaft mixers exist and may be placed into two separate groups, change pan and fixed pan. The main hazards associated with change pan mixers are caused by the rotating shaft and impeller blade. In some instances the rotating shaft can be protected by an outer casing, but when this is not possible the entire mixer should be protected by a metallic mesh safety cage. The safety cage should be fitted with an electro micro safety switch that will prevent the machine operating until all safety doors are securely closed and locked.

In recent years modern types of fixed pan mixers have been introduced which, because of their design and construction, allow more advanced safety techniques to be employed. The safety principle of the mixer is very simple, i.e. the mixing pan vessel protects the stirrer and mixing blade and cannot be turned on unless it is in the correct position and locked down. Similarly other rotating parts are suitably protected by micro switches and do not rotate or move unless the whole of the machine is secured in the correct working position.

Ball mills

Ball mills vary in size and capacity and must be positioned in such a way to allow safe rotation and be completely protected by a fixed safety cage. Access to the ball mill at ground floor level will be essential for maintenance and discharging. The entrance gate through the ball mill cage must be protected in such a way that the mill cannot rotate and is in a fixed position by a safe brake system. For the larger capacity ball mills it would be essential to have a mezzanine floor erected above the ball mill. Access to the platform would be by a stairway and the platform suitably protected by a safety barrier. A manhole in the ball mill is essential for loading, discharge is often performed using a separate valve system. Both these systems must be protected in such a way that they cannot be left in the open position when the mill is operated. To prevent build-up of internal pressure in the mill it should be fitted with a vent valve. The mill must be vented before either the manhole or discharge valve are open.

On occasions it will be necessary for maintenance engineers to enter the ball mill. A permit to work system must be employed for any workman entering into an enclosed space and the vapour concentration inside the ball mill must be measured and tested as safe by a competent person. The mill must be disconnected from the electrical supply at the main isolator, e.g. fuse removed and a notice placed on the switch stating 'Do not switch on – workman working on mill.'

Bead mills

Bead mills may have the dispersion cylinder in a vertical or horizontal position. Bead mills are moderately modern in design and the machine does not give a high hazard while in operation. Any exposed moving or rotating parts should be well protected.

Multi-roll mills

The most common multi-roll mill in common use is the three- or triple-roll mill. The hazards associated with a three-roll mill are the doctor blade and the nip of the middle and furthest roller from the blade. The nip between these two rollers must be protected by a safety bar which is interlocked with the mill. The doctor blade is extremely sharp and is mounted in the apron and fits onto the first roller. Great care must be taken when replacing the doctor blade and a blade guard should be employed to prevent cuts. Should a foreign object be caught in the nip of the rollers, the mill should be switched off, the pressure released and the object withdrawn.

Permit to work

To ensure that maintenance staff and contractors are aware of the correct procedures that are necessary for safe working, a 'permit-to-work' should be issued. Such a document is illustrated in Figs. 17.8 and 17.9.

Section 2(2)a of the Health and Safety at Work, etc. Act 1974 requires '. . . the provision and maintenance of . . . systems of work that, so far as reasonably practicable, are safe and without risk to health.' Employers need to consider whether the implementation of a permit-to-work system would reduce the risks associated with their production processes and/or provide an adequate control measure under COSHH.

Safety audits

Safety audits or inspections provide an excellent means of ensuring the workplace, machinery, apparatus and raw materials are safe and of satisfactory standard or condition. The audit list could be prepared in conjunction with a local health and safety committee and different lists used in different departments or laboratories. The audit should have listed the manager/supervisor responsible for the location and the nominated safety officer or representative.

Notice to Contractors working at   Co Ltd

This is a Chemical Works and contains substances which must be handled with caution, some being highly flammable, and therefore it is imperative that the works rules for Health & Safety are adhered to.

The essential rules are listed below. If in doubt the Site Engineering Department or the Safety Officer of the working location will advise.

1. A Permit-to-Work must be obtained from the Site Engineering Department before commencement of work in any location on the   site.

2. No Smoking except in permitted areas.

3. Any drilling, cutting, burning, welding must be authorised in writing by the Site Engineering Department.

4. Do not enter buildings, interfere with machinery, electrical wire systems or pipe systems unless authorised by the Site Engineering Department.

5. In the event of fire or any emergency, all contractors' employees should make their way to the main gate and report to the gatekeeper.

6. Private cars are not allowed on site. If tools and materials are brought in cars, then after unloading the vehicle must be taken to the car park.

7. The Speed Limit on this site is 10 m.p.h.

8. Contractors must arrange for the removal of rubbish, and for ensuring that the work area is left in a clean condition.

Fig. 17.8 Notice to contractors.

PD 9-2-1 SERIAL Nº 5001

COATES BROTHERS PLC

SOUTH WALES FACTORY

PERMIT-TO-WORK CERTIFICATE

1 LOCATION ....

2 PERMIT VALID (MAXIMUM OF 8 HOURS)

FROM.............. TO.............. DATE(S)..............

3 WORK TO BE CARRIED OUT ....

....

4 (a) ISOLATION OF PLANT/WORK AREA (cross out items not applicable)
Precautions have been taken for the isolation of the following :— STEAM/WATER/GASES/FLUID HEAT MEDIUM/MECHANICAL DRIVES/PIPE LINES. SLUDGES AND DEPOSITS HAVE BEEN REMOVED. APPROPRIATE SAFETY TICKETS HAVE BEEN ATTACHED.

SIGNED.............................. DEPT. MANAGER. DATE & TIME..............................

(b) ISOLATION OF ELECTRICAL POWER. I certify that all electrical power has been isolated to allow work to be safely carried out.

SIGNED.............................. AUTHORISED PERSON DATE & TIME..............................

5 ATMOSPHERIC TEST CERTIFICATE. I certify that the atmosphere in the work area/vessel has been examined for the presence of toxic, asphyxiating or flammable gases having regard to the previous contents and is considered to be safe for entry (or the commencement of work).
NOTE Where gases or fumes may be present, entry without breathing apparatus must not be permitted.

SIGNED.............................. QUALIFIED TESTER. DATE & TIME..............................

AN ADEQUATE SUPPLY OF FRESH AIR IS UNNECESSARY/NECESSARY .... DEPT. MANAGER

6 NAKED FLAME CERTIFICATE. Naked flame or other sources of ignitions ARE/ARE NOT permitted.
TYPE PERMITTED (i.e. Burning, welding, grinding, drilling)..............................

SIGNED.............................. DEPT. MANAGER

7 PRECAUTIONS TO BE TAKEN. The following protection/precautions must be used during the work (State TYPE against each item required and State NONE against each item not required).

(1) Eye protection, type ....

(2) Gloves, type..............................

(3) Protective clothing, type..............................

(4) Breathing apparatus, type..............................
(Duration of cylinders to be made known to user)

(5) Respirators, type ....

(6) Fire extinguishers, type ....

(7) Flameproof lamps..............................

(8) Safety belt and line ....

(9) Safety man ....

(10) Vessel .... COLD/HOT

(11) Rescue equipment in good order and available..............................

(12) Surrounding areas are safe..............................

OTHER PRECAUTIONS..............................

SIGNED.............................. DEPT. MANAGER

8 CERTIFICATION: I certify that the precautions in section 4 have been taken and the work may commence subject to the requirements of SECTIONS 5, 6, and 7.

SIGNED .... DEPT MANAGER DATE ....

If the certified conditions change, work must be suspended immediately and the certifier informed.

9 ACCEPTANCE. I have read the permit and understand the nature of the work and the safety precautions to be taken.

SIGNED (1).............................. C/NO ....

SIGNED (2).............................. C/NO ....

SIGNED (3).............................. C/NO ....

10 WORK COMPLETED/SUSPENDED.
All persons assisting in work, tools, and gear have been withdrawn

SIGNED .... DATE..............

11 ACCEPTED AND PERMIT CANCELLED

SIGNED .... DATE & TIME ....

Fig. 17.9 Permit to work.

The audit team should submit a dated safety report to the manager of the department who will give instructions for any faults to be remedied. The audit team must not take on the job of repairing any fault.

The preparation of a safety audit list could be formulated using the following topics.

- Structural damage – fittings which may cause a safety hazard, safe means of escape, noise level, extraction ventilation.
- Electrics – all wiring plugs and connections earthing points and clips.
- Fixed equipment and machinery – all guards to be securely in position and giving protection from all dangerous moving parts. Safety equipment such as mill bars and interlocks should receive special attention.
- Portable equipment – pumps, motors, generators, etc. should be inspected for mechanical and electrical features.
- Laboratory glassware and reagents – ensure cracked or dangerous glassware is not in use and broken glassware is disposed by a safe method. Laboratory reagents inspected to ensure correct storage with special attention to flammability and toxicity.

Safety equipment

Fire extinguishers shall be regularly inspected. Ensure all fire and safety signs, fire precautions and notices are exhibited.

Personal protection

Ensure ear protectors, eye protectors, first aid box, washing and emergency showers come up to an acceptable standard of hygiene.

Stores and working areas

Ensure a good standard of housekeeping is achieved. Remember to inspect stores, dark rooms and instrument rooms. If ladders or steps are used to reach shelves, ensure they are in a safe condition.

Safety committee

Section 2 (7) of the Health and Safety at Work, etc. Act states that an employer must establish a safety committee[96b,97] if requested by at least two safety representatives appointed by a recognized trade union. The membership of the safety committee should be established to include representatives from all major departments of a factory or site, and this should be a matter for management and union to decide. The committee should meet at regular intervals, chaired by a senior manager and minutes should be recorded.

The objectives of a safety committee are to bring together employer and employee and promote all aspects of health and safety. It would be beneficial for a safety committee to draw up a statement of intent so that all members were aware of its aims and objectives.

The health and safety committee should include the following items for its meetings:

- study accident reports and offer recommendations for accident prevention;
- fire alarm procedure;
- report on fire drills and fire training;
- health and safety training;
- study health and safety data sheets for raw materials and products;
- new health and safety legislation.

Other items relevant to health and safety should be included as necessary.

Safety representatives

A safety representative is appointed by a recognized trade union [96b–98] and, as far as is reasonably practicable, should have 2 years' experience in the same or similar employment. An employer shall permit a safety representative to take time off with pay during working hours to receive necessary training at a recommended establishment centre.

The major functions of a safety representative are:

- to attend safety committee meetings;
- submit reports to the safety committee following investigations and safety audits;
- investigate potential hazards and dangerous occurrences;
- investigate complaints relating to employees' health, safety and welfare;
- conduct safety and fire preventions audits.

It is common practice for the factory inspector to send the safety representative a copy of all correspondance sent to the management.

Safety policies

A company employing five or more people must produce a written statement of its health and safety policy. The policy will spell out the implications of HSW and other health and safety constraints relevant to the work carried out. It is important to stress the need for co-operation from the workforce and good communication at all levels. The policy should be signed by a senior director of the company.

Employees should be able to determine from the policy statement how they fit into the health and safety structure, what their own duties are and where to go for advice, where to report an accident or potential hazard and how to obtain first-aid or other assistance. The written policy must detail the management structure responsible for implementing the policy.

The HSC has published a free leaflet providing advice to employers [99]. The HSE reviewed the experience of some of their staff on policy matters in an earlier publication [100]. More recently, the HSE has issued guidance with the small business in mind [101].

General guidance to printers

A general guide to health and safety issues for both employers and employees in small printing firms has been prepared by PIAC and published by the HSE [102] as part of a series of advisory booklets aimed at printers.

17.5 SPECIFIC PRINTING INK APPLICATIONS

Food wrapper inks

Extensive use of wrapping and packaging materials for food continues to ensure good hygiene during handling and increases sales appeal. Printing inks are applied to such packaging materials to provide directions for use, to identify the package contents and to improve sales appeal. However, printing inks cannot be made from truly edible materials, except in very special circumstances, as the range of suitable components is, in general, too limited. For this reason printing inks must not be used in such a way that they come into contact with food.

In the UK there are no specific regulations relating to the use of printing inks for the printing of food wrappers or packaging provided the printed material is not intended to come into contact with food. The SPBIM has promoted recommendations[103] that its members adopt to ensure that printing on food wrappers does not pose any health hazard and can be relied upon with confidence.

For a wrapper which will come into direct contact with food the ink must be applied to the outside of the wrapper. The wrapper itself must form a functional barrier between the printed surface and must prevent the ink from migrating or bleeding through the barrier and contaminating the foodstuff. The ink should be printed in such a way that set-off is avoided, ensuring that the wrapping surface in contact with the foodstuff is free of printing ink. An ink film is extremely thin (Chapter 2 gives indications of typical film thicknesses). Consequently the total quantity of ink involved is extremely small.

To ensure maximum safety, inks for immediate food wrappers should be formulated on non-toxic raw materials and should not contain materials listed on the SBPIM exclusion list. The application of an overprint varnish or lacquer to provide a functional barrier between the ink and food stuff will not necessarily provide a suitable barrier to prevent migration and therefore cannot be accepted as a suitable material to ensure contamination does not occur. The suitability of an overprint varnish or lacquer as an impermeable barrier should be assessed before finally deciding on a particular formulation.

If ink films or their byproducts were to come into contact with food, then the provisions of the 'Materials and Articles in Contact with Food Regulations' would apply to the printed composite. The printing process employed will be governed by the type of stock being printed but will generally be flexo, gravure or litho. Dependent upon the process, inks dry by evaporation or oxidation and, in the absence of strict quality assurance controls, there exists the possibility of adverse taint and odour effects on the packaged food.

The odorous volatile solvents coming from the inks are chemically identifiable and well known. Trained testing panels can readily identify both the taint and the odour. However it is possible that, for example, a trace of fatty acid in a coating could react with an additive or intermediate and finally break down to odorous byproducts such as aldehydes or ketones which in turn could introduce alien taint or odour to the food. The term 'organoleptic' refers to odour and taint characteristics and can only be perceived by nose and tongue. Threshold taste values for various substances have been reported as low as 5 ppb and the odour threshold for butyric acid (rancid butter odour) is 1 ppb.

There is a wide range of statutory instruments [104–107] under the various Food and Drugs Acts in the UK, which specify permitted food ingredients under such headings as emulsifiers, solvents, anti-oxidants, colouring matter and preservatives. In addition, one set of regulations [108] referred to above specifically controls the marketing and use of materials and articles which are intended to come into (direct) contact with food.

The regulations require that materials which are in their finished state and are intended to come into contact with food will, *inter alia*, be labelled 'for food use'. Such materials will not endanger human health or bring about a deterioration in the organoleptic characteristics of the food. Clearly, wet inks are not 'in their finished state'. The Food Contaminants Branch of the Ministry of Agriculture Fisheries and Food have confirmed that, in their view, printing inks manufactured in accordance with SBPIM's policy would not conflict with the provisions of the Materials and Articles in Contact with Food Regulations 1978.

In the absence of direct UK legislative guidance, it is common in the UK for print customers to seek reassurance in respect of so-called 'FDA Approval', for example. It is now well established that, in the USA, the Food and Drug Administration (FDA) has no authority under the food additives provisions of the US Federal Food and Cosmetic Act to 'approve' specific products, and that no food additive regulations on the general subject of printing inks exist. The FDA are on record [109] as believing that printing inks applied to the outer surface of food packaging materials are not food additives within the meaning of section 201 (3) of that Act, provided that the packaging material serves as a functional barrier between the printing ink and the food.

This said, it is common for UK converters to seek from their ink suppliers assurances as to the safety of the coatings to be applied, by reference to FDA regulations. This is understandable in the absence of direct guidance from our own legislators.

Under such circumstances, it is necessary for printing ink manufacturers to understand the relevant sections of the Federal Regulations [110], but at the same time to ensure the enquirers are aware of the real scope of those regulations. The sections that are relevant in this context are:

21CFR174	Indirect food additives: General
21CFR175	Indirect food additives: Adhesives and components of coatings
21CFR176	Indirect food additives: Paper and paperboard components

21CFR177	Indirect food additives: Polymers
21CFR182	Substances generally recognized as safe
21CFR186	Indirect food substances affirmed as generally recognized as safe.

In addition, parts of two further sections are of some relevance:

21CFR178	Adjuvants, production aids and sensitizers
21CFR181	Prior sanctioned food ingredients.

21CFR175.300, covering resinous and polymeric coatings for use as food contact surfaces, is the most useful guide in this context.

In broad terms, these regulations describe the components of coatings, for example, which may become added to food as a result of their use on articles that contact food (packaging material, for example).

Substances that, under conditions of good manufacturing practice, may be safely used as components of articles that contact food, are:

- substances generally recognized as safe in, or on, food;
- substances generally recognized as safe for their intended use in food packaging;
- substances used in accordance with a prior sanction or approval;
- substances permitted for use by specific regulations (such as 21CFR175.300, for example).

It should be noted, however, that coatings are described not only in terms of their permitted ingredients but also in terms of the level of extractibles present in the coating in the finished form, i.e. migration level into defined simulants from the coated substrate.

Their selection of simulants for testing of the components of food coating has been reviewed in the light of 'recent studies' which demonstrated that migration into certain aqueous foods is higher than would be predicted into water or 3% acetic acid. Additionally, it has been shown that migration into heptane predicts an exaggerated migration into fatty foods (but the official simulant for the testing of the finished coating as listed in 21CFR175 is still heptane).

In 21CFR175.300, for example, coatings intended for one-time use on containers having a capacity in excess of 1 gallon, will not yield an extractable content of greater than 1.8 milligrammes per square inch under the extraction conditions (solvent type and time/temperature regime) appropriate to the food type in question (for example, dairy products, dry solids or acidic foods).

For the time being, the 'Plastic materials and articles intended to come into contact with foodstuffs' directive[111] specifically excludes some coatings from the definition of plastic (Article 1(3)) and notes that Annex II, Sections A and B 'do not yet include monomers and other starting substances used only in the manufacture of . . . printing inks' (Article 3(5)).

Thus, inkmakers must take direct responsibility for the formulation of indirect food wrapper inks themselves, using such individual guidance as is available to them. The responsibility is greatest in the selection of pigments, since colouring materials listed in various states' positive lists are very limiting. For members of the Society, this means relying on the published recommendation[112].

Nevertheless, it is the ultimate responsibility of the packer to ensure that the final package composite does not adversely affect the contents and this dictates a high level of co-operation between all those in the processing chain from inkmaker to food packer.

Coatings for toys, pencils and graphic instruments

Toys bearing an EC mark must satisfy the essential safety requirements of the 1989 Toys (Safety) Regulations [113]. Of particular concern to inkmakers are the constraints on chemical properties. In general, toys (and that includes the packaging if it possesses intended play value) must not present health hazards or risks of physical injury by ingestion, inhalation or contact with skin, mucous tissues or eyes. Specific constraints are applied to the bio-availability of eight heavy metals and 'bio-availability' is interpreted to mean the soluble extract having toxicological significance.

The approved method for measuring the soluble extract is described in EN71 Part 3, December 1988, which was published as BS5665: Part 3:1989 [114]. This European method specifically addressed the problem of analysing for heavy metals in circumstances where the coating film cannot be removed from its substrate. However, in practice, the new methods have not been without their problems and certainly the reproducibility (probably associated with sample preparation) has given rise to concern since the method was issued. A further revision is under consideration and its consultation period ended in January 1993. The proposed migration level of barium would be raised from 500 mg/kg of the toy coating material to 1000 mg/kg and much greater attention would be paid to sample preparation and the extraction procedure for paper and board would be altered.

The same range of heavy metals is controlled in the Pencils and Graphic Instruments Regulations[115] but with variation in the conditions and

Table 17.6 Recommended maximum toxicity levels

Substance	Pencils and graphic instruments (ppm)	Toys (ppm)
Antimony (soluble)	250	60
Arsenic (soluble)	100	25
Barium (soluble)	500	500
Cadmium (soluble)	100	75
Chromium (soluble)	Not specified	60
Chromium (hexavalent soluble)	100	Not specified
Lead (soluble)	250	90
Mercury (soluble)	100	60
Selenium	Not specified	500

concentrations for lead and chromium.

The Society has issued a recommendation to members relating to the formulation of inks for toys. This recommendation relies on the specific exclusion of heavy metal components (except barium sulphate) rather than relying on an analysis of the dried film. It is important to note that regulations apply to the dried film, and it is thus the responsibility of the printer or toymaker to satisfy himself that the dry coating meets the requirements of the regulations. No written guarantee can be given by the ink manufacturer that inks 'conform' to the Toy Regulations (Table 17.6).

Printing inks for use on children's comics

The SPBIM considers that, as the print on children's comics could affect the health of children, the same limitations should be applied to printing inks for children's comics as are applied to printing inks for use on immediate food wrappers.

SBPIM general exclusion list

The SBPIM have consulted with their members and agreed on a list of exclusion materials which should not be used because of possible health effects. This list, known as the general exclusion list, includes known carcinogens, cumulative poisons or other toxic substances that should not be used in printing ink manufacture. Additions are made to this list as new knowledge becomes available.

General exclusion list

The following materials must be excluded from any printing ink or varnish:

- compounds of lead (except where specific exemption has been given by the SBPIM);
- compounds of cadmium;
- zinc chromate;
- pigments and dyestuffs based on selenium, arsenic, antimony and mercury;
- CI Solvent Blue 7 50400 (Induline);
- toluene di-isocyanate (TDI);
- 2-nitropropane (2-NP);
- polychlorinated biphenyls (PCBs);
- tricresyl phosphate;
- 4,4-tetramethyl diamino benzophenone (Michlers ketone);
- 2,4-dimethyl-6-*t*-butylphenol (Topanol A);
- diamino stilbene and derivatives;
- auramine;
- acrylates with a Draize figure greater than 3.4 on the 0 to 8 scale. The Draize figure for any acrylate to be determined by a recognized research institute. (Specific acrylates to be excluded are listed in a separate exclusion list – see next section.)

Ultra-violet and electron-beam curable inks

Ultra-violet inks, varnishes and lacquers are formulated on polyfunctional acrylates and other materials which can be polymerized to a hard film when exposed to high intensity UV light. The Printing Industry Advisory Committee (PIAC), on which the SBPIM is represented, has made recommendations[116] and laid down guidelines on safety in the manufacture and use of inks, varnishes and lacquers cured by UV light.

A revised edition of this booklet was published at the beginning of 1993[144]. Some functional acrylate/methacrylate monomers can cause skin irritation and others in addition cause skin sensitization. Sensitized persons must be removed from any further exposure to these materials and transferred to another job. Extraction ventilation must always be employed when handling UV products.

Electron-beam curable inks use similar raw materials to those employed in UV inks and similar handling and formulation techniques should be applied.

The Primary Irritation Index (PII), or the Draize rating, is determined from a patch test performed on animals and measures the degree of irritation imparted by a substance. The scale ranges from 0 to 8; 0 is low and 8 is high irritancy. The SBPIM has recommended to its members that only materials with a Draize rating of less than 3.5 and with no sensitizing effect be used in UV-cured inks. The Draize test has changed little since it was developed in 1944 and has aroused protests which can be justified on both scientific and humanitarian grounds. Work is in progress to develop alternative testing with the intention that animal testing can then be drastically reduced or better still, no longer carried out.

The practice of the application of classification criteria for acrylate monomers and oligomers is currently under active review by the raw material suppliers. Hopefully, the temporary differences in interpretation will be rapidly resolved.

The following acrylates are currently recommended for exclusion from from UV-curable ink formulations and are taken from the PIAC booklet as updated by SBPIM in their exclusion list:

2-Hydroxyethyl acrylate	Phenoxy ethyl acrylate (phenyl cellosolve acrylate)
2-Hydroxypropyl acrylate	Dicyclopentenyl acrylate
Methyl carbamyl ethyl acrylate	2-Ethyl hexyl acrylate (2EHA)
1,3-Butylene glycol diacrylate	Diethylene glycol diacrylate (DEGDA)
Butyl glycidyl ether acrylate	Tetra-ethylene glycol diacrylate (TEGDA)
1,6-Hexanediol diacrylate	Mixtures of pentaerythritol tri and tetra-acrylates (PETA)
Phenyl glycidyl ether acrylate	Trimetholpropane triacrylate (TMPTA)
Acrylate of glycidyl ester of C_{9-11}-carboxylic acid	Butanediol diacrylate (BDDA)
Neopentylglycol diacrylate	

Under the Health and Safety (Emissions into the Atmosphere) Regulations[46], acrylates are controlled under some circumstances. Thus:

> Works in which acrylates are . . . stored in fixed tanks with aggregate capacity exceeding 20 tonnes have a statutory duty to use best practical means to both prevent acrylate emissions into the atmosphere and render harmless and inoffensive such emissions that do occur.

Normally we would expect acrylates to be stored in 200-litre barrels within ink factories.

17.6 SOME INTERNATIONAL CONSTRAINTS

Throughout this chapter we have deliberately concentrated on UK legislation. Nevertheless, since the UK joined the Common Market, UK legislation has increasingly reflected European practice (though in some cases we were rather slow to adopt Community directives). O'Neil prepared [117], and has regularly updated [118, 119], an excellent review covering health and safety and environmental legislation of some 20–30 countries with special attention to the UK, USA, EC and to some extent Japan. This has now been replaced by the monthly *Coatings, Regulations, and the Environment* [120].

Chemical inventories

US regulations [121] require the Environmental Protection Agency to 'identify, compile, keep current and publish' a list of chemical substances (as defined) which are manufactured, imported or processed for commercial purposes in the USA. The initial inventory was published in 1979; the current edition [122] contains 63 000 chemical substances. In contrast to EC's EINECS, TSCA Inventory is a growing list of chemical substances and as such provides a rolling definition of 'new'. Neither lists toxic substances: they simply list commercially existing chemical substances of their respective territories.

The TSCA list itself is globally well known and indirectly carries much authority internationally; it has been used as the basis for several other national lists. However, the EPA has been beset with criticism for its enforcement of the Act itself and became bogged down in legal tangles in its attempt to gather toxicological information on which to base its risk evaluations [123].

In contrast, the European approach to prioritization seems to have attracted very much less controversy.

The Australian Core Inventory (ACOIN) has now been finalized as the Australian Inventory of Chemical Substances (AICS) [124]. This latter serves the same purpose as EINECS and the TSCA inventory.

The UN has published[125] a list of recommendations and legislation of various countries and international organizations on a selected range of potentially toxic chemicals. Of particular interest are the various labelling requirements, suitability for food contact and occupational exposure limits.

In the latter context the International Labour Office has published [126] a list of some 2100 substances with their occupational limits as assigned by countries and includes a useful review of the countries covered.

In contrast to the TSCA Inventory, the Registry of Toxic Effects of Chemical Substances (RTECS) [127] provides details of the known toxic data of some 80 000 chemical substances, though NIOSH estimate that the number of unique substances for which toxicity data may be available is approximately 250 000. The ultimate intention of the registry is to identify all known toxic substances and to provide pertinent data on the toxic effects from known doses entering an organism by any route described. It has long been the single most significant source of LD_{50} data in the world.

The data must not be considered as providing definitive values for describing 'safe' versus 'toxic' doses for human exposure. Rather it is one source of data for critical evaluation by experienced investigators. Indeed NIOSH themselves conduct critical reviews of occupational exposure hazards and the resulting criteria documents are used as the basis for recommendations to US Department of Labor, Occupational Safety and Health Administration, for US occupational exposure standards.

The last complete printed edition of RTECS was 1982 with a supplement in 1983. For non-US residents it is probably more convenient to use the microfiche edition which, in any event, is more up to date, being republished with amendments on a quarterly basis.

The Japanese Ministry of Labour has promulgated [128] lists of chemical substances which required both regular air monitoring and labelling. Associated enforcement regulations [129] classify solvents according to the level of ventilation required for their safe use (and also specifies the personal protection required):

- Class 2 local ventilation required, e.g. toluene, 2-ethoxyethanol;
- Class 3 general ventilation required, e.g. petroleum distillates.

The Japanese Industrial Safety and Health Association (JISHA) published [130] an extensive review of the original Industrial Safety and Health Law 1972 and its amendments up to 1988, together with their related ordinances up to October 1990.

New chemicals need to be notified [131] to the Ministry of International Trade and Industry (MITI) by reference to a definitive list of existing substances.

The inclusion of a limited CAS Registry Number Index in the 5th edition [132] made this handbook very much easier to use. Substantial improvements are planned for its successor *Current ENCS Listed Chemicals* scheduled for publication across the first quarter of 1993. This will include, in separate volumes, a CAS Number Index, Chemical Name Index (alphabetical) and a Molecular Formula Index as well as an ENCS Number Index, but at a considerably increased cover price.

US data communication

Under the provisions of OSHA's Hazards Communications Standard [133] a greater emphasis is placed on the use of labels to communicate hazards to employees than is the case under analogous UK requirements.

Published in November 1985 and coming into effect in May 1986, the OSHA standard requires ink manufacturers to:

- review their material inventories (raw materials, intermediates and finished goods);
- establish a hazardous materials list;
- prepare materials safety data sheets and labels (for all materials hazardous or not);
- prepare a written hazard communications program;
- provide employees with information and training on all hazardous materials in their workplace.

The National Association of Printing Ink Manufacturers (NAPIM) have developed a range of documentation to enable their members to discharge their communications responsibilities. NAPIM first published their material safety data sheets (MSDS) in 1971. These have been revised and a guide published describing in detail the requirements of each section of the new MSDS.

They have also published [134] a guide to a Hazardous Materials Identification System (HMIS) based on a system developed by National Paint and Coatings Association (NPCA) and the National Fire Protection Association (NFPA) for example. HMIS is a visual system using colour coding and simple numerical ratings (on a scale from 0 to 4; 0, low hazard and 4, high hazard) and is designed to communicate risk information to workers with respect to health hazard (blue), flammability hazard (red) and reactivity (yellow). Recommendations for personal protection equipment is indicated by a two-letter code and by symbol (Figs. 17.10 and 17.11).

Health hazard ratings are assessed from acute oral, dermal and inhalation toxicity data, generally with a tenfold difference between labels. On this basis 1H corresponds to the middle of the European 'Harmful' range and 2H embraces the bottom end of 'Toxic' and the top end of 'Harmful'. Flammability criteria bear some resemblance to European criteria; thus 4F corresponds, approximately, to extremely flammable, and 3F corresponds, approximately, to highly flammable. Thereafter the two sets of criteria diverge.

Food wrapper coatings

In 1982 the UN's Chemical Industry Committee of the Economic Commission for Europe decided to undertake a feasibility study in the area of food additives legislation. Its aim was to provide a condensed survey of then current legislation in various countries and publish a short description of the coverage of that legislation. Twenty-three European countries and the USA replied to the questionnaire and their responses were published in 1984 [135]. The level of detail varied considerably from country to country.

In 1985 CEPE published a review of European national legislation and practice for coatings for metal substates in direct contact with food. A

HEALTH	**1**
FLAMMABILITY	**3**
REACTIVITY	**0**
Personal Protection	**SF**

SF – Employee must wear gloves and splash goggles

Fig. 17.10 Recommendations for personal protective equipment.

revised version was published in 1991 [136].

The compilation was intended to assist producers and users of coatings compositions, particularly bearing in mind the needs of countries which have no national legislation of their own. The document goes some way to compensating for the continuing absence of a European directive in this area.

EC directives and drafts on packaging in contact with food have so far concentrated on plastic substrates and, as noted in the section on food wrapper ink, p. 943 have specifically excluded surface coatings such as inks and paints from their scope. Difficulties with overall migration limits have slowed progress on the directive but it now seems likely that the Single European Act may have the effect of overcoming delays (by introducing qualified majority agreement to replace unanimity in the decision procedures) and at the same time hasten progress on subsequent daughter-directives dealing with paper and metal substrates and associated coatings.

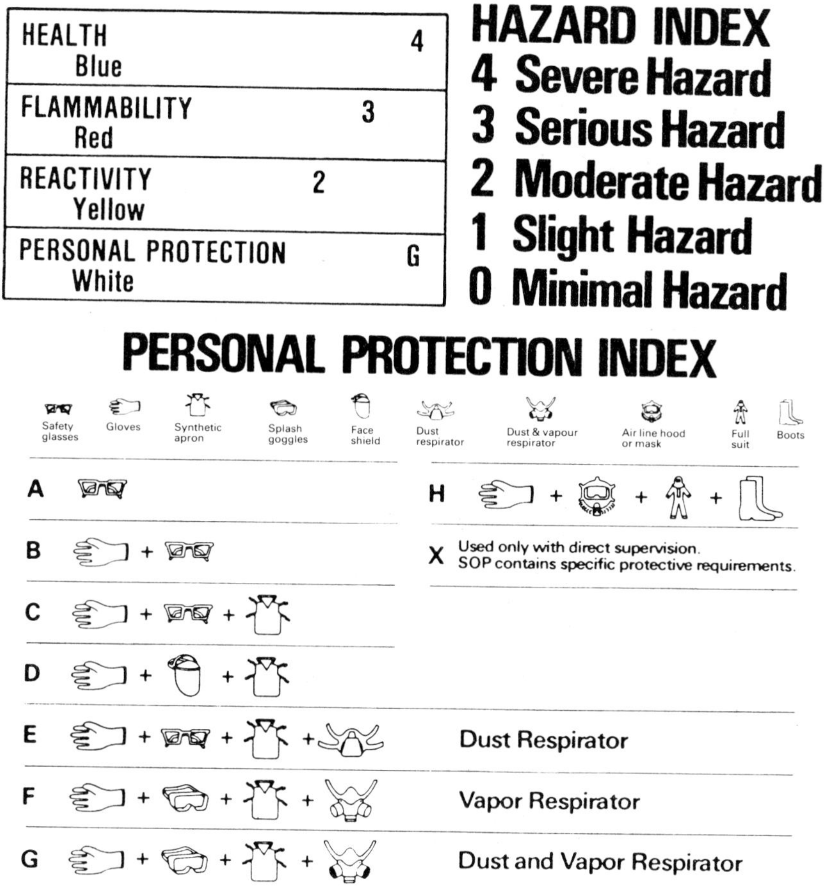

Fig. 17.11 Examples of a type of hazard label (adapted from the American Industrial Hygiene Association).

In the meantime various trade bodies, such as the Association of Plastics Manufacturers in Europe (APME) for example, have prepared 'temporary' guidance either for the use of their members or as discussion documents for EC consideration. APME have prepared such guidance covering colourants for food contact plastics.

EC drafts have drawn heavily on West German and Dutch sources. In the absence of more territorially relevant guidance these two sources can go some way to providing positive guidance in the UK. In particular, Dutch legislation has a growing section on global and specific test regimes for migration of coatings components. An English translation [137] of the fifth version of the draft packaging and food utensils regulation was prepared by the Netherlands Plastics Federation. West German legislation [138] presents a positive list of materials, in the main subdivided by chemical type, which may be used for food packaging applications. Chapter XL is entitled 'Lacquers and Varnishes for Food Containers and

Packs'. Permitted ingredients include pigments among the auxiliary materials.

In the UK frequent appeal to the US Food, Drug and Cosmetic Act for guidance is encountered. For this reason, and despite their non-territorial relevance, FDA constraints have been dealt with in the section on food wrapper inks (p. 944).

Transport

There are separate international recommendations for the transport of dangerous goods multimodally, by sea, by air, by road and by rail. These continue to be developed but there is evidence of increasing harmonization with UN recommendations[72]. Mostyn has compiled a series of guides bringing together the various route requirements for the main UN classes of dangerous goods[139].

As already noted, ADR is currently being substantially revised prior to its adoption in the UK to control inland road journeys for dangerous goods.

Light-gauge metal packaging, made of steel or tinplate (thickness less than 0.5 mm) is extensively used in the conveyance of printing inks. From 1989 such packaging will be subject to regulations for sea, air, and international road/rail transport and will be required to pass performance tests and bear an official mark as evidence of compliance[140].

Constraints on packaging waste

United States of America

The Source Reduction Council of CONEG (Coalition of North Eastern Governors) produced some Model Toxics Legislations, the final draft of which was dated 3 January 1990.

The intention was to provide to the individual US States a model on which to base a control regime for four named heavy metals as they appear in packaging components. In the definition of 'packaging component' ink was specifically mentioned.

Section 4 of the model states that '. . . no later than two years after the adoption of (the) act, no package or packaging component shall be offered for sale or for promotional purposes by its manufacturer or distributor in the state of XXXX which includes, in the package itself or in any packaging component, inks, dyes, pigments . . . or any other additives, any lead, cadmium, mercury or hexavalent chromium which has been *intentionally introduced* as an element during manufacturing or distribution, as opposed to the *incidental presence* of any of these elements.'

Somewhat confusingly, it then goes on to lay down a limit regime for intentionally introduced heavy metals (but does not explain how an analyst could distinguish between 'intentionally present' and 'incidentally present' metal components.

The sum concentration of the four specified metals shall not exceed:

- 600 ppm effective two years after adoption;
- 250 ppm effective three years after adoption;

- 100 ppm effective four years after adoption.

Section 6 requires that manufacturers will supply a certificate of compliance to the purchaser.

Individual State-implementations of the model exhibit a range of interpretations. A useful review of the situation has been published by Fitzpatrick [141].

Europe (EC)

The topic is likely to be of increasing importance to Europeans, since the EC draft directive on packaging and packaging waste [142] adopts the same constraints on the composition of 'packaging and packaging components' and has the same timescale range for implementation (with respect to the adoption date of the directive).

Europe (Germany)

Germany has tackled the control from a different standpoint [143], which could be described as administrative or financial in character. Manufacturers and retailers are obliged to take responsibility for the packaging used (they are defined to be the originators, in a waste disposal context). Recycling is given absolute priority over thermal recovery. Returned packaging material must be either:

- reused
- submitted for recovery **outside** the public waste disposal system.

Systems to cope with various sources of packaging waste were brought into effect across a timescale December 1991 to January 1993:

1 December 1991	Transportation packaging (collected and reclaimed).
1 April 1992	Consumer allowed to leave secondary packaging at the point of sale.
1 January 1993	Primary (sales) packaging collected at the retail store; material must be reclaimed.
1 January 1993	Mandatory deposit scheme (0.50 DM/container) will be applied to: • beverage containers; • soap and detergent containers; • emulsion paint containers.

Manufacturers/distributors have the option to replace the collection and mandatory deposit system with their own regular household collection scheme (but will still be required to recycle the collected material).

In the event that private collection schemes are initiated, then at least 50% recovery will be required from 1 January 1993 to 30 June 1995. This will rise to at least 80% recovery specifically for glass, paper, metals, plastics and mixed packages from 1 July 1995.

As from 1 July 1995 sorting quotas will also be imposed: 90% for glass, tinplate and aluminium, and 80% for other packaging.

All useful, separated material must be recycled; incineration of useful material is not permitted. Documentary proof that sorting quotas are being met will be required.

The decree explicitly acknowledges that the effect sought is the reduction of packing to that deemed absolutely necessary for the protection of the contents.

Under Article 2(1), the provisions of the ordinance apply (*inter alia*) to any business undertaking which brings into circulation '. . . packaged products at any commercial level, unless the packaging has been contaminated by hazardous substances or preparations or those that present an environmental risk'. While many printing inks are likely either to be hazardous or to present an environmental risk, it is equally clear that pallets, shrink wrapping and cardboard outer cartons are unlikely to be contaminated and, as commonly encountered, are likely to be within the scope of 'transport packaging'. But note that under Article 4 packaging, which is both transport and sales packaging, '. . . shall be treated as sales packaging'.

GLOSSARY OF ABBREVIATIONS

ACGIH	American Conference of Governmental Industrial Hygienists
ACOIN	Australian Core Inventory
ACTS	Advisory Committee on Toxic Substances (to HSE)
ADR	Accord Européen Relatif au Transport International des Merchandises Dangereuses par Route
AICS	Australian Inventory of Chemical Substances
APME	Association of Plastics Manufacturers in Europe
CEFIC	European Council of Chemical Manufacturers' Federations
CEPE	European Confederation of Paint, Printing Ink and Artists' Colour Manufacturers Associations
CHIP	(UK) Chemical (Hazard Information and Packaging) Regulations [1993]
CIA	Chemical Industries Association
CIMAH	Control of Industrial Major Accident Hazards
COPA	Control of Pollution Act (1974)
COSHH	Control of Substances Hazardous to Health (1988)
CPL	Classification, Packaging and Labelling of Dangerous Substances Regulations (SI 1984, No. 1244, as amended)
ECETOC	European Chemical Industry Ecology and Toxicology Centre
ECOIN	European Core Inventory
EINECS	European Inventory of Existing Chemical Substances
ELINCS	European List of New Chemical Substances
ENCS	Existing and New Chemical Substances (Japan, MITI)
EPA	(UK) Environmental Protection Act 1990
EPA	(US) Environmental Protection Agency
FDA	(US) Federal Drug Administration
HASAWA	see HSW
HAZCHEM	a CIA-initiated scheme for vehicle marking – Carrying Dangerous Goods
HFLR	Highly Flammable Liquids Regulations (1972)
HMIP	Her Majesty's Inspectorate of Pollution
HMIS	Hazardous Material Identification System
HSC	Health and Safety Commission
HSE	Health and Safety Executive
HSW	Health and Safety at Work, etc. Act 1974
IATA	International Air Transport Association
IMDG	International Maritime Dangerous Goods (code)
IMO	International Maritime Organisation
IPC	Integrated Pollution Control
MAK	Maximale Arbeitsplatzkonzentrationen (Germany)
MEL	(UK) Maximum Exposure Limit
MSDS	Material Safety Data Sheet
NAPIM	(US) National Association of Printing Ink Manufacturers
NFPA	(US) National Fire Protection Association
NIHHS	Notification of Installations Handling Hazardous Substances Regulations (1982)
NIOSH	(US) National Institute for Occupational Safety and Health (Dept of Health and Human Services)
NPCA	(US) National Paint and Coatings Association
OEL	Occupational Exposure Limits
OES	(UK) Occupational Exposure Standard
OSHA	(US) Occupational Safety and Health Administration (Dept of Labor)

OTS	(US) Office of Toxic Substances (EPA)
PIAC	Printing Industry Advisory Committee (to HSE)
PII	Primary Irritation Index
PMA	Paintmakers Association (of Great Britain)
PPE	Personal Protective Equipment
PRA	Paint Research Association (Teddington)
RIDDOR	Reporting of Injuries, Diseases and Dangerous Occurrences Regulations
RTECS	Registry of Toxic Effects of Chemical Substances (NIOSH)
SBPIM	Society of British Printing Ink Manufacturers
SI	Statutory Instrument (UK)
TLV	Threshold Limit Values (established by ACGIH)
TREMCARD	Transport Emergency Card (Trademark CEFIC)
TSCA	(US) Toxic Substances Control Act
TWA	Time-weighted average (applied to OELs)

(Note the Statutory Instruments quoted are the principal regulations in each case; many have subsequently been amended.)

REFERENCES

1. Lord Robens (1972) *Safety and Health at Work: Report of the Committee 1970–1972* Command Publication 5034, HMSO. (ISBN 010 150340 7)
2. *Health and Safety at Work, etc. Act 1974,* Chapter 37. (ISBN 010 543774 3)
3. Health and Safety Commission HSC3, *Health and Safety at Work, etc. Act 1974. Advice to Employers,* reprinted 1985; HSC2, *Health and Safety at Work. etc. Act 1974. The Act Outlined,* reprinted 1985.
4. Health and Safety Commission Guidance on the Act L1, *A Guide to the Health and Safety at Work, etc. Act 1974,* 4th edn, 1990. (ISBN 011 885555 7)
5. Statutory Instrument 1992 No. 2051. *The Management of Health and Safety at Work Regulations 1992.* (ISBN 011 025051 6); Health and Safety Commission Approved Code of Practice, *Management of Health and Safety at Work,* 1992. (ISBN 011 886330 4)
6. *Environmental Protection Act 1990,* Chapter 43. (ISBN 010 544390 5)
7. *Control of Pollution Act 1974,* Chapter 40, (ISBN 010 544074 4)
8. Department of the Environment. Secretary of State's Guidance PG6/11, *Manufacture of Printing Ink* (February 1992). (ISBN 011 752596 0)
9. Statutory Instrument 1992 No.588, *The Controlled Waste Regulations 1992.* (ISBN 011 023588 6)
10. Statutory Instrument 1980 No.1709, *Control of Pollution (Special Waste) Regulations 1980.* (ISBN 011 007709 1)
11. Official Journal of the European Communities **34** L337 31.12.91. *Council Directive of 12.12.91 on Hazardous Waste,* 91/689/EEC. (ISBN 011 969556 1)
12. Statutory Instrument 1991 No.2839, *The Environmental Protection (Duty of Care) Regulations 1991.* (ISBN 011 015853 9)
13. *The Sale of Goods Act 1979.* Chapter 54. (ISBN 010 545479 6)
14. Statutory Instrument 1989 No.1275, *The Toys (Safety) Regulations 1989.* (ISBN 011 097275 9)
15. Statutory Instrument 1974 No.226, *The Pencils and Graphic Instruments (Safety) Regulations 1974.* (ISBN 011 040226 X)
16. *The Consumer Protection Act 1987,* Chapter 43. (ISBN 010 544387 5)
17. Commission of the European Communities, *Constructing EINECS: Basic Documents* Vols I–IV, 1981. (ISBN 92 8252454 X)
18. *Official Journal of the European Communities* **33** C146A 15.06.90.
19. Health and Safety Commission, *Draft Proposal for the Chemicals (Hazard Information and Packaging) Regulations and Associated Approved Documents.* Consultative Document CD48 (Parts I and II), 1992.
20. Statutory Instrument 1982 No.1496, *The Notification of New Substances Regulations 1982.* (ISBN 011 027496 2)
21. Health and Safety Executive Health and Safety Booklet HS(R)14, *A Guide to the Notification of New Substances Regulations 1982,* 1982. (ISBN 011 883660 9)
22. Commission Directive of 25.04.84, 84/449/EEC. *Official Journal of the European Communities* **27** L251 19.09. 84.
23. Commission Communication Publication of ELINCS, 93/C130/02. *Official Journal of the European Communities* **36** C130 10.05.93.
24. Health and Safety Executive Guidance Note EH40/93, *Occupational Exposure Limits 1993,* (1993)(ISBN 011 882030 6)
25. Health and Safety Executive Guidance Note EH44. *Dust in the Work Place General Principles of Protection, 1991. (ISBN 011 885595 6)*

26. Health and Safety Executive MDHS14, *General Methods for the Gravimetric Determination of Respirable and Total Inhalable Dust*, 1989. (ISBN 011 885954 4)
27. American Conference of Governmental Industrial Hygienists, *Threshold Limit Values for Chemical Substances and Physical Agents and Biological Exposure Indices for 1991–1992. (ISBN 0 93671292 9)*
28. American Conference of Governmental Industrial Hygienists, *Documentation of the Threshold Limit Values and Biological Exposure Indices*, 5th edn, revised 1990.
29. Statutory Instrument 1983 No.1649, *The Asbestos (Licensing) Regulations 1983*. (ISBN 011 037649 8)
30. Health and Safety Executive Guidance Note EH10, *Asbestos – Exposure Limits and Measurements of Airborne Dust Concentrations*, 1990. (ISBN 011 885552 2)
31. Statutory Instrument 1988 No.1657, *The Control of Substances Hazardous to Health Regulations 1988*. (ISBN 011 087657 1)
32. Health and Safety Executive Health and Safety Series Booklet HS(G)37, *An Introduction to Local Exhaust Ventilation*, 1987. (ISBN 011 883954 3)
33. Health and Safety Executive, Health and Safety Series Booklet HS(G)54, *The Maintenance, Examination and Testing of Local Exhaust Ventilation*, 1990. (ISBN 011 885438 0)
34. Health and Safety Executive Health and Safety Series Booklet HS(G)53, *Respiratory Protective Equipment. A Practical Guide for Users*, 1990. (ISBN 011 885522 0)
35. Health and Safety Executive, *Respiratory Equipment; Legislative Requirements and List*, 3rd edn, 1992. (ISBN 011 886382 7)
36. Statutory Instrument 1992 No.2966, *The Personal Protective Equipment at Work Regulations 1992*. (ISBN 011 025832 0)
37. Health and Safety Commission Approved Code of Practice L5, *Control of Substances Hazardous to Health* (General ACoP Third Edition), and *Control of Carcinogenic Substances* (Carcinogens ACoP Third Edition) 1991. (ISBN 011 885698 7)
38. Health and Safety Executive, *COSHH Assessments. A Step-by-step Guide to Assessment and the Skills Needed for it*, 1988. (ISBN 011 885470 4)
39. Health and Safety Executive, *Health Surveillance under COSHH. Guidance to Employers*, 1990. (ISBN 011 885447 X)
40. Health and Safety Executive Health and Safety Series Booklet HS(G)77, *COSHH and Peripatetic Workers*, 1992. (ISBN 011 885733 9)
41. Health and Safety Executive Health and Safety Series Booklet HS(G)70, *The Control of Legionellosis, including Legionnaire's Disease*, 1992. (ISBN 011 885660 X)
42. Health and Safety Commission Approved Code of Practice L8, *The Prevention or Control of Legionellosis (including Legionnaires' Disease)*, 1992. (ISBN 011 885659 6)
43. Health and Safety Commission Consultative Document CD50. Draft Approved Code of Practice, *Control of Respiratory Sensitisers*, 1992.
44. Statutory Instrument 1980 No.1248, *The Control of Lead at Work Regulations 1980*. (ISBN 011 007248 0)
45. Health and Safety Commission Approved Code of Practice, *Control of Lead at Work*, 2nd edn, (1985). (ISBN 011 883780 X)
46. Statutory Instrument 1983 No.943, *The Health and Safety (Emissions into the Atmosphere) Regulations 1983*. (ISBN 011 036943 2)
47. Health and Safety Executive Guidance Note EH28, *Control of Lead: Air Sampling Techniques and Strategies*, 1981. (ISBN 011 883393 6)
48. *Fire Precautions Act 1971*, Chapter 40. (ISBN 010 544071 X)
49. Home Office, *Guide to the Fire Precautions in Existing Places of Work that Require*

a Fire Certificate, 1989 (ISBN 011 340906 0)

50. Health and Safety Commission. Printing Industry Advisory Committee, *Fire Safety in the Printing Industry*, 1992. (ISBN 011 886375 4)
51. Statutory Instrument 1976 No.2003, *Fire Certificates (Special Premises) Regulations 1976*. (ISBN 011 602003 8)
52. *The Fire Safety and Safety of Places of Sport Act 1987*, Chapter 27. (ISBN 010 542787 X)
53. Statutory Instrument 1972 No.917, *The Highly Flammable Liquids and Liquefied Petroleum Gases Regulations 1972*. (ISBN 011 020917 6)
54. Department of Employment, *Highly Flammable Liquids and Liquefied Petroleum Gases: Guide to the Regulations*, 1973. (ISBN 011 360904 3)
55. Health and Safety Executive Health and Safety Series Booklet HS(G)50, *The Storage of Flammable Liquids in Fixed Tanks (up to 10,000 m^3 total capacity)*, 1990. (ISBN 011 885532 8)
56. Health and Safety Executive Health and Safety Series Booklet HS(G)51, *The Storage of Flammable Liquids in Containers*, 1990. (ISBN 011 885537 6)
57. Health and Safety Executive Health and Safety Series Booklet HS(G)4, *Highly Flammable Liquids in the Paint Industry*, 1978. (ISBN 011 883219 0)
58. Health and Safety Executive Health and Safety Series Booklet HS(G)22, *Electrical Apparatus for Use in Potentially Explosive Atmospheres*, 1984. (ISBN 011 883746 X)
59. Health and Safety Executive Health and Safety Series Booklet HS(R)15 rev, *Administrative Guidance on the European Community 'Explosive Atmospheres' Directive (76/117/EEC) and 79/196/EEC) and Related Directives*, 2nd edn, (1987). (ISBN 011 883880 6)
60. Health and Safety Commission, *Information Approved for the Classification, Packaging and Labelling of Dangerous Substances for Supply and Conveyance by Road*, 3rd edn, Authorised and Approved List, 1990. (ISBN 011 885542 5)
61. Statutory Instrument 1984 No.1244, *Classification, Packaging and Labelling of Dangerous Substances Regulations 1984*. (ISBN 011 047244 6)
62. Health and Safety Executive, Health and Safety Series Booklet HS(R)22, *A Guide to the Classification, Packaging and Labelling of Dangerous Substances Regulations 1984*, 1985. (ISBN 011 883794 X)
63. Health and Safety Executive Health and Safety Series Booklet HS(G)27. 3rd edn, *Substances for Use at Work: The Provision of Information*, 1989. (ISBN 011 885458 5)
64. Statutory Instrument 1984 No.1902, *The Control of Industrial Major Accident Hazard Regulations 1984*. (ISBN 011 047902 5)
65. Health and Safety Executive Health and Safety Series Booklet HS(R)21, *A Guide to the Control of Industrial Major Accidental Hazards Regulations 1984*, revised 1990. (ISBN 011 885579 4)
66. Health and Safety Executive Health and Safety Series Booklet HS(G)25, *The Control of industrial Major Accidental Hazards Regulations 1984 (CIMAH): Further Guidance on Emergency Plans*, 1985. (ISBN 011 883831 8)
67. Statutory Instrument 1990 No.304, *The Dangerous Substances (Notification and Marking of Sites) Regulations 1990*. (ISBN 011 003304 3)
68. *The Planning (Hazardous Substances) Act 1990*, Chapter 10. (ISBN 010 541090 X)
69. Statutory Instrument 1992 No.743, *The Road Traffic (Carriage of Dangerous Goods in Road Tankers and Tank Containers) Regulations 1992*. (ISBN 011 023743 9)
70. Statutory Instrument 1992 No.742, *The Road Traffic (Carriage of Dangerous Substances in Packages etc) Regulations 1992* (ISBN 011 023742 0)
71. Statutory Instrument 1992 No.744, *The Road Traffic (Training of Drivers of Vehicles Carrying Dangerous Goods) Regulations 1992*. (ISBN 011 023744 7)

72. United Nations, *Transport of Dangerous Goods: Recommendations of the Committee of Experts on the Transport of Dangerous Goods*, 7th edn, 1991. (ISBN 92 1139 035 4)
73. International Maritime Organisation, *IMDG Code, 1992 Consolidated Edition*, Vols I–IV. (Incorporating amendment 26–91.)
74. International Air Transport Association, *Dangerous Goods Regulations*, 33rd edn, 1992. (ISBN 92 9035 355 4)
75. Department of Transport, *European Agreement Concerning the International Carriage of Dangerous Goods by Road ADR*, 1993 edn, 1992. (ISBN 011 551114 8)
76. Home Office/National Chemical Emergency Centre, *HAZCHEM LIST No.7 Emergency Action Codes and Supplementary Information for Dealing with Incidents Involving Dangerous Substances Conveyed in Bulk by Road or Rail*, 1992.
77. National Chemical Emergency Centre, *CEFIC Tremcards 1991. Reference Edition*, CEFIC Transport Emergency Cards, Single Substance Cards, Volumes 1 and 2, 1991.
78. National Chemical Emergency Centre, *CEFIC Tremcards 1991. Reference Edition*, CEFIC Transport Emergency Cards, Group Cards, 1991.
79. Statutory Instrument 1981 No.917, *The Health and Safety (First Aid) Regulations 1981*. (ISBN 011 016917 4)
80. Health and Safety Commission Code of Practice 42, *First Aid at Work: Health and Safety (First Aid) Regulations 1981 and Guidance*, 1990. (ISBN 011 885536 0)
81. Statutory Instrument 1985 No.2023, *The Reporting of Injuries, Diseases and Dangerous Occurrences Regulations 1985*. (ISBN 011 058023 0)
82. Health and Safety Executive Health and Safety Series Booklet HS(R)23; *A Guide to the Reporting of Injuries and Dangerous Occurrences Regulations 1985*, 1986. (ISBN 011 883858 X)
83. Health and Safety Executive HSE 11 (Rev), *Reporting an Injury or a Dangerous Occurrence*, 1986.
84. Health and Safety Executive HSE 17, *Reporting a Case of a Disease*, 1986.
85. Statutory Instrument 1992 No.2793, *The Manual Handling Operations Regulations 1992*. (ISBN 011 025920 3)
86. Health and Safety Executive Guidance on Regulations L23, *Manual Handling: Manual Handling Operations Regulations 1992*, (1992). (ISBN 011 886335 5)
87. Department of Employment Code of Practice, *Reducing the Exposure of Employed Persons to Noise*, 1972. (ISBN 011 880340 9)
88. Statutory Instrument 1989 No.1790, *The Noise at Work Regulations 1989*. (ISBN 011 097790 4)
89. Health and Safety Executive, *Noise at Work. Noise Guides Nos. 1 and 2*, 1989. (ISBN 011 885512 3)
90. Health and Safety Executive Health and Safety Series Booklet HS(G)56, *Noise at Work*, 1990. (ISBN 011 885430 5)
91. Health and Safety Commission, Printing Industry Advisory Committee, *Noise Reduction at Web-fed Presses*, 1988. (ISBN 011 883972 1)
92. Health and Safety Executive Health and Safety Series Booklet HS(G)38, *Lighting at Work*, 1991. (ISBN 011 883964 0)
93. Statutory Instrument 1992 No.2792, *The Health and Safety (Display Screen Equipment) Regulations 1992*. (ISBN 011 025919 X)
94. Health and Safety Executive Guidance on Regulations L26, *Display Screen Equipment Work*, 1992. (ISBN 011 886331 2)
95. Health and Safety Executive Approved Code of Practice L24, *Workplace Health, Safety and Welfare*, 1992. (ISBN 011 886333 9)

96a. Statutory Instrument 1992 No.2932, *The Provision and Use of Work Equipment 1992*. (ISBN 011 025849 5)
96b. Statutory Instrument 1977 No.500, *The Safety Representatives and Safety Committees Regulations 1977*. (ISBN 011 070500 9)
97. Health and Safety Commission, *Safety Representatives and Safety Committees*, 1977. (ISBN 011 880335 2)
98. Health and Safety Commission HSC9, *Time Off for the Training of Safety Representatives*, 1978.
99. Health and Safety Commission HSC6 (rev), *Writing a Safety Policy Statement. Advice to Employers*, 1985.
100. Health and Safety Executive, *Effective Policies for Health and Safety. A Review Drawn from Previous Work and Experience of the Accident Prevention Unit of HM Factory Inspectorate*, 1980. (ISBN 011 883254 9)
101. *Our Health and Safety Policy Statement*, (revised 1989). (ISBN 011 885510 7)
102. Health and Safety Commission Printing Industry Advisory Committee, *Health and Safety for Small Firms in the Printing Industry*, 1986. (ISBN 011 883851 2)
103. Society of British Printing Ink Manufacturers, *Guide to Materials and Substances for Exclusion from Printing Inks and Varnishes*, 1990.
104. Statutory Instrument 1978 No.108, *The Antioxidants in Food Regulations 1978*.
105. Statutory Instrument 1973 No.1340, *The Colouring Matter in Food Regulations 1973*.
106. Statutory Instrument 1989 No.876, *The Emulsifiers and Stabilisers in Food Regulations 1989*. (ISBN 011 096876 X)
107. Statutory Instrument 1967 No.1582, *The Solvents in Food Regulations 1967*.
108. Statutory Instrument 1987 No.1523, *The Materials and Articles in Contact with Food Regulations 1987*. (ISBN 011 077523 6)
109. Editorial, *American Ink Maker* **58** (9) (1980) 21 (September 1980).
110. Code of Federal Regulations Title 21, *Food and Drug Administration*, Office of the Federal Register, National Archives and Records Administration (published annually).
111. Official Journal of the European Communities *33* L349, 13.12.90. pp26–48. *Commission Directive of 23 February 1990 Relating to Plastic Materials and Articles Intended to Come into Contact with Foodstuffs 90/128/EEC*. [Corrigendum to *33* L75, 21.3.90. pp. 19–40.]
112. Society of British Printing Ink Manufacturers, *Printing Inks for Use on Food Wrappers and Packages*, 1988.
113. Statutory Instrument 1989 No.1275, *The Toys (Safety) Regulations 1989*. (ISBN 011 097275 9)
114. British Standard BS5665: Part 3: 1989. EN71: Part 3: 1988, *Safety of Toys, Part 3 Specification for Migration of Certain Elements*. (ISBN 058 017312 7)
115. Statutory Instrument 1974 No.226, *The Pencils and Graphic Instruments (Safety) Regulations 1974*. (ISBN 011 040226 X)
116. Health and Safety Commission, Printing Ink Advisory Committee, *Safety in the Use of Inks, Varnishes and Lacquers Cured by Ultraviolet Light*, (1983). (ISBN 011 883678 1)
117. O'Neill, L.A. (1981) *Health and Safety, Environmental Pollution and the Paint Industry*, 2nd edn, Paint Research Association.
118. O'Neill, L.A. (1986) *Health and Safety, Environmental Pollution and the Paint Industry*, supplement to the 2nd edn of 1981, Paint Research Association.
119. O'Neill, L.A, *Survey of Hazards, Pollution and Legislation in the Coatings Field*. Quarterly from May 1986 to 1992. Paint Research Association.
120. O'Neill, L.A. and Cé, *Coating Regulations and the Environment (CORE)*, Monthly from June 1992, Paint Research Association.
121. *Toxic Substances Control Act 1976*.

122. (US) Environmental Protection Agency, Office of Toxic Substances, *Toxic Substances Control Act (TSCA) Chemical Substances Inventory* Vols I-V (January 1986) and *1990 Supplement to the 1985 Edition of the TSCA Inventory* (June 1990). EPA560/7–85–002 and EPA560/7–90–003.
123. Plamond, Israel and Scoville. *Recent Developments under the Toxic Substances Control Act*. RadTech 92 North America, Volume II, pp. 722–725, Boston, April 1992.
124. Worksafe Australia, National Occupational Health and Safety Commission, *Australian Inventory of Chemical Substances*, 1992. (microfiche)
125. United Nations Environment Program IRPTC Data Profile No.7, *International Register of Potentially Toxic Chemicals Legal File 1986*, 1987. (ISBN 928 071163 6)
126. International Labour Office, *Occupational Exposure Limits for Airborne Toxic Substances*, 3rd edn, 1991. (ISBN 922 107293 2)
127. (US) Department of Health and Human Services. National Institute for Occupational Safety and Health, *Registry of Toxic Effects of Chemical Substances*. (Microfilm edition, reissued with update twice per year.)
128. (Japanese) Ministry of Labour, *Industrial Safety and Health Law* (Law No.57 of 1972); *Ordinance on Prevention of Hazards due to Specified Chemical Substances* (Ordinance No.39 of 1972); *Ordinance on Prevention of Hazards due to Dust* (Ordinance No.18 of 1979).
129. (Japanese) Ministry of Labour, *Ordinance on the Prevention of Organic Solvent Poisoning* (Ordinance No.36 of 1972).
130. Japanese Industrial Safety and Health Association, *Industrial Safety and Health Law and Related Legislation of Japan*, 1991. (ISBN 480 590392 9)
131. (Japanese) Ministry of International Trade and Industry, *Chemical Substances Inspection and Control Law*. (Law No.117, October 1973)
132. The Chemical Daily Co., Tokyo, *Handbook of Existing and New Chemical Substances*, 5th edn, 1992.
133. (US) *29 Code of Federal Regulations 1910.1200*, 1985.
134. National Association of Printing Ink Manufacturers. *HMIS: Guidelines for a Hazardous Materials Identification System of Raw Materials*, 1984.
135. United Nations Economic Commission for Europe, *Regulations and Legislation on Food Additives and Chemicals for Food Packaging*, ECE/CHEM/54 (1984).
136. Confédération Europeénne des Associations de Fabricants de Peintures, d'Encres, d'Imprimerie et de Coleurs d'Art (CEPE), *Survey of Current Regulations in Europe Concerning Printing Ink for Food Packaging*, 1991.
137. Nederlandse Vereniging – Federatie voor Kunststoffen. Translation of the 5th version of *The draft packaging and food utensils regulations (food laws)*, 1975.
138. *Kunststoffe im Lebensmittelverkehr*, Carl Hetmans Verlag, 1984.
139. Mostyn, H. (1986) *Packaging Dangerous Goods for Transport. General Guide, and Guide to Class 3 Inflammable Liquids*, Aurigny Ltd.
140. PIRA Packaging Division (1987) *Light gauge metal packaging for transport of dangerous goods*, PK/DE/D24.
141. Fitzpatrick, H.F. (1992) Legislative and Regulatory Issues facing the Colour Industry. *American Ink Maker*, **70**(6), 15, 18–20, 25, 62, 64, 66, 70, 74.
142. EC Draft Directive COM(92) 278/7. Packaging and Waste (July 1992) 92/C263/01. *Official Journal of the European Coimmunities*, **35** C263, 19.
143. Bundesgesetzblatt Part 1 1991, *The Avoidance of Packaging Waste*, pp. 1234–38. (Verpackungsverordnung 12.06.91).
144. Health and Safety Commission, Printing Ink Advisory Committee, Safety in Use of Inks, Varnishes and Lacquers Cured by Ultraviolet light or Electron Beam Techniques (1993). (ISBN 011 882045 1).

Index